CELLULAR AND MOLECULAR BIOLOGY OF BONE

CELLULAR AND MOLECULAR BIOLOGY OF BONE

Edited by
Masaki Noda
Department of Molecular Pharmacology
Division of Functional Disorder Research
Medical Research Institute
Tokyo Medical and Dental University
Tokyo, Japan

ACADEMIC PRESS, INC.
A Division of Harcourt Brace & Company
San Diego New York Boston London Sydney Tokyo Toronto

Cover photo: Longitudinal section through neonatal rat femur stained with Masson trichrome. Photo courtesy of R. Tracy Ballock, M.D.

This book is printed on acid-free paper. ∞

Academic Press, Inc.
1250 Sixth Avenue, San Diego, California 92101-4311

United Kingdom Edition published by
Academic Press Limited
24–28 Oval Road, London NW1 7DX

Library of Congress Cataloging-in-Publication Data

Cellular and molecular biology of bone / edited by Masaki Noda,
p. cm.
Includes bibliographical references and index.
ISBN 0-12-520225-3 (hardcover)
1. Bone cells. 2. Bones--Molecular aspects. I. Noda, Masaki.
QP88.2.C46 1993
599'.087--dc20 93-14783
CIP

PRINTED IN THE UNITED STATES OF AMERICA
93 94 95 96 97 98 QW 9 8 7 6 5 4 3 2 1

CONTENTS

4 BONE MORPHOGENETIC PROTEINS AND THEIR GENE EXPRESSION

John M. Wozney

5 OUR UNDERSTANDING OF INHERITED SKELETAL FRAGILITY AND WHAT THIS HAS TAUGHT US ABOUT BONE STRUCTURE AND FUNCTION

Jeffrey Bonadio and Steven A. Goldstein

CONTRIBUTORS

Numbers in parentheses indicate the pages on which the authors' contributions begin.

Abdul-Badi Abou- Samra (321), Massachusetts General Hospital and Harvard Medical School, Boston, Massachusetts 02114

Jane E. Aubin (1), Medical Research Council Group in Periodontal Physiology, University of Toronto, Toronto, Ontario M5S 1A8

R. Tracy Ballock (97), Laboratory of Chemoprevention, National Institute of Health, National Cancer Institute, Bethesda, Maryland 20892

Roland Baron (445), Departments of Orthopedics, Cell Biology, and Cell and Molecular Physiology, Yale University School of Medicine, New Haven, Connecticut 06510

Jeffrey Bonadio (169), Department of Pathology, Howard Hughes Medical Institute, University of Michigan, Ann Arbor, Michigan 48109

Myles A. Brown (257), Departments of Medicine, Dana Farber Cancer Institute and Harvard Medical School, Boston, Massachusetts 02115

Munmun Chakraborty (445), Departments of Orthopedics, Cell Biology, and Cell and Molecular Physiology, Yale University School of Medicine, New Haven, Connecticut 06510

Diptendu Chatterjee (445), Departments of Orthopedics, Cell Biology, and Cell and Molecular Physiology, Yale University School of Medicine, New Haven, Connecticut 06510

Gilbert J. Cote (343), Departments of Medicine and Cell Biology, Baylor College of Medicine and VA Medical Center and Section of Endocrinology, M.D. Anderson Cancer Center, University of Texas, Houston, Texas 77030

Marie Demay (321), Massachusetts General Hospital, and Harvard Medical School, Boston, Massachusetts 02114

Randall L. Duncan (413), Renal Division, Jewish Hospital/Washington University, St. Louis, Missouri 63110

Gregor Eichele (287), V. and M. McLean Department of Biochemistry, Baylor College of Medicine, Houston, Texas 77030

Robert F. Gagel (343), Departments of Medicine and Cell Biology, Baylor College of Medicine and VA Medical Center and Section of Endocrinology, M.D. Anderson Cancer Center, University of Texas, Houston, Texas 77030

Christopher K. Glass (257), Division of Cellular and Molecular Medicine and, Center for Molecular Genetics, University of California, San Diego, La Jolla, California 92093

Steven A. Goldstein (169), Orthopedic Research Laboratories, Section of Orthopedic Surgery, University of Michigan, Ann Arbor, Michigan 48109

Agamemnon E. Grigoriadis (497), Research Institute of Molecular Pathology, A-1030, Vienna, Austria

Anne-Marie Heegaard (191), Bone Research Branch, National Institute of Dental Research, National Institutes of Health, Bethesda, Maryland 20892

Johan N.M. Heersche (1), Medical Research Council Group in Periodontal Physiology, University of Toronto, Toronto, Ontario M5S 1A8

William Horne (445), Departments of Orthopedics, Cell Biology, and Cell and Molecular Physiology, Yale University School of Medicine, New Haven, Connecticut 06510

Keith A. Hruska (413), Renal Division, Jewish Hospital of St. Louis, St. Louis, Missouri 63110

Kyomi Ibaraki (191), Bone Research Branch, National Institute of Dental Research, National Institutes of Health, Bethesda, Maryland 20892

Harald Jüppner (321), Massachusetts General Hospital and Harvard Medical School, Boston, Massachusetts 02114

Sandra A. Kerner (235), Departments of Pediatrics and Cell Biology, Baylor College of Medicine, Houston, Texas 77030 and Ligand Pharmaceuticals, Inc., San Diego, California 92121

Janet M. Kerr (191), Bone Research Branch, National Institute of Dental Research, National Institutes of Health, Bethesda, Maryland 20892

Robert A. Kesterson (235), Departments of Pediatrics and Cell Biology, Baylor College of Medicine, Houston, Texas 77030 and Ligand Pharmaceuticals, Inc., San Diego, California 92121

Seong-Jin Kim (97), Laboratory of Chemoprevention, National Institutes of Health, National Cancer Institute, Bethesda, Maryland 20892

Henry Kronenberg (321), Massachusetts General Hospital and Harvard Medical School, Boston, Massachusetts 02114

Jane B. Lian (47), Department of Cell Biology, University of Massachusetts Medical Center, Worcester, Massachusetts 01655

Sergio Line[1] (539), National Institute of Dental Research, National Institutes of Health, Bethesda, Maryland 20892

Abderrahim Lomri (445), Departments of Orthopedics, Cell Biology, and Cell and Molecular Physiology, Yale University School of Medicine, New Haven, Connecticut 06510

Meetha Medhora (413), Renal Division, Jewish Hospital/Washington University, St. Louis, Missouri 63110

Lynn Neff (445), Departments of Orthopedics, Cell Biology, and Cell and Molecular Physiology, Yale University School of Medicine, New Haven, Connecticut 06510

Keiichi Ozono (235), Departments of Pediatrics and Cell Biology, Baylor College of Medicine, Houston, Texas 77030 and Ligand Pharmaceuticals, Inc., San Diego, California 92121

Sara Peleg (343), Department of Medical Specialities, Section of Endocrinology, M. D. Anderson Cancer Center, University of Texas, Houston, Texas 77030

J. Wesley Pike (235), Departments of Pediatrics and Cell Biology, Baylor College of Medicine, Houston, Texas 77030 and Ligand Pharmaceuticals, Inc., San Diego, California 92121

Jan-Hindrik Ravesloot (445), Departments of Orthopedics, Cell Biology, and Cell and Molecular Physiology, Yale University School of Medicine, New Haven, Connecticut 06510

Craig Rhodes (539), National Institute of Dental Research, National Institutes of Health, Bethesda, Maryland 20892

Felice Rolnick (413), Renal Division, Jewish Hospital/Washington University, St. Louis, Missouri 63110

Gino Segre (321), Massachusetts General Hospital and Harvard Medical School, Boston, Massachusetts 02114

Susan M. Smith (287), Department of Nutritional Sciences, University of Wisconsin, Madison, Wisconsin 53706

Teruki Sone (235), Departments of Pediatrics and Cell Biology, Baylor College of Medicine, Houston, Texas 77030 and Ligand Pharmaceuticals, Inc., San Diego, California 92121

Gary S. Stein (47), Department of Cell Biology, University of Massachusetts Medical Center, Worcester, Massachusetts 01655

Christina Thaller (287), V. and M. McLean Department of Biochemistry, Baylor College of Medicine, Houston, Texas 77030

---

[1] *Present address:* Faculdade de Odontolgia de Piracicabe-UNICAMP, Av. Limeira s/n, Caixa Postal 52, 13400 Piracicaba-São Paulo, Brazil.

Kursad Turksen[2] (1), Medical Research Council Group in Periodontal Physiology, University of Toronto, Toronto, Ontario M5S 1A8

Erwin F. Wagner (497), Research Institute of Molecular Pathology, A-1030 Vienna, Austria

Zhao-Qi Wang (497), Research Institute of Molecular Pathology, A-1030 Vienna, Austria

John M. Wozney (131), Genetics Institute, Cambridge, Massachusetts 02140

Yoshihiko Yamada (539), National Institute of Dental Research, National Institutes of Health, Bethesda, Maryland 20892

Kensuke Yamakawa (413), Renal Division, Jewish Hospital/Washington University, St. Louis, Missouri 63110

Toshiyuki Yoneda[3] (375), Department of Medicine, University of Texas Health Science Center, Division of Endocrinology and Metabolism, San Antonio, Texas 78284

Marian F. Young (191), Bone Research Branch, National Institute of Dental Research, National Institutes of Health, Bethesda, Maryland 20892

---

[2]*Present address:* Howard Hughes Medical Institute, University of Chicago, Chicago, Illinois 60637.

[3]*Present address:* Division of Molecular Cell Biology, Medical Research Institute, Tokyo Medical and Dental University, Tokyo, Japan 101.

PREFACE

The study of bone cell biology has been undertaken by multidisciplinary groups whose fields cover basic sciences such as developmental biology, molecular endocrinology, genetics, physiology, pharmacology, and biochemistry, as well as clinical sciences such as endocrinology, orthopedics, and dental medicine. Bone biology has been regarded as one of the applied sciences; however, recent progress in bone biology suggests that it could serve as a leading model in each of the above-mentioned disciplines. In the past few years, research in this field has advanced rapidly because of the availability of new technologies and recent developments in biology. Molecular and cellular biological techniques have been the most successful in bone biology research. This rapid progress, however, increases the knowledge gap between researchers—even between those working within the same field of bone biology.

To accomplish the goals of bone biology research most efficiently, it is necessary to compile ground-breaking information into a concise format. The purpose of this book is to introduce forefront research in bone biology, where powerful molecular and cellular biological techniques are successfully utilized, as well as to give a comprehensive picture of the most up-to-date information in bone biology. This book offers a concise state-of-the-art view of bone biology and will provide a resource not only for experts in the field, but also for undergraduate students, newcomers, and practitioners.

Masaki Noda

1

OSTEOBLASTIC CELL LINEAGE

JANE E. AUBIN, KURSAD TURKSEN, and JOHAN N. M. HEERSCHE

Cellular and Molecular Biology of Bone

I. INTRODUCTION

Bone formation takes place in the organism during embryonic development, growth, remodeling, and fracture repair and when induced experimentally—for example, by the implantation of decalcified bone matrix or purified or recombinant members of the bone morphogenetic protein family (Reddi, 1985; Urist, 1989; Wozney *et al.*, 1990; Wozney, 1992). There is clearly a large reservoir of cells in the body capable of osteogenesis throughout life. During the past decade, new methods have been developed to study the cell biology of bone and gain insight into the various cell types important in bone function (for reviews, see Rodan and Rodan, 1984; Nijweide *et al.*, 1986, 1988; Heersche and Aubin, 1990; Aubin *et al.*, 1990b). It has also become increasingly clear that the metabolic activities of bone are under the control of a large number of systemic and local factors (Martin *et al.*, 1987; Stern, 1988; Marcus, 1988; Martin, 1989; Mundy, 1989). Despite these advances, many questions remain. For example, detailed knowledge of the lineage of the osteoblast, including identification of transitional steps from stem cell to committed osteoprogenitor to osteoblast, interactions of cells within the lineage, and identification and regulation of stem cells and different levels of committed progenitors, is largely lacking. This chapter provides a review of the osteoblast lineage and possibilities for recognizing stages of differentiation or maturity. As such, it reviews current concepts of the origin, lineage, and differentiation of osteoblasts and currently available tools to study them, with emphasis on *in vitro* model systems.

II. CELLS OF THE OSTEOBLAST LINEAGE

A. General Morphological and Histological Definition

Based on morphological and histological studies, osteoblastic cells are categorized in a presumed linear sequence progressing from osteoprogenitor to preosteoblasts, to osteoblasts, and then to lining cells or osteocytes (Nijweide *et al.*, 1986; Martin *et al.*, 1987; Marks and Popoff, 1988; Bonucci, 1990; Wlodarski, 1990). Earlier morphological definitions of the active osteoblast as a cuboidal, polar, basophilic cell lining the bone matrix at sites of active matrix formation (Cameron, 1968; Holtrop, 1975) have been supplemented more recently by elucidation of their specific products—for example, type I collagen (Leblond, 1989), osteocalcin (Hauschka *et al.*, 1989), osteopontin (SPP1) (Butler, 1989), and bone sialoprotein (Sodek *et al.*, 1992a,b). Active osteoblasts give a strong histochemical reaction for alkaline phosphatase (APase) that disappears when cells cease their synthetic activity (Doty and Schofield, 1976) or become embedded in matrix as osteocytes (Holtrop, 1975). However, a number of criteria, such as morphology (Marotti, 1976; Villaneuva *et al.*,

1981), biosynthetic activity detected by biochemical analysis (Otawara and Price, 1986), immunohistochemistry (Mark *et al.*, 1988) or *in situ* hybridization (Heersche *et al.*, 1992), suggest that newly differentiated osteoblasts (cuboidal, osteocalcin low or negative) differ from more mature osteoblasts later in their secretory lifetime (more flattened osteocalcin high). Thus, maturational stage—not only stage of differentiation—ultimately will have to be elucidated with these and other markers (see later).

In a region where osteoblasts are laying down bone matrix, the cuboidal cells directly behind them have been called preosteoblasts (Pritchard, 1952; Luk *et al.*, 1974). Based on kinetic studies, it has been suggested that preosteoblasts are the precursors of the osteoblast in the regions of growing bone (Owen, 1963, 1967; Kember, 1971). Preosteoblasts morphologically resemble the osteoblast and show some markers of the osteoblast (e.g., APase activity [Doty and Schofield, 1976]), but they are clearly recognizably different from osteoblasts in not expressing others (see later). Other possibly earlier precursor cells may reside in the heterogeneous layer of proliferating cells behind the osteoblast/preosteoblast layer (Pritchard, 1972a,b). These earlier cells may be osteoprogenitor cells (Young, 1962). In addition to their position in the tissue near bone surfaces, osteoprogenitors are fibroblastic or spindle-shaped with oval or elongated nuclei and notable glycogen content (Scott, 1967). Their appearance and location in the tissues are the main criteria to define their presence. It seems likely, but there is no proof, that the farther away from the bone surface the osteogenic cell is the less differentiated it will be (Scott, 1967).

The osteocyte is considered the most mature or terminally differentiated cell of the osteoblast lineage (Jande and Belanger, 1973; Holtrop, 1975). Osteocytes are embedded in bone matrix occupying spaces (*lacuanae*) in the interior of bone and are connected to adjacent cells by cytoplasmic projections within channels (*canaliculi*) through the mineralized matrix (Menton *et al.*, 1984). The presence of gap junctions between the cytoplasmic projections is thought to allow these cells to communicate. Some, but not all, of the biochemical features of the osteoblast are expressed in the osteocyte (see also later).

In the adult, the majority of bone surfaces are occupied by another cell type with a distinct phenotype, the bone lining cell, which displays a flat and highly elongated cell shape with a spindle-shaped nucleus (Menton *et al.*, 1984). Bone lining cells are usually designated as part of the osteoblast lineage because they are believed to be derived from osteoblasts that have ceased their activity and flattened out on bone surfaces that are undergoing neither formation nor resorption (Luk *et al.*, 1974; Miller and Jee, 1987). They have fewer organelles than the active osteoblast (Cameron, 1968), further suggesting that they may be largely inactive cells. Little is known about their function: Because they are joined to their neighbors and to nearby osteocytes by gap junctional complexes, it has been hypothesized that bone lining cells may function as a selective barrier between bone and other extracellular fluid compartments and may contribute to mineral homeostasis by regulating the fluxes of calcium and phosphate in and out of bone fluids and to control of the growth of bone crystals by maintaining a suitable microenvironment (Miller *et al.*, 1980; Miller and Jee, 1987).

B. Osteoblast Phenotypic Expression and Osteoblast Markers

In addition to the general morphological criteria already outlined, cells within the osteoblast lineage exhibit several characteristics that have proven useful as aids in their identification. Of these, high APase activity, as already indicated, parathyroid hormone (PTH) binding and a PTH-stimulated adenylate cyclase, and the ability to synthesize a number of noncollagenous bone matrix proteins are important features. A brief description of some of these aspects of the osteoblast and/or its matrix follows. This description, rather than being comprehensive for any one molecule, is slanted toward how these may be used to help define osteoblast differentiation and lineage. However, it should be pointed out that virtually none of these markers is unique to the osteoblast although they represent major components of osteoblast expression and development. Moreover, it is possible that the differentiation of the future osteoblast could be detected earlier if suitable early markers could be found. This tissue will be discussed later.

1. Alkaline Phosphatase (APase EC 3.1.3.1)

Alkaline phosphatase is an ectoenzyme that can hydrolyze monophosphate esters at a high pH optimum (for general review, see Wuthier and Register, 1984; Harris, 1989; Whyte, 1989). Structural studies have indicated that APase interacts with specific phospholipids on the osteoblast plasma membrane (for review, see Cross, 1987; Ferguson and Williams, 1988; Low, 1989a,b). Alkaline phosphatase belongs to a growing class of cell surface proteins (including Thy-1 and N-CAM) that are covalently bound to phosphatidyl inositol (PI) phospholipid complexes in the plasma membrane. Thus, membrane-bound APase can be released from cells (e.g., the ROS and UMR cell lines) by PI-specific phospholipase C (Noda *et al.*, 1987; Turksen and Aubin, 1991).

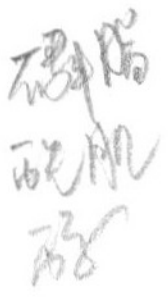

Although to date the precise physiological role of this enzyme in bone is not known, APase activity is present when cells become recognized as preosteoblast and osteoblast, but it is absent from the osteocyte (Doty and Schofield, 1976). It is generally accepted that as the specific activity of APase in a population of bone cells increases there is a corresponding shift to a more differentiated state, and the level of osteoblastic APase has been routinely used in *in vitro* experiments as a relative marker of osteoblast differentiation (Rodan and Rodan, 1984). Recently, a number of studies have suggested that PI-linked proteins may be involved in transmembrane signaling (for review, see Saltier *et al.*, 1989; Low, 1989a,b). Therefore, it is interesting to speculate that APase may play a role in the regulation of osteoblast differentiation. A more commonly ascribed function for APase is in mineralization; however, controversy over its precise role in this process still exists (for a recent discussion, see Heersche *et al.*, 1990). In keeping with the histological observation that APase is already present on preosteoblastic cells prior to their assuming the cuboidal shape typically associated with the mature osteoblast, a variety of recent studies also suggest that APase expression appears in differentiating osteoblastic cells prior to expression of the matrix molecule osteocalcin (Bronckers *et al.*, 1987; Aronow *et al.*, 1990; Owen *et al.*, 1990).

2. Hormone and Growth Factor Response

The response of osteoblastic cells to a variety of hormones acting through adenylate cyclase has been studied extensively (for review, see Peck and Woods, 1988) and details of the usually complex effects are beyond the scope of this chapter. However, many mixed cell populations and clonal populations, derived from either osteosarcomas or normal tissues, have been characterized as osteoblastic based on their having a PTH response (i.e., a PTH-stimulatable adenylate cyclase). Earlier, PTH receptors were demonstrated on the osteoblast and its immediate precursors (Silve *et al.*, 1982; Rao *et al.*, 1983), but some recent studies suggest that the highest number of receptors is on a relatively undifferentiated cell, with relatively few on the mature osteoblast itself (Rouleau *et al.*, 1988, 1990). These latter data may be difficult to reconcile with data from some studies *in vitro* in which a PTH-stimulated adenylate cyclase response was acquired along with APase in a clonal, osteoprogenitorlike cell line, RCT-1, induced to mature with retinoic acid (Heath *et al.*, 1989). It is possible that RCT-1 cells proceed along a differentiation sequence but become blocked before terminal differentiation (see later) or that the magnitude of the adenylate cyclase response to PTH does not correlate directly with the number of receptors a cell possesses. However, Bernier *et al.* (1991) used UMR 106 cells to suggest that, although some cells in the population bound both epidermal growth factor (EGF) and PTH, the cells binding the highest levels of PTH were different from those binding the highest levels of EGF, with the former relatively quiescent (possibly more mature) compared to the latter. Other recent studies in which high levels of PTH receptor messenger RNA (mRNA) were localized by *in situ* hybridization in mature functional osteoblasts in the primary spongiosa of rat long bones also suggest that the mature osteoblast may contain relatively large numbers of PTH receptors (Abou-Samra *et al.*, 1992; Urena *et al.*, 1992). These kinds of studies point to the discrepancies and voids in our understanding of when and on what cells markers like the PTH receptors first appear and/or disappear during the osteoblast lineage (see also later).

Prostaglandins (PGs) are low molecular weight lipids synthesized from arachidonic acid and have important biological effects in a variety of cell systems including bone (for review, see Curtis-Prior, 1988). PGs mediate the activities of a number of other factors and influence expression of many aspects of the osteoblast phenotype (for review, see Harvey, 1988). Different osteoblast populations have been shown to synthesize and/or respond to PGE_2 (e.g., ROS 17/2.8 vs. UMR 106.06; Partridge *et al.*, 1980; Aubin *et al.*, 1982). It has been suggested that the ability to respond to PGE_2 is a marker of a less differentiated osteoblastic cell, but confirmation of such a hypothesis awaits further proof.

Many other hormones have effects on osteoblastic cells, including steroid hormones (for reviews, see Martin *et al.*, 1987), but it can be relatively difficult to discriminate stage of lineage based on effects of these hormones and to discriminate between changes in a maturational or differentiation state versus an up- or down-regulation of particular genes without concomitant changes in differentiation program. This is true also for a variety of growth factor effects. Among growth factors with marked effects on cells of the osteoblast lineage are insulinlike growth factors I and II, EGF, fibroblast growth factor, platelet-derived growth factor, and transforming growth factor β (TGF-β); other molecules with potent activities are cytokines such as interleukin 1 (e.g., Krane *et al.*, 1988; Canalis *et al.*, 1989a,b). A common theme (and often common problem) in analyzing the effects of these agents on osteoblastic cells is the heterogeneity (stimulatory, inhibitory, no effect) of the responses measured by different people in different model systems (for a general discussion, see Nijweide *et al.*, 1986, 1988; Heersche and Aubin, 1990). In addition, many of the agents tested have biphasic effects, suggesting that their ability to regulate osteoblast properties and differentiation is complex and may be dependent on timing, concentration, and whether other agents are present. However, an increasing body of data suggests that some factors may elicit different effects on osteoblastic cells as they mature from osteoprogenitor to more mature cells (e.g., EGF [Antosz *et al.*, 1987; Canalis *et al.*, 1989a,b], TGF-β [Antosz *et al.*, 1989; Noda, 1989]), whereas others may recognize only particular subpopulations within the lineage (Nijweide *et al.*, 1986; Heersche and Aubin, 1990). The apparent inconsistencies in effects of many agents in different osteoblast systems, or even in the same model system at different stages of culture, could be sorted out, at least in part, by being able to define more precisely the subpopulation makeup of the cells or tissues being used and to sort the heterogeneity observed into physiologically meaningful classes. Several possible approaches exist to begin to conquer these difficulties. One is to map receptor location to particular cell types resident in tissues. In this way, Martineau-Doize *et al.* (1988), for example, recently identified the major EGF binding cells in rat trabecular bone to be not the osteoblast but a less differentiated precursor cell. The problem remains, however, that some osteoblast populations thought to be more mature by certain criteria also respond to EGF (Canalis *et al.*, 1989a,b; Antosz *et al.*, 1989), underscoring the need for a battery of other markers and other approaches (see later).

3. Bone Matrix Proteins

Our understanding of bone and osteoblast products has been advanced by new procedures for isolating bone matrix molecules (for reviews, see Heinegard and Oldberg, 1989; Boskey, 1989; Robey, 1989). At least some of these bone matrix molecules may also serve as differentiation markers in the osteoblast lineage.

The most abundant protein of the organic matrix of bone is type I collagen, but the noncollagenous proteins, which presumably lend unique properties to bone tissue and influence their ability to mineralize, are now drawing much attention (for general discussion, see Boskey, 1989; Simkiss and Wilbur, 1989; Lowenstam and Weiner, 1989; Glimcher, 1989). One of these, osteocalcin, is abundant in adult bone, comprising 10–20% of the noncollagenous protein of bone matrix, but is rare in embryonic bone. Osteocalcin is found only in bone, teeth, and mineralizing cartilage. Its distinguishing feature is the presence of three residues of γ-carboxyglutamic acid per molecule added via posttranscriptional modifications in a vitamin K-dependent process (for reviews, see Price, 1988; Hauschka *et al.*, 1989). Its exact physiological role is not know, but *in vitro* osteocalcin serves as an inhibitor of crystal growth (Price, 1983) and it has been postulated to play a role in osteoclast recruitment and bone resorption (Mundy and Poser, 1983; Lian *et al.*, 1984; Glowacki *et al.*, 1991). In rat bone, it has been shown to be secreted after the onset of mineralization, but its presence is not essential for mineralization to occur although it accumulates in the mineralized bone. In normal lamellar bone, osteocalcin has been localized in the osteocyte where the arborizing cytoplasmic processes were intensely stained. In newly apposed bone, the matrix did not show any osteocalcin staining, suggesting that osteocalcin deposition may be a relatively late event in the process of new bone formation (Groot *et al.*, 1986;

Bronckers *et al.*, 1987; Vermeulen *et al.*, 1989). This is supported by the developmental appearance of osteocalcin, which is present in the osteoblasts and chondrocytes in the hypertrophic zone, but not in the preosteoblasts (Bronckers *et al.*, 1987). The observation, made with *in situ* hybridization, that osteocalcin mRNA becomes detectable in osteoblasts in the primary spongiosa of long bones after mineralization has begun also supports this view (Heersche *et al.*, 1992).

Osteonectin, despite its name and its abundance in bone (Termine *et al.*, 1981; for review, see Tracy *et al.*, 1988; Schulz and Jundt, 1989), is not exclusively an osteoblast product. It is synthesized by many cells (e.g., fibroblasts of various sites [Wasi *et al.*, 1984; Tung *et al.*, 1985; Zung *et al.*, 1986], parietal endoderm during early mouse development (SPARC [secreted protein acidic and rich in cysteine; Mason *et al.*, 1986; Sage *et al.*, 1989), the basement membrane producing EHS tumor [BM-40; Mann *et al.*, 1987]), and it is homologous to the M_r 43,000 culture shock protein produced by bovine endothelial cells cultured for prolonged periods with repeated passages (Sage, 1986; Sage *et al.*, 1984, 1986). The distribution of osteonectin mRNA determined in both embryonic and adult tissues suggests that it is expressed in tissues undergoing remodeling and/or morphogenesis (Howe *et al.*, 1988; Nomura *et al.*, 1988; Sage *et al.*, 1989).

At least two sialoproteins are osteoblast products. Osteopontin (also termed bone phosphoprotein, bone sialoprotein I, secreted phosphoprotein 1, SPP, 2ar, η-1,pp69) has been isolated from bone (Prince *et al.*, 1987; for review, see Senger *et al.*, 1989; Butler, 1989) and is an Arg–Gly–Asp–Ser (RGDS)-containing, phosphorylated, sialic acid-rich, Ca-binding protein (Oldberg *et al.*, 1986). Osteopontin is present in the bone matrix, osteoblasts, osteocytes, and stromal cells in marrow but is also present in the proximal convoluted tubules of kidney, neurons, and sensory and secretory cells in the internal ear (Mark *et al.*, 1987, 1988; Butler, 1989) and is also expressed by hypertrophic chondrocytes (Franzen *et al.*, 1989) and chondrocytes in mineralizing cartilage (Chen *et al.*, 1991a,b). 2ar, which codes for mouse osteopontin, is known to be expressed by transformed cells and is inducible by tumor promoters and growth factors in a variety of cell lines (Craig *et al.*, 1988; Nomura *et al.*, 1988; Smith and Denhardt, 1987), including T lymphocytes (η-1, early T-lymphocyte activating factor; Patarca *et al.*, 1989). The role of osteopontin in bone is not known, although it may mediate the attachment and spreading of osteoblasts and osteoclasts (Reinholt *et al.*, 1990), and its phosphorylation and sulfation could have roles in biomineralization (Glimcher, 1989; Addadi *et al.*, 1987). Although osteopontin is synthesized by various nonosteogenic and osteogenic cell populations, osteoblasts produce a highly phosphorylated and sulfated form of the protein (Nagata *et al.*, 1989). Osteopontin has been localized within the developing bone cells in embryonic femur during early stages of osteogenesis and appeared before mineralization, earlier than osteocalcin. Therefore, osteopontin is thought to be an earlier marker than osteocalcin, possibly identifying osteoblast precursor cells (i.e., preosteoblasts; Mark *et al.*, 1987). Bone sialoprotein (BSP) has a higher content of sialic acid and glutamic acid and is more highly sulfated than osteopontin (Oldberg *et al.*, 1988; Ecarot-Charrier *et al.*, 1989). BSP contains an RGD cell binding sequence and binds to cells via an integrin, the vitronectin receptor (Oldberg *et al.*, 1988). BSP is present in the mineralized matrix of bone but is also present in the mineralized zone of hypertrophic cartilage. Osteoblasts, osteocytes, and chondrocytes in the mineralized cartilage label intensely for BSP (Oldberg *et al.*, 1988; Chen *et al.*, 1991b). Interestingly, both osteopontin and BSP mRNA levels are highest in new bone and 21-day fetal bone and decline thereafter. Osteopontin mRNA localized by *in situ* hybridization in calvarial bone is reported to be high in cells of fibrous periosteum and endosteum but highest in cuboidal osteoblasts on the periosteal surface of newly forming bone (Weinreb *et al.*, 1990; Sodek *et al.*, 1992a,b). BSP mRNA appears restricted to active osteoblasts and is especially high at sites of *de novo* bone formation and in developing sutures (Sodek *et al.*, 1992a,b). In long bones, osteopontin is expressed in chondrocytes of hypertrophic cartilage and in stromal cells of marrow in addition to osteoblasts at sites of bone formation, whereas BSP mRNA is restricted to sites of new mineralized tissue formation (bone and calcifying cartilage) with less expression in older bone (Chen *et al.*, 1991a). Thus, again, certain matrix molecules may help elucidate maturation stages of the osteoblast or its progenitors besides marking a different differentiation stage.

A variety of other proteins associated with bone are being investigated, but in most cases there is as yet little direct information of when during a differentiation sequence they begin to be expressed. For example, thrombospondin, originally purified from platelet granules (for review, see Mosher, 1990), has been identified in bone and as an osteoblast product (Robey *et al.*, 1989), but when during differentiation osteogenic cells first express thrombospondin is not yet known. Tenascin is another RGD-containing extracellular matrix component (Bourdon and Rouslahti, 1989) originally identified in muscle (Chiquet and Fambraught, 1984; for review, see Chiquet-Ehrismann, 1990) but reported to be localized around the osteogenic cells invading the cartilage model in bones that form by endochondral ossification and in the condensing mesenchyme of bones that form by intramembranous ossification (Mackie *et al.*, 1987). It has been localized in mature bone matrix but persists on periosteal and endosteal surfaces. The proteoglycans of bone, notably biglycan (CS-PGI), decorin (CS-PGII) (Fisher *et al.*, 1989), and CS-PGIII (Goldberg *et al.*, 1988) also deserve more attention as potential lineage markers, but as yet it is difficult to use any of these molecules as precise osteogenic differentiation or maturational markers.

III. ORIGIN AND LINEAGE OF THE OSTEOBLAST

A. Mesenchymal Stem Cells and Multipotential and Restricted Progenitors

The ontogeny of bone cells and the developmental relationship between these cells pose a major problem in bone biology (for reviews, see Friedenstein, 1976; Hall, 1987; Nijweide *et al.*, 1986; Owen and Friedenstein, 1988; Owen, 1988; Peck and Woods, 1988). Understanding the origin and differentiation of osteoblastic cells has been hampered both by the lack of a detailed understanding of osteoblast functions and by the fact that unequivocal criteria for identification of osteoblast cells at early developmental stages have not been established. Markers such as those discussed earlier—APase activity, collagen and noncollagen protein synthesis, and hormone response—may generally reflect cells already relatively mature (for a general discussion, see Rodan and Rodan, 1984; Nijweide *et al.*, 1986). Clearly, in comparison to certain other cell types (e.g., the hemopoietic system, which has contributed much to the knowledge of osteoclast lineage) little is known about the osteogenic cell system.

Studies using [^{3}H]-thymidine autoradiography have supported the concept that osteoprogenitors are undifferentiated mesenchymal cells residing in the stromal tissues surrounding bone marrow and in other connective tissue compartments (e.g., the periosteal layer of bone [Young, 1962; Owen, 1963, 1967; for detailed review and critique of early data, see Owen, 1970; Kember, 1971; Krukowski *et al.*, 1983]). However, the number or series of steps leading from mesenchymal stem cell to committed osteoprogenitor to osteoblast and related cells is not known. Based in part on labeling studies, studies with populations inserted into diffusion chambers, and other osteogenic models (Friedenstein, 1976; Urist *et al.*, 1983; Nijweide *et al.*, 1986; for a recent review, see Owen 1988), and lineage models for the hemopoietic system (Lajtha, 1979; Till and McCulloch, 1980; Ogawa *et al.*, 1983; for a recent discussion of the concepts, see Wolpert, 1988; Torok-Storb, 1988; Hall and Watt, 1989), a model for osteoblast cell differentiation was proposed by Owen (1985). According to that model, cells of the stromal fibroblastic system of bone and marrow contain pluripotent stem cells that are able to generate several cell lines including the osteogenic line (for recent review, see Owen, 1988; Beresford, 1989).

Friedenstein first demonstrated that bone marrow stroma contains cells that have the capacity to form bone when transplanted *in vivo* in diffusion chambers (Friedenstein *et al.*, 1968). Subsequently, he and others, notably Owen and colleagues, demonstrated that, in addition to bone, cartilage, marrow adipocytes, and fibrous tissue also formed (Ashton *et al.*, 1980; Friedenstein, 1980) and that all the tissues could arise from single clones or fibroblast colony-forming units (CFU-F; Friedenstein, 1980; Friedenstein *et al.*, 1987). Direct *in vitro* cellular evidence for the existence of a multipotential mesenchymal stem cell was lacking until the observation that 5-azacytidine treatment induced the differentiation of the mouse embryo fibroblast line C3H 10T1/2 clone 8 (10T1/2) (Reznikoff *et al.*, 1973). 5-Azacytidine induced stem cells within the 10T1/2 population to differentiate into muscle cells, adipocytes, and, at a low frequency, chondrocytes (Constantinides *et al.*, 1977; Taylor and Jones, 1979, 1982b). Single colony and subclone analyses detecting end-stage (recognizable tissue) phenotypes demonstrated that the clonal 10T1/2 cell line contained subpopulations of progenitor cells that could differentiate into one, two, or three cell types (Taylor and Jones, 1982a) but that only the subclones restricted to a single lineage (specifically myogenic and adipogenic) were stable in culture (Konieczny and Emerson, 1984). This work was extended when a rat multipotential mesenchymal progenitor line was isolated by limiting dilution of cells derived from 21-day fetal rat calvaria. Among the clones isolated with expression of various properties associated with the osteoblast phenotype (Aubin *et al.*, 1982, 1990b), one line (RCJ 3.1), when cultured under conditions favoring bone formation *in vitro* and in the presence of dexamethasone (DEX), was capable of differentiating into muscle, fat, cartilage, and bone (Grigoriadis *et al.*, 1988). By using colony assays and isolation of subclones of various restriction profiles, the types of progenitor cells present in the RCJ 3.1 population were analyzed along with their minimum differentiation potential and the frequency with which they appear in the population, allowing analysis of the kinds of lineage decisions such a clonal multipotential cell line could make. It was concluded that within the mesenchymal multilineage hierarchy in which osteoblasts reside, multipotential cells (e.g., the original parental cell giving rise to the RCJ 3.1 population) comprise a low frequency of the total population, with tri- and bipotential cells making up a larger proportion (i.e., ~2, 10, and 20%, respectively, of the total population; Grigoriadis *et al.*, 1990). The differentiated progeny resulting from restricted monopotential cells comprise the majority of the population, resulting in a hierarchical schema in which the rare mesenchymal stem cells are related to their more abundant descendents, including the cells giving rise to muscle, fat, cartilage, bone, and probably fibroblasts (for discussion on the difficulties in determining the fibroblast lineage, see Grigoriadis *et al.*, 1990) (see Fig. 1 and Aubin *et al.*, 1990b; Grigoriadis *et al.*, 1988, 1990).

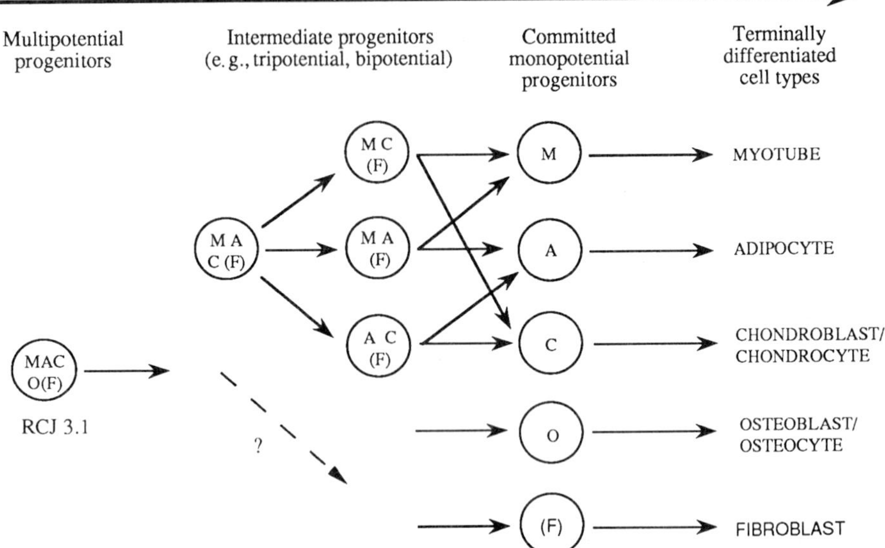

FIGURE 1 Proposed lineage diagram based on differentiation *in vitro* of the clonal cell line RCJ 3.1 and some of its subclones. The dashed line indicates that conclusions could not be drawn on this part of the lineage from the RCJ 3.1 studies, in which bone colonies formed at low frequency. For further details and discussion, see Aubin *et al.* (1990b), and Grigoriadis *et al.* (1988, 1990).

Several caveats affecting interpretation of the results are important. First, such approaches in which only end-stage phenotypes are analyzed under a given set of conditions will normally not allow discrimination of alternate pathways within the hierarchy because not all stem or progenitor cells may express their proliferation and differentiation capacities. Second, since the end-stage phenotypes may be removed from the original multipotential progenitor by several generations, it may not be possible to make unambiguous conclusions regarding the factors and mechanisms involved in commitment and restriction of different cell types to specific lineages. However, combining colony analyses with subclone isolation and isolation of progenitors of intermediate and increasingly restricted differentiation potential allows some of this to be addressed.

It is clear that DEX can regulate both the earlier multipotential and bipotential cells as well as the restricted monopotential cells (see Grigoriadis *et al.*, 1988, 1990), but it will be important to identify the other factors and hormones that may regulate commitment and/or differentiation in this multilineage cascade (see later). For example, a comparison of the RCJ 3.1 cell line with the well-characterized 10T1/2, for which subclone information is also available, reveals that there are many differences between the two lines, including differences in their basal expression of different phenotypes and their responsiveness to different inducers. In contrast to the 10T1/2 cell line, RCJ 3.1 appears not to respond to 5-azacytidine, nor does 10T1/2 appear to respond to DEX with changes in the commitment–differentiation program (Grigoriadis *et al.*, 1988). While RCJ 3.1 expresses osteoblast properties and forms bone nodules at a low frequency without addition of culture additives other than ascorbic acid and β-glycerophosphate, 10T1/2 does not. However, Katagiri *et al.* (1990) showed that treatment with BMP-2 caused 10T1/2 to increase expression of APase and synthesize osteocalcin, suggesting that these may be conditions under which the osteoblast program may be turned on in 10T1/2. While it has not been achieved from 10T1/2, whose bipotential subclones are unstable and cease to differentiate after only one passage, it has been possible to subclone stable tri-, bi-, and monopotential clonal lines, including those restricted to chondrocytic development, from RCJ 3.1 (Grigoriadis *et al.*, 1990). At least three types of bipotential cell were detectable in RCJ 3.1 populations; these expressed different combinations of muscle cells, fat cells, and cartilage cells that are clearly stable in the presence of DEX for at least five to six passages (~20 doublings). The ability to identify these different progenitors and their inherently different capacities to commit to different lineages and differentiate (i.e., low or high producers of differentiated progeny) suggest that commitment in the RCJ 3.1 cell line is random. This is consistent with the stochastic model of cell differentiation proposed many years ago for hemopoietic cells (Till *et al.*, 1964) and quite widely supported by others for hemopoietic and nonhemopoietic systems (e.g., O'Neill and Stockdale, 1972; Nakahata *et al.*, 1982; Bennet, 1983; Leary *et al.*, 1984; Lim *et al.*, 1984; Ogana, 1989) but is in contrast to the hypothesis that a stem cell undergoes a fixed predetermined sequence of events that leads to lineage restriction (e.g., Nicola and Johnson, 1982; Temple and Raff, 1986).

A number of other clonal multipotential cell populations have now been isolated by different labs, and these express markedly different capacities to differentiate along the various mesenchymal pathways. It will be increasingly important to attempt to determine the basis for the nonidentical phenotypes observed *in vitro*. For example, C26, a line derived from newborn rat calvaria cells, expresses osteoblastic properties, makes myotubes, gives rise to mineralized tissue when cultured in collagen gels, and produces adipocytes when treated with DEX (Yamaguchi and Kahn, 1991). Notably, when treated with BMP-2, C26 cells lose the capacity to make muscle but increase expression of APase, synthesize osteocalcin, and respond better to PTH (Yamaguchi *et al.*, 1991). Although extremely interesting, a detailed analysis of which cells (multipotential or more restricted or both) within the population are responding to the BMP-2 with changes in their expression of differentiated properties and possibly with changes in commitment has not yet been reported. While use of the mixed populations may allow commitment and differentiation modifiers to be identified, it is only detailed subclone analyses and possibly isolation of progenitors with more or less restricted phenotypes that will allow elucidation of the molecular effects of a variety of regulators on discrete cell types that cannot be ascertained unambiguously by study of mixed multipotential cells. It is well to be cautious of the marked heterogeneity of expression of differentiated progeny and the conditions under which they are expressed in both the stem cells and their progeny in 10T1/2, RCJ 3.1, and C26. Even given the probability of stochastic expression, it is currently unclear how much of this heterogeneity represents a true heterogeneity of the stem cells and their ability to commit to different mesenchymal lineages or is artifact of long-term culturing with concomitant changes in the rodent genome, known to undergo relatively frequent chromosomal alterations. Large-scale chromosomal rearrangements detectable by banding changes can and do occur with passage of rodent cells, and probably more frequent but more difficult to detect are the small more subtle but potentially as damaging rearrangements, deletions, amplifications, and point mutations that occur to the genetic record in culture (Farber and Liskay, 1974; Sasaki and Kodama, 1987; Loo *et al.*, 1989). What role these may play in the heterogeneity observed to date remains to be determined.

The existence of progenitor populations with potential to differentiate along more than one lineage pathway is becoming increasingly evident in the field of osteogenesis, primarily because of the advancement of culture conditions favoring expression of, and/or allowing detection of, diverse phenotypes. The identification of the multipotentiality of RCJ 3.1 (and C26) would not have been possible without growing the cells under the conditions favoring bone formation *in vitro* (ascorbic acid and β-glycerophosphate) and the inclusion of DEX in the culture medium (which had previously been shown to stimulate bone formation [Tenenbaum and Heersche, 1985; Bellows *et al.*, 1987, 1989, 1990]). Kellerman and colleagues prepared an SV-40-immortalized clonal line (C1) from mouse mesodermal cells, which they reported to be osteogenic (Kellerman *et al.*, 1990), but which on analysis under different growth conditions (medium also supplemented with DEX or DEX and insulin) was found to be capable also of making cartilage and fat, respectively (Forest *et al.*, 1992). Owen and colleagues have identified in rabbit bone marrow colonies of cells in which adipocytic or osteogenic (bone–cartilage) expression can be discerned (Bennet *et al.*, 1991). While we have been able to identify single colonies simultaneously containing different, discrete, and recognizable progeny within the same colony in medium with ascorbic acid, β-glycerophosphate, and DEX, osteogenesis in the rabbit marrow colonies was not seen *in vitro* and was detectable only *in vivo* in diffusion chambers; similarly, C1 has not been reported to express more than one phenotype simultaneously under one culture condition. Experiments to distinguish the molecular mechanisms underlying the ability of cells to express multipotentiality, commit to a restricted phenotype, and/or display plasticity are now crucial (Blau and Baltimore, 1991).

B. Osteo-Chondroprogenitors

It is now clear that populations of cells from bone marrow stroma (e.g., Maniatopoulos *et al.*, 1988) and mouse (Ecarot-Charrier *et al.*, 1983, 1988) and rat (Nefussi *et al.*, 1985; Bellows *et al.*, 1986a) calvariae contain committed osteoprogenitors cells that, under appropriate conditions, will differentiate into osteoblasts forming bone (see later). However, other cell types are identifiable within these cultures, and depending on the culture conditions these too can be made to express their differentiated functions. Thus, in mixed fetal rat calvaria populations chondrogenic precursors can be identified that are distinctly different from the osteogenic precursors (Bellows *et al.*, 1989). It is important to note that, in contrast to the multipotential cells already referred to, these progenitors appear monopotential or restricted, because they do not form mixed colonies, and colonies with single, terminally differentiated cell types occur in spatially well-separated parts of culture dishes, often in temporal sequences. None of these observations speak to a detectable common bipotential progenitor for cartilage and bone. Certain observations suggest that a restricted bipotential progenitor for bone- and cartilage-forming cells (*osteo-chondroprogenitor* [Hall, 1970, 1978] or *skeletoblasts* (Stutzmann and Petrovic, 1982]) exists. The fact that cartilage appears in the fracture callus during fracture repair, the observations that tension and pressure applied to the embryonic chick mandible *in vitro* stimulate the periosteum to make cartilage rather than bone (Hall, 1978), and the corticosteroid-induced expression of the cartilage phenotype in organ cultures of chick periosteum (Heersche *et al.*, 1984) all support, but do not prove, this. In addition, cells isolated from embryonic chick limb bud have the capacity to differentiate into either bone or cartilage depending on the density at which the cells are plated (Osdoby and Caplan, 1979). All of these observations are as consistent with the presence of separate committed monopotential progenitors being present in mixed cell populations as a bipotential progenitor for the two tissue types. However, a clonal population isolated from differentiating teratocarcinoma cells *in vitro* was found to give rise to tumors containing both cartilage and bone when injected into syngeneic mice (Nicolas *et al.*, 1980). More recently, Manduca *et al.* (1992) reported a clonal cell line derived from embryonic chick tibial cells that expressed bipotentiality for bone and cartilage. These latter studies suggest that such a common progenitor exists, but proof will await a variety of other manipulations (see earlier). Furthermore, in terms of these two phenotypes, it seems appropriate to consider other possible interpretations, the most notable being transdifferentiation or a direct phenotypic switch from one cell type to another (for general review of the transdifferentiation concept, see Yamada, 1977; Eguchi, 1986; Okada, 1986; Nathanson, 1989; Beresford, 1990; for review and critique of transdifferentiation in bone, see Hall, 1970, 1978; Beresford, 1981). There is ultrastructural evidence indicating that hypertrophic chondrocytes do not die (Hunziker *et al.*, 1987; Yoshioka and Yagi, 1988) and that they may transdifferentiate to osteoblasts in mouse metatarsal bone organ cultures (Thesingh and Scherft, 1986; Thesingh *et al.*, 1991), chick embryonic femura (Roach, 1992), and mandibular condyles (Strauss *et al.*, 1990). The recent observations by Moskalewski and Malejczyk (1989) that interrenal transplantation of isolated chondrocytes results in bone formation also provide evidence for a transdifferentiation in which cells expressing a cartilage phenotype switch to ones expressing bone associated markers. In another recent study, Cancedda *et al.* (1992) showed that hypertrophic chondrocytes from chick tibiae undergo "further differentiation" *in vitro* to express the osteoblast phenotype including the formation of a mineralized matrix. In this regard, it is striking that normal hypertrophic chondrocytes express a variety of markers also expressed by osteoblasts, including BSP, osteopontin, osteonectin, PTH receptor, and APase (see earlier and Cancedda *et al.*, 1992); therefore, whether a hypertrophic chondrocyte "transdifferentiates" to an osteoblast or goes through a further "maturational progression" to an osteoblast, both events leading to formation of a mineralized bonelike matrix, requires further analysis.

IV. OSTEOBLAST HETEROGENEITY: SUBPOPULATIONS, STAGES OF DIFFERENTIATION, OR ABERRANT EXPRESSION *IN VITRO?*

The preceding outlined studies expanded our understanding of the mesenchymal cell hierarchy and provided some insight into recognizable stages in the restricted osteoblast lineage, but they did not help to identify how many steps are in the pathway (see Fig. 2), how cells at each step differ, and what each kind of cell responds to. One major problem continues to be the inability to identify cells at different stages of osteoblast differentiation due to the paucity of markers specific for the different stages, especially earlier stages. As already mentioned, up to now preosteoblasts, osteoblasts, and osteocytes *in situ* have been charac-

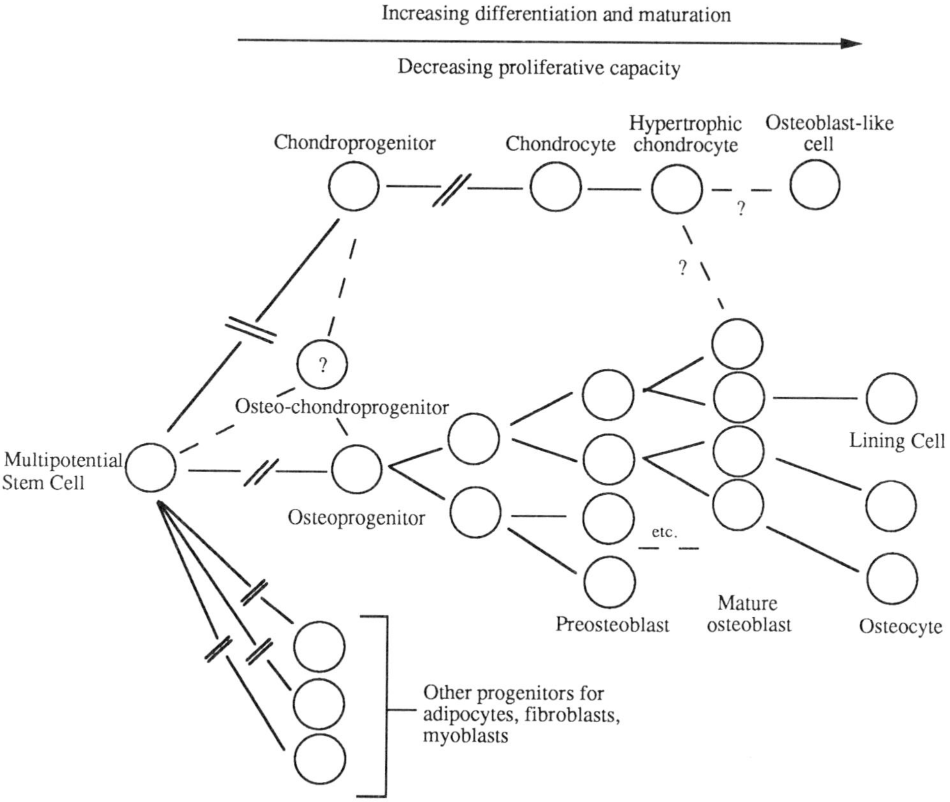

FIGURE 2 Postulated steps in the osteoblast lineage implying recognizable stages of differentiation. Updated from Aubin *et al.* (1990b).

terized predominantly by morphological and histochemical criteria. To identify cells earlier than the preosteoblast in the osteoblast lineage has been difficult, because the majority of evidence suggests that earlier progenitor cells do not differ in morphological characteristics from the fibroblasts of the fibrous layer of periosteum. Biochemical characterization of isolated cell populations—that is, analysis of properties such as responsiveness to hormones (e.g., PTH), high levels of APase, and, more recently, synthesis of certain noncollagenous proteins—have been used to distinguish osteoblasts from other cell types (Aubin *et al.*, 1982; Rodan and Rodan, 1984; Guenther *et al.*, 1988). Other properties, such as synthesis of type I collagen and responsiveness to PGE_2 and EGF, are observed (Cohn and Wong, 1979; Aubin *et al.*, 1982; Sodek *et al.*, 1985; Heersche *et al.*, 1985) but are not unique to cells of the osteoblast lineage, although they may help to identify cells as more or less mature (see also later). When comparisons are made among the cells from several of the most commonly used bone cell systems (e.g., cells from rat or mouse calvaria, cell lines from rat and human osteosarcoma), marked heterogeneity is observed for virtually every characteristic investigated (e.g., hormone response, both kind and magnitude; synthesis of matrix molecules; APase activity) (Aubin *et al.*, 1982; Grigoriadis *et al.*, 1985, 1986; Heersche *et al.*, 1985; Sodek *et al.*, 1985; Bellows *et al.*, 1986b; Aubin *et al.*, 1988; Guenther *et al.*, 1989) (Table I). The origin of this heterogeneity is not known, partly because it can be difficult to correlate certain biochemical or hormonal features with a particular kind of bone cell (i.e., progenitor, preosteoblast, osteoblast or other mesenchymal cell). While such heterogeneity may reflect specialized subpopulations, another possibility is that cells with different responses may represent osteoblastic cells at different stages of differentiation or maturation.

To address this latter possibility, several kinds of studies are of interest. In one approach, less mature more proliferative progenitor cells from periosteal tissue have been compared to those derived from the periosteal free bone (e.g., Canalis, 1980). Sequential digestion of cells from rodent calvaria has yielded populations with lower (populations released earlier) or higher (populations released later) expression of the previously noted osteoblast properties (Rao *et al.*, 1977, Wong, 1980). However, while extremely powerful in advancing the understanding of what properties contribute to the osteoblast phenotype, these populations are all still mixed, comprising cells of multiple lineages, at diverse stages of differentiation or maturation (Nijweide *et al.*, 1986; Aubin *et al.*, 1990; Heersche and Aubin, 1990). Clonal cell lines have been isolated as a further step to overcoming this aspect of the problem. Osteoblastlike osteosarcoma lines from rat (e.g., ROS 17/2.8 [Majeska *et al.*, 1978] UMR 106 [Partridge *et al.*, 1983] and human (e.g., SaOS-2 [Rodan *et al.*, 1987], MG-63 [Franceschi *et al.*, 1988]) and clonal lines derived from normal bone cell populations (e.g., mouse [MC3T3-E1; Kodama *et al.*, 1982], rat calvariae [Aubin *et al.*, 1982; Guenther *et al.*, 1988]) and mouse bone marrow stroma (Benayahu *et al.*, 1991) have helped to solidify the correlation of particular properties with osteoblast and/or other cell types (Rodan and Rodan, 1984; Nijweide *et al.*, 1986; Heersche and Aubin, 1990). Particularly notable in terms of lineage are attempts to correlate progressive changes in particular osteoblast-associated properties with a progression in the state of differentiation of cells. For example, in the osteoblastlike osteosarcoma line ROS 17/2.8, APase activity increases as cells remain in culture for prolonged periods (i.e., enter the post-confluence stage of the cell cycle), and certain hormones (e.g., DEX) facilitate this transition, consistent with the cells acquiring a more differentiated phenotype (Majeska and Rodan, 1985). However, in at least some of the lines derived from osteosarcoma, including ROS 17/2.8, the regulation between proliferation and expression of differentiated phenotype may be aberrant compared to their normal counterparts (Stein *et al.*, 1990). In another recent series of studies, retinoic acid has been shown to induce an increased expression of APase activity in the clonal cell line UMR 201 (Ng *et al.*, 1988) and in the line RCT-1, in which not only APase but also PTH response was increased (Heath *et al.*, 1989). In none of these studies, however, is it possible to order unambiguously the clonal cell lines (i.e., in a unidirectional lineage sequence using basal or induced expression of current marker molecules) or correlate them without exception to data from other sorts of approaches (i.e., ligand binding or immunocytochemistry *in situ*). It is possible that some of the heterogeneity is the result of culture artifact (i.e., the chromosomal alterations as already outlined or aberrant expression *in vitro* due to continuous stimulation by the presence of serum). On the other hand, true functional heterogeneity representative of a comparable heterogeneity *in vivo* cannot be excluded.

These kinds of outstanding problems have necessitated consideration of still other approaches, two of which—the use of novel new assays and the identification of new marker molecules for the osteoblast lineage—will be detailed below.

V. INDIRECT IDENTIFICATION OF THE OSTEOPROGENITOR CELL

A. Bone Formation *in Vitro*

Given that direct identification of certain kinds or stages of osteoprogenitor cells has been difficult, indirect functional assays may be applied to identify an osteoprogenitor based on its capacity to produce recognizable differentiated progeny cells (i.e., osteoblasts) without requiring isolation of the progenitor or even knowing any of its precise biochemical features. Such an approach has been employed to recognize various hemopoietic progenitors (for review, see Metcalf, 1988, 1989), the progenitors for cells of nerve (Raff, 1989; Raff and Lillien, 1988) and fibroblast (Bayreuther *et al.*, 1988), and those for muscle, fat, and cartilage (Jones *et al.*, 1983; Harrington and Jones, 1988) and muscle, fat, cartilage, and bone (Grigoriadis *et al.*, 1988; see also earlier). This kind of approach to study the committed osteoprogenitor has been developed and characterized in detail over several years. After Tenenbaum and Heersche (1982) demonstrated that folded chick periostea formed mineralized bone *in vitro* more reproducibly in the presence of ascorbic acid and β-glycerophosphate, Ecarot-Charrier *et al.* (1983) showed that nonenzymatically isolated murine calvaria cells cultured with the same two additives produced a bonelike material *in vitro*. These two additives are now being widely used in a variety of isolated bone cell systems, with

TABLE I Comparisons among Cells from Several of the Most Commonly Used Bone Cell Systems

Osteoblast characteristic	Cell line[a]						
	RCJ 1.20	RCJ 3.1	ROS 17/2.8	UMR 106.06	MC3T3-E1	SaOS-2	MG-63
Osteogenesis *in vivo*	−[1]	ND	+[10]	+[20]	ND	+[34]	ND
Mineralized bone formation *in vitro*	−[1]	+[7,8]	−	−	+[27]	ND	−[41][b]
Cuboidal morphology *in vitro*	+[1]	+[8]	+[10]	+[20]	+[27]	−[34]	−[41]
Alkaline phosphatase activity	+[1]	+[8]	+[11]	+[21]	+[30]	+[36]	−[43]
Phosphatidylinositol-phospholipase C releasable alkaline phosphatase	ND	ND	+[19]	+[26]	ND	+[40]	−
Parathyroid hormone-stimulated cyclic adenosine monophosphate	+[1]	+[7]	+[10]	+[21]	+[28]	+[34]	−[36]
Parathyroid hormone-related protein stimulated cyclic adenosine monophosphate	ND	ND	+[18]	+[25]	+[33]	ND	ND
Parathyroid hormone receptor	ND	ND	+[17]	+[24]	+[27]	+[36]	−[36]
ProstaglandinE_2-stimulated cyclic adenosine monophosphate	+[1]	+[7]	−[1]	+[21]	+[29]	+[35]	+
Type I collagen synthesis	+[1]	+[8]	+[1]	+[20]	+[30]	ND	+[42]
Type III collagen synthesis	+[1]	+[8]	+[1]	−[47]	+		+
Osteocalcin (bone gla protein) synthesis	−[2]	ND	+[12]	−[22]	+[31]	−[37]	−[44]
Osteonectin synthesis	+[3]	ND	+[13]	ND	ND	+[38]	ND

Osteopontin synthesis	ND	ND	+	ND	+[32]	ND	ND
Glucocorticoid response/receptor	+[4]	+[8]	+[14]	+[22]	ND	+[39]	+[42]
1,25(OH)2D3 response/receptor	+[5]	+[9]	+[15]	+[21]	+[29]	+[34]	+[45]
Epidermal growth factor receptor	+[6]	ND	−[16]	+[23]	+[29]	+[46]	+[46]

[a]Original papers that reported the isolation of these cell lines are the following. RCJ 1.20: Aubin *et al.*, *J. Cell Biol.* **92;** 452, 1982; RCJ 3.1: Aubin *et al.*, *J. Cell Biol.* **92,** 452, 1982; ROS 17/2: Majeska *et al.*, *Endocrinology* **107,** 1494, 1980; UMR 106: Partridge *et al.*, *FEBS Lett.* **115,** 139, 1980; MC3T3-E1 Kodama *et al.*, *Jpn J. Oral Biol.* **23,** 899, 1982; SaOS-2: Ponten and Saksela, *Int. J. Cancer* **2** 434, 1967; MG-63; Billiau *et al.*, *Antimicrob. Agents Chemother.* **12,** 11, 1977.

1. Aubin *et al.* (1982); 2. Price, Aubin, and Heersche, unpublished observation; 3. Domenicucci and Sodek, personal communications; 4. Millar *et al.* (1982); 5. Petkovitch *et al.* (1983); 6. Petkovich *et al.* (1987); 7. Aubin *et al.* (1988); 8. Grigoriadis *et al.* (1988); 9. Grigoriadis *et al.* (1986); 10. Majeska *et al.* (1980); 11. Majeska and Rodan, (1982); 12. Nishimoto and Price, (1980); 13. Otsuka and Sodek, unpublished observations; 14. Haussler *et al.* (1980); 15. Monolagas *et al.* (1980); 16. Imai *et al.* (1988); 17. Newman *et al.* (1989); 18. Donahue *et al.* (1990); 19. Noda (1987); 20. Martin *et al.* (1976); 21. Patridge *et al.* (1980); 22. Martin, unpublished observation; *see also* Bernier *et al.* (1991) Fraser *et al.* (1988). *J. Bio Chem,* **263,** 911. 23. Ng *et al.* (1983); 24. Mitchell *et al.* (1990); 25. Civitelli *et al.* (1985); 26. Turksen and Aubin, (1991); 27. Suda *et al.* (1983); 28. Harrison *et al.* (1990); 29. Kumegawa *et al.* (1984); 30. Kodama *et al.* (1982); 31. Casser-Bette *et al.* (1990); 32. Noda *et al.* (1990); 33. Yamada *et al.* (1989); 34. Rodan *et al.* (1987); 35. Shapiro *et al.* (1990); 36. Fukayama and Tashjian, (1941, 1990); 37. Mohonen *et al.* (1990); 38. Howe *et al.* (1990); 39. Sutherland *et al.* (1990); 40. Fedde *et al.* (1990); 41. Dedhar *et al.* (1987); 42. Lajeunesse *et al.* (1990); 43. Franceschi *et al.* (1985); 44. Francceschi *et al.* (1988); 45. Valaja *et al.* (1990); 46. Shupnik and Tashjian, (1982); 47. Blair *et al.* (1986).

[b]However, an Arn–Gly–Asp-resistant variant (PRV) did form mineralized matrix *in vitro.*

ND = not determined.

cells from animals of diverse ages from embryonic to adult and from numerous species (chicken, rat, bovine, human) to elicit deposition of bonelike material (e.g., human [Robey and Termine, 1985], chick [Nijweide *et al.*, 1982; Gerstenfeld *et al.*, 1988], mouse [Ecarot-Charrier *et al.*, 1983], mouse teratocarcinoma [Kellermann *et al.*, 1990]). When cells enzymatically isolated from fetal rat calvaria are grown in media supplemented with the same additives, discrete three-dimensional nodular structures are formed (Nefussi *et al.*, 1985; Bellows *et al.*, 1986; Zimmermann *et al.*, 1988; Aronow *et al.*, 1990; Owen *et al.*, 1990). The nodules resemble woven bone histologically and are covered by an APase-positive cuboidal layer of cells, and immunolabeling has documented the presence of bone matrix proteins, such as type I collagen and osteonectin (Bellows *et al.*, 1986). By transmission electron microscopy, it has been verified that the overall morphology of the nodule, the cell ultrastructure analogous to that of the osteoblast and osteocyte, the relationship of the cells to the highly organized, dense collagenous matrix, and matrix-mineral characteristics are all reminiscent of a true osseous structure. Electron probe and electron and X-ray diffraction analysis verified the mineral to be hydroxyapaptite. Mineral is deposited only in the nodular areas of the culture and not in the fibroblast areas (Bhargava *et al.*, 1988; Ecarot-Charrier *et al.*, 1988).

B. A Colony Assay for the Quantification of Osteoprogenitor Cells and Assessment of Their Proliferative and Self-Renewal Capacity

The number of bone nodules formed in isolated rat calvaria cell populations bears a linear relationship to the total number of cells plated over a very broad range of plating densities (Bellows *et al.*, 1986a, 1987). However, a more detailed analysis by limiting dilution indicated a "single-hit" phenomenon (i.e., only one cell type was limiting for nodule formation, a cell present at a low frequency under standard culture conditions [Bellows and Aubin, 1989]). The data are consistent with the limiting cell type being the osteoprogenitor itself, with one osteoprogenitor giving rise to one bone nodule under normal plating conditions, and with such cells being present at a measurable but low frequency (i.e., $<1\%$) of the total population under standard isolation and culture conditions (Bellows and Aubin, 1989). Thus, the numbers of nodules or colonies capable of forming bone can be counted unambiguously for an assessment of osteoprogenitor numbers recoverable from calvarial or bone marrow (Aubin *et al.*, 1990a,b; McCulloch *et al.*, 1991; for reviews, see Friedenstein, 1990; Aubin *et al.*, 1992). Certain hormones and growth factors influence the number of osteoprogenitors that divide and differentiate to express formation of bone nodules, a phenomenon addressed later.

By measuring the average size and cellularity of nodules, it has been estimated that an osteoprogenitor goes through 6–7 doublings *in vitro* to yield functional bone-forming osteoblasts. However, the osteoprogenitors appear to have a limited capacity for self-renewal and survive in the populations up to population doublings of approximately 10. Dexamethasone, however, is one agent that can extend this lifetime up to 16–18 population doublings and increase the proliferative capacity of at least some of the osteoprogenitors (see also later). Also of note is the observation that osteoprogenitor cells appear to begin to cycle later *in vitro* than the rest of the population (Bellows *et al.*, 1990). This is in agreement with other studies that suggest that the CFU-F giving rise to bone in diffusion chambers are also quiescent when first placed in culture, not entering S phase until 28–60 hr after being placed in culture (Keilis-Borok *et al.*, 1971). Such behavior supports the hypothesis that the earliest cells in a lineage are relatively quiescent and have a long G0 time, a low turnover rate, and a low probability of triggering into cycle (i.e., are protected in an environmental niche by the tissue structure) (Potten *et al.*, 1979). These studies also point to the interest in detecting normal osteoprogenitors in primary cultures. Similar analyses are difficult, if not impossible, in rodent lines even when of clonal origin, because many of these spontaneously immortalize on long-term growth up to large cell numbers, nor are they achievable in lines obtained by viral transformation. In both of these cases, proliferative lifetime and self-renewal capacity cannot unambiguously be assessed compared to normal cell counterparts, nor is proliferation–differentiation apparently so tightly coupled. Thus, it seems appropriate for analyses of both kinds of cell systems to proceed.

C. Regulation of the Osteoprogenitors

Morphologically recognizable osteoblasts and the bone nodules appear in the long-term rat calvaria cell cultures at predictable and reproducible periods after plating. Thus, the cultures go through a period of log-phase growth, and only 2–3 days subsequent to confluence do cuboidal osteoblasts appear (Nefussi *et al.*, 1985; Bellows *et al.*, 1986a). In a further more detailed analyses of the relationships between proliferation and expression of differentiated expression, Stein, Lian, and their colleagues have defined a differentiation sequence in which temporally ordered periods are detectable, which they have called growth (proliferation) and extracellular matrix biosynthesis; extracellular matrix development, maturation, and organization; and extracellular matrix mineralization (for a recent review, see Lian and Stein, 1992). Agents affecting nodule numbers might elicit their effects at different times of the culture period, possibly representing their abilities to influence cells at different stages of this proliferation–differentiation sequence. That this does appear to be the case is evident from the findings that among a variety of agents tested many have complex biphasic effects on bone nodule numbers and expression of osteoblast-associated markers (for summaries, see Aubin *et al.*, 1992; Lian and Stein, 1992). Such complex, biphasic effects on osteoprogenitor versus osteoblast expression are in agreement with a variety of other approaches including primary cell, tissue, and organ cultures and a variety of established, transformed, and/or tumorigenic cells lines (for reviews, see Rodan and Rodan, 1984; Heersche and Aubin, 1990).

One class of hormones of particular interest with respect to bone metabolism is glucocorticoids. While glucocorticoid excess *in vivo* is generally associated with net bone loss due to a decrease in bone formation and an increase in bone resorption, *in vitro* glucocorticoids at near physiological concentrations generally stimulate parameters associated with bone formation (for discussion, see Bellows *et al.*, 1987; Heersche and Aubin, 1990). Differences in effects on proliferation and differentiation of certain osteoprogenitor cells have been observed depending on the culture systems used. In organ culture systems, glucocorticoids inhibit periosteal or progenitor cell proliferation (Chyun *et al.*, 1984) and/or stimulate proliferation of a population of cells that subsequently differentiate into osteoblasts but inhibit proliferation of possibly less differentiated progenitor cells (Tenenbaum and Heersche, 1985; McCulloch and Tenenbaum, 1986). In nodule-forming primary cultures, however, both mature and immature osteoprogenitor cells appear to be stimulated to proliferate and differentiate by DEX: The number and size of bone nodules formed increases, and the self-renewal capacity of the bone-nodule forming cell increases (Bellows *et al.*, 1987, 1990). These mass population studies and limiting dilution analyses (Bellows and Aubin, 1989) suggested that a subpopulation of cells depends on DEX or natural glucocorticoids for expression of bone nodule formation. The additional nodules formed in the presence of DEX could result from increasing the proliferative capacity of more mature osteoprogenitors enabling these cells to achieve sufficient numbers of cell divisions to form a visible nodule. Alternatively, DEX could be acting on a separate subpopulation of osteoprogenitors requiring glucocorticoids to proliferate and/or differentiate along an osteogenic pathway and all of the data are consistent with both views and suggested that both occurred. Immunoselection with antibodies against rat APase demonstrated that the osteoprogenitors present in rat calvaria cultures could be separated into two "classes": one capable of expressing bone formation without exogenous DEX stimulation (APase-positive), and one expressing bone formation only in its presence (APase-negative) (Turksen and Aubin, 1991). Thus, the data are consistent with the hypothesis that these two populations of progenitor cells represent different stages of differentiation of cells in the osteoblast lineage, with the latter representing a less mature cell than the former and requiring different regulatory signals for its expression than the latter (Turksen and Aubin, 1991). This possibility is of particular interest, and the challenge is to identify these immature cells in mixed populations and intact tissues and to isolate them for further analysis. How these immature APase-negative osteoprogenitors in primary culture relate to those defined in clonal, established, or immortalized cell lines that also start out APase-negative but are stimulated to differentiate (but not terminally differentiate) *in vitro* by retinoic acid (Ng *et al.*, 1988; Heath *et al.*, 1989) is currently not clear. Nevertheless, these kinds of studies suggest that it will be of interest and importance to attempt to determine what kinds of markers these immature osteoprogenitors do express so that other approaches can be used for their identification and isolation.

On the basis of the immunoselection and functional assay for bone formation described in this section, and the detection of the known osteoblastic markers by a variety of approaches, as summarized earlier in this chapter, we propose in Figure 3 a lineage scheme in which less mature, APase-negative, glucocorticoid-requiring osteoprogenitors are separate from a more mature APase-positive osteoprogenitor, and we have tentatively placed these cells along a lineage path encompassing cells expressing other osteoblastic markers. Clearly, much remains to be learned about the molecular and biochemical features of cells earlier than the preosteoblast.

VI. MONOCLONAL ANTIBODIES FOR IDENTIFICATION OF CELLS IN THE OSTEOBLAST LINEAGE

The preparation of monoclonal antibodies recognizing either specific cell surface or cytoplasmic antigens or secreted matrix molecules and their use in both immunohistochemical and biochemical studies would seem to be a promising approach to help resolve some of the problems already outlined.

The generally accepted concept is that differentiating cells will be found to possess cell surface antigens unique to different stages of maturation (e.g., Metcalf, 1989). For example, cell surface composition clearly differs between cells of differing functions (e.g., lymphocytes vs. neurons), between the same cell at distinct stages of its maturity (e.g., an erythroblast vs. an erythrocyte), and between different surface regions of a single specialized cell (e.g., on the neuron, the synaptic region vs. cell body; the apical vs. basal side of epithelial cells). The molecular complexity of the cell surface as displayed on the surface of differentiating cells and on tumor cells (Nicolson, 1984) presumably reflects the complex role of the cell surface in cellular behavior (e.g., receptors) and

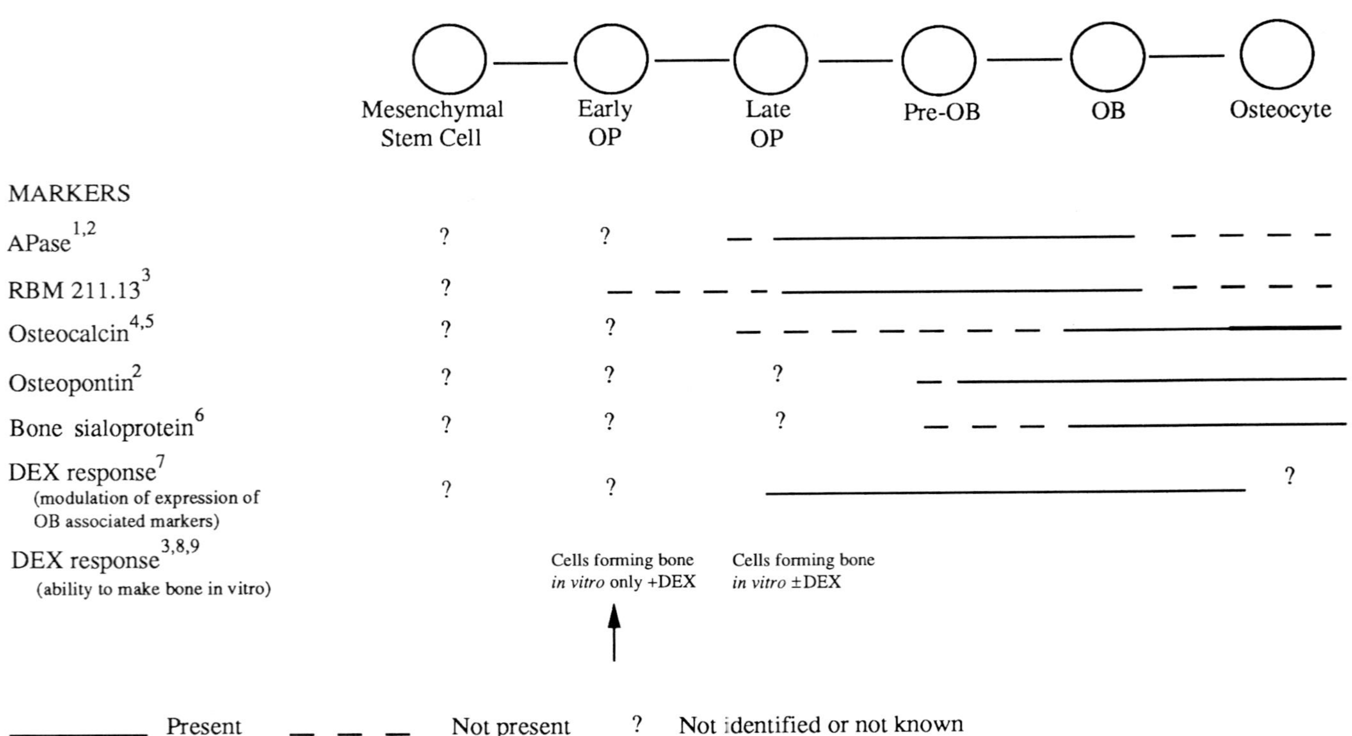

Mesenchymal Stem Cell
Early OP
Late OP
Pre-OB
OB
Osteocyte
MARKERS
APase[1,2]
RBM 211.13[3]
Osteocalcin[4,5]
Osteopontin[2]
Bone sialoprotein[6]
DEX response[7]
(modulation of expression of OB associated markers)
DEX response[3,8,9]
(ability to make bone in vitro)
Cells forming bone *in vitro* only +DEX
Cells forming bone *in vitro* ±DEX
Present
Not present
? Not identified or not known

nutrition (transport proteins). In the past, a few naturally occurring probes such as toxins (e.g., α-bungarotoxin, tetanus toxin; Bayne *et al.*, 1981; Raff *et al.*, 1979) and lectins (Lis and Sharon, 1986; Sharon, 1984) have been used to study cell surface components. However, such specific but naturally occurring probes appear to be rare. In addition, specialized cells may also express specific cytoplasmic molecules. With the development of hybridoma technology (Kohler and Milstein, 1975), reagents specific for different cell types have been generated through immunization (for review, see Milstein and Lennox, 1980; Fujita, 1987; for examples and discussions, see McKay *et al.*, 1981; Eisenbarth, 1981; Valentino *et al.*, 1985; Nicholas, 1986). For example, monoclonal antibodies have been used to study the hemopoietic (Nicola, 1982; Sprangrude *et al.*, 1988), myogenic and chondrogenic lineages (e.g., Sasse *et al.*, 1984). In neurobiology, monoclonal antibodies have been used to demonstrate the antigenic diversity of neuronal subpopulations in the leech (Zipser and McKay, 1981) and glial subclasses in rat optic nerve (Raff, 1989) and to reveal the differences in the cell surface characteristics of *Drosophila* and rat retinal cells (Zipursky *et al.*, 1984; Barnstable, 1987).

From these studies and others, the most striking impression is that different types of cells have a quite different surface composition, and there is every reason to believe that bone would be similar, because, as already outlined, both morphological and biochemical properties indicate heterogeneity among osteoblastic cells. While antibodies against tumor antigens expressed on osteogenic sarcoma cells were first described several years ago, none of these apparently recognized antigens expressed on the normal osteoblast counterparts (e.g., see Embleton, 1984; Nakamura *et al.*, 1987; Hosoi *et al.*, 1982; Heiner *et al.*, 1987; Bruland *et al.*, 1988; Tsai *et al.*, 1990). However, the possibility that cells at different stages of the osteoblast lineage might express antigenic determinants unique or distinctive to these stages has recently been made

FIGURE 3 A proposed lineage diagram with differentiation steps for cells of the osteoblast lineage. Immunoselection with antibodies against alkaline phosphatase (APase) allowed discrimination of an APase-positive osteoprogenitor (OP) making bone *in vitro* without dexamethasone (DEX) (late OP) from an earlier APase-negative OP (early OP), which *in vitro* proliferates and/or differentiates only in the presence of DEX. We have superimposed on this scheme where we think these types of OP cells reside in comparison to cells expressing other osteoblast (OB) markers. For conciseness and ease of comparison, this figure is based primarily on immunocytochemical evidence for the expression of OB-associated markers. However, other biochemical approaches support the data referenced (for reviews, see Rodan and Rodan, 1984; Nijweide *et al.*, 1986; Martin *et al.*, 1987; Butler, 1989; Section II.B). 1. Doty and Schofield, 1976; 2. Mark *et al.*, 1987a; 3. Turksen and Aubin, 1991; 4. Bronckers *et al.*, 1985; 5. Mark *et al.*, 1987b; 6. Chen *et al.*, 1991b; 7. Rodan and Rodan, 1984; Martin *et al.*, 1987; 8. Tennenbaum and Heersche, 1985; 9. Bellows *et al.*, 1987; Bellows and Aubin, 1989. ———, present; — — —, not present; ?, not identified or not known.

compelling by several reports. Using a chick osteoblastic cell model, Nijweide and Mulder (1986) reported a monoclonal antibody reacting only with osteocytes, and Bruder and Caplan (1989, 1990a,b) recently have reported a series of antibodies also apparently labeling different subpopulations or differentiation–maturational stages of chick osteoblastic cells. Recently, monoclonal antibodies raised against rat osteoblastic populations from calvaria or bone marrow stroma expressed osteoblast-associated staining both *in vivo* and *in vitro* and limited ability to label other tissues (Turksen *et al.*, 1992).

A major challenge in the field of osteoblast differentiation is to recognize unambiguously cells at different stages of differentiation or maturation. Of the antibodies raised against injected cells and reported to date, those against APase appear quite commonly (Bruder and Caplan, 1990a,b; Turksen and Aubin, 1991; Nakamura *et al.*, 1987). Among the others reported, however, several different patterns of reactivity and specificity are evident. Individual cells within osteoblastic clonal cell lines stain with different intensities, and staining is completely absent in other well-characterized lines by these different antibodies; comparing these patterns to staining patterns of bone tissue should help in discriminating stages in this lineage. Furthermore, several of the cell and tissue distributions detectable are unique compared to all other known osteoblast markers outlined earlier. Both cell surface and cytoplasmic staining has been seen. Although such data on expression of cell surface or cytoplasmic antigens do not in themselves constitute proof of a cell lineage relationship, it would appear that the system will be a useful one to investigate further aspects of osteoblast differentiation (Turksen *et al.*, 1992). Interestingly, however, the majority reported up to now recognize the mature osteoblasts, osteocytes, and some both osteoblasts and chondrocytes. Whether the earlier progenitors in the osteoblast lineage will be detectable by these approaches remains to be seen.

VII. CONCLUDING REMARKS

Throughout this chapter, data are discussed using osteoblastic cells, cell lines, and clonally derived populations, many of which derived from different tissues or organs and different anatomical sites. Many of these express properties of progenitor populations, and some of these have been made to progress through some differentiation or maturation steps *in vitro,* although an unambiguous and detailed lineage sequence awaits further analysis. In this regard, while it is true that many of the commonly used markers of the late-stage cells of the osteoblast lineage reside in osteogenic cells irrespective of their locations, a growing body of data suggests that differences will be found among the differentiated phenotypes and maturational states achieved by these different cells.

Given the paucity of definitive information, we have not addressed in detail here the germ layer of origin (i.e., ectodermal vs. mesodermal), in discussion of the cells, clones, and tissues already described, nor dwelled on inductive versus conductive mechanisms of bone formation (for a recent review, see Hall, 1988). Nevertheless, these issues will become increasingly important as the lineage maps derived for osteoblastic cells become more detailed.

REFERENCES

Abou-Samra, A.-B., Juppner, H., Force, T., Freeman, M. W., Kong, X.-F., Schipani, E., Urena, P., Richards, J., Bonventre, J. V., Potts, J. T., Kronenberg, H. M., and Segre, G. V. (1992). Expression cloning of a common receptor for parathyroid hormone-related peptide from rat osteoblast-like cells: A single receptor stimulates intracellular accumulation of both cAMP and inositol triphosphates and increases intracellular free calcium. *Proc. Natl. Acad. Sci. USA* **89,** 2732–2736.

Addadi, L., Moradian, J., Shay, E., Maroudas, N. G., and Weiner, S. (1987). A chemical model for the cooperation of sulfates and carboxylates in calcite crystal nucleation: Relevance to biomineralization. *Proc. Natl. Acad. Sci. USA* **84,** 2732–2736.

Antosz, M. E., Bellows, C. G., and Aubin, J. E. (1987). Biphasic effects of epidermal growth factor on bone nodule formation by isolated rat calvaria cells *in vitro. J. Bone Miner. Res.* **2,** 385–393.

Antosz, M. E., Bellows, C. G., and Aubin, J. E. (1989). Effects of transforming growth factor beta and epidermal growth factor on cell proliferation and the formation of bone nodules in isolated fetal rat calvaria cells. *J. Cell. Physiol.* **140,** 386–395.

Aronow, M. A., Gerstenfeld, L. C., Owen, T. A., Tassinari, M. S., Stein, G. S., and Lian, J. B. (1990). Factors that promote progressive development of the osteoblast phenotype in cultured fetal rat calvaria cells. *J. Cell Physiol.* **143,** 213–221.

Ashton, B. A., Allen, T. D., Howlett, C. R., Eaglesom, C. C., Hattori, A., and Owen, M. (1980). Formation of bone and cartilage by marrow stromal cells in diffusion chambers in vivo. *Clin. Orthop. Relat. Res.* **151,** 294–307.

Aubin, J. E., Heersche, J. N. M., Merrilees, M. J., and Sodek, J. (1982). Isolation of bone cell clones with differences in growth hormone responses and extracellular matrix production. *J. Cell Biol.* **92,** 452–462.

Aubin, J. E., Tertinegg, I., Ber, R., and Heersche, J. N. M. (1988). Consistent patterns of changing hormone responsiveness during continuous culture of cloned rat calvaria cells. *J. Bone Miner. Res.* **3,** 333–339.

Aubin, J. E., Fung, S.-W., and Georgis, W. (1990a). The influence of non-ostegenic hemopoietic cells on bone formation by bone marrow stromal populations. *J. Bone Miner. Res.* **5,** (suppl. 2), S81.

Aubin, J. E., Heersche, J. N. M., Bellows, C. G., and Grigoriadis, A. E. (1990b). Osteoblast lineage analysis in fetal rat calvaria cells. *In* "Calcium Regulation and Bone Metabolism" (D. V. Cohn, F. H. Glorieux, and T. J. Martin, eds.), pp. 362–370. Elsevier, Amsterdam.

Aubin, J. E., Bellows, C. G., Turksen, K., Liu, F., and Heersche, J. N. M. (1992). Analysis of the osteoblast lineage and regulation of differentiation. *In* "Chemistry and Biology of Mineralized Tissues" (H. Slavkin and P. Price, eds.), pp. 267–276. Excerpta Medica, Amsterdam.

Barnstable, C. (1987). Immunological studies of the diversity and development of the mammalian visual system. *Immunological Rev.* **100,** 47–78.

Bayne, E. K., Gardner, J., and Fambrough, D. M. (1981). Monoclonal antibodies to extracellular matrix antigens in chicken skeletal muscle. *In* "Monoclonal Antibodies to Neural Antigens" (R. McKay *et al.*, eds.), pp. 259–270. Cold Spring Harbor Laboratories, Cold Spring, New York.

Bayreuther, K., Rodemann, H. P., Francz, P. I., and Maier, K. (1988). Differentiation of fibroblast stem cells. *J. Cell Sci.* **10,** 115–130.

Bellows, C. G., and Aubin, J. E. (1989). Determination of numbers of osteoprogenitors present in isolated fetal rat calvaria cells in vitro. *Dev. Biol.* **133,** 8–13.

Bellows, C. G., Aubin, J. E., Heersche, J. N. M., and Antosz, M. E. (1986a). Mineralized bone nodules formed in vitro from enzymatically released rat calvaria cell populations. *Calcif. Tissue Int.* **36,** 143–154.

Bellows, C. G., Sodek, J., Yao, K.-L., and Aubin, J. E. (1986b). Phenotypic differences in subclones and long term cultures of clonally derived rat bone cell lines. *J. Cell. Biochem.* **31,** 153–169.

Bellows, C. G., Aubin, J. E., and Heersche, J. N. M. (1987). Physiological concentrations of glucocorticoids stimulate formation of bone nodules from isolated rat calvaria cells in vitro. *Endocrinology* **121,** 1985–1992.

Bellows, C. G., Heersche, J. N. M., and Aubin, J. E. (1989). Effects of dexamethasone on expression and maintenance of carrtilage in serum-containing cultures of calvaria cells. *Cell Tissue Res.* **256,** 145–151.

Bellows, C. G., Heersche, J. N. M., and Aubin, J. E. (1990). Determination of the capacity for proliferation and differentiation of osteoprogenitor cells in the presence and absence of dexamethasone. *Dev. Biol.* **140,** 132–138.

Benayahu, D., Fried, A., Zipori, D., and Weintroub, S. (1991). Subpopulations of marrow stromal cells share a variety of osteoblastic markers. *Calcif. Tissue Int.* **49,** 202–207.

Bennett, D. C. (1983). Differentiation in mouse melanoma cells initial reversibility and an on-off stochastic model. *Cell* **34,** 445–453.

Bennett, J. H., Joyner, C. J., Triffit, J. T. and Owen, M. E. (1991). Adipocyte cells cultured from marrow have osteogenic potential. *J. Cell Sci.* **99,** 131–139.

Beresford, J. N. (1989). Osteogenic stem cells and the stromal system of bone and marrow. *Clin. Orthop. Relat. Res.* **240,** 270–280.

Beresford, W. A. (1981). "Chondroid Bone Secondary Cartilage and Metaplasia." Urban and Schwarzenberg, Baltimore, Maryland.

Beresford, W. A. (1990). Direct transdifferentiation: Can cells change their phenotype without dividing? *Cell Differ. Dev.* **29,** 81–93.

Bernier, S. M., Rouleau M. F., and Goltzman, D. (1991). Biochemical and morphological analysis of the interaction of epidermal growth factor and parathyroid hormone with UMR 106 osteosarcoma cells. *Endrocrinology* **128,** 2752–2760.

Bhargava, U., Bar-Lev, M., Bellows, C. G., and Aubin, J. E. (1988). Ultrastructured analysis of bone nodules formed in vitro by isolated fetal rat calvaria cells. *Bone* **9,** 155–163.

Blair, H. C., Kahn, A. J., Crouch, E. C., Jeffrey, J. J., and Teitelbaum, S. L. (1986). Isolated osteoclasts resorb the organic and inorganic components of bone. *J. Cell Biol.* **102,** 1164–1172.

Blau, H. M., and Baltimore, D. (1991). Differentiation requires continuous regulation. *J. Cell Biol.* **112,** 781–783.

Bonucci, E. (1990). The histology, histochemistry and ultrastructure of bone. *In* "Bone Regulatory Factors" (A. Pecile and B. de Bernard, eds.), pp. 15–39. Plenum Press, New York.

Boskey, A. L. (1989). Noncollagenous matrix proteins and their role in mineralization. *Bone Mineral* **6,** 111–123.

Bourdon, M. A., and Rouslahti, E. (1989). Tenascin mediates cell attachment through an RGD dependent receptor. *J. Cell Biol.* **108,** 1149–1155.

Bronckers, A. L. J. J., Gay, S., Dimuzio, M. T., and Butler, W. T. (1985). Immunolocalization of γ-carboxyglutamic acid-containing proteins in developing rat bones. *Collagen Rel. Res.* **5,** 273–281.

Bronckers, A. L. J. J., Gay, S., Frinkelman, R. D., and Butler, W. T. (1987). Developmental appearance of Gla proteins (osteocalcin) and alkaline phosphatase in tooth germs and bones of the rat. *Bone Mineral* **2,** 361–373.

Bruder, S. P., and Caplan, A. I. (1989). First bone formation and the dissection of an osteogenic lineage in the embryonic chick tibia is revealed by monoclonal antibodies against osteoblast. *Bone* **10,** 359–375.

Bruder, S. P., and Caplan, A. I. (1990a). A monoclonal antibody against the surface of osteoblasts recognizes alkaline phosphatase isoenzyme in bone liver kidney and intestine. *Bone* **11,** 133–139.

Bruder, S. P., and Caplan, A. I. (1990b). Osteogenic cell lineage analysis is facilitated by organ cultures of embryonic chick periosteum. *Dev. Biol.* **141,** 319–329.

Bruland, O. S., Fodstad, O., Stanwig, A. E., and Phil, A. (1988). Expression and characterization of a novel human osteosarcoma-associated cell surface antigen. *Cancer Res.* **48,** 5302–5308.

Butler, W. T. (1989). The nature and significance of osteopontin. *Connec. Tissue Res.* **23,** 123–136.

Cameron, D. A. (1968). The Golgi apparatus in bone and cartilage cells. *Clin. Orthop.* **58,** 191–211.

Canalis, E. (1980). Effect of insulin-like growth factor I on DNA and protein synthesis in cultured rat calvaria. *J. Clin. Invest.* **66,** 709–713.

Canalis, E., McCarthy, T. L., and Centrella, M. (1989a). The regulation of bone formation by local growth factors. *Bone Miner. Res.* **6,** 27–56.

Canalis, E., McCarthy, T. L., and Centrella, M. (1989b). The role of growth factors in skeletal remodeling. *Endo. Metabol. Clinics North Am.* **18,** 903–918.

Cancedda, F. D., Gentili, C., Manduca, P., and Cancedda, R. (1992). Hypertrophic chondrocytes undergo further differentiation in culture. *J. Cell Biol.* **117,** 427–435.

Casser-Bette, M., Murray, A. B., Closs, E. I., Erfle, V., and Schmidt, J. (1990). Bone formation by osteoblast-like cells in a three-dimensional cell culture. *Calc. Tiss. Internat.* **46,** 46–56.

Chen, J.-K., Shapiro, H. S., Wrana, J. L., Reimers, S., Heersche, J. N. M., and Sodek, J. (1991a). Localization of bone sialoprotein (BSP) expression to the site of mineralized tissue formation in fetal rat tissue by in situ hybridization. *Matrix* **11,** 133–143.

Chen, J.-K., Zhang, Q., McCulloch, C. A. G., and Sodek, J. (1991b). Immunohistochemical localization of bone sialoprotein (BSP) in fetal porcine bone tissues: Comparisons with secreted phosphoprotein 1 (SPP1, osteopontin) and SPARC (osteonectin). *Histochem. J.* **23,** 281–289.

Chiquet, M., and Fambraught, D. M. (1984). Chick myotendinous antigen. I. A monoclonal antibody as a marker for tendon and muscle morphogenesis. *J. Cell Biol.* **98,** 1226–1936.

Chiquet-Ehrismann, R. (1990). What distinguishes tenascin from fibronectin?. *FASEB J.* **4,** 2598–2604.

Chyun, Y. S., Kream, B. E., and Raisz, L. G. (1984). Cortisol decreases bone formation by inhibiting periosteal cell proliferation. *Endocrinology* **114,** 477–480.

Cohn, D. V., and Wong, G. L. (1979). Isolated bone cells. *In* "Skeletal Research: An Experimental Approach" (D. J., Simmons and A. S. Kunin, eds.), pp. 3–20. Academic Press, New York.

Constantinides, P. G., Jones, P. A., and Gevers, W. (1977). Functional striated muscle cells from non-myoblast precursors following 5-azacytidine treatment. *Nature (London)* **267,** 364–366.

Craig, A. F., Nemir, M., Mukherjee, B. B., Chambers, A. F., and Denhardt, D. T. (1988). Identification of the major phosphoprotein secreted by many rodent cell lines as 2AR/osteopontin: Enhanced expression in H-RAS transformed 3T3 cells. *Biochem. Biophys. Res. Commun.* **157,** 166–173.

Craig, A. M., Nemir, M., Mukherjee, B. B., Chambers, A. F., and Denhardt, D. T. (1989). Osteopontin, a transformation-associated cell adhesion phosphoprotein, is induced by 12-O-tetradecanoylphorbol 13-acetate in mouse epidermis. *J. Biol. Chem.* **264,** 9682–9689.

Cross, G. A. M. (1987). Eukaryotic protein modification and membrane attachment via phosphatidylinositol. *Cell* **48,** 179–181.

Curtis-Prior, P. B. (eds.). (1988). "Prostaglandins: Biology and Chemistry of Prostaglandins and Related Eicosanoids." Churchill Livingstone, Edinburgh.

Dedhar, S., Argraves, W. S., Suzuki, S., Ruoslahti, E., and Pierschbacher, M. D. (1987). Human osteosarcoma cells resistant to detachment by an Arg–Gly–Asp-containing peptide overproduce the fibronectin receptor. *J. Cell Biol.* **105,** 1175–1182.

Donahue, H. J., Fryer, M. J., and Heath III, Hunter. (1990). Structure-function relationships for full-length recombinant parathyroid hormone-related peptide and its amino-terminal fragmants: Effects on cytosolic calcium ion mobilization and adenylate cyclase activation in rat osteoblast-like cells. *Endocrinology,* **126,** 1471–1477.

Doty, S. B., and Schofield, B. H. (1976). Enzyme histochemistry of bone and cartilage cells. *Prog. Histochem. Cytochem.* **8,** 1–38.

Ecarot-Charrier, B., Glorieux, F. H., van der Rest, M., and Pereira, G. (1983). Osteoblasts isolated from mouse calvaria initiate matrix mineralization in culture. *J. Cell Biol.* **96,** 639–643.

Ecarot-Charrier, B., Shepard, N., Charette, G., Grynpas, M., and Glorieux, F. H. (1988). Mineralization in osteoblast cultures: A light and electron microscopic study. *Bone* **9,** 147–154.

Ecarot-Charrier, B., Bouchard, F., and Delloye, C. (1989). bone sialoprotein II synthesized by cultured osteoblasts contains tyrosine sulfate. *J. Biol. Chem.* **264,** 20049–20053.

Eguchi, G. (1986). Instability in cell commitment of vertebrate pigmented epithelial cells and their transdifferentiation into lens cells. *Curr. Topics Dev. Biol.* **20,** 21–37.

Eisenbarth, G. S. (1981). Application of monoclonal antibody techniques to biochemical research. *Anal. Biochem.* **111,** 1–16.

Embleton, M. J. (1984). Monoclonal antibodies to osteogenic sarcoma antigens. *In* "Monoclonal Antibodies and Cancer" (G. L. Wright, ed.), pp. 181–189. Marcel Dekker, New York.

Farber, F. A., and Liskay, R. M. (1974). Karyotypic analysis of a near-diploid established mouse cell line. *Cytogenet. Cell Genet.* **13,** 384–396.

Fedde, K. N., Lane, C. C., and Whyte, M. P. (1988). Alkaline phosphatase is an ectoenzyme that acts on micromolar concentrations of natural substrates at physiological pH in human osteosarcoma (SAOs-2) cells. *Arch. Biochem. Biophys.* **264,** 400–409.

Ferguson, M. A. J., and Williams, A. F. (1988). Cell surface anchoring of proteins via glycosyl-phophatidylinositol structures. *Annu. Rev. Biochem.* **57,** 285–320.

Fisher, L. W., Termine, J. D., and Young, F. M. (1989). Deduced protein sequence of bone small proteoglycan I (biglycan) shows homology with proteoglycan II (decorin) and several nonconnective tissue proteins in a variety of species. *J. Biol. Chem.* **264,** 4571–4576.

Forest, C., Poliard, A., Lamblin, D., Nifuji, A., Marie, P. J., Buc-Caron, M. H., and Kellerman, O. (1992). A committed mesodermal clone derived from mouse teratocarcinoma which differentiates into bone, cartilage or fat depending upon the inducers. *J. Bone Miner. Res.* **7**(suppl. 1), S123.

Franceschi, R. T., James, W. M., and Zerlauth, G. (1985). 1α,25-dihydroxyvitamin D_3 specific regulation of growth, morphology, and fibronectin in a human osteosarcoma cell line. *J. Cell. Physiol.* **123,** 401–409.

Franceschi, R. T., Romano, P. R., and Park, K. Y. (1988). Regulation of type I collagen synthesis by 1,25-dihydroxyvitamin D_3 in human osteosarcoma cells. *J. Biol. Chem.* **263,** 18938–18945.

Franzen, A., Oldberg, A., and Solursh, M. (1989). Possible recruitment of osteoblastic precursor cells from hypertropic chondrocytes during initial osteogenesis in cartilaginous limbs of young rats. *Matrix* **9,** 261–265.

Fraser, J. D., Otawara, Y., and Price, P. A. (1988). 1,25-dihydroxyvitamin D_3 stimulates the synthesis of matrix γ-carboxyglutamic acid protein by osteosarcoma cells. *J. Biol. Chem.* **263,** 911–916.

Friedenstein, A. J. (1976). Precursor cells of mechanocytes. *Int. Rev. Cytol.* **47,** 327–359.

Friedenstein, A. J. (1980). Stromal mechanisms of bone marrow: Cloning in vitro and retransplantation in vivo. *In* "Immunobiology of Bone Marrow Transplantation" (S. Theinfelder, ed.), pp. 19–29. Springer-Verlag, Berlin.

Friedenstein, A. J. (1990). Osteogenic stem cells in the bone marrow. *In* "Bone and Mineral Research" (J. N. M. Heersche and J. A. Kanis, eds.), Vol. 7, pp. 243–272. Elsevier, Amsterdam.

Friedenstein, A. J., Petrakova, K. V., Kurolesova, A. I., and Frolova, G. P. (1968). Heterotopic transplants of bone marrow. Analysis of precursor cells for osteogenic and hemtopoietic tissues. *Transplantation* **6,** 230–247.

Friedenstein, A. J., Chailalhyan, R. K., and Gerasimov, U. V. (1987). Bone marrow osteogenic stem cells: In vitro cultivation and transplantation in diffusion chambers. *Cell Tissue Kinetics* **20,** 263–272.

Fujita, S. C. (1987). Monoclonal antibody approaches to neurogenesis. *Curr. Top. Dev. Biol.* **21,** 255–275.

Fukayam S., and Tashjian, A. H. (1990). Stimulation by parathyroid hormone of $^{45}Ca^{2+}$ uptake in osteoblast-like cells: Possible involvement of alkaline phosphatase. *Endocrinology* **126,** 1941–1949.

Gerstenfeld, L. C., Chipman, S. D., Kelly, C. M., Hodgens, K. J., and Lee, D. D. (1988). Collagen expression, ultrastructural assembly, and mineralization in cultures of chick embryo osteoblasts. *J. Cell Biol.* **106,** 979–989.

Glimcher, M. J. (1989). Mechanisms of calcification in bone: Role of collagen fibrils and collagen–phosphoprotein complexes in vitro and in vivo. *Anat. Rec.* **224,** 139–153.

Glowacki, J., Rey, C., Glimcher, M. J., Cox, K. A., and Lian, J. (1991). A role for osteocalcin in osteoclast differentiation. *J. Cell. Biochem.* **45,** 292–302.

Goldberg, H. A., Domenicucci, C., Pringle, G., and Sodek, J. (1988). Mineral-binding proteoglycans of fetal porcine calvarial bone. *J. Biol. Chem.* **263,** 12092–12101.

Grigoriadis, A. E., Petkovich, P. M., Ber, R., Aubin, J. E., and Heersche, J. N. M. (1985). Subclone heterogeneity in a clonally-derived osteoblast-like cell line. *Bone* **6,** 249–256.

Grigoriadis, A. E., Petkovich, P. M., Rosenthal, E. E., and Heersche, J. N. M. (1986). Modulation by retinoic acid of 1,25-dihydroxyvitamin D_3 effects on alkaline phosphatase activity and parathyroid hormone responsiveness in an osteoblast-like osteosarcoma cell line. *Endocrinology* **119,** 932–939.

Grigoriadis, A. E., Heersche, J. N. M., and Aubin, J. E. (1988). Differentiation of muscle fat cartilage and bone from progenitor cells present in a bone derived clonal cell population: Effects of dexamethasone. *J. Cell Biol.* **106,** 2139–2151.

Grigoriadis, A. E., Heersche, J. N. M., and Aubin, J. E. (1990). Continuously growing bipotential and monopotential myogenic, adipogenic and chondrogenic subclones isolated from the multipotential RCJ 3.1 clonal cell line. *Dev. Biol.* **142,** 313–318.

Groot, C. G., Danes, J. K., Blok, J., Hoogendijk, A., and Hauschka, P. V. (1986). Light and electron microscopic demonstration of osteocalcin antigenicity in embryonic and adult rat bone. *Bone Miner.* **1,** 379–385.

Guenther, H. L., Cecchini, M. G., Elford, P. R., and Fleisch, H. (1988). Effects of transforming growth factor type beta upon bone cell populations grown either in monolayer of semisolid medium. *J. Bone Miner. Res.* **3,** 269–278.

Guenther, H. L., Hofstetter, W., Stutzer, A., Muhlbauber, R., and Fleisch, H. (1989). Evidence for heterogeneity of the osteoblastic phenotype determined with clonal rat bone cells established from transforming growth factor β induced cell colonies grown anchorage independently in semisolid medium. *Endocrinology* **125,** 2092–2102.

Hall, B. K. (1970). Cellular differentiation in skeletal tissues. *Biol. Rev.* **45,** 455–484.

Hall, B. K. (1978). "Developmental and Cellular Skeletal Biology." Academic Press, New York.

Hall, B. K. (1987). Earlier evidence of cartilage and bone development in embryonic life. *Clin. Orthop. Relat. Res.* **225,** 255–272.

Hall, B. K. (1988). The embryonic development of bone. *Am. Sci.* **76,** 174–181.

Hall, P. A., and Watt, F. M. (1989). Stem cells: The generation and maintenance of cellular diversity. *Development* **106,** 619–633.

Harrington, M. A., and Jones, P. A. (1988). Mesodermal determination genes: Evidence from DNA methylation studies. *Bioessays* **8,** 100–103.

Harris, H. (1989). The human alkaline phosphatases: What we know and what we don't know. *Clin. Chim. Acta* **186,** 133–150.

Harrison, J. R., Vargas, S. J., Peterson, D. N., Lorenzo, J. A., and Kream, B. E. (1990). Interleukin 1α and phorbol ester inhibit collagen synthesis in osteoblastic MC3T3E1 cells by a transcriptional mechanism. *Mol. Endocrinol.* **4,** 184–190.

Harvey, W. (1988). Source of prostaglandins and their influnce on bone resorption and formation. *In* "Prostaglandins in Bone Resorption" (W. Harvey and A. Bennet, eds.), pp. 27–41. CRC Press, Baco Raton, Florida.

Hauschka, P. V., Lian, J. B., Cole, D. E. C., and Gundberg, C. M. (1989). Osteocalcin and matrix gla protein: Vitamin K-dependent proteins in bone. *Physiol. Rev.* **69,** 990–1046.

Haussler, M. R., Manolagas, S. C., and Deftos, L. J. (1980). Glucocorticoid receptor in clonal osteosarcoma cell lines: A novel system for investigating bone-active hormones. *Biochem. Biophys. Res. Commun.* **94,** 373–380.

Heath, J. K., Rodan, S. B., Yoon, K. G., and Rodan, G. A. (1989). Rat calvarial cell lines immortalized with SV-40 large T-antigen—Constitutive and retinoic acid-inducible expression of osteoblastic features. *Endocrinology* **124,** 3060–3068.

Heersche, J. N. M., and Aubin, J. E. (1990). Regulation of cellular activity of osteoblasts. *In* "Bone: A Treatise. Vol. I: The Osteoblast and Osteocyte" (B. K. Hall, ed.), pp. 327–349. Telford Press, Caldwell, New Jersey.

Heersche, J. N. M., Pitaru, S., Aubin, J. E., and Tenenbaum, H. C. (1984). Corticosteroid induced expression of cartilage phenotype in cultured membrane bone periosteum. *In* "Endocrine Control of Bone and Calcium Metabolism" (D. V. Cohn, T. Fujita, J. T. Potts Jr., and R. V. Talmage, eds.), pp. 147–150. Elsevier, Amsterdam.

Heersche, J. N. M., Aubin, J. E., Grigoriadis, A., and Moriya, Y. (1985). Hormone responsiveness of bone cell populations. *In* "The Chemistry and Biology of Mineralized Tissues" (W. T. Butler, ed.), pp. 287–295. EBSCO Media, Birmingham, Alabama.

Heersche, J. N. M., Tenenbaum, H. C., Tam, C. S., Bellows, C. G., and Aubin, J. E. (1990). The role of cells in the calcification process. *In* "Bone Regulatory Factors" (A. Pecile and B. de Bernard, eds.), pp. 69–78. Plenum Press, New York.

Heersche, J. N. M., Reimers, S. M., Wrana, J. L., Waye, M. M. Y., and Gupta, A. K. (1992). Changes in expression of alpha 1 type I collagen and osteocalcin mRNA in osteoblasts and odontoblasts at different stages of maturity as shown by in situ hybridization. *Proc. Finn. Dent. Soc.* **88**(suppl. 1), 173–182.

Heinegard, D., and Oldberg, A. (1989). Structure and biology of cartilage and bone matrix noncollagenous macromolecules. *FASEB J.* **3,** 2042–2051.

Heiner, J. P., Miraldi, F., Kallick, S., Makley, J., Neely, J., Smith-Mensah, W. H., and Cheng, N. K. V. (1987). Localization of GD2-specific monoclonal antibody 3F8 in human osteosarcoma. *Cancer Res.* **47,** 5377–5384.

Holtrop, M. E. (1975). The ultrastructure of bone. *Ann. Clin. Lab. Sci.* **5,** 264–271.

Hosoi, S., Nakamura, T., Higashi, S., Yamamuro, T., Toyama, S., Shinomiya, K., and Mikawa, H. (1982). Detection of human osteosarcoma associated antigen by monoclonal antibodies. *Cancer Res.* **42,** 654–661.

Howe, C. C., Overton, G. C., Sawicki, J., Solter, D., Stain, P., and Strickland, S. (1988). Expression of SPARC/osteonectin transcript in murine embryos and gonads. *Differentiation* **37,** 20–25.

Howe, C. C., Kath, R., Mancianti, M. L., Herlyn, M., Mueller, S. and Cristofalo, V. (1990). Expression and structure of human SPARC transcripts: SPARC mRNA is expressed by human cells involved in extracellular matrix production and some of these cells show an unusual expression. *Exp. Cell Res.* **188,** 185–191.

Hunziker, E. B., Schenk, R. K., and Cruz-Orive, L.-M. (1987). Quantitation of chodrocyte performance in growth-plate cartilage during longitudinal bone growth. *J. Bone Joint Surg.* **69A,** 164–173.

Imai, Y., Rodan, S. B., and Rodan, G. A. (1988). Effects of retinoic acid on alkaline phosphatase messanger ribonucleic acid, catecholamine receptors, and G proteins in ROS 17/2.8 cells. *Endocrinology* **122,** 456–463.

Jande, S. S., and Belanger, L. F. (1973). The life cycle of the osteocyte. *Clin. Orthop.* **94,** 281–305.

Jones, P. A., Taylor, S. M., and Wilson, V. (1983). DNA modification, differentiation and transformation *J. Exp. Zoll.* **228,** 287–295.

Katagiri, T., Yamaguchi, A., Ikeda, T., Yoshiki, S., Wozney, J. M., Rosen, V., Wang, E. A., Tanaka, H., Omura, S., and Suda, T. (1990). The non-osteogenic pluripotent cell line 10T1/2 is induced to differentiate into osteoblastic cells by recombinant human bone morphogenetic protein-2. *Biochem. Biophys. Res. Commun.* **172,** 295–299.

Keilis-Borok, I. V., Latzinik, N. V., Epichina, S. Y., and Friedenstein, A. J. (1971). Dynamics of the formation of fibroblast colonies in monolayer cultures of bone marrow, according to 3H-thymidine incorporation experiments. *Cytologia* **13,** 1402–1409.

Kellerman, O., Buc-Caron, M. H., Marie, P. J., Lamblin, D., and Jacob, F. (1990). An immortalized osteogenic cell line derived from mouse teratocarcinoma is able to mineralize in vivo and in vitro. *J. Cell Biol.* **110,** 123–132.

Kember, N. F. (1971). Cell population kinetics of bone growth: The first ten years of autoradiographic studies with tritiated thymidine. *Clin. Orthop. Relat. Res.* **76,** 213–230.

Kodama, H., Amagai, Y., Sudo, H., Kasai, S., and Yamamoto, S. (1982). Establishment of a clonal osteogenic cell line from newborn mouse calvaria. *Jpn. J. Oral Biol.* **23,** 899–904.

Kohler, G., and Milstein, C. (1975). Continous cultures of fused cells secreting antibody of predefined specificity. *Nature (London)* **256,** 495–497.

Konieczny, S. F., and Emerson, C. P. (1984). 5-Azacytidine induction of stable mesodermal stem cell lineages from 10T1/2 cells: Evidence for regulatory genes controlling determination. *Cell* **38,** 791–800.

Krane, S. M., Goldring, M. B., and Goldring, S. R. (1988). Cytokines. *In* "Cell and Molecular Biology of Vertebrate Hard Tissues" (D. Evered and S. Harnett, eds., Vol. 136, pp. 239–257. Chichester Ciba Foundation Symposium. Wiley and Sons, New York.

Krukowski, M., Simmons, D. J., and Kahn, A. J. (1983). Cell lineage studies. *In* "Skeletal Research: An Experimental Approach" (A. S. Kunin and D. J. Simmons, eds.), Vol. 2, pp. 89–117. Academic Press, New York.

Lajeunesse, D., Leclerc, M., and Brunette, M. (1990). Regulation of osteocalcin and alkaline phosphatase synthesis by the human osteosarcoma cell line MG-63. *J. Bone Miner. Res.* **5,** S92.

Lajtha, L. G. (1979). Stem cell concepts. *Differentiation* **14,** 23–34.

Leary, A. G., Ogawa, M., Strauss, L. C., and Civin, C. I. (1984). Single cell origin of multilineage colonies in culture: Evidence that differentiation of multipotent progenitors and restriction or prolifertive potential of monopotent progenitors are stochastic processes. *J. Clin. Invest.* **74,** 2193–2197.

Leblond, C. P. (1989). Synthesis and secretion of collagen by cells of connective tissue bone and dentin. *Anat. Rec.* **224,** 123–128.

Lian, J. B., and Stein, G. S. (1992). Concepts of osteoblast growth and differentiation: Basis for modulation of bone cell development and tissue formation. *Crit. Rev. Oral Biol. Med.* **3,** 269–305.

Lian, J. B., Tassinari, M., and Glowacki, J. (1984). Resorption of implanted bone from normal and warfarin-treated rats. *J. Clinic. Invest.* **73,** 1223–1226.

Lim, B., Jamal, N., and Messner, H. A. (1984). Flexible association of hemopoietic differentiation programs in multilineage colonies. *J. Cell. Physiol.* **121,** 291–297.

Lis, H., and Sharon, N. (1986). Lectins as molecules and as tools. *Annu. Rev. Biochem.* **55,** 35–67.

Loo, D., Rawson, C., Ernst, T., Shirahata, S., and Barnes, D. (1989). Primary and multipassage culture of mouse embryo cells in serum-containing and serum-free media. *In* "Cell Growth and Division. A Practical Approach" (R. Baserga, ed.), pp. 17–35. IRL Press, New York.

Low, M. G. (1989a). The glycosyl-phosphatidylinositol anchor of membrane proteins. *Biochim. Biophys. Acta* **988,** 427–454.

Low, M. G. (1989b). Glycosyl-phosphatidyinositol: A versatile anchor for cell surface proteins. *FASEB J.* **3,** 1600–1608.

Lowenstam, H. A., and Weiner, S. (1989). "On Mineralization." Oxford University Press, New York.

Luk, S. C., Nopajaroonsri, C., and Simon, G. T. (1974). The ultrastructure of endosteum: A topographic study in young adult rabbits. *J. Ultrastruct. Res.* **46,** 165–183.

Mackie, E. J., Thesleff, I., and Chiquet-Ehrismann, R. (1987). Tenascin is associated with chondrogenic and osteogenic differentiation in vivo and promotes chondrogensis in vitro. *J. Cell Biol.* **105,** 2569–2579.

Mahonen, A., Pirskanen, A., Keinanen, R., and Maenpaa, P. (1990). Effect of $1,25(OH)_2D_3$ on its receptor mRNA levels and osteocalcin synthesis in human osteosarcoma cells. *Biochim. Biophys. Acta.* **1048,** 30–37.

Majeska, R. J., and Rodan, G. A. (1985). Culture and activity of osteoblasts and osteoblast-like cells. *In* "The Chemistry and Biology of Mineralized Tissues" (W. T. Butler, ed.), pp. 279–285. EBSCO Media, Birmingham, Alabama.

Majeska, R. J., Rodan, S. B., and Rodan, G. A. (1978). Maintenance of parathyroid hormone response in clonal rat osteosarcoma lines. *Exp. Cell Res.* **111,** 465–468.

Majeska, R. J., Rodan, S. B., and Rodan, G. A. (1980). Parathyroid hormone-responsive clonal cell lines from rat osteosarcoma. *Endocrinology* **107,** 1494–1503.

Manduca, P., Cancedda, F. D., and Cancedda, R. (1992). Chondrogenic differentiation in chick embryo osteoblast cultures. *Eur. J. Cell Biol.* **57,** 193–201.

Maniatopoulos, C., Sodek, J., and Melcher, A. H. (1988). Bone formation in vitro by stromal cells obtained from bone marrow of young adult rats. *Cell Tissue Res.* **254,** 317–330.

Mann, K., Deutzmann, R., Paulsson, M., and Timpl, R. (1987). Solubilization of protein BM-40 from a basement membrane tumor with chelating agents and evidence for its identity with osteonectin and SPARC. *FEBS Lett.* **218,** 167–171.

Manolagas, S. C., Spiess, Y. H., Burton, D. W., and Deftos, L. J. (1983). Mechanism of action of 1,25 dihydroxyvitamin D_3-induced stimulation of alkaline phosphatase in cultured osteoblast-like cells. *Mol. Cell. Endocrinol.* **33,** 27–36.

Marcus, R. (1988). Endocrine control of bone and mineral metabolism. *In* "Metabolic Bone and Mineral Disorders" (S. C. Manolagas and J. M. Olefsky, eds.), pp. 13–32. Churchill Livingstone, New York.

Mark, M. P., Prince, C. W., Osawa, T., Gay, S., Bronckers, A. L., and Butler, W. T. (1987a). Immunohistochemical demonstration of a 44-KD phosphoprotein in developing rat bones. *J. Histochem. Cytochem.* **35,** 707–715.

Mark, M. P., Prince, C. W., Gay, S., Austin, R. L., and Butler, W. T. (1988). 44-kDa bone phosphoprotein (osteopontin) antigenicity at ectopic sites in newborn rats: Kidney and nervous tissues. *Cell Tissue Res.* **251,** 23–30.

Mark, M. P., Prince, C. W., Gay, S., Austin, R. L., Bhowen, M., Finkelman, R. D., and Butler, W. T. (1987b). A comparative immunocytochemical study on the subcellular distributions of 44-kD bone phosphoprotein and bone γ-carboxyglutamic acid (gla)-containing protein in osteoblasts. *J. Bone Miner. Res.* **2,** 337–346.

Marks, S. C., and Popoff, S. N. (1988). Bone cell biology: The regulation of development structure and function in the skeleton. *Am. J. Anat.* **183,** 1–44.

Marotti, G. (1976). Decrement in volume of osteoblasts during osteon formation and its effect on the corresponding osteocytes, *Bone Histomorphomtery. Proc. 2nd. International Workshop, Armour Montagu, Paris,* pp. 385–393.

Martin, T. J. (1989). Bone cell physiology. *Endo. Metabol. Clinics North Am.* **18,** 833–858.

Martin, T. J., Ingleton, P. M., Underwood, J. C. E., Michelangeli, V. P., Hunt, N. H., and Melick, R. A. (1976). Parathyroid-hormone responsive adenylate cyclase in induced transplantable osteogenic rat sarcoma. *Nature.* **260,** 436–438.

Martin, T. J., Raisz, L. G., and Rodan, G. A. (1987). Calcium regulation and bone metabolism. *In* "Clinical Endrocrinology of Calcium Metabolism" (T. J. Martin and L. G. Raisz, eds.), pp. 1–52. Marcel Dekker, New York.

Martineau-Doize, B., Lai, W. H., Warshawsky, H., and Bergeron, I. (1988). In vivo demonstration of cell types in bone that harbor epidermal growth factor receptors. *Endocrinology* **123,** 841–858.

Mason, I. J., Taylor, A., Williams, J. C., Sage, H., and Hogan, B. L. M. (1986). Evidence from molecular cloning that SPARC, a major product of mouse embryo parietal endoderm, is related to an endothelial cell culture shock glycoprotein of Mr 43 000. *EMBO J.* **5,** 1465–1472.

McCulloch, C. A. G., and Tenenbaum, H. C. (1986). Dexamethasone induces proliferation and terminal differentiation of osteogenic cells in tissue culture. *Anat. Rec.* **215,** 397–402.

McCulloch, C. A. G., Struguresco, M., Hughes, F., Melcher, A. H., and Aubin, J. E. (1991). Osteogenic progenitor cells in rat bone marrow stromal populations exhibit self-renewal in cultre. *Blood* **77,** 1906–1911.

McKay, R., Raff, M. C., and Reichardt, L. F. (1981). Introduction. *In* "Monoclonal Antibodies to Neural Antigens" (R. McKay, M. C. Raff, and L. F. Reichardt, eds.), pp. 1–13. Cold Spring Harbor Laboratories, Cold Spring, New York.

Menton, D. N., Simmons, D. J., Chang, S. L., and Orr, B. Y. (1984). From bone lining cell to osteocyte–An SEM study. *Anat. Rec.* **209,** 29–39.

Metcalf, D. (1989). The molecular control of cell division differentiation commitment and maturation in haemopoietic cells. *Nature (London)* **339,** 27–30.

Millar, S., Heersche, J. N. M., and Aubin, J. E. (1982). Regulation of alkaline phosphatase activity in cloned bone cell populations. *J. Dent. Res.* **61,** 256a.

Miller, S. C., and Jee, W. S. S. (1987). The bone lining cell: A distinct phenotype? *Calcif. Tissue Int.* **41,** 1–5.

Miller, S. C., Bowman, B. M., Smith, J. M., and Jee, W. S. S. (1980). Characterization of endosteal bone-lining cells from fatty marrow bone sites in adult beagles. *Anat. Rec.* **198,** 163–173.

Milstein, C., and Lennox, E. (1980). The use of monoclonal antibody techniques in the study of developing cell surfaces. *Curr. Top. Dev. Biol.* **14,** 1–32.

Mosher, D. F. (1990). Physiology of thrombospondin. *Annu. Rev. Med.* **41,** 85–97.

Moskalewski, S., and Malejczyk, J. (1989). Bone formation following intrarenal transplantation of isolated murine chondrocytes: Chondrocyte–bone transdifferentiation? *Development* **107,** 473–480.

Mundy, G. R. (1989). "Calcium Homeostasis: Hypercalcimia and Hypocalcemia." Martin Dunitz, London.

Mundy, G. R., and Poser, J. W. (1983). Chemotactic activity of γ-carboxyglutamic containing protein in bone. *Calcif. Tissue Int.* **35,** 164–168.

Nagata, T., Todescan, R., Goldberg, H. A., Zhang, Q., and Sodek, J. (1989). Sulfation of secreted phosphoprotein I (SPP1, osteopontin) is associated with mineralized tissue formation. *Biochem. Biophys. Res. Commun.* **165,** 234–240.

Nakahata, T., Gross, A. J., and Ogawa, M. (1982). A stochastic model of self-renewal and commitment to differentiation of the primitive hemopoietic stem cells in culture. *J. Cell Physiol.* **113,** 455–458.

Nakamura, T., Gross, M., Yamamuro, T., and Liao, S.-K. (1987). Identification of a human osteosarcoma-associated glycoprotein with monoclonal antibodies: Relationship with alkaline phosphatase. *Biochem. Cell Biol.* **65,** 1091–1097.

Nathanson, M. A. (1989). Differentiation of muscular tissues. *Int. Rev. Cytology* **116,** 89–164.

Nefussi, J.-R., Boy-Lefevre, M. L., Boulekbache, H., and Forest, N. (1985). Mineralization in vitro of matrix formed by osteoblasts isolated by collagenease digestion. *Differentiation* **29,** 160–168.

Newman, W., Beall, L. D., Levine, M. A., Cone, J. L., Randhawa, Z. I., and Bertolini, D. R. (1989). Biotinylated parathyroid as a probe for the parathyroid hormone receptor. *J. Biol. Chem.* **264,** 16359–16366.

Ng, K. W., Partridge, N. C., Niall, M., and Martin, T. J. (1983). Epidermal growth factor receptors in clonal lines of a rat osteogenic sarcoma and in osteoblast-rich rat bone cells. *Calc. Tiss. Int.* **35,** 298–303.

Ng, K. W., Gummer, P. R., Michelangeli, V. P., Baterman, J. F., Mascara, T., Cole, W. G., and Martin, T. J. (1988). Regulation of alkaline phosphatase expression in a neonatal rat calvarial cell strain by retinoic acid. *J. Bone Miner. Res.* **3,** 53–61.

Nicholas, R. (1986). "Hybridoma Technology: An Annotated Listing of Key Papers 1975–1985." Mansell Pub. Ltd., London.

Nicola, N. A. (1982). Electronic cell sorting of hemopoietic progenitor cells. *Cell Separation: Methods and Selected Applications* **1,** 191–221.

Nicola, N. A. and Johnson, G. R. (1982). The production of committed hemopoietic colony-forming cells from multipotential precursor cells *in vitro*. *Blood* **60,** 1019–1029.

Nicolas, J. F., Gaillard, J., Jacob, H., and Jacob, F. (1980). Bone forming cell line derived from embryonal carcinoma cells. *Nature (London)* **286,** 716–718.

Nicolson, A. L. (1984). Cell surface molecules and tumor metastasis. Regulation of metastasis phenotypic diversity. *Exp. Cell Res.* **150,** 3–22.

Nijweide, P. J., and Mulder, R. J. P. (1986). Identification of osteocytes in osteoblast-like cultures using a monoclonal antibody specifically directed against osteocytes. *Histochemistry* **84,** 343–350.

Nijweide, P. J., van Iperen-van Gent, A. S., Kawilarang-de Haas, E. W. M., van der Plas, A., and Wassenaar, A. M. (1982). Bone formation and calcification by isolated osteoblast-like cells. *J. Cell Biol.* **93,** 318–323.

Nijweide, P. J., Burger, E. H., and Feyen, J. H. M. (1986). Cells of bone: Proliferation differentiation and hormonal regulation. *Physiol. Rev.* **66,** 855–886.

Nijweide, P. J., Van der Plas, A., and Olthof, A. A. (1988). Osteoblastic differentiation. *In* "Cell and Molecular Biology of Vertebrate Hard Tissues" (D. Evered and S. Harnett, eds.), pp. 61–72. Ciba Foundation Sym. 136, John Wiley and Sons, New York.

Nishimoto, S. K., and Price, P. A. (1980). Secretion of the vitamin K-dependent protein of bone by rat osteosarcoma cells. *J. Biol. Chem.* **255,** 6579–6583.

Noda, M., Vogal, R. L., Hasson, D. M., and Rodan, G. A. (1990). Leukemia inhibitory factor suppresses proliferation, alkaline phosphatase activity, and type I collagen messenger ribonucleic acid level and enhances osteopontin mRNA level in murine osteoblast-like (MC3T3E1) cells. *Endocrinology* **127,** 185–190.

Noda, M. (1989). Transcriptional regulation of osteocalcin production by transforming growth factor β in rat osteoblast like cells. *Endocrinology* **124,** 612–617.

Noda, M., Yoon, K., Rodan, G. A., and Koppel, D. E. (1987). High lateral mobility of endogenous and transfected alakaline phosphatase: A phosphatidylinositol-anchored membrane protein. *J. Cell Biol.* **105,** 1671–1677.

Nomura, S., Wills, A. J., Edwards, D. R., Heath, J. K., and Hogan, B. L. M. (1988). Developmental expression of 2ar (osteopontin) and SPARC (osteonectin) RNA as revealed by in situ hybridization. *J. Cell Biol.* **106,** 441–450.

Ogawa, M. (1989). Hemopoietic stem cells: Stochastic differentiation and humoral control of proliferation. *Environ. Health Persp.* **80,** 199–207.

Ogawa, M., Porter, P. N., and Nakahat, T. (1983). Renewal and commitment to differentiation of haemopoietic stem cell (an interpretive review). *Blood* **61,** 823–830.

Okada, T. S. (1986). Transdifferentiation in animal cells: Fact or artifact? *Dev. Growth Differ.* **28,** 213–221.

Oldberg, A., Franzen, A., and Heinegard, D. (1986). Cloning and sequence analysis of rat bone sialoprotein (osteopontin) cDNA reveals an Arg-Gly-Asp cell binding sequence *Proc. Natl. Acad. Sci. USA* **83,** 8819–8823.

Oldberg, A., Frazen, A., Heinegard, D., Pierschbacher, M., and Rouslahti, E. (1988). Identification of a bone sialoprotein receptor in osteosarcoma cells. *J. Biol. Chem.* **263,** 19433–19436.

O'Neill, M. C., and Stockdale, F. E. (1972). A kinetic analysis of myogenesis in vitro. *J. Cell Biol.* **52,** 52–65.

Osdoby, P., and Caplan, A. I. (1979). Osteogenesis in cultures of limb mesenchymal cells. *Dev. Biol.* **73,** 84–102.

Otawara, Y., and Price, P. A. (1986). Developmental appearance of matrix Gla protein during calcification in the rat. *J. Biol. Chem.* **261,** 10828–10835.

Owen, M. (1963). Cell population kinetics of an osteogenic tissue. *J. Cell Biol.* **19,** 19–32.

Owen, M. (1967). Uptake of [^{3}H]-uridine into precursor pools and RNA in osteogenic cells. *J. Cell Sci.* **2,** 39–56.

Owen, M. (1970). The origin of bone cells. *Int. Rev. Cytol.* **28,** 213–238.

Owen, M. (1985). Lineage of osteogenic cells and their relationship to the stromal system. *In* "Bone and Mineral Research" (W. A. Peck, ed.), Vol. 3, pp. 1–25. Elsevier, Amsterdam.

Owen, M. (1988). Marrow stromal stem cells. *J. Cell Sci.* **S10,** 63–76.

Owen, M., and Friedenstein, A. J. (1988). Stromal stem cells: Marrow-derived osteogenic precursors. *In* "Cell and Molecular Biology of Vertebrate Hard Tissues" Chichester Ciba Foundation Symposium (D. Evered and S. Harnett, eds.), Vol. 136, pp. 42–60. Wiley, New York.

Owen, T. A., Aronow, M., Shalhoub, V., Barone, L. M., Wilming, L., Tassinari, M. S.,

Kennedy, M. B., Pockwinse, S., Lian, J. B., and Stein, G. S. (1990). Progressive development of the rat osteoblast phenotype in vitro: Reciprocal relationships in expression of genes associated with osteoblast proliferation and differentiation during formation of the bone extracellular matrix. *J. Cell Physiol.* **143,** 420–430.

Partridge, N. C., Frampton, R. J., Eisman, J. A., Michelangeli, V. P., Elms, E., Bradley, T. R., and Martin, T. J. (1980). Receptors for 1,25$(OH)_2$ vitamin D_3 enriched cloned osteoblast-like rat osteogenic sarcoma cells. *FEBS Lett.* **115,** 139–142.

Partridge, N. C., Frampton, R. J., Eisman, J. A., Michelangeli, V. P., Elms, E., Bradley, T. R., and Martin, T. J. (1980). Receptors for 1,25$(OH)_2$ vitamin D_3 enriched in cloned osteoblast-like rat osteogenic sarcoma cells. *FEBS Lett.* **115,** 139–142.

Partridge, N. C., Alcorn, D., Michelangeli, V. P., Ryan, G., and Martin, T. J. (1983). Morphological and biochemical characterization of four clonal osteogenic sarcoma cells of rat origin. *Cancer Res.* **43,** 4308–4314.

Patarca, R., Freeman, G. J., Singh, R. P., Wei, F.-Y., Durfee, T., Blattner, F., Regnier, D. C., Kozak, C. A., Mock, B. A., Morse, H. C., Jerrells, T. R., and Cantor, H. (1989). Structural and functional studies of the early T lymphocyte activation (ETA-1) gene. *J. Exp. Med.* **170,** 145–161.

Peck, W. A., and Woods, W. L. (1988). The cells of bone. *In* "Osteoporosis" (B. L. Riggs and L. J. Melton, eds.), pp. 1–44. Raven Press, New York.

Petkovich, P. M., Heersche, J. N. M., Tinker, D. O., and Jones, G. (1984). Retinoic acid stimulates, 125 dihydroxyvitamin D_3 binding in rat osteosarcoma cells. *J. Biol. Chem.* **259,** 8274–8280.

Petkovich, P. M., Wrana, J. L., Grigoriadis, A. E., Heersche, J. N. M., and Sodek, J. (1987). 1,25-dihydroxyvitamin D_3 increases epidermal growth factor receptors and transforming growth factor-like activity in a bone-derived cell line. *J. Biol. Chem.* **262,** 13424–13428.

Potten, C. S., Schofield, R., and Lajtha, L. G. (1979). A comparison of cell replacement in bone marrow, testis and three regions of surface epithelium. *Biochim. Biophys. Acta* **560,** 281–299.

Price, P. A. (1983). Osteocalcin. *In* "Bone and Mineral Research" (W. A. Peck, ed.), Vol. 1, pp. 157–190. Excerpta Medica, Amsterdam.

Price, P. A. (1988). Role of vitamin K-dependent proteins in bone metabolism. *Annu. Rev. Nutr.* **8,** 565–583.

Prince, C. W., Oosawa, T., Butler, W. T., Tomana, M., Bhown, A. J., Bhown, M., and Schrohenloher, R. E. (1987). Isolation characterization and biosynthesis of a phosphorylated glycoprotein from rat bone. *J. Biol. Chem.* **262,** 2900–2907.

Pritchard, J. J. (1972a). Growth and differentiation of bone and connective tissue. *In* "Differentiation and Growth of Cells in Vertebrate Tissues" (G. Goldspink, ed.), pp. 101–128. Chapman and Hall, London.

Pritchard, J. J. (1972b). The osteoblast. *In* "The Biochemistry and Physiology of Bone" (G. H. Bourne, ed.), Vol. I, pp. 21–43. Academic Press, New York.

Raff, M. C. (1989). Glial cell diversification in the rat optic nerve. *Science* **243,** 1450–1455.

Raff, M. C., and Lillien, L. E. (1988). Differentiation of a bipotential glial progenitor cell—What controls the timing and the choice of developmental pathway. *J. Cell Sci.* **S10,** 77–83.

Raff, M. C., Fields, K. L., Hakamori, S., Mirsky, R., Pruss, R. M., and Winter, J. (1979). Cell type specific markers for distinguishing and studying neurons and the major classes of glial cells in culture. *Brain Res.* **174,** 283–294.

Rao, L. G., Ng, B., Brunette, D., and Heersche, J. N. M. (1977). Parathyroid hormone and prostaglandin E_1 response in a selected population of bone cells after repeated subculture and storage at −80°C. *Endocrinology* **100,** 1233–1241.

Rao, L. G., Murray, T. M., and Heersche, J. N. M. (1983). Immunohistochemical demonstration of parathyroid hormone binding to specific cell types in fixed rat bone tissues. *Endocrinology* **113,** 805–810.

Reddi, A. H. (1985). Regulation of bone differentiation by local and systemic factors. *In* "Bone and Mineral Research" (W. A. Peck, ed.), Vol. 3, pp. 27–47. Elsevier, New York.

Reinholt, F. P., Hultenby, K., Oldberg, A., and Heinegard, D. (1990). Osteopontin—A possible anchor of osteoclasts to bone. *Proc. Natl. Acad. Sci. USA* **87,** 4473–4475.

Reznikoff, C. A., Brankow, D. W., and Heidelberger, C. (1973). Establishment and characterization of a cloned line of C3H mouse embryo cells sensitive to postconfluence inhibition of division. *Cancer Res.* **33,** 3231–3238.

Roach, H. I. (1992). Transdifferentiation of hypertrophic chondrocytes into cells capable of producing a mineralized bone matrix. *Bone Miner.* **19,** 1–20.

Robey, P. G. (1989). The biochemistry of bone. *Endo. Metabol. Clinics North Am.* **18,** 859–902.

Robey, P. G., and Termine, J. D. (1985). Human bone cells in vitro. *Calcif. Tissue Int.* **37,** 453–460.

Robey, P. G., Young, M. F., Fisher, L. W., and McClain, T. D. (1989). Thrombospondin is an osteoblast-derived component of mineralized extracellular matrix. *J. Cell Biol.* **108,** 719–727.

Rodan, G. A., and Rodan, S. B. (1984). Expression of the osteoblastic phenotype. *In* "Bone and Mineral Research" (W. A. Peck, ed.), Vol. 2, pp. 244–285. Elsevier, Amsterdam.

Rodan, S. B., Imai, Y., Thiede, M. A., Wesolowski, G., Thompson, D., Bar-Shavit, Z., Shull, S., Mann, K., and Rodan, G. A. (1987). Characterization of a human osteosarcoma cell line (SaOS-2) with osteoblastic properties. *Cancer Res.* **47,** 4961–4966.

Rouleau, M. F., Mitchell, J., and Goltzman, D. (1988). In vivo distribution of parathyroid hormone receptors in bone: Evidence that a predominant osseous target cell is not the mature osteoblasts. *Endocrinology* **123,** 187–191.

Rouleau, M. F., Mitchell, J., and Goltzman, D. (1990). Characterization of the major parathyroid hormone target cell in the endosteal metaphysis of rat long bones. *J. Bone Miner. Res.* **5,** 1043–1053.

Sage, H. (1986). Culture shock selective uptake and rapid release of a novel serum protein by endothelial cells in vitro. *J. Biol. Chem.* **261,** 7082–7092.

Sage, H., Johnson, C., and Bornstein, P. (1984). Characterization of a novel serum albumin binding glycoprotein secreted by endothelial cells in culture. *J. Biol. Chem.* **259,** 3993–4007.

Sage, H., Pritzl, P., and Bramson, R. (1986). Endothelial cell injury in vitro is associated with increased secretion of an Mr 43000 glycoprotein ligand. *J. Cell. Physiol.* **127,** 373–387.

Sage, H., Vernon, R. B., Decker, J., Funk, S., and Iruelaarispe, M. L. (1989). Distribution of the calcium-binding protein Sparc in tissues of embryonic and adult mice. *J. Histochem. Cytochem.* **37,** 819–829.

Saltier, A. R., Osterman, D. G., Darnell, J. C., Chan, B. L., and Sorbara-Cazan, L. R. (1989). The role of glycosylphosphoinosides in signal transduction. *Recent Prog. Hormon. Res.* **45,** 353–382.

Sasaki, M. S., and Kodama, S. (1987). Establishment and some mutational characteristics of 3T3-like near-diploid mouse cell line. *J. Cell. Physiol.* **131,** 114–122.

Sasse, J., Horwitz, A., Pacifici, M., and Holtzer, H. (1984). Seperation of precursor myogenic and chondrogenic cells in early limb bud mesenchyme by a monoclonal antibody. *J. Cell Biol.* **99,** 1856–1866.

Schulz, A., and Jundt, G. (1989). Immunohistological demonstration of osteonectin in normal bone tissue and in bone tumors. *In* "Biological Characterization of Bone Tumors" (A. Roessner, ed.), pp. 31–54. Springer-Verlag, Berlin.

Scott, B. L. (1967). Thymidine ^{3}H electron microscope radioautography of osteogenic cells in the fetal rat. *J. Cell Biol.* **35,** 115–126.

Senger, D. R., Peruzzi, C. A., and Papadopoulos, A. (1989). Elevated expression of secreted phosphoprotein I (osteopontin, 2ar) as a consequence of neoplastic transformation. *Anticancer Res.* **9,** 1291–1300.

Shapiro, S., Tatakis, D. N., and Dziak, R. (1990). Effects of tumor necrosis factor α on parathyroid hormone-induced increases in osteoblastic cell cyclic AMP. *Calc. Tiss. Internat.* **46,** 60–62.

Shupnik, M. A., and Tashjian, A. H. (1982). Epidermal growth factor and phorbol ester actions on human osteosarcoma cells. *J. Biol. Chem.* **257,** 12161–12164.

Sharon, N. (1984). Use of lectins for separation of cells. *Cell Separation* **3,** 13–46.

Silve, C. M., Hradek, G. T., Jones, A. L., and Arnaud, C. D. (1982). Parathyroid hormone receptor in intact embryonic chicken bone: Characterization and cellular localization. *J. Cell Biol.* **94,** 379–386.

Simkiss, K., and Wibur, K. M. (1989). "Biomineralization: Cell Biology and Mineral Deposition." Academic Press, San Diego.

Smith, J. H., and Denhardt, D. T. (1987). Molecular cloning of a tumor promoter inducible mRNA found in JB6 mouse epidermal cells: Induction is stable at high not at low cell densities. *J. Cell. Biochem.* **34,** 13–22.

Sodek, J., Bellows, C. G., Aubin, J. E., Limeback, H., Otsuka, K., and Yao, K.-L. (1985). Differences in collagen gene expression in subclones and long-term cultures of clonal derived rat bone cell populations. *In* "The Chemistry and Biology of Mineralized Tissues" (W. T. Butler, ed.), pp. 303–306. EBSCO Media, Birmingham, Alabama.

Sodek, J., Chen, J.-K., Kasugai, S., Nagata, T., Zhang, Q., McKee, M. D., and Nanci A. (1992a). Elucidating the functions of bone sialoprotein and osteopontin in bone formation. *In* "Chemistry and Biology of Mineralized Tissues" (H. Slavkin and P. Price, eds.), pp. 297–306. Excerpta Medica, Amsterdam.

Sodek, J., Chen, J.-K., Kasugai, S., Nagata, T., Zhang, Q., Shapiro, H. S., Wrana, J. L., and Goldberg, H. A. (1992b). Sialoproteins in bone remodelling. *In* "The Biological Mechanisms of Tooth Movement and Craniofacial Adaptation" (Z. Davidovitch, ed.), pp. 127–136. EBSCO Media, Birmingham, Alabama.

Sprangrude, G. J., Heimfeld, S., and Weissman, I. L. (1988). Purification and characterization of mouse hematopoietic stem cells. *Science* **241,** 58–62.

Stein, G. S., Lian, J. B., and Owen, T. A. (1990). Relationship of cell growth to the regulation of tissue-specific gene expression during osteoblast differentiation. *FASEB J.* **4,** 3111–3123.

Stern, P. H. (1988). Cellular Physiology of bone. *In* "Metabolic Bone and Mineral Disorders" (S. C. Manolagas and J. M. Olefsky, eds.), pp. 1–13. Churchill Livingstone, New York.

Strauss, P. G., Cross, E. I., Schmidt, J., and Erfle, V. (1990). Gene expression during osteogenic differentiation in mandibular condyles in vitro. *J. Cell Biol.* **110,** 1369–1378.

Stutzmann, J. J., and Petrovic, A. G. (1982). Bone cell histogenesis: The skeletoblast as a stem cell for preosteoblasts and for secondary type prechondroblats. *In* "Factors and Mechanisms Influencing Bone Growth," (A. D. Dixon and B. G. Samat, eds.), pp. 29–43. Alan Liss, New York.

Sudo, H., Kodama, H., Amagai, Y., Yamamoto, S., and Kasai, S. (1983). *In vitro* differentiation and calcification in a new clonal osteogenic cell line derived from newborn mouse calvaria. *J. Cell Biol.* **96,** 191–198.

Taylor, S. M., and Jones, P. A. (1979). Multiple new phenotypes induced in 10T1/2 and 3T3 cells treated with 5-azacytidine. *Cell* **17,** 771–779.

Taylor, S. M., and Jones, P. A. (1982a). Changes in phenotypic expression in embryonic and adult cells treated with 5-azacytidine. *J. Cell. Physiol.* **111,** 187–194.

Taylor, S. M., and Jones, P. A. (1982b). Mechanisms of action of eukaryotic DNA methyltransferase. Use of 5-azacytidine-containing DNA. *J. Mol. Biol.* **162,** 679–692.

Temple, S., and Raff, M. C. (1985). Differentiation of a bipotential glial progenitor cell in single cell microculture. *Nature (London)* **313,** 223–225.

Temple, S., and Raff, M. C. (1986). Clonal analysis of oligodendrocyte development in culture: Evidence for a developmental clock that counts cell divisions. *Cell* **44,** 773–779.

Tenenbaum, H. C., and Heersche, J. N. M. (1982). Differentiation of osteoblasts and formation of mineralized bone in vitro. *Calcif. Tissue Int.* **34,** 76–79.

Tenenbaum, H. C., and Heersche, J. N. M. (1985). Dexamethasone stimulates osteogenesis in chick periosteum in vitro. *Endocrinology* **117,** 2211–2217.

Termine, J. D., Kleinman, H. K., Whitson, S. W., Conn, M. L., McGarvey, M. L., and Martin, G. R. (1981). Osteonectin: A bone specific protein linking mineral to collagen. *Cell* **26,** 99–105.

Thesingh, C. W., and Scherft, J. P. (1986). Bone matrix formation by transformed chondrocytes in organ cultures of stripped embryonic metatarsalia. *In* "Cell Mediated Calcification and Matrix Vesicles" (Y. Ali, ed.), pp. 309–314. Elsevier, Amsterdam.

Thesingh, C. W., Groot, C. G., and Wassenaar, A. M. (1991). Transdifferentiation of hypertrophic chondrocytes in murine metatarsal bones, induced by co-cultured cerebrum. *Bone Miner.* **8,** 25–40.

Till, J. E., and McCulloch, E. A. (1980). Hemopoietic stem cell differentiation. *Biochim. Biophys. Acta* **605,** 431–464.

Till, J. E., McCulloch, E. A., and Siminovitch, L. (1964). A stochastic model of stem cell proliferation based on the growth of spleen colony forming cells. *Proc. Natl. Acad. Sci. USA* **51,** 29–36.

Torok-Storb, B. (1988). Cellular interactions. *Blood* **72,** 373–385.

Tracy, R. P., Shull, S., Riggs, B. L., and Mann, K. G. (1988). The osteonectin family of proteins. *Int. J. Biochem.* **20,** 653–660.

Tsai, C. C., McGuire, M. H., Mellitt, R. J., Ritter, R. A., Xu, J., Litwicki, D. J., and Roodman, S. T. (1990). Monoclonal antibody to human osteosarcoma: A novel Mr 26000 protein recognized by murine hybridoma TMMR-2. *Cancer Res.* **50,** 152–161.

Tung, P. S., Domenicucci, C., Wasi, S., and Sodek, J. (1985). Specific immunohistochemical localization of osteonectin and collagen type I and III in fetal and adult porcine dental tissues. *J. Histochem. Cytochem.* **33,** 531–540.

Turksen, K., and Aubin, J. E. (1991). Positive and negative immunoselection of two classes of osteoprogenitor cells. *J. Cell Biol.* **114,** 373–384.

Turksen, K., Bhargava, U., Moe, H. K., and Aubin, J. E. (1992). Isolation of monoclonal antibodies recognizing rat bone-associated molecules in vitro and in vivo. *J. Histochem. Cytochem.* **40,** 1339–1352.

Urena, P., Lee, K., Weaver, D., Kong, X. F., Brown, D., Bond, A. T., Abou-Samra, A. B., and Segre, G. (1992). PTH/PTHrP receptor mRNA expression as assessed by Northern blot and in situ hybridization. *J. Bone Miner. Res.* **7**(suppl. 1), S119.

Urist, M. (1989). Bone morphogenetic protein bone regeneration heterotopic ossification and bone–bone marrow consortium. *In* "Bone and Mineral Research" (W. A. Peck, ed.), Vol. 6, pp. 37–112. Elsevier, Amsterdam.

Urist, M., Delange, R. J., and Finerman, G. A. M. (1983). Bone cell differentiation and growth factors. *Science* **220,** 680–686.

Valaja, T., Mahonen, A., Pirskanen, A., and Maenpaa, P. H. (1990). Affinity of 22-oxa-1,25$(OH)_2D_3$ for 1,25-dihydroxyvitamin D receptor and its effects on the synthesis of osteocalcin in human osteosarcoma cells. *Biochem. Biophys. Res. Commun.* **169,** 629–635.

Valentino, K. L., Winter, J., and Reichardt, L. F. (1985). Applications of monoclonal antibodies to neuroscience research. *Annu. Rev. Neurosci.* **8,** 199–232.

Vermeulen, A. H., Vermeer, C., and Bosman, F. T. (1989). Histochemical detection of osteocalcin in normal and pathological human bone. *J. Histochem. Cytochem.* **37,** 1503–1508.

Villaneuva, A. R., Mathews, C. H. E., and Parfitt, A. M. (1981). Relationship between the size and shape of osteoblasts and the width of the osteoid seams in bone, *Proc. 2nd. Workshop on Bone Morphometry, Niigata University School of Medicine, Niigata,* pp. 191–199.

Wasi, S., Otsuka, K., Yao, K.-L., Tung, P. S., Aubin, J. E., Sodek, J., and Termine, J. D. (1984). An osteonectin-like protein in porcine periodontal ligament and its synthesis by periodontal ligament fibroblasts. *Can. J. Biochem. Cell Biol.* **2,** 470–478.

Weinreb, M., Shinar, D., and Rodan, G. A. (1990). Different pattern of alkaline phosphatase, osteopontin and osteocalcin in developing rat bone visualized by in situ hybridization. *J. Bone Miner. Res.* **5,** 831–842.

Whyte, M. P. (1989). Alkaline phosphatase: Physiological role explored in hypophosphatasia. *In* "Bone and Mineral Research" (W. A. Peck, ed.), Vol. 6, pp. 175–218. Elsevier, Amsterdam.

Wlodarski, K. H. (1990). Properties and origin of osteoblasts. *Clin. Orthop. Relat. Res.* **252,** 277–293.

Wolpert, L. (1988). Stem cells: A Problem in asymmetry. *J. Cell Sci.* **10,** 1–19.

Wong, G. (1980). Bone cell cultures as an experimental model. *Arthritis and Rheumatism* **23,** 1081–1086.

Wozney, J. M. (1992). The bone morphogenetic protein family and osteogenesis. *Mol. Reproduction Dev.* **32,** 160–167.

Wozney, J. M., Rosen, V., Byrne, M., Celeste, A. J., Moutsatsos, I., and Wang, E. A. (1990). Growth factors influencing bone development. *J. Cell Sci.* **13,** 149–156.

Wuthier, R. E., and Register, T. C. (1984). Role of alkaline phosphatase, a polyfunctional enzyme, in mineralizing tissues. *In* "The Chemistry and Biology of Mineralized Tissues" (W. T. Butler, ed.), pp. 114–124. EBSCO Media, Birmingham, Alabama.

Yamada, H., Tsutsumi, M., Fukase, M., Fujimori, A., Yamamoto, Y., Miyauchi, A., Fujii, Y., Noda, T., Fujii, N., and Fujita, T. (1989). Effects of human PTH-related peptide and human PTH omn cyclic AMP production and cytosolic free calcium in an osteoblastic cell clone. *Bone and Mineral.* **6,** 45–54.

Yamada, T. (1977). "Control Mechanisms in Cell-Type Conversion in Newt Lens Regeneration." S. Karger, Basel.

Yamaguchi, A., and Kahn, A. J. (1991). Clonal osteogenic cell lines express myogenic and adipocytic developmental potential. *Calcif. Tissue Int.* **49,** 221–225.

Yamaguchi, A., Katagiri, T., Ikeda, T., Wozney, J. M., Rosen, V., Want, E. A., Kahn, A. J., Suda, T., and Yoshiki, S. (1991). Recombinant human bone morphogenetic protein-2 stimulates osteoblastic maturation and inhibits myogenic differentiation in vitro. *J. Cell Biol.* **113,** 681–687.

Yoshioka, C., and Yagi, T. (1988). Electron microscopic observations on the fate of hypertropic chodrocytes in condylar cartilage of rat mandible. *J. Craniofacial Genetics Dev. Biol.* **8,** 253–264.

Young, R. W. (1962). Regional differences in cell generation time in growing rat tibia. *Exp. Cell Res.* **26,** 562–567.

Zimmerman, B., Wachtel, H. C., Somogyi, H., Merker, H.-J., and Bernimoulin, J.-P. (1988). Bone formation by rat calvaria cell growth at high density in organoid culture. *Cell Differen. Develop.* **25,** 145–154.

Zipser, B., and McKay, R. (1981). Monoclonal antibodies distinguish identifiable neurons in the leech. *Nature (London)* **289,** 549–554.

Zipursky, S. L., Venkattesh, T. R., Teplow, D. B., and Benzer, S. (1984). Neuronal development in the *Drosophila* retina monoclonal antibodies as molecular probes. *Cell* **36,** 15–26.

Zung, P., Domenicucci, C., Wasi, S., Kuwata, F., and Sodek, J. (1986). Osteonectin is a minor component of mineralized connective tissues in rat. *Biochem. Cell Biol.* **64,** 356–362.

2

MOLECULAR MECHANISMS MEDIATING DEVELOPMENTAL AND HORMONE-REGULATED EXPRESSION OF GENES IN OSTEOBLASTS: An Integrated Relationship of Cell Growth and Differentiation

GARY S. STEIN and JANE B. LIAN

Cellular and Molecular Biology of Bone

I. INTRODUCTION

A functional relationship between cell growth and the initiation and progression of events associated with differentiation has been viewed as a fundamental question by developmental biologists for more than a century. In the case of bone, as observed with other systems (Fig. 1), the relationship of growth and differentiation must be maintained and stringently regulated, both during development and throughout the life of the organism to support tissue remodeling.

For many years, bone was defined anatomically and examined largely in a descriptive manner by ultrastructural analysis and by biochemical and histochemical methods (Pechak *et al.*, 1986; Bruder and Caplan, 1989; Nijweide and Mulder, 1986). These studies provided the basis for our understanding of bone tissue organization and orchestration of the progressive recruitment, proliferation, and differentiation of the various cellular components of bone tissue. Now, complemented by an increased knowledge of molecular mechanisms that are associated with

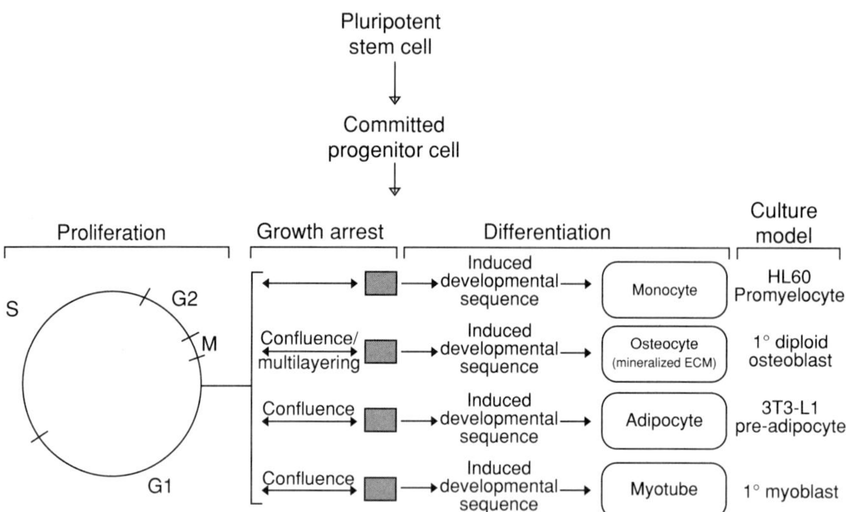

FIGURE 1 Schematic representation of the events associated with progressive expression of monocyte, granulocyte, osteoblast, adipocyte, and myotube phenotypes. Initially, the cells (HL-60 promyelocytic leukemia cells, primary cultures of embryonic calvarial osteoblasts, 3T3-L1 preadipocytes, or myoblasts) actively proliferate, expressing cell cycle and cell growth-regulated genes as well as genes encoding extracellular matrix (ECM) proteins. After growth arrest, a developmental sequence involving the sequential and selective expression of genes that result in the differentiated cell and tissue phenotype occurs. Completion of the proliferation period marks an important transition point where expression of tissue-specific genes, often functionally coupled to the down-regulation of proliferation, is initiated.

and regulate expression of genes encoding phenotypic components of bone and those that may control the progressive development and maturation of the bone cell phenotype, our understanding of bone cell and tissue differentiation is rapidly expanding.

In this chapter, we present an overview of approaches that we have been pursuing to address experimentally the proliferation–differentiation relationship during progressive development of the osteoblast phenotype. We will focus on several longstanding and fundamental questions which provide a conceptual framework to postulate mechanisms that can serve as a basis for further investigating the manner in which cell growth supports initial events associated with biosynthesis of the bone extracellular matrix while suppressing genes expressed postproliferatively. The implications of aberrations in growth control for deregulation of genes associated with bone cell differentiation in transformed osteoblasts and osteosarcoma cells will also be considered.

II. THE OSTEOBLAST DEVELOPMENTAL SEQUENCE: A FUNCTIONALLY COUPLED, INTEGRATED RELATIONSHIP BETWEEN EXPRESSION OF CELL GROWTH AND OSTEOBLAST-RELATED GENES

Over the past two decades, a foundation for our understanding of bone cell differentiation has been established, first by the identification and characterization of proteins secreted by osteoblasts as well as those associated with the bone extracellular matrix (structural proteins and growth factors) and more recently by analysis of cellular messenger RNA (mRNA) transcripts. It has been recognized, however, that defining mechanisms operative in the development and maintenance of the osteoblast phenotype necessitates comprehensive knowledge of cell growth and tissue-specific genes expressed with regard to (1) the time course and extent of expression, (2) temporal versus functionally coupled relationships, (3) control at both transcriptional and a series of posttranscriptional levels, and (4) signaling pathways and macromolecules that integrate the activities of physiological regulatory mediators. The principal concepts that have emerged include, but are not restricted to, (1) proliferation supporting initial stages of bone cell-related gene expression while restricting expression of postproliferative genes, and (2) down-regulation of proliferation contributing to induction of genes expressed following completion of osteoblast proliferative activity. This fundamental relationship between proliferation and differentiation has defined both a biological problem and an experimental approach that is being examined in a broad spectrum of cells and tissues undergoing differentiation.

A. The Ordered Expression of Genes during Development of the Osteoblast Phenotype

The development of methods for culture of normal diploid calvarial-derived osteoblasts under conditions that support development of a tissuelike organization (Bellows *et al.*, 1986; Bhargava *et al.*, 1988) similar to embryonic bone (Yoon *et al.*, 1987) provided the basis for experimentally addressing the proliferation–differentiation relationship as reflected by structural, biochemical, and molecular parameters within the context of physiological regulation (Owen *et al.*, 1990a).

By trypsin-collagenase digestion of fetal rat calvaria, we obtain a population of cells from the third sequential digest that express significant levels of osteoblast phenotypic markers including alkaline phosphatase (APase), osteopontin, and osteocalcin mRNA (Owen *et al.*, 1990a; Aronow *et al.*, 1990a). When placed into culture, these bone cell genes are down-regulated, and every plated cell initiates proliferation, as indicated by [^{3}H]thymidine autoradiography (Fig. 2). The cells multilayer, forming nodules. When proliferation ceases, APase levels increase sufficiently for histochemical detection. An ordered deposition of mineral initiates within the extracellular matrix of these nodules, resulting in the development of a bone tissue-like organization, with an orthogonal arrangement of collagen layers between osteoblasts and a mineralized extracellular matrix.

The sequential expression of cell growth and tissue-specific genes has been mapped by Northern blot analysis (Owen *et al.*, 1990a) and *in situ* hybridization (Pockwinse *et al.*, 1992) during progressive development of the bone cell phenotype from a proliferating cell to a mature osteocytic cell in a mineralized type I collagen extracellular matrix (Aronow *et al.*, 1990a; Owen *et al.*, 1990a) (Fig. 3). This temporal sequence of gene expression has defined three distinct periods: growth (proliferation) and extracellular matrix biosynthesis; extracellular matrix development, maturation, and organization; and extracellular matrix mineralization. The biological relevance of the osteoblast culture system as a model for bone cell differentiation is supported by a sequence of gene expression (cell growth and osteoblast-related genes) that is similar to the pattern of gene expression and tissue distribution determined by *in situ* hybridization in neonatal bones (Weinreb *et al.*, 1990; Sandberg *et al.*, 1988; Lyons *et al.*, 1989; Nomura *et al.*, 1988) and during fetal calvarial development *in vivo* (Yoon *et al.*, 1987).

During the first 10–12 days of culture following isolation of fetal rat calvarial-derived osteoblasts, a period of active proliferation is reflected by mitotic activity with expression of cell cycle- (e.g., histone) and cell growth- (e.g., c-*myc*, c-*fos*, c-*jun*) regulated genes. These genes encode proteins that support proliferation by functioning as transactivation fac-

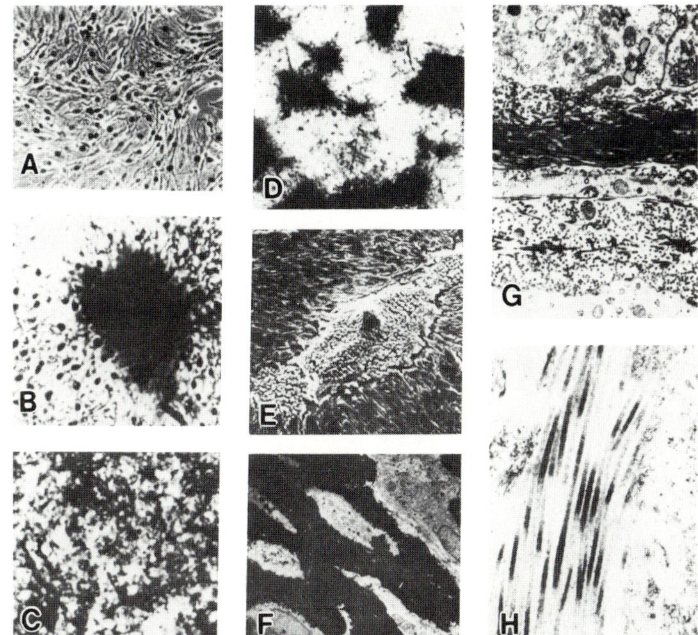

FIGURE 2 (A–D) Morphology of normal rat diploid osteoblast cultures. (A and B) Light microscopy of cultured osteoblasts isolated from 21-day-old fetal calvaria bone by sequential collagenase digestion (Aronow *et al.*, 1990a). Cells labeled with [^{3}H]-thymidine show more than 95% of the cells are proliferating after plating (A, day 5). The cells multilayer, forming nodules (B, day 12), and the onset of expression of the osteoblast phenotype is indicated by alkaline phosphatase staining of cells where proliferation has ceased in the center of the nodule. (C) Day 18 shows alkaline phosphatase histochemistry in more than 90% of the cells. (D) Day 28 shows mineralized nodules throughout the dish. Von Kossa silver staining (A–D) photographed at 100×. (E) Scanning electron micrograph of a nodule on day 28 at 20×. (F–H) Transmission electron microscopy. (F) Mineralized matrix enveloping osteoblasts. (G) Orthogonal organization of the collagen matrix in thick layers between the osteoblasts *in vitro*. (H) Higher magnification of the collagen bundles shows mineral deposition within the collagen fibrils and absence of intracellular calcification.

tors in the case of c-*myc* and c-*fos* and as proteins that play a primary role in packaging newly replicated DNA into chromatin in the case of histones. Modifications in expression of proliferation-related genes (e.g., c-*fos* or c-*sarc* expression *in vivo* [in transgenic animals]) results in altered bone formation (Rüther *et al.*, 1987; Soriano *et al.*, 1991), indicating the importance of their regulated expression during the proliferative period for control of cell growth as well as for control of genes later during osteoblast development that directly involves bone cell differentiation.

During this proliferation period, and fundamental to development of the bone cell phenotype, several genes associated with formation of

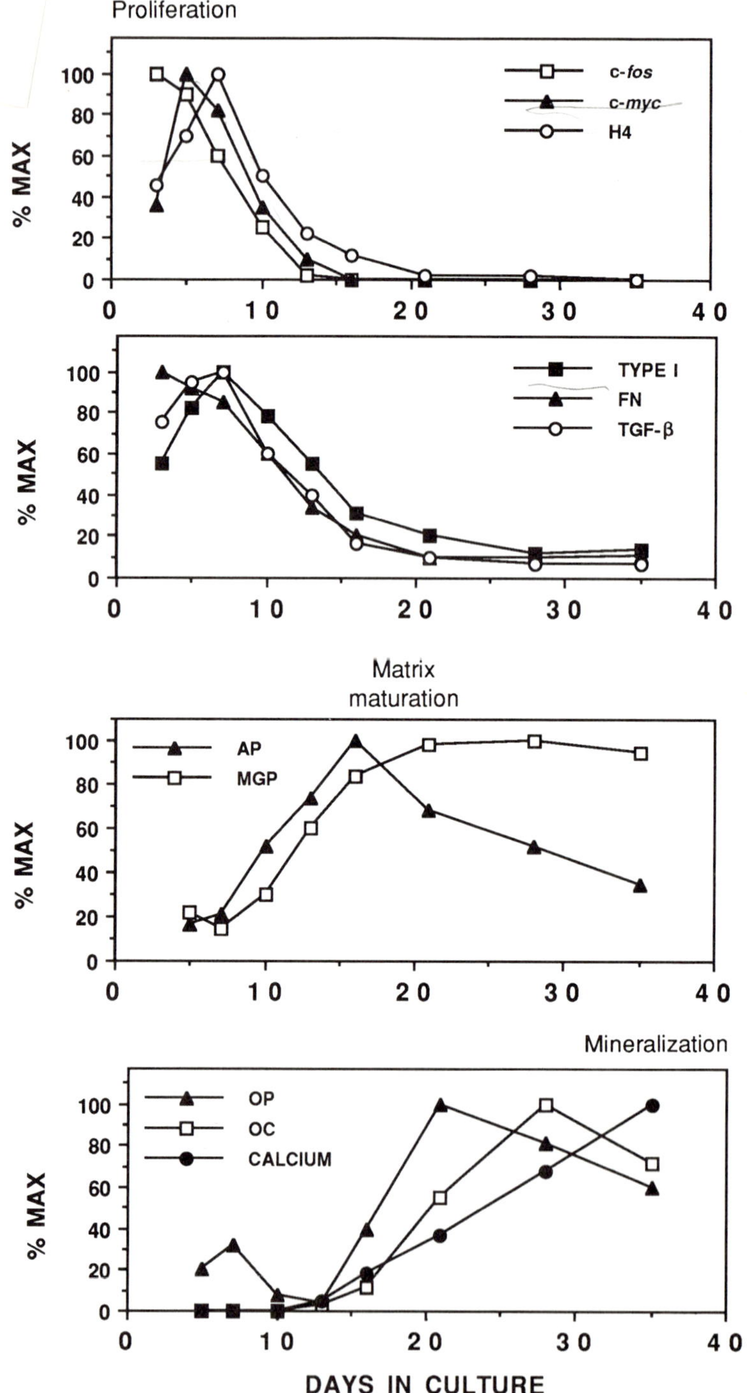

Proliferation
c-fos
c-myc
H4
% MAX
TYPE I
FN
TGF-β
Matrix
maturation
AP
MGP
Mineralization
OP
OC
CALCIUM
DAYS IN CULTURE

the extracellular matrix (type I collagen, fibronectin, and transforming growth factor β [TGF-β]) are actively expressed and then gradually down-regulated with collagen mRNA being maintained at a low basal level during subsequent stages of osteoblast differentiation. The parallel relationship of high TGF-β levels with type I collagen gene expression in cultured osteoblasts (Owen *et al.*, 1990a) is similar to the pattern of expression observed during endochondral bone formation *in vivo* (Carrington *et al.*, 1988; Bortell *et al.*, 1990; Nomura *et al.*, 1988; Lyons *et al.*, 1989) and is consistent with a major role for TGF-β in regulating extracellular matrix biosynthesis (Sporn *et al.*, 1983; Joyce *et al.*, 1990).

Immediately following the down-regulation of proliferation, proteins associated with bone cell phenotype are detected. For example, APase enzyme activity and mRNA are increased more than 10-fold. At this time, a differentiation-specific histone gene is expressed (Shalhoub *et al.*, 1989; Collart *et al.*, 1991). During the immediate postproliferative period (from 12 to 18 days), the extracellular matrix undergoes a series of modifications in composition and organization that renders it competent for mineralization. As the cultures progress into the mineralization stage, all cells become APase-positive histochemically (Fig. 2). In heavily mineralized cultures, cellular levels of APase mRNA decline (Fig. 3). At this time, maximal levels of collagenase gene expression are observed and may relate to a remodeling of the matrix to support tissue organization and mineral deposition (Shalhoub *et al.*, in press).

Several bone-synthesized proteins associate with the mineralized matrix *in vivo* (Fisher *et al.*, 1987; Franzen and Heinegard, 1985; Hauschka *et al.*, 1989; Whitson *et al.*, 1984). With the onset of mineralization, other bone-expressed genes are induced; for example, bone sialoprotein

FIGURE 3 Temporal expression of cell growth and osteoblast phenotype-related genes during the development of *in vitro* formed bonelike tissue by normal diploid rat osteoblasts. Isolated primary cells were initially cultured in modified Eagle's medium with 10% fetal calf serum (FCS) and then, after confluence, in BGJb medium supplemented with 10% FCS and 50 μg/ml ascorbic acid. Expression of genes associated with proliferation (*upper 2 panels*), matrix maturation (*third panel*), and extracellular matrix mineralization in normal diploid osteoblast cultures are indicated. Cellular RNA was isolated at the times indicated (3, 5, 7, 10, 12, 16, 21, 28, and 35 days) during the differentiation time course and assayed for the steady-state levels of various transcripts by Northern blot analysis. The resulting blots were quantitated by scanning densitometry and the results plotted relative to the maximal expression of each transcript. Three periods of gene expression are represented. During proliferation, genes are H4 histone reflecting DNA synthesis, c-*myc*, and c-*fos*. Genes expressed for formation of the bone extracellular matrix are type I collagen (TYPE I), fibronectin (FN) and transforming growth factor-β (TFG-β). During matrix maturation, alkaline phosphatase (AP) is expressed postproliferatively and associated with extracellular matrix maturation. During mineralization, genes represented that are induced to high levels with onset of extracellular matrix mineralization are osteopontin (OP) and osteocalcin (OC). Calcium (Ca^{2+}) accumulation is indicated.

(Nagata *et al.*, 1991), osteopontin, and osteocalcin (bone gla protein) (Owen *et al.*, 1990a; Gerstenfeld *et al.*, 1987, 1990) are increased, paralleling the accumulation of mineral. Osteopontin is expressed during the period of active proliferation (at 25% of maximal levels), decreases postproliferatively, and then exhibits induction at the onset of mineralization, achieving peak levels of expression during mineralization (Days 16–20). This pattern of osteopontin expression has been observed in other osteoblastic developmental systems (Strauss *et al.*, 1990; Gerstenfeld *et al.*, 1990) and may reflect multiple functional properties. Increased osteopontin expression can occur following serum stimulation of quiescent fibroblasts, oncogene transformation, or phorbol ester treatment of fibroblasts (Craig *et al.*, 1989). Here one can speculate that proliferation or tumorigenic-related functions of osteopontin may be related to control of relationships between cells and extracellular matrices, particularly in light of the Arg–Gly–Asp-containing sequence that mediates cell attachment (Oldberg *et al.*, 1986). However, the levels of osteopontin mRNA observed during the proliferation period may in part reflect mRNA transcribed *in vivo* in osteoblasts undergoing matrix mineralization prior to the isolation from fetal calvaria. Consistent with high levels of osteopontin expression later in the osteoblast developmental sequence are the calcium-binding properties of this acidic glycoprotein containing O-phosphoserine (Glimcher, 1989). Additionally, recent results suggest that osteopontin may be preferentially interacting with osteoclasts implicating this phosphoprotein in the resorption process (Reinholt *et al.*, 1990; Miyauchi *et al.*, 1991).

The vitamin K-dependent protein, osteocalcin (Lian and Friedman, 1978), in contrast to osteopontin, is expressed only postproliferatively with the onset of nodule formation. This 5.7-kDa calcium-binding protein, characterized by 3-γ-carboxy-glutamic acid residues (Gla) that bind tightly to hydroxyapatite, is maximally expressed with mineralization of the extracellular matrix *in vivo* (Hauschka *et al.*, 1989) and *in vitro* (Owen *et al.*, 1990a). A high correlation (0.92) of osteocalcin mRNA and synthesis in relation to calcification of the extracellular matrix occurs *in vitro* in rat osteoblast cultures (Aronow *et al.*, 1990a), similar to expression (Weinreb *et al.*, 1990) and protein accumulation in the bone extracellular matrix *in vivo* (Hauschka *et al.*, 1989). Several modifications of culture conditions (Owen *et al.*, 1990a) demonstrate that induction of high osteocalcin and osteopontin mRNA levels depend on formation of a mineralized extracellular matrix. These relationships are not found for another vitamin K-dependent protein in bone, matrix Gla protein (MGP) (Fraser and Price, 1988). MGP is characterized by the presence of five Gla residues in a 10-kDa molecule, is present in proliferating osteoblasts and throughout the culture period, is increased in relation to accumulation of collagen, but is not functionally coupled with extracellular matrix mineralization (Barone *et al.*, 1991). Osteocalcin has been shown to contribute to regulation of the mineral phase in bone, both *in vitro* as a potential inhibitor of mineral nucleation (Boskey *et al.*, 1985; Romberg *et al.*, 1986) and *in vivo* as a bone matrix signal that promotes osteoclast differentiation and activation (Glowacki and Lian, 1987). Thus, expression late in the osteoblast development sequence suggests that osteocalcin is a marker of the mature osteoblast, which is consistent with a possible role for the synthesis and binding of osteocalcin to mineral in the coupling of bone formation to resorption. Within the context of involvement of osteocalcin in bone resorption, osteocalcin mRNA levels and accumulated protein are significantly decreased in bone of the TL and OP osteopetrotic rats compared to normal littermates that have reduced numbers of osteoclasts (Lian and Marks, 1990; Shalhoub *et al.*, 1991).

B. Two Transition-Restriction Points during Development of the Osteoblast Phenotype Characterize the Proliferation–Differentiation Relationship

The sequential and stringently regulated expression of genes that defines three principal periods of osteoblast phenotype development (proliferation, extracellular matrix development and maturation, and mineralization) is schematically illustrated in Fig. 4 as a reciprocal and functionally coupled relationship between proliferation and differentiation. Two transition points are key elements of this temporal expression of genes that support cell growth and differentiation: The first transition point is at the completion of the proliferation period, when genes for cell cycle and cell growth control are down-regulated and expression of genes encoding proteins for extracellular matrix maturation and organization is initiated, and the second is at the onset of extracellular matrix mineralization. These transitions have been experimentally established (Owen *et al.*, 1990a) and functionally defined as restriction points during osteoblast differentiation, to which developmental expression of genes can proceed but cannot pass, without additional cellular signaling. Thus, establishing the basis of cellular competency for progression toward development of the mature osteoblast phenotype necessitates identification of the signaling pathways operative at the developmental transition points by which genes are selectively activated and/or suppressed. Additionally, the regulatory mechanisms that serve as the rate-limiting steps at these strategic points during osteoblast differentiation must be characterized. Understanding biological regulatory mechanisms that support bone cell development within the context of the proliferation–differentiation relationship provides a viable basis for addressing skele-

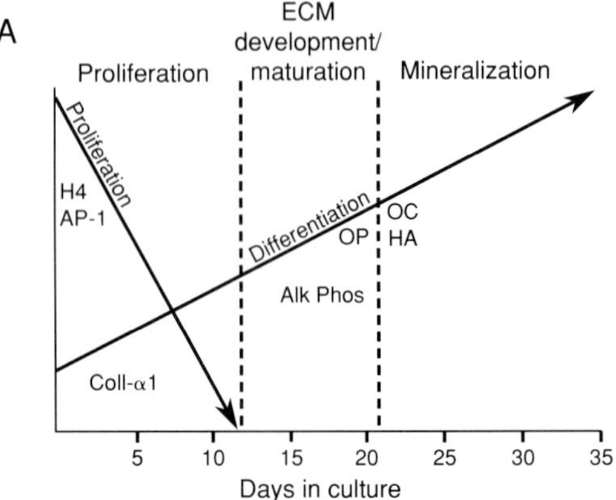
A
ECM development/ maturation
Proliferation
Mineralization
Proliferation
H4
AP-1
Differentiation
OP
OC
HA
Alk Phos
Coll-α1
5
10
15
20
25
30
35
Days in culture
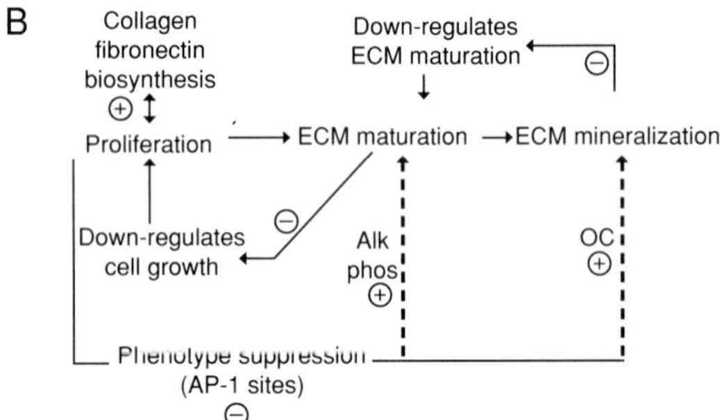
B
Collagen fibronectin biosynthesis
Down-regulates ECM maturation
Proliferation
ECM maturation
ECM mineralization
Down-regulates cell growth
Alk phos
OC
(AP-1 sites)
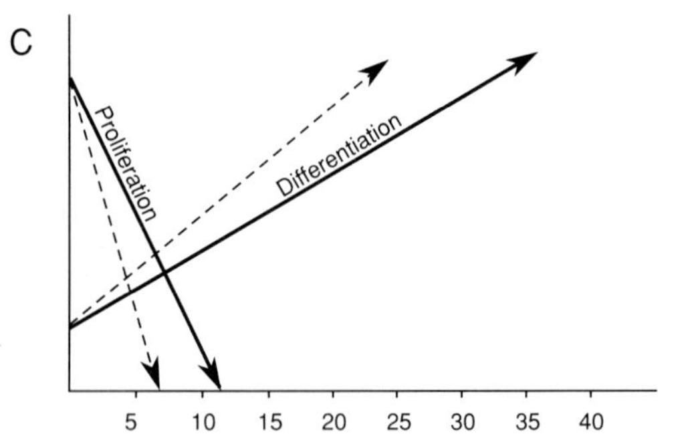
C
Proliferation
Differentiation
5
10
15
20
25
30
35
40
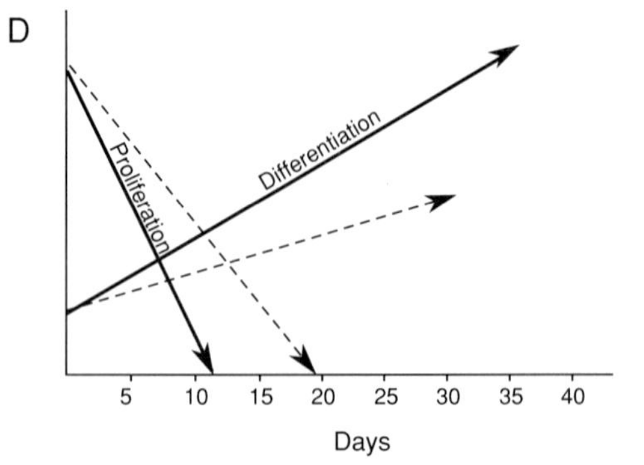
D
Proliferation
Differentiation
5
10
15
20
25
30
35
40
Days

tal disorders and, particularly, bone cell cancer, where aberrations in the proliferation–differentiation relationship may facilitate both diagnosis and evaluation of targeted therapy.

The relationship between cell growth and expression of osteoblast-related genes can be addressed experimentally, as schematically illustrated in Figure 4, by abbreviating or extending the proliferation period of the osteoblast developmental sequence and determining which genes associated with extracellular matrix biosynthesis, maturation, or mineralization are prematurely up-regulated or delayed in initiation of expression. The growth–differentiation relationship can be further approached by studies in which the onset and progression of differentiation are systematically modified and by observing consequent changes in both expression of genes during proliferation and the influence on proliferative activity. Table I and several recent reviews (Lian *et al.*, 1992; Stein *et al.*, 1992a) summarize examples of this experimental approach. Indicated is how we have used both biochemical interventions and hormonal modulation of the osteoblast developmental sequence to distinguish the extent to which expression of specific genes are components of a

FIGURE 4 Reciprocal and functionally coupled relationship between cell growth and differentiation-related gene expression. (A) These relationships are schematically illustrated as arrows representing changes in expression of cell cycle- and cell growth-regulated genes. The three principal periods of the osteoblast developmental sequence are designated within broken vertical lines (proliferation, matrix development and maturation, and mineralization). These broken lines indicate the two experimentally established principal transition points in the developmental sequence exhibited by normal diploid osteoblasts during the progressive acquisition of the bone cell phenotype—the first at the completion of proliferation when genes associated with matrix development and maturation are up-regulated, and the second at the onset of ECM mineralization. (B) A series of signaling mechanisms are illustrated whereby the proliferation period supports the synthesis of a type I collagen–fibronectin ECM, which continues to mature and mineralize. The formation of this matrix down-regulates proliferation, and matrix mineralization down-regulates the expression of genes associated with the ECM maturation period. The occupancy of AP-1 sites in the osteocalcin (OC) and the alkaline phosphatase (ALK PHOS) gene promoters by Fos–Jun and/or related proteins are proposed to suppress both basal 10 m*M* β-glycerol phosphate. (C and D) An experimental approach to assess the influence of proliferation on expression of genes associated with osteoblast differentiation by determining consequences of growth inhibition on differentiation parameters (i.e., establishing which are prematurely induced and/or exhibit elevated levels of expression [C]). This will facilitate distinguishing changes in the levels of expression of osteoblast (differentiation) genes that are coupled to the proliferative state of the cell versus those genes that become temporally expressed, dependent on the formation and organization of the ECM following the completion of proliferative activity. (D) Potential consequences of extending the proliferation period or delaying the onset of the differentiation. In this way, one can further establish which differentiation-related genes are expressed dependent on the shut-down of proliferation (i.e., which proliferation-related events depend on formation of the bone ECM. OP, osteopontin.

TABLE I Experimental Results Supporting a Functional Relationship between Proliferation and Differentiation during Progressive Development of the Osteoblast Phenotype

Biological modification	Modifier	Influence on osteoblast phenotype parameter
1. Proliferation inhibition	Hydroxyurea	Up-regulation of alkaline phosphatase and osteopontin; no effect on osteocalcin. Induction to, but not past, onset of extracellular matrix mineralization.
2. Delayed extracellular matrix mineralization	Absence of β-glycerol phosphate in culture media	No change in gene expression during proliferation or extracellular matrix mineralization/organization; delayed extracellular matrix mineralization and osteocalcin gene expression
3. Accelerated differentiation	Dexamethasone	Increased number of nodules. Accelerated up-regulation of postproliferative osteoblast phenotype genes
4. Block differentiation	Chronic vitamin D treatment/ hydroxyurea treatment	Inhibition of collagen, mineralized matrix formation, and subsequent expression of osteoblast related genes
5. Decrease extracellular matrix biosynthesis	Decreased concentration or elimination of ascorbic acid	Increased proliferation and decreased expression of collagen and post-proliferative osteoblast-related genes
6. Increase representation of collagen matrix	Plate cells on/in type I collagen	Increased expression of alkaline phosphatase and osteocalcin
7. Promote differentiation	Acute vitamin D treatment	Gene expression reflects the mature osteoblast phenotype (e.g., decreased collagen and alkaline phosphatase gene expression and increased osteopontin and osteocalcin gene expression)

temporal sequence and those that are functionally coupled (i.e., causally related).

The biochemical modifications presented in Table I indicate how inhibition of proliferation supports up-regulated expression of genes that are expressed only during progression of the osteoblast developmental sequence up to the stage where mineralization is initiated (e.g., APase and osteopontin but not osteocalcin). Additionally, there is evidence for the concept that expression of at least a second set of genes is not directly coupled to the down-regulation of proliferation but, rather, to development of the more differentiated osteoblast in a mineralized matrix. In the absence of mineralization, osteopontin and osteocalcin are not induced to high levels and APase does not decline. Expression of some bone cell genes are not effected, such as collagen, MGP, and osteonectin.

Although unquestionably a simplification of an extremely complex series of biological interactions, the temporal pattern of expression suggests a working model for the relationship between growth and differentiation whereby genes involved in the production and deposition of the extracellular matrix must be expressed during the proliferative period for differentiation to occur. One can postulate that proliferation is functionally related to the synthesis of a bone-specific extracellular matrix and that the maturation and organization of the extracellular matrix contributes to the shutdown of proliferation, which then promotes expression of genes that render the matrix competent for mineralization, a final process that is essential for complete expression of the mature osteoblast phenotype (Fig. 4). The onset of extracellular matrix mineralization and/or events earlier during the mineralization period may be responsible for the down-regulation of genes expressed during extracellular matrix maturation and organization. Clearly, in this model the development of an extracellular matrix is integrally related to the differentiation stages. This relationship is evident from a series of studies in which cells were cultured at various concentrations of ascorbic acid (Owen *et al.*, 1990a). With resulting higher levels of collagen synthesis and accumulation of the extracellular matrix, proliferation ceases at a lower cell density, and, coordinately, APase mRNA levels and enzyme activity per cell are greater. Thus, a contribution of signals from the extracellular matrix promotes progressive differentiation of the osteoblast, and this is indicated by findings in other cell culture systems (Quarles *et al.*, 1992; Franceschi and Young, 1990; Aronow *et al.*, 1990a).

A complementary experimental approach substantiating the support and/or inductive effect of the collagen extracellular matrix in osteoblast differentiation is culturing primary diploid osteoblasts on type I collagen film or in a collagen gel. Here, we observe an accelerated progression of osteoblast phenotype development, indicated by early and enhanced expression of APase and osteocalcin, which is in agreement with a recent report by Reddi and co-workers (Vukicevic *et al.*, 1990). Thus, our working model provides a basis for addressing whether particular stages of osteoblast differentiation exhibit selective responsiveness to actions of hormones and other physiological factors that influence osteoblast activity as well as other questions related to molecular mechanisms associated with bone formation.

III. HORMONE MODIFICATIONS ON DEVELOPMENT OF THE OSTEOBLAST PHENOTYPE

Hormonal effects on the growth–differentiation relationship have provided a viable experimental approach to assess the consequences of accelerated or delayed differentiation on expression of cell growth and tissue-specific genes related to development of the osteoblast phenotype. Glucocorticoids promote differentiation of progenitor cells to the osteoblast phenotype (Leboy *et al.*, 1991). Chronic treatment of osteoblast cultures with dexamethasone, a synthetic glucocorticoid, increases the number of mineralized bone nodules in primary fetal rat calvarial osteoblast cultures (Bellows *et al.*, 1987). In subcultivated cells, the effect is even more pronounced since initiation of the osteoblast developmental sequence in the absence of 10^{-7} *M* dexamethasone is significantly delayed compared to progression of the developmental sequence in primary osteoblast cultures (Aronow *et al.*, 1990a). The pattern of gene expression in passaged cells is modified by dexamethasone in that glucocorticoids promote expression of the differentiation parameters to levels observed in primary cell cultures (Fig. 5). Osteopontin, for example, is increased by dexamethasone during the proliferation period, whereas APase and osteocalcin are up-regulated parallel to an increase in the number of bone nodules and earlier induction of mineralization compared to non-dexamethasone-treated cultures (Fig. 5). Thus, expression of these osteoblast phenotypic markers reflects the normal development temporal sequence of bone nodule formation. Whether some of the relationships between genes that are coupled or functionally related have been altered due to direct modulation by glucocorticoids or whether the accelerated differentiation is secondary to other modifications in osteoblast developmental parameters remains to be established. However, the presence of several glucocorticoid-responsive genes (Stromstedt *et al.*, 1991; Heinrichs *et al.*, 1992; Canalis, 1983) expressed during osteoblast growth and differentiation and the effects of glucocorticoids on mRNA stability (Shalhoub *et al.*, in press) are consistent with the observed modifications in physiological control of the osteoblast phenotype.

Hormone modulation of osteoblast growth and differentiation is also illustrated by the effects of vitamin D. From an historical perspective, it has been well known for a number of years that vitamin D anabolically and catabolically modulates bone cell metabolic activities (Chendal, 1983; Marie and Travers, 1983; Marie *et al.*, 1985; Wronski *et al.*, 1986), and more recently it has become apparent that this occurs through selective expression of a series of vitamin D-responsive genes (Majeska and Rodan, 1982; Kim and Chen, 1989; Harrison *et al.*, 1989; Prince and Butler, 1987; Noda *et al.*, 1990; Owen *et al.*, 1991; Franceschi *et al.*, 1988). Vitamin D treatment of osteoblasts alters levels of gene expression to one that reflects a more mature differentiated cell (Fig. 6). For example, with short-term exposure (up to 48 hr), proliferation is down-regulated with collagen and APase mRNA levels, whereas osteocalcin, MGP, and osteopontin are up-regulated. This hormone-responsive pro-

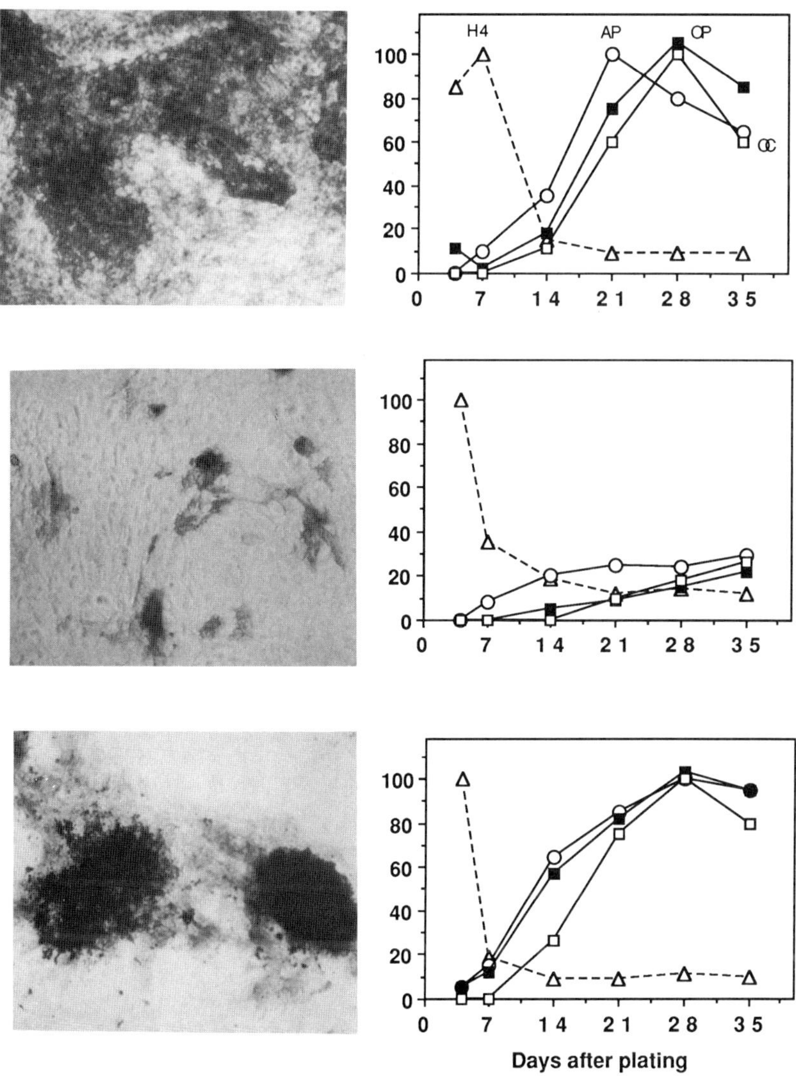

FIGURE 5 Glucocorticoid influence on osteoblast growth and differentiation. (Left) Histochemical staining of cultures for alkaline phosphatase and mineral (von Kossa) to show promotion of bone nodule formation by dexamethasone. Top panel shows primary cultures, middle panel secondary osteoblast cultures, and lower panel secondary cultures treated continuously with 10^{-7} *M* dexamethasone. All cultures shown are 28 days after plating stained. (Right) Corresponding levels of gene expression in the primary cultures second passage rat osteoblasts in the absence (middle) or presence of 10^{-7} *M* dexamethasone (lower). The mRNA levels of the genes in passaged cells were quantitated relative to the dexamethasone-treated cells from densitometric readings normalized to rRNA. Note the earlier down-regulation of proliferation (H4 histone) in second-passage cells and up-regulation of alkaline phosphatase (AP), osteopontin (OP), and osteocalcin (OC) by dexamethasone.

file of gene expression is similar to that observed in the mature osteoblasts in a mineralized matrix. However, duration of exposure of osteoblasts to vitamin D is another consideration of hormone action on growth and differentiation. Chronic treatment of osteoblasts with 10^{-8} *M* 1,25$(OH)_2D_3$, if initiated during the proliferation period, can in fact block differentiation (Fig. 6). Here the down-regulation of collagen and APase gene expression by vitamin D prevents formation and mineralization of the bone extracellular matrix. Thus, there is a consequential absence of expression of the mature osteoblast phenotype genes such as osteocalcin and osteopontin, which are coupled to mineral deposition. When continuous vitamin D treatment is initiated after the onset of mineralization, when basal expression of osteocalcin is occurring, the osteocalcin gene is then up-regulated (Fig. 6) during continuous exposure to vitamin D. Thus, at least *in vitro,* differentiation to the mature osteoblast phenotype in rat fetal-derived cells selectively depends on formation of the bonelike extracellular matrix.

Not only do hormones influence the osteoblast growth and differentiation relationship, but hormone responsiveness also is a function of the developmental stage of the osteoblast. This is illustrated by several findings, including (1) differential stimulation of APase by vitamin D dependent upon the basal level of expression in subclones of the ROS 17/2.8 cell (Speiss *et al.*, 1986; Fraser and Price, 1990), and (2) reports of either stimulatory or inhibiting effects of vitamin D on collagen synthesis in various cell lines and culture systems (Kyeyume-Nyombi *et al.*, 1989; Franceschi *et al.*, 1988; Canalis and Lian, 1985). In the normal diploid rat osteoblast cultures at different stages of phenotype development, from the proliferation period to the mature osteocytelike cell in the mineralized matrix, we observed pleiotropic effects on several vitamin D-regulated genes that depend on the stage of osteoblast differentiation or maturation (Owen *et al.*, 1991). For example, histone H4, collagen and APase are down-regulated in proliferating osteoblasts, but in mature osteoblasts in a mineralized matrix, an up-regulation is observed. This latter effect of 1,25$(OH)_2D_3$ *in vitro* is similar to the action of this hormone *in vivo* (Wronski *et al.*, 1986; Hock *et al.*, 1986). Additionally, hormone responsiveness of a particular gene depends on its basal level of expression, as illustrated by a requirement of basal expression for hormone stimulation and the differences in extent of hormone regulation. Osteocalcin and osteopontin are stimulated 20-fold greater when basal levels are low. Confirmation of this concept of developmental responsiveness of the osteoblast to vitamin D and glucocorticoid has recently been provided in two studies. Chick osteoblasts were isolated at different embryonic ages, where early embryonic cells were promoted to differentiate by vitamin D and late embryonic cells with mature osteoblast phenotype properties showed an inhibition of these functions in re-

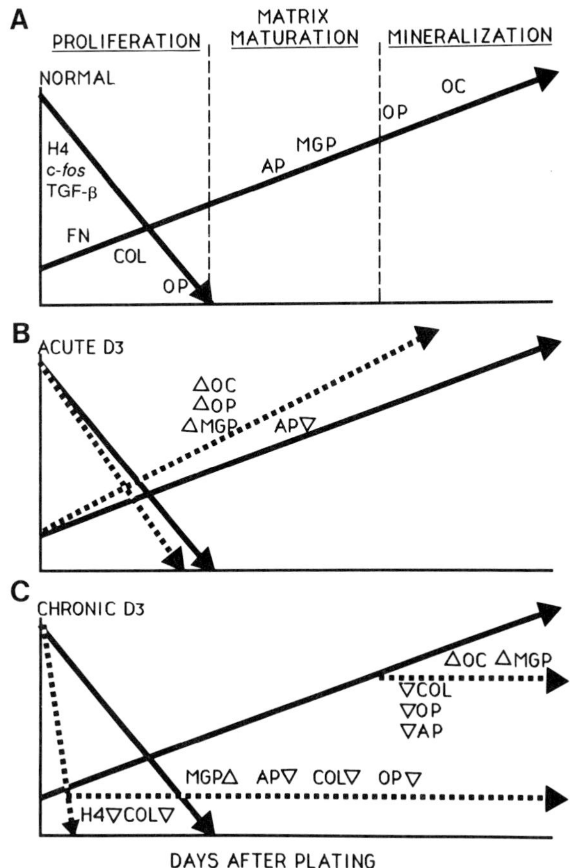

FIGURE 6 Vitamin D-related modifications of the proliferation–differentiation relationship in normal diploid cells during the rat osteoblast developmental sequence. (A) The proliferation–differentiation relationships are schematically illustrated for normal developmental stages as described in Fig. 4 for comparison to the lower panels. (B) Schematically illustrates the acceleration of the relationship between growth and differentiation by vitamin D in short-term treatment (up to 48 hr). Hormone accelerates proliferation and differentiation (broken arrows) by changing levels of gene expression in one period to that found in the following period. (C) The inhibitory effects of vitamin D on differentiation of osteoblasts *in vitro* (broken arrows) are illustrated for continuous hormone treatment initiated on day 5 or day 20 reflected by inhibition (day 5) or cessation (from day 20) of the formation and growth of mineralized nodules. AP, alkaline phosphatase; COL, type ∝I collagen; FN, fibronectin; H4, histone; MGP, matrix Gla protein; OC, osteocalcin; OP, osteopontin; TGF-β, transforming growth factor β.

sponse to hormone (Broess *et al.*, submitted). Studies by Turksen and Aubin (1991) examining glucocorticoid effects in promoting bone nodule formation showed 30-fold greater effect on APase-negative osteoprogenitor cells compared to mature APase-positive osteoblasts.

IV. MOLECULAR MECHANISMS OPERATIVE IN DEVELOPMENTAL EXPRESSION OF A CELL GROWTH AND BONE-SPECIFIC GENE DURING OSTEOBLAST DIFFERENTIATION

Two striking examples of genes in which expression is controlled at multiple levels and modified with respect to competency for responsiveness to physiological regulatory signals at various stages of osteoblast differentiation are the cell cycle-regulated histone genes at the proliferation–differentiation transition point and the osteocalcin gene at the onset of extracellular matrix mineralization. Transcriptionally, these genes are controlled by promoters with modular organizations of positive and negative regulatory elements that interact in a sequence-specific manner with a diverse series of physiological mediators. Additional options exercised by the osteoblast during differentiation for modulating transcriptional control of the histone and osteocalcin genes include the recruitment of transcription factors and the extent to which sequence-specific DNA binding proteins are phosphorylated (Roesler *et al.*, 1988). Messenger RNA stability also accounts for changes in the extent to which the histone and osteocalcin genes are expressed during the osteoblast developmental sequence where cell cycle variations during proliferation and effects of steroid hormones influence the rates of histone and osteocalcin mRNA turnover.

A. Histone Gene Regulation at the Proliferation–Differentiation Transition Point

As indicated in Fig. 7, both transcriptional control (two- to threefold enhancement) and regulation of histone mRNA stability contribute to coupling histone gene expression with DNA replication, accounting for the restriction of histone protein synthesis and the cellular representation of histone mRNA to the S phase of the cell cycle (Owen *et al.*, 1990a; Plumb *et al.*, 1983; Ramsey-Ewing *et al.*, submitted; Marashi *et al.*, 1982; Stein *et al.*, 1975; Holthuis *et al.*, 1990). Then, at the completion of the proliferation period, when osteoblasts cease to traverse the cell cycle, histone gene expression is completely down-regulated transcriptionally (and histone mRNA selectively destabilized) (Morris *et al.*, 1991; Sierra *et al.*, 1983; Baumbach *et al.*, 1987; Owen *et al.*, 1990c; Stein *et al.*, 1992b). We confirmed the contribution of transcriptional control to histone gene expression in transgenic animals, carrying a histone gene promoter-CAT fusion gene construct where cell growth-related expression was observed in calvaria, long bone, and calvarial-derived osteoblast cultures (Gerbaulet *et al.*, 1992). We identified regulatory elements in the histone gene promoter designated Sites I–IV (Fig. 8) by deletion and site-specific

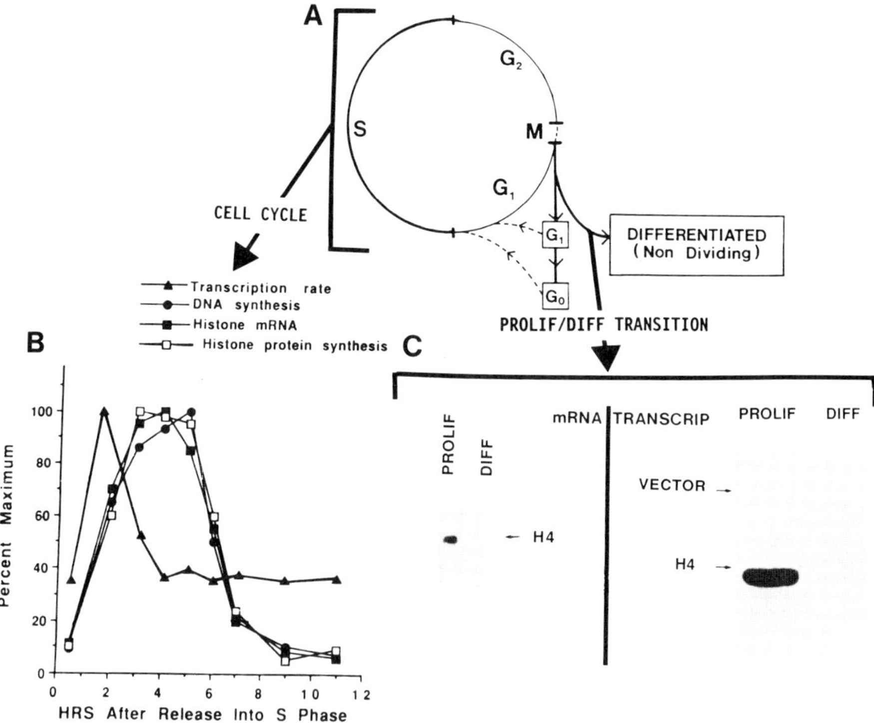

FIGURE 7 Regulation of the histone H4 gene in osteoblasts. (A) Schematic representation of the cell cycle (G_1, S, G_2, mitosis), indicating the pathway associated with the post-proliferative onset of differentiation initiated following completion of mitosis (proliferation–differentiation [PROLIF/DIFF] transition). (B) Data defining the principal biochemical parameters of histone gene expression, indicating the restriction of histone protein synthesis and the representation of histone messenger RNA (mRNA) to S-phase cells (DNA synthesis). Constitutive transcription of histone genes is evident throughout the cell cycle with an enhanced transcriptional level during the initial 2 hr of S phase. These results establish the combined contribution of transcription and mRNA stability to the S phase-specific regulation of histone biosynthesis in proliferating cells with histone mRNA levels as the rate-limiting step. (C) In contrast, the completion of proliferative activity at the onset of differentiation is mediated by transcriptional down-regulation of histone gene expression, supported by a parallel decline in rate of transcription (nuclear run-on analysis) and cellular mRNA levels (Northern blot analysis).

mutagenesis that contribute independently and cooperatively to transcription, both *in vitro* (cell-free) (Sierra *et al.*, 1982) and *in vivo* in intact cells (Ramsey-Ewing *et al.*, submitted; Kroeger *et al.*, 1987) and in transgenic animals (van Wijnen *et al.*, 1991a; Gerbaulet *et al.*, 1992). These sequences modulate promoter activity during the cell cycle, the down-regulation of transcription that accompanies the onset of osteoblast differentiation and the subsequent postproliferative repression of histone

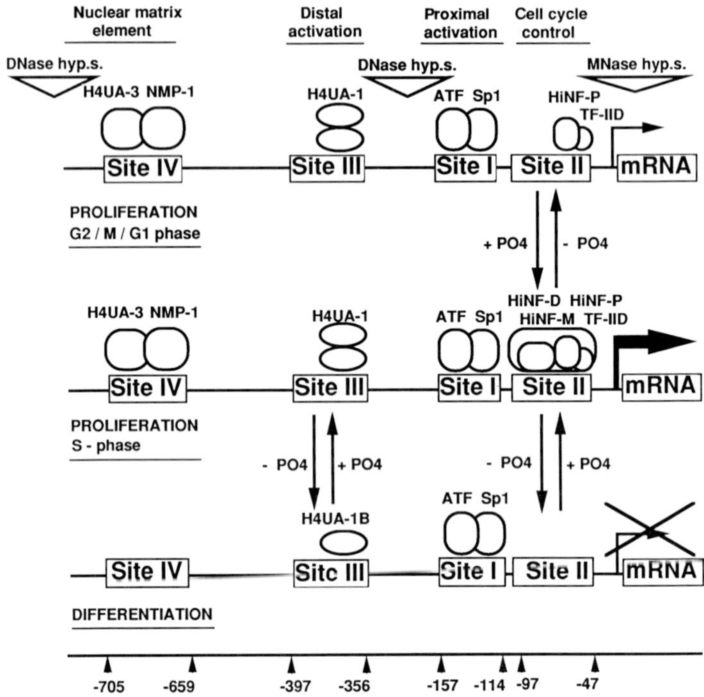

FIGURE 8 Regulatory organization of the H4-F0108 gene promoter. Each line represents the overall modular organization of the H4-F0108 gene promoter (top panel), which is reflected by distal (sites IV and III) and proximal (sites I and II) domains (boxes) that are interspersed with several nuclease hypersensitive regions (inverted triangles). Multiple promoter-binding proteins interact with each site (squares, rectangles, and ovals) and determine the extent and timing of histone gene transcription (arrow indicates transcriptional start site). Sites I and II are *in vitro* protein–DNA interaction domains established in the intact cell by genomic DNaseI footprinting and genomic dimethylsulfate fingerprinting. Each domain spans a series of sequence motifs and coincide with several *in vivo* and *in vitro* transcriptional elements that influence cap-site initiation of mRNA, basal levels of transcription, and/or the cell cycle periodicity of histone gene transcription. Interactions with sequence-specific *trans*-acting factors have been established. Sites III and IV are distal activating protein–DNA interaction domains, with site IV being a component of a putative nuclear matrix attachment site. Lines 1–3 represent the regulation of H4-histone gene transcription by occupancy of Site II by HiNF-D, HiNF-M, HiNF-P, and TF-IID during the S and G2/M/G1 phases of the cell cycle, resulting in basal (thin arrow) or maximal levels (thick arrow) of transcription. Shutdown of transcription at the cessation of proliferation (thin arrow covered by x) coincides with modifications of protein–protein interactions at site III and down-regulation of site II occupancy by cognate factors. The contribution of phosphorylation to protein–DNA interactions at site II and protein–protein interactions at site III are also indicated.

genes when expression of genes associated with development and maintenance of the osteoblast phenotype is ongoing. By a systematic examination of protein–DNA interactions in the proximal histone gene regulatory elements *in vivo* and *in vitro*, we have defined at single nucleotide resolution, a complex series of transcription factors that interact with Sites I–IV (Fig. 8) (van der Houven van Oordt *et al.*, 1992; Pauli *et al.*, 1987; van Wijnen *et al.*, 1989; Stein *et al.*, 1989). Protein–DNA interactions, together with protein–protein associations at these regulatory elements, exhibit modifications that accompany and appear to be functionally related to modifications in histone gene promoter activity in proliferating osteoblasts and at the proliferation–differentiation transition point.

At Site II, where we have demonstrated that sequences responsible for cell cycle control of this gene reside (Ramsey-Ewing *et al.*, submitted), we identified four transcription factors. H4TFII and HiNF-P exhibit constitutive binding throughout the cell cycle while HiNF-D and HiNF-M are stringently cell cycle-regulated with occupancy at Site II restricted to S phase (Pauli *et al.*, 1987; van Wijnen *et al.*, 1989; Stein *et al.*, 1989). Transcription factors interacting at Site I, ATF and Sp1, remain constitutive during the cell cycle. Similarly, promoter factor interactions at Sites III (H4UA-1 and H4UA-2) and IV (NMP-1) are retained during G1, S, G2, and mitosis. However, at the onset of differentiation, the down-regulation of histone gene transcription is accompanied by complete loss of transcription factor occupancy at Site II and accelerated electrophoretic mobility of a protein–DNA complex at Site III (H4UA-b) (Owen *et al.*, 1990c; Holthuis *et al.*, 1990; van Wijnen *et al.*, 1991b; van der Houven van Oordt *et al.*, 1992; van den Ent *et al.*, in press). Occupancy of Site I by ATF and Sp1 persists in differentiated osteoblasts, despite the absence of proliferation. We have initiated studies to pursue the mechanism by which factor-regulatory element binding is controlled. Recently, we have obtained evidence supporting the involvement of transcription factor phosphorylation in modulating interactions of DNA-binding proteins at Site II of the histone gene promoter in osteoblasts during the cell cycle and at the onset of differentiation when histone gene transcription is down-regulated (Fig. 8) (van Wijnen *et al.*, 1991a). These findings suggest a requirement of Site II transcription factor phosphorylation for mediating the up-regulation of histone gene transcription during the S phase of the cell cycle and a requirement for Site II transcription factor dephosphorylation for the termination of HiNF-D and HiNF-M–Site II interactions at the completion of S phase and the decline in transcription of histone genes. Additionally, these results provide an indication of a possible mechanism for regulating the activity of a rate-limiting histone gene transcription factor while extending the problem of transcriptional regulation during the cell cycle in proliferating osteoblasts and at the proliferation–differentiation transition in the osteoblast developmental sequence to the cellular signaling pathways that control protein phosphorylation. Other results suggest that the transition from a low mobility to a high mobility protein–DNA complex at Site III with the onset of differentiation is due to dephosphorylation of the H4UA-1 transcription factor that leads to the low mobility H4UA-1b factor interaction at Site III. Here it should be noted that both high-mobility H4UA-1 and low-mobility H4UA-2 complexes exhibit identical protein–DNA interactions at single nucleotide resolution consistent with a protein–protein association that is abrogated by dephosphorylation when histone gene transcription is repressed postproliferatively (van der Houven van Oordt *et al.*, 1992).

B. Transcriptional Control of the Osteocalcin Gene Expression at Multiple Levels at the Extracellular Matrix Mineralization Transition Point

Osteocalcin is a bone-specific protein. While the functions of osteocalcin remain to be definitively established, it is reasonable to anticipate that the protein may have multiple biological activities. This is reflected by changes in bone and serum osteocalcin levels in response to and/or causally related to a broad spectrum of physiologic circumstances where homeostatic mechanisms controlling calcium metabolism are operative. The calcitropic hormones (e.g., vitamin D, parathyroid hormone) participate in regulation of cellular osteocalcin levels and the extent to which osteocalcin is secreted. The picture that is emerging is one of a protein where regulation must be mediated by multiple factors and signaling mechanisms to accommodate a diverse series of physiological requirements (reviewed in Lian and Gundberg, 1988).

1. Organization of the Osteocalcin Gene

Sequence analysis of the osteocalcin gene provides an indication of the complexity of transcriptional and post-transcriptional regulation that mediates the extent to which the osteocalcin gene is expressed during development of the osteoblast phenotype. The sequence organization of the mRNA coding region of the gene reflects the biochemical events associated with processing of the initially synthesized 10-kDa precursor (Fig. 9). In contrast to other bone-related genes (e.g., type I collagen [Lichtler *et al.*, 1989], APase [Weiss *et al.*, 1988; Zernik *et al.*, 1990; Knoll *et al.*, 1988; Matsuura *et al.*, 1990]), only a single mRNA transcript has been observed from the osteocalcin gene (Celeste *et al.*, 1986; Lian *et al.*, 1989). It appears that one transcription initiation site is utilized, and the four splicing events to fuse exons 1–4 into an mRNA encoding the osteo-

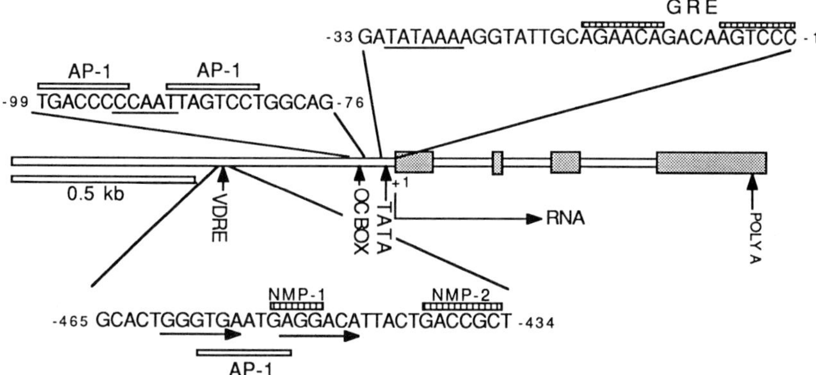

FIGURE 9 Rat osteocalcin (OC) gene promoter indicating sequences for physiological regulatory elements. Structural organization of the rat OC gene. Sequences of those regulatory elements in the proximal promoter that have been defined and partially characterized. These include the vitamin D-responsive element (VDRE); the osteocalcin box (OC box), which is a primary proximal transcription regulatory element containing the CCAAT motif as a central core; and the TATA motif contiguous to a high-affinity glucocorticoid response element (GRE) defined in the rat gene (Heinrichs *et al.*, 1992). Within the OC box and VDRE elements are also found active AP-1 sites that bind the oncogene-encoded Fos–Jun protein complex. The solid circles above or below G residues indicate vitamin D receptor protein–DNA interactions defined at single nucleotide resolution within the VDRE (Markose *et al.*, 1990). Also indicated are AP-1 sites (Owen *et al.*, 1990) and nuclear matrix protein-binding sites designated NMP-1 and NMP-2. The intron and exon (shaded boxes) organization is illustrated.

calcin prepropeptide do not vary as a function of known biological parameters. The presence of only a single osteocalcin mRNA additionally supports one functional poly(A) site and a fixed length to the 3′-poly(A) sequence. It should be noted that although regulation of osteocalcin expression does not appear to be modulated by changes in the organization of the mRNA transcripts, this does not preclude the presence of sequences in the transcribed region of the osteocalcin gene that contribute to the regulation of transcription. There is evidence indicating that sequences in the first exon suppress transcription of osteocalcin in ROS 17/2.8 cells (Frenkel *et al.*, 1992; Demay *et al.*, 1991).

The representation of consensus sequences in the promoter for regulatory elements that are responsive to steroid hormones, basal regulatory factors, and tissue-specific transactivation factors contributes to our understanding of regulatory mechanisms and biological activity. Identification and characterization of proteins that interact in a sequence-specific manner with their cognate regulatory elements in the osteocalcin gene promoter further establishes the physiological mediators of osteocalcin gene transcription and the circumstances under which they participate in regulation. Residing in the 5′ flanking sequences of osteocalcin gene promoter are sequences bearing homology to steroid receptor binding complexes (e.g., retinoic acid, vitamin D, estrogen, glucocorticoids) and to regulatory factors such as cyclic adenosine monophosphate and γ-interferon (Nanes *et al.*, 1990). These steroid hormones and other physiologic mediators have been shown to influence osteocalcin mRNA or transcription levels in osteoblastic cells (Nishimoto *et al.*, 1987; Beresford *et al.*, 1984; Theofan and Price, 1989; Noda *et al.*, 1988; Noda, 1989; Aronow *et al.*, 1990b; Lian *et al.*, 1989). As with many genes transcribed by RNA polymerase II, a TATA motif is located at −28 to −31 in the rat gene promoter and a CCAAT element at −76 to −99, and AP-1 and AP-2 sites are present at several sites in the osteocalcin gene promoter. The overlap of regulatory domains, as illustrated by the TATA/GRE, AP1/CCAAT, and AP1/VDRE (−466 to −437) provide a basis for combined activities of physiological mediators.

In the osteocalcin gene promoter, to date only four regulatory sequences, which are schematically illustrated in Fig. 9, have been established by specific criteria: (1) influence on transcription by deletion and mutational analyses and (2) identification and characterization of sequence-specific regulatory element occupancy by cognate transcription factors. The osteocalcin box, a 24-nucleotide element with a CCAAT motif as the central core, is highly conserved between the human and rat genes and an essential basal regulatory sequence for expression (Lian *et al.*, 1989). The vitamin D-responsive element (Markose *et al.*, 1990; Demay *et al.*, 1990; Terpening *et al.*, 1991; Kerner *et al.*, 1989; Morrison *et al.*, 1989) and a GRE associated with the TATA domain have been directly shown to modulate steroid hormone effects on osteocalcin gene (Morrison *et al.*, 1989; Schepmoes *et al.*, 1991; Stromstedt *et al.*, 1991; Heinrichs *et al.*, 1992) transcriptional activity. Both the GRE/TATA and VDRE elements exhibit protein–DNA interactions that influence vitamin D-mediated control of osteocalcin gene transcription (Bortell *et al.*, 1992). In normal diploid cells, the steroid hormones, vitamin D and glucocorticoid, function as enhancers modulating osteocalcin transcription postproliferatively only when the gene is transcribed at basal levels (Owen *et al.*, 1991). The complexity of this developmentally mediated control of osteocalcin transcription is reflected by differences in promoter factor–DNA complexes at the major regulatory elements and as a function of osteoblast differentiation (Fig. 10). Synergism between these osteocalcin gene promoter elements is reflected by enhanced transcription and associated increased representation of transcription factor complexes at the TATA/GRE when normal diploid rat osteoblasts are treated with glucocorticoid and vitamin D (Fig. 10E) (Owen *et al.*, 1993; Bortell *et al.*, 1993). Interaction of the glucocorticoid receptor with other transcription factors is known (Strähle *et al.*, 1988) as well as interactions between the GRE and other hormone-response elements (Ankenbouer *et al.*, 1988; Grange *et al.*, 1989).

It is the VDRE of the osteocalcin gene that was the first vitamin D receptor promoter binding sequence to be identified by deletion and mutational analyses of promoter segments (Demay *et al.*, 1990; Terpening *et al.*, 1991; Kerner *et al.*, 1989; Morrison *et al.*, 1989) as well as by defining the binding of vitamin D receptor complexes (Markose *et al.*, 1990) at single nucleotide resolution. Understanding the structural and functional properties of the osteocalcin gene VDRE and the complexity of regulatory events at this transcriptional control element is becoming increasingly apparent. This is providing insight into the involvement of vitamin D as a mediator of osteocalcin gene expression within the context of a broad spectrum of physiological responses of the osteoblast. For example, consensus sequences for another steroid hormone (retinoic acid) (Schüle *et al.*, 1990a) and for the nuclear protooncogene-encoded Fos and Jun proteins (Lian *et al.*, 1989) are found within the VDRE. Together with results supporting the ability of retinoic acid (Schüle *et al.*, 1990a; Nishimoto *et al.*, 1987) and the Fos–Jun complex (Owen *et al.*, 1990c) to modulate osteocalcin gene transcription, a functional basis for synergistic and/or antagonistic control of multiple regulatory activities is provided within the VDRE. Furthermore, an explanation is in part provided for positive or negative activity of a single regulatory sequence under different biological conditions on the basis of variations in the representation of factors that have the potential for binding.

2. Phenotype Suppression: Oncogene-Mediated Suppression of the Bone-Specific Osteocalcin Gene in Proliferating Osteoblasts

Several experimental results support the concept of coordinate occupancy by Fos–Jun protein complexes at the VDRE and OC box regulatory elements providing a potential molecular mechanism to account for the absence of osteocalcin gene expression in proliferating osteoblasts. Expression of c-*fos* and c-*jun* have been shown to occur primarily during the proliferative period of the osteoblast developmental sequence (Owen *et al.*, 1990a; Shalhoub *et al.*, 1989). AP-1 binding activity is observed primarily in proliferating osteoblasts and dramatically decreases after the down-regulation of proliferation and the initiation of extracellular matrix maturation and mineralization, at which time osteocalcin gene transcription is initiated (Owen *et al.*, 1993) (Fig. 10). We have shown that protein–DNA contacts do not occur at G residues within the AP-1 consensus sequences of the VDRE and OC box when the gene is actively transcribed (Markose *et al.*, 1990). Furthermore, we have demonstrated sequence-specific interactions of these promoter sequences with a stable heterodimeric Fos–Jun complex and that mutations in the vitamin D receptor binding domain of the VDRE and in the CCAAT motif of the OC box promote Fos–Jun binding in the absence of element-specific factor interactions (Owen *et al.*, 1993) (Fig. 11). The latter results support

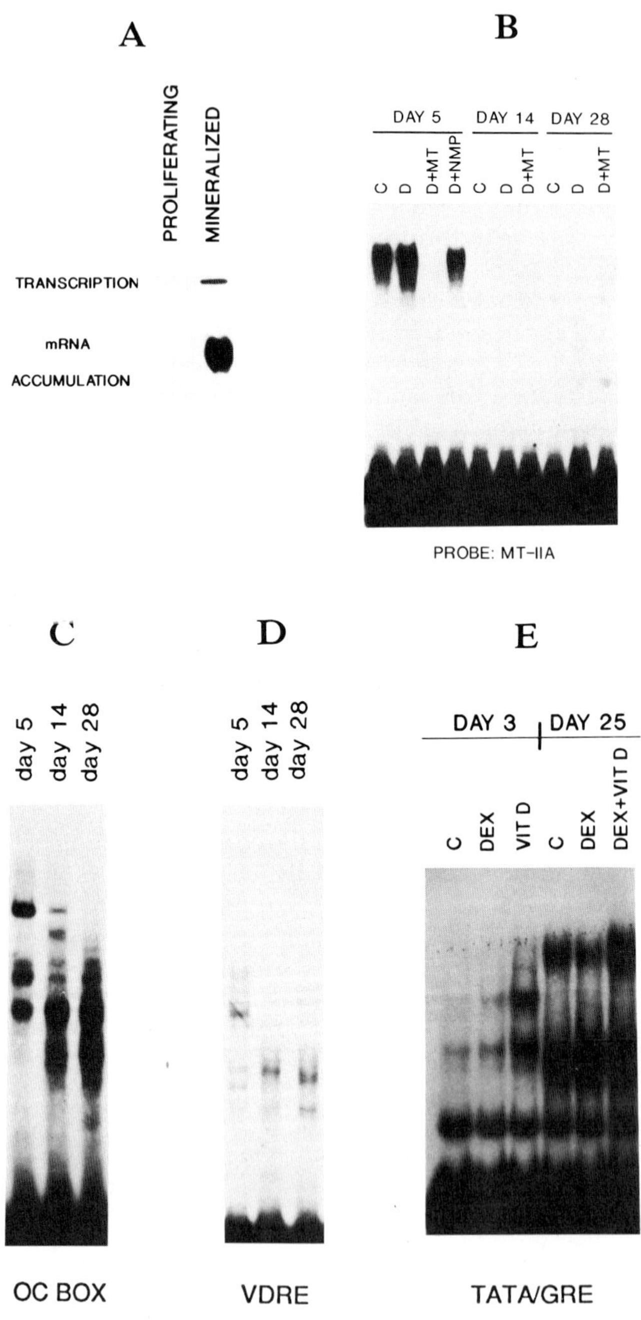
A
PROLIFERATING
MINERALIZED
TRANSCRIPTION
mRNA
ACCUMULATION
B
DAY 5
DAY 14
DAY 28
C
D
D+MT
D+NMP
C
D
D+MT
C
D
D+MT
PROBE: MT-IIA
C
day 5
day 14
day 28
OC BOX
D
day 5
day 14
day 28
VDRE
E
DAY 3
DAY 25
C
DEX
VIT D
C
DEX
DEX+VIT D
TATA/GRE

mutually exclusive binding of the Fos–Jun complex and VDRE as well as OC box factors. Additionally, experiments in which transfection of c-*fos* and c-*jun* into cells expressing osteocalcin resulted in the down-regulation of osteocalcin gene transcription further supports an oncogene-mediated suppression of the osteocalcin gene (Schüle *et al.*, 1990a). These results are consistent with a model in which coordinate occupancy of the AP-1 sites in the VDRE and osteocalcin box in proliferating osteoblasts may suppress both basal level and vitamin D-enhanced osteocalcin gene transcription, a phenomenon described as *phenotype suppression* (Owen *et al.*, 1990c; Lian *et al.*, 1991).

A remaining question is the mechanism by which the osteocalcin gene is rendered transcribable and vitamin D responsive following the down-regulation of proliferation. Here, the possibilities include (1) release of the Fos–Jun complex from the AP-1 sites to permit occupancy by the vitamin D receptor complex and/or by tissue-specific osteocalcin box transcription factors, or (2) modifications of the Fos–Jun complex that facilitates binding of activation-related factors. With respect to the latter possibility, binding of the Fos–Jun complex may pleiotropically play a dual positive and negative role in the regulation of transcription. Vitamin D differentially stimulates expression of various members of the *fos* and *jun* family (Candeliere *et al.*, 1991) providing the possibility for variations in the AP-1 complex under different biological conditions. The human osteocalcin gene VDRE, in contrast to the VDRE of the rat osteocalcin gene, has an additional upstream AP-1 site (solid underline in Fig. 11). While the internal AP-1 site may function in suppression of transcription, the upstream AP-1 site in the human VDRE has been shown to *enhance* vitamin D-stimulated gene transcription but is not necessary for vitamin D (Ozono *et al.*, 1990) regulation. This suggests that responsive-

FIGURE 10 Protein–DNA interactions in the osteocalcin gene promoter during the osteoblast developmental sequence. (A) Osteocalcin gene transcription and mRNA levels in proliferating day 5 diploid rat osteoblasts and from cells harvested from a mineralized matrix at day 28. (B–E) Nuclear-factor proteins were prepared from primary osteoblast cultures harvested during active proliferation on day 5 and postproliferatively on day 14 or from mineralized cultures on day 28. Gel retardation assays were carried out using either the AP-1 metallothionein II_a (MTIIA) or consensus sequence (Rauscher *et al.*, 1990; Lee *et al.*, 1987) as probe under conditions that maximize for Fos–Jun protein binding (Owen *et al.*, 1990c). (B) Lanes labeled D are from cells treated with 10^{-8} *M* 1,25-dihydroxyvitamin D_3 24 hr prior to harvest. Competition is shown with MTII or a nonhomologous sequence (NMP). (C) shows protein–DNA interactions with the probe spanning the OC box; (D) the vitamin D_3-responsive element (VDRE) probe. For the TATA/GRE domain probe shown in (E), extracts are compared from both 10^{-7} *M* DEX and vitamin D-treated (VIT D) cellular extracts harvested on day 3 or day 25. Very different protein–DNA interactions are found at the VDRE and OC box in extracts from proliferating cells when AP-1 activity is found and the OC gene is not transcribed compared to day 14 and 28-day cell extracts.

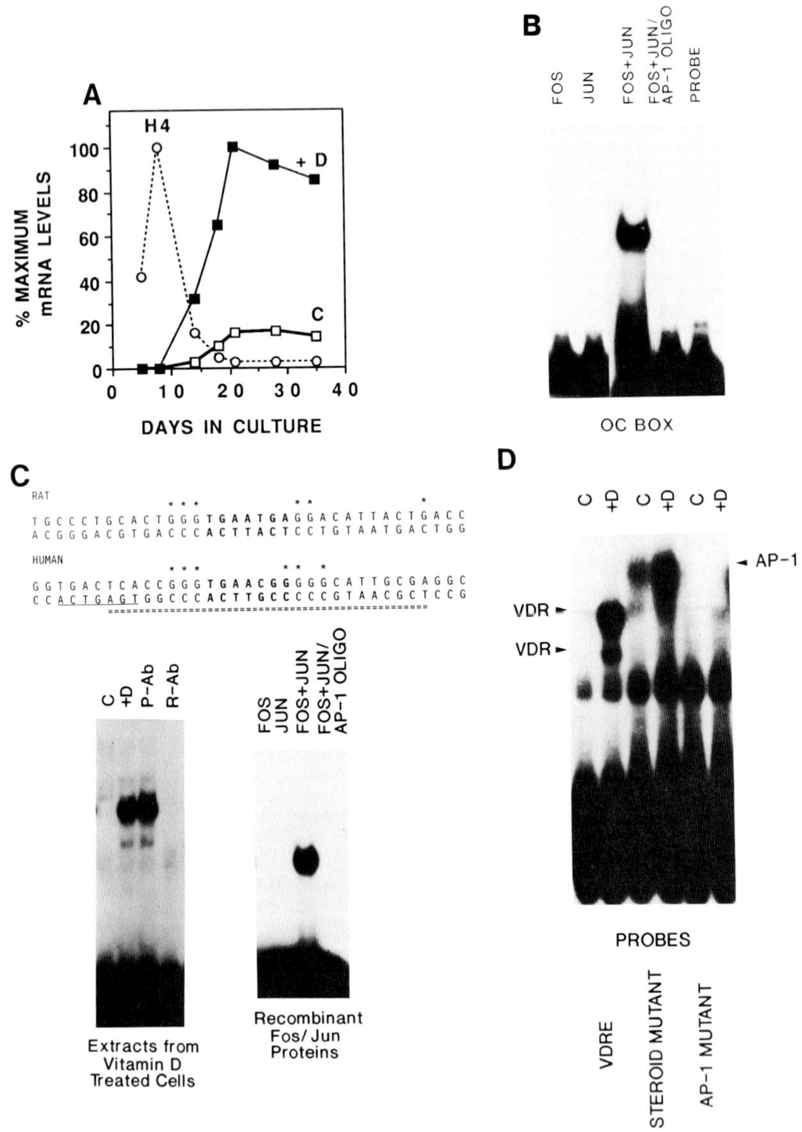

FIGURE 11 Fos–Jun protein complex binding to the OC box and human and rat vitamin D_3-responsive element (VDRE) sequences. (A) Transcriptional control of osteocalcin gene expression and vitamin D modulation during the osteoblast developmental sequence. Vitamin D stimulation of osteocalcin mRNA and synthesis does not occur in normal proliferating rat osteoblasts. H4 histone mRNA is an indication of proliferative activity (○). Osteocalcin mRNA levels were determined in control cultures (□) and those treated for 24 hr with 10^{-8} M 1,25dihydroxyvitamin D_3 (■). (B) Recombinant fos and jun proteins bind to the OC box as a sequence-specific heterodimeric complex (Owen *et al.*, 1991). (C) Fos–Jun protein complex binding to the human VDRE sequences. Nuclear extracts from 48-hr

ness of the human and rat osteocalcin genes to factors that influence AP-1 activity may differ. Yet, the similar organization of the internal AP-1 sites of the VDRE and the identical AP-1 organization in the osteocalcin box for the human and rat osteocalcin genes suggest that there are shared functional properties of the promoter elements, perhaps in regulating expression in relation to the proliferative state of the cells. The Fos–Jun complex may suppress osteocalcin gene transcription when proliferation is ongoing by directly or indirectly modulating sequence-specific interactions at the vitamin D receptor binding domain. Then postproliferatively, the Fos–Jun complex may facilitate vitamin D receptor binding to support the sequential up-regulation of osteocalcin and other vitamin D-responsive genes.

One must define the mechanism by which (1) the VDRE is rendered refractory to binding of the regulatory complex in immature osteoblasts not expressing osteocalcin or in nonosteoblastic cells and (2) the requirements for competency of the VDRE for modulating transcription. Several studies indicate that both protein–DNA interactions and protein–protein interactions are involved (Markose *et al.*, 1990). The requirement of an accessory factor for binding of the vitamin D receptor complex to the osteocalcin gene VDRE (Liao *et al.*, 1990; Ross *et al.*, 1992) and the findings that establish vitamin D receptor phosphorylation as essential for hormone–receptor complex formation (Brown and DeLuca, 1990; Hsieh *et al.*, 1991) reflect the complexity of events associated with activity of the VDRE. The recently characterized RXR retinoic acid receptor is one such accessory factor that contributes to the vitamin D receptor complex and transcriptional activity (Kliewer *et al.*, 1992).

Further experimental results are necessary to determine whether the organization of steroid receptor binding domains and AP-1 sites within

---

vitamin D-treated ROS cells were used in gel mobility shift assays (left panel) with an oligonucleotide probe to the human VDRE, as indicated by the dotted underline in the top panel sequence. The vitamin D receptor complex is blocked by preincubation with antibody recognizing the rat receptor (R-Ab) but not the pig-specific antibody (P-Ab) (provided by H. DeLuca, Madison, Wisconsin). The probe used for these protein–DNA interactions has the classic AP-1 consensus deleted (solid underline) but retains the AP-1 site homologous to the rat VDRE (bold letters) that resides within the core of this regulatory element. The right panel demonstrates that this is an active AP-1 site by binding of only the Fos–Jun protein complex, which can be competed by the AP-1 oligonucleotide. (D) Mutational analysis of the VDRE establishing mutual exclusion of VDR and AP-1 binding proteins to the VDRE. Nuclear protein extracts from control (C) and 24 hr 10^{-8} *M* 1,25$(OH)_2D_3$-treated (+D) ROS 17/2.8 osteosarcoma cells expressing osteocalcin were compared for sequence-specific protein–DNA binding using the wild-type VDRE sequence, which binds vitamin D receptor (VDR); a probe with a mutation in the steroid half element, which binds Fos–Jun but not VDR; and a probe with a mutated AP-1 site. Mutation of the AP-1 sequence that overlaps one of the steroid half elements prevents both VDR and Fos–Jun binding. AP-1 activity was confirmed by competition studies (data not shown).

promoters of genes that are hormone responsive can provide a general mechanism for the phenotype suppression and/or activation in different periods of cell and tissue differentiation. However, support for such a model is provided by the association of AP-1 sites within consensus sequences for other steroid responsive elements—for example, glucocorticoids (Diamond *et al*, 1990; Schüle *et al.*, 1990a,b). Undoubtedly the phenotype suppression model represents a simplification of an extremely complex series of protein–DNA interactions whereby multiple physiological signals are transduced to the nucleus, resulting in modifications in transcription that progressively alters phenotypic properties of cells leading to structural and functional events associated with differentiation. However, the strength of such a model is that it provides a basis for experimentally addressing the functional significance of organization of regulatory elements for phenotypic marker genes of cell differentiation within the context of gene expression associated with proliferation.

C. Involvement of Chromatin Structure and Nucleosome Organization in the Three-Dimensional Properties of the Histone and Osteocalcin Gene Promoters That Integrate Activities at Basal and Enhancer Regulatory Elements

A basic question with respect to transcriptional regulation of histone and osteocalcin genes during development of the osteoblast phenotype within the nucleus of an intact cell is how can transcription of specific genes be selectively initiated with a limited representation of the regulatory factors and the regulatory elements? We have examined transcriptional regulation of the histone and osteocalcin genes within the three-dimensional context of nuclear architecture, addressing the combined potential involvement of chromatin structure and nucleosome organization in integrating the activities at multiple independent regulatory sequences and the nuclear matrix in the concentration and localization of promoter regulatory elements and sequence-specific transcription factors.

1. Chromatin Structure and Nucleosome Organization

Several features of chromatin structure provide a basis for developmental modifications in competency of regulatory sequences for transactivation factor binding, both independently and by functional cooperativity between the multiple basal and enhancer elements of the histone and osteocalcin gene promoters.

The presence of nucleosomes in the H4 histone gene promoter

(Chrysogelos *et al.*, 1985; Moreno *et al.*, 1986) provides the possibility for increasing the proximity of independent regulatory elements that support synergistic and/or antagonistic cooperative interactions between histone gene DNA binding activities. Involvement of chromatin structure with transcriptional regulation as related to growth control is consistent with variations in nucleosome organization as a function of the cell cycle progression (Moreno *et al.*, 1986), which may enhance and/or restrict accessibility of transcription factors and modulate the extent to which DNA-bound factors are phosphorylated. Cell cycle and growth-related modifications in chromatin organization of the histone gene promoter include modifications in nucleosome spacing as well as protein–protein and protein–DNA interactions both within nucleosomes and in the internucleosomal sequences. This is reflected by accessibility to micrococcal nucleus, DNaseI, S1 nucleus, and a series of restriction enzymes (Pauli *et al.*, 1988).

Results indicating the presence of nucleosomes (Bortell *et al.*, 1992), each encompassing approximately 180 basepairs within the osteocalcin gene promoter sequences spanning the vitamin D-responsive element and the proximal basal regulatory region, reduces the potential distance between multiple promoter regulatory domains. This raises the possibilities of functional cooperativity between a series of basal and enhancer elements that are responsive to physiological mediators of osteocalcin gene expression. Additionally, reduced distances between the VDRE and basal elements of the osteocalcin gene 5′ regulatory region provides a basis for interactions of vitamin D receptor and proximal promoter transactivation factors bound at elements that enhance transcriptional activity. Such possibilities are supported by our recent demonstration that sequence-specific modifications in chromatin structure of the osteocalcin gene promoter, indicated by DNaseI hypersensitivity of the VDRE and proximal basal regulatory elements (OC box and GRE/TATA), reflect the level of transcription. Nuclease hypersensitivity is tissue-specific (bone cell-restricted) and related to the level of basal transcription and vitamin D responsiveness (Montecino *et al.*, 1992).

2. The Nuclear Matrix

The nuclear matrix is operationally defined as a network of polymorphic, anastomosing filaments within the nucleus (Capco *et al.*, 1982; Fey and Penman, 1988). Structurally and functionally, the nuclear matrix undergoes modifications in both composition and organization that are responsive to and support cellular requirements for chromatin structure and gene expression (Schaack *et al.*, 1990; Stein *et al.*, 1991; Fey *et al.*, 1991; Stief *et al.*, 1989; Zeitlin *et al.*, 1987; Zenk *et al.*, 1990). Evidence for

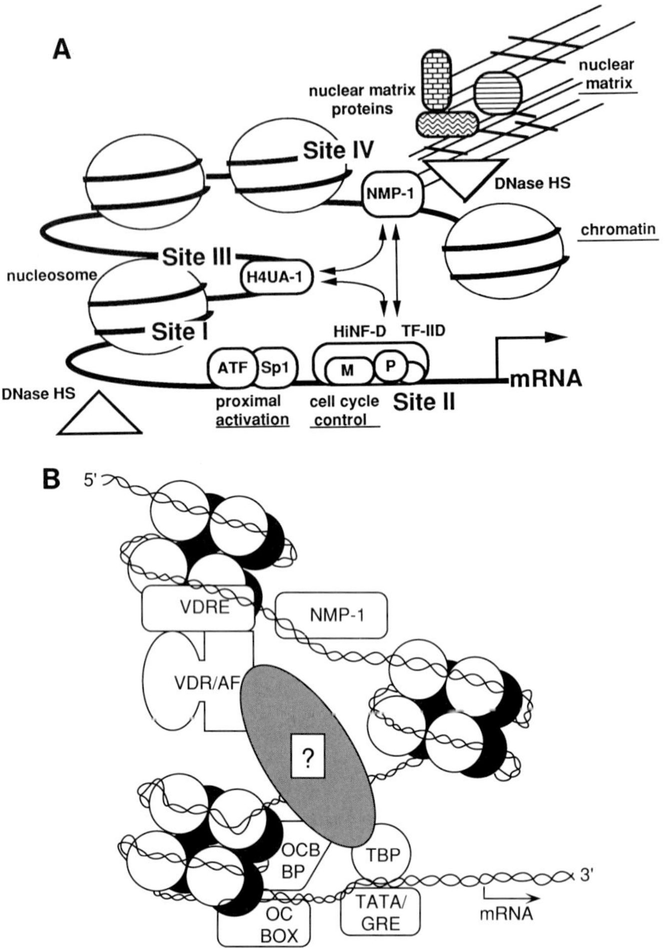

FIGURE 12 Model for three-dimensional configurations of the histone and osteocalcin gene promoters showing potential interactions between independent regulatory elements is shown. (A) Spatial integration of intra- and extracellular signals modulating H4-F0108 gene transcription by reversible alterations in chromatin structure and nucleosomal organization of the promoter is shown. Relationship of distal and proximal protein–DNA interaction sites where H4-F0108 promoter DNA (solid black line) is packaged into nucleosomes (open ovals) is depicted. Indicated are possible cooperative and/or mutually exclusive higher-order nucleoprotein interactions (thin arrows) between various DNA-bound *trans*-acting factors. The presence of a putative attachment site (site IV) with the nuclear matrix (network of lines) containing matrix-associated sequence-specific DNA-binding proteins (rounded boxes filled with alternative symbols) provides a basis for restricted mobility of the promoter to a confined position within the nucleus as well as for the concentration and localization of transcription factors. (B) Three-dimensional organization of the rat osteocalcin (OC) gene promoter showing interactions between the vitamin D-responsive element (VDRE) and the basal regulatory OC box and TATA/GRE elements. Schematically illustrated are nuclear matrix protein–DNA interactions near the VDRE that could serve to structurally anchor this region of the promoter to impose conformational constraints on chromatin organization and/or to concentrate transcriptional factors that facilitate VDR binding to the VDRE. Association of the OC gene promoter with the nuclear

participation of the nuclear matrix in transcriptional control during the developmental sequence associated with osteoblast growth and differentiation is provided by changes in nuclear matrix protein composition that are consistent with each of the three principal periods of osteoblast differentiation and dramatic modifications at the two key transition points—at completion of the proliferation period and at the onset of extracellular matrix mineralization (Dworetzky *et al.*, 1990).

More direct evidence linking the nuclear matrix with transcriptional control during osteoblast phenotype development is provided by a relationship of the nuclear matrix to activity of the histone and osteocalcin genes. The H4 histone gene is associated with the nuclear matrix only during the proliferation period of the osteoblast developmental sequence when actively transcribed, at which time the distal promoter factor NMP-1, a unique 84-kDa ATF transcription factor that binds in a sequence-specific manner to a strong positive regulatory element residing between −589 and −730, is a nuclear matrix component (Dworetzky *et al.*, 1992). A dual role for the nuclear matrix binding domain in the H4 histone gene promoter is therefore suggested: contributing to the up-regulation of histone gene transcription by functioning as a nuclear matrix attachment site and serving to concentrate and localize a 84-kDa ATF DNA binding protein (Dworetzky *et al.*, 1992). Not to be dismissed is the potential for gene–nuclear matrix association of the NMP-1 site to impose constraints on chromatin structure. Taken together with transcription-related modifications in the chromatin organization of the histone gene promoter, we have proposed a model incorporating a relationship between chromatin structure and gene expression, which is schematically presented in Figure 12.

Further support for involvement of the nuclear matrix in transcriptional control is our recent observation that a homologous NMP-1 site resides in the osteocalcin gene promoter. We have also shown that sequence-specific DNA binding proteins are associated with the nuclear matrix when the osteocalcin gene is expressed (Bortell *et al.*, 1992) and when physiological mediators modify the level of osteocalcin gene transcription (Bidwell *et al.*, 1991). Identification of an NMP-1 site in the osteocalcin gene promoter adjacent to the vitamin D responsive element serves as a basis for postulating a model for the three-dimensional organization of the 5′ regulatory sequences that is consistent with synergistic interactions of multiple regulatory elements that contribute to up-

---

matrix together with the presence of nucleosomes between the VDRE, OC box, and TATA/GRE reduces the distances between these regulatory elements, thereby potentially modulating cooperative interactions that enhance OC gene expression. The question mark indicates the potential involvement of auxiliary proteins in mediating the interactions between these regulatory complexes. AF, accessory factor; OCB BP, OC box binding proteins; TBP, TATA/GRE binding proteins.

regulation of osteocalcin gene transcription (Bortell *et al.*, 1992) (Fig. 12). Here, in addition to gene and transcription factor localization, a potential mechanism for nuclear matrix-mediated structural constraints and a conformation of the osteocalcin gene promoter that facilitates vitamin D responsiveness may be operative.

V. CONSEQUENCE OF THE ABROGATION OF GROWTH CONTROL ON DEREGULATION OF DIFFERENTIATION GENES IN TUMOR GROWTH CELLS

In several transformed osteoblasts and osteosarcoma cell lines that have been extensively examined (e.g., in the ROS 17/2.8 rat osteosarcoma cell line [Majeska *et al.*, 1985]), there is a deregulation of the sequential pattern of gene expression observed in diploid osteoblasts. As shown in Figure 13, a relaxation of control mechanisms permits sequentially expressed genes in diploid osteoblasts to be expressed simultaneously during proliferation, as reflected by concomitant expression of the histone H4 gene and tissue-specific genes including APase, osteopontin, and osteocalcin (Stein *et al.*, 1990). It should be emphasized that while in normal diploid osteoblasts, some genes (e.g., osteocalcin, osteopontin) are expressed or induced to high levels only at the onset of extracellular matrix mineralization, in osteosarcoma cells these genes are expressed in the absence of a bonelike extracellular matrix. Perturbations are implicated in the signaling mechanisms that interface the down-regulation of cell growth and induction of genes that support extracellular matrix maturation and specialization with those that control gene expression associated with extracellular matrix mineralization. Our understanding of the manner in which gene responsiveness to regulators of osteoblast growth and differentiation is compromised in osteosarcoma cells can thereby provide insight into molecular mechanisms that are deregulated.

A. Abrogation of Growth Control: Deregulation of Histone H4 Transcription

Transcriptional control of the H4 histone gene is strikingly modified in osteosarcoma cells. An abrogation of the cell cycle and proliferation-regulated changes in transcription factor binding to the Site II and Site III H4 histone promoter elements by nuclear extracts from tumor cells is observed (Fig. 14) (Holthuis *et al.*, 1990; van Wijnen *et al.*, 1991b). A similar deregulation of growth-related transcription factor interactions occurs in an H3 histone gene promoter (van Wijnen *et al.*, 1991b). Enzymatic dephosphorylation of nuclear extracts results in loss of HiNF-D–Site II interactions and increased electrophoretic mobility of the Site III

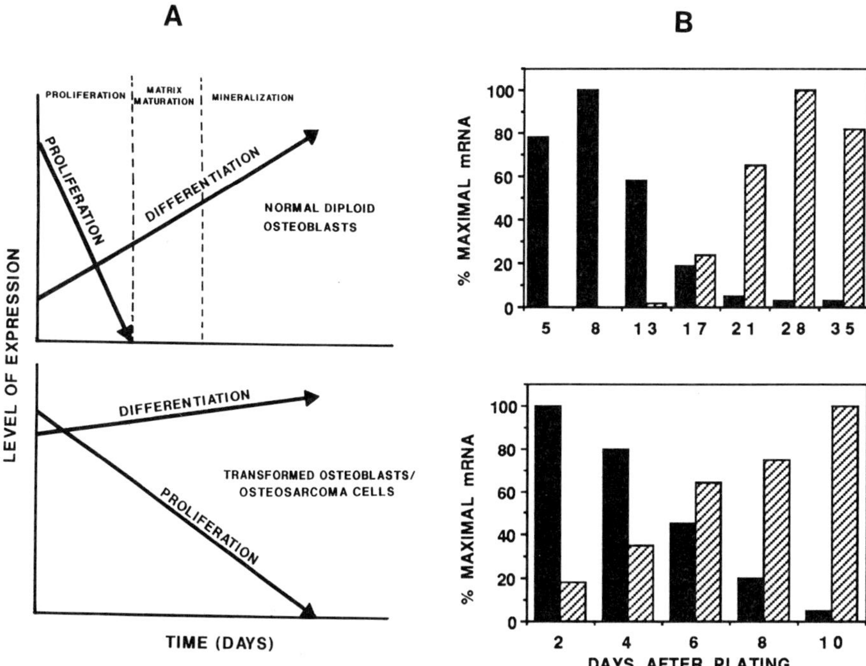

FIGURE 13 Model of the reciprocal relationship between proliferation and differentiation in normal diploid cells during the rat osteoblast developmental sequence and in osteosarcoma (transformed) cells. (A) Top panel illustrates, for comparison to the lower panel, the growth–differentiation relationship operative in normal diploid osteoblasts in contrast to the lower panel. The lower panel schematically illustrates the deregulation of the relationship between growth and differentiation in transformed osteoblasts or osteosarcoma cells. The proliferation vector reflects the continuous expression of cell growth and expression of cell cycle- and proliferation-related genes. In contrast to normal diploid cells, the constitutive expression of osteoblast differentiation phenotype markers in transformed cells reflects the absence of the two developmentally important transition points observed in normal diploid cells. In osteosarcoma cells, cell growth and tissue-specific gene expression occur concomitantly; thus, the relationship between growth and differentiation is deregulated. (B) Illustrates the differences in expression of a cell growth-regulated histone gene (H4, solid bar) and the differentiation-specific osteocalcin gene (hatch bar) for rat osteoblasts (top) and ROS 17/2.8 osteosarcoma cells (bottom).

protein–DNA complex in normal cells. Therefore, involvement of phosphorylation at these two histone gene promoter elements is implicated (Kleinsmith *et al.*, 1976). Indeed, specific phosphorylation events (kinase activities) play a key role in cell cycle progression (e.g., cdc2 regulation of the cyclin gene [Nurse, 1990]). Thus, for a tumor cell, continual traverse of the cell cycle may be associated with modifications in the level of transcription factor phosphorylation or the utilization of specific phosphorylation sites. The observations are not unique to the ROS 17/2.8 cell

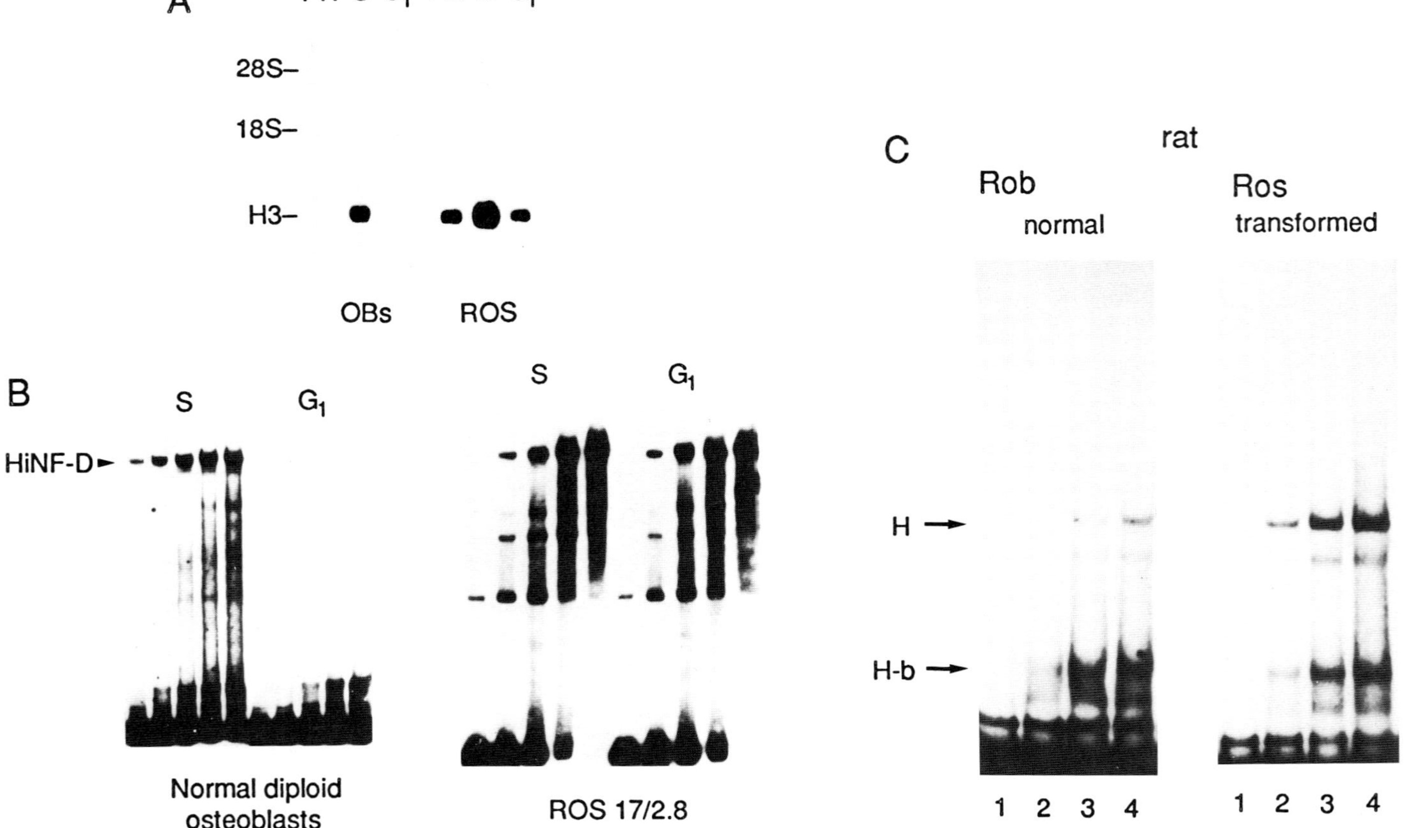
A
PR S G1 PR S G1
28S–
18S–
H3–
OBs
ROS
B
S
G1
HiNF-D
Normal diploid osteoblasts
S
G1
ROS 17/2.8
C
rat
Rob
normal
Ros
transformed
H
H-b
1 2 3 4
1 2 3 4

but a comparison of several tumor cell lines revealed similar differences to their normal diploid counterpart. The identical relationship has been shown in WI38 (normal fibroblasts) compared to the transformed HeLa cell line (van der Houven van Oordt *et al.*, 1992).

B. Relaxation of Transcriptional Regulation of the Osteocalcin Gene in Tumor Cell Lines

The postproliferative and developmental expression of osteocalcin in normal diploid osteoblasts and its regulation by steroid hormones only when osteocalcin expression is ongoing postproliferatively has been well documented. We have proposed that AP-1 sites that overlap the basal regulatory CAAT-containing element, the OC Box, and the vitamin D-responsive element bind the nuclear protooncogene encoded fos–jun protein complex as a mechanism by which osteocalcin expression and vitamin D modulation is suppressed in normal actively dividing cells (Owen *et al.*, 1990b; Lian *et al.*, 1991). In ROS 17/2.8 cells, osteocalcin expression occurs in the presence of AP-1 activity in contrast to the diploid cells.

To understand osteocalcin expression in proliferating tumor cells, we have examined the transcription factor complexes at basal regulatory sequences and at the VDRE of the osteocalcin gene promoter. In normal diploid osteoblasts, nuclear extracts from proliferating cells and in post-proliferative osteocalcin expressing cells exhibit differences in protein–DNA interactions. In contrast, the tumor cell nuclear extracts from proliferating or confluent cultures are similar and support protein–DNA interactions that are strikingly different from those that characterize the normal osteoblasts. These differences are found at each of the basal elements and the VDRE.

Analysis of nuclear extracts from normal osteoblasts and proliferating as well as confluent ROS cells revealed different forms of the vitamin D receptor complex (Fig. 15). In normal proliferating cells, both high and low mobility complexes are observed with equal representation. In the expressing differentiated osteoblasts, only the rapid-mobility form of the vitamin D receptor complex is found. In contrast, in tumor cells, interactions at the VDRE are identical between proliferating and confluent

FIGURE 14 Deregulation of histone gene promoter expression in osteosarcoma cells. (A) Histone H4 messenger RNA levels in synchronized osteoblasts (OBs) and ROS 17/2.8 osteosarcoma cells. PR, prerelease period of the cell cycle. (B) Site II protein–DNA interactions during the S and G1 periods of the cell cycle. Note that HiNF-D–Site II interactions are restricted to normal diploid osteoblasts while these interactions are constitutive in ROS 17/2.8 cells. (C) Site III protein–DNA interactions also show increased representation of the slower mobility form in the osteosarcoma cells (ROS 17/2.8) compared to normal osteoblasts.

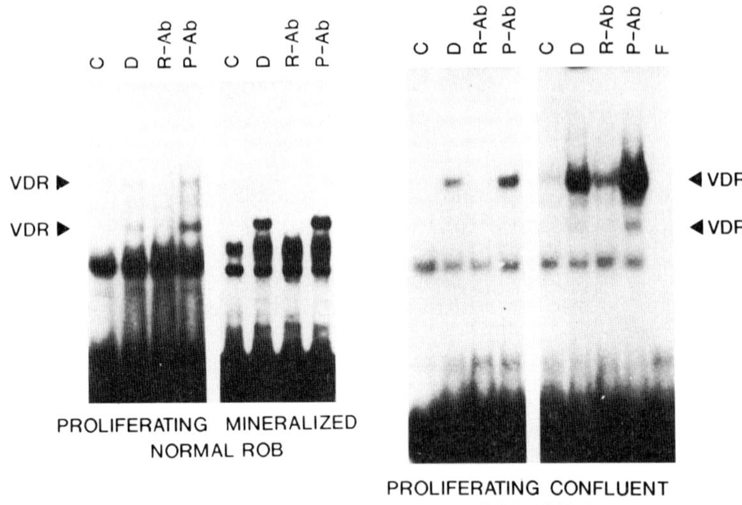

FIGURE 15 Different properties of the vitamin D receptor (VDR) complex between normal osteoblasts and ROS 17/2.8 cells. Protein–DNA interactions at the rat osteocalcin vitamin D-responsive element are compared using nuclear proteins from untreated (C) or 10^{-8} *M* vitamin-D treated (D) confluent (day 8) ROS 17/2.8 cells or mature (day 32) normal diploid osteoblasts. Gel mobility shift assays show enhancement of protein–DNA interactions with a ^{32}P-labeled rat osteocalcin VDRE oligonucleotide DNA probe following treatment of the cells with vitamin D (VDR; indicated by arrows). Monoclonal antibodies against the porcine VDR and that detect only the porcine vitamin D receptor (P-Ab) or that cross-react with the rat VDR (R-Ab) were preincubated with the nuclear proteins for 20 min prior to DNA probe addition. These antibodies demonstrate that the vitamin D-enhanced protein–DNA interactions observed in both cell types are dependent on the binding of the VDR.

cultures, and here the lower mobility complex predominates (approx. 10 to 1). Both forms are competed by an antibody (Bortell *et al.*, 1992) that recognizes the endogenous rat vitamin D receptor. A modification of the accessory protein–vitamin D receptor–DNA interaction may account for the higher mobility. Preliminary studies indicate that the higher mobility form is phosphorylated. By treatment of nuclear extracts with phosphatase, the higher molecular weight form can be converted to lower molecular weight forms. Thus, we see a second example of the predominance of phosphorylated forms of the transcription factor complexes in tumor cells compared to normal diploid differentiated osteoblasts. The dominance of the one form of the vitamin D receptor complex in ROS 17/2.8 may allow for vitamin D regulation of the osteocalcin gene in proliferating cells and in the presence of AP-1 activity.

We can speculate that the deregulation of growth control in osteosarcoma cells may reflect modifications in the activity of tumor suppressor genes (reviewed by Weinberg, 1989; Levine, 1990). Alternatively, or together with the loss of stringent growth control, there may be modifications in the regulatory sequences and/or in the factors that control the progressive expression of tissue-specific genes and their response to cell growth or morphogenic regulatory factors. While the specific molecular mechanisms remain to be established, by further understanding the deregulation of growth in bone tumors, we can anticipate gaining additional insight into control of the tightly coupled relationship between proliferation and development of the osteoblast phenotype.

ACKNOWLEDGMENTS

Studies reported from the authors' laboratories were supported by grants from the National Institutes of Health (AR33920, AR35166, AR39588, GM32010), the March of Dimes Birth Defects Foundation, and the Northeast Osteogenesis Imperfecta Society. The authors thank Ms. Christine Dunshee for photographic assistance.

REFERENCES

Ankenbauer, W., Strähle, U., and Schütz, G. (1988). Synergistic action of glucocorticoid and estradiol responsive elements. *Proc. Natl. Acad. Sci. USA* **85**, 7526–7530.

Aronow, M. A., Gerstenfeld, L. C., Owen, T. A., Tassinari, M. S., Stein, G. S., and Lian, J. B. (1990a). Factors that promote progressive development of the osteoblast phenotype in cultured fetal rat calvaria cells. *J. Cell. Physiol.* **143**, 213–221.

Aronow, M. A., Owen, T. A., Stein, G. S., and Lian, J. B. (1990b). Estrogen inhibition of osteocalcin mRNA expression occurs in cultured rat osteoblasts only after formation of a mineralized extracellular matrix. *J. Bone Miner. Res.* **5**, S273.

Baker, A. R., McDonnell, D. P., Hughes, M., Crisp, T. M., Mangelsdorf, D. J., Haussler, M. R., Pike, J. W., Shine, J., and O'Malley, B. W. (1988). Cloning and expression of full-length cDNA encoding human vitamin D receptor. *Proc. Natl. Acad. Sci. USA* **85**, 3294–3298.

Barone, L. M., Owen, T. A., Tassinari, M. S., Bortell, R., Stein, G. S., and Lian, J. B. (1991). Developmental expression and hormonal regulation of the rat matrix gla protein (MGP) gene in chondrogenesis and osteogenesis. *J. Cell. Biochem.* **46**, 351–365.

Barrack, E. R., and Coffey, D. S. (1983). *In* "Gene Regulation by Steroid Hormones, II" (A. K. Roy and J. H. Clark, eds.), pp. 239–266. Springer-Verlag. New York, N.Y.

Baumbach, L. L., Stein, G. S., and Stein, J. L. (1987). Regulation of human histone gene expression: Transcriptional and post-transcriptional control in the coupling of histone messenger RNA stability with DNA replication. *Biochemistry* **26**, 6178–6187.

Bellows, C. G., Aubin, J. E., Heersche, H. N. M., and Antosz, M. E. (1986). Mineralized bone nodules formed *in vitro* from enzymatically released rat calvaria cell populations. *Calcif. Tissue Int.* **38**, 143–154.

Bellows, C. G., Aubin, J. E., and Heersche, J. N. M. (1987). Physiological concentrations of glucocorticoids stimulate formation of bone nodules from isolated rat calvaria cells *in vitro*. *Endocrinology* **121**, 1985–1992.

Beresford, J. N., Gallagher, J. A., Poser, J. W., and Russell, R. G. G. (1984). Production of osteocalcin by human bone cells in vitro. Effects of 1,25$(OH)_2D_3$, 24,25$(OH)_2D_3$, parathyroid hormone, and glucocorticoids. *Metab. Bone Dis. Relat. Res.* **5**, 229–234.

Beresford, J. N., Gallagher, J. A., and Russell, R. G. G. (1986). 1,25-Dihydroxyvitamin D_3 and human bone-derived cells *in vitro:* Effects on alkaline phosphatase, type I collagen and proliferation. *Endocrinology* **119,** 1176–1785.

Bhargava, U., Bar-Lev, M., Bellows, C. G., and Aubin, J. E. (1988). Ultrastructural analysis of bone nodules formed in vitro by isolated fetal rat calvaria cells. *Bone* **9,** 155–163.

Bidwell, J. P., Fryer, M. J., Firek, A. F., Donahue, H. J., Heath III, H. (1991). Desensitization of rat osteoblast-like cells (ROS 17/2.8) to parathyroid hormone uncouples the adenosine 3′, 5′-monophosphate and cytosolic ionized calcium response limbs. *Endocrinology* **128,** 1021–1028.

Bortell, R., Barone, L. M., Tassinari, M. S., Lian, J. B., and Stein, G. S. (1990). Gene expression during endochondral bone development: Evidence for coordinate expression of transforming growth factor β and collagen type I. *J. Cell. Biochem.* **44,** 81–91.

Bortell, R., Owen, T. A., van Wijnen, A. J., Bidwell, J. P., Gavazzo, P., Breen, E., DeLuca, H., Stein, G. S., and Lian, J. B. (1992). Vitamin D responsive protein/DNA interactions at multiple promoter regulatory elements that contribute to the level of rat osteocalcin gene expression. *Proc. Natl. Acad. Sci. USA* **89,** 6119–6123.

Bortell, R., Owen, T. A., Shalhoub, V., van Wijnen, A. J., Aronow, M. A., Heinrichs, A., Stein, J. L., Lian, J. B., and Stein, G. S. (1993). Constitutive transcription of the osteocalcin gene in osteosarcoma cells is reflected by altered protein/DNA interactions at promoter regulatory elements. *Proc. Natl. Acad. Sci., USA.* **90,** 2300–2304.

Boskey, A. L. (1989). Noncollagenous matrix proteins and their role in mineralization. *Bone Miner.* **6,** 111–123.

Boskey, A. L., Wians, F. H., Jr., and Hauschka, P. V. (1985). The effect of osteocalcin on *in vitro* lipid-induced hydroxyapatite formation and seeded hydroxyapatite growth. *Calcif. Tiss. Int.* **37,** 57–62.

Brown, T. A., and DeLuca, H. F. (1990). Phosphorylation of the 1,25-dihydroxyvitamin D_3 receptor: A primary event in 1,25-dihydroxyvitamin D_3 action. *J. Biol. Chem.* **265,** 10025–10029.

Bruder, S. P., and Caplan, A. I. (1989). Discrete stages within the osteogenic lineage are revealed by alterations in the cell surface architecture of embryonic bone cells. *Conn. Tissue Res.* **20,** 73–79.

Canalis, E. (1983). Effect of glucocorticoids on type I collagen synthesis, alkaline phosphatase activity and deoxyribonucleic acid content in cultured rat calvariae. *Endocrinology* **112,** 931–939.

Canalis, E., and Lian, J. B. (1985). $1,25(OH)_2D_3$ effects on collagen and DNA synthesis in periosteum and periosteum-free calvariae. *Bone* **6,** 457–460.

Candeliere, G. A., Prud'homme, J., and St-Arnaud, R. (1991). Differential stimulation of fos and jun family members by calcitriol in osteoblastic cells. *Mol. Endrocinol.* **12,** 1780–1788.

Capco, D. G., Wan, K. M., and Penman, S. (1982). The nuclear matrix: Three-dimensional architecture and protein composition. *Cell* **29,** 847–858.

Carrington, J. L., Roberts, A. B., Flanders, K. C., Roche, N. S., and Reddi, A. H. (1988). Accumulation, localization and compartmentation of transforming growth factor β during endochondral bone development. *J. Cell Biol.* **107,** 1969–1975.

Celeste, A. J., Rosen, V., Buecker, J. L., Kriz, R., Wang, E. A., and Wozney, J. M. (1986). Isolation of the human gene for bone Gla protein utilizing mouse and rat cDNA clones. *EMBO J.* **5,** 1885–1890.

Chen, T. L., Cohn, C. M., Morey-Holton, E., and Feldman, D. (1983). 1α,25-Dihydroxyvitamin D_3 receptors in cultured rat osteoblast-like cells. *J. Biol. Chem.* **258,** 4350–4355.

Chrysogelos, S., Riley, D. E., Stein, G., and Stein, J. (1985). A human histone H4 gene exhibits cell cycle dependent changes in chromatin structure that correlate with its expression. *Proc. Natl. Acad. Sci. USA* **82,** 7535–7539.

Collart, D., Ramsey-Ewing, A., Bortell, R., Lian, J., Stein, J., and Stein, G. (1991). Isolation and characterization of a cDNA from a human histone H2B gene which is reciprocally expressed in relation to replication-dependent H2B histone genes during HL60 cell differentiation. *Biochemistry* **30,** 1610–1617.

Craig, A. M., Smith, J. H., and Denhardt, D. T. (1989). Osteopontin, a transformation-associated cell adhesion phosphoprotein, is induced by 12-O-tetradecanoylphorbol 13-acetate in mouse epidermis. *J. Biol. Chem.* **264,** 9682–9689.

Demay, M. B., Gerardi, J. M., DeLuca, H. F., and Kronenberg, H. M. (1990). DNA sequences in the rat osteocalcin gene that bind the 1,25-dihydroxyvitamin D_3 receptor and confer responsive to 1,25-dihyroxyvitamin D_3. *Proc. Natl. Acad. Sci. USA* **87,** 369–373.

Demay, M. B., DeLuca, H., and Korenenberg, H. M. (1991). Identification of sequences in the human parathyroid hormone gene that bind the 1,25-dihydroxyvitamin D_3 receptor. *J. Bone Miner. Res.* **6,** S238.

Diamond, M. I., Miner, J. N., Yoshinaga, S.K., and Yamamoto, K. R. (1990). Transcription factor interactions: Selectors of positive or negative regulation from a single DNA element. *Science* **249,** 1266–1272.

Dworetzky, S. I., Wright, K. L., Fey, E. G., Penman, S., Lian, J. B., Stein, J. L., and Stein, G. S. (1992). Sequence-specific DNA binding proteins are components of a nuclear matrix attachment site. *Proc. Natl. Acad. Sci. USA* **89,** 4178–4182.

Dworetzky, S. I., Fey, E. G., Penman, S., Lian, J. B., Stein, J. L., and Stein, G. S. (1990). Progressive changes in the protein composition of the nuclear matrix during rat osteoblast differentiation. *Proc. Natl. Acad. Sci. USA* **87,** 4605–4607.

Ecarot-Charrier, B., Glorieux, F. H., van der Rest, M., and Pereira, G. (1983). Osteoblasts isolated from mouse calvaria initiate matrix mineralization in culture. *J. Cell Biol.* **96,** 639–643.

Ecarot-Charrier, B., Shepard, N., Charette, G., Grynpas, M., and Glorieux, F. H. (1988). Mineralization in osteoblast cultures: A light and electron microscopic study. *Bone* **9,** 147–154.

Fey, E. G., and Penman, S. (1988). Nuclear matrix proteins reflect cell type of origin in cultured human cells. *Proc. Natl. Acad. Sci. USA* **85,** 121–125.

Fey, E. G., Bangs, P., Sparks, C., and Odgren, P. (1991). The nuclear matrix: Defining structural and functional roles. *Crit. Rev. Eukaryotic Gene Exp.* **1,** 127–143.

Fisher, L. W., Hawins, G. R., Turcoss, N., and Termine, J. D. (1987). Purification and partial characterization of small proteoglycans I and II, bone sialoproteins I and II, and osteonectin from the mineral compartment of developing human bone. *J. Biol. Chem.* **262,** 9702–9708.

Franceschi, R. T., and Young, J. (1990). Regulation of alkaline phosphatase by 1,25-dihydroxyvitamin D_3 and ascorbic acid in bone-derived cells. *J. Bone Miner. Res.* **5,** 1157–1167.

Franceschi, R. T., Romano, P. R., and Park, K. Y. (1988). Regulation of type I collagen synthesis by 1,25-dihydroxyvitamin D_3 in human osteosarcoma cells. *J. Biol. Chem.* **263,** 18938–18945.

Franzen, A., and Heinegard, D. (1985). Isolation and characterization of two sialoproteins present only in bone calcified matrix. *Biochem. J.* **232,** 715–724.

Fraser, J. D., and Price, P. A. (1988). Lung, heart, and kidney express high levels of mRNA for the vitamin K-dependent matrix Gla protein: Implications for the possible functions of matrix Gla protein and for the tissue distribution of the γ-carboxylase. *J. Biol. Chem.* **263,** 11033–11036.

Fraser, J. D., and Price, P. A. (1990). Induction of matrix Gla protein synthesis during prolonged 1,25-dihydroxyvitamin D_3 treatment of osteosarcoma cells. *Calcif. Tissue Int.* **46,** 270–279.

Frenkel, B., Mijnes, J., Aronow, M. A., Zambetti, G., Banerjee, C., Stein, J. L., Lian, J. B., and Stein, G. S. (1992). Repression of promoter activity mediated by a silencer within the transcribed sequence of the rat osteocalcin gene. *J. Bone and Miner. Res.* **7,** 5107.

Gerbaulet, S. P., van Wijnen, A. J., Aronin, N., Tassinari, M., Lian, J. B., Stein, J. L., and Stein, G. S. (1992). Downregulation of histone H4 gene transcription during post-natal development in transgenic mice and transgenically derived calvarial osteoblast cultures. *J. Cell Biochem.* **49,** 137–147.

Gerstenfeld, L. C., Chipman, S. D., Glowacki, J., and Lian, J. B. (1987). Expression of differentiated function by mineralizing cultures of chicken osteoblasts. *Dev. Biol.* **122,** 49–60.

Gerstenfeld, L. C., Gotoh, Y., McKee, M. D., Nanci, A., Landis, W. J., and Glimcher, M. J. (1990). Ultrastructural immunolocalization of a major 66 kDa phosphoprotein synthesized by chicken osteoblasts during mineralization in vitro. *Anat. Rec.* **228,** 93–103.

Gerstenfeld, L. C., Broess, M., Bruder, S., Caplan, A., and Landis, W. J. (1992). Regulation of osteoblast extracellular matrix formation: Post-translational regulation of extracellular matrix deposition and relationship between embryonic development and hormonal response. *Connect. Tiss. Res.* (in press).

Glimcher, M. J. (1989). Mechanism of calcification: Role of collagen fibrils and collagen–phosphoprotein complexes in vitro and in vivo. *Anat. Rec.* **224,** 139–153.

Glowacki, J., and Lian, J. (1987). Impaired recruitment of osteoclast progenitors in deficient osteocalcin-depleted bone implants. *Cell Differ.* **21,** 247–254.

Grange, T., Roux, J., Rigeaud, G., and Pictet, R. (1989). Two remote glucocorticoid responsive units interact cooperatively to promote glucocorticoid induction of rat tyrosine aminotransferase gene expression. *Nucleic Acids Res.* **17,** 8695–8709.

Harrison, J. R., Peterson, D. N., Lichtler, A. C., Mador, A. T., Rowe, D. W., and Kream, B. E. (1989). 1,25-Dihydroxyvitamin D_3 inhibits transcription of type I collagen genes in the rat osteosarcoma cell line ROS 17/2.8. *Endocrinology* **125,** 327–333.

Hauschka, P. V., Lian, J. B., and Gallop, P. M. (1975). Direct identification of the calcium-binding amino acid, γ-carboxyglutamate, in mineralized tissue. *Proc. Natl. Acad. Sci. USA* **72,** 3925–3929.

Hauschka, P. V., Lian, J. B., Cole, D. E. C., and Gundberg, C. M. (1989). Osteocalcin and matrix Gla protein: Vitamin K-dependent proteins in bone. *Physiologic Rev.* **69,** 990–1047.

Heinrichs, A. A., Bortell, R., Stein, J. L., Litwack, G., Lian, J. B., and Stein, G. S. (1992). Identification of multiple glucocorticoid receptor binding sites in the proximal rat osteocalcin gene promoter. *J. Bone Miner. Res.* **7,** 5103.

Hock, J. M., Gunness-Hey, M., Poser, J., Olson, H., Bell, N. H., and Raisz, L. G. (1986). Stimulation of undermineralized matrix formation by 1,25 dihydroxyvitamin D_3 in long bones of rats. *Calcif. Tissue Int.* **28,** 79–86.

Holthuis, J., Owen, T. A., van Wijnen, A. J., Wright, K. L., Ramsey-Ewing, A., Kennedy, M. B., Carter, R., Cosenza, S. C., Soprano, K. J., Lian, J. B., Stein, J. L., and Stein, G. S. (1990). Tumor cells exhibit deregulation of the cell cycle histone gene promoter factor HiNF-D. *Science* **247,** 1454–1457.

Hsieh, J.-C., Jurutka, P. W., Galligan, M. A., Terpening, C. M., Haussler, C. A., Samuels, D. S., Shimizu, Y., Shimizu, N., and Haussler, M. R. (1991). Human vitamin D receptor is selectively phosphorylated by protein kinase C on serine 51, a residue crucial to its trans-activation function. *Proc. Natl. Acad. Sci. USA* **88,** 9315–9319.

Joyce, M. E., Roberts, A. B., Sporn, M. B., and Bolander, M. E. (1990). Transforming growth factor-β and the initiation of chondrogenesis and osteogenesis in the rat femur. *J. Cell Biol.* **110,** 2195–2207.

Kerner, S. A., Scott, R. A., and Pike, J. W. (1989). Sequence elements in the human osteocalcin gene confer basal activation and inducible response to hormonal vitamin D_3. *Proc. Natl. Acad. Sci. USA* **86,** 4455–4459.

Kim, H. T., and Chen, T. L. (1989). 1,25-Dihydroxyvitamin D_3 interaction with dexamethasone and retinoic acid: Effects of procollagen messenger ribonucleic acid levels in rat osteoblast-like cells. *Mol. Endocrinol.* **3,** 97–104.

Kleinsmith, L. J., Stein, J., and Stein, G. (1976). Dephosphorylation of nonhistone proteins specifically alters the pattern of gene transcription in reconstituted chromatin. *Proc. Natl. Acad. Sci. USA* **73,** 1174–1178.

Kliewer, S. A., Umesono, Kazuhiko, Mangelsdorf, D. J., and Evans, R. M. (1992). Retinoid X receptor interacts with nuclear receptors in retinoic acid, thyroid hormone and vitamin D_3 signalling. *Nature (London)* **355,** 446–450.

Knoll, B. J., Rothblum, K. N., and Longley, M. (1988). Nucleotide sequence of the human placental alkaline phosphatase gene: Evolution of the 5′ flanking region by deletion/substitution. *J. Biol. Chem.* **263,** 12020–12027.

Kroeger, P., Stewart, C., Schaap, T., van Wijnen, A., Hirshman, J., Helms, S., Stein, G., and Stein, J. (1987). Proximal and distal regulatory elements that influence *in vivo* expression of a cell cycle-dependent human H4 histone gene. *Proc. Natl. Sci. USA* **84,** 3982–3986.

Kyeyune-Nyombi, E., Lau, W. K.-H., Baylink, D. J., and Strong, D. D. (1989). Stimulation of cellular alkaline phosphatase activity and its messenger RNA level in a human osteosarcoma cell line by 1,25-dihydroxyvitamin D_3. *Arch. Biochem. Biophys.* **275,** 363–370.

Leboy, P. S., Beresford, J. N., Devlin, C., and Owen, M. E. (1991). Dexamethasone induction of osteoblast mRNAs in rat marrow stromal cell cultures. *J. Cell. Physiol.* **146,** 370–378.

Levine, A. J. (1990). Tumor suppressor genes. *Bioessays* **12,** 60–66.

Lian, J. B., and Friedman, P. A. (1978). The vitamin K-dependent synthesis of gamma-carboxyglutamic acid by bone microsomes. *J. Biol. Chem.* **253,** 6623–6626.

Lian, J. B., and Gundberg, C. M. (1988). Osteocalicin: Biochemical considerations and clinical applications. *Clin. Orthop. Relat. Res.* (M. Urist, ed.) **226,** 267–291.

Lian, J. B., and Stein, G. S. (1992). Transcriptional control of vitamin D regulated proteins. *J. Cell. Biochem.* **49,** 37–45.

Lian, J. B., and Marks, S. C. (1990). Osteopetrosis in the rat: Coexistence of reduction in the osteocalcin and bone resorption. *Endocrinology* **126,** 955–962.

Lian, J., Stewart, C., Puchacz, E., Mackowiak, S., Shalhoub, V. Collart, D., Zambetti, G., and Stein, G. (1989). Structure of the rat osteocalcin gene and regulation of vitamin D-dependent expression. *Proc. Natl. Acad. Sci. USA* **86,** 1143–1147.

Lian, J. B., Stein, G. S., Bortell, R., and Owen, T. A. (1991). Phenotype suppression: A postulated molecular mechanism for mediating the relationship of proliferation and differentiation by fos/jun interactions at AP-1 sites in steroid responsive promoter elements of tissue-specific genes. *J. Cell. Biochem.* **45,** 9–14.

Lian, J. B., Stein, G. S., Owen, T. A., Tassinari, M. S., Aronow, M., Collart, D., Shalhoub, V., Peura, S., Dworetzky, S., and Pockwinse, S. (1992). Gene expression during development of the osteoblast phenotype: An integrated relationship of cell growth to differentiation. *In* "Molecular and Cellular Approaches to the Control of Proliferation and Differentiation." (G. S. Stein and J. B. Lian, eds.), pp. 165–223. Academic Press, San Diego.

Liao, J., Ozono, K., Sone, T., McDonnell, D. P., and Pike, J. W. (1990). Vitamin D receptor interaction with specific DNA requires a nuclear protein and 1,25-dihydroxyvitamin D_3. *Proc. Natl. Acad. Sci. USA* **87,** 9751–9755.

Lichtler, A., Stover, M. L., Angilly, J., Kream, B., and Rowe, J. (1989). Isolation and characterization of the rat α-1 (I) collagen promoter. *J. Biol. Chem.* **264,** 3072–3077.

Lyons, K. M., Pelton, R. W., and Hogan, B. L. M. (1989). Patterns of expression of murine Vgr-1 and BMP-2a RNA suggest that transforming growth factor-β-like genes coordinately regulate aspects of embryonic development. *Genes Dev.* **1,** 1657–1668.

Majeska, R. J., and Rodan, G. A. (1982). The effect of $1,25(OH)_2D_3$ on alkaline phosphatase in osteoblastic osteosarcoma cells. *J. Biol. Chem.* **257,** 3362–3365.

Majeska, R. J., Nair, B. C., and Rodan, G. A. (1985). Glucocorticoid regulation of alkaline phosphatase in the osteoblastic osteosarcoma cell line ROS 17/2.8. *Endocrinology* **116,** 170–179.

Marashi, F., Baumbach, L., Rickles, R., Sierra, F., Stein, J. L., and Stein, G. S. (1982). Histone proteins in HeLa S_3 cells are synthesized in a cell cycle stage specific manner. *Science* **215,** 683–685.

Marie, P. J., and Travers, R. (1983). Continuous infusion of 1,25-dihydroxyvitamin D_3 stimulates bone turnover in the normal young mouse. *Calcif. Tissue Int.* **35,** 418–425.

Marie, P. J., Hott, M., and Garba, M.-T. (1985). Contrasting effects of 1,25-dihydroxyvitamin D_3 on bone matrix and mineral appositional rates in the mouse. *Metabolism* **34,** 777–783.

Markose, E. R., Stein, J. L., Stein, G. S., and Lian, J. B. (1990). Vitamin D-mediated modifications in protein–DNA interactions at two promoter elements of the osteocalcin gene. *Proc. Natl. Acad. Sci. USA* **87,** 1701–1705.

Matsuura, S., Kishi, F., and Kajii, T. (1990). Characterization of a 5′-flanking region of the human liver/bone/kidney alkaline phosphatase gene: Two kinds of mRNA from a single gene. *Biochem. Biophys. Res. Commun.* **168,** 993–1000.

Miyauchi, A., Alvarez, J., Greenfield, E. M., Teti, A., Grano, M., Colucci, S., Zambonin-Zallone, A., Ross, F. P., Teitelbaum, S. L., Cheresh, D., and Hruska, K. A. (1991). Recognition of osteopontin and related peptides by an $\alpha_v\beta_3$ integrin stimulates immediate cell signals in osteoclasts. *J. Biol. Chem.* **266,** 20369–20374.

Montecino, M., Breen, E., Lian, J., Stein, G., and Stein, J. (1992). Changes in chromatin structure of the osteocalcin gene promoter associated with basal and vitamin D enhanced transcription. *J. Bone Miner. Res.* **7,** 5167.

Moreno, M. L., Chrysogelos, S. A., Stein, G. S., and Stein, J. L. (1986). Reversible changes in the nucleosomal organization of a human H4 histone gene during the cell cycle. *Biochemistry* **25,** 5364–5370.

Morris, T. D., Weber, L. A., Hickey, E., Stein, G. S., and Stein, J. L. (1991). Changes in the stability of a human H3 histone mRNA during the HeLa cell cycle. *Mol. Cell. Biol.* **11,** 544–553.

Morrison, N. A., Shine, J., Fragonas J.-C., Verkest, V., McMenemy, L., and Eisman, J. A. (1989). 1,25-Dihydroxyvitamin D-responsive element and glucocorticoid repression in the osteocalcin gene. *Science* **246,** 1158–1161.

Nagata, T., Bellows, C. G., Kasugai, S., Butler, W. T., and Sodek, J. (1991). Biosynthesis of bone proteins [SPP-1 (secreted phosphoprotein-1, osteopontin), BSP (bone sialoprotein) and SPARC (osteonectin)] in association with mineralized-tissue formation by fetal-rat calvarial cells in culture. *J. Biochem.* **274,** 513–520.

Nanes, M. S., Rubin, J., Titus, L., Hendy, G. N., and Catherwood, B. D. (1990). Interferon-γ inhibits 1,25-dihydroxyvitamin D_3-stimulated synthesis of bone GLA protein in rat osteosarcoma cells by a pretranslational mechanism. *Endocrinology* **127,** 588–594.

Nijweide, P. J., and Mulder, R.J.P. (1986). Identification of osteocytes in osteoblast-like cell cultures using a monoclonal antibody specifically directed against osteocytes. *Histochemistry* **84,** 342–347.

Nishimoto, S. K., Salka, C., and Nimni, M. E. (1987). Retinoic acid and glucocorticoids enhance the effect of 1,25-dihydroxyvitamin D_3 on bone γ-carboxyglutamic acid protein synthesis by rat osteosarcoma cells. *J. Bone Miner. Res.* **2,** 571–577.

Noda, M. (1989). Transcriptional regulation of osteocalcin production by transforming growth factor-β in rat osteoblast-like cells. *Endocrinology* **124,** 612–617.

Noda, M., Yoon, K., and Rodan, G. A. (1988). Cyclic AMP-mediated stabilization of osteocalcin mRNA in rat osteoblast-like cells treated with parathyroid hormone. *J. Biol. Chem.* **263,** 18574–18577.

Noda, M., Vogel, R. L., Craig, A. M., Prahl, J., DeLuca, H. F., and Denhardt, D. T. (1990). Identification of a DNA sequence responsible for binding of the 1,25-dihydroxyvitamin D_3 receptor and 1,25-dihydroxyvitamin D_3 enhancement of mouse secreted phosphoprotein 1 (Spp-1 or osteopontin) gene expression. *Proc. Natl. Acad. Sci. USA* **87,** 9995–9999.

Nomura, S., Wills, A. J., Edwards, D. R., Heath, J. K., and Hogan, B. L. M. (1988). Developmental expression of 2ar (osteopontin) and SPARC (osteonectin) RNA as revealed by in situ hybridization. *J. Cell Biol.* **106,** 441–450.

Nurse, P. (1990). Universal control mechanism regulating onset of M-phase. *Nature (London)* **344,** 503–508.

Oldberg, Å., Franzén, A., and Heinegård, D. (1986). Cloning and sequence analysis of rat bone sialoprotein (osteopontin) cDNA reveals an Arg-Gly-Asp cell-binding sequence. *Proc. Natl. Acad. Sci. USA* **83,** 8819–8823.

Owen, T. A., Aronow, M., Shalhoub, V., Barone, L. M., Wilming, L., Tassinari, M. S., Kennedy, M. B., Pockwinse, S., Lian, J. B., and Stein, G. S. (1990a). Progressive development of the rat osteoblast phenotype *in vitro:* Reciprocal relationships in expression of genes associated with osteoblast proliferation and differentiation during formation of the bone extracellular matrix. *J. Cell. Physiol.* **143,** 420–430.

Owen, T. A., Bortell, R., Yocum, S. A., Smock, S. L., Zhang, M., Abate, C., Shalhoub, V., Aronin, N., Wright, K. L., van Wijnen, A. J., Stein, J. L., Curran, T., Lian, J. B., and Stein, G. S. (1990b). Coordinate occupancy of AP-1 sites in the vitamin D responsive and CCAAT box elements by fos-jun in the osteocalcin gene: A model for phenotype suppression of transcription. *Proc. Natl. Acad. Sci. USA* **87,** 9990–9994.

Owen, T. A., Holthuis, J., Markose, E., van Wijnen, A. J., Wolfe, S. A., Grimes, S. Lian, J. B., and Stein, G. S. (1990c). Modifications of protein–DNA interactions in the proximal promoter of a cell growth-regulated histone gene during the onset and progression of osteoblast differentiation. *Proc. Natl. Acad. Sci. USA* **87,** 5129–5133.

Owen, T. A., Aronow, M. A., Barone, L. M., Bettencourt, B., Stein, G., and Lian, J. B. (1991). Pleiotropic effects of vitamin D on osteoblast gene expression are related to the proliferative and differentiated state of the bone cell phenotype: Dependency upon basal levels of gene expression, duration of exposure and bone matrix competency in normal rat osteoblast cultures. *Endocrinology* **128,** 1496–1504.

Owen, T. A., Holthuis, J., Markose, E., van Wijnen, A. J., Wolfe, S. A., Grimes, S., Lian, J. B., and Stein, G. S. (1990c). Modifications of protein–DNA interactions in the proximal promoter of a cell growth-regulated histone gene during the onset and progression of osteoblast differentiation. *Proc. Natl. Acad. Sci. USA* **87,** 5129–5133.

Ozkaynak, E., Rueger, D. C., Drier, E. A., Corbett, C., Ridge, R. J., Sampath, T. K., and Oppermann, H. (1990). OP-1 cDNA encodes an osteogenic protein in the TGF-β family. *EMBO J.* **9,** 2085–2093.

Ozono, K., Liao, J., Kerner, S. A., Scott, R. A., and Pike, J. W. (1990). The vitamin D-responsive element in the human osteocalcin gene. *J. Biol. Chem.* **265,** 21881–21888.

Pauli, U., Chrysogelos, S., Stein, G., Stein, J., and Nick, H. (1987). Protein–DNA interactions *in vivo* upstream of a cell cycle regulated human H4 histone gene. *Science* **236,** 1308–1311.

Pauli, U., Chrysogelos, S., Stein, J., and Stein, G. (1988). Native genomic blotting: High-resolution mapping of DNase I-hypersensitive sites and protein–DNA interactions. *Proc. Natl. Acad. Sci. USA* **85,** 16–20.

Pauli, U., Chrysogelos, S., Nick, H., Stein, G., and Stein, J. (1989). *In vivo* protein binding sites and nuclease hypersensitivity in the promoter region of a cell cycle regulated human H3 histone gene. *Nucleic Acids Res.* **17,** 2333–2350.

Pechak, D. G., Kujawa, J. J., and Caplan, A. I. (1986). Morphological and histochemical events during first bone formation in embryonic chick limbs. *Bone* **7,** 441–458.

Plumb, M., Stein, J., and Stein, G. (1983). Coordinate regulation of multiple histone mRNAs during the cell cycle in HeLa cells. *Nucleic Acids Res.* **11,** 2391–2410.

Pockwinse, S., Wilming, L., Conlon, D., Stein, G. S., and Lian, J. B. (1992). Expression of cell growth and bone specific genes at single cell resolution during development of bone tissue-like organization in primary osteoblast cultures. *J. Cell Biochem.* **49,** 310–323.

Prince, C. W., and Butler, W. T. (1987). 1,25-Dihydroxyvitamin D_2 regulated the biosynthesis of osteopontin, a bone-derived cell attachment protein, in clonal osteoblast-like osteosarcoma cells. *Collagen Rel. Res.* **7,** 305–313.

Quarles, L. D., Yohay, D. A., Lever, L. W., Caton, R., and Wenstrup, R. J. (1992). Distinct proliferative and differentiated stages of murine MC3T3-E1 cells in culture: An *in vitro* model of osteoblast development. *J. Bone Miner. Res.* **7,** 683–692.

Raisz, L. G., and Kream, B. E. (1983a). Regulation of bone formation (part 1). *New Eng. J. Med.* **309,** 29–35.

Raisz, L. G., and Kream, B. E. (1983b). Regulation of bone formation (part 2). *New Eng. J. Med.* **309,** 83–89.

Ramsey-Ewing, A., van Wijnen, A., Stein, G. S., and Stein, J. L. Delineation of a human histone H4 cell cycle element in vivo: The master switch for H4 gene transcription.

Reinholt, F. P., Hultenby, K., Oldberg, Å., and Heinegård, D. (1990). Osteopontin—A possible anchor of osteoclasts to bone. *Proc. Natl. Acad. Sci. USA* **87,** 4473–4475.

Rodan, G. A., and Noda, M. (1991). Gene expression in osteoblastic cells. *Crit. Rev. Eukaryotic Gene Expression* **1,** 85–98.

Roesler, W. J., Vandenbark, G. R., and Hanson, R. W. (1988). Cyclic AMP and the induction of eukaryotic gene transcription. *J. Biol. Chem.* **263,** 9063–9066.

Romberg, R. W., Werness, P. G., Riggs, B. L., and Mann, K. G. (1986). Inhibition of hydroxyapatite crystal growth by bone-specific and other calcium-binding proteins. *Biochemistry* **25,** 1176–1180.

Ross, T. K., Moss, V. E., Prahl, J. M., and DeLuca, H. F. (1992). A nuclear protein essential for binding of rat 1,25-dihydroxyvitamin D_3 receptor to its response elements. *Proc. Natl. Acad. Sci. USA* **89,** 256–260.

Rüther, U., Garber, C., Komitowski, D., Müller, R., and Wagner, E. F. (1987). Deregulated c-*fos* expression interferes with normal bone development in transgenic mice. *Nature (London)* **325,** 412–416.

Sandberg, M., Autio-Harmainen, H., and Vuorio, E. (1988). Localization of the expression of types I, III, and IV collagen, TGF-β1 and c-fos genes in developing human calvarial bones. *Dev. Biol.* **130,** 324–334.

Schaack, J., Ho, W. Y.-W., Freimuth, P., and Shenk, T. (1990). Adenovirus terminal protein mediates both nuclear matrix association and efficient transcription of adenovirus DNA. *Genes Dev.* **4,** 1197–1208.

Schepmoes, G., Breen, E., Owen, T. A., Aronow, M. A., Stein, G. S., and Lian, J. B. (1991). Influence of dexamethasone on the vitamin D-mediated regulation of osteocalcin gene expression. *J. Cell. Biochem.* **47,** 184–196.

Schüle, R., Muller, M., Kaltschmidt, C., and Renkawitz, R. (1988). Many transcription factors interact synergistically with steroid receptors. *Science* **241,** 1418–1420.

Schüle, R., Kazuhiko, U., Mangelsdorf, D. J., Bolado, J., Pike, J. W., and Evans, R. M. (1990a). Jun-Fos and receptors for vitamins A and D recognize a common response element in the human osteocalcin gene. *Cell* **61,** 497–504.

Schüle, R., Rangarajan, P., Kliewer, S., Ransome, L. J., Bolado, J., Yang, N., Verma, I. M., and Evans, R. M. (1990b). Functional antagonism between oncoprotein c-Jun and the glucocorticoid receptor. *Cell* **62,** 1217–1226.

Shalhoub, V., Gerstenfeld, L. C., Collart, D., Lian, J. B., and Stein, G. S. (1989). Downregulation of cell growth and cell cycle regulated genes during chick osteoblast differentiation with the reciprocal expression of histone gene variants. *Biochemistry* **28,** 5318–5322.

Shalhoub, V., Jackson, M. E., Lian, J. B., Stein, G. S., and Marks, S. C., Jr. (1991). Gene expression during skeletal development in three osteopetrotic rat mutations: Evidence for osteoblast abnormalities. *J. Biol. Chem.* **266,** 9847–9856.

Shalhoub, V., Conlon, D., Tassinari, M., Quinn, C., Partridge, N., Stein, G. S., and Lian, J. B. (1992). Glucocorticoids promote development of the osteoblast phenotype by selectively modulating expression of cell growth and differentiation associated genes. *J. Cell. Biochem.* **50,** 425–440.

Sierra, F., Lichtler, A., Marashi, F., Rickles, R., Van Dyke, T., Clark, S., Wells, J., Stein, G., and Stein, J. (1982). Organization of human histone genes. *Proc. Natl. Acad. Sci. USA* **79,** 1795–1799.

Sierra, F., Stein, G., and Stein, J. (1983). Structure and *in vitro* transcription of a human H4 histone gene. *Nucleic Acids Res.* **11,** 7069–7086.

Soriano, P., Montgomery, C., Geske, R., and Bradley, A. (1991). Targeted disruption of the c-src proto-oncogene leads to osteopetrosis in mice. *Cell* **64,** 693–702.

Speiss, Y. H., Price, P. A., Deftos, J. L., and Manolagas, S. C. (1986). Phenotype-associated changes in the effects of 1,25-dihydroxyvitamin D_3 on alkaline phosphatase and bone Gla-protein of rat osteoblastic cells. *Endocrinology* **118,** 1340–1346.

Sporn, M. B., Roberts, A. B. Shull, J. H., Smith, J. M., Ward, J. M., and Sodek, J. (1983). Polypeptide transforming growth factor isolated from bovine sources and used for wound healing in vivo. *Science* **219,** 1329–1331.

Stein, G., Park, W., Thrall, C., Mans, R., and Stein, J. (1975). Regulation of cell cycle stage-specific transcription of histone genes from chromatin by non-histone chromosomal proteins. *Nature (London)* **257,** 764–767.

Stein, G., Lian, J., Stein, J., Briggs, R., Shalhoub, V., Wright, K., Pauli, U., and van Wijnen, A. J. (1989). Altered binding of human histone gene transcription factors during the shutdown of proliferation and onset of differentiation in HL-60 cells. *Proc. Natl. Acad. Sci. USA* **86,** 1865–1869.

Stein, G. S., Lian, J. B., and Owen, T. A. (1990). Relationship of cell growth to the regulation of tissue-specific gene expression during osteoblast differentiation. *FASEB J.* **4,** 3111–3123.

Stein, G. S., Lian, J. B., Dworetzky, S. I., Owen, T. A., Bortell, R., Bidwell, J. P., and van Wijnen, A. J. (1991). Regulation of transcription-factor activity during growth and differentiation: Involvement of the nuclear matrix in concentration and localization of promoter binding proteins. *J. Cell. Biochem.* **47,** 300–305.

Stein, G. S., Lian, J. B., Owen, T. A., Holthuis, J., Bortell, R., and van Wijnen, A. J. (1992a). Molecular mechanisms that mediate a functional relationship between proliferation and differentiation. *In* "Molecular and Cellular Approaches to the Control of Proliferation and Differentiation." (G. S. Stein and J. B. Lian, eds.). pp. 299–341. Academic Press, San Diego.

Stein, G. S., Stein, J. L., van Wijnen, A. J., and Lian, J. B. (1992b). Regulation of histone gene expression. *Curr. Opinion Cell Biol.* **4,** 166–173.

Stief, A., Winter, D. M., Strätling, W. H., and Sippel, A. E. (1989). A nuclear DNA attachment element mediates elevated and position-independent gene activity. *Nature (London)* **341,** 343–345.

Strähle, U., Schmid, W., and Schütz, G. (1988). Synergistic action of the glucocorticoid receptor with transcription factors. *EMBO J.* **7,** 3389–3395.

Strauss, P. G., Closs, E. I., Schmidt, J., and Erfle, V. (1990). Gene expression during osteogenic differentiation in mandibular condyles in vitro. *J. Cell Biol.* **110,** 1369.

Stromstedt, P. E., Poellinger, L., Gustafsson, J. A., and Cralstedt-Duke, J. (1991). The glucocorticoid receptor binds to a sequence overlapping the TATA box of the human osteocalcin promoter: A potential mechanisms for negative regulation. *Mol. Cell. Biol.* **11,** 3379–3383.

Terpening, C. M., Haussler, C. A., Jurutka, P. W., Galligan, M. A., Komm, B. S., and Haussler, M. R. (1991). The vitamin D-responsive element in the rat bone Gla protein gene is an imperfect direct repeat that cooperates with other cis-elements in 1,25-dihydroxyvitamin D_3-mediated transcriptional activation. *Mol. Endocrinol.* **5,** 373–385.

Theofan, G., and Price, P. A. (1989). Bone gla protein messenger ribonucleic acid is regulated by both 1,25-dihydroxyvitamin D_3 and 3′,5′-cyclic adenosine monophosphate in rat osteosarcoma cells. *Mol. Endocrinol.* **3,** 36–43.

Turksen, K., and Aubin, J. E. (1991). Positive and negative immunoselection for enrichment of two classes of osteoprogenitor cells. *J. Cell Biol.* **114,** 373–384.

van den Ent, F. M. I., van Wijnen, A. J., Last, T. J., Bortell, R., Stein, J. L., Lian, J. B., and Stein, G. S. (1993). Concerted control of multiple histone promoter factors during reciprocal regulation of cell cycle control and bone-related genes in cell-density inhibited osteosarcoma ROS 17/2.8 cells. *Cancer Research,* (in press).

van der Houven van Oordt, C. W., van Wijnen, A. J., Carter, R., Soprano, K., Lian, J. B., Stein, G. S., and Stein, J. L. (1992). Protein–DNA interactions at the H4-site III upstream transcriptional element of a cell cycle regulated histone H4 gene: Differences in normal versus tumor cells. *J. Cell. Biochem.* **49,** 93–110.

van Wijnen, A. J., Wright, K. L., Lian, J. B., Stein, J. L., and Stein, G. S. (1989). Human H4 histone gene transcription requires the proliferation specific nuclear factor HiNF-D: Auxilliary roles for HiNF-C (Sp1-Like) and HiNF-A (high mobility group-like). *J. Biol. Chem.* **264,** 15034–15042.

van Wijnen, A. J., Choi, T. K., Owen, T. A., Wright, K. L., Lian, J. B., Jaenisch, R., Stein, J. L., and Stein, G. S. (1991a). Involvement of the cell cycle regulated nuclear factor HiNF-D in cell growth control of a human H4 histone gene during hepatic development in transgenic mice. *Proc. Natl. Acad. Sci. USA* **88,** 2573–2577.

van Wijnen, A. J., Ramsey-Ewing, A., Bortell, R., Owen, T. A., Lian, J. B., Stein, J. L., and Stein, G. S. (1991b). Transcriptional element H4-site II of cell cycle regulated human H4 histone genes is a multipartite protein/DNA interaction site for factors HiNF-D, HiNF-M and HiNF-P: Involvement of phosphorylation. *J. Cell. Biochem.* **46,** 174–189.

Vukicevic, S., Luyten, F. P., Kleinman, H. K., and Reddi, A. H. (1990). Differentiation of canalicular cell processes in bone cells by basement membrane matrix components: Regulation by discrete domains of laminin. *Cell* **63,** 437–445.

Weinberg, R. A. (1989). Oncogenes, antioncogenes, and the molecular bases of multistep carcinogenesis. *Cancer Res.* **49,** 3713–3721.

Weinreb, M., Shinar, D., and Rodan, G. A. (1990). Different pattern of alkaline phosphatase, osteopontin, and osteocalcin expression in developing rat bone visualized by in situ hybridization. *J. Bone Miner. Res.* **5,** 831–842.

Weiss, M. J., Ray, K., Henthorn, P. S., Lamb, B., Kadesch, T., and Harris, H. (1988). Structure of the human liver/bone/kidney alkaline phosphatase gene. *J. Biol. Chem.* **263,** 12002–12010.

Whitson, S. W., Harrison, W., Dunlap, M. K., Bowers, D. E., Jr., Fisher, L. W., Robey, P. G., and Termine, J. D. (1984). Fetal bovine cells synthesize bone specific matrix proteins. *J. Cell Biol.* **88,** 607–614.

Wronski, T. J., Halloran, B. P., Bikle, D. D., Globus, R. K., and Morey-Holton, E. R. (1986). Chronic administration of 1,25-dihydroxyvitamin D_3: Increased bone but impaired mineralization. *Endocrinology* **119,** 2580–2585.

Yoon, K., Buenaga, R., and Rodan, G. A. (1987). Tissue specificity and developmental expression of osteopontin. *Biochem. Biophys. Res. Commun.* **148,** 1129–1136.

Yoon, K., Rutledge, S. J., Buenaga, R. F., and Rodan, G. A. (1988). Characterization of the rat osteocalcin gene: Stimulation of promoter activity by 1,25-dihydroxyvitamin D_3. *Biochemistry* **27,** 8521–8526.

Zeitlin, S., Parent, A., Silverstein, S., and Efstratiadis, A. (1987). Pre-mRNA splicing and the nuclear matrix. *Mol. Cell. Biol.* **7,** 111–120.

Zenk, D. W., Ginder, G. D., and Brotherton, T. W. (1990). A nuclear matrix protein binds very tightly to DNA in the avian β-globin gene enhancer. *Biochemistry* **29,** 5221–5226.

Zernik, J., Thiede, M. A., Twarog, K., Stover, M. L., Rodan, G. A., Upholt, W. B., and Rowe, D. W. (1990). Cloning and analysis of the 5′ region of the rat bone/liver/kidney/placenta alkaline phosphatase gene a dual-function promoter. *Matrix* **10,** 38–47.

3

CELLULAR AND MOLECULAR BIOLOGY OF TRANSFORMING GROWTH FACTOR β

SEONG-JIN KIM and R. TRACY BALLOCK

Cellular and Molecular Biology of Bone

I. INTRODUCTION

In recent years, there has been an exponential increase in understanding of the chemistry and biology of the family of peptides called transforming growth factor β (TGF-β). The original narrow definition of TGF-β as an agent inducing the transformed phenotype in mesenchymal cells (for a recent review, see Roberts and Sporn, 1990) has now been supplanted by the knowledge that this protein affects many functions in nearly every tissue and cell type, not only mesenchymal cells but also lymphoid cells, monocytes, and neural cells. It is now known that TGF-β plays essential roles in embryogenesis, particularly during periods of morphogenesis, and that some of the same embryological mechanisms are reiterated in the adult during the normal processes of tissue remodeling and repair. These mechanisms may also function aberrantly in various pathological processes, including carcinogenesis. Its broad spectrum of cellular targets and its multifunctional actions suggest that TGF-β has a pivotal control function in many physiological and pathological processes. The finding of five distinct, highly conserved, yet functionally similar TGF-βs presents a challenge to unravel the specific roles of each of these closely related, developmentally relevant peptides.

II. CHEMISTRY OF TRANSFORMING GROWTH FACTOR β

TGF-β, now called TGF-β1, was originally purified to homogeniety from human platelets (Assoian *et al.*, 1983), human placenta (Frolik *et al.*, 1983), and bovine kidney (Roberts *et al.*, 1983) and identified as a homodimeric peptide with a molecular weight of 25,000. The human complementary DNA (cDNA) clone sequence indicates that the monomer is synthesized as the COOH-terminal 112 amino acids of a 390-amino acid precursor (Derynck *et al.*, 1985) (Fig. 1). A few years later, a second form of the peptide, TGF-β2, was purified from several sources including platelets, bovine bone, and cultured cells (Seyedin *et al.*, 1985; Cheifetz *et al.*, 1987; Wrann *et al.*, 1987). Although only 71% homologous, TGF-β2 is interchangeable with TGF-β1 in most biological assays. More recently, three new forms of the peptide, called TGF-β3, TGF-β4, and TGF-β5, have been identified by screening of cDNA libraries (ten Dijke *et al.*, 1988; Jakowlew *et al.*, 1988a, b; Kondaiah *et al.*, 1990). All five TGF-βs are 64–83% homologous and share essential structural features such as synthesis from a long precursor and conservation of all nine cysteine residues in the processed peptide (Fig. 1). To the extent that it has been

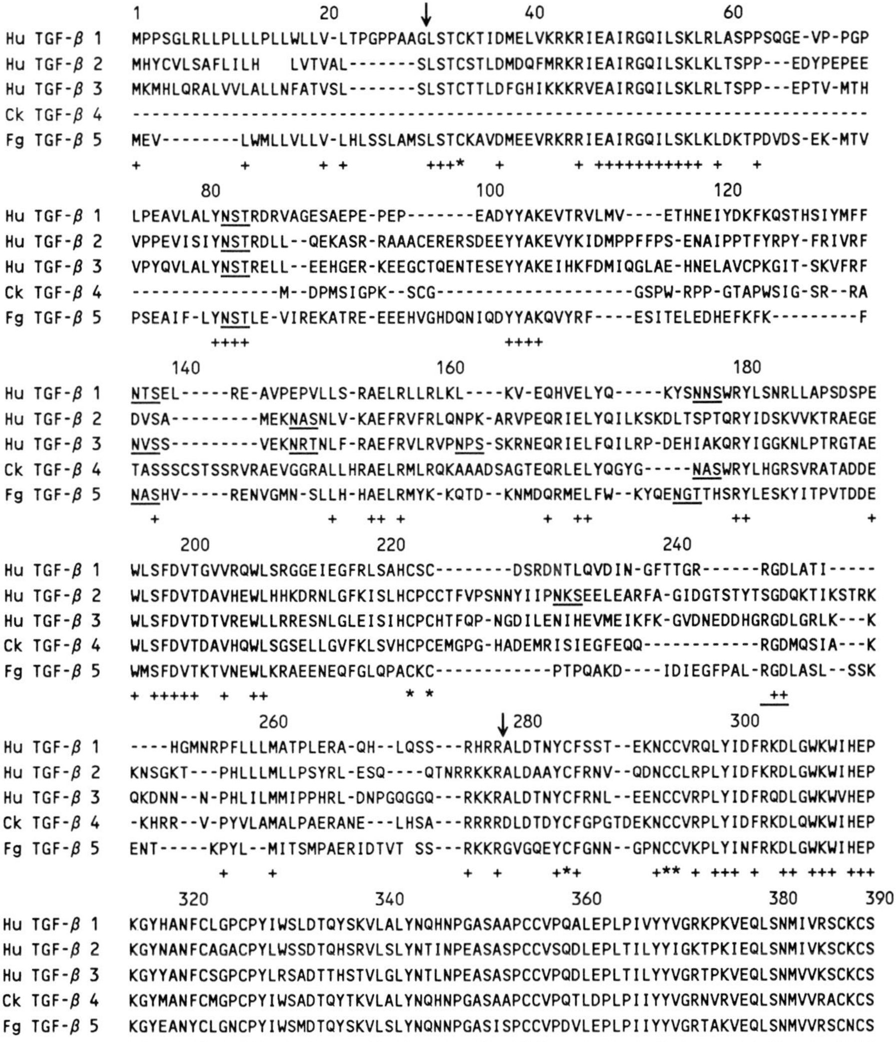

FIGURE 1 Amino acid sequences of the transforming growth factor β (TGF-β) precursors. Arrows indicate the position of proteolytic processing resulting in cleavage of the signal peptide of TGF-β1 and of the mature TGF-βs. N-linked glycosylation sites are underlined, as is the integrin cellular recognition sequence, RGD. +, amino acids that are conserved in all TGF-βs; *, cysteines that are conserved in all TGF-βs. Recent data suggests that TGF-β4 has a typical precursor sequence beginning with a signal peptide (Burt and Jakowlew, 1992).

examined, each of these different TGF-βs is more than 98% conserved among species; thus, for example, TGF-β1 is identical in humans, monkey, pig, cow, and chicken, and TGF-β3 has only one conservative amino acid substitution between humans and chicken.

In addition to the TGF-βs, many other proteins have now been found to belong to the TGF-β supergene family by virtue of amino acid homologies, particularly with respect to conservation of seven of the nine cysteine residues of TGF-β (see Roberts and Sporn, 1990). These proteins are each only 30–40% homologous to the TGF-βs and are functionally distinct. A unifying feature of the biology of all of the peptides of the supergene family is their ability to regulate developmental processes. Thus, Müllerian inhibiting substance induces regression of the female rudiments of the developing male reproductive system. The inhibins and activins regulate the activity of the gonadotropin, follicle-stimulating hormone (FSH). The activins exert the mesoderm-inducing activity in the *Xenopus* embryo. The bone morphogenetic proteins are thought to play a role in the formation of cartilage and bone *in vivo*. The putative product of the decapentaplegic gene complex in *Drosophila* directs dorsal–ventral patterning in the developing fly embryo, and Vg1, an amphibian gene expressed in frog oocytes, is postulated to be involved in the process of induction of mesoderm from ectoderm during gastrulation in the amphibian embryo. The chemical and biological relatedness of these proteins and the TGF-βs raises the possibility that this superfamily of proteins diverged early in evolution from a common ancestral gene encoding a protein essential to the development of very primitive organisms.

Expression of recombinant TGF-β3 and isolation of TGF-β5 from medium conditioned by *Xenopus* tadpole cells (Roberts *et al.*, 1989) has demonstrated that these two TGF-βs also bind to TGF-β receptors and have activity equivalent to TGF-β1 and TGF-β2 in a standard assay of growth inhibition (Danielpour *et al.*, 1989). In addition, like TGF-β1 and TGF-β2, TGF-β3 and TGF-β5 are secreted from cells in a biologically latent form, which must first be activated before the peptide can bind to signaling receptors. The latent form of TGF-β is a noncovalent complex of the processed peptide and the remainder of the TGF-β precursor (Wakefield *et al.*, 1988). TGF-β4 has not yet been expressed. It is unique among the TGF-βs in that it lacks a signal peptide sequence, has a shorter precursor (308 amino acids compared to 382–412 for the other TGF-βs), and has an insertion of 2 amino acids in the processed coding region (Jakowlew *et al.*, 1988b). Whether TGF-β4 might play a unique intracellular role is currently being investigated.

III. MULTIPLE ACTIONS OF TRANSFORMING GROWTH FACTOR β

When the multifunctionality of TGF-β was first discovered, it was thought to be a property unique to this particular peptide; however, it is now realized that many peptide growth factors are multifunctional and are capable of eliciting a very broad range of unrelated responses in many different types of cell (Tucker *et al.*, 1984; Roberts *et al.*, 1985; Massague *et al.*, 1986; Kehrl *et al.*, 1986a, b). There are now many different actions of TGF-β that have been discovered in various cell culture systems. These actions can arbitrarily be categorized as either proliferative, antiproliferative, or unrelated to proliferation (Fig. 2).

TGF-β was originally discovered in an assay that measured its ability to cause a proliferative effect, namely, its ability to stimulate the anchorage-independent growth of fibroblasts. There are many examples of cells of mesenchymal origin whose proliferation is enhanced by TGF-β (Roberts *et al.*, 1981). This assay clearly identifies TGF-β as a bona fide growth factor.

TGF-β-like activities are produced by bone cells, and high levels of TGF-β are found in the extracellular bone matrix, suggesting an important physiological function of TGF-βs in this tissue (Centrella *et al.*, 1987). TGF-β stimulates cell replication and collagen production in cultured fetal rat bone cells (Centrella *et al.*, 1986, 1987, 1988) and induces chondrogenesis in embryonic rat mesenchymal cells (Seyedin *et al.*, 1986). Recently, it has been shown that TGF-β3 appeared to be three- to

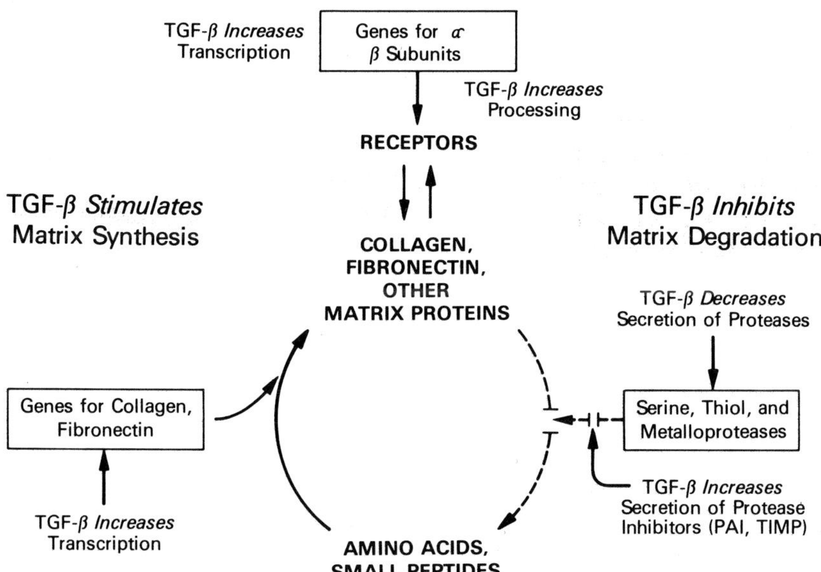

FIGURE 2 Mechanisms of transforming growth factor β (TGF-β) enhancement of both the accumulation of extracellular matrix proteins and the integrins, including enhancement of synthesis of matrix proteins, inhibition of the degradation of matrix proteins, and enhanced synthesis of the receptors for matrix proteins. PAI, plasminogen activator inhibitor; TIMP, tissue inhibitor of metalloendoproteases.

fivefold more potent than TGF-β1 in a direct comparison of functions associated with bone formation (i.e., mitogenesis, collagen synthesis, and alkaline phosphatase activity) (ten Dijke *et al.*, 1990). An important cell type for which TGF-β is mitogenic in monolayer culture is the osteoblast (Centrella *et al.*, 1987; Robey *et al.*, 1987); these cells not only respond to TGF-β with a growth response, but they also produce high levels of TGF-β, a finding suggesting that this peptide might exert some type of autocrine growth control in bone.

TGF-β also has strong mitogenic activity for Schwann cells of the peripheral nervous system, a cell type for which there are few known mitogens. Because the Schwann cell is believed to play an important role in regeneration of peripheral nerve after injury, this finding raises the intriguing possibility that TGF-β might function as an intrinsic mediator of repair of peripheral neuronal damage. Other data strongly implicate TGF-β in promotion of wound healing and connective tissue repair.

TGF-β is a potent inhibitor of the proliferation of many cells *in vitro*, particularly of epithelial cells (Tucker *et al.*, 1984; Moses *et al.*, 1985; Masui *et al.*, 1986). Most recently, it has been shown that TGF-β is a strong inhibitor of proliferation in many primary or secondary cell cultures, including hepatocytes (Hayashi and Carr, 1985), embryonic fibroblasts (Anzano *et al.*, 1986), keratinocytes (Shipley *et al.*, 1986), and bronchial epithelial cells (Masui *et al.*, 1986). It has long been known that serum may inhibit growth of many epithelial cells in culture. TGF-β, which is found in high concentrations in serum (Childs *et al.*, 1982) and inhibits proliferation of most epithelial cells, now appears to be a principal mediator of this effect. TGF-β is present in platelets (Assoian *et al.*, 1983) in an amount equivalent to the platelet-derived growth factor (PDGF) and is released from α granules of platelets when blood clots (Assoian and Sporn, 1986).

TGF-β has many effects, unrelated to proliferation, on many different cell types. This is hardly surprising considering the nearly universal occurrence of TGF-β receptors on all types of cells (Cheifetz *et al.*, 1987; Segarini *et al.*, 1989). A specific action for TGF-β is the enhancement of the formation of extracellular matrix and inhibition of matrix degradation. This is undoubtedly of major importance with respect to embryogenesis and to repair of tissue injury. A primary effect of TGF-β is its strong enhancement of the formation of both collagen and fibronectin in fibroblasts of human, rat, mouse, and chicken origin (Ignotz and Massague, 1986; Roberts *et al.*, 1986; Wrana *et al.*, 1986). The effect is specific for TGF-β; epidermal growth factor (EGF) and PDGF are ineffective. More than one mechanism may contribute to the enhancement of the synthesis of matrix proteins by TGF-β. Several studies have shown that TGF-β increases messenger RNA (mRNA) levels for type I, III, and V collagen and fibronectin (Varga *et al.*, 1987; Rossi *et al.*, 1988; Madri *et al.*, 1988).

Raghow *et al.* (1987) suggested that the increased mRNA levels result from stabilization of the message rather than increased synthesis. This is based on the observation that TGF-β had no effect on transcription of collagen mRNA using a nuclear runoff assay. However, other experiments have shown that TGF-β can directly stimulate the activity of the mouse a2(I) collagen promoter (Rossi *et al.*, 1988). The data suggest that TGF-β has direct effects on the transcriptional activity of the promoter and, moreover, that its stimulatory effects are mediated through a nuclear factor 1 (NF1) binding site. The NF1 binding site was first identified as a critical element in the TGF-β-dependent activation of a2(I) collagen transcription by deletion analysis of the promoter (Rossi *et al.*, 1988). These experiments illustrated that deletion of an NF1 binding site abolished the TGF-β inducibility of the promoter.

A second action of TGF-β directly related to enhanced formation of extracellular matrix is its ability to inhibit the proteolytic degradation of newly formed matrix proteins. This occurs by two distinct but coordinated mechanisms. The first mechanism involves an increase in the formation and secretion of protease inhibitors, and the second mechanism involves a decrease in the secretion of protease themselves (Chiang and Nilsen-Hamilton, 1986; Laiho *et al.*, 1986; Lund *et al.*, 1987; Edwards *et al.*, 1987; Overall *et al.*, 1989; Redini *et al.*, 1988; Matrisian *et al.*, 1986).

The stimulatory effects of TGF-β on function of other cell types are diverse and cell-specific. As examples, TGF-β enhances prostaglandin release and mobilization of calcium from calvarial organ cultures (Tashijian *et al.*, 1985), whereas in nondividing ovarian granulosa cells it markedly potentiates the ability of FSH to induce the production of progesterone and estrogen (Adashi and Resnick, 1986) and alters the formation of receptors for luteinizing hormone induced by FSH (Knecht *et al.*, 1986). Several reports have demonstrated that TGF-β can also control the direction of cellular differentiation. Low concentration of TGF-β, which have no effect on cell proliferation, block the fusion of various types of myoblasts into myotubes (Massague *et al.*, 1986). The adipogenic differentiation of 3T3-L1 cells, induced by insulin and dexamethasone, is also blocked by TGF-β without affecting mitogenesis (Ignotz and Massague, 1985).

IV. TRANSFORMING GROWTH FACTOR β IN THE REGULATION OF BONE FORMATION AND REPAIR

Although nearly all cells synthesize and respond to TGF-β, the highest levels of this peptide growth factor have been found in bone and cartilage (Seyedin *et al.*, 1986). TGF-β has also been found to be an important regulator of extracellular matrix synthesis and degradation, suggesting a potentially critical function of this growth factor in the regulation of bone formation and repair.

A. Transforming Growth Factor β Is Present at Sites of Bone Formation and Repair

Using polyclonal antibodies raised against a synthetic peptide corresponding to the amino terminus of the mature TGF-β molecule, it has been possible to localize TGF-β to areas of bone formation and repair *in vivo*. Two different polyclonal anti-TGF-β antibodies have been reported to date, both raised against a similar amino acid sequence (Ellingsworth *et al.*, 1986; Flanders *et al.*, 1988). Interestingly, one antibody seems to recognize only intracellular TGF-β, while the second antibody recognizes only extracellular peptides. The reason for this divergent pattern of localization has not been elucidated. Immunohistochemical studies of 11- to 18-day-old mouse embryos using anti-TGF-β antibodies have demonstrated positive staining for TGF-β in tissues of mesodermal origin, such as connective tissue, cartilage, and bone (Heine *et al.*, 1987). In particular, intense staining for TGF-β was observed in tissues actively participating in cartilage and bone formation. These include not only areas of intramembranous ossification, such as the developing calvarium, but also areas of endochondral ossification, including vertebral bodies and long bones (Fig. 3) (Heine *et al.*, 1987). TGF-β immunostaining has also been demonstrated in the chondrocytes of the growth plates of more mature animals, which increase the length of long bones through the endochondral mechanism (Jingushi *et al.*, 1990).

TGF-β is also present at sites of bone repair. Immediately following fracture, TGF-β staining can be appreciated in the area of the fracture hematoma and proliferating periosteal mesenchymal cells (Joyce *et al.*, 1990a). At later stages of fracture healing, TGF-β is evident within osteoblasts forming the intramembranous hard callus as well as within chondroprogenitor cells and chondrocytes comprising the soft callus. It is hypothesized that TGF-β, and perhaps other growth factors released from degranulating platelets in the fracture hematoma, initiates the cascade of cellular events resulting in bone and cartilage formation during fracture repair.

B. Osteoblasts Synthesize and Respond to Transforming Growth Factor β *in Vitro*

Both primary osteoblasts and transformed osteoblasts synthesize and respond to TGF-β in culture. The synthesis and release of TGF-β from these cells can also be modulated by systemic hormones known to regulate bone formation and resorption. Primary osteoblast cultures have

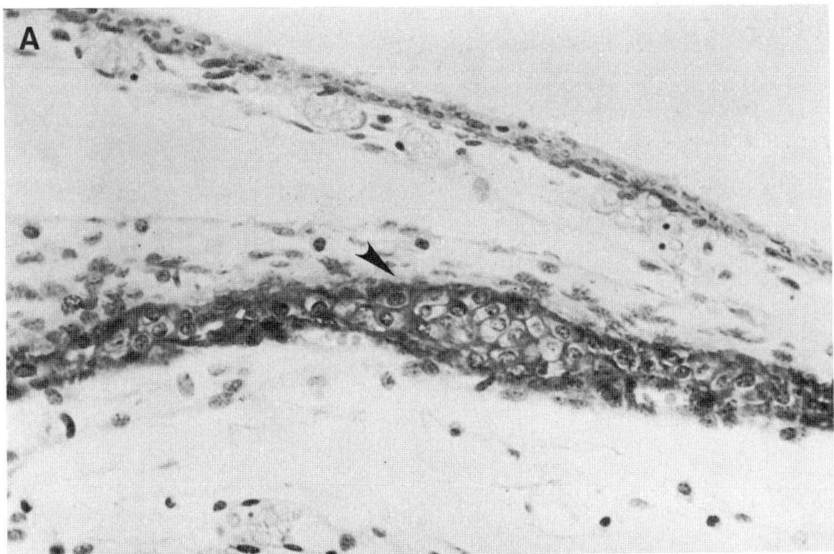

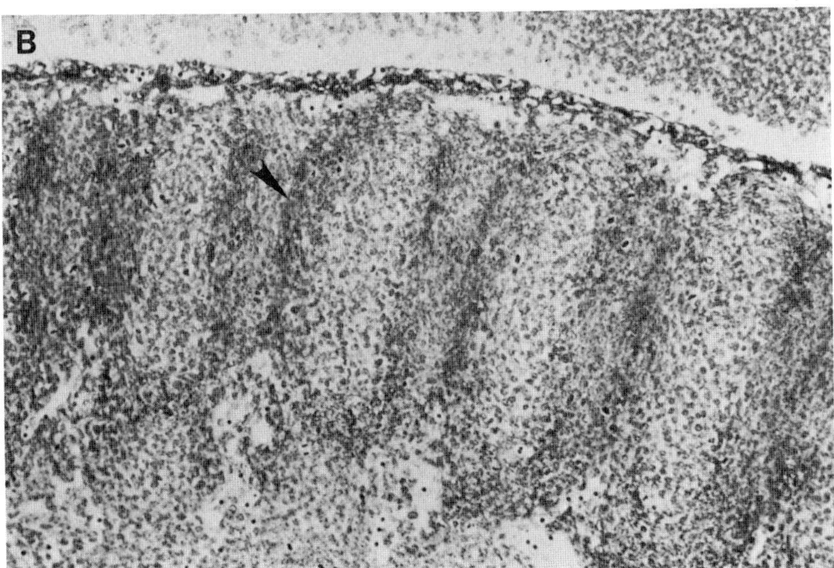

FIGURE 3 Immunohistochemical localization of transforming growth factor (TGF) at sites of bone formation in the mouse embryo. (A) Developing calvarium at day 15 of mouse embryogenesis demonstrates positive staining for TGF-β in osteoblasts participating in intramembranous ossification. The arrowhead points to positive cells. (B) Sclerotomes in the caudal spinal column of a day 13 embryo show marked staining in areas that demarcate the future vertebral bodies. The arrowhead points to one of the areas of positive cells. Further details in Heine *et al.*, 1987. Reprinted with permission.

been used to demonstrate that these cells transcribe mRNAs for TGF-β, translate this message at high levels, and secrete TGF-β protein in its latent form into the culture medium (Robey *et al.*, 1987).

Hormones involved in the systemic regulation of mineral metabolism also modulate the synthesis of TGF-β by osteoblasts. Factors that are known to increase serum calcium by stimulating bone resorption, such as parathyroid hormone and 1,25-dihydroxyvitamin D_3, increase the TGF-β activity in the culture medium conditioned by osteoblasts (Pfeilschifter and Mundy, 1987). Estradiol has also been shown to increase the transcription of TGF-β mRNA by osteoblasts, although levels of TGF-β protein synthesis were not measured (Komm *et al.*, 1988).

TGF-β has multiple actions in almost every cell type studied, and bone-forming cells are no exception. As with other cells, these effects depend on the stage of differentiation of the target cells, the culture conditions, and the presence or absence of other growth factors. In osteoblastic cells, conflicting results have been reported on the effect of TGF-β depending on whether the target cells were primary osteoblasts or transformed osteoblasts. In general, however, TGF-β has been found to have a biphasic mitogenic effect on cultured osteoblasts and stimulates synthesis of collagen and noncollagenous extracellular matrix proteins. In addition, TGF-β inhibits some feature of the fully differentiated osteoblast phenotype, such as alkaline phosphatase activity and osteocalcin synthesis.

Although TGF-β inhibits the growth of many cell types, it appears to be a biphasic stimulator of mitogenic activity by primary osteoblasts and MC3T3 cells (Centrella *et al.*, 1987). At low concentrations, TGF-β stimulates osteoblast chemotaxis as well as DNA synthesis and cell division (Pfeilschifter *et al.*, 1990). At higher concentrations, however, this mitogenic activity is decreased. In addition to its effects on cell growth and migration, TGF-β also promotes the synthesis of bone matrix proteins by primary and transformed osteoblasts. Osteoblast synthesis of the noncollagenous matrix proteins osteonectin and osteopontin is stimulated by TGF-β, and collagen synthesis is increased two- to threefold by concentrations of TGF-β as low as 1 ng/ml (Noda and Rodan, 1989). This increase in collagen synthesis appears to be both dose-, and time-dependent.

Although TGF-β appears to promote osteoblastic synthesis of several of the proteins found in bone matrix, other markers of the terminally differentiated osteoblast are inhibited. These include synthesis of osteocalcin, which is believed to be secreted late in osteoblast development, as well as alkaline phosphatase activity, which is necessary for matrix mineralization (Noda, 1989; Vukicevic *et al.*, 1990).

Whether TGF-β exerts its effect on bone matrix synthesis by direct mechanisms is controversial. Although increases in gene transcription for type I collagen and other matrix proteins have been observed, post-transcriptional and post-translational regulatory events have not been ruled out (Centrella *et al.*, 1991). Robey *et al.* (1987) also suggested that the increase in collagen accumulation seen in primary bovine osteoblast cultures in response to TGF-β results from mitotic expansion of the population of collagen-producing cells. It is possible that multiple direct and indirect control mechanisms may be utilized in the regulation of bone matrix synthesis by TGF-β *in vivo.*

Although TGF-β stimulates the growth of primary osteoblasts and promotes collagen synthesis by these cells, conflicting results have been obtained using transformed osteoblasts as the target cell population. In ROS $^{17}/_{2}$ and UMR 106 cells, TGF-β has been found to inhibit cellular proliferation and stimulate expression of the differentiated osteoblastic phenotype (Pfeilschifter *et al.*, 1987). The biological significance of these conflicting results is unclear.

C. Transforming Growth Factor β Is a Potential Coupling Factor Linking Bone Formation and Resorption

Indirect evidence is accumulating to implicate TGF-β as a potential coupling factor during bone remodeling, coordinately regulating the processes of matrix synthesis and matrix resorption.

TGF-β is stored in the mineralized bone matrix in its latent form and can be activated by exposure to a low pH environment similar to that generated by the osteoclast during bone resorption. It has been demonstrated that latent TGF-β added to osteoclastic cell cultures is activated to produce the biologically active peptide *in vitro* (Oreffo *et al.*, 1989).

TGF-β has also been shown to promote bone matrix synthesis and inhibit matrix degradation. The positive effect of TGF-β on osteoblast chemotaxis, proliferation, and matrix synthesis has already been outlined. TGF-β has also been shown to inhibit bone resorption by inhibiting both the formation and activity of osteoclastic cells. In several studies of osteoclast differentiation, TGF-β blocks the conversion of circulating monocytes to multinucleated osteoclasts (Chenu *et al.*, 1988). TGF-β also directly suppresses the bone-resorbing activity of osteoclasts in both avian and rodent models (Pfeilschifter *et al.*, 1988).

D. Transforming Growth Factor β Induces Bone and Cartilage Formation *in Vivo*

The conflicting data on the effect of TGF-β on cultured osteoblasts has necessitated *in vivo* experiments to clarify further the role of this peptide in bone formation. Several reports now demonstrate that TGF-β injected into the area of the periosteum will result in bone formation by osteoblasts *in vivo* (Joyce *et al.*, 1990a; Noda and Camilliere, 1989; Marcelli *et al.*, 1990). This induction of bone formation is apparent 3 days after two daily injections of TGF-β in doses ranging from 50 ng to 5 μg. The bone that is formed in response to injection of TGF-β appears to mineralize normally and persist for at least 40 days following injection.

The pathway of bone formation resulting from TGF-β injection depends on the type of bone injected. Injection of flat bones results in intramembranous ossification, whereas injection of long bones results in both intramembranous and endochondral ossification (Fig. 4). It is interesting to note that these are the same pathways of bone formation involved in the development and repair of these respective types of bone. In fact, when TGF-β is injected into long bone periosteum for 14 days, the resulting bone and cartilage tissue bears a striking histological resemblance to the pattern of bone and cartilage formation associated with fracture healing. This suggests that TGF-β may be an important regulator of bone repair by initiating the sequence of cellular events that lead to bone formation following injury.

Although TGF-β effectively induces bone formation by periosteal cells *in vivo*, it does not induce bone formation by the endochondral route when implanted subcutaneously or intramuscularly with demineralized bone matrix (Celeste *et al.*, 1990). This assay has been used to demonstrate osteogenic activity by a number of novel proteins, most of which are members of the bone morphogenetic protein class of the TGF-β superfamily (Celeste *et al.*, 1990).

V. TRANSCRIPTIONAL CONTROL OF EXPRESSION OF TRANSFORMING GROWTH FACTOR βs

A. *In Vivo* and *in Vitro* Regulation of Transcription of Transforming Growth Factor βs

An up-regulation of TGF-β1 transcription has been reported for several *in vivo* systems involving injury and repair such as phorbol ester treatment of skin (Akhurst *et al.*, 1988), experimental myocardial infaction (Thompson *et al.*, 1988), bone fracture healing (Joyce *et al.*, 1990a), and liver regeneration (Braun *et al.*, 1988). These reports suggest that primary signals of stress or injury or secondary repair responses result in activation of TGF-β1 transcription, consistent with the well-documented role of the protein in tissue repair and bone formation.

Much less is known about the *in vivo* regulation of transcription of TGF-β2, TGF-β3, TGF-β4, and TGF-β5. However, substantial evidence now indicates that expression of TGF-βs2, TGF-β3, and TGF-β5 is under strict developmental control. For example, in the chicken heart, expres-

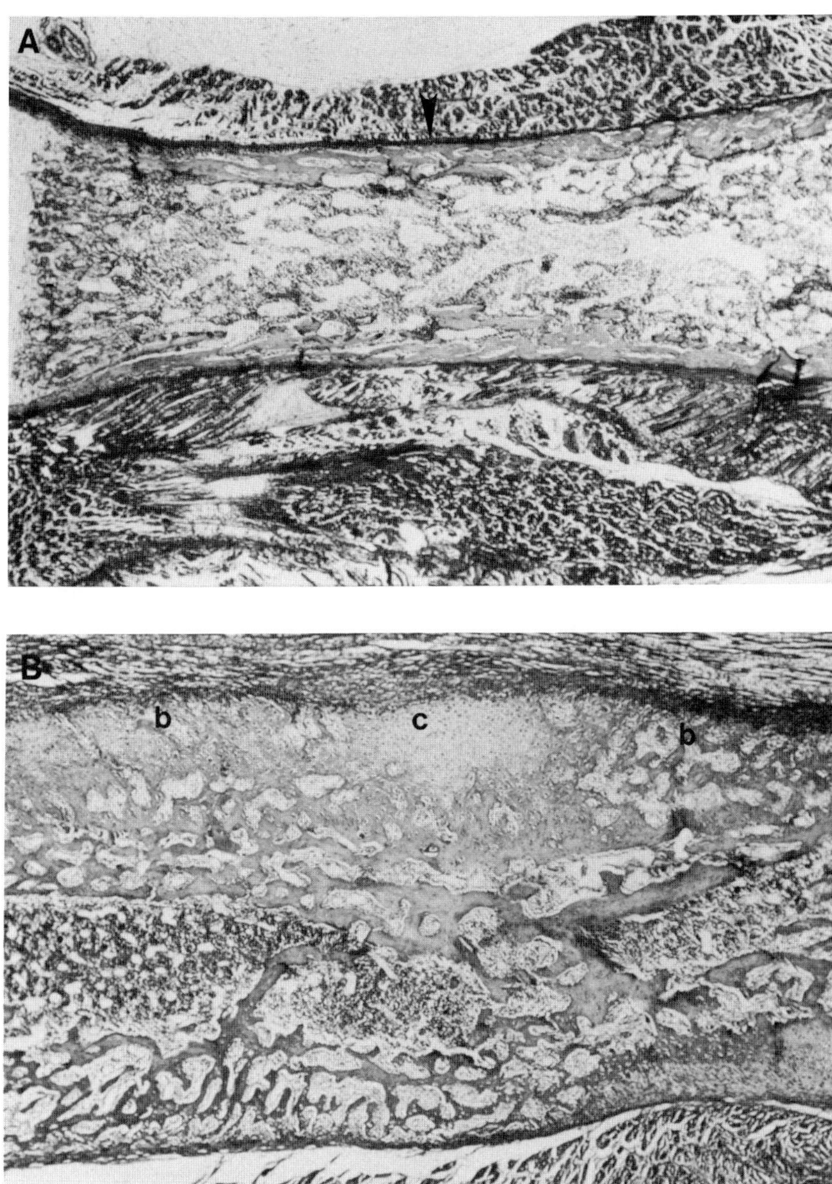

FIGURE 4 Periosteal injection of transforming growth factor β (TGF-β) *in vivo* induces bone and cartilage formation. (A) Longitudial section of uninjected neonatal rat femur. The arrowhead points to the site of TGF-β injection in the injected specimens. (B) Neonatal rat femur 3 days after two daily injection of 200 ng of TGF-β into the periosteum, demonstrating cartilage formation at the site of injection and intramembranous bone formation on either side of the injection site. b, intramembranous bone; c, cartilage.

sion of TGF-β2 is maximal at Day 6, whereas that of TGF-β3 increases steadily until 18 days of development. It should also be noted that TGF-β1 transcripts, which are prominently expressed throughout gestation of the mouse embryo (Heine *et al.*, 1987), are not detected in development of the chicken. Expression of TGF-β5 is also tightly regulated during development of the frog, first appearing at the early neurula stage and remaining high through the tadpole stage and into adulthood (Kondaiah *et al.*, 1990).

Control of TGF-β1 transcription has been demonstrated in several *in vitro* systems including viral transformation (Jakowlew *et al.*, 1988c; Birchenall-Roberts *et al.*, 1990; Geiser *et al.*, 1991), activation of lymphocytes (Kehrl *et al.*, 1986b), hormone treatment (Komm *et al.*, 1988), and autoinduction (Van Obberghen-Schilling *et al.*, 1988; Bascom *et al.*, 1989). In each of these, increased levels of TGF-β1 mRNA can be correlated with increased amounts of TGF-β1 secreted into the medium. Recent studies show that expression of the different TGF-β isoforms is independently regulated in cells *in vitro* (Bang *et al.*, 1992). For example, when mouse keratinocytes are induced to differentiate by shifting the calcium concentration of the medium from 0.05 to 1.4 m*M*, TGF-β2 mRNA levels are increased 10-fold, whereas TGF-β1 mRNA levels are reduced 3- to 5-fold. This is accompanied by a 50-fold increase in secreted TGF-β2 over a 48-hr period (Glick *et al.*, 1990). Moreover, data with v-*ras*^{H}-transformed cells respond to changes in calcium concentrations in the same manner as the primary keratinocytes, but the cells secrete exclusively TGF-β1 at both low and high calcium concentrations. These studies demonstrate not only independent control of TGF-β1 and TGF-β2 transcription, but also that viral transformation can alter the type of TGF-β secreted by a cell and that translational control can be independent of transcriptional control.

Recently, it has been shown that expression of the mRNAs for TGF-β1, TGF-β2, and TGF-β3 could mediate some of the effects of retinoic acid *in vivo* and *in vitro*. Retinoic acid induced the expression of TGF-β2 both in mouse keratinocytes in culture and in the intact epidermis *in vivo* (Glick *et al.*, 1989), suggesting that this mechanism could reflect a more global interaction between retinoids and the TGF-β family in other cell types. Using polyclonal antipeptide antibodies to the different TGF-β isoforms, Glick *et al.* (1991) demonstrated that retinoic acid regulates the expression of TGF-β in the vitamin A-deficient rat. Basal expression of TGF-β2 diminished under conditions of vitamin A deficiency. Systemic retinoic acid induced expression of all TGF-β isoforms in the epidermis and of TGF-β2 and TGF-β3 in the respiratory epithelium as well as in the intestinal mucosa and lamina propria. Of the three isoforms, TGF-β2 appears to be most sensitive to tissue retinoid levels. In tissue such as the bronchial and intestinal epithelium, where TGF-β2 was detected in the normal controls, depletion of tissue retinoids resulted in the loss of this basal expression.

Recent studies have shown that retinoic acid can induce TGF-β2 and TGF-β3 in a variety of cells *in vitro* (Glick *et al.*, 1990, 1991). Danielpour *et al.* (1991) demonstrated that EGF induces secretion of TGF-β1 and not TGF-β2, whereas retinoic acid induces secretion of TGF-β2 and not TGF-β1 in NRK-49F normal rat kidney fibroblasts and A549 human lung carcinoma cells. Moreover, treatment with EGF diminishes the levels of TGF-β2, whereas retinoic acid decreases the levels of TGF-β1 in both cell lines, suggesting a specific retinoic acid effect on the expression of TGF-β isoforms. Other studies show the cell type or tissue-specific regulation of TGF-β isoforms by retinoic acid. A recent study showed that retinoic acid causes an induction of TGF-β2 and TGF-β3 mRNAs in chicken chondrocytes and an inhibition of these same TGF-β mRNAs in cultured myocytes; effects on secretion of TGF-β2 parallel those of the mRNAs. These studies indicate that the interaction between retinoid receptors and members of the TGF-β superfamily is probably very complex *in vivo*.

B. Expression of Transforming Growth Factor β in Bone Culture by Osteotropic Hormones

It has been recently suggested that osteotropic hormones and growth factors may influence release of TGF-β from the bone matrix, its expression and synthesis by bone cells, binding to its cellular receptor, or its biological activity in models of bone formation. Several reports demonstrated that parathyroid hormone, 1,25-dihydroxyvitamin D_3, interleukin 1, and estrogen, all factors that stimulate bone resorption, induce the activity of TGF-β in culture medium in intact bone organ cultures, osteoblastlike osteosarcoma cells, and osteoblastlike cells from fetal rats (Petkovich *et al.*, 1987; Pfeilschifter and Mundy, 1987). TGF-β1 mRNA levels were enhanced in cultured human osteoblastlike osteosarcoma (HOS TE85) cells treated with estradiol. As has recently been shown for induction of TGF-α mRNA by its respective ligand, TGF-β1 treatment increases the level of its own mRNA in a variety of both fibroblastic and epithelial cells (Van Obberghen-Schilling *et al.*, 1988; Bascom *et al.*, 1989). In rat osteosarcoma cells (ROS $^{17}/_{2}$), an increase of TGF-β mRNA can be detected within 3 hr following TGF-β1 addition. TGF-β1 and type I collagen mRNAs are coordinately modulated in rat kidney (NRK) and 3T3 cells by TGF-β, providing important clues to the mechanistic aspects of TGF-β1 gene regulation. Autoinduction of TGF-β1 in fibroblastic cells may be of significance in wound healing, whereas autoinduction in bone cells may also be important in the regulation of bone formation and repair.

VI. CHARACTERIZATION OF THE PROMOTERS FOR TRANSFORMING GROWTH FACTOR β1, β2, AND β3

To begin to understand the mechanism of the differential regulation of expression of the TGF-β isoforms, an analysis was undertaken of the 5′ flanking regions of the human genes for TGF-β1, TGF-β2, and TGF-β3 (Kim *et al.*, 1989b; Noma *et al.*, 1991; Lafyatis *et al.*, 1990). A summary of the main features of each of these promoter sequences is presented in Figure 5. Although the TGF-β1, TGF-β2, and TGF-β3 amino acid sequences are highly conserved, further analysis has revealed that there is substantial divergence among the nucleotide sequences of their respective gene promoters. The TGF-β2 and TGF-β3 genes contain classic TATA box elements found 20–30 nucleotides upstream of the transcriptional start sites, whereas the 5′ flanking region of the TGF-β1 gene lacks a TATA box and contains high GC-rich regions containing multiple Sp1 binding sites. Moreover, a variety of experimental approaches have identified activator protein 1 (AP-1) sites, which bind the Jun–Fos complex, as the major positive regulatory sequences involved in up-regulation of TGF-β1 expression (Kim *et al.*, 1989a, b, c, 1990a, b). Both the human TGF-β2 gene and the human TGF-β3 gene contain an upstream CRE/ATF element that is required for basal and induced levels of gene expression.

A. Characterization of the Promoter Regions of the Transforming Growth Factor β1 Gene

To understand the mechanisms underlying the regulation of expression of TGF-β1, and to define the DNA sequences essential for its constituitive and induced expression, the promoter region of the human and mouse TGF-β1 genes has been characterized. Analysis of the transcriptional start sites of human TGF-β1 mRNAs by S1 nuclease protection assay revealed two major start sites 271 nucleotides from one another; several minor sites were also identified. The TGF-β1 promoter contains neither a TATA box nor a CAAT box, is very GC-rich, and contains 11 putative Sp1 binding sites. To determine the location of sites that may be important for the function of the TGF-β1 promoter, progressively deleted constructs of the 5′ flanking region of the TGF-β1 promoter linked to the chloramphenicol acetyltransferase (CAT) gene were transfected into various cell lines. The pattern of expression of the deletion mutants was similar in all of the cell lines examined. Sequences responsible for both promotion and inhibition of transcription were located in the region extending from 1400 to 300 basepairs (bp) upstream of the first major TGF-β1 transcriptional start site. The 130-bp fragment located between 453 and 323 bp upstream of the first start site contains positive

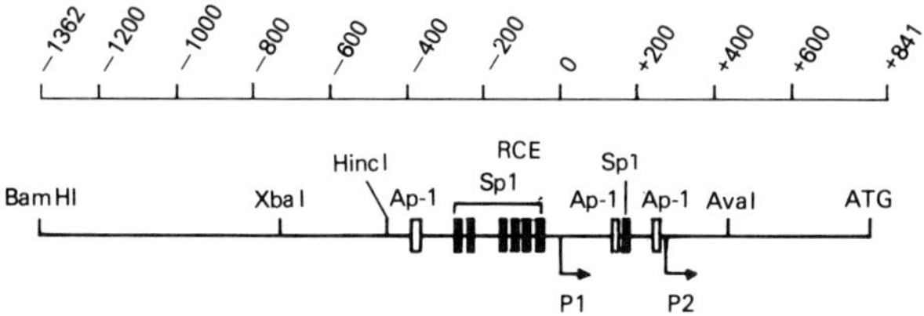

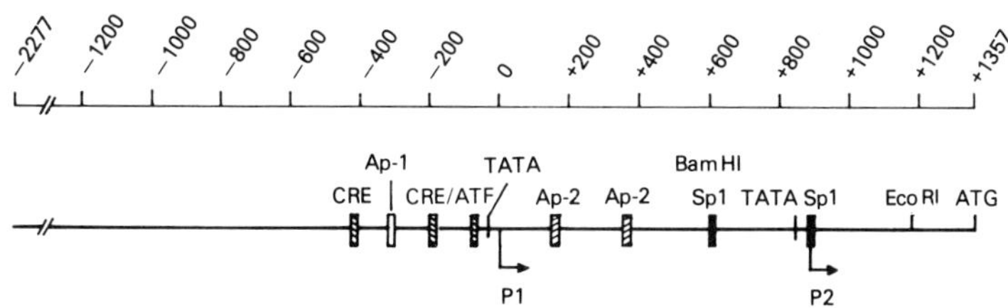

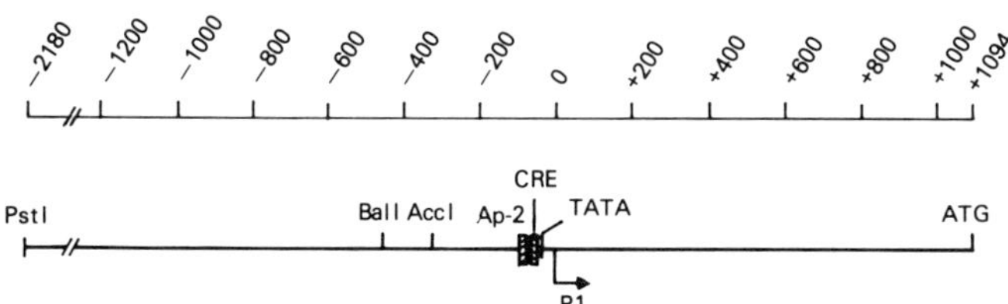

FIGURE 5 Comparison of the major features of the promoter regions of the human transforming growth factor β1 (TGF-β1), TGF-β2, and TGF-β3 genes. Transcriptional start sites are designated as P1 and P2. The concensus transcription factor binding sites are indicated.

regulatory activity in all cells tested. The regulatory elements found in the second promoter of the human TGF-β1 gene were also characterized. This analysis showed that the second promoter contains two positive regulatory elements, one in a 23-bp fragment between nucleotides +150 and +173 and the other in the region between nucleotides +247 and +267.

1. Promoter Sequences of the Human Transforming Growth Factor β1 Gene Responsive to Transforming Growth Factor β1 Autoinduction

As already described, several growth factors including PDGF, TGF-α, and TGF-β1 (Van Obberghen-Schilling *et al.*, 1988; Kim *et al.*, 1990a)

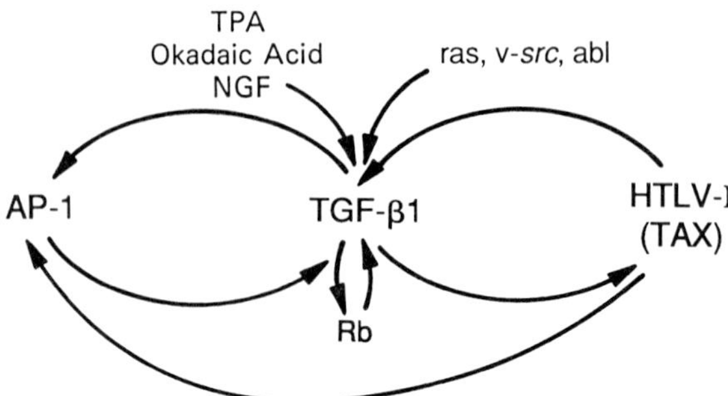

FIGURE 6 Schematic illustration of the pathways for transcriptional regulation of transforming growth factor β1 (TGF-β1). HTLV-I, human T-lymphotropic virus type I. TPA, 12-0-tetradecanoylphorbol-13-acetate; NGF, nerve growth factor; Rb, retinoblastoma.

autoregulate expression of their mRNAs, resulting in increased secretion of the respective peptides. Such autoinduction can amplify responses to these growth factors during normal development or in pathological processes such as carcinogenesis. To investigate the mechanism of autoinduction of TGF-β1, the promoter regions of the TGF-β1 gene responsive to autoinduction were examined (Fig. 6). Two distinct regions of the TGF-β1 promoter were found to be responsive to autoregulation. Sequences located between nucleotides −453 and 323 and between the two major transcriptional start sites displayed positive regulatory activities and were induced by TGF-β1. These regions contain multiple putative binding sites for the AP-1 complex. To confirm the possibility that AP-1 binding sites play a pivotal role in TGF-β1 autoregulation, DNase-1 footprinting assays using affinity-purified HeLa cell AP-1 or a bacterially expressed c-Jun fusion protein were performed. The putative AP-1 binding site located between positions −353 and −359 in the promoter region was protected. To confirm the possibility that the AP-1 binding site is involved in TGF-β1 autoinduction, deletion mutants of the TGF-β1/CAT fusion gene were co-transfected into F9 embryonal carcinoma cells (harboring an inducible c-*fos* gene) with RSV–c-*jun* or RSV–mut-*jun* vectors, which specify the production of either the wild-type Jun or a mutant protein that can no longer bind DNA. Both promoter–CAT chimeric genes were transactivated by Jun in F9 cells. These results strongly suggest that the AP-1 binding sites mediate the autoregulation of TGF-β1 and that expression of both the Jun and Fos components of the AP-1 complex is required. TGF-β1 treatment rapidly induces the expression of c-*jun* mRNA in A-549 cells and also increases its own message. Both of these events, as well as c-*jun* autoinduction, are mediated by binding of the AP-1 complex to specific sequences. These data suggest that the ability of growth factors and nuclear protooncogenes–transcription factors to control cellular growth, differentiation, and development may depend, in part, on their ability to regulate each other's expression.

2. Transcriptional Regulation of the Transforming Growth Factor β1 Gene by the Viral or Oncogene Transformation of the Cells

Several studies have revealed that viral or oncogene transformation of cells leads to an increase in TGF-β1 message. Harvey sarcoma virus-transformed NIH 3T3 cells and Moloney sarcoma virus-transformed NRK cells show increased expression of the mRNA for TGF-β1 (Jakowlew *et al.*, 1988c). Geiser *et al.* (1991) showed that the *ras*-induced increase in TGF-β1 mRNA is due in part to transcriptional activation. The interleukin-3-dependent myeloid precursor cell line (32D-123) with retroviruses containing the *src* oncogene also expressed significantly higher levels of TGF-β1 mRNA than the parental 32D-123 cell line (Birchenall-Roberts *et al.*, 1990). Analysis of different TGF-β1 promoter constructs regulated by $pp60^{v\text{-}src}$ indicated that sequences responsive to high levels of *src* induction contain binding sites for AP-1 (Fig. 6). In 32D-123 cells, serum is required for the induction of c-*fos*, c-*jun*, and, consequently, TGF-β1. These results favor the idea that $pp60^{v\text{-}src}$ replaces serum in directly regulating the intracellular pathways that activate c-*fos* and c-*jun*. It has been proposed that deregulated secretion of growth factors such as TGF-β1 by tumor cells may stimulate tumor growth (Furstenberger *et al.*, 1989; Roberts *et al.*, 1988). The abnormal regulation of *jun* by v-*src*, and the resulting constituitive expression of TGF-β1, could be involved in carcinogenesis. Recently, Cartwright *et al.* (1990) identified the abnormal expression of tyrosine kinases as an early event in the genesis of human colon carcinoma. Aberrant expression of tyrosine kinases consequently deregulates cellular responses (e.g., to TGF-β) that control growth.

Cells transformed by retroviruses have been shown to secrete increased levels of TGF-β1. Recently, it has been reported that fresh leukemic cells from adult T-cell leukemia patients constituitively produce high levels of TGF-β1 mRNA and secrete TGF-β1 but not TGF-β2 into the culture medium (Niitsu *et al.*, 1988; Kim *et al.*, 1990b). Co-transfection of deletion constructs of the TGF-β1 promoter/CAT chimeric gene with the expression vector of the human T-lymphotropic virus type I (HTLV-I) *tax* gene revealed regions homologous with AP-1 binding sites that were required for Tax-induced transactivation of the TGF-β1 expression *in vivo*, the patterns of TGF-β1 expression in transgenic mice carrying the HTLV-I *tax* gene under the control of the viral long terminal repeat (LTR)

(Kim *et al.*, 1991b) were examined. Tumors from these mice and other tissues, such as submaxillary glands and skeletal muscle, which express high levels of *tax* mRNA selectively, express high levels of TGF-β1 mRNA and its protein. These data suggest that TGF-β might mediate certain biologic effects seen in these transgenic mice. For example, animals with significant tumor burdens have extensive peripheral granulocytosis (Kim *et al.*, 1991b). The localization of tumors on the ears and tails of the transgenic animals, sites of frequent wounding resulting from aggressive behavior, suggests that local release of TGF-β at these sites may serve to amplify further the growth of the tumor. Interestingly, TGF-β1 treatment induces the expression of *tax* mRNA in PX-1 and PX-2 cells derived from neurofibromas of *tax*-transgenic mice and also increases the level of its own mRNA. Previously, it has been demonstrated that another reciprocal set of interactions between TGF-β1 and Jun, namely, autoinduction of TGF-β1, is mediated by the AP-1 (Jun–Fos) complex, and that TGF-β1 treatment of cells resulted in prolonged induction of *jun* mRNA mediated through the AP-1 site in the c-*jun* promoter. Jeang *et al.* (1990) recently reported that Jun also strongly activates the expression of the HTLV-I LTR. These results suggest that induction of *tax* mRNA by the AP-1 complex binding site within the HTLV-I LTR and that the increased production of TGF-β1 mediated by Tax supports increased replication of HTLV-I. Taken together, these data, which demonstrate the interaction between TGF-β1 and HTLV-I, suggest that the ability of viruses to induce the expression of growth factors may be important in the pathogenesis of a variety of diseases.

3. Regulation of Transforming Growth Factor β1 Gene Expression by the Product of the Retinoblastoma Gene

TGF-βs regulate cell differentiation, cell growth, and many aspects of cell function and are the most potent growth-inhibitory polypeptides known for a wide variety of cell types including most normal and transformed epithelial, endothelial, fibroblast, lymphoid, and hematopoietic cells (Massague *et al.*, 1986; Masui *et al.*, 1986; Tucker *et al.*, 1984; Roberts *et al.*, 1985; Kehrl *et al.*, 1986a). Although TGF-βs arrest growth in the late G1 phase of the cell cycle, the mechanism by which they inhibit cell proliferation is virtually unknown (Russel *et al.*, 1988; Silberstein and Daniel, 1987). In certain cells, such as mouse keratinocytes, TGF-β1 has been shown to reduce the expression of growth-related genes, including c-*myc*, by inhibiting transcriptional initiation (Pietenpol *et al.*, 1990). This inhibitory activity was abrogated by various viral transforming proteins, including the E7 protein of human papilloma virus type 16, the E1A protein of adenovirus type 5, and the large tumor (T) antigen of simian virus 40, all of which have been demonstrated to associate with the retinoblastoma gene product RB (DeCaprio *et al.*, 1988; Dyson *et al.*, 1989). Although the RB gene product is synthesized in all phases of the cell cycle, it also has properties of a cell cycle regulatory protein. Whereas underphosphorylated forms of RB are the primary species seen in the G0 and G1 phases of the cell cycle, the protein undergoes phosphorylation at multiple sites as cells traverse the G1–S boundary (DeCaprio *et al.*, 1989; Buchkovich *et al.*, 1989; Chen *et al.*, 1989; Mihara *et al.*, 1989). Treatment of cells with TGF-β appears to prevent phosphorylation of RB and retain RB in the underphosphorylated, growth-suppressive state (Laiho *et al.*, 1990). Recently, the existence of the reciprocal process has been shown, namely, regulation of TGF-β1 expression by RB (Fig. 6). Sequences homologous to a previously defined RB-responsive element, the RCE in c-*fos* promoter (Robbins *et al.*, 1990), mediate the regulation of human TGF-β1 promoter activity by the RB protein. In agreement with the recent report demonstrating that RB down-regulates c-*fos* transcription in NIH 3T3 cells, we have shown that RB also down-regulates TGF-β1 transcription in NIH 3T3 cells. Interestingly, depending on the particular cells assayed, RB can also up-regulate the TGF-β1 gene. Thus, RB positively regulates transcription of both the TGF-β1 and c-*fos* genes in CCL-64 and A-549 cells (Kim *et al.*, 1991a, 1992a). Induction of TGF-β1 transcription by RB is presumably mediated by underphosphorylated RB, since the basal level of transcription of the TGF-β1 promoter is greater in quiescent cells.

Previous studies have suggested that RB might mediate the effects of TGF-β on growth and that these two peptides might share a common pathway. Laiho *et al.* (1990) proposed that TGF-β might regulate the state of phosphorylation of RB, enhancing its underphosphorylated, growth-inhibitory form. Pietenpol *et al.* (1990) showed that the down-regulation of c-*myc* expression associated with inhibition of the growth of keratinocytes by TGF-β might also be mediated by RB. However, data from cells lacking either TGF-β responsivity or RB suggest that links between TGF-β and RB are not obligatory and that each peptide independently can inhibit cell growth (Ong *et al.*, 1990).

B. Characterization of Human Transforming Growth Factor β2 Promoter

TGF-β2 has been isolated from various cells and tissues (Seyedin *et al.*, 1985; Wrann *et al.*, 1987). Although their functional significance has not been completely elucidated, there are multiple TGF-β2 mRNA transcripts of approximately 2.8, 3.8, 4.0, 5.1, and 5.8 kb (Madisen *et al.*, 1988; O'Reilly *et al.*, 1992). Isolation and sequencing of the human and mouse TGF-β2 cDNAs have revealed multiple potential polyadenylation signals (Miller *et al.*, 1989b), and it is possible that differential utilization of these signals in the 3′ untranslated region (UTR) could generate some of the heterogeneity in transcript size (Madisen *et al.*, 1988). Moreover, the 5.1-kb transcript has been shown to contain an additional 84 nucleotides in the amino-terminal precursor coding region. These nucleotides encode an additional in-frame 29 amino acids, presumably derived from alternative splicing of a separate exon (Webb *et al.*, 1988). The steady-state level of each transcript varies among different cell lines and the 2.8- and 5.1-kb transcripts are not detected in some murine papilloma cell lines. Collectively, these studies suggest that regulation of TGF-β2 expression occurs at multiple levels, including transcription, alternative splicing of exons, alternative utilization of poly(A)$^+$ sites, and post-transcriptional stabilization of the mRNA.

In situ hybridization, Northern analyses, and immunohistochemical studies (Pelton *et al.*, 1989; Millan *et al.*, 1991) have all shown that TGF-β2 is expressed in a developmentally specific pattern that overlaps but is distinct from those of TGF-β1 and TGF-β3. Moreover, the TGF-β gene promoters have been cloned and sequence analysis has revealed striking differences in their structure, suggesting that these differences in expression may be mediated through tissue-specific gene transcription (Kim *et al.*, 1989b; Lafyatis *et al.*, 1990; Noma *et al.*, 1991).

Recently, the promoter region for the human TGF-β2 gene was isolated and characterized (Noma *et al.*, 1991; O'Reilly *et al.*, 1992). Sequence analysis of the upstream DNA identified a consensus TATA box 1387 nucleotides from the start site of the coding region. A transcriptional initiation site was identified by S1 nuclease analysis to be 30 nucleotides downstream from the TATA box, as is characteristic of many eukaryotic promoters. Promoter function was demonstrated using a series of CAT constructs containing 5′ serially deleted fragments of genomic DNA upstream from the initiation site. Sequences responsible for putative enhancer and silencer regions were identified between −778 and −40 relative to the transcription initiation site. The presence of an upstream cyclic acid responsive element (CRE/ATF)-like element at −74 resulted in a 5- to 10-fold increase in CAT activity over a minimal construct extending to −62. TGF-β2 is expressed from multiple promoters, one of which is regulated, in part, through a CRE/ATF-like element (O'Reilly *et al.*, 1992; Kim *et al.*, 1992c). Characterization of multiple promoters regulating the expression of distinct TGF-β2 transcripts is presently in progress.

C. Characterization of the Transforming Growth Factor β3 Promoter

TGF-β3 has been cloned from humans, chicken, and mice (ten Dijke *et al.*, 1988; Jakowlew *et al.*, 1988a; Miller *et al.*, 1989a; Denhez *et al.*, 1990). Although the specific *in vivo* roles of this form of TGF-β are unknown, the pattern of embryonic and tissue-specific expression of TGF-β3 suggest that it is involved in embryogenesis and cell differentiation.

Transcription from the TGF-β3 promoter is regulated by multiple upstream regions. The cyclic adenenosine monophosphate (cAMP)-responsive element at −47 was found to mediate both basal and cAMP-induced activity of the promoter. Immediately upstream of the CRE in the TGF-β3 promoter is a consensus AP-2 binding site. AP-2 binding sites have been found in a number of genes, including the enhancer regions of the SV40 and the human metallothionein IIA genes (Imagawa *et al.*, 1987; Williams *et al.*, 1988). The AP-2 site in the human metallothionein IIA gene enhances promoter activity in response to both forskolin and phorbol ester. Retinoic acid-induced differentiation of human embryonal carcinoma cells was found to increase expression of the AP-2 gene (Williams *et al.*, 1988), suggesting that AP-2 may have a role in regulating gene expression during embryogenesis. Since TGF-β3 is expressed during mouse embryonic development, the AP-2 site in the TGF-β3 promoter might be important in embryonic TGF-β3 expression.

The region upstream from the CRE and AP-2 site, between −60 and −221, induced markedly increased transcriptional activity in the cell line A375 (Lafyatis *et al.*, 1990). This cell line expresses TGF-β3 mRNA at levels much greater than in other cell lines tested. It is therefore likely that this region contains the enhancer element.

A significant level of expression of TGF-β3 mRNA has been demonstrated in chicken chondrocytes, cardiac myocytes, and the skeletal myocytes (Jakowlew *et al.*, 1992). Addition of TGF-β1, TGF-β2, or TGF-β3 in cultured sternal chondrocytes and cardiac myocytes resulted in an increase in expression of TGF-β3 mRNA. Retinoic acid induces TGF-β3 mRNA in chicken chondrocytes and an inhibition of this same mRNA in cultured myocytes. It is not clear whether the effects of retinoic acid occur by a post-transcriptional mechanism. Previous studies demonstrated that both mouse skeletal muscle tissue and the mouse myoblast cell line C2C12 express high levels of TGF-β3 mRNA (Lafyatis *et al.*, 1990). The differentiation of myoblasts to myotubes has been studied in a variety of myoblast cell lines. TGF-β1 is known to completely inhibit differentiation of myoblasts (Massague *et al.*, 1986). Transcripts for TGF-β1, TGF-β2, and TGF-β3 were seen in both unfused and fused C2C12 cells. TGF-β1 transcripts decreased after fusion, while TGF-β2 and TGF-β3 transcripts increased. TGF-β3 expression in myoblasts is stimulated through several different regions of the TGF-β3 promoter from −301 to −47 and in the long 5′ UTR (Lafyatis *et al.*, 1991). These regions are mostly distinct from the region between −499 and −221 responsible for the increased expression of TGF-β3 stimulated after myoblast fusion. The increased expression of TGF-β3 during muscle differentiation might be due to the binding of the regulatory factor(s) to the promoter region between −301 and −47 and/or the 5′ UTR during some stage in the commitment of mesodermal progenitor cells to a myoblast lineage. The binding of additional factor(s) to the region between −499 and −221 may stimulate high levels of expression in the mature myotube. Understanding the regulation of the TGF-β3 promoter in the context of regulation of the promoters of muscle-specific genes will require further study. The functional significance of the high-level expression of TGF-β3 in skeletal and cardiac muscles is unclear. During development, a potential role for TGF-β3 could be in directing some aspect(s) of embryogenesis through intercellular signaling to induce or inhibit the differentiation of cells in adjacent tissues.

VII. POST-TRANSCRIPTIONAL REGULATION OF TRANSFORMING GROWTH FACTOR β ISOFORMS

In addition to transcriptional regulation, numerous reports suggest that expression of TGF-βs may also be regulated post-transcriptionally. For example, in PC3 and BSC-1 cells, which secrete predominantly TGF-β2 but relatively little TGF-β1 protein, the TGF-β1 mRNA level is higher than that of TGF-β2. Assoian *et al.* (1987) reported that TGF-β1 mRNA is expressed at similar levels in unstimulated monocytes and in monocytes activated to become macrophages; however, TGF-β1 protein is secreted only by activated macrophages, suggesting that expression is controlled at the level of translation. Members of the steroid hormone superfamily may play an important role in the expression of TGF-β isoforms *in vitro* in a target-specific manner. The inhibitory effects of the antiestrogens and gestodene on breast carcinoma cells and of retinoids on keratinocytes are partially reversed by neutralizing antibodies to TGF-β (Colletta *et al.*, 1990a, b). This suggests that the induction of TGF-β subtypes has functional significance in the mechanism of action of steroids and retinoids. Colletta *et al.* (1990b) demonstrated that antiestrogens induce the production of TGF-β1 in mammary carcinoma cells and fetal fibroblasts with little or no change in TGF-β1 mRNA levels. This suggests that the antiestrogens and gestodene may be affecting the efficiency of translation of the TGF-β1 mRNA. TGF-β1 transcription was increased after treatment of keratinocytes with retinoic acid, but there was no accompanying increase in secretion of the TGF-β1 peptide, whereas retinoic acid specifically induces TGF-β2 in these cells. The induction of TGF-β2 by retinoids is accompanied by an increase in TGF-β2 mRNAs, but little change in transcription rates, suggesting an effect of retinoids on message stability. Retinoids also induce the TGF-β3 expression in chicken chondrocytes with little change in transcription rate.

The TGF-β1 mRNA contains an unusually long 5′ untranslated sequence, which is rich in GC content (Kim *et al.*, 1989b). Experiments in which a cDNA fragment of the 5′ UTR were inserted upstream of the coding region of a human growth hormone reporter gene indicated that the 5′ UTR has a potent inhibitory effect on the translation of the reporter gene in certain cell backgrounds. Computer analysis of this region of the TGF-β1 5′ UTR indicated the possible existence of a stem-loop structure (Kim *et al.*, 1992b). *In vitro* gel retardation assays using radiolabeled RNA probes from this region of the TGF-β1 5′ UTR demonstrate the binding of a specific protein.

Recently, it has been reported that when the 5′ UTR of TGF-β3 was introduced upstream of the coding sequence of CAT, *in vitro* translation was inhibited (Arrick *et al.*, 1991). The 5′ UTR of TGF-β3 mRNA contained 11 ATG triplets. The two upstream open reading frames closest to the initiator codon for the TGF-β3 coding sequence also decreased translational efficiency, since mutation of either ATG resulted in increased translation.

VIII. CONCLUSION

The past few years have witnessed an explosion of interest in and understanding of TGF-β. This multifunctional growth factor is of critical importance in the regulation of bone formation and repair. TGF-β is produced during key steps of intramembranous and endochondral ossification and may initiate the process of bone repair following fracture.

Much work remains to be done, however, in understanding the mechanisms by which TGF-β regulates cell growth and differentiation. The areas of transcriptional regulation, post-transcriptional regulation, receptor cloning and signal transduction, interaction with hormones and growth factors, and their role in physiology and disease will be the most intensively studied. Gleaning answers to these questions will challenge current available technologies but will be critically important in our attempts to understand the complexities of cell function and proliferation.

ACKNOWLEDGMENTS

The authors thank their scientific colleagues who aided in the preparation of this review by providing preprints or by allowing us to refer to their unpublished work. We also wish to thank Drs. Michael Sporn and Anita Roberts for their continuous support for the projects presented here and critical review of the manuscript.

REFERENCES

Adashi, E. Y., and Resnick, C. E. (1986). Antagonistic interactions of transforming growth factors in the regulation of granulosa cell differentiation. *Endocrinology* **119,** 1879–1881.

Akhurst, R. J., Fee, F., and Balmain, A. (1988). Localized production of TGF-β mRNA in tumor promoter-stimulated mouse epidermis. *Nature (London)* **331,** 363–365.

Anzano, M. A., Roberts, A. B., and Sporn, M. B. (1986). Anchorage-independent growth of primary rat embryo cells is induced by platelet-derived growth factor and inhibited by type-beta transforming growth factor. *J. Cell. Physiol.* **126,** 312–318.

Arrick, B. A., Lee, A. L., Grendell, R. L., and Derynck, R. (1991). Inhibition of translation of transforming growth factor-β3 mRNA by its 5′ untranslated region. *Mol. Cell. Biol.* **11,** 4306–4313.

Assoian, R. K., and Sporn, M. B. (1986). Type-beta transforming growth factor in human platelets: release during platelet degranulation and action on vascular smooth muscle cells. *J. Cell. Biol.* **102,** 1217–1223.

Assoian, R. K., Komoriya, A., Heyers, C. A., Miller, D. M., and Sporn, M. B. (1983). Transforming growth factor-β in human platelets. *J. Biol. Chem.* **258,** 7155–7160.

Assoian, R. K., Fleurdelys, B. E., Stevenson, H. C., Miller, P. J., Madtes, D. K., Raines, E. W., Ross, R., and Sporn, M. B. (1987). Expression and secretion of type β transforming growth factor by activated human macrophages. *Proc. Natl. Acad. Sci. USA* **84,** 6020–6024.

Bang, Y.-J., Kim, S.-J., Danielpour, D., O'Reilly, M. A., Kim, K. Y., Myers, C. E., and Trepel, J. B. (1992). Molecular mechanism of cyclic AMP-induced growth arrest of PC-3 prostate carcinoma cells: Evidence that cyclic AMP acts through regulation of transforming growth factor β. *Proc. Natl. Acad. Sci. USA* **89,** 3556–3560.

Bascom, C. C., Wolfshohl, J. R., Coffey, R. J., Jr., Madisen, L., Webb, N. R., Purchio, A. R., Derynck, R., and Moses, H. L. (1989). Complex regulation of transforming growth factor β1, β2, and β3 mRNA expression in mouse fibroblasts and keratinocytes by transforming growth factors β1 and β2. *Mol. Cell. Biol.* **9,** 5508–5515.

Birchenall-Roberts, M. C., Ruscetti, F. W., Kasper, J., Lee, H.-D., Friedman, R., Geiser, A. G., Sporn, M. B., Roberts, A. B., and Kim, S.-J. (1990). Transcriptional regulation of the transforming growth factor β1 promoter by v-src gene products is mediated through the AP-1 complex. *Mol. Cell. Biol.* **10,** 4978–4983.

Braun, L., Mead, J. E., Panzica, M., Mikumo, R., Bell, G. I., and Fausto, N. (1988). Transforming growth factor-β mRNA increases during liver regeneration: A possible paracrine mechanism of growth regulation. *Proc. Natl. Acad. Sci. USA* **85,** 1539–1543.

Buchkovich, K., Duffy, L. A., and Harlow, E. (1989). The retinoblastoma protein is phosphorylated during specific phases of the cell cycle. *Cell* **58,** 1097–1105.

Burt, D. W., and Jakowlew, S. B. (1992). Correction: A new interpretation of a chicken transforming growth factor-β4 complementary DNA. *Mol. Endocrinol.* **6,** 989–992.

Cartwright, C. A., Meisler, A. I., and Eckhart, W. (1990). Activation of the $pp60^{c\text{-}src}$ protein kinase is an early event in colonic carcinogenesis. *Proc. Natl. Acad. Sci. USA* **87,** 558–562.

Celeste, A. J., Iannazzi, J. A., Taylor, R. C., Hewick, R. M., Rosen, V., Wang, E. A., and Wozney, J. M. (1990). Identification of transforming growth factor β family members present in bone-inductive protein purified from bovine bone. *Biochemistry* **87,** 9843–9847.

Centrella, M., Massague, J., and Canalis, E. (1986). Human platelet-derived transforming growth factor-β stimulates parameters of bone growth in fetal rat calvariae. *Endocrinology* **119,** 2306–2312.

Centrella, M., McCarty, T. L., and Canalis, E. (1987). Transforming growth factor is a bifunctional regulator of replication and collagen synthesis in osteoblast-enriched cell cultures from fetal rat bone. *J. Biol. Chem.* **262,** 2869–2874.

Centrella, M., McCarthy, T. L., and Canalis, E. (1988). Parathyroid hormone modulates transforming growth factor β activity and binding in osteoblast-enriched cell cultures from fetal rat parietal bone. *Proc. Natl. Acad. Sci. USA* **85,** 5889–5893.

Centrella, M., McCarthy, T. L., and Canalis, E. (1991). Transforming growth factor-β and remodeling of bone. *J. Bone Joint Surg.* **73-A,** 1418–1428.

Cheifetz, S., Weatherbee, J. A., Tsang, M. L. S., Anderson, J. K., Mole, J. E., Lucas, R., and Massague, J. (1987). The transforming growth factor beta system, a complex pattern of cross-reactive ligands and receptors. *Cell* **48,** 409–415.

Chen, P.-L., Scully, P., Shew, J.-Y., Wang, J. Y. J., and Lee, W.-H. (1989). Phosphorylation of the retinoblastoma gene product is modulated during the cell cycle and cellular differentiation. *Cell* **58,** 1193–1198.

Chenu, C., Pfeilschrifter, J., Mundy, G. R., and Roodman, G. D. (1988). Transforming growth factor-β inhibits formation of osteoclast-like cells in long-term human marrow cultures. *Proc. Natl. Acad. Sci. USA* **85,** 5683–5687.

Chiang, C. P., and Nilsen-Hamilton, M. (1986). Opposite and selective effects of epidermal growth factor and human platelet transforming growth factor-β on the production of secreted proteins by murine 3T3 cells and human fibroblasts. *J. Biol. Chem.* **261,** 10478–10481.

Childs, C. B., Proper, J. A., Tucker, R. F., and Moses, H. L. (1982). Serum contains a platelet-derived transforming growth factor. *Proc. Natl. Acad. Sci. USA* **79,** 5312–5316.

Colletta, A. A., Wakefield, L. M., Howell, F. V., Van Roozendaal, K. E. P., Danielpour, D., Ebbs, S. R., Sporn, M. B., and Baum, M. (1990). Antiestrogens induce the secretion of active transforming growth factor β from human fibroblasts. *Br. J. Cancer* **62,** 405–409.

Colletta, A. A., Wakefield, L. M., Howell, F. V., Danielpour, D., Baum, M., and Sporn, M. B. (1991). The growth inhibition of human breast cancer cells by a novel synthetic progestin is partly mediated by the induction of transforming growth factors β. *J. Clin. Invest.* **87,** 277–283.

Danielpour, D., Dart, L. L., Flanders, K. C., Roberts, A. B., and Sporn, M. B. (1989). Immunodetection and quantitation of the two forms of transforming growth factor-beta (TGF-β1 and TGF-β2) secreted by cells in culture. *J. Cell. Physiol.* **138,** 79–86.

Danielpour, D., Kim, K. Y., Winokur, T. S., and Sporn, M. B. (1991). Differential expression of the expression of transforming growth factor βs 1 and 2 by retinoic acid, epidermal growth factor and dexamethasone in NRK-49F and A549 cells. *J. Cell. Physiol.* **148,** 235–244.

DeCaprio, J. A., Ludlow, J. W., Figge, J., Shew, J.-Y., Huang, C.-M., Lee, W.-H., Marsilio, E., Paucha, E., and Livingston, D. M. (1988). SV40 large tumor antigen forms a specific complex with the product of the retinoblastoma susceptibility gene. *Cell* **54,** 275–283.

DeCaprio, J. A., Ludlow, J. W., Lynch, D., Fukukawa, Y., Griffin, J., Piwnica-Worms, H., Huang, C.-M., and Livingston, D. M. (1989). The product of the retinoblastoma susceptibility gene has properties of a cell cycle regulatory element. *Cell* **58,** 1085–1095.

Denhez, F., Latyatis, R., Kondaiah, P., Roberts, A. B., and Sporn, M. B. (1990). Cloning by polymerase chain reaction of a new TGF-β, mTGF-β3. *Growth Factors* **3,** 139–146.

Derynck, R., Jarrett, J. A., Chen, E. Y., Eaton, D. H., Bell, J. R., Assoian, R. B., Roberts, A. B., Sporn, M. B., and Goeddel, D. V. (1985). Human transforming growth factor-beta cDNA sequence and expression in tumor cell lines. *Nature (London)* **316,** 701–705.

Dyson, N., Howley, P. M., Munger, K., and Harlow, E. (1989). The human papilloma virus-16 E7 oncoprotein is able to bind to the retinoblastoma gene product. *Science* **243,** 934–937.

Edwards, D. R., Murphy, G., Reynolds, J. J., Whitman, S. E., Docherty, A. J. P., Angel, P., and Heath, J. K. (1987). Transforming growth factor beta modulates the expression of collagenase and metalloproteinase inhibitor. *EMBO J.* **6,** 1899–1904.

Ellingsworth, L. R., Brennan, J. E., Fok, K., Rosen, D. M., Bentz, H., Piez, K. A., and Seyedin, S. M. (1986). Antibodies to the N-terminal portion of cartilage-inducing factor A and transforming growth factor beta. *J. Biol. Chem.* **261,** 12362–12367.

Flanders, K. C., Roberts, A. B., Ling, N., Fleurdelys, B. E., and Sporn, M. B. (1988). Antibodies to peptide determinants in transforming growth factor-beta and their application. *Biochemistry* **27,** 739–746.

Frolik, C. A., Dart, L. L., Meyers, C. A., Smith, D. M., and Sporn, M. B. (1983). Purification and initial characterization of a type beta transforming growth factor from human placenta. *Proc. Natl. Acad. Sci. USA* **80,** 3676–3680.

Furstenberger, G., Rogers, M., Schnapke, R., Bauer, G., Holter, P., and Marks, F. (1989). Stimulatory role of transforming growth factors in multistage skin carcinogenesis: Possible explanation for the tumor-inducing effect of wounding in initiated NMRI mouse skin. *Int. J. Cancer* **43,** 915–921.

Geiser, A. G., Kim, S.-J., Roberts, A. B., and Sporn, M. B. (1991). Characterization of the mouse transforming growth factor-β1 promoter and activation by the Ha-ras oncogene. *Mol. Cell. Biol.* **11,** 84–92.

Glick, A. B., Flanders, K. C., Danielpour, D., Yuspa, S. H., and Sporn, M. B. (1989). Retinoic acid induces transforming growth factor β2 in cultured keratinocytes and mouse epidermis. *Cell Regul.* **1,** 87–97.

Glick, A. B., Danielpour, D., Morgan, D. L., Sporn, M. B., and Yuspa, S. H. (1990). Induction of transforming growth factor β2 during terminal differentiation of primary mouse keratinocytes. *Mol. Endocrinol.* **4,** 46–52.

Glick, A. B., McCune, B. K., Abdulkarem, N., Flanders, K. C., Lumadue, J. A., Smith, J. M., and Sporn, M. B. (1991). Complex regulation of TGF-β expression by retinoic acid in the vitamin A-deficient rat. *Development* **111,** 1081–1086.

Hayashi, I., and Carr, B. I. (1985). DNA synthesis in rat hepatocytes: Inhibition by a platelet factor and stimulation by an endogenous factor. *J. Cell. Physiol.* **125,** 82–88.

Heine, U., Munoz, E. F., Flanders, K. C., Ellingsworth, L. R., Lam, H. Y., Thompson, N. L., Roberts, A. B., and Sporn, M. B. (1987). Role of transforming growth factor-β in the development of the mouse embryo. *J. Cell. Biol.* **105,** 2861–2876.

Ignotz, R. A., and Massague, J. (1985). Type β transforming growth factor controls the adipogenic differentiation of 3T3 fibroblasts *Proc. Natl. Acad. Sci. USA* **82,** 8530–8534.

Ignotz, R. A., and Massague, J. (1986). Transforming growth factor-beta stimulates the expression of fibronectin and collagen and their incorporation into the extracellular matrix. *J. Biol. Chem.* **261,** 4337–4345.

Imagawa, M., Chiu, R., and Karin, M. (1987). Transcription factor AP-2 mediates induction by two different signal-transduction pathways: Protein kinase C and cAMP. *Cell* **51,** 251–260.

Jakowlew, S. B., Dillard, P. J., Kondaiah, P., Sporn, M. B., and Roberts, A. B. (1988a). Complementary deoxyribonucleic acid cloning of a novel transforming growth factor-β messenger ribonucleic acid from chick embryo chondrocytes. *Mol. Endocrinol.* **2,** 747–755.

Jakowlew, S. B., Dillard, P. J., Sporn, M. B., and Roberts, A. B. (1988b). Complementary deoxyribonucleic acid cloning of an mRNA encoding transforming growth factor-beta 4 from chick embryo chondrocytes. *Mol. Endocrinol.* **2,** 1186–1195.

Jakowlew, S. B., Kondaiah, P., Flanders, K. C., Thompson, N. L., Dillard, P. J., Sporn, M. B., and Roberts, A. B. (1988c). Increased coordinate expression of growth factor mRNA accompanies viral transformation of rodent cells. *Oncogene Res.* **2,** 135–148.

Jakowlew, S. B., Cubert, J., Danielpour, D., Sporn, M. B., and Roberts, A. B. (1992).

Differential regulation of the expression of transforming growth factor-β mRNAs by growth factors and retinoic acid in chicken embryo chondrocytes, myocytes, and fibroblasts. *J. Cell. Physiol.* **150,** 377–385.

Jeang, K.-T., Chiu, R., Santos, E., and Kim, S.-J. (1990). Oncoprotein Jun is a transcriptional activator of the HTLV-I long terminal repeat. *Virology* **181,** 218–227.

Jingushi, S., Joyce, M. E., Flanders, K. C., Hjelmeland, L., Roberts, A. B., Sporn, M. B., Muniz, O., Howell, D., Dean, D., Ryan, U. and Bolander, M. E. (1990). Distribution of acidic fibroblast growth factor, basic growth factor, and transforming growth factor-β1 in rat growth plate. *In* "Calcium Regulation and Bone Metabolism" (D. V. Cohn, F. H. Glorieux, and T. J. Martin, eds.), pp. 298–303. Elsevier Science Publishers, New York.

Joyce, M. E., Jingushi, S., and Bolander, M. E. (1990a). Transforming growth factor-β in the regulation of fracture repair. *Orthop. Clin. North Am.* **21,** 199–209.

Joyce, M. E., Roberts, A. B., Sporn, M. B., and Bolander, M. E. (1990b). Transforming growth factor-beta and the initiation of chondrogenesis and osteogenesis in the rat femur. *J. Cell. Biol.* **110,** 2195–2207.

Kehrl, J. H., Roberts, A. B., Wakefield, L. M., Jakowlew, S. B., Sporn, M. B., and Fauci, A. S. (1986a). Transforming growth factor beta is an important immunomodulatory protein for human B-lymphocytes. *J. Immunol.* **137,** 3855–3860.

Kehrl, J. H., Wakefield, L. M., Roberts, A. B., Jakowlew, S. B., Alvarez-Mon, M., Derynck, R., Sporn, M. B., and Fauci, A. S. (1986b). Production of transforming growth factor beta by human T lymphocytes and its potential role in the regulation of T cell growth. *J. Exp. Med.* **163,** 1037–1050.

Kim, S.-J., Denhez, F., Kim, K.-Y., Holt, J. T., Sporn, M. B., and Roberts, A. B. (1989a). Activation of the second promoter of the TGF-β1 gene by TGF-β1 and phorbol ester occurs through the same target sequences. *J. Biol. Chem.* **264,** 19373–19378.

Kim, S.-J., Glick, A. B., Sporn, M. B., and Roberts, A. B. (1989b). Characterization of the promoter region of the human transforming growth factor-β1 gene. *J. Biol. Chem.* **264,** 402–408.

Kim, S.-J., Jeang, K.-T., Glick, A. B., Sporn, M. B., and Roberts, A. B. (1989c). Promoter sequences of the human transforming growth factor-β1 gene responsive to transforming growth factor-β1 autoinduction. *J. Biol. Chem.* **264,** 7041–7045.

Kim, S.-J., Angel, P., Lafyatis, R., Hattori, K., Kim, K. Y., Sporn, M. B., Karin, M., and Roberts, A. B. (1990a). Autoinduction of transforming growth factor β1 is mediated by the AP-1 complex. *Mol. Cell. Biol.* **10,** 1492–1497.

Kim, S.-J., Kehrl, J. H., Burton, J., Tendler, C. L., Jeang, K.-T., Danielpour, D., Thevenin, C., Kim, K. Y., Sporn, M. B., and Roberts, A. B. (1990b). Transactivation of the transforming growth factor β1 (TGF-β1) gene by human T-lymphotropic virus type 1 Tax: A potential mechanism for the increased production of TGF-β1 in adult T-cell leukemia. *J. Exp. Med.* **172,** 121–129.

Kim, S.-J., Lee, H.-D., Robbins, P. D., Busam, K., Sporn, M. B., and Roberts, A. B. (1991a). Regulation of transforming growth factor β1 gene expression by the product of the retinoblastoma-susceptibility gene. *Proc. Natl. Acad. Sci. USA* **88,** 3052–3056.

Kim, S.-J., Winokur, T. S., Lee, H.-D., Danielpour, D., Kim, K. Y., Geiser, A. G., Chen, L.-S., Sporn, M. B., Roberts, A. B., and Jay, G. (1991b). Overexpression of transforming growth factor-β in transgenic mice carrying the human T-cell lymphotropic virus type I tax gene. *Mol. Cell. Biol.* **11,** 5222–5228.

Kim, S.-J., Onwuta, U. S., Lee, Y. I., Li, R., Botchan, M. R., and Robbins, P. D. (1992a). The retinoblastoma gene product regulates Sp1-mediated transcription. *Mol. Cell. Biol.* **12,** 2455–2463.

Kim, S.-J., Park, K., Koeller, D., Kim, K. Y., Wakefield, L. M., Sporn, M. B., and Roberts,

A. B. (1992b). The 5′ untranslated region of the human transforming growth factor-β1 gene exerts an inhibitory effects on translation. *J. Biol. Chem.* **267,** 13702–13707.

Kim, S.-J., Wagner, S., Liu, F., O'Reilly, M. A., Robbins, P. D., and Green, M. R. (1992c). The retinoblastoma gene product activates transcription of the human TGF-β2 gene through ATF-2. *Nature* (London). **358,** 331–334.

Knecht, M., Feng, P., and Catt, K. (1986). TGF-beta regulates the expression of luteinizing hormone receptors in ovarian granulosa cells. *Biochem. Biophys. Res. Commun.* **139,** 800–807.

Komm, B. S., Terpening, C. M., Benz, D. J., Graeme, K. A., Gallegos, A., Korc, M., Greene, G. L., O'Malley, B. W., and Haussler, M. R. (1988). Estrogen binding, receptor mRNA, and biological response in osteoblast-like osteosarcoma cells. *Science* **241,** 81–84.

Kondaiah, P., Sands, M. J., Smith, J. M., Fields, A., Roberts, A. B., Sporn, M. B., and Melton, D. A. (1990). Identification of a novel transforming growth factor-β mRNA in *Xenopus laevis*. *J. Biol. Chem.* **265,** 1089–1093.

Lafyatis, R., Lechleider, R., Kim, S.-J., Jakowlew, S. B., Roberts, A. B., and Sporn, M. B. (1990). Structural and functional characterization of the transforming growth factor-β3 promoter: A cAMP responsive element regulates basal and induced transcription. *J. Biol. Chem.* **265,** 19128–19136.

Lafyatis, R., Lechleider, R., Roberts, A. B., and Sporn, M. B. (1991). Secretion and transcriptional regulation of transforming growth factor-β3 during myogenesis. *Mol. Cell. Biol.* **11,** 3795–3803.

Laiho, M., Saksela, O., Andreasen, P. A., and Keski-Oja, J. (1986). Enhanced production and extracellular deposition of the endothelial-type plasminogen activator inhibitos in cultured human lung fibroblasts by transforming growth factor-β. *J. Cell. Biol.* **103,** 2403–2410.

Laiho, M., DeCaprio, J. A., Ludlow, J. W., Livington, D. M., and Massague, J. (1990). Growth inhibition by TGF-β linked to suppression of retinoblastoma protein phosphorylation. *Cell* **62,** 175–185.

Lund, L. R., Riccio, A., Andreasen, P. A., Nielsen, L. S., Kristensen, P., Laiho, M., Blasi, F., and Dano, K. (1987). Transforming growth factor-β is a strong and fast acting positive regulator of the level of type-1 plasminogen activator inhibitor mRNA in WI-38 human lung fibroblasts. *EMBO J.* **6,** 1281–1286.

Madisen, L., Webb, N. R., Rose, T. M., Marquardt, H., Ikeda, T., Twardzik, D., Seyedin, S., and Purchio, A. F. (1988). Transforming growth factor-β2: cDNA cloning and sequence analysis. *DNA* **7,** 1–8.

Madri, J. A., Pratt, B. M., and Tucker, A. (1988). Phenotypic modulation of endothelial cells by transforming growth factor-β depends upon the composition and organization of the extracellular matrix. *J. Cell. Biol.* **106,** 1375–1384.

Marcelli, C., Yates, A. J., and Mundy, G. R. (1990). *In-vivo* effects of human recombinant transforming growth factor β on bone turnover in normal mice. *J. Bone Miner. Res.* **5,** 1087–1095.

Massague, J., Cheifetz, S., Endo, T., and Nadal-Ginard, B. (1986). Type β transforming growth factor is an inhibitor of myogenic differentiation. *Proc. Natl. Acad. Sci. USA* **83,** 8206–8210.

Masui, T., Wakefield, L. M., Lechner, J. F., LaVeck, M. A., Sporn, M. B., and Harris, C. C. (1986). Type β transforming growth factor is the primary differentiation inducing serum factor for normal human bronchial epithelial cells. *Proc. Natl. Acad. Sci. USA* **83,** 2438–2442.

Matrisian, L. M., Leroy, P., Ruhlmann, C., Gesnel, M.-C., and Breathnach, R. (1986). Isolation of the oncogene and epidermal growth factor-induced transin gene: Complex control in rat fibroblasts. *Mol. Cell. Biol.* **6,** 1679–1686.

Mihara, K., Cao, X.-R., Yen, A., Chandler, S., Driscoll, B., Murphree A. L., T'Ang, A., and

Fung, Y.-K. T. (1989). Cell cycle-dependent regulation of phosphorylation of the human retinoblastoma gene product. *Science* **246,** 1300–1303.

Millan, F. A., Denhez, F., Kondaiah, P., and Akhurst, R. J. (1991). Embryonic gene expression patterns of TGF-β1, β2 and β3 suggest different developmental functions *in vivo. Development* **111,** 131–144.

Miller, D. A., Lee, A., Matsui, Y., Chen, E. Y., Moses, H. L., and Derynck, R. (1989a). Complementary DNA cloning of the murine transforming growth factor-beta3 precursor and the comparative expression of TGF-beta3 and TGF-beta1 messenger RNA in murine and adult tissues. *Mol. Endocrinol.* **3,** 1926–1934.

Miller, D. A., Lee, A., Pelton, R. W., Chen, E. Y., Moses, H. L., and Derynck, R. (1989b). Murine transforming growth factor-β2 cDNA sequence and expression in adult tissues and embryos. *Mol. Endocrinol.* **3,** 1108–1114.

Moses, H. L., Tucker, R. F., Leof, E. B., Coffey, R. J., Halper, J., and Shipley, G. D. (1985). Type beta transforming growth factor is a growth stimulator and a growth inhibitor. *In* "Cancer Cells," Vol. 3, pp. 65–71. Cold Spring Harbor, New York.

Niitsu, Y., Urushizaki, Y., Koshida, Y., Terui, K., Mahara, Y., Kohgo, Y., and Urushizaki, I. (1988). Expression of the TGF-β gene in adult T cell leukemia. *Blood* **71,** 263–267.

Noda, M. (1989). Transcriptional regulation of osteocalcin production by transforming growth factor-β in rat osteoblast-like cells. *Endocrinology* **124,** 612–617.

Noda, M., and Camilliere, J. J. (1989). *In vivo* stimulation of bone formation by transforming growth factor-β. *Endocrinology* **125,** 2991–2994.

Noda, M., and Rodan, G. A. (1989). Type β transforming growth factor regulates expression of genes encoding bone matrix proteins *Connect. Tissue Res.* **21,** 71–75.

Noma, T., Glick, A. B., Geiser, A. G., O'Reilly, M. A., Miller, J., Roberts, A. B., and Sporn, M. B. (1991). Molecular cloning and structure of the human transforming growth factor-β2 gene promoter. *Growth Factors* **4,** 247–255.

Ong, G., Sikora, K., and Gullick, W. J. (1990). Inactivation of the retinoblastoma gene does not lead to loss of TGF-β receptors or response to TGF-β in breast cancer cell lines. *Oncogene* **6,** 761–763.

Oreffo, R. O., Mundy, G. R., Seyedin, S. M., and Bonewald, L. F. (1989). Activation of the bone-derived latent TGF-β complex by isolated osteoblasts. *Biochem. Biophys. Res. Commun.* **158,** 817–823.

O'Reilly, M. A., Geiser, A. G., Kim, S.-J., Bruggeman, L., Luu, A. X., Roberts, A. B., and Sporn, M. B. (1992). Identification of a CRE/ATF element in the human transforming growth factor-β2 gene that is essential for promoter activity **267,** 19938–19943.

Overall, C. M., Wrana, J. L., and Sodek, J. (1989). Independent regulation of collagenase, 72 KDa-progelatinase, and metalloendoproteinase inhibitor (TIMP) expression in human fibroblasts by transforming growth factor-β. *J. Biol. Chem.* **264,** 1860–1869.

Pelton, R. W., Nomura, S., Moses, H. L., and Hogan, B. L. M. (1989). Expression of transforming growth factor-β2 RNA during murine embryogenesis. *Development* **106,** 759–767.

Petkovich, P. M., Wrana, J. L., Grigoriadis, A. E., Heersche, J. N. M., and Sodek, J. (1987). 1,25-Dihydroxyvitamin D_3 increases epidermal growth factor receptors and transforming growth factor β-like activity in a bone-derived cell line. *J. Biol. Chem.* **262,** 13424–13428.

Pfeilschifter, J., and Mundy, G. R. (1987). Modulation of type β transforming growth factor activity in bone cultures by osteotropic hormones. *Proc. Natl. Acad. Sci. USA* **84,** 2024–2028.

Pfeilschifter, J., Souza, D. S. N., and Mundy, G. R. (1987). Effects of transforming growth factor-β on osteoblastic osteosarcoma cells. *Endocrinology* **121,** 212–218.

Pfeilschifter, J., Seyedin, S. M., and Mundy, G. R. (1988). *In-vivo* effects of human recombinant transforming growth factor β on bone resorption in fetal rat long bone cultures. *J. Clin. Invest.* **82,** 680–685.

Pfeilschifter, J., Wolf, O., Naumann, A., Minne, H. W., Mundy, G. R., and Ziegler, R. (1990). Chemotactic response of osteoblast-like cells to transforming growth factor beta. *J. Bone. Miner. Res.* **5,** 825–830.

Pietenpol, J. A., Stein, R. W., Moran, E., Yaciuk, P., Schlegel, R., Lyons, R. M., Pittelkow, M. R., Munger, K., Howley, P. M., and Moses, H. L. (1990). TGF-β1 inhibition of c-myc transcription and growth in keratinocytes is abrogated by viral transforming proteins with pRB binding domains. *Cell* **61,** 777–785.

Raghow, R., Postlethwaite, A. E., Keski-Oja, J., Moses, H. L., and Kang, A. H. (1987). Transforming growth factor-β increases steady state levels of type I procollagen and fibronectin messenger RNAs post-transcriptionally in cultured human dermal fibroblasts. *J. Clin. Invest.* **79,** 1285–1288.

Redini, F., Lafuma, C., Pujol, J.-P., Robert, L., and Hornebeck, W. (1988). Effect of cytokines and growth factors on the expression of elastase activity by human synoviocytes, dermal fibroblasts and rabbit articular chondrocytes. *Biochem. Biophys. Res. Commun.* **155,** 786–793.

Robbins, P. D., Horowitz, J. M., and Mulligan, R. C. (1990). Negative regulation of human *c-fos* expression by the retinoblastoma gene product. *Nature (London)* **346,** 668–671.

Roberts, A. B., and Sporn, M. B. (1990). The transforming growth factors-βs. *In* "Handbook of Experimental Pharmacology," Vol. 95/1. "Peptide Growth Factors and Their Receptors," pp. 419–472. Springer-Verlag, Heidelberg.

Roberts, A. B., Anzano, M. A., Lamb, L. C., Smith, J. M., and Sporn, M. B. (1981). New class of transforming growth factors potentiated by epidermal growth factor. *Proc. Natl. Acad. Sci. USA* **78,** 5339–5343.

Roberts, A. B., Anzano, M. A., Meyers, C. A., Wideman, J., Blacher, R., Pan, Y.-C., Stein, S., Lehrman, S. R., Smith, J. M., Lamb, L. C., and Sporn, M. B. (1983). Purification and properties of a type beta transforming growth factor from bovine kidney. *Biochemistry* **22,** 5692–5698.

Roberts, A. B., Anzano, M. A., Wakefield, L. M., Roche, N. S., Stern, D. F., and Sporn, M. B. (1985). Type beta transforming growth factor: A bifunctional regulator of cellular growth. *Proc. Natl. Acad. Sci. USA* **82,** 119–123.

Roberts, A. B., Sporns, M. B., Assoian, R. K., Smith, J. M., Roche, N. S., Wakefield, L. M., Heine, U. I., Liotta, L. A., Falanga, V., Kehrl, J. H., and Fauci, A. S. (1986). Transforming growth factor type-beta: Rapid induction of fibrosis and angiogenesis *in vivo* and stimulation of collagen formation *in vitro*. *Proc. Natl. Acad. Sci. USA* **83,** 4167–4171.

Roberts, A. B., Thompson, N. L., Heine, U., Flanders, K. C., and Sporn, M. B. (1988). Transforming growth factor-β: Possible roles in carcinogenesis. *Br. J. Cancer* **57,** 594–600.

Roberts, A. B., Rosa, F., Roche, N. S., Coligan, J., Garfield, M., Rebbert, M. L., Kondaiah, P., Danielpour, D., Kehrl, J. H., Wahl, S. M., Dawid, I. B., and Sporn, M. B. (1990). Isolation and characterization of TGF-β2 and TGF-β5 from medium conditioned by *Xenopus* XTC cells. *Growth Factors* **3,** 277–286.

Robey, P. G., Young, M. F., Flanders, K. C., Roche, N. S., Kondaiah, P., Reddi, A. H., Termine, J. D., Sporn, M. B., and Roberts, A. B. (1987). Osteoblasts synthesize and respond to TGF-β *in vitro*. *J. Cell. Biol.* **105,** 457–463.

Rossi, P., Karsenty, G., Roberts, A. B., Roche, N. S., Sporn, M. B., and de Crombrugghe, B. (1988). A nuclear factor 1 binding site mediates the transcriptional activation of a type I collagen promoter by transforming growth factor-β. *Cell* **52,** 405–414.

Russel, W. E., Coffey, R. J., Ouellette, A. J., and Moses, H. L. (1988). Transforming growth factor beta reversibly inhibits the early proliferative response tp partial hepatectomy in the rat. *Proc. Natl. Acad. Sci. USA* **85,** 5126–5130.

Segarini, P. R., Rosen, D. M., and Seyedin, S. M. (1989). Binding of TGF-β to cell surface proteins varies with cell type. *Mol. Endocrinol.* **3,** 261–272.

Seyedin, S. M., Thomas, T. C., Thomson, A. Y., Rosen, D. M., and Piez, K. A. (1985). Purification and characterization of two cartilage-inducing factors from bovine demineralized bone. *Proc. Natl. Acad. Sci. USA* **82,** 2267–2271.

Seyedin, S. M., Thompson, A. Y., Bentz, H., Rosen, D. M., McPherson, J. M., Conti, A., Siegel, N. R., Galluppi, G. R., and Piez, K. A. (1986). Cartilage-inducing factor-A. *J. Biol. Chem.* **261,** 5693–5695.

Shipley, G. D., Pittelkow, M. R., Wille, J. J., Scott, R. E., and Moses, H. L. (1986). Reversible inhibition of normal human prokeratinocyte proliferation by type β transforming growth factor-growth inhibitor in serum-free medium. *Cancer Res.* **46,** 2068–2071.

Silberstein, G. B., and Daniel, C. W. (1987). Reversible inhibition of mammary gland growth by transforming growth factor-β. *Science* **237,** 291–293.

Tashijian, A. H., Voelkel, E. F., Lazzaro, M., Singer, F. R., Roberts, A. B., Derynck, R., Winkler, M. E., and Levine, L. (1985). Human transforming growth factors alpha and beta stimulate prostaglandin production and bone resorption in cultured mouse calvaria. *Proc. Natl. Acad. Sci. USA* **82,** 4535–4538.

ten Dijke, P., Hanson, P., Iwata, K. K., Pieler, C., and Foulkes, J. G. (1988). Identification of a new member of the transforming growth factor-β gene family. *Proc. Natl. Acad. Sci. USA* **85,** 4715–4719.

ten Dijke, P., Iwata, K. K., Goddard, C., Pieler, C., Canalis, E., McCarthy, T. L., and Centrella, M. (1990). Recombinant transforming growth factor type β3: Biological activities and receptor-binding properties in isolated bone cells. *Mol. Cell. Biol.* **10,** 4473–4479.

Thompson, N. L., Bazoberry, F., Speir, E. H., Casscells, W., Ferrans, V. J., Flanders, K. C., Kondaiah, P., Geiser, A. G., and Sporn, M. B. (1988). Transforming growth factor-β1 in acute myocardial infarction in rats. *Growth Factors* **1,** 91–99.

Tucker, R. F., Shipley, G. D., Moses, H. L., and Holley, R. W. (1984). Growth inhibitor from BSC-1 cells closely related to platelet type beta transforming growth factor. *Science* **226,** 705–707.

Van Obberghen-Schilling, E., Roche, N. S., Flanders, K. C., Sporn, M. B., and Roberts, A. B. (1988). Transforming growth factor β1 positively regulates its own expression in normal and transformed cells. *J. Biol. Chem.* **263,** 7741–7746.

Varga, J., Rosenbloom, J., and Jimenez, S. A. (1987). Transforming growth factor-β (TGF-β) causes a persistent increase in steady-state amounts of type I an II and type III collagen and fibronectin mRNAs in normal human dermal fibroblasts. *Biochem. J.* **247,** 597–604.

Vukicevic, S., Luyten, F. P., and Reddi, A. H. (1990). Osteogenin inhibits proliferation and stimulates differentiation in mouse osteoblast-like cells (MC 3T3-E1). *Biochem. Biophys. Res. Commun.* **166,** 750–756.

Wakefield, L. M., Smith, D. M., Flanders, K. C., and Sporn, M. B. (1988). Latent transforming growth factor-β from human platelets. *J. Biol. Chem.* **263,** 7646–7654.

Webb, N. R., Madison, L., Rose, T. M., and Purchio, A. F. (1988). Structural and sequence of TGF-β2 cDNA clones predicts two different precursor proteins produced by alternative mRNA splicing. *DNA* **7,** 493–497.

Williams, T., Admon, A., Luscher, B., and Tijan, R. (1988). Cloning and expression of AP-2, a cell-type-specific transcription factor that activates inducible enhancer elements. *Genes Dev.* **2,** 1557–1569.

Wrana, J. L., Sodek, J., Ber, R. L., and Bellows, C. G. (1986). The effects of platelet-derived transforming growth factor-β on normal human diploid gingival fibroblasts. *Eur. J. Biochem.* **159,** 69–76.

Wrann, M., Bodmer, S., de Martin, R., Siepl, C., Hofer-Warbinek, R., Frei, K., Hofer, E., and Fontana, A. (1987). T cell suppressor from human glioblastoma cells is a 12.5 KD protein closely related to transforming growth factor-beta. *EMBO J.* **6,** 1633–1636.

4

BONE MORPHOGENETIC PROTEINS AND THEIR GENE EXPRESSION

JOHN M. WOZNEY

Cellular and Molecular Biology of Bone

I. INTRODUCTION

The search for the molecule or molecules responsible for the bone- and cartilage-inductive activity present in bone and other tissue extracts has led to the discovery of a novel set of molecules called the bone morphogenetic proteins (BMPs). The structures of seven proteins, BMP-1 through BMP-7, have been elucidated by molecular cloning. BMP-2 through BMP-7 form a unique subfamily within the transforming growth factor β (TGF-β) superfamily, and evidence indicating that these proteins are responsible for bone-inductive activity isolated from these tissue extracts has accumulated. The unique inductive activities of these proteins along with their presence in bone suggests that they are important regulators of bone repair processes and may well be involved in the normal maintenance of bone tissue. Multiple therapeutic uses in the wide variety of settings where bone has been lost through physiological or traumatic processes are potential clinical indications for BMPs. Recent studies on the cellular activities of the BMPs indicate that, as expected from their activities in animal systems, the BMPs essentially act as differentiation factors, causing induction and increased expression of multiple differentiated phenotypes in mesenchymal cells. Investigations in several areas indicate that members of the BMP family are also involved in a variety of developmental processes; they are likely to be important signaling molecules during formation of the skeleton in vertebrates. In addition, they probably play some role in the development of other organ and tissue systems that form via mesenchymal–epithelial interactions and possibly function to deliver or interpret positional information in a wide variety of organisms.

II. BONE MORPHOGENETIC PROTEIN FAMILY

A. Structure of BMP-2 through BMP-7

Seven molecules, which have been named BMP-1 through BMP-7, have been identified and their corresponding molecular clones obtained using fragmentary amino acid sequence information from bone-inductive extracts derived from bovine bone (Wozney *et al.*, 1988; Wozney, 1989; Özkaynak *et al.*, 1990; Celeste *et al.*, 1990). These amino acid sequences allowed the design of multiple oligonucleotide probes, which were initially used to obtain bovine genomic or complementary DNA (cDNA) clones for each of the BMPs (Wozney, 1990). These bovine clones were used to isolate human cDNA clones that contained the entire coding sequences for each of the individual BMPs. The human cDNAs for BMP-1, BMP-2, BMP-4, BMP-5, and BMP-7 were each derived from libraries constructed from the human osteoblastlike osteosarcoma cell line

U-2 OS, the BMP-3 cDNAs were derived from a cDNA library of the H128 small cell lung carcinoma cell line, and clones for BMP-6 were initially derived from U-2 OS cDNA libraries and subsequently from human placenta and brain cDNA libraries. Other reports have indicated the cloning of BMP-7, also called OP-1, from hippocampus and placenta cDNA libraries (Özkaynak *et al.*, 1990).

From the primary amino acid sequences of the human BMPs derived from these molecular clones, six out of the seven BMPs were found to be related to each other and to be members of the TGF-β superfamily. An alignment of these amino acid sequences (Fig. 1) indicates that significant amino acid identity exists among all the BMPs in the carboxy-terminal region of the proteins (identical residues are indicated by asterisks). This region contains seven cysteine residues, whose presence and relative positions are conserved among all reported members of the TGF-β superfamily. Similar to other members of the TGF-β family, the BMPs are synthesized within the cell in a precursor form (see Fig. 2). All six BMPs contain hydrophobic secretory leader sequences and substantial propeptide regions. The mature portion of each molecule resides at the carboxy-terminus of the prepropeptide and includes the seven-cysteine domain. Each mature, active BMP is a dimeric molecule containing two of these polypeptides. In this chapter, a molecule such as "BMP-2" refers to a homodimer of two BMP-2 chains. It should be pointed out that potential exists for formation of heterodimeric forms of the BMPs, either by dimerization of two different BMPs or one BMP with another member of the TGF-β superfamily. Such heterodimeric proteins are known to exist within the superfamily and may possess different specific activities, receptor binding characteristics, or entirely novel activities relative to the homodimeric forms. For example, TGF-β1.2 and TGF-β2.3 have been shown to exist naturally and to have similar activities to the TGF-β homodimeric forms, although their specific activities vary in particular assay systems (Cheifetz *et al.*, 1988; Ogawa *et al.*, 1992). In contrast, the activity of activin (e.g., an inhibin β_A homodimer) is opposite to that of inhibin (an inhibin α–inhibin β_A heterodimer) in every assay system examined to date (Hsueh *et al.*, 1987).

Analysis of the primary amino acid sequences of the six related BMPs allows them to be grouped into three separate sets. One set consists of BMP-2 and BMP-4, which are 92% identical in the seven-cysteine region; in fact, BMP-4 was originally isolated by cross-hybridization to a BMP-2 probe. BMP-5, BMP-6, and BMP-7 are also closely related to one another, possessing an average of 89% amino acid identity in their corresponding regions, and define a second BMP subset. BMP-3 is the sole member of the third subset identified to date. The BMP-2/BMP-4 and BMP-5/BMP-6/BMP-7 subsets are more closely related to one another

```
BMP-2                  MVAGTRCLLALLLPQVLLGGA..AGLVPELGRRKFA...AASSGRPSSQPSDEVL.....SEFE
BMP-4                  MIPGNRMLMVVLLCQVLLGGASHASLIPETGKKKVAEIQGHAGGRRSGQ.SHELL.....RDFE
BMP-5           MHLTVFLLKGIVGF..LWSCWVLVGYAKGGLG................DNHVHSSFIYRRLRNHERREIQ
BMP-6  MPGLGRRAQWLCWWWGLLCSCCGPPPLRPPLPAAAAAAAGGQLLGDGGSPGRTEQPPPSPQSSSGFLYRRLKTQEKREMQ
BMP-7           MHVRSLRAAAPHSFVALWAPLFLLRSALADFSL...............DNEVHSSFIHRRLRSQERREMQ
BMP-3                   MAGASRLLFLWLGCFCVSLAQGERPKPPFPELRKAVPGDRTAGGGPDSELQPQDKVSEHMLRL
                                           *

BMP-2  LRLLSMFGLKQRPTP...............................SRDAVVPPYMLDLYRRHSGQPG.....SPAPDHR
BMP-4  ATLLQMFGLRRRPQP...............................SKSAVIPDYMRDLYRLQSGEEEEEQIHSTGLEYP
BMP-5  REILSILGLPHRPRPF...........................SPGKQASSAPLFMLDLYNAMTNEENPEESEYSVR...
BMP-6  KEILSVLGLPHRPRPLHGLQQPQPPALRQQEEQQQQQLPRGEPPPGRLKSAPLFMLDLYNALSADNDEDGASEGERQQS
BMP-7  REILSILGLPHRPRPH...........................LQGKH.NSAPMFMLDLYNAMAVEEGGGPGG.......
BMP-3  YDRYSTVQAARTPGSLEGGSQPWRPRLLREGNTVRSFRAAAAETLERK....GLYIFNLTSLTKSENILSATLYFCIGE.
                   *                                             *

BMP-2  LERAA.............................................SRANTVRSFHHEESLEELPETSGKTTRRFF
BMP-4  .ERPA.............................................SRANTVRSFHHEEHLENIPGTSENSAFRFL
BMP-5  .ASLAEETRGARKGYPASPNGYPRRIQLSRTTPLTTQSPPLASLHDTNFLNDADMVMSFVNLVERDKDFSHQRRHYKEFR
BMP-6  WPHEAASSSQRRQPPPGAAHPLNRKSLLAPGSG.SGGASPLTSAQDSAFLNDADMVMSFVNLVEYDKEFSPRQRHHKEFK
BMP-7  ............QGF.....SYPYKAVFS......TQGPPLASLQDSHFLTDADMVMSFVNLVEHDKEFFHPRYHHREFR
BMP-3  LGNISLSCPVSGGCSHHAQRKHIQIDLSAWTLKFSRNQSQLLGHLSVDMAKSHRDIMSW..LSKDITQFLRKAKENEEFL
                                                                *                     *

BMP-2  FNLSSIPTEEFITSAELQVFREQMQDALGNNSSFHHRINIYEIIK.PATANSKFPVTRLLDTRLVNQNASRWESFDVTPA
BMP-4  FNLSSIPENEVISSAELRLFREQVDQGPDWERGF.HRINIYEVMKPPAFVVPGHLITRLLDTRLVHHNVTRWETFDVSPA
BMP-5  FDLTQIPHGEAVTAAEFRIYKDRSNNRFENET...IKISIYQIIKEYTNRDADLF...LLDTRKAQALDVGWLVFDITVT
BMP-6  FNLSQIPEGEVVTAAEFRIYKDCVMGSFKNQT...FLISIYQVLQEHQHRDSDLF...LLDTRVVWASEEGWLEFDITAT
BMP-7  FDLSKIPEGEAVTAAEFRIYKDYIRERFDNET...FRISVYQVLQEHLGRESDLF...LLDSRTLWASEEGWLVFDITAT
BMP-3  IGFNITSKGRQLPKRRLP.FPEPYILVYANDAAISEPESVVSSLQGHRNFPTG..............TVPKWDSHIRAAL
                                                                              *

BMP-2  VMRWTAQGHANHGFVVEVAHLEEKQGVSKRHVRISRSLHQDEHSWSQIRPLLVTF.......GHDGKGHPL..HKREKRQ
BMP-4  VLRWTREKQPNYGLAIEVTHLHQTRTHQGQHVRISRSLPQGSGNWAQLRPLLVTF.......GHDGRGHALTRRRRAKRS
BMP-5  SNHWVINPQNNLGLQLCAETGDGRSINVKSAGLVGRQGPQSKQ......PFMVAFFKASEVLLRSVR.AANKRKNQNRNK
BMP-6  SNLWVVTPQHNMGLQLSVVTRDGVHVHPRAAGLVGRDGPYDKQ......PFMVAFFKVSEVHVRTTRSASSRRRQQSRNR
BMP-7  SNHWVVNPRHNLGLQLSVETLDGQSINPKLAGLIGRHGPQNKQ......PFMVAFFKATEVHFRSIRSTGSKQRSQNRSK
BMP-3  SIERRKKRSTGVLLPLQNNELPGAEYQYKKDEVWEERKP................YKTLQAQAPEKSKNKKKQRKGPHRK

BMP-2  AK.HKQRKRLKS.S..........CKRHPLYVDFSDVGWNDWIVAPPGYHAFYCHGECPFPLADHLNSTNHAIVQTLVNS
BMP-4  PKHHSQRARKKNKN..........CRRHSLYVDFSDVGWNDWIVAPPGYQAFYCHGDCPFPLADHLNSTNHAIVQTLVNS
BMP-5  SSSHQDSSRMSSVGDYNTSEQKQACKKHELYVSFRDLGWQDWIIAPEGYAAFYCDGECSFPLNAHMNATNHAIVQTLVHL
BMP-6  STQSQDVARVSSASDYNSSELKTACRKHELYVSFQDLGWQDWIIAPKGYAANYCDGECSFPLNAHMNATNHAIVQTLVHL
BMP-7  TPKNQEALRMANVAENSSSDQRQACKKHELYVSFRDLGWQDWIIAPEGYAAYYCEGECAFPLNSYMNATNHAIVQTLVHF
BMP-3  SQTLQFDEQTLKKARRKQWIEPRNCARRYLKVDFADIGWSEWIISPKSFDAYYCSGACQFPMPKSLKPSNHATIQSIVRA
                               C    * * * * **  **  *    * *C * C **        ***  *  *

BMP-2  VN.SK.IPKACCVPTELSAISMLYLDENEKVVLKNYQDMVVEGCGCR
BMP-4  VN.SS.IPKACCVPTELSAISMLYLDEYDKVVLKNYQEMVVEGCGCR
BMP-5  MF.PDHVPKPCCAPTKLNAISVLYFDDSSNVILKKYRNMVVRSCGCH
BMP-6  MN.PEYVPKPCCAPTKLNAISVLYFDDNSNVILKKYRNMVVRACGCH
BMP-7  IN.PETVPKPCCAPTQLNAISVLYFDDSSNVILKKYRNMVVRACGCH
BMP-3  VGVVPGIPEPCCVPEKMSSLSILFFDENKNVVLKVYPNMTVESCACR
              *  CC *      * *  *    * ** *  * *  C C
```

FIGURE 1 Alignment of the amino acid sequences of the bone morphogenetic protein (BMP) family of proteins. Asterisks indicate positions where an amino acid residue is conserved among all six proteins. Residues that are conserved among BMP-2/BMP-4/BMP-5/BMP-6/BMP-7 are shaded. The mature regions of the proteins are indicated with bold lettering. The processing positions for the mature regions were determined as indicated in the text.

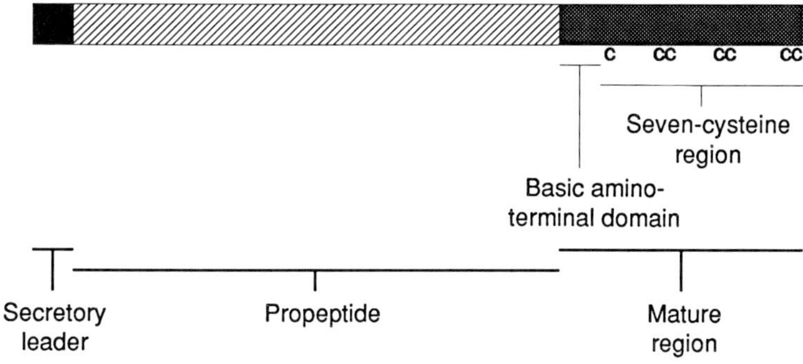

BMP	Additional names	Amino acids (preproprotein)	Amino acids (mature monomer)	Glycosylation sites (mature monomer)
BMP-2	BMP-2A	396	114	1
BMP-4	BMP-2B	408	116	2
BMP-5		454	138	3
BMP-6	Vgr-1	513	139	3
BMP-7	OP-1	431	139	3
BMP-3	Osteogenin	472	?	1

FIGURE 2 Schematic structure of the bone morphogenetic proteins (BMPs). Nomenclature and biochemical characteristics of the BMPs are given in the bottom half.

than they are to BMP-3; amino acid residues that are common to these five molecules are shaded in Figure 1. It is interesting to note that blocks of homology in these molecules persist even into the propeptide region.

This subdivision of the BMPs into three groups extends to other biochemical features of the proteins (Fig. 2, bottom). All the mature BMPs have amino-terminal domains that are basic (the domain amino-terminal to the seven-cysteine regions). The size of this domain, and thus the size of the mature proteins, is similar for BMP-2 and BMP-4; BMP-5, BMP-6, and BMP-7 are also almost identical in size. It should be noted that the site of propeptide processing is not clear for all of the BMPs. In cases where the amino-terminus is known, either from natural sources or from recombinant systems, processing occurs at the consensus sequence Arg–X–X–Arg to yield the mature peptide. The numbers in Fig. 2 are derived from amino-terminal sequencing of natural BMPs, from the major amino-terminus detected in BMPs expressed in recombinant systems (see later), or by analogy with other BMPs. Multiple processing sites have been detected in recombinant systems, and it is possible that multiple forms exist *in vivo.* Each of the BMPs is glycosylated, which is unlike the TGF-βs. BMP-2 contains one asn-linked glycosylation site, BMP-4 contains two (one in the corresponding position to that in BMP-2), and BMP-5, BMP-6, and BMP-7 each contain three in closely related positions. BMP-3 contains a single glycosylation site, at a unique position relative to the other BMPs.

The amino acid sequences of several of the BMPs have also been derived from several nonhuman species and, in general, there is striking conservation of the primary amino acid sequences among very divergent species. The murine and human BMP-2 sequences are identical in the mature region (Dickinson *et al.,* 1990); comparison of this sequence with the *Xenopus* sequence indicates two conservative (Lys to Arg) changes in the basic amino-terminal region and only two differences within the seven-cysteine domain (Plessow *et al.,* 1991). Similarly, the murine and human BMP-4 sequences are identical in the seven-cysteine domain, with only two changes in the amino-terminus. Two *Xenopus* BMP-4 cDNAs have been identified, one of which is identical to the human sequence in the seven-cysteine region, and one of which contains a single amino acid difference (Köster *et al.,* 1991). The murine BMP-6 (Vgr-1) and BMP-7 sequences are also quite similar to those of the human forms with only one (BMP-7) or two (BMP-6) amino acid differences in the seven-cysteine region; BMP-6 has five differences in the basic amino-terminal domain, while BMP-7 has only two (Lyons *et al.,* 1989a; Özkaynak *et al.,* 1991).

B. Relationship to the Transforming Growth Factor β Superfamily

The BMPs form a subset of growth and differentiation factors within the TGF-β superfamily. The relationship among the amino acid sequences in the seven-cysteine domains of the published TGF-β family members is shown in Figure 3. The most closely related TGF-β superfamily members to BMP-2 and BMP-4 are *decapentaplegic* (*dpp*) and Vg1. *Decapentaplegic* is a *Drosophila* protein (Padgett *et al.,* 1987) that was shown by genetic methods to be requisite for embryonic development (see Section IV.D). Vg1 is a *Xenopus* factor whose messenger RNA (mRNA) is localized to the vegetal hemisphere of the oocyte; although its exact function has remained elusive, it too is believed to be involved in embryonic development (Weeks and Melton, 1987). BMP-5, BMP-6, BMP-7, and a recently identified new member of this subset, called BMP-8 (Celeste *et al.* 1992), are closely related to *60A*, a *Drosophila* protein of unknown function expressed in the early embryo (Wharton *et al.,* 1991; Doctor *et al.,* 1992). Although BMP-3 lies outside of these two groups of proteins, it is the next most closely related TGF-β superfamily member. Due to the confusing nomenclature for the BMPs (Section III.B) and the fact that they have

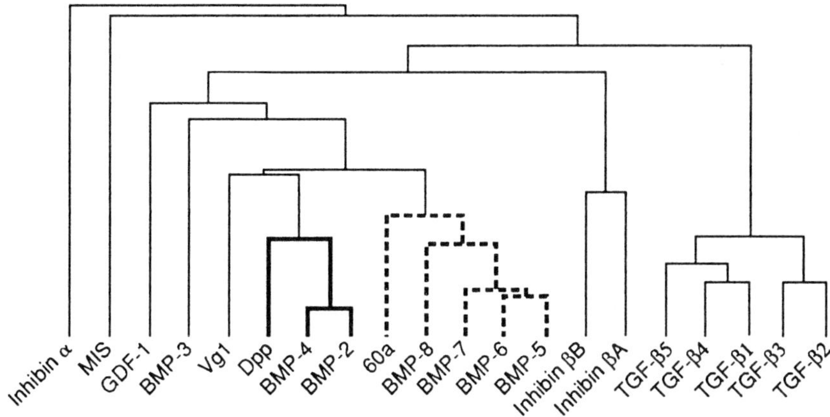

FIGURE 3 Relationship among the transforming growth factor β (TGF-β) superfamily members. This figure was generated by comparison of the seven-cysteine regions of the proteins by the GCG program PileUp. Note that this figure is representative of an analysis generated by pairwise comparison of all members of the family and is not purported to represent their evolutionary relationships. MIS, Müllerian inhibiting substance.

additional functions other than bone formation, it has been proposed that this family of factors be renamed the DVR (for *dpp* Vg1-related) family (Lyons *et al.*, 1991). However, inclusion of BMP-3 in this family necessitates inclusion of GDF-1 because it is also closely related, and given the importance of activins (inhibin β_A and inhibin β_B homodimers and heterodimers) in developmental processes these may also actually reside within the DVR family. Therefore, criteria still need to be clearly established for inclusion into the DVR family.

Other TGF-β superfamily members clearly lie outside of the BMP family. Five TGF-βs have been identified: TGF-β1, TGF-β2, and TGF-β3 in humans, TGF-β4 in chicken, and TGF-β5 in *Xenopus;* all of these factors are more closely related to one another than to the BMPs and, in fact, are quite distantly related, showing an average of only about 37% amino acid identity in the seven-cysteine region to the BMP molecules. Müllerian inhibiting substance and inhibin α are also quite distantly related.

C. BMP-1

BMP-1 was originally identified and cloned as a component of purified bone-inductive extracts along with BMP-2 through BMP-8. It is not a TGF-β superfamily member and, thus, not related by sequence to the other BMPs. Its amino acid sequence, as derived from the cDNA, indicates that it consists of multiple domains (Fig. 4). It has a hydrophobic secretory leader sequence at its amino-terminus, suggesting that it is a

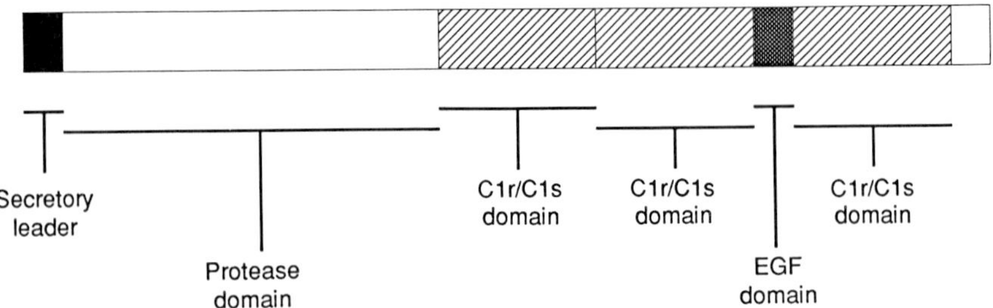

FIGURE 4 Schematic domain structure of bone morphogenetic protein 1. EGF, epidermal growth factor.

secreted factor. Interestingly, BMP-1 has a domain homologous to a protease found in *Astacus* (Titani *et al.*, 1987). This domain has now also been identified in a number of proteins leading to the proposal that this family of metalloendoproteinases be termed the Astacin family (Dumermuth *et al.*, 1991). BMP-1 also has three internal repeat domains, which show homology to the C1s and C1r components of complement. The second and third of these domains are separated by an epidermal growth factor (EGF) domain, which is present in numerous proteins.

Several proteins with postulated roles in development have now been found that show homology to BMP-1 and contain similar domain structures. Perhaps the most interesting is *tolloid*, a *Drosophila* gene that is similar to, but less severe than, *dpp* in its mutant phenotype (Shimell *et al.*, 1991). The *tolloid* gene product contains an Astacin family protease domain, five (rather than the three in BMP-1) C1r/C1s domains, and two (rather than the one in BMP-1) EGF domains. Because the motif of two C1r/C1s domains separated by an EGF domain appears to be important in protein–protein interactions, it has been suggested that *tolloid* may physically interact with other proteins, the obvious candidate for which is *dpp*. Two proteins have been found in sea urchin embryos that also consist of the same three domain types as BMP-1 (Lepage *et al.*, 1992; Reynolds *et al.*, 1992). SpAN, isolated from *Strongylocentrotus purpuratus*, and BP10, isolated from *Paracentrotus lividus*, display 76% amino acid identity overall and may represent the same protein isolated from different sea urchin species. Both are transiently expressed in the blastula of early sea urchin embryos, and their mRNAs are distributed in a spatially restricted manner. The presence of homologs of all the BMP proteins in a variety of widely divergent species suggest their importance in multiple types of developmental and other tissue processes.

D. BMP Genes

Most of the BMP genes have been assigned to specific chromosomes in both the mouse and human genomes (Table I) and, in some cases, are

TABLE I Chromosomal Localizations of the Bone Morphogenetic Proteins

	Mouse		Human	
BMP	Chromosome	Possible mutant association	Chromosome	Possible disease association
BMP-1	14		8	Multiple hereditary osteochondromatosis
BMP-2	2	Tightskin (*tsk*)	20p12	Holt-Oram syndrome
BMP-3	5		4p14–q21	Dentinogenesis imperfecta type II
BMP-4	14	Pugnose (*pn*)	14	Holt-Oram syndrome
BMP-5			6	
BMP-6	13	Congenital hydocephalus (*ch*)	6	
BMP-7			20	Holt-Oram syndrome

candidate genes for developmental anomalies (Dickinson *et al.*, 1990; Ceci *et al.*, 1990). In the mouse, the BMP-2 gene is localized to chromosome 2, where it is a candidate gene for the tight skin (*tsk*) mutant. This mutant is characterized by excess collagen deposition in the subcutaneous connective tissue as well as by increased growth of cartilage, bone, and tendons. The BMP-4 gene is localized to chromosome 14 and may be a candidate for the pugnose (*pn*) locus, which results in abnormalities in skull bone development. A second BMP-4-related gene has been localized to the X chromosome, but no human homolog of this gene has been found, suggesting that the mouse sequence might be a pseudogene. The BMP-6 gene is present on chromosome 13, possibly at the congenital hydrocephalus (*ch*) locus, which is associated with abnormalities in the growth and differentiation of the skeletal system and kidney.

In the human, the BMP-2 and BMP-7 genes are both located on chromosome 20, and BMP-2 has been sublocalized to 20p12 (Tabas *et al.*, 1991, 1992; Rao *et al.*, 1992; Hahn *et al.*, 1992). Sublocalization data on BMP-7 are necessary before it can be determined if these genes are linked. Both BMP-5 and BMP-6 are on chromosome 6. Again, it is not known if these two BMP genes are linked. The BMP-3 gene is located on chromosome 4 between p14 and q21. Interestingly, dentinogenesis imperfecta type II, a disease of tooth development, has been associated with this chromosome. Both chromosomes 2 and 20 have been implicated in Holt-Oram syndrome (HOS) based on a break point and inversion in single individuals, so that BMP-2, BMP-4, or BMP-7 might be involved. HOS is characterized by flaws in cardiac and skeletal development, resulting in septal defects and upper limb deformities. Given the osteogenic activities of the BMPs, any of their genes could be candidates for involvement in fibrodysplasia ossificans progressiva (FOP), a rare genetic disorder characterized by successive ossification of the skeletal muscles and abnormalities in the development of the toes. As distribution of the FOP lesions follows a gradient similar to that observed with mutations of *dpp,* the proposal has been made that either BMP-2 or BMP-4 might be involved in this disease process (Kaplan *et al.*, 1990). The disease could result from either an error in the regulation of the BMP genes resulting in their over- or misexpression or a BMP receptor defect causing oversensitivity to the signalling molecule.

Limited information is available on the structure of the BMP genes. Both the BMP-2 and BMP-4 genes have very similar structures to that of the *dpp* gene, including a single intron within the coding region placed at an exactly corresponding position in all three genes (St. Johnston *et al.*, 1990; Wozney, unpublished). In addition, all three have an intron placed closely 5′ to the initiator methionine codon. The BMP-7 gene structure is quite different, consisting of seven exons (E. Özkaynak and H. Oppermann, personal communication). Even though the TGF-β genes also have seven exons, they are not delimited at the corresponding positions of the BMP genes.

E. BMP mRNAs

The mRNAs for the BMPs have now been found in a variety of cell and tissue types (Wozney *et al.*, 1990; Özkaynak *et al.*, 1991; Lyons *et al.*, 1989a; Gazit *et al.*, 1991; Bonewald *et al.*, 1991; Bentley *et al.*, 1992; Graveley *et al.*, 1992). Table II details where they have been found from published reports, and some attempt has been made where possible to order the sources according to the relative abundance of the BMP mRNA. While bone and bone cells are major sources for some of the BMP mRNAs, many tissues outside of bone have substantial amounts of some of the BMP mRNAs. For example, a major site of BMP-3 expression is the lung, and the major source for BMP-7 mRNA is the kidney. The fact that multiple tissue and cell types synthesize at least the mRNAs for BMPs may explain the multiple tissue sources for osteoinductive activity (Section III.A) and also suggests that the BMPs play roles in addition to that of bone formation in the adult organism.

Several cell or tissue types synthesize multiple BMP mRNAs. The osteosarcoma cell line U-2 OS, for example, synthesizes the mRNAs for BMP-2, BMP-4, BMP-5, BMP-6, and BMP-7. Primary osteosarcomas are known to make BMP activity, and this activity has been suggested to be

Table II Bone Morphogenetic Protein mRNA tissue and cell line sources

BMP	Tissue	Cell line
BMP-2	Bone spleen, liver, brain, lung, kidney, heart, placenta	U-2 OS (osteosarcoma) MG-63 (osteosarcoma) F9 (embryonal carcinoma)
BMP-3	Lung, brain	PC-3 (prostate) H128 (small cell lung carcinoma)
BMP-4	Bone, lung, kidney, brain, speen, liver, heart, placenta	U-2 OS (osteosarcoma) MG-63 (osteosarcoma) PC-3 (prostate) DU-145 (prostate) F9 (embryonal carcinoma)
BMP-5	placenta	U-2 OS (osteosarcoma) MG-63 (osteosarcoma)
BMP-6	Calvaria lung, brain, placenta, kidney, uterus, muscle, skin	U-2 OS (osteosarcoma) F-9 (+ RA + cyclic adenosine monophosphate) PC-3, DU-145, PAIII (prostate carcinomas)
BMP-7	Kidney, placenta, brain, calvaria, spleen, lung, heart, liver, adrenal, bladder	U-2 OS (osteosarcoma)

correlated with the prognostic outcome of the disease (Yoshikawa *et al.*, 1985, 1988a, b). Whether this represents production of a particular set of BMPs and whether expression of these specific BMPs represents a particular stage of differentiation of the osteoblast with osteosarcomas, remain unclear. Likewise, interest in production of BMP mRNAs by prostate carcinoma cells has been driven by the ability of these cells to metastasize to bone. Several BMP mRNAs are made by prostate tumor cells (Harris *et al.*, 1991b; Hart *et al.*, 1991; Bentley *et al.*, 1992), and while it seems likely that their expression may relate to the osteogenesis observed at the site of metastasis, the existence of a unique pattern of BMP mRNA expression in prostate carcinoma cells is not clear.

III. *IN VIVO* ACTIVITIES

A. Sources and Assay Systems

The BMP proteins were originally identified by their ability to induce bone when implanted subcutaneously in rodents. Transitional epithelium and various epithelial cell lines were some of the first tissue and cell types shown to have bone-inductive capacities in this type of assay (Neuhof, 1917; Huggins, 1931; Anderson *et al.*, 1964; Anderson and Coulter, 1968; Wlodarski, 1969; Wlodarski *et al.*, 1971; Anderson, 1976;

Hall and van Exan, 1982). Demineralized bone and extracts of demineralized bone also induce bone formation in this model and comprise the most extensively studied systems for derivation of BMP activity (Urist, 1965; Urist *et al.*, 1973, 1979b; Reddi and Anderson, 1976). In addition, bone-inductive activity has been found in extracts of various mouse and human osteosarcomas (Urist *et al.*, 1977, 1979a; Takaoka *et al.*, 1980, 1982, 1989; Hanamura *et al.*, 1980; Bauer and Urist, 1981; Yoshikawa *et al.*, 1984; Tsuda *et al.*, 1989), tooth matrix, or dentine (de Groot and Deshmukh, 1975; Butler *et al.*, 1977).

Most of the assay systems used for detecting bone-inductive activity consist of the implantation of materials at either a subcutaneous or an intramuscular site in mice or rats. Depending of the amounts of material to be implanted, the material may be implanted directly or combined with a matrix. The use of a matrix appears to enhance the reproducibility and sensitivity of the bone-inductive response. A system using demineralized, extracted rat bone matrix as a carrier for BMP materials has been described (Sampath and Reddi, 1981). In this assay system, the carrier matrix is derived from the diaphyseal portion of rat long bones that have been ground into particles of a specific size and subsequently demineralized (Sampath and Reddi, 1984). The endogenous BMP activity present is removed by repeated extractions with guanidine and discarded. The remaining carrier consists primarily of bone collagen, with minor components of bone proteoglycans and other noncollagenous proteins with no osteoinductive activity. The BMP material to be assayed is then deposited onto this matrix by precipitation with ethanol, by dialysis against water, or by lyophilization. This BMP–matrix combination is then implanted subcutaneously into a rat for a period of days, and then the implants are removed for histological examination focused on the presence of newly formed cartilage and/or bone. Using this system, it has been demonstrated that there is some species specificity for the matrix component (i.e., the use of xenogeneic matrix materials decreases or eliminates the bone-inductive activity of bone extracts), presumably because of interference by the elicitation of an immune response (Sampath and Reddi, 1983). On the other hand, bone-inductive activity is not species-dependent; bone extracts from a variety of species do contain inductive activity when assayed with rat matrix in this system.

Temporal histological analysis of bone induction in this assay system suggests that bone extracts and other BMP materials result in bone formation through an endochondral series of events (Reddi and Huggins, 1972, 1975; Reddi, 1981). Morphologically undifferentiated cells appear to migrate to the area of implantation where they subsequently proliferate. These cells then begin to differentiate into chondrocytes, which proceed to mature and calcify. A vascular network at this point can be observed to invade the implant site, and bone begins to be deposited as the calcified cartilage and rat bone matrix is removed. Finally, bone marrow elements are observed at the site, and the end result is a piece of living bone tissue. Various biochemical markers of the different stages of this process have also been detailed. While surrogate end points for bone formation such as alkaline phosphatase activity or calcium content may be used to compare the potencies of BMP preparations, the complexity of this bone-inductive process advocates histological analysis as the most meaningful parameter for study.

B. Ectopic Bone Induction

Most of the biochemical analysis of bone-inductive materials has come from the purification from extracts of bone matrix. Early purification work on extracts of bovine and human bone resulted in impure preparations of materials and the ascription of bone-inductive activity to various proteins with biochemical characteristics now known to be inconsistent with those of the true inductive molecules (Urist *et al.*, 1983, 1984, 1987a; Sampath *et al.*, 1987). The first substantial purification of this activity from bovine bone (Wang *et al.*, 1988) used a process that has resulted in the identification and cloning of BMP-1 through BMP-8. This purification scheme produced a mixture of the BMP molecules after a 300,000-fold purification with respect to the initial guanidine extract of bone. In addition, this work incorporated biochemical characterization of the BMP activity, correlating the presence of basic proteins of approximately 30 kDa with the bone-inductive activity. Yields from this purification were difficult to calculate due to the inability to assay accurately crude bovine bone extracts in the rat ectopic assay system, but information from this and subsequent purifications suggests that the amount of BMP proteins present in bone matrix is approximately 1 μg/kg mineralized bone.

Subsequent purifications of bone-inductive activity have substantiated the claim that BMP family members are responsible for this activity. One purification of similar magnitude from bovine bone suggests that bone-inductive activity is accounted for by a mixture of BMP-2 and BMP-7 (or OP-1) polypeptides (Sampath *et al.*, 1990). The majority of material in this preparation was found to be homodimeric, suggesting that it was a mixture of the BMP-2 and BMP-7 homodimers, although the presence of small amounts of heterodimer could not be excluded. Another purification from bovine bone indicated the presence of the single molecule BMP-3, also called osteogenin (Luyten *et al.*, 1989). Additional purifications of osteogenic activity from bone using a different carrier matrix (bovine dermal collagen and ceramic) yielded material whose major component was reported to be a novel protein called osteoinductive factor (OIF); this material has subsequently been found to contain BMP-2 and BMP-3 along with OIF (Bentz *et al.*, 1989, 1990, 1991). In this system, TGF-β2 enhances the osteoinductive activity of this mixture of proteins. OIF, now renamed osteoglycin, is itself an interesting molecule, being a proteoglycan with a leucine-rich repeat contained in the other small proteoglycans, decorin, biglycan, and fibromodulin (Madisen *et al.*, 1990). It shows substantial similarity to a small proteoglycan specifically expressed in epiphyseal calcifying cartilage (Shinomura and Kimata, 1992). Finally, purification of bone-inductive activity from a murine osteosarcoma has revealed the presence of BMP-4 (Takaoka *et al.*, 1992; Matsui *et al.*, 1992). Taken together, these results indicate that the inductive components from all of these purifications consist of members of the BMP family. It should be noted that lack of detection of a specific BMP protein does not prove that it is not present; BMPs other than those originally reported could exist in some of these preparations.

The availability of recombinant BMPs has allowed proof of the osteoinductive capacity of individual BMP proteins as well as detailed characterization of their *in vivo* activities. Recombinant human BMP-2 (rhBMP-2) was the first molecule studied in detail (Wang *et al.*, 1990). Recombinant human BMP-2 was produced using a mammalian cell expression system (Chinese hamster ovary [CHO] cells). Purification and biochemical characterization of the molecules from cells that overexpressed the rhBMP-2 protein indicated that this material was glycosylated, dimerized, and processed, similar to the BMP molecules found in bone (Israel *et al.*, 1992). Implantation of rhBMP-2 in the rat ectopic assay system revealed that this molecule was osteoinductive. Histological examination of the results from implanting different doses for varying amounts of time indicated that the bone formation process seen with rhBMP-2 was indistinguishable from that of bone-derived extracts. Thus, implantation of rhBMP-2 resulted in cartilage formation at early times; the cartilage matured and was removed while being replaced by bone tissue that ultimately included bone marrow elements. Therefore, a single BMP molecule by itself is capable of initiating the bone formation process *in vivo*. This study also revealed that the amount of BMP implanted affects the rate at which bone formation occurs, such that increasing amounts of rhBMP-2 resulted in earlier bone formation. Bone formation could be seen to occur as early as 5 days postimplantation when relatively large amounts of rhBMP-2 are used; in this case, bone formation appeared to occur concurrently with cartilage formation, suggesting rhBMP-2 may be able to influence both the endochondral bone induction pathway and direct bone formation. Nevertheless, in all cases rhBMP-2-induced bone formation appeared to be associated with cartilage formation, though the cartilage formed was always replaced at later times with bone. Additionally, the bone formed in response to rhBMP-2 was always spatially limited to the matrix material with which the protein was implanted: Increasing the amount of rhBMP-2 implanted beyond a certain point did not increase the volume of bone formed. The amounts of rhBMP-2 necessary for a particular amount of bone formation were noted to be approximately 10-fold greater than that of bone-derived BMPs, possibly due to the presence of multiple different BMP polypeptides in the latter material.

Subsequent studies have indicated that other recombinant BMPs are also chondrogenic and osteogenic when implanted as single molecules in this assay system. BMP-4 has been synthesized in CHO cells (LaPan *et al.*, 1991) as well as in another mammalian cell expression system, 293 cells, a human embryonic kidney cell line (Hammonds *et al.*, 1991). In the latter case, use of the BMP-2 propeptide region rather than the BMP-4 propeptide region resulted in increased levels of BMP-4 expression. Recombinant human BMP-4 appears to have very similar activity to BMP-2 *in vivo,* inducing cartilage and bone in the rat ectopic assay system. BMP-6 (E. Wang, personal communication) and BMP-7 (Sampath *et al.*, 1991, 1992) have also been expressed in CHO cells and, like BMP-2 and BMP-4, are chondro- and osteoinductive. It is difficult to compare directly the relative activities of all of these molecules, because different researchers have used slightly different matrix carrier systems and used different scoring systems to report the activity. However, it is known that rhBMP-5, while it is osteoinductive, clearly has a lower specific activity in the rat ectopic assay system than does rhBMP-2 (D'Alessandro *et al.*, 1991; Cox *et al.*, 1991). In summary, multiple individual BMPs unequivocally induce both cartilage and bone formation in this assay system, the end point of implantation of any BMP examined to date is bone, and different BMPs may have different quantitative activities.

C. Matrix Interactions

For the local induction of bone by BMPs, the inductive protein is immobilized at the site by combination with a carrier. In the preceding assay system, guanidine-extracted allogeneic bone matrix is used as the carrier system. While it is not known with what components of bone matrix the BMPs interact, it is known that BMP-3 (osteogenin) interacts with type IV and I collagen and that BMP-2 can be sequestered into the extracellular matrix (Paralkar *et al.*, 1990; Israel *et al.*, 1992). However, bone matrix does not provide an additional component necessary for bone induction, as a variety of other carrier systems have been used. For example, when partially purified bone extracts have been combined with bovine dermal collagen, fibrin, hydroxyapatite, tricalcium phosphate, calcium sulfate, or synthetic matrices such as poly(lactic acid), poly(glycolic acid), or copolymers, they retain their ability to induce bone formation (Kawamura *et al.*, 1987; Yamazaki *et al.*, 1988; Takaoka *et al.*, 1988, 1991; Kawamura and Urist, 1988; Desilets *et al.*, 1990; Ripamonti, 1991; Sato *et al.*, 1991). In fact, rhBMP-2 alone when implanted at high doses can still induce bone (Wang *et al.*, 1990; Wozney *et al.*, 1990). However, the presence of a matrix material both influences the volume of bone formed and decreases the amount of rhBMP-2 needed to induce bone formation. Presumably the role of the matrix in this system is to immobilize the inductive molecule at the implantation site for a period of time sufficient to influence the responding cells; whether the matrix results in slow release of the BMP such that a chemotactic gradient is created, or the cells respond to the molecule as it is bound to the matrix, remains unclear.

D. Animal Models of Bone Replacement

1. Bone-Derived Bone Morphogenetic Proteins

Partially purified bone extracts have been used in a variety of animal model systems to test the ability of osteoinductive proteins to heal bony defects. Trephine defects in the skulls of rats (Doll *et al.*, 1990; Mark *et al.*, 1990), rabbits (Moore *et al.*, 1990), dogs (Sato and Urist, 1985; Urist *et al.*, 1987b; Nilsson and Urist, 1991), sheep (Lindholm *et al.*, 1988), and monkeys (Ferguson *et al.*, 1987) have been successfully healed using partially purified allogeneic or xenogeneic (usually bovine) purified bone extracts. A variety of carrier matrices for the inductive extracts have been used in these studies including extracted allogeneic bone matrix, tricalcium phosphate, polylactic–polyglycolic acid copolymers, collagen, and hydroxyapatite. These studies indicate that bone extracts containing BMPs can successfully induce bone that will heal bony defects in animals ranging from the rat to nonhuman primates. Similar results have been obtained in animal models employing long bone defects in the rabbit ulna and femur (Bolander and Balian, 1986; Sato *et al.*, 1991) and in the canine ulna (Nilsson *et al.*, 1986; Johnson *et al.*, 1989). Canine models of long bone nonunions (Heckman *et al.*, 1991) and spinal fusions (Lovell *et al.*, 1989) have been successfully treated with purified bone extracts. Extracts of bone can also be used to create bony structures of a specific shape by treatment of a muscle flap inside a silastic mold (Khouri *et al.*, 1991). BMP-containing extracts have also been used to induce the formation of osteodentine in amputated dental pulp (Nakashima, 1990), suggesting a role for the BMPs in the healing of mineralized tissues other than bone. All of these studies indicate that BMP molecules will be useful in the treatment of numerous clinical indications where the induction of bone is desired (see Einhorn, 1992).

2. Recombinant Bone Morphogenetic Proteins

Several studies using recombinant BMP molecules to treat bony defects in animal models have been reported. Recombinant human BMP-2 has been used in three models of bone formation in different species (see Schaub and Wozney, 1991). In each model, a critical-sized bone defect was created that, if left untreated, would not be able to heal by osteoconduction, thereby providing a stringent test for the osteoinductive capacity of rhBMP-2. In one study, rhBMP-2 combined with extracted inactive rat bone matrix was shown to heal a 0.5-cm segmental defect in rat femurs (Yasko *et al.*, 1992b). Evaluated radiographically, the number of animals demonstrating union across the defect site increased with time from 3 to 9 weeks. Biomechanical testing indicated integration of the newly formed bone with the existing bone, resulting in stable union across the defect. These studies have been extended using several synthetic and collagenous matrices indicating comparable bone induction without the use of the bone matrix carrier (Yasko *et al.*, 1991, 1992a).

Recombinant human BMP-2 has also demonstrated efficacy in a similar model in the sheep (Gerhart *et al.*, 1992), showing the ability to heal bony defects in a higher animal whose bone structure and size more closely approximate those of humans. In this model, 2.5-cm femoral defects were implanted with rhBMP-2 and sheep bone matrix carrier and blood. Healing of the defect observed radiographically at 12 weeks was comparable to that observed with autogenous bone graft in control animals. Though the sample numbers were not large, biomechanical testing suggested the rhBMP-2-treated femurs had strengths comparable to the contralateral (unoperated) femurs after the 12-week treatment period. Several of the rhBMP-2-treated animals were maintained in pasture for 1 year prior to sacrifice. Analysis of the treated femurs at this time revealed that the new bone had begun normal remodeling under load-bearing consistent with the model that BMP-2 triggers the normal process of bone formation. A third study has demonstrated the efficacy of rhBMP-2 in the healing of large canine mandibular defects as assessed by histologic, radiographic, and functional criteria, suggesting potential utility in cranio-/maxillofacial and dental indications as well (Toriumi *et al.*, 1991). BMP-7 (OP-1) has also been used to heal large segmental defects (Cook *et al.*, 1991, 1992). In this study, 1.5-cm defects in rabbit ulnae were treated with BMP-7 and rabbit-extracted bone matrix. Similar to the results with rhBMP-2, radiographic union beginning at 3 weeks postsurgery was observed in these animals.

E. Clinical Use

The clinical utility of osteoinductive compositions has been recognized for many years. Demineralized bone matrix containing BMP activity has been used to treat humans in a variety of settings including both long bone and craniofacial defects. Several studies using partially purified extracts of human bone in the treatment of bony defects have been reported (Johnson *et al.*, 1988a, b, 1990). Human bone extracts combined with various matrices including bone noncollagenous proteins, polylactic acid and polyglycolic acid polymers, and autologous bone graft have been successfully implanted to treat nonunions or segmental defects in human femurs and tibiae. Unfortunately, the inability to perform controlled studies in these cases prevents clear evaluation of the efficacy of the treatments. The promising studies with the recombinant BMPs in multiple animal models of bone replacement suggest that the availability of pathogen-free osteoinductive proteins combined with resorbable matrices for clinical use is imminent.

IV. MECHANISMS OF ACTION

A. Activities on Bone and Cartilage Cells *in Vitro*

BMP-induced cartilage and bone formation *in vivo* is clearly a complex multistage process and undoubtedly involves the activities of multiple locally produced growth factors and systemically available hormones. Investigation of the individual steps in this process by examining the effects of the BMPs on various cell types *in vitro* has allowed us to begin to elucidate the activities of these factors at the cellular level. Studies on the roles of individual BMP molecules have been complicated by two facts: First, few direct comparisons of different BMPs have been made in the same cell systems, and, second, some of these studies have been performed with purified bone-derived BMPs, which are likely to contain several different BMP molecules. However, some general trends are emerging.

1. Control of Cell Proliferation

Multiple cell types respond to the BMPs by increasing or decreasing their proliferation as monitored by ^{3}H-thymidine incorporation into DNA or cell number (Table III). Osteoblastic or osteoprogenitor cells, in general, respond to treatment with the BMPs by increases in cell proliferation. BMP-4 and BMP-7 have been shown to be mitogenic when tested on primary fetal rat calvarial cells (Chen *et al.*, 1991b; Maliakal *et al.*, 1991); BMP-2 treatment increased proliferation of the C26 rat calvarial osteoprogenitor cells (Yamaguchi *et al.*, 1991). Primary calvarial-derived periosteal cells initially respond to osteogenin with a decrease in proliferative rate but ultimately by an increase in cell number (Vukicevic *et al.*, 1989). BMP-7 has been shown to be mitogenic for the human osteosarcoma cell line TE85 (Knutsen *et al.*, 1991) while BMP-2 appears to have no effect on these cells (Wozney *et al.*, 1992). In contrast to the proliferative effects of the BMPs, MC3T3-E1 cells, a cell line believed to be representative of an early differentiated osteoblast, are strongly inhibited in their growth by BMP-4 and osteogenin treatment (Vukicevic *et al.*, 1990a; Paralkar *et al.*, 1991). Mature chondrocytes show a small increase in their growth when treated with osteogenin, whereas no effect has been observed on chondroprogenitor cells. These results taken together indicate that multiple cell types in the cartilage and bone lineages are able to respond to the BMPs; thus, these factors may be involved in many of the regulatory steps of cartilage and bone growth and maintenance.

2. Control of Cell Differentiation

BMP treatment results in the differentiation of mesenchymal cells into several phenotypes and also increases the expression of markers associated with these phenotypes in committed cells. For example, C26 cells, a rat calvarial-derived multipotential cell line is known to differentiate into osteoblasts, muscle cells, and adipocytes (Yamaguchi and Kahn, 1991). When cultured with BMP-2, C26 cells respond by differentiating into cells of the osteoblast phenotype (Yamaguchi *et al.*, 1991). BMP-2 not only increases expression of alkaline phosphatase and parathyroid hormone (PTH) receptors in these cells but also induces expression of osteocalcin (bone gla protein), a specific marker for differentiated osteoblasts. BMPs are the only known protein factors that are capable of inducing expression of this osteoblast marker. The level of expression of BMP-2-induced osteocalcin is greatly enhanced by the addition of 1,25-dihydroxyvitamin D_3 (1,25$(OH)_2D_3$). BMP-2 also decreases expression of the myogenic phenotype in C26 cells, as shown by the inhibition of the production of desmin-positive myotubes in these cell cultures. Similarly, osteoprogenitor cells from the bone marrow stroma will differentiate into osteoblastlike cells after treatment with BMP-2. The mouse stromal cell line W-20-17, which can differentiate into both osteoblasts and adipocytes, increases expression of alkaline phosphatase and PTH receptors and also is induced to express osteocalcin when treated with BMP-2 (Thies *et al.*, 1992a). Interestingly, 1,25$(OH)_2D_3$ decreases the expression of osteocalcin in these cells (Thies *et al.*, 1992b). BMP-2, BMP-4, BMP-5, BMP-6, and BMP-7 all induce some alkaline phosphatase expression in W-20-17 cells but vary in specific activity. For example, BMP-5 exhibits much lower specific activity than BMP-2 in this assay system, while BMP-4 is significantly more active. The quantitative difference in activity between BMP-2 and BMP-4 can be obviated by removal of their amino-terminal domains with trypsin (the regions that are most different between these closely related molecules), suggesting a role for this domain in receptor binding or ligand presentation (LaPan *et al.*, 1991). Primary fetal rat calvarial cells, probably a mixture of osteoprogenitor and committed osteoblasts, also increase expression of markers of the osteoblast

Table III *In Vitro* Activities of the Bone Morphogenetic Proteins

Cell type	BMP	Effect on proliferation	Effect on phenotype [marker(s)]	References
C26 (multipotent)	BMP-2	↑	↑ OB (ALP, PTH, BGP) ↓ MU (desmin (+) myotubes)	Yamaguchi *et al.*, 1991
C20 (osteoblast)	BMP-2	↓ (slight)	↑ OB (PTH only)	Yamaguchi *et al.*, 1991
MC3T3-E1 (osteoblast)	BMP-2	None[b]	↑ OB (ALP)	Takuwa *et al.*, 1991
	BMP-4	↓		Paralkar *et al.*, 1991
	BMPs[b]	↓	↑ OB (ALP, PTH, type I)	Vukicevic *et al.*, 1990a; Hiraki *et al.*, 1991
W-20-17 (multipotent)	BMP-2	None	↑ OB (ALP, PTH, BGP)	Thies *et al.*, 1992a
	BMP-4	None	↑ OB (ALP)	Thies *et al.*, 1992a; LaPan *et al.*, 1991
1° calvarial (mixed osteoblast/osteoprogenitor)	BMPs[b]	↑	↑ OB (ALP, PTH, type I)	Vukicevic *et al.*, 1989
	BMP-4	↑	↑ OB (ALP, type I)	T. L. Chen *et al.*, 1991b
	BMP-7	↑	↑ OB (ALP, PTH, BGP, type I)	Maliakal *et al.*, 1991
Periosteal (osteoprogenitor)	BMPs[b]	↓ Then ↑	↑ OB (ALP, PTH, type I)	Vukicevic *et al.*, 1989
CFK1 (osteoprogenitor)	BMP-2	↑	↑ OB (ALP)	Bernier and Goltzman, 1992
TE85 (osteoblast/osteosarcoma)	BMP-7	↑	↑ OB (ALP)	Knutsen *et al.*, 1991
	BMP-2	None		Wozney *et al.*, 1992
Fetal rat chondroblasts	BMPs[b]	↑	↑ CB ($^{35}SO_4$)	Vukicevic *et al.*, 1989

Rabbit articular chondrocytes	BMPs[c]	↑	↑ CB ($^{35}SO_4$, type II,[c] PG[c])	Vukicevic *et al.*, 1989; Harrison *et al.*, 1991[d]; Hiraki *et al.*, 1991
Bovine articular cartilage	BMPs[c], BMP-4		↑ CB (PG)	Luyten *et al.*, 1992
Chink limb bud (multipotent)	BMPs[c]	None	↑ CB ($^{35}SO_4$, type II)	Carrington *et al.*, 1991
	BMP-2			Roark and Kosher, 1991
	BMP-4		↑ CB ($^{35}SO_4$, alcian, ALP)	Chen *et al.*, 1991a
Mouse limb bud (multipotent)	BMP-2		↑ CB ($^{35}SO_4$, alcian, ALP)	
			↑ OB (ALP, BGP)	Rosen *et al.*, 1991
C3H10T1/2 (multipotent)	BMP-2		↑ CB, ↑ OB, ↑ AP	Katagiri *et al.*, 1990b; E. A. Wang *et al.*, 1992; Yamaguchi *et al.*, 1992

[a]Alcian, alcian blue staining focil/nodules; ALP, alkaline phosphatase; AP, adipocyte; BGP, BGP or osteocalcin; CB, chondroblast; MU, muscle; OB, osteoblast; PG, cartilage proteglycan; PTH, cyclic adenosine monophosphate response to parathyroid hormone; $^{35}SO_4$, sulfate incorporation into extracellular matrix; type I(II), type I(II) collagen.

[b]Probably due to lower specific activity.

[c]"Osteogenin"; bone-derived BMPs.

[d]Colony formation in agrarose.

phenotype when treated with BMP-4, BMP-7, or osteogenin, as do primary periosteal fibroblasts (Vukicevic *et al.*, 1989; Chen *et al.*, 1991b; Maliakal *et al.*, 1991). BMPs can also increase expression of osteoblast markers in differentiated osteoblastic cells such as MC3T3-E1 and C20 cells (Vukicevic *et al.*, 1990a; Takuwa *et al.*, 1991; Yamaguchi *et al.*, 1991). The magnitude of the observed increase apparently depends on the constitutive levels of these proteins that the cells already express.

BMPs also have effects on chondroblasts, chondrocytes, and chondroprogenitor cells. Osteogenin has been shown to increase sulfate incorporation into the large aggregating proteoglycans (specific for cartilage) in fetal rat chondroblasts and rabbit articular chondrocytes (Vukicevic *et al.*, 1989; Harrison *et al.*, 1991; Hiraki *et al.*, 1991), whereas BMP-4 and osteogenin have similar effects on bovine articular cartilage *in vitro* (Luyten *et al.*, 1992). In the latter case, BMPs are as effective in maintaining the differentiated state of cartilage tissue explants as are serum, insulinlike growth factor I, or TGF-β. BMP-2, BMP-4, and osteogenin have all been demonstrated to induce expression of various markers of differentiated chondrocytes in embryonic chick limb bud cells (Carrington *et al.*, 1991; Roark and Kosher, 1991; P. Chen *et al.*, 1991a), whereas mouse limb bud cells have been shown to increase expression of both chondroblast and osteoblast phenotype markers when treated with BMPs (Rosen *et al.*, 1991). The ability of the BMPs to enhance both osteoblast and chondroblast phenotypes is consistent with a role for the BMPs in development of the embryonic skeleton (see later). In addition, the embryonic mesenchymal cell line C3H10T1/2, which has been shown to be capable of differentiating into adipocytes, muscle cells, and chondroblasts on the addition of azacytidine, differentiates into osteoblastic, chondroblastic, and adipocytic cells upon treatment with BMP-2 (Katagiri *et al.*, 1990b; Wang *et al.*, 1992; Yamaguchi *et al.*, 1992), again suggesting an important role for the BMPs during embryonic development.

Although the BMPs are members of the TGF-β superfamily, the BMPs and TGF-βs have few common effects on cells *in vitro*. For example, TGF-β inhibits production of alkaline phosphatase by a number of cell types including C26, W-20-17, and MC3T3-E1 cells (Noda and Rodan, 1986; Katagiri *et al.*, 1990a), while the BMPs increase its expression in each cell system. Also in contrast to the BMPs, TGF-β suppresses sulfate incorporation into proteoglycan in limb bud cultures and cultured chondrocytes. Some of the TGF-β effects such as inhibition of sulfate incorporation in chick limb bud cell cultures and inhibition of alkaline phosphatase in MC3T3-E1 cells can be reversed by the BMPs, whereas the converse can also be true: TGF-β inhibits BMP-2-induced alkaline phosphatase in W-20-17 cells, and both TGF-β and inhibin antagonize BMP-4-stimulated alkaline phosphatase in mk90 cells (Dudley *et al.*, 1991). On the other hand, a few effects are common to the TGF-βs and BMPs. For example, TGF-β is also a mitogen for primary calvarial cells, although this effect is biphasic (Centrella *et al.*, 1986). In addition, both TGF-β and BMPs inhibit proliferation of MC3T3-E1 cells (Katagiri *et al.*, 1990a).

The BMPs also have chemotactic effects on mesenchymal cells. Purified bone-derived BMP has been shown to be a chemoattractant for both the osteoblastlike ROS 17/2 cells (Padley *et al.*, 1991) and the primary muscle-derived cells from neonatal rats (Landesman and Reddi, 1986). In addition, rhBMP-2 is chemotactic for the osteoblastlike MC3T3-E1 cells, whereas it has no chemotactic effect on Swiss 3T3 cells (Hughes, 1992). Similar to the chemotactic effects of other growth factors, the optimal concentration needed (0.25 ng/ml) is significantly lower than that needed to induce phenotypic changes in these cells.

The BMPs *in vitro* therefore have significant effects on cells during multiple stages in the endochondral bone formation process observed upon implantation of a BMP *in vivo*. Through chemotaxis, BMPs may bring cells to the implantation site; BMPs may induce mesenchymal progenitors to differentiate into cartilage-forming and bone-forming cells; and BMPs may affect proliferation of these cells at various stages of the development process. The most profound effects observed for the BMPs, and those that are most specific to the BMPs, are the differentiation of mesenchymal progenitor cells into cell types including chondroblasts and osteoblasts. While it is unlikely that the implanted BMP remains at the site long enough to direct all of these processes *in vivo*, the expression of various BMP molecules by a variety of cell types including osteoblasts (see Section II.E) suggests that a cascade of the expression of multiple BMPs may regulate, in part, the bone formation process initiated by a single implanted BMP.

B. Activities on Other Cell Types *in Vitro*

The BMPs also have effects on cells outside the cartilage and bone lineages. For example, BMPs appear to be involved in the differentiation of F9 embryonal carcinoma cells into endodermal cells (Rogers *et al.*, 1992). During the differentiation of F9 cells into parietal endoderm induced by treatment with retinoic acid and cyclic adenosine monophosphate, BMP-2 mRNA expression is increased more than 10-fold, while BMP-4 mRNA expression is decreased more than 10-fold. It is interesting to note that in this system these closely related BMPs (BMP-2 and BMP-4) are independently regulated, as they appear to be in MG-63 cells (Harris *et al.*, 1991a). In addition, BMP-2 alters the growth, morphology, and gene expression of the retinoic acid-treated F9 cells. For example, BMP-2 treatment increases retinoic acid receptor-β and decreases cellular retinoic acid binding protein I mRNA expression in these cells, suggesting that BMP-2 may be involved in regulating the responses of the cells to retinoic acid. BMP-2 has also been shown to modulate the TGF-β effects on astrocytes (Schluesener, 1991; Schluesener and Meyermann, 1991). Whereas multidrug transport in rat astrocyte cell lines is decreased by TGF-β1, TGF-β1.2, and TGF-β2 treatment, this effect is antagonized by the addition of BMP-2. Thus, it is clear that the BMPs can affect cells either directly or by modulating the activities of other growth factors (e.g., TGF-β) or regulatory agents (e.g., retinoic acid) on these cells.

C. Receptors

A single report describes the presence of high-affinity binding sites for BMP-4 on MC3T3-E1 and NIH3T3 cells (Paralkar *et al.*, 1991). By cross-linking experiments, binding proteins of 200 and 70 kDa are present on MC3T3-E1 cells, whereas proteins of 200 and 90 kDa are present on NIH3T3 cells. It is perhaps surprising that there are receptors of different sizes for the same protein on two cell types of the same species. No competition by TGF-β was found for the BMP-4 binding proteins, nor was competition by BMP-4 for the TGF-β receptors observed (Attisano *et al.*, 1992), consistent with most of the activity data on cells *in vitro*. While BMP-3–osteogenin binding sites have been localized to developing cartilage and bone (Vukicevic *et al.*, 1990b), this may represent binding to extracellular matrix proteins. The cloning of the TGF-β type II receptor and several activin receptors indicates that they are related members of a family of serine-threonine kinase receptor molecules (Mathews and Vale, 1991; Wang *et al.*, 1991; Attisano *et al.*, 1992). The BMP receptors are therefore predicted to also be members of this family. The detailed characterization of BMP receptors, and the determination of how many different receptors exist for the BMP family of proteins, will await their cloning.

D. Role in Embryonic Development

Several lines of evidence suggest that the BMPs are involved in multiple developmental processes during embryogenesis. Homologs of many of the BMPs have been identified in a wide variety of species, even in species such as the fruit fly, which do not contain cartilage or bone tissue, and these proteins appear to be expressed in a developmentally regulated manner. For example, mRNAs for several of the BMPs, including BMP-4 and BMP-7, have been found to be present in the early *Xenopus* embryo. At this stage of *Xenopus* development, activin A, another TGF-β superfamily member, is a potent mesoderm inducer in animal cap explants (Smith *et al.*, 1990; Ueno *et al.*, 1990). Although initial experiments indicated that supernatants derived from COS cells expressing BMP-4 contained mesoderm-inducing activity in the animal cap assay (Köster *et al.*, 1991), it is now clear that this activity is not due to BMP-4, as purified recombinant BMP-4 does not reproducibly induce mesodermal structures. In fact, recent evidence indicates that BMP-4 antagonizes the effects of activin in this system and may act after the initial mesoderm induction event to play a role in the specification of ventral mesoderm (Dale *et al.*, 1992; Jones *et al.*, 1992).

Two TGF-β superfamily members have been detected in *Drosophila*. One of these, *dpp*, is most closely related to BMP-2 and BMP-4; the other, *60A*, is most closely related to BMP-5, BMP-6, BMP-7, and BMP-8. The powerful genetic systems available in *Drosophila* have indicated that *dpp* is required early in embryogenesis for dorsoventral patterning, being necessary for the specification of dorsal ectoderm. Later in development it is required for the proximal–distal organization of the imaginal disks (e.g., for correct distal outgrowth of the appendages). Its expression pattern indicates that it is expressed at the anterior–posterior boundary within the disks (Masucci *et al.*, 1990). *Decapentaplegic* is also necessary for differentiation of the midgut (Panganiban *et al.*, 1990; Immerglück *et al.*, 1990). In this system, experimental evidence suggests that *dpp* serves as a signaling molecule between the mesoderm and endoderm and is both controlled by and controls expression of homeobox genes. As noted in Section II.B, *dpp* is likely to be the *Drosophila* homolog of BMP-2 and BMP-4, based on the high degree of amino acid homology in the mature regions of the molecules and on their structures, in which a single intron with the coding region resides in an analogous position in all three genes. The *dpp* gene is subject to complex regulation, with both multiple alternatively spliced 5′ exons and a large complex 3′ regulatory region (St. Johnston *et al.*, 1990; Blackman *et al.*, 1991). The suggestion has been made that this complex regulation allows a single gene to function in a manner analogous to both the BMP-2 and BMP-4 genes in mammalian species. The supposition that BMP-2, BMP-4, and *dpp* serve analogous functions is strengthened by the demonstration that the ligand region of BMP-4 can replace that of *dpp* and fully rescue the *dpp* mutant phenotype (Padgett *et al.*, 1993).

Less is known about the function of the *60A* gene (Wharton *et al.*, 1991; Doctor *et al.*, 1992), which was detected by homology to the BMP family rather than by genetic analysis. However, it is known to be expressed throughout development, with specific temporal peaks of expression. *60A* expression is first detected at the beginning of gastrulation, mostly in the mesoderm of the extending germ band. Expression later in development is primarily in the foregut and hindgut. Because the only two *Drosophila* TGF-β family members that exist are probably descended from the same ancestral genes as the BMP-2/BMP-4 and

BMP-5/BMP-6/BMP-7/BMP-8 subgroups, and because these *Drosophila* genes are known to be required for or implicated in development, the BMPs are strong candidates for developmentally important genes in mammalian systems.

It is interesting to note that *Drosophila tolloid,* which is similar in sequence and domain structure to BMP-1, exhibits a similar mutant phenotype to that of *dpp.* It, like *dpp,* is required for formation of dorsal structures in the developing embryo. Further evidence that these two diverse proteins genetically interact comes from the observation that extra copies of the *dpp* gene can compensate for *tolloid* mutations. Recent evidence suggests that *tolloid* generates a gradient of *dpp* activity (Ferguson and Anderson, 1992) and, thus, in some way may activate it. The suggestion has been made that this may be through cleavage of a precursor or inactive form by the *tolloid* protease activity (Shimell *et al.*, 1991). By analogy, it is possible that BMP-1 similarly increases the activities of the other BMPs in some manner; however, this has not been demonstrated.

Much of our knowledge of the possible roles of BMPs in mammalian development comes from localization of BMP mRNAs in the developing mouse embryo by *in situ* hybridization (Rosen *et al.*, 1989; Lyons *et al.*, 1989b, 1990; Pelton *et al.*, 1990). Specific localization patterns in the limb bud suggest roles for the BMPs in the developing skeletal system as well as their known roles in the induction of cartilage and bone in the adult animal. As early as 9.5 day post coitum (p.c.), BMP-2 and BMP-4 mRNA expression can be detected in the thickened epithelium of the ventral surface of the limb bud, and at 10.5 days p.c. they are expressed specifically in the apical ectodermal ridge. Since the apical ectodermal ridge is known to be necessary for outgrowth of the limb bud through creation of the progress zone, these BMPs may be involved in influencing the proliferation or differentiation state of the mesodermal cells in the underlying progress zone. A low level of BMP-4 mRNA is also found in the mesenchyme of the limb bud, distributed in an anterior to posterior and distal to proximal gradient. By 12.5 days p.c., neither BMP-2 nor BMP-4 transcripts can be found in the epithelium; rather, they can be found in the interdigital mesenchyme surrounding the areas of mesenchymal condensations that will develop into the cartilaginous structures of the limbs. In contrast, BMP-6 expression is not found in the limb until later in development, where it is localized to the mature hypertrophic cartilage. Together, these results suggest specific roles for each of the BMPs in limb development.

BMP mRNAs can also be detected in a variety of locations outside of the developing skeletal system. BMP-2 and BMP-4 can be found in many tissues where mesenchymal–epithelial inductive interactions occur. These include the heart, tooth buds, craniofacial processes, and developing whisker follicles. Specific sites of BMP-6 expression can also be found in the embryonic heart and skin (but not in the whisker follicles) and also in the central nervous system and brain as they develop. BMP-7 protein expression has also been localized to multiple tissue types during development (Vukicevic *et al.*, 1992). These specific localizations of the different BMP mRNAs, coupled with their expression by a variety of cell types and tissues in the adult organism (see Section II.E), suggest that the BMPs are involved in a variety of developmental processes in addition to the induction of cartilage and bone in the embryo and in the adult (Rosen and Thies, 1992). Much work is still needed to elucidate the mechanisms by which the different BMPs act in controlling or inducing these processes.

ACKNOWLEDGMENTS

Many thanks to Vicki Rosen, Steve Clark, and Carolyn Pratt for helpful comments on the text, to Sue Avery for her help with the literature search, and to the many people on the BMP team at Genetics Institute who have made research on this family of proteins possible.

REFERENCES

Anderson, H. C. (1976). Osteogenetic epithelial–mesenchymal cell interactions. *Clin. Orthop. Relat. Res.* **119,** 211–224.

Anderson, H. C., and Coulter, P. R. (1968). Bone-inducing capability of cultured human cells (FL and HeLa) compared to that of various types of injury. *Fed. Proc.* **27,** 475.

Anderson, H. C., Merker, P. C., and Fogh, J. (1964). Formation of tumors containing bone after intramuscular injection of transformed human amnion cells (FL) into cortisone-treated mice. *Am. J. Pathol.* **44,** 507–519.

Attisano, L., Wrana, J. L., Cheifetz, S., and Massagué, J. (1992). Novel activin receptors: Distinct genes and alternative mRNA splicing generate a repertoire of serine-threonine kinase receptors. *Cell* **68,** 97–108.

Bauer, F. C. H., and Urist, M. R. (1981). Human osteosarcoma-derived soluble bone morphogenetic protein. *Clin. Orthop. Relat. Res.* **154,** 291–295.

Bentley, H., Hart, K. A., Seid, J. M., Black, D., Hamdy, F., and Russell, R. G. G. (1992). Expression of the bone morphogenetic proteins (BMPs) in human bone, bone derived cell lines and other tissues. *Calcif. Tissue Int.* **50S,** A3.

Bentz, H., Nathan, R. M., Rosen, D. M., Armstrong, R. M., Thompson, A. Y., Segarini, P. R., Mathews, M. C., Dasch, J. R., Piez, K. A., and Seyedin, S. M. (1989). Purification and characterization of a unique osteoinductive factor from bovine bone. *J. Biol. Chem.* **264,** 20805–20810.

Bentz, H., Chang, R.-J., Thompson, A. Y., Glaser, C. B., and Rosen, D. M. (1990). Amino acid sequence of bovine osteoinductive factor. *J. Biol. Chem.* **265,** 5024–5029.

Bentz, H., Thompson, A. Y., Armstrong, R., Chang, R., Piez, K. A., and Rosen, D. M. (1991). Transforming growth factor-β2 enhances the osteoinductive activity of a bovine bone-derived fraction containing bone morphogenetic protein-2 and 3. *Matrix* **11,** 269–275.

Bernier, S. M., and Goltzman, D. (1992). Effect of protein and steroidal osteotropic agents on differentiation and epidermal growth factor-mediated growth of the CFK1 osseous cell line. *J. Cell. Physiol.* **152,** 317–327.

Blackman, R. K., Sanicola, M., Raftery, L. A., Gillvet, T., and Gelbart, W. M. (1991). An extensive 3′ *cis*-regulatory region directs the imaginal disk expression of *decapentaplegic,* a member of the TGF-β family in *Drosophila. Development* **111,** 657–665.

Bolander, M. E., and Balian, G. (1986). The use of demineralized bone matrix in the repair of segmental defects: Augmentation with extracted matrix proteins and a comparison with autologous grafts. *J. Bone Joint Surg.* **68-A,** 1264–1274.

Bonewald, L. F., Mundy, G. R., Kester, M. B., Harris, M. A., and Harris, S. E. (1991). Transforming growth factor beta (TGFβ) and bone morphogenetic protein (BMP) 4 expression in human MG-63 bone cells is enhanced by TGF-β. *J. Bone Miner. Res.* **6,** S258.

Butler, W. T., Mikulski, A., and Urist, M. R. (1977). Noncollagenous proteins of a rat dentin matrix possessing bone morphogenetic activity. *J. Dent. Res.* **56,** 228–232.

Carrington, J. L., Chen, P., Yanagishita, M., and Reddi, A. H. (1991). Osteogenin (bone morphogenetic protein-3) stimulates cartilage formation by chick limb bud cells *in vitro. Dev. Biol.* **146,** 406–415.

Ceci, J. D., Kingsley, D. M., Silan, C. M., Copeland, N. G., and Jenkins, N. A. (1990). An interspecific backcross linkage map of the proximal half of mouse chromosome 14. *Genomics* **6,** 673–678.

Celeste, A. J., Iannazzi, J. A., Taylor, R. C., Hewick, R. M., Rosen, V., Wang, E. A., and Wozney, J. M. (1990). Identification of transforming growth factor β family members present in bone-inductive protein purified from bovine bone. *Proc. Natl Acad. Sci. USA* **87,** 9843–9847.

Celeste, A. J., Taylor, R., Yamaji, N., Wang, J., Ross, J., and Wozney, J. (1992). Molecular cloning of BMP-8: A protein present in bovine bone which is highly related to the BMP-5/6/7 subfamily of osteoinductive molecules. *J. Cell. Biochem. Suppl.* **16F,** 100.

Centrella, M., Massagué, J., and Canalis, E. (1986). Human platelet-derived transforming growth factor-beta stimulates parameters of bone growth in fetal rat calvariae. *Endocrinology* **119,** 2306–2312.

Cheifetz, S., Bassols, A., Stanley, K., Ohta, M., Greenberger, J., and Massagué, J. (1988). Heterodimeric transforming growth factor β: Biological properties and interaction with three types of cell surface receptors. *J. Biol. Chem.* **263,** 10783–10789.

Chen, P., Carrington, J. L., Hammonds, R. G., and Reddi, A. H. (1991a). Stimulation of chondrogenesis in limb bud mesoderm cells by recombinant human bone morphogenetic protein 2B (BMP-2B) and modulation by transforming growth factor β1 and β2. *Exp. Cell Res.* **195,** 509–515.

Chen, T. L., Bates, R. L., Dudley, A., Hammonds, R. G., Jr., and Amento, E. P. (1991b). Bone morphogenetic protein-2b stimulation of growth and osteogenic phenotypes in rat osteoblast-like cells: Comparison with TGF-β1. *J. Bone Miner. Res.* **6,** 1387–1393.

Cook, S. D., Baffes, G. C., Sampath, T. K., and Rueger, D. C. (1991). Healing of segmental defects using recombinant human osteogenic protein (rhOP-1). *J. Bone Miner. Res.* **6,** S154.

Cook, S. D., Baffes, G. C., Sampath, T. K., and Rueger, D. C. (1992). Healing of large segmental defects using recombinant human osteogenic protein (rhOP-1). *Trans. Orthop. Res. Soc.* **17,** 581.

Cox, K., Holtrop, M., D'Alessandro, J. S., Wang, E. A., Wozney, J. M., and Rosen, V. (1991). Histological and ultrastructural comparison of the *in vivo* activities of rhBMP-2 and rhBMP-5. *J. Bone Miner. Res.* **6,** S155.

Dale, L., Howes, G., Price, B. M. J., and Smith, J. C. (1992). Bone morphogenetic protein 4: A ventralizing factor in early *Xenopus* development. *Development* **115,** 573–585.

D'Alessandro, J. S., Cox, K. A., Israel, D. I., LaPan, P., Moutsatsos, I. K., Nove, J., Rosen, V., Ryan, M. C., Wozney, J. M., and Wang, E. A. (1991). Purification, characterization and activities of recombinant bone morphogenetic protein 5. *J. Bone Miner. Res.* **6,** S153.

de Groot, K., and Deshmukh, A. (1975). The subcutaneous implantation of xenogeneic decalcified teeth. *J. Periodontol.* **46,** 78–81.

Desilets, C. P., Marden, L. J., Patterson, A. L., and Hollinger, J. O. (1990). Development of synthetic bone-repair materials for craniofacial reconstruction. *J. Craniofac. Surg.* **1,** 150–153.

Dickinson, M. E., Kobrin, M. S., Silan, C. M., Kingsley, D. M., Justice, M. J., Miller, D. A., Ceci, J. D., Lock, L. F., Lee, A., Buchberg, A. M., Siracusa, L. D., Lyons, K. M., Derynck, R., Hogan, B. L. M., Copeland, N. G., and Jenkins, N. A. (1990). Chromosomal localization of seven members of the murine TGF-β superfamily suggest close linkage to several morphogenetic mutant loci. *Genomics* **6,** 505–520.

Doctor, J. S., Jackson, P. D., Rashka, K. E., Visalli, M., and Hoffmann, F. M. (1992). Sequence, biochemical characterization, and developmental expression of a new member of the TGF-β superfamily in *Drosophila melanogaster. Dev. Biol.* **151,** 491–505.

Doll, B. A., Towle, H. J., Hollinger, J. O., Reddi, A. H., and Mellonig, J. T. (1990). The osteogenic potential of two composite graft systems using osteogenin. *J. Periodontol.* **61,** 745–750.

Dudley, A. T., Morris, S. G., and Hammonds, R. G. (1991). TGF-β1 and inhibin A antagonize BMP2b-stimulated alkaline phosphatase levels in mk90 cells. *J. Cell Biol.* **115,** 449a.

Dumermuth, E., Sterchi, E. E., Jiang, W., Wolz, R. L., Bond, J. S., Flannery, A. V., and Beynon, R. J. (1991). The Astacin family of metalloendopeptidases. *J. Biol. Chem.* **266,** 21381–21385.

Einhorn, T. A. (1992). Clinical applications of recombinant gene technology: Bone and cartilage repair. *Cells Materials* **2,** 1–11.

Ferguson, D., Davis, W. L., Urist, M. R., Hurt, W. C., and Allen, E. P. (1987). Bovine bone morphogenetic protein (bBMP) fraction-induced repair of craniotomy defects in the Rhesus monkey. *Clin. Orthop. Relat. Res.* **219,** 251–290.

Ferguson, E. L., and Anderson, K. V. (1992). Localized enhancement and repression of the activity of the TGF-β family member, *decapentaplegic,* is necessary for dorsal–ventral pattern formation in the *Drosophila* embryo. *Development* **114,** 583–597.

Gazit, D., Kahn, A., and Derynck, R. (1991). Regulation of expression of various members of the TGF-β superfamily in clonal cell lines with osteogenic potential. *J. Bone Miner. Res.* **6,** S258.

Gerhart, T. N., Kirker-Head, C. A., Kriz, M. J., Holtrop, M. E., Hennig, G. E., and Wang, E. A. (1992). Healing segmental femoral defects in sheep using recombinant human bone morphogenetic protein (rhBMP-2). *Clin. Orthop. Relat. Res.* (in press).

Graveley, R. M., Rahman, S., Bentley, H., Seid, J. M., Wishart, W., Fuchs, S., Nordmann, R., and Russell, R. G. G. (1992). Analysis of cytokine expression in human trabecular (osteoblast-like) bone cells in culture by PCR. *Calcif. Tissue Int.* **50S,** A8.

Hahn, G. V., Cohen, R. B., Wozney, J. M., Levitz, C. L., Zasloff, M. A., and Kaplan, F. S. (1992). A bone morphogenetic protein subfamily: Chromosomal localization of human genes for BMP5, BMP6, and BMP7 *Genomics* **14,** 759–762.

Hall, B. K., and van Exan, R. J. (1982). Induction of bone by epithelial cell products. *J. Embryol. Exp. Morph.* **69,** 37–46.

Hammonds, R. G., Jr., Schwall, R., Dudley, A., Berkemeier, L., Lai, C., Lee, J., Cunningham, N., Reddi, A. H., Wood, W. I., and Mason, A. J. (1991). Bone-inducing activity of mature BMP-2b produced from a hybrid BMP-2a/2b precursor. *Mol. Endocrinol.* **5,** 149–155.

Hanamura, H., Higuchi, Y., Nakagawa, M., Iwata, H., Nogami, H., and Urist, M. R. (1980). Solubilized bone morphogenetic protein (BMP) from mouse osteosarcoma and rat demineralized bone matrix. *Clin. Orthop. Relat. Res.* **148,** 281–290.

Harris, S. E., Bonewald, L. F., Harris, M. A., Sabatini, M., and Mundy, G. R. (1991a). Retinoic acid (RA) regulates expression of early growth response genes, BMP 4, and BMP 2 mRNA in human osteoblastic cells. *J. Bone Miner. Res.* **6,** S199.

Harris, S. E., Harris, M. A., Mahy, P., Sabatini, M., Dunn, J., Boyce, B., and Mundy, G. R. (1991b). Expression of BMP 4 and BMP 3-like mRNA by human prostate carcinoma cells. *J. Bone Miner. Res.* **6,** S193.

Harrison, E. T., Jr., Luyten, F. P., and Reddi, A. H. (1991). Osteogenin promotes reexpression of cartilage phenotype by dedifferentiated articular chondrocytes in serum-free medium. *Exp. Cell Res.* **192,** 340–345.

Hart, K. A., Bentley, H., Seid, J., Russell, R. G. G., Oreffo, R. O. C., and Johnstone, D. (1991). Detection of bone morphogenetic protein (BMP) expression in human prostatic tissue and cell lines. *J. Bone Miner. Res.* **6,** S260.

Heckman, J. D., Boyan, B. D., Aufdemorte, T. B., and Abbott, J. T. (1991). The use of bone morphogenetic protein in the treatment of non-union in a canine model. *J. Bone Joint Surg.* **73A,** 750–764.

Hiraki, Y., Inoue, H., Shigeno, C., Sanma, Y., Bentz, H., Rosen, D. M., Asada, A., and Suzuki, F. (1991). Bone morphogenetic proteins (BMP-2 and BMP-3) promote growth and expression of the differentiated phenotype of rabbit chondrocytes and osteoblastic MC3T3-E1 cells *in vitro*. *J. Bone Miner. Res.* **6,** 1373–1385.

Hsueh, A. J. W., Dahl, K. D., Vaughan, J., Tucker, E., Rivier, J., Bardin, C. W., and Vale, W. (1987). Heterodimers and homodimers of inhibin subunits have different paracrine action in the modulation of luteinizing hormone-stimulated androgen biosynthesis. *Proc. Natl. Acad. Sci. USA* **84,** 5082–5086.

Huggins, C. B. (1931). The formation of bone under the influence of epithelium of the urinary tract. *Arch. Surg.* **22,** 377–408.

Hughes, F. J. (1992). Chemotactic effects of BMP2 on osteoblastic and fibroblastic cell lines. *J. Dent. Res.* **71,** 730.

Immerglück, K., Lawrence, P. A., and Bienz, M. (1990). Induction across germ layers in *Drosophila* mediated by a genetic cascade. *Cell* **62,** 261–268.

Israel, D. I., Nove, J., Kerns, K. M., Moutsatsos, I. K., and Kaufman, R. J. (1992). Expression and characterization of bone morphogenetic protein-2 in Chinese hamster ovary cells. *Growth Factors* **7,** 139–150.

Johnson, E. E., Urist, M. R., and Finerman, G. A. M. (1988a). Bone morphogenetic protein augmentation grafting of resistant femoral nonunions: A preliminary report. *Clin. Orthop. Relat. Res.* **230,** 257–265.

Johnson, E. E., Urist, M. R., and Finerman, G. A. M. (1988b). Repair of segmental defects of the tibia with cancellous bone grafts augmented with human bone morphogenetic protein. *Clin. Orthop. Relat. Res.* **236,** 249–257.

Johnson, E. E., Urist, M. R., Schmalzried, T. P., Chotivichit, A., Huang, H. K., and Finerman, G. A. M. (1989). Autogeneic cancellous bone grafts in extensive segmental ulnar defects in dogs. *Clin. Orthop. Relat. Res.* **243,** 254–265.

Johnson, E. E., Urist, M. R., and Finerman, G. A. M. (1990). Distal metaphyseal tibial nonunion: Deformity and bone loss treated by open reduction, internal fixation, and human bone morphogenetic protein (hBMP). *Clin. Orthop. Relat. Res.* **250,** 234–240.

Jones, C. M., Lyons, K. M., LaPan, P. M., Wright, C. V. E., and Hogan, B. L. M. (1992). DVR-4 (bone morphogenetic protein-4) as a posterior-ventralizing factor in *Xenopus* mesoderm induction. *Development* **115,** 639–647.

Kaplan, F. S., Tabas, J. A., and Zasloff, M. A. (1990). Fibrodysplasia ossificans progressiva: A clue from the fly? *Calcif. Tissue Int.* **47,** 117–125.

Katagiri, T., Lee, T., Takeshima, H., Suda, T., Tanaka, H., and Omura, S. (1990a). Transforming growth factor-β modulates proliferation and differentiation of mouse clonal osteoblastic MC3T3-E1 cells depending on their maturation stages. *Bone Miner.* **11,** 285–293.

Katagiri, T., Yamaguchi, A., Ikeda, T., Yoshiki, S., Wozney, J. M., Rosen, V., Wang, E. A., Tanaka, H., Omura, S., and Suda, T. (1990b). The non-osteogenic mouse pluripotent cell line, C3H10T1/2, is induced to differentiate into osteoblastic cells by recombinant human bone morphogenetic protein-2. *Biochem. Biophys. Res. Commun.* **172,** 295–299.

Kawamura, M., and Urist, M. R. (1988). Human fibrin is a physiologic delivery system of bone morphogenetic protein. *Clin. Orthop. Relat. Res.* **235,** 302–310.

Kawamura, M., Iwata, H., Sato, K., and Miura, T. (1987). Chondroosteogenetic response to crude bone matrix proteins bound to hydroxyapatite. *Clin. Orthop. Relat. Res.* **217,** 281–292.

Khouri, R. K., Koudsi, B., and Reddi, H. (1991). Tissue transformation into bone *in vivo. J. Am. Med. Assoc.* **266,** 1953–1955.

Knutsen, R., Mohan, S., Wergedal, J., Sampath, K., and Baylink, D. J. (1991). Osteogenic protein-1 (OP-1) stimulates proliferation and differentiation of human bone cells (HBC) *in vitro. J. Bone Miner. Res.* **6,** S141.

Köster, M., Plessow, S., Clement, J. H., Lorenz, A., Tiedemann, H., and Knöchel, W. (1991). Bone morphogenetic protein 4 (BMP-4), a member of the TGF-β family, in early embryos of *Xenopus laevis:* Analysis of mesoderm inducing activity. *Mech. Dev.* **33,** 191–200.

Landesman, R., and Reddi, A. H. (1986). Chemotaxis of muscle-derived mesenchymal cells to bone-inductive proteins of rat. *Calcif. Tissue Int.* **39,** 259–262.

LaPan, P., Bauduy, M., Cox, K. A., D'Alessandro, J. S., Israel, D. I., Nove, J., Rosen, V., Wozney, J. M., Moutsatsos, I. K., and Wang, E. A. (1991). Purification, characterization and activities of recombinant human bone morphogenetic protein 4. *J. Bone Miner. Res. 6,* S153.

Lepage, T., Ghiglione, C., and Gache, C. (1992). Spatial and temporal expression pattern during sea urchin embryogenesis of a gene coding for a protease homologous to the human protein BMP-1 and to the product of the *Drosophila* dorsal–ventral patterning gene *tolloid. Development* **114,** 147–164.

Lindholm, T. C., Lindholm, T. S., Alitalo, I., and Urist, M. R. (1988). Bovine bone morphogenetic protein (bBMP) induced repair of skull trephine defects in sheep. *Clin. Orthop. Relat. Res.* **227,** 265–268.

Lovell, T. P., Dawson, E. G., Nilsson, O. S., and Urist, M. R. (1989). Augmentation of spinal fusion with bone morphogenetic protein in dogs. *Clin. Orthop. Relat. Res.* **243,** 266–275.

Luyten, F. P., Cunningham, N. S., Ma, S., Muthukumaran, N., Hammonds, R. G., Nevins, W. B., Wood, W. I., and Reddi, A. H. (1989). Purification and partial amino acid sequence of osteogenin, a protein initiating bone differentiation. *J. Biol. Chem.* **264,** 13377–13380.

Luyten, F. P., Yu, Y. M., Yanagishita, M., Vukicevic, S., Hammonds, R. G., and Reddi, A. H. (1992). Natural bovine osteogenin and recombinant human bone morphogenetic protein-2B are equipotent in the maintenance of proteoglycans in bovine articular cartilage explant cultures. *J. Biol. Chem.* **267,** 3691–3695.

Lyons, K., Graycar, J. L., Lee, A., Hashmi, S., Lindquist, P. B., Chen, E. Y., Hogan, B. L. M., and Derynck, R. (1989a). *Vgr-1,* a mammalian gene related to *Xenopus Vg-1,* is a member of the transforming growth factor β gene superfamily. *Proc. Natl. Acad. Sci. USA* **86,** 4554–4558.

Lyons, K. M., Pelton, R. W., and Hogan, B. L. M. (1989b). Patterns of expression of murine Vgr-1 and BMP-2a RNA suggest that transforming growth factor-β-like genes coordinately regulate aspects of embryonic development. *Genes Dev.* **3,** 1657–1668.

Lyons, K. M., Pelton, R. W., and Hogan, B. L. M. (1990). Organogenesis and pattern formation in the mouse: RNA distribution patterns suggest a role for bone morphogenetic protein-2A (BMP-2A). *Development* **109,** 833–844.

Lyons, K. M., Jones, C. M., and Hogan, B. L. M. (1991). The DVR gene family in embryonic development. *Trends Genet.* **7,** 408–412.

Madisen, L., Neubauer, M., Plowman, G., Rosen, D., Segarini, P., Dasch, J., Thompson, A., Ziman, J., Bentz, H., and Purchio, A. F. (1990). Molecular cloning of a novel bone-forming compound: Osteoinductive factor. *DNA* **9,** 303–309.

Maliakal, J. C., Hauschka, P. V., and Sampath, T. K. (1991). Recombinant human osteogenic protein (hOP-1) promotes the growth of osteoblasts and stimulates expression of the osteoblast phenotype in culture. *J. Bone Miner. Res.* **6,** S251.

Mark, D. E., Hollinger, J. O., Hastings, C., Jr., Chen, G., Marden, L. J., and Reddi, A. H. (1990). Repair of calvarial nonunions by osteogenin, a bone-inductive protein. *J. Plast. Reconstr. Surg.* **86,** 623–630.

Masucci, J. D., Miltenberger, R. J., and Hoffmann, F. M. (1990). Pattern-specific expression of the *Drosophila decapentaplegic* gene in imaginal disks is regulated by 3′ *cis*-regulatory elements. *Genes Dev.* **4,** 2011–2023.

Mathews, L. S., and Vale, W. W. (1991). Expression cloning of an activin receptor, a predicted transmembrane serine kinase. *Cell* **65,** 973–982.

Matsui, M., Oikawa, S., Hashimoto, J., Yoshikawa, H., Miyamoto, S., Suzuki, S., Tsuruoka, N., Katayama, T., Tawaragi, Y., and Takaoka, K. (1992). Gene cloning and expression of a murine bone morphogenetic protein derived from osteosarcoma. *Calcif. Tissue Int.* **50S,** A14.

Moore, J. C., Matukas, V. J., Deatherage, J. R., and Miller, E. J. (1990). Craniofacial osseous restoration with osteoinductive proteins in a collagenous delivery system. *Int. J. Oral Maxillofac. Surg.* **19,** 172–176.

Nakashima, M. (1990). The induction of reparative dentine in the amputated dental pulp of the dog by bone morphogenetic protein. *Arch. Oral Biol.* **35,** 493–497.

Neuhof, H. (1917). Fascia transplantation into visceral defects. *Surg. Gynecol. Obstet.* **24,** 383–427.

Nilsson, O. S., and Urist, M. R. (1991). Immune inhibition of repair of canine skull trephine defects implanted with partially purified bovine morphogenetic protein. *Int. Orthop.* **15,** 257–263.

Nilsson, O. S., Urist, M. R., Dawson, E. G., Schmalzried, T. P., and Finerman, G. A. M. (1986). Bone repair induced by bone morphogenetic protein in ulnar defects in dogs. *J. Bone Joint Surg.* **68B,** 635–642.

Noda, M., and Rodan, G. A. (1986). Type-β transforming growth factor inhibits proliferation and expression of alkaline phosphatase in murine osteoblast-like cells. *Biochem. Biophys. Res. Commun.* **140,** 56–65.

Ogawa, Y., Schmidt, D. K., Dasch, J. R., Chang, R., and Glaser, C. B. (1992). Purification and characterization of transforming growth factor-β2.3 and -β1.2 heterodimers from bovine bone. *J. Biol. Chem.* **267,** 2325–2328.

Özkaynak, E., Rueger, D. C., Drier, E. A., Corbett, C., Ridge, R. J., Sampath, T. K., and Oppermann, H. (1990). OP-1 cDNA encodes an osteogenic protein in the TGF-β family. *EMBO J.* **9,** 2085–2093.

Özkaynak, E., Schnegelsberg, P. N. J., and Oppermann, H. (1991). Murine osteogenic protein (OP-1): High levels of mRNA in the kidney. *Biochem. Biophys. Res. Commun.* **179,** 116–123.

Padgett, R. W., St. Johnston, R. D., and Gelbart, W. M. (1993). A transcript from a *Drosophila* pattern gene predicts a protein homologous to the transforming growth factor-β family. *Nature (London)* **325,** 81–84.

Padgett, R. W., Wozney, J. M., and Gelbart, W. M. (1992). Human BMP sequences can confer normal dorsal-ventral patterning in the *Drosophila* embryo *Proc. Natl. Acad. Sci. U.S.A.* **90,** 2905–2909.

Padley, R. A., Cobb, C. M., Killoy, W. J., Newhouse, N. L., and Boyan, B. D. (1991). *In vitro* chemotactic response of osteoblast-like osteosarcoma cells to a partially purified protein extract of demineralized bone matrix. **62,** 15–20.

Panganiban, G. E. F., Reuter, R., Scott, M. P., and Hoffmann, F. M. (1990). A *Drosophila* growth factor homolog, *decapentaplegic,* regulates homeotic gene expression within and across germ layers during midgut morphogenesis. *Development* **110,** 1041–1050.

Paralkar, V. M., Nandedkar, A. K. N., Pointer, R. H., Kleinman, H. K., and Reddi, A. H. (1990). Interaction of osteogenin, a heparin binding bone morphogenetic protein, with type IV collagen. *J. Biol. Chem.* **265,** 17281–17284.

Paralkar, V. M., Hammonds, R. G., and Reddi, A. H. (1991). Identification and characterization of cellular binding proteins (receptors) for recombinant human bone morphogenetic protein 2B, an initiator of bone differentiation cascade. *Proc. Natl. Acad. Sci. USA* **88,** 3397–3401.

Pelton, R. W., Dickinson, M. E., Moses, H. L., and Hogan, B. L. M. (1990). *In situ* hybridization analysis of TGFβ3 RNA expression during mouse development: Comparative studies with TGFβ1 and β2. *Development* **110,** 609–620.

Plessow, S., Köster, M., and Knöchel, W. (1991). cDNA sequence of *Xenopus laevis* bone morphogenetic protein 2 (BMP-2). *Biochim. Biophys. Acta* **1089,** 280–282.

Rao, V. V. N. G., Löffler, C., Wozney, J. M., and Hansmann, I. (1992). The gene for bone morphogenetic protein 2A (BMP2A) is localized to human chromosome 20p12 by radioactive and nonradioactive *in situ* hybridization (ISH & FISH) *Hum. Genet.* **90,** 299–302.

Reddi, A. H. (1981). Cell biology and biochemistry of endochondral bone development. *Collagen Rel. Res.* **1,** 209–226.

Reddi, A. H., and Anderson, W. A. (1976). Collagenous bone matrix-induced endochondral ossification and hemopoiesis. *J. Cell Biol.* **69,** 557–572.

Reddi, A. H., and Huggins, C. (1972). Biochemical sequences in the transformation of normal fibroblasts in adolescent rats. *Proc. Natl. Acad. Sci. USA* **69,** 1601–1605.

Reddi, A. H., and Huggins, C. B. (1975). Formation of bone marrow in fibroblast-transformation ossicles. *Proc. Natl. Acad. Sci. USA* **72,** 2212–2216.

Reynolds, S. D., Angerer, L. M., Palis, J., Nasir, A., and Angerer, R. C. (1992). Early mRNAs, spatially restricted along the animal–vegetal axis of sea urchin embryos, include one encoding a protein related to tolloid and BMP-1. *Development* **114,** 769–786.

Ripamonti, U. (1991). The morphogenesis of bone in replicas of porous hydroxyapatite obtained from conversion of calcium carbonate exoskeleton of coral. *J. Bone Joint Surg.* **73-A,** 692–703.

Roark, E. F., and Kosher, R. A. (1991). Differential effects of transforming growth factor-βs and bone morphogenetic protein-2 on chick limb chondrogenesis. *J. Cell Biol.* **115,** 449a.

Rogers, M. B., Rosen, V., Wozney, J. M., and Gudas, L. J. (1992). Bone morphogenetic proteins-2 and -4 are involved in the retinoic acid-induced differentiation of embryonal carcinoma cells. *Mol. Biol. Cell* **3,** 189–196.

Rosen, V., and Thies, R. S. (1992). The BMP proteins in bone formation and repair. *Trends Genet.* **8,** 97–102.

Rosen, V., Wozney, J. M., Wang, E. A., Cordes, P., Celeste, A., McQuaid, D., and Kurtzberg, L. (1989). Purification and molecular cloning of a novel group of BMPs and localization of BMP mRNA in developing bone. *Connec. Tissue Res.* **20,** 313–319.

Rosen, V., Bauduy, M., McQuaid, D., Donaldson, D., Thies, S., and Wozney, J. (1991). Expression of osteoblast-like phenotype in mouse embryo limb bud cell lines cultured in BMP-2 and retinoic acid. *Proc. Endocrine Soc.* **73,** 57.

Sampath, T. K., and Reddi, A. H. (1981). Dissociative extraction and reconstitution of extracellular matrix components involved in local bone differentiation. *Proc. Natl. Acad. Sci. USA* **78,** 7599–7603.

Sampath, T. K., and Reddi, A. H. (1983). Homology of bone-inductive proteins from human, monkey, bovine, and rat extracellular matrix. *Proc. Natl. Acad. Sci. USA* **80,** 6591–6595.

Sampath, T. K., and Reddi, A. H. (1984). Importance of geometry of the extracellular matrix in endochondral bone differentiation. *J. Cell Biol.* **98,** 2192–2197.

Sampath, T. K., Muthukumaran, N., and Reddi, A. H. (1987). Isolation of osteogenin, an extracellular matrix-associated, bone-inductive protein, by heparin affinity chromatography. *Proc. Natl. Acad. Sci. USA* **84,** 7109–7113.

Sampath, T. K., Coughlin, J. E., Whetsone, R. M., Banach, D., Corbett, C., Ridge, R. J., Özkaynak, E., Oppermann, H., and Rueger, D. C. (1990). Bovine osteogenic protein is composed of dimers of OP-1 and BMP-2A, two members of the transforming growth factor-β superfamily. *J. Biol. Chem.* **265,** 13198–13205.

Sampath, T. K., Özkaynak, E., Jones, W. K., Sasak, H., Tucker, R., Tucker, M., Kusmik, W., Lightholder, J., Pang, R., Corbett, C., Oppermann, H., and Rueger, D. C. (1991). Recombinant human osteogenic protein (hOP-1) induces new bone formation with a specific activity comparable to that of natural bovine osteogenic protein. *J. Bone Miner. Res.* **6,** S155.

Sampath, T. K., Özkaynak, E., Jones, W., Sasak, H., Tucker, R., Tucker, M., Kusmik, W., Lightholder, J., Pang, R., Corbett, C., Oppermann, H., and Rueger, D. C. (1992). Recombinant human osteogenic protein (hOP-1) induces new bone formation with a specific activity comparable to that of natural bovine OP. *Trans. Orthop. Res. Soc.* **17,** 72.

Sato, K., and Urist, M. R. (1985). Induced regeneration of calvaria by bone morphogenetic protein (BMP) in dogs. *Clin. Orthop. Relat. Res.* **197,** 301–311.

Sato, T., Kawamura, M., Sato, K., Iwata, H., and Miura, T. (1991). Bone morphogenesis of rabbit bone morphogenetic protein-bound hydroxyapatite-fibrin composite. *Clin. Orthop. Relat. Res.* **263,** 254–262.

Schaub, R. G., and Wozney, J. M. (1991). Novel agents that promote bone regeneration. *Cur. Opin. Biotechnol.* **2,** 868–871.

Schluesener, H. J. (1991). Transforming growth factors type β inhibit multidrug transport in rat astrocyte cell lines. *Autoimmunity* **9,** 269–275.

Schluesener, H. J., and Meyermann, R. (1991). Spontaneous multidrug transport in human glioma cells is regulated by transforming growth factors type β. *Acta Neuropathol.* **81,** 641–648.

Shimell, M. J., Ferguson, E. L., Childs, S. R., and O'Connor, M. B. (1991). The *Drosophila* dorsal–ventral patterning gene *tolloid* is related to human bone morphogenetic protein 1. *Cell* **67,** 469–481.

Shinomura, T., and Kimata, K. (1992). Proteoglycan-Lb, a small dermatan sulfate proteoglycan expressed in embryonic chick epiphyseal cartilage, is structurally related to osteoinductive factor. *J. Biol. Chem.* **267,** 1265–1270.

Smith, J. C., Price, B. M. J., Van Nimmen, K., and Huylebroeck, D. (1990). Identification of a potent *Xenopus* mesoderm-inducing factor as a homologue of activin A. *Nature (London)* **345,** 729–731.

St. Johnston, R. D., Hoffman, F. M., Blackman, R. K., Segal, D., Grimaila, R., Padgett, R. W., Irick, H. A., and Gelbart, W. M. (1990). Molecular organization of the *decapentaplegic* gene in *Drosophila melanogaster*. *Genes Dev.* **4,** 1114–1127.

Tabas, J. A., Zasloff, M., Wasmuth, J. J., Emanuel, B. S., Altherr, M. R., McPherson, J. D., Wozney, J. M., and Kaplan, F. S. (1991). Bone morphogenetic protein: Chromosomal localization of human genes for BMP-1, BMP-2A, and BMP-3. *Genomics* **9,** 283–289.

Tabas, J. A., Hahn, G. V., Cohen, R. B., Seaunez, H. N., Modi, W. S., Wozney, J. M., Zasloff, M., and Kaplan, F. S. (1992). Bone morphogenetic protein (BMP): Chromosomal assignment of the human gene for BMP4. *Clin. Orthop. Relat. Res.* (in press).

Takaoka, K., Ono, K., Amitani, K., Kishimoto, R., and Nagata, Y. (1980). Solubilization and concentration of a bone-inducing substance from a murine osteosarcoma. *Clin. Orthop. Relat. Res.* **148,** 274–280.

Takaoka, K., Yoshikawa, H., Shimizu, N., Ono, K., Amitani, K., and Nakata, Y. (1982). Partial purification of bone-inducing substances from a murine osteosarcoma. *Clin. Orthop. Relat. Res.* **164,** 265–270.

Takaoka, K., Nakahara, H., Yoshikawa, H., Masuhara, K., Tsuda, T., and Ono, K. (1988). Ectopic bone induction on and in porous hydroxyapatite combined with collagen and bone morphogenetic protein. *Clin. Orthop. Relat. Res.* **234,** 250–254.

Takaoka, K., Yoshikawa, H., Masuhara, K., Sugamoto, K., Tsuda, T., Aoki, Y., Ono, K., and Sakamoto, Y. (1989). Establishment of a cell line producing bone morphogenetic protein from a human osteosarcoma. *Clin. Orthop. Relat. Res.* **244,** 258–264.

Takaoka, K., Koezuka, M., and Nakahara, H. (1991). Telopeptide-depleted bovine skin collagen as a carrier for bone morphogenetic protein. *J. Orthop. Res.* **9,** 902–907.

Takaoka, K., Yoshikawa, H., Hashimoto, J., Masuhara, K., Miyamoto, S., Suzuki, S., and Matsui, M. (1992). Purification and characterization of a bone-inducing factor from a murine osteosarcoma. *Calcif. Tissue Int.* **50S,** A19.

Takuwa, Y., Ohse, C., Wang, E. A., Wozney, J. M., and Yamashita, K. (1991). Bone morphogenetic protein-2 stimulates alkaline phosphatase activity and collagen synthesis in cultured osteoblastic cells, MC3T3-E1. *Biochem. Biophys. Res. Commun.* **174,** 96–101.

Thies, R. S., Bauduy, M., Ashton, B. A., Kurtzberg, L., Wozney, J. M., and Rosen, V. (1992a). Recombinant human bone morphogenetic protein-2 induces osteoblastic differentiation in W-20-17 stromal cells. *Endocrinology* **130,** 1318–1324.

Thies, R. S., Song, J., Wozney, J. M., and Rosen, V. (1992b). Vitamin D inhibits bone morphogenetic protein-2 induction of osteocalcin in W-20-17 bone marrow stromal cells. *Proc. Endocrine Soc.* **74,** 443.

Titani, K., Torff, H.-J., Hormel, S., Kumar, S., Walsh, K. A., Rödl, J., Neurath, H., and Zwilling, R. (1987). Amino acid sequence of a unique protease from the crayfish *Astacus fluviatilis. Biochemistry* **26,** 222–226.

Toriumi, D. M., Kotler, H. S., Luxenberg, D. P., Holtrop, M. E., and Wang, E. A. (1991). Mandibular reconstruction with a recombinant bone-inducing factor. *Arch. Otolaryngol. Head Neck Surg.* **117,** 1101–1112.

Tsuda, T., Masuhara, K., Yoshikawa, H., Shimizu, N., and Takaoka, K. (1989). Establishment of an osteoinductive murine osteosarcoma clonal cell line showing osteoblastic phenotypic traits. *Bone* **10,** 195–200.

Ueno, N., Nishimatsu, S., and Murakami, K. (1990). Activin as a cell differentiation factor. *Prog. Growth Factor Res.* **2,** 113–124.

Urist, M. R. (1965). Bone: Formation by autoinduction. *Science* **150,** 893–899.

Urist, M. R., Iwata, H., Ceccotti, P. L., Dorfman, R. L., Boyd, S. D., McDowell, R. M., and Chien, C. (1973). Bone morphogenesis in implants of insoluble bone gelatin. *Proc. Natl. Acad. Sci. USA* **70,** 3511–3515.

Urist, M. R., Nakata, N., Felser, J. M., Nogami, H., Hanamura, H., Miki, T., and Finerman, G. A. M. (1977). An osteosarcoma cell and matrix retained morphogen for normal bone formation. *Clin. Orthop. Relat. Res.* **124,** 251–266.

Urist, M. R., Grant, T. T., Lindholm, T. S., Mirra, J. M., Hirano, H., and Finerman, G. A. M. (1979a). Induction of new-bone formation in the host bed by human bone-tumor transplants in athymic nude mice. *J. Bone Joint Surg.* **61A,** 1207–1210.

Urist, M. R., Mikulski, A., and Lietze, A. (1979b). Solubilized and insolubilized bone morphogenetic protein. *Proc. Natl. Acad. Sci. USA* **76,** 1828–1832.

Urist, M. R., DeLange, R. J., and Finerman, A. M. (1983). Bone cell differentiation and growth factors. *Science* **220,** 680–686.

Urist, M. R., Huo, Y. K., Brownell, A. G., Hohl, W. M., Buyske, J., Lietze, A., Tempst, P., Hunkapiller, M., and DeLange, R. J. (1984). Purification of bovine bone morphogenetic protein by hydroxyapatite chromatography. *Proc. Natl. Acad. Sci. USA* **81,** 371–375.

Urist, M. R., Chang, J. J., Lietze, A., Huo, Y. K., Brownell, A. G., and DeLange, R. J. (1987a). Preparation and bioassay of bone morphogenetic protein and polypeptide fragments. *Methods Enzymol.* **146,** 294–312.

Urist, M. R., Nilsson, O., Rasmussen, J., Hirota, W., Lovell, T., Schmalzreid, T., and Finerman, G. A. M. (1987b). Bone regeneration under the influence of a bone morphogenetic protein (BMP) beta tricalcium phosphate (TCP) composite in skull trephine defects in dogs. *Clin. Orthop. Relat. Res.* **214,** 295–304.

Vukicevic, S., Luyten, F. P., and Reddi, A. H. (1989). Stimulation of the expression of osteogenic and chondrogenic phenotypes *in vitro* by osteogenin. *Proc. Natl. Acad. Sci. USA* **86,** 8793–8797.

Vukicevic, S., Luyten, F. P., and Reddi, A. H. (1990a). Osteogenin inhibits proliferation and stimulates differentiation in mouse osteoblast-like cells (MC3T3-E1). *Biochem. Biophys. Res. Commun.* **166,** 750–756.

Vukicevic, S., Paralkar, V. M., Cunningham, N. S., Gutkind, J. S., and Reddi, A. H. (1990b). Autoradiographic localization of osteogenin binding sites in cartilage and bone during rat embryonic development. *Dev. Biol.* **140,** 209–214.

Vukicevic, S., Chen, P., Reddi, A. H., and Sampath, T. K. (1992). Localization of the osteogenic protein-1 during early human embryonic development. *Calcif. Tissue Int.* **50S,** A31.

Wang, E. A., Rosen, V., Cordes, P., Hewick, R. M., Kriz, M. J., Luxenberg, D. P., Sibley, B. S., and Wozney, J. M. (1988). Purification and characterization of other distinct bone-inducing factors. *Proc. Natl. Acad. Sci. USA* **85,** 9484–9488.

Wang, E. A., Rosen, V., D'Alessandro, J. S., Bauduy, M., Cordes, P., Harada, T., Israel, D., Hewick, R. M., Kerns, K., LaPan, P., Luxenberg, D. P., McQuaid, D., Moutsatsos, I., Nove, J., and Wozney, J. M. (1990). Recombinant human bone morphogenetic protein induces bone formation. *Proc. Natl. Acad. Sci. USA* **87,** 2220–2224.

Wang, E. A., Gerhart, T. N., and Toriumi, D. M. (1992). Bone morphogenetic proteins and development, *Proc. 4th International Symposium on Chemistry and Biology of Mineralized Tissues,* San Diego, CA.

Wang, X., Lin, H. Y., Ng-Eaton, E., Downward, J., Lodish, H. F., and Weinberg, R. A. (1991). Expression cloning and characterization of the TGF-β type III receptor. *Cell* **67,** 797–805.

Weeks, D. L., and Melton, D. A. (1987). A maternal mRNA localized to the vegetal hemisphere in *Xenopus* eggs codes for a growth factor related to TGF-β. *Cell* **51,** 861–867.

Wharton, K. A., Thomsen, G. H., and Gelbart, W. M. (1991). *Drosophila* 60A gene, another transforming growth factor β family member, is closely related to human bone morphogenetic proteins. *Proc. Natl. Acad. Sci. USA* **88,** 9214–9218.

Wlodarski, K. (1969). The inductive properties of epithelial established cell lines. *Exp. Cell Res.* **57,** 446–448.

Wlodarski, K., Poltorak, A., and Koziorowska, J. (1971). Species specificity of osteogenesis induced by WISH cell line and bone induction by Vaccinia virus transformed human fibroblasts. *Calcif. Tissue Res.* **7,** 345–352.

Wozney, J. M. (1989). Bone morphogenetic proteins. *Prog. Growth Factor Res.* **1,** 267–280.

Wozney, J. M. (1990). Using purified protein to clone its gene. *Methods Enzymol.* **182,** 738–751.

Wozney, J. M., Rosen, V., Celeste, A. J., Mitsock, L. M., Whitters, M. J., Kriz, R. W., Hewick, R. M., and Wang, E. A. (1988). Novel regulators of bone formation: Molecular clones and activities. *Science* **242,** 1528–1534.

Wozney, J. M., Rosen, V., Byrne, M., Celeste, A. J., Moutsatsos, I., and Wang, E. A. (1990). Growth factors influencing bone development. *J. Cell Sci. Suppl.* **13,** 149–156.

Wozney, J. M., McQuaid, D., Thies, S., and Rosen, V. (1992). The BMP family and bone cell differentiation, *Proc. Osteosarcoma Research Conference 1991* Pittsburg, PA. (J. Novak, ed.). Hogrefe and Huber (in press).

Yamaguchi, A., and Kahn, A. J. (1991). Clonal osteogenic cell lines express myogenic and adipocytic developmental potential. *Calcif. Tissue Int.* **49,** 221–225.

Yamaguchi, A., Katagiri, T., Ikeda, T., Wozney, J. M., Rosen, V., Wang, E. A., Kahn, A. J., Suda, T., and Yoshiki, S. (1991). Recombinant human bone morphogenetic protein-2 stimulates osteoblastic maturation and inhibits myogenic differentiation *in vitro. J. Cell Biol.* **113,** 681–687.

Yamaguchi, A., Ikeda, T., Katagiri, T., Suda, T., and Yoshiki, S. (1992). BMP-2 induces differentiation of a non-osteogenic fibroblastic cell line (C3H10T1/2) into both osteoblasts and chondroblasts *in vitro. Bone Miner.* **17S,** 191.

Yamazaki, Y., Oida, S., Akimoto, Y., and Shioda, S. (1988). Response of the mouse femoral muscle to an implant of a composite of bone morphogenetic protein and plaster of Paris. *Clin. Orthop. Relat. Res.* **234,** 240–249.

Yasko, A. W., Lane, J. M., Waller, S., Fellinger, E. J., Tomin, M., Ron, E., and Wang, E. A. (1991). Biologic and synthetic matrix carriers for recombinant human BMP-2: Orthotopic osteoinduction. *J. Bone Miner. Res.* **6,** S155.

Yasko, A. W., Fellinger, E. J., Waller, S., Tomin, A., Peterson, M., Wang, E. A., and Lane, J. M. (1992a). Comparison of biological and synthetic carriers for recombinant human BMP (rhBMP-2) induced bone formation. *Trans. Orthop. Res. Soc.* **17,** 71.

Yasko, A. W., Lane, J. M., Fellinger, E. J., Rosen, V., Wozney, J. M., and Wang, E. A. (1992b). The healing of segmental bone defects, induced by recombinant human bone morphogenetic protein (rhBMP-2). A radiographic, histological, and biomechanical study in rats. *J. Bone Joint Surg.* **74A,** 659–670.

Yoshikawa, H., Takaoka, K., Shimizu, N., and Ono, K. (1984). Solubility of a bone-inducing substance from a murine osteosarcoma. *Clin. Orthop. Relat. Res.* **182,** 231–235.

Yoshikawa, H., Takaoka, K., Hamada, H., and Ono, K. (1985). Clinical significance of bone morphogenetic activity in osteosarcoma. *Cancer* **56,** 1682–1687.

Yoshikawa, H., Hashimoto, J., Masuhara, K., Takaoka, K., and Ono, K. (1988a). Inhibition by tumor necrosis factor of induction of ectopic bone formation by osteosarcoma-derived bone-inducing substance. *Bone* **9,** 391–396.

Yoshikawa, H., Takaoka, K., Masuhara, K., Ono, K., and Sakamoto, Y. (1988b). Prognostic significance of bone morphogenetic activity in osteosarcoma tissue. *Cancer* **61,** 569–573.

5

OUR UNDERSTANDING OF INHERITED SKELETAL FRAGILITY AND WHAT THIS HAS TAUGHT US ABOUT BONE STRUCTURE AND FUNCTION

JEFFREY BONADIO and STEVEN A. GOLDSTEIN

Cellular and Molecular Biology of Bone

I. INTRODUCTION

This chapter reviews current ideas about the molecular pathogenesis of the human inherited disease osteogenesis imperfecta (OI).

First described in the eighteenth century, OI refers to a group of inherited connective tissue diseases, the hallmarks of which are bone and soft connective tissue fragility. Males and females are affected equally, and the overall incidence of disease, based on ascertainment studies, from around the world, currently is estimated to be 1 in 5000–14,000 live births. OI is not known to be endemic to a particular race, nationality, or region of the world. All known cases of OI have resulted from mutations in the extracellular matrix (ECM) molecule type I collagen, a prominent constituent of the connective tissue of bone, tendon, ligament, tooth, skin dermis, the wall of large blood vessels, and sclera.

II. MAPPING THE OSTEOGENESIS IMPERFECTA LOCUS

The OI locus was mapped by V. McKusick, G. Martin, and co-workers in 1974. Their hypothesis—that type I collagen genes would harbor mutations that cause OI—was based in part on previous histological studies describing abnormalities in the collagen of affected individuals. To establish this hypothesis, the investigators chose to examine procollagen biosynthesis by OI cells in culture. McKusick had obtained OI skin tissues from which primary fibroblast cultures were established. Martin's laboratory had been studying the intermediate steps in procollagen biosynthesis by cultured fibroblasts (Layman *et al.*, 1971) and, thus, could analyze the ability of the OI cell strains to synthesize and secrete procollagen.

Several abnormalities in type I procollagen biosynthesis by OI cells were eventually described (Penttinen *et al.*, 1975), and the various OI cell strains were segregated into three groups. The first group secreted an abnormal ratio of collagen types I and III, probably because of a selective defect in type I collagen production. In the second group, abnormalities in the type I : type III collagen ratio could be detected only after proteolytic digestion of experimental samples with pepsin, which suggested a qualitative change in type I procollagen structure. The third group was composed of OI cells for which no detectable biosynthetic abnormality could be demonstrated. The interpretation of these results provided an important foundation for much of the subsequent analysis of OI. Thus, McKusick, Martin, and co-workers (1) established a relationship between the OI phenotype and biosynthetic alterations in either the production or structure of type I procollagen, (2) suggested that genetic and biochemical heterogeneity may be responsible for the well-established clinical heterogeneity associated with OI, and (3) suggested, in an attempt to unify their concept of disease, that alterations in collagen production and structure could lead to a decrease in tissue collagen content and the generalized tissue fragility characteristic of OI.

III. CLINICAL CLASSIFICATION

In 1979, Sillence *et al.*, 1979 established a clinical classification scheme that divided OI patients according to their natural history and mode of inheritance. Sillence and co-workers recognized four types of OI (Table I). OI type I is a mild disorder characterized by bone fracture without deformity, blue sclerae, normal or near-normal stature, and autosomal dominant inheritance (Sillence, 1981). The incidence is now estimated to be 1 in 5–20,000 live births, and males and females are affected equally. Osteopenia is associated with an increased rate of long bone fracture upon ambulation. For reasons not well understood, fracture frequency decreases dramatically at puberty and during young adult life but increases once again in late middle age (Paterson *et al.*, 1984). Hearing loss, often beginning in the second or third decade, is a feature of this disease in about half the families. The hearing loss can progress despite the general decline in fracture frequency (Pedersen, 1985). Conductive or mixed hearing loss is said to be more common than sensorineural hearing loss alone. Dentinogenesis imperfecta is observed in a small subset of these individuals. Given its mild manifestations, OI type I can remain undocumented in many families. In several cases, this has led to the unfortunate misdiagnosis of child abuse.

In contrast, OI types II–IV represent a spectrum of more severe disorders associated with a shortened life-span. OI type II, the perinatal

TABLE I The Sillence Classification of Osteoporosis Imperfecta[a]

I	II	III	IV
"Mild, dominant" sclerae	"Perinatal lethal"	"Progressively deformed"	Mild, normal
Matrix fragility, mild	Matrix fragility, extreme	Matrix fragility, moderately severe	Matrix fragility, moderate
Dominant inheritance	Dominant inheritance	Dominant inheritance	Dominant inheritance
Normal life-span	Lethal outcome	Shortened life-span	Shortened life-span

[a]Refer to the text for additional discussion of phenotypic manifestations. Note that the original classification scheme suggested that osteogenesis imperfecta types I–IV were associated with a recessive pattern of inheritance. A current view is that the great majority of all osteogenesis imperfecta types results from collagen mutations that act in a dominant fashion.

lethal form, is characterized by short stature, a soft calvarium, blue sclerae, fragile skin, a small chest, floppy-appearing lower extremities (due to external rotation and abduction of the femurs), fragile tendons and ligaments, bone fracture with severe deformity, and death in the perinatal period due to respiratory insufficiency (Sillence *et al.*, 1984). Radiographic signs of bone weakness include compression of the femurs, bowing of the tibiae, broad and beaded ribs (other appearances have been noted; e.g., see Byers, 1989), and calvarial thinning. OI type III is characterized by short stature, a triangular facies, severe scoliosis, and bone fracture with moderate deformity. Scoliosis can lead to emphysema and a shortened life-span due to respiratory insufficiency. OI type IV is characterized by normal sclerae, bone fracture with mild to moderate deformity, tooth defects, and a natural history that essentially is intermediate between OI type II and OI type I.

IV. EVIDENCE OF GENETIC AND BIOCHEMICAL HETEROGENEITY

The Sillence classification has served the genetics community well, providing a basis for clear and informative counseling of OI families. It has also contributed to our understanding of disease. For example, in 1981–1982 Byers and co-workers established the genetic heterogeneity of OI by providing evidence that structural and null mutations cause distinctive alterations in procollagen biosynthesis and distinctive phenotypic manifestations. This work provided evidence that unique biochemical defects in procollagen biosynthesis segregated with different types of OI as defined by Sillence.

Barsh and Byers (1981) initially showed that the OI type II phenotype was associated with structural mutations in type I procollagen. In one OI cell strain, they found one population of proα1(I) chains (proα1$(I)_f$) that migrated faster than normal on sodium dodecyl sulfate–polyacrylamide gel electrophoresis (SDS–PAGE) and a second population (proα1$(I)_s$) that migrated more slowly than normal. To explain these findings, Barsh and Byers suggested that the proα1$(I)_f$ and proα1$(I)_s$ chains were separate gene products. The proα1$(I)_f$ chain was shown by cyanogen bromide (CNBr) peptide mapping to have a defect in its primary structure, perhaps a deletion of amino acids to account for its abnormal electrophoretic migration, whereas proα1$(I)_s$ was thought to be excessively modified. Additionally, the mutation was associated with a block in type I procollagen intracellular transport, leading to a delay in procollagen secretion and an overall reduction in the amount of procollagen in the culture medium.

In contrast to their study of OI type II, Barsh et al. (1982) showed that the OI type I phenotype was associated with a defect in type I procollagen production. They studied primary fibroblast cell lines from three unrelated individuals with OI type I. Total collagen production by all three cell lines was decreased to almost 50% of normal. Type III collagen was produced at normal levels, and the altered type I : type III collagen ratio was therefore due to a selective decrease in type I collagen production. Type I procollagen secretion by the OI cells and the structure of proα1(I) and proα2(I) chains were normal, suggesting that the defect in type I procollagen production occurred during an early step in the procollagen biosynthetic pathway. In fact, the investigators could find no evidence of defective procollagen assembly, but they did find indirect evidence of decreased proα1(I) chain synthesis to about 50% of normal. The rate of proα2(I) chain synthesis appeared to be normal. (The decrease in proα1(I) chain synthesis apparently created an "excess" of proα2(I) chains; these chains, rather than assembling into triple helical molecules, appeared to be degraded inside the cultured cells.)

V. CHARACTERIZATION OF THE FIRST MOLECULAR DEFECT IN A COLLAGEN GENE

Investigative efforts next focused on the probability that structural mutations in the triple helical domain of type I collagen were associated with OI type II. In this regard, the laboratories of Byers and Prockop established an important experimental strategy for localizing defects within the triple helical domain of procollagen molecules. The approach involved fragmentation of the triple helical domain by chemical or enzymatic cleavage and analysis of the constituent peptides by SDS-PAGE. As already described, Byers and co-workers used CNBr cleavage, while Prockop and co-workers achieved the same result with fibroblast collagenase (Fig. 1).

Williams and Prockop (1983), studying the same OI type II cell strain as Barsh and Byers, also identified a population of fast- and slow-migrating proα1(I) collagen chains and provided evidence that the slow- and fast-migrating proα1(I) chains were incorporated into disulfide-linked procollagen trimers. They described the synthesis of three populations of procollagen molecules: One population had three normal-length proα chains, a second had one shortened and two normal-length chains, and a third had two shortened and one normal-length chains. Molecules containing three shortened chains were not described. The abnormal conformation of procollagen molecules incorporating the mutant proα chain was elegantly demonstrated by thermal denaturation studies that utilized limited proteolytic digestion by trypsin and chymotrypsin. Williams and Prockop localized the deletion within the triple helical domain of fast-migrating proα1(I) collagen chains following asymmetrical cleavage by the enzyme fibroblast collagenase.

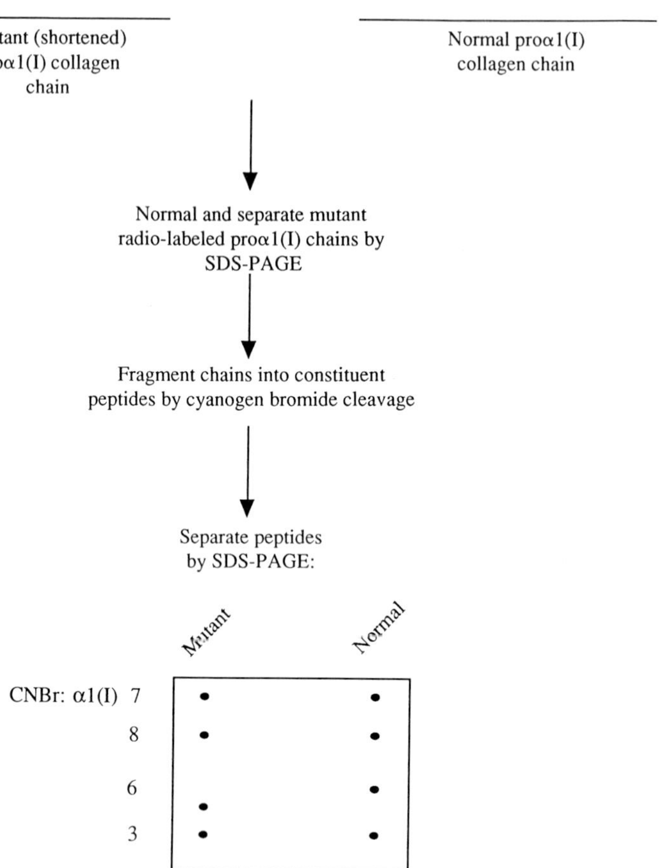

FIGURE 1 Schematic view of the strategy to localize a deletion of amino acid sequence from an α1(I) collagen chain. Radiolabeled α1(I) chains are separated by sodium dodecyl sulfate–polyacrylamide gel electrophoresis (SDS–PAGE) and fragmented in the gel with cyanogen bromide (CNBr), and the constituent peptides are separated by electrophoresis in a second-dimension gel. In the example shown, the deletion results in the abnormally fast migration of the α1(I) CNBr6 peptide in the second-dimension gel.

The molecular defect harbored by these OI type II cells was subsequently characterized by Southern blot hybridization and an S1 nuclease protection assay. Ramirez, Prockop, and co-workers initially demonstrated that the proband was heterozygous for a deletion of ~500 basepairs from a proα1(I) collagen gene (Chu *et al.*, 1983). Two groups then sequenced across the deletion breakpoints (Chu *et al.*, 1985; Barsh *et al.*, 1985), demonstrating that three exons were deleted from a proα1(I) collagen gene. The deletion was most likely generated by a process of recombination between nonhomologous introns. The deleted exons encoded a total of 84 amino acids from the triple helical domain of mutant proα1(I) chains. The mutation did not shift the frame of the coding sequence; therefore, the mutant gene was able to encode the fast-migrating proα1(I)$_f$ chain observed previously. The deletion encompassed a Met residue, which explained the loss of one peptide and the appearance of a novel fusion peptide following CNBr digestion of mutant procollagen molecules secreted by the cultured OI fibroblasts (Barsh *et al.*, 1985).

VI. MAPPING SINGLE AMINO ACID SUBSTITUTIONS: THE PREDOMINANT STRUCTURAL MUTATION ASSOCIATED WITH OSTEOGENESIS IMPERFECTA II–IV

While it was clear from screening studies that most mutations associated with OI type II altered the structure of the genetic sequence encoding the type I collagen triple helical domain, it also became apparent that a majority of mutations did not result in large deletions or insertions of amino acid sequence. Investigators therefore turned their attention to the problem of determining the other classes of structural mutation associated with OI. While relatively large, the loci for the α1(I) and α2(I) chains are similar in that each is interrupted by more than 50 introns (for a review, see Sandell and Boyd, 1990) and a majority of the α1(I) and α2(I) exons encode perfect Gly–X–Y amino acid repeats consisting of either 18 or 36 amino acids. Small deletions or insertions could have resulted therefore from the loss or gain of a small number of exons. Alternatively, it was considered possible that single amino acid substitutions caused most cases of the lethal form of OI.

Mutations had to be characterized at the sequence level in order to discriminate between these possibilities. An important first step in this process was to develop a system to map subtle structural alternations in mutant type I procollagen molecules. An initial strategy to map subtle mutations in the triple helical domain of type I collagen was based on the observation that abnormal procollagen molecules from different OI cell strains were not always overmodified (hydroxylated and glycosylated) to the same extent along the triple helix (Steinmann *et al.*, 1984; Bateman *et al.*, 1984; Bonadio *et al.*, 1985; Bonadio and Byers, 1985). Overmodification of mutant procollagen was often asymmetrically distributed within this domain, suggesting that triple helix formation proceeded at an appropriate rate until the region of the mutation was reached. Thereafter the rate of formation of a stable triple helix was delayed and all chains in abnormal molecules were available to the post-translational modifying enzymes for a longer period of time than normal. Similar to the way in which a large deletion was localized within the collagen triple helix, the extent of overmodification within the triple helical domain was ascertained by fragmentation of the triple helical domain into constituent peptides and analysis by SDS–PAGE. The model predicted that OI mutations would be localized at the COOH terminus of the overmodified domain.

The mapping strategy specifically assumed an orderly process of normal triple helix assembly. In general, type I procollagen biosynthesis resembles that of other secreted glycoproteins (Kuhn, 1987). The α1(I) and α2(I) messenger RNAs (mRNAs) appear to be translated in a 2:1 ratio. The initial translation products, preproα1(I) and preproα2(I) chains, have amino-terminal signal sequences that allow them to be translocated into the lumen of the rough endoplasmic reticulum. The signal sequence is cleaved from the chain as a co-translational event. In contrast to many globular proteins, nascent procollagen chains are completely translated before they assemble and achieve a stable quaternary structure. Proα1(I) and proα2(I) chains associate in a 2:1 ratio via amino acid sequences within the carboxy-terminal propeptides, the association is stabilized by the formation of interchain disulfide bonds, and assembly of the triple-stranded helix takes place soon thereafter. Several important co- and post-translational modifications take place on proα chains before a stable triple helical conformation is achieved. High mannose oligosaccharides are added to the carboxy-terminal propeptide domains of proα1(I) and proα2(I) chains. The oligosaccharide apparently is not processed further because mannose residues on secreted procollagen molecules are susceptible to cleavage by endoglycosidase H. Certain prolyl residues within the triple helical domain are hydroxylated. Hydroxyproline residues make a direct contribution to the stability of the triple helix by participating in the formation of interchain hydrogen bonds. Certain lysyl residues are also hydroxylated, and glucose and galactose are added to a small number of hydroxylysyl residues via O-linked glycosidic bonds. These events in collagen biosynthesis all take place in the rough endoplasmic reticulum of connective tissue cells, and they are regulated in part by the rate at which a stable triple helical conformation is achieved. Structural mutations appeared to alter the process of triple helical assembly, resulting in unstable, overmodified, poorly secreted procollagen molecules. Because assembly normally proceeds in a linear fashion from the COOH to the NH_2 terminus, it was predictable that mutations would be localized at one end (the COOH terminus) *of the overmodified region* (Fig. 2).

With a mapping strategy in hand, investigators next turned to the problem of cloning and sequencing structural mutations in the triple helical domain of collagen proα chains. Given the accumulated evidence regarding the abnormal structure of procollagen produced by OI cells *in vitro,* it was reasonable to assume that a majority of structural collagen mutations would be localized to coding sequence. Given the size and complexity of the α1(I) and α2(I) collagen genes and the knowledge that

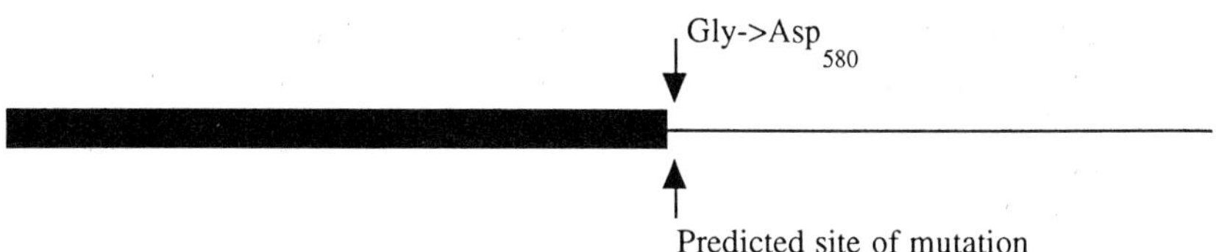

FIGURE 2 The predicted correlation between the extent of overmodification within the triple helical domain of a mutant procollagen molecule and the location of a single amino acid substitution within the domain. The type I collagen triple helix is shown schematically, with the overmodified region shown as a black rectangle. In this example, the COOH-terminus of the overmodified region and the location of the mutation, a Gly → Asp substitution at position 580 of the α1(I) collagen chain produced by cells from an individual with OI type II, were tightly correlated (E. Smiley, D. Bole, P. H. Byers, and J. Bonadio, unpublished data).

collagen exons were generally smaller than collagen introns, it was desirable to approach this problem at the level of mRNA rather than at the level of the collagen gene. This strategy became possible with the availability of the polymerase chain reaction technique. Patterson *et al.* (1989) applied the polymerase chain reaction to the problem of analyzing a single amino acid substitution in the triple helical domain of a procollagen α chain (Fig. 3). Total RNA was isolated from a small number of OI cells in culture, complementary DNA (cDNA) was synthesized using

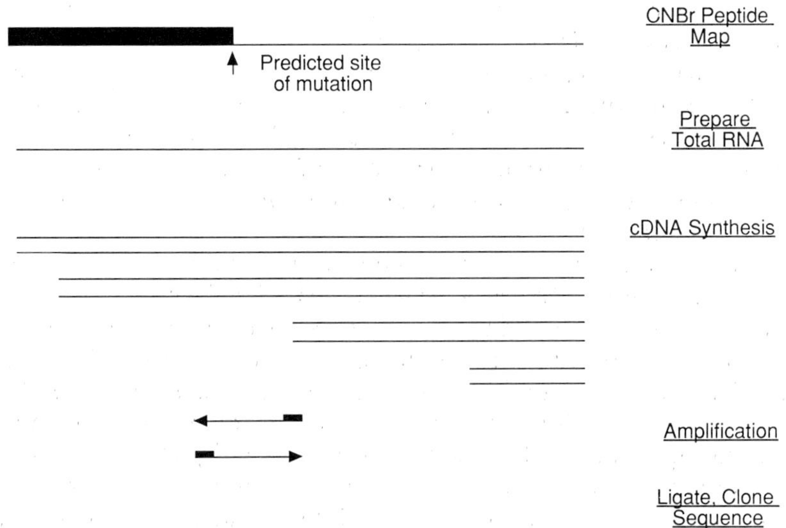

FIGURE 3 Schematic depicting the strategy developed by Patterson *et al.* (1989) to characterize osteogenesis imperfecta mutations at the level of messenger RNA. cDNA, complementary DNA; CNBr, cyanogen bromide.

collagen-specific primers, and fragments of the α1(I) and α2(I) coding sequence targeted by the CNBr-peptide map were amplified and sequenced. Sequence analysis of the amplified cDNA revealed a new, heterozygous, Gly → Val substitution at residue 256 of the triple helical domain of α1(I) chains produced by the OI cells. The polymerase chain reaction technique was efficient because it allowed one to quickly and simultaneously prepare targeted regions from cDNAs encoding the α1(I) and α2(I) collagen chains.

Sophisticated new techniques that allow the precise, allele-specific localization of mutations in collagen cDNA molecules have more recently become available (Genovese *et al.*, 1989; Marini *et al.*, 1989; Lamande *et al.*, 1989), while automated technologies have facilitated the sequence analysis (Kuivaniemi *et al.*, 1991). The convergence of these technologies and the polymerase chain reaction has shortened the process of mutation analysis to a few weeks and resulted in a rapid increase in our understanding of structural mutations that cause OI. Approximately 90 OI mutations have been characterized since 1989 alone, so that information now exists on the sequence of more than 100 OI mutations in all. The majority are heterozygous, single amino acid substitutions for conserved glycine residues in the triple helical domain. The substitutions have shown no preference for a particular region of the triple helix, which suggests that mutation of any conserved glycine would result in a clinically significant phenotype. Structural mutations have been localized to the α1(I) and α2(I) collagen chains of patients with OI types II–IV. A number of deletions and insertions (recombinatorial errors or mRNA processing defects) have also been identified. About half of the latter structural mutations have left the frame of the coding sequence intact while the other half have resulted in a frameshift. Finally, there is one reported case of a single amino acid substitution outside of the triple helical domain of an α1(I) collagen chain, a Gly → Cys substitution in the third amino acid beyond the COOH terminus of the triple helical domain (Cohn *et al.*, 1988).

That a correlation exists between overmodification and the location of mutations within the triple helical domain has been verified repeatedly. To our knowledge, mutations have occurred at or near the COOH terminus of the overmodified domain, in every case in which the location of the mutation and the extent of overmodification have been documented. Moreover, this correlation has been extended to mutations in other collagen types with a domain structure and pattern of assembly similar to that of type I procollagen (e.g., type III procollagen molecules synthesized by EDS IV cells in culture [Superti-Furga *et al.*, 1989], type II procollagen molecules in cartilage from infants with achondrogenesis ([Cohn and Byers, 1988; Vissing *et al.*, 1989; Tiller *et al.*, 1990]).

VII. INHERITANCE

Early on, only the mild form of OI, as first defined by Sillence (OI type I), was considered to be an autosomal dominant disease. Most forms of OI were thought to be recessive because pedigree analysis of the typical OI family showed no evidence of an inherited connective tissue disorder, the parents of the proband were unaffected, and multiple affected siblings had been well documented in several families. However, the analysis of procollagen biosynthesis by OI fibroblasts in culture and the molecular genetic characterization of a large number of OI mutations clearly demonstrated that the vast majority of OI mutations are inherited as heterozygous alleles that act in a dominant fashion, suggesting that most OI mutations were in fact new rather than inherited (Fig. 4A). Single case reports of recessive inheritance due to consanguinity (Philajaniemi *et al.*, 1984), of suspected recessive inheritance due to uniparental disomy (Bonadio *et al.*, 1990a) and of compound heterozygosity (de Wet *et al.*, 1983), have also been published.

The observation that most OI mutations were new dominant alleles did not, however, fit well with certain OI pedigrees in which half-siblings (i.e., affected siblings that have only one parent in common)

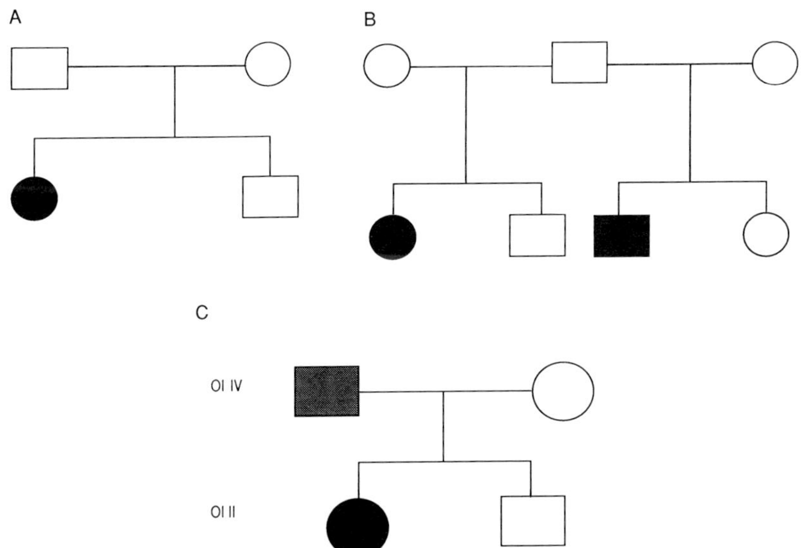

FIGURE 4 Common patterns of inheritance of osteogenesis imperfecta (OI) mutant alleles. (A) New mutation acting in a dominant fashion. (B) Gonadal mosaicism. Illustration adapted from Cohn *et al.* (1990). (C) Somatic mosaicism. Illustration adapted from Wallis *et al.* (1990).

with the same OI phenotype were documented or other OI pedigrees in which one parent with a mild form of OI gave rise to an affected infant with the lethal form. Molecular genetic analysis of one of these families, in which an unaffected father sired two such half-siblings with the perinatal lethal form of OI, established that the father's germ cells and the dermal fibroblasts from the two infants had the same Gly → Cys single amino acid substitution in an α1(I) collagen gene (Cohen *et al.*, 1990). Moreover, the study of additional families with more than one affected sib have shown that the parent carrying the mutant allele is mosaic not only in the germ cell lineage but also in selected somatic cells (Wallis *et al.*, 1990) (Fig. 4B and C). Thus, it appears as though both sporadic and recurrent cases of OI can be accounted for by mosaicism, which would be consistent with similar findings with several other human disorders including pseudoachondroplasia, Duchenne muscular dystrophy, and Marfan syndrome. Furthermore, the phenotype of the parent carrying the mutation may vary from normal to mild OI, apparently depending on the extent of mosaicism in connective tissues, and the probability of recurrence in these families will depend on the extent of germ-line mosaicism. These observations are of great importance because they suggest that, in addition to the qualitative aspects of the collagen mutation, the phenotypic expression of OI (and perhaps other forms of inherited connective tissue disease) may depend on the relative number of cells expressing the mutation and their tissue distribution.

A. Pathogenesis I: Structural Mutations

Structural collagen mutations appear to disrupt the normal assembly of three proα chains into a highly ordered procollagen molecule. Like many ECM molecules, type I collagen has an extended rather than globular conformation with dimensions of approximately 3000 × 15Å (Piez, 1984). The molecule is a heteropolymer in mature connective tissues, consisting of constituent α1(I) and α2(I) chains in a 2:1 stoichiometry. Each α chain is made up of more than 1000 amino acid residues. The amino acid sequence of the triple helical domain of an α chain can be represented as an invariant Gly–X–Y repeat. The molecule has a high degree of order. X-ray diffraction studies have shown that three α chains polymerize to form a right-handed triple helix, the α chains are staggered with respect to one another by one amino acid, every third residue of an α chain is folded into the center of the triple helix, and glycine is the only residue small enough to be folded in this manner and yet maintain the normal geometry of the triple-stranded collagen helix. As such, the repeated glycine residues are thought to form a central element along the length of the collagen molecule. Substitutions for conserved glycine residues appear to introduce an element of disorder into the procollagen molecule in that they delay procollagen assembly and alter procollagen structure (Fig. 5). Consequently, molecules incorporating one or more mutant chains are unstable, overmodified, and poorly secreted from their site of assembly in the lumen of the rough endoplasmic reticulum. Similar effects have been observed with mutations that alter the length of the triple helical domain (i.e., deletions or insertions of amino acid sequence).

The phenotypic manifestations of OI reflect the widespread distribution of type I collagen in connective tissues. Therefore, OI can and should be thought of as a systemic disorder of type I collagen rather than simply as an inherited form of bone fragility (as its name might imply). How then does a change in the amino acid sequence in a single population of collagen proα chain cause connective tissue fragility? First, we consider the normal arrangement and function of collagen. Then, we discuss how the effect of a structural mutation may be propagated within the ECM in a way that compromises connective tissue integrity.

Upon secretion, procollagen molecules normally are processed to collagen and assembled (with other macromolecules) to form collagen fibrils (Piez, 1984). Although still under investigation, the packing of collagen molecules within a fibril is known to be precise. Type I collagen molecules are parallel and in register with respect to one another (separated by regular gaps estimated to be 36 nm) and they actually assemble into paracrystalline fibrillar arrays in many tissues. They are thermodynamic considerations, the surface chemistry of the type I collagen molecule, the formation of covalent cross-links between lysyl and hydroxylysyl residues between nearby molecules within fibrils, and the

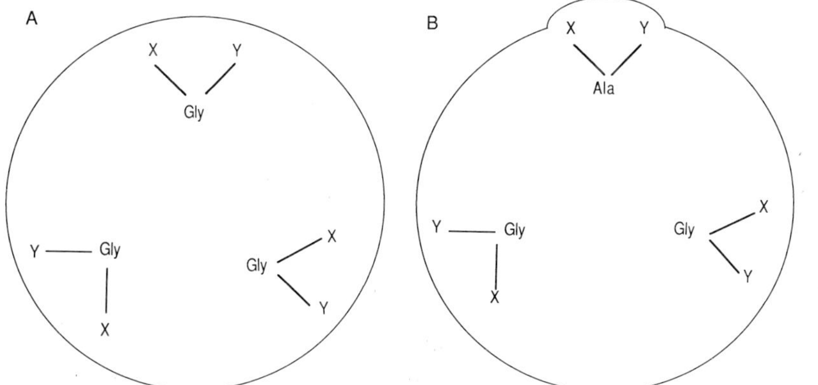

FIGURE 5 Transverse view illustrating the relationship of Gly-X-Y repeats from three collagen α chains that have assembled into a triple helix. (A) The normal relationship. Illustration adapted from Piez (1984). (B) The abnormal relationship caused by a single amino acid substitution for glycine. Illustration taken from Traub and Steinmann (1986).

interaction between collagen and other matrix components are thought to govern fibril formation. Finally, the fibril is assembled, often with precise spatial and temporal organization, into a three-dimensional ECM consisting of collagens, noncollagenous glycoproteins, and proteoglycans.

Both the ordered structure of type I collagen molecules and fibrils and the ordered arrangement of fibrils within the ECM contribute directly to the material and structural properties of connective tissue (Agarwal and Broutman, 1980). It is convenient to think of the ECM of most connective tissues as a protein multimer comprised of collagens, proteoglycans, noncollagenous glycoproteins, and water. The constituents of this multimer assemble in time and space. If the assembly process is accomplished correctly, the ECM will have certain functional attributes that allow it to make an important contribution to processes as diverse as cell adhesion and migration, tissue morphogenesis, wound healing, mechanical function (e.g., locomotion), and protection from injury. The ECM may vary considerably from one tissue to the next, and differences can be seen in the type, relative amount, and organization of the constituent macromolecules that make up individual matrices. These are important considerations because the function of connective tissue ultimately reflects the biochemical constituents of the ECM, the manner in which these constituents are organized, and the ability of connective tissue cells to sense their environment and respond appropriately.

Multiprotein structures (protein heteropolymers) such as muscle, the mitotic apparatus, the spliceosome, and the transcription apparatus are sensitive to mutations that result in the incorporation of a defective, or structurally altered, protein constituent. Given the protein multimer analogy, structural collagen mutations would be expected to have dominant negative effects (i.e., a change in the structure and/or functional capacity of collagen would be expected to disrupt the function of the entire ECM). Thus, outside of cells, the altered conformation of mutant molecules may disrupt the normal assembly of collagen molecules into fibrils. Although not well investigated, it is assumed that the altered conformation of collagen fibrils ultimately leads to alternations in the structure and function of the ECM and, consequently, the phenotypic manifestations of OI (Fig. 6). It is noteworthy that these views are based, for the most part, on the study of skin fibroblasts in culture taken from affected individuals, and that we do not yet know how closely the lessons learned from these *in vitro* studies apply to OI *in vivo*. A critically important experiment in this regard was performed by Jaenisch, Bateman, Cole, and co-workers using transgenic mice (Stacey *et al.*, 1988). These investigators used the technique of site-directed mutagenesis to create a mouse homolog of a previously documented Gly → Cys single amino acid substitution at position 988 of the $\alpha1(I)$ collagen chain, which

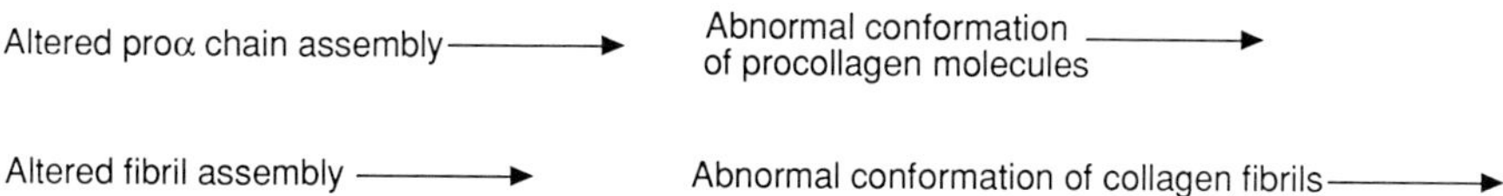

FIGURE 6 Schematic depicting a personal view of the molecular pathogenesis of the connective tissue fragility that characterizes osteogenesis imperfecta.

was associated with OI type II. Mice expressing the collagen transgene also demonstrated a phenotype similar to OI type II including a lethal outcome, a reduced amount of type I collagen in tissues, and an astonishing number of fractures.

The collagen triple helix contains a number of subdomains that ultimately facilitate procollagen assembly into the ECM. Examples of these subdomains include those that mediate chain assembly and chain propagation into a triple-stranded helix and others that mediate the interaction between collagen and fibronectin, collagen and proteoglycans, collagen and the turnover enzyme fibroblast collagenase, and collagen and the cell surface. At present, little is known about the relationship between the structure of these subdomains and their function. However, their existence raises several interesting questions related to the clinical heterogeneity of OI. For example, what are the phenotypic consequences of a glycine substitution that changes the conformation of the collagen–fibronectin binding site as opposed to one that changes the collagen–cell surface interaction site? How might the consequences differ if the mutation at these sites was a Gly → Cys substitution as opposed to a Gly → Val substitution? Similar kinds of questions may be asked for each of the subdomains of the collagen molecule and for each of the possible amino acid substitutions for glycine (Ala, Arg, Asp, Cys, Trp, Tyr, and Val). Moreover, differences may be encountered depending on whether the mutation is localized to the α1(I) or α2(I) collagen chain (Wenstrup *et al.*, 1990). Finally, it should be kept in mind that mutations in one part of the collagen molecule may disrupt the function of subdomains far removed from the site of mutation (Vogel *et al.*, 1988). Ultimately, an improved understanding of these structure–function relationships may come from the continued investigation of mutant collagen alleles, and this knowledge may provide important insights into the phenotypic variability known to exist within the OI patient population.

B. Pathogenesis II: Null Mutations

Barsh and Byers originally hypothesized that heterozygous null α1(I) collagen alleles produce the OI type I phenotype, suggesting that a principal effect of the null allele would be to decrease collagen production by ~50%. In fact, this finding has been observed consistently, and it represents an important biochemical correlate of disease. However, regulatory mutations potentially represent a broad class of mutation that could disrupt collagen production at the level of either transcription or translation. As already described, the collagen loci are complex and Southern blotting of genomic DNA from more than 20 affected individuals has shown that these mutations do not commonly arise by large genetic rearrangements (for discussion, see Willing *et al.*, 1990). Therefore, regulatory collagen mutations likely are subtle in character. Preliminary mapping of these mutations is required to characterize them efficiently, however, mapping strategy has yet to be developed, and only one naturally occurring mutation associated with OI type I, a small deletion near the 3′ end of a proα1(I) allele that interferes with the molecular assembly and/or stability of procollagen molecules that incorporate this chain (Willing *et al.*, 1990), has been characterized to date. Consequently, the nature of mutations responsible for the OI type I phenotype has remained poorly understood.

In the absence of more sequence information, an experimental model of this form of OI would be helpful. A step toward this end was achieved by work demonstrating that a transgenic mouse strain, Mov13, models OI type I. The Mov13 strain was generated by exposing mouse embryos to Moloney murine leukemia virus (Schnieke *et al.*, 1983; Harbers *et al.*, 1984). Genetic and molecular evidence indicated that a single copy of the provirus integrated into the first intron of the α1(I) collagen gene. The proviral insert is associated with a change in chromatin conformation and *de novo* methylation of the gene, and it prevents initiation of transcription (Hartung *et al.*, 1986; Jahner and Jaenisch, 1985; Breindl *et al.*, 1984). Mov13 mice bred to have one defective allele produce 50% less α1(I) collagen mRNA than normal (Lohler *et al.*, 1984). Given the nature of the mutation harbored by Mov13 mice, it was hypothesized that mice with one defective allele fit the Barsh and Byers prediction and, thus, would serve as a model of OI type I.

Experiments were conducted at three levels to establish the phenotype of Mov13 mice (Bonadio *et al.*, 1990b). Standard collagen quantitation techniques were used to provide initial evidence that the α1(I) collagen transcription defect affected nonmineralized connective tissue (skin dermis). Mov13 mice were shown to have early-onset progressive hearing loss in physiological tests (evoked auditory brainstem responses) of hearing sensitivity. Biomechanical tests were used to quantify the structural properties of long bone. By this method, the femurs and humeri of Mov13 mice were shown to have significantly reduced bending strength.

Thus, the heterozygous null mutation was associated with dominant morphological and functional defects in both mineralized and nonmineralized connective tissue and with early-onset progressive hearing loss. As a model, Mov13 mice provide strong support for the original hypothesis of Barsh and Byers. Moreover, a reduced amount of type I collagen apparently has adverse consequences for collagen fiber structure and the organization and structural integrity of connective tissue. The Mov13 mutation probably represents an example of a null allele that reduces the amount of one component of a multiprotein structure, in this case the ECM, disrupting its integrity and function because of the imbalance. Although it is somewhat unusual for a heterozygous null allele to be associated with a dominant phenotype, the ordered assembly of the collagen molecule into fibrils and the abundance and ordered assembly of fibrils within the ECM may be relevant to consider in this regard.

OI type I is similar in some respects to osteoporosis. Involutional osteoporosis is a significant health care problem causing more than 1 million fractures per year in the United States (Riggs and Melton, 1986). Two forms of the disease have been described. Type I osteoporosis primarily affects women at some time after the menopause, whereas type II is age-related and occurs in both sexes. Insufficient accumulation of skeletal mass in young adulthood may contribute to both types of osteoporosis. Normally, bone tissue content increases until age 30 and then declines. It is thought that if bone density drops below a threshold level, fractures occur following minor trauma. Type I osteoporosis appears to be triggered by an accelerated loss of bone associated with falling estrogen levels in women at the time of menopause. Prockop and co-workers recently described the case of a 52-year-old postmenopausal woman who ". . . developed acute mid-thoracic back pain following a severe jolt while driving a truck" (Spotila *et al.*, 1991, p. 5424). X-ray examination of the spine revealed an anterior vertebral compression fracture and ". . . generalized demineralization of the spinal column consistent with osteoporosis" (ibid., 5424). Her sclerae had a bluish cast, she had a mild high frequency hearing loss, and her teeth and her height were judged to be normal. She had suffered five previous fractures, three before puberty and two during young adulthood. The investigators identified and characterized an apparently new heterozygous single amino acid substitution, Gly → Ser, at position 661 of the α2(I) collagen chain. The glycine substitution was associated with overmodification of the triple helical domain of type I procollagen molecules produced by the individual's skin fibroblasts in culture. The investigators concluded that ". . . there may be phenotypic and genotypic overlap between osteogenesis imperfecta type I and postmenopausal osteoporosis and that a subset of women with postmenopausal osteoporosis may have mutations in the genes for type I procollagen." (ibid., 5423). Of course, this conclusion is largely a matter of semantics (or nosology), because it could just as well have been concluded that this case represented an example of a mild form of OI that remained undetected until late in life.

VIII. SUMMARY

Information regarding the molecular pathogenesis of OI has accumulated rapidly over the last few years. A disease classification based on natural history, the molecular basis of the disease, a description of the mechanism of the inheritance, and several model systems have been reported. In particular, the recent development of new model systems to investigate the effect of mutations at the level of collagen fibril assembly (Kadler *et al.*, 1988) and at the level of bone structure and function (Stacey *et al.*, 1988; Bonadio *et al.*, 1990b) represents a future avenue of investigation for OI research. This work has the potential of providing insight into the pathogenesis of OI at the level of connective tissue and, thus, providing insight into rational forms of therapy for skeletal and nonmineralized connective tissue fragility.

ACKNOWLEDGMENTS

This work was supported in part by grants from the NIH. The authors acknowledge the important contributions made by our colleagues J. Kuhn, M. Wong, K. Jepsen, T. Pangilinian, and E. Smiley.

REFERENCES

Agarwal, B. D., and Broutman, L. J. (1980). "Analysis and Performance of Fiber Composites." J. Wiley and Sons, New York.

Barsh, G. S., and Byers, P. H. (1981). Reduced secretion of structurally abnormal type I procollagen in a form of osteogenesis imperfecta. *Proc. Natl. Acad. Sci. USA* **78,** 5142–5146.

Barsh, G. S., David, K. E., and Byers, P. H. (1982). Type I osteogenesis imperfecta: A nonfunctional allele for proα1(I) chains of type I collagen. *Proc. Natl. Acad. Sci. USA* **79,** 3838–3842.

Barsh, G. S., Roush, C. L., Bonadio, J., Byers, P. H., and Gelinas, R. E. (1985). Intro-mediated recombination may cause a deletion in an α1 type 1 collagen chain in a lethal form of osteogenesis imperfecta. *Proc. Natl. Acad. Sci. USA* **82,** 2870–2874.

Bateman, J. F., Mascara, T., Chan, D., and Cole, W. G. (1984). Abnormal type I collagen metabolism by cultured fibroblasts in lethal perinatal osteogenesis imperfecta. *Biochem. J.* **217,** 103–107.

Bonadio, J. F., and Byers, P. H. (1985). Subtle structural alterations in the chains of type I procollagen produce osteogenesis imperfecta type II. *Nature (London)* **316,** 363–366.

Bonadio, J., Ramirez, F., and Barr, M. (1990a). An intron mutation in the human α1(I) collagen gene alters the efficiency of pre-mRNA splicing and is associated with osteogenesis imperfecta type II. *J. Biol. Chem.* **265,** 2262–2268.

Bonadio, J., Saunders, T. L., Tsai, E., Goldstein, S. A., Morris-Wiman, J., Brinkley, L., Dolan, D. F., Altschuler, R. A., Hawkins, J. E., Bateman, J. F., Mascara, T., and Jaenisch, R. (1990b). Transgenic mouse model of the mild dominant form of osteogenesis imperfecta. *Proc. Natl. Acad. Sci. USA* **87,** 7145–7149.

Bonadio, J. F., Holbrook, K. A., Gelinas, R. E., Jacob, J., and Byers, P. H. (1985). Altered triple helical structure of type I procollagen in lethal perinatal osteogenesis imperfecta. *J. Biol. Chem.* **260,** 1734–1742.

Breindl, M., Harbers, K., and Jaenisch, R. (1984). Retrovirus-induced lethal mutation in collagen I gene of mice is associated with an altered chromatin structure. *Cell* **38,** 9–16.

Byers, P. H. (1989). Inherited disorders of collagen gene structure and expression. *Am. J. Hum. Genet.* **34,** 72–80.

Chu, M.-L., Williams, C. J., Pepe, G., Hirsch, J. L., Prockop, D. J., and Ramirez, F. (1983). Internal deletion in a collagen gene in a perinatal lethal form of osteogenesis imperfecta. *Nature (London)* **304,** 78–80.

Chu, M.-L., Gargiulo, V., Williams, C. J., and Ramirez, F. (1985). Multiexon deletion in an osteogenesis imperfecta variant with increased type III collagen mRNA. *J. Biol. Chem.* **260,** 691–694.

Cohn, D. H., Apone, S., Eyre, D. R., Starman, B. J., Andreassen, P., Charbonneau, H., Nicholls, A. C., Pope, F. M., and Byers, P. H. (1988). Substitution of cysteine for glycine within the carboxyl-terminal telopeptide of the α1 chain of type I collagen produces mild osteogenesis imperfecta. *J. Biol. Chem.* **263,** 14605–14607.

Cohn, D. H., Starman, B. J., Blumberg, B., and Byers, P. H. (1990). Recurrence of lethal osteogenesis imperfecta due to parental mosaicism for a dominant mutation in a human type I collagen gene (COL1A1). *Am. J. Hum. Genet.* **46,** 591–601.

de Wet, W. J., Pihlajaniemi, T., Myers, J., Kelly, T. E., and Prockop, D. J. (1983). Synthesis of a shortened pro-α2(I) chain and decreased synthesis of pro-α2(I) chains in a proband with osteogenesis imperfecta. *J. Biol. Chem.* **258,** 7721–7728.

Genovese, C., Brufsky, A., Shapiro, J., and Rowe, D. (1989). Detection of mutations in human type I collagen mRNA in osteogenesis imperfecta by indirect RNase protection. *J. Biol. Chem.* **264,** 9632–9637.

Harbers, K., Kuehn, M., Delius, H., and Jaenisch, R. (1984). Insertion of retrovirus into the first intron of α1(I) collagen gene leads to embryonic lethal mutation in mice. *Proc. Natl. Acad. Sci. USA* **81,** 1504–1508.

Hartung, S., Jaenisch, R., and Breindl, M. (1986). Retrovirus insertion inactivates mouse α1(I) collagen gene by blocking initiation of transcription. *Nature (London)* **320,** 365–367.

Jahner, D., and Jaenisch, R. (1985). Retrovirus-induced *de novo* methylation of flanking host sequences correlates with gene inactivity. *Nature (London)* **315,** 594–597.

Kadler, K. E., Hojima, Y., and Prockop, D. J. (1988). Assembly of type I collagen fibrils *de novo*. *J. Biol. Chem.* **263,** 10517–10523.

Kuhn, K. (1987). The Classical Collagens: Type I, II, and III. *In* "Structure and Function of Collagen Types" (R. Mayne and R. E. Burgeson, eds.), pp. 1–42. Academic Press, Orlando, Florida.

Kuivaniemi, H., Tromp, G., and Prockop, D. J. (1991). Mutations in collagen genes: cause of rare and some common diseases in humans. *FASEB J.* **5,** 2052–2060.

Lamande, S. R., Dahl. H.-H., Cole, W. G., and Bateman, J. F. (1989). Characterization of point mutations in the collagen COL1A1 and COL1A2 genes causing lethal perinatal osteogenesis imperfecta. *J. Biol. Chem.* **264,** 15809–15812.

Layman, D. L., McGoodwin, E. B., and Martin, G. R. (1971). The nature of the collagen synthesized by cultured human fibroblasts. *Proc. Natl. Acad. Sci. USA* **68,** 454–458.

Lohler, J., Timpl, R., and Jaenisch, R. (1984). Embryonic lethal mutation in a mouse collagen I gene causes rupture of blood vessels and is associated with erythropoietic and mesenchymal cell death. *Cell* **38,** 597–607.

Marini, J. C., Grange, D. K., Gottesman, G. S., Lewis, M. B., and Koeplin, D. A. (1989). Osteogenesis imperfecta type IV. Detection of a point mutation in one α1(I) collagen allele (COL1A1) by RNA/RNA hybrid analysis. *J. Biol. Chem.* **264,** 11893–11900.

Paterson, C. R., McAllion, S., and Stellman, J. L. (1984). Osteogenesis imperfecta after the menopause. *N. Engl. J. Med.* **310,** 1694–1969.

Patterson, E., Smiley, E., and Bonadio, J. (1989). RNA sequence analysis of a perinatal lethal osteogenesis imperfecta mutation. *J. Biol. Chem.* **264,** 10083–10087.

Pedersen, U., (1985). "Osteogenesis Imperfecta Clinical Features, Hearing Loss and Stapedectomy." Almqvist and Wiksell Periodical Company, Stockholm.

Penttinen, R. P., Lichtenstein, J. R., Martin, G. R., and McKusick, V. A. (1975). Abnormal collagen metabolism in cultured cells in osteogenesis imperfecta. *Proc. Natl. Acad. Sci. USA* **72,** 586–589.

Piez, K. A. (1982). Structure and assembly of the native collagen fibril. *Connect. Tissue Res.* **10,** 25–36.

Piez, K. A. (1984), In "Extracellular Matrix Biochemistry" K. A. Piez and A. H. Reddu (Eds.), pp. 1–39 New York: Elsevier Science.

Pihlajaniemi, T., Dickson, L. A., Pope, F. M., Korhonen, V. R., Nicholls, A., Prockop, D. J., and Myers, J. C. (1984). Osteogenesis imperfecta: Cloning of a pro-α2(I) collagen gene with a frameshift mutation. *J. Biol. Chem.* **259,** 12941–12944.

Riggs, D. L., and Melton, L. J. (1986). Involutional osteoporosis. *N. Engl. J. Med.* **314,** 1676–1686.

Sandell, L. J., and Boyd, C. D. (1990). Conserved and divergent sequence and functional elements within collagen genes. *In* "Extracellular Matrix Genes." (pp. 1–56). Academic Press, San Diego.

Schnieke, A., Harbers, K., and Jaenisch, R. (1983). Embryonic lethal mutation in mice induced by retrovirus insertion into the α1(I) collagen gene. *Nature (London)* **304,** 315–320.

Sillence, D. O. (1981). Osteogenesis imperfecta: An expanding panorama of variants. *Clin. Orthop. Relat. Res.* **159,** 11–25.

Sillence, D. O., Senn, A., and Danks, D. M. (1979). Genetic heterogeneity in osteogenesis imperfecta. *J. Med. Genet.* **16,** 101–106.

Sillence, D. O., Barlow, K. K., Garber, A. P., Hall, J. G., and Rimoin, D. L. (1984). Osteogenesis imperfecta type II: Delineation of the phenotype with reference to genetivc heterogeneity. *Am. J. Med. Genet.* **17,** 407–423.

Spotila, L. D., Constantinou, C. D., Sereda, L., Ganguly, A., Riggs, B. L., and Prockop, D. J. (1991). Mutation in a gene for type I procollagen (COL1A2) in a woman with postmenopausal osteoporosis: Evidence for phenotypic and genotypic overlap with mild osteogenesis imperfecta. *Proc. Natl. Acad. Sci. USA* **88,** 5423–5427.

Stacey, A., Bateman, J., Choi, T., Mascara, T., Cole, W., and Jeanisch, R. (1988). Perinatal lethal osteogenesis imperfecta in transgenic mice bearing an engineered mutant proα1(I) collagen gene. *Nature (London)* **332,** 131–136.

Steinmann, B., Rao, V. H., Vogel, A., Bruckner, P., Gitzelmann, R., and Byers, P. H. (1984). Cysteine in the triple-helical domain of one allelic product of the α1(I) gene of type I collagen produces a lethal form of osteogenesis imperfecta. *J. Biol. Chem.* **259,** 11129–11138.

Superti-Furga, A., Steinmann, B., Ramirez, F. H., and Byers, P. H. (1989). Molecular defects of type III procollagen in Ehlers-Danlos syndrome type IV. *Hum. Genet.* **82,** 104–108.

Tiller, G. E., Rimoin, D. L., Murray, L. W., and Cohn, D. H. (1990). Tandem duplication within a type II collagen gene (*COL2A1*) exon in an individual with spondyloepiphyseal dysplasia. *Proc. Natl. Acad. Sci. USA* **87,** 3889–3893.

Traub, W., and Steinmann, B. (1986). Structural study of mutant type I collagen from a patient with lethal osteogenesis imperfecta containing an intramolecular disulfide bond in the triple-helical domain. *FEBS Lett.* **198,** 213–216.

Vissing, H., D'Alessio, M., Lee, B., Ramirez, F., Godfrey, M., and Hollister, D. W. (1989). Glycine to serine substitution in the triple-helical domain of pro-α1(II) collagen results in a lethal perinatal form of short-limbed dwarfism. *J. Biol. Chem.* **264,** 18265–18267.

Vogel, B. E., Doelz, R., Kadler, K. E., Hojima, Y., Engel, J., and Prockop, D. J. (1988). A substitution of cysteine for glycine 748 of the α1 chain produces a kink at this site in the procollagen I molecule and an altered *N*-proteinase cleavage site over 225 nm away. *J. Biol. Chem.* **263,** 19249–19255.

Wallis, G. A., Starman, B. J., Zinn, A. B., and Byers, P. H. (1990). Variable expression of osteogenesis imperfecta in a nuclear family is explained by somatic mosiacism for a lethal point mutation in the α1(I) gene (COL1A1) of type I collagen in a parent. *Am. J. Hum. Genet.* **46,** 1034–1040.

Wenstrup, R. J., Willing, M. C., Starman, B. J., and Byers, P. H. (1990). Distinct biochemical phenotypes predict clinical severity in nonlethal variants of osteogenesis imperfecta. *Am. J. Hum. Genet.* **46,** 975–982.

Williams, C. J., and Prockop. D. J. (1983). Synthesis and processing of a type I procollagen containing shortened pro-α1(I) chains by fibroblasts from a patient with osteogenesis imperfecta. *J. Biol. Chem.* **258,** 5915–5921.

Willing, M. C., Cohn, D. H., and Byers, P. H. (1990). Frameshift mutations near the 3′ end of the COL1A1 gene of type I collagen predicts an elongated proα1(I) chain and results in osteogenesis imperfecta type I. *J. Clin. Invest.* **85,** 282–290.

6

MOLECULAR AND CELLULAR BIOLOGY OF THE MAJOR NONCOLLAGENOUS PROTEINS IN BONE

MARIAN F. YOUNG, KYOMI IBARAKI, JANET M. KERR, and ANNE-MARIE HEEGAARD

Cellular and Molecular Biology of Bone

I. INTRODUCTION

The matrix of bone is composed of a multitude of proteins that give it unique characteristics and function. To determine the precise role of these proteins, they were isolated from the mineralized compartment and subsequently analyzed by classical biochemical procedures. Using antibodies prepared against the purified proteins, complementary DNA (cDNA) clones were then isolated and complete protein sequences determined. These cDNAs have since been instrumental in defining the cell that produce the proteins as well as the mechanisms that control the expression of the genes at the genomic level. The goal of this chapter is to describe the new (and sometimes surprising) findings that have unfolded from these analyses.

This chapter is divided into three major sections. The first section is exclusively devoted to the structure and expression of a protein called osteonectin. As the name implies, it was once believed to have bone-specific functions, namely, the initiation of bone mineralization. By cDNA cloning and subsequent protein sequencing, it has become apparent that osteonectin is widely distributed in the body and likely to have multiple functions in mineralized and nonmineralized tissues as well. Despite its wide tissue distribution, it is, comparatively, highly expressed in bone (see Table I). For this reason, the osteonectin gene has been used as a model to study transcriptional control in human- and bovine-derived cultured bone cells. In these cultures, the nuclear regulation of osteonectin is quite complex and still the focus of ongoing research.

The second section describes the two cell attachment bone matrix proteins, osteopontin (OPN) and bone sialoprotein (BSP). While they both share the same small cell attachment motif (Arg–Gly–Asp [RGD]), they are otherwise distinct in structure and in their extensive post-translational modifications. The transcriptional control of the two genes also appears to be distinct and varies considerably with direct hormone/vitamin treatment as well as during development and with cellular transformation (see Tables II and III). This differential regulation is probably related to differences in the DNA sequences of the two gene promoters. The final section discusses the two small proteoglycans decorin and biglycan, which are, apparently, part of a gene "family." Like

TABLE I Tissue Distribution of Osteonectin/SPARC[a]

High	Low	Very low
Extraembryonic tissues		
Parietal endoderm	Yolk sac	Neuroectoderm
Placenta	Trophoblast	
	Amnion	
Newborn mouse tissues		
Skeletal bone	Lung	Brain—most of neuronal cells
Calvaria	Cartilage	Liver
Odontoblast	Heart	Stomach—secretory cells
Skin—dermis, connective tissue, sheath of whisker follicle	Ameloblast	Skin—epidermis
Adult tissues		
Lung	Skeletal muscle	Heart
Adrenal gland	Kidney	Bladder
Ovary—granulosa cells		
Testis—Leydig cells		

[a]High, low, and very low refer to the relative expression of osteonectin/SPARC mRNA as determined by *in situ* hybridization (Holland *et al.*, 1987; Nomura *et al.*, 1988) and Northern analysis (Mason *et al.*, 1986b, Nomura *et al.*, 1988) of mouse tissues.

OPN and BSP, the small proteoglycans are likely to have distinct functions in mineralized and nonmineralized tissues as well.

Clearly the bone matrix contains many additional noncollagenous proteins that are not discussed in detail in this chapter including, for example, osteocalcin, matrix Gla protein, fibronectin, thrombospondin, and alkaline phosphatase; some of these proteins will be the topic of separate chapters in this volume. We have selected osteonectin/SPARC (Secreted Protein, Acidic and Rich in Cysteine) BSP, OPN, and the two small proteoglycans biglycan and decorin because they are well characterized biochemically and genetically. In general, each section of this chapter will focus on molecular questions related to (1) protein structure and function and (2) the mechanisms that control the expression of these genes at the nuclear level.

II. OSTEONECTIN/SPARC/BM-40

A. Osteonectin: An Extracellular Matrix Component Identical to SPARC, the 43-kDa "Culture Shock" Protein, and BM-40

Osteonectin was originally isolated from fetal bovine mineralized bone matrix as a relatively abundant noncollagenous component (Termine *et al.*, 1981a). It has a molecular weight of 32,000 based on molecular sieve column chromatography and properties of a phosphoglycoprotein containing 3% sialic acid, 1.6% hexosamines, and 0.5% bound phosphate. The name osteonectin was given because of (1) its ability to bind to Ca^{2+}, hydroxyapatite, and native type I collagen and (2) its localization in bone and another mineralized tissue, dentin (Termine *et al.*, 1981a, b). It was proposed, therefore, to be a nucleator of mineralization. Later, a 43-kDa serum albumin-binding glycoprotein was purified from bovine aortic endothelial (BAE) cell culture media and was found to be secreted by a wide spectrum of normal and transformed cells in culture (Sage *et al.*, 1984). Interestingly, production of this protein increased when the BAE cells were exposed to endotoxin or mechanical stress in response to sparse plating density (this condition was termed by Sage *et al.* (1986) as a "culture shock"). In a separate study, Mason *et al.* (1986a) carried out differential screening of both parietal endoderm and differentiated F9 mouse embryonal teratocarcinoma cDNA libraries and characterized a cDNA for a protein called SPARC. F9 teratocarcinoma have the ability to differentiate *in vitro* by addition of retinoic acid and dibutyryl cyclic adenosine monophosphate (cAMP) and subsequently begin to make basement membrane matrix components (Strickland and Mahdavi, 1978). During the process of F9 cell differentiation, expression of SPARC increased 20-fold (Mason *et al.*, 1986b). The identity of SPARC as the 43-kDa culture shock protein was revealed by strong homology between the deduced amino acid sequence from SPARC cDNAs and partial amino acid sequence of the 43-kDa culture shock protein (Mason *et al.*, 1986a). Similarities between these two proteins were also demonstrated by immunocross-reactivity and by tryptic peptide mapping (Mason *et al.*, 1986a). In parallel, a third laboratory described a protein called BM-40 (M_r 40,000). This protein was purified from mouse Engelbreth-Holm-Swarm (EHS) tumor and was found in tissue extracts of Reichert's membrane, culture media from mouse teratocarcinoma PYS cells, F9 cells, and Schwann cells (Dziadek *et al.*, 1986).

Final proof that SPARC, culture shock protein, osteonectin, and BM-40 (Mann *et al.*, 1987) were identical was revealed from the strong homology of directly determined amino acid sequences (bovine adult bone, Romberg *et al.*, 1985; BM-40, Mann *et al.*, 1987; bovine and porcine bone, Domenicucci *et al.*, 1988) and amino acid sequences deduced from cDNA sequences (bovine fetal bone, Young *et al.*, 1986; Bolander *et al.*, 1988; human placenta, Swaroop *et al.*, 1988; human osteosarcoma cell line SaOS-2, Villarreal *et al.*, 1989; human bone, Young *et al.*, 1990a; mouse and human BM-40, Lankat-Buttgereit *et al.*, 1988). Predicted amino acid sequences of human SPARC, osteonectin, and BM-40 revealed a complete agreement among these three transcripts from different sources. Figure 1 shows high homology of osteonectin/SPARC/BM-40 among several vertebrate species from frog (Damjanovski *et al.*, 1992) to

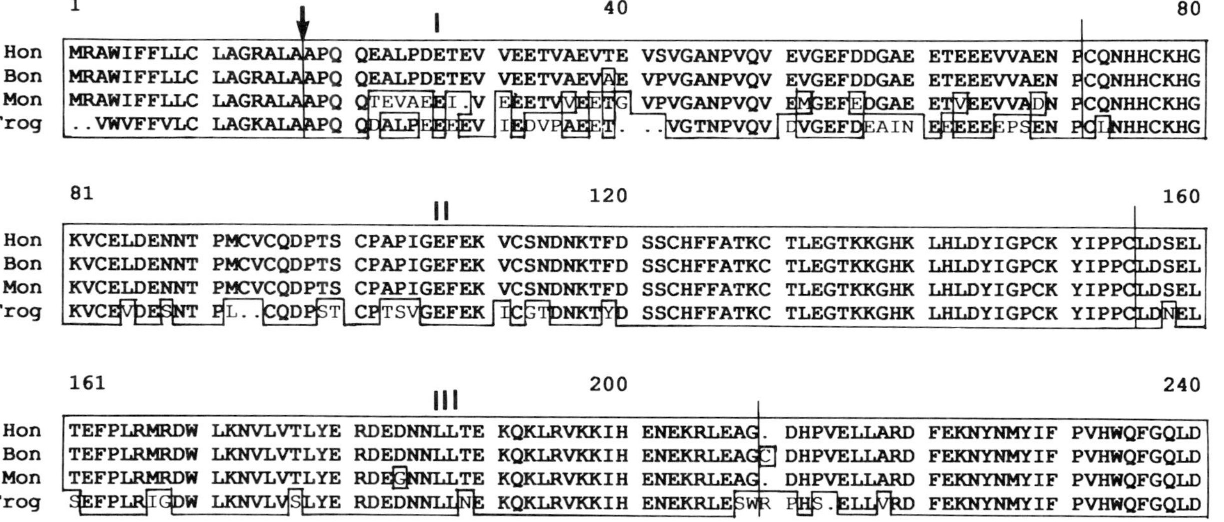

FIGURE 1 Osteonectin/SPARC/BM-40 is a protein highly conserved among vertebrates. Amino acid sequences predicted from osteonectin/SPARC/BM-40 cDNA sequences. Abbreviations Hon, Bon, Mon and Frog indicate predicted amino acid sequences from human (Swaroop *et al.*, 1988), bovine (Bolander *et al.*, 1988), mouse (Mason *et al.*, 1986a), and *Xenopus laevis* (Damjanovski *et al.*, 1992), respectively. Regions labeled I, II, III, and IV are predicted structural and potentially functional domains in osteonectin based on structure analysis. Boxed sequences are identical to human osteonectin.

human. Osteonectin is highly conserved (92% homology between mouse and human), suggesting that function of this protein may have been essential during evolution.

B. Osteonectin: Highly Conserved Structural Domains during Evolution

The entire amino acid sequence of human osteonectin predicted from the cDNA nucleotide sequence contains 303 amino acids (bovine, 304; mouse, 302) including a 17-residue signal peptide (see Fig. 1), giving it a predicted molecular weight of ~33,000 (Young *et al.*, 1990a). DNA sequence analysis of SPARC/BM40/osteonectin reveals four important domains in the predicted protein molecule, which are conformationally and functionally distinct (Mason *et al.*, 1986a; Engel *et al.*, 1987; Swaroop *et al.*, 1988; Bolander *et al.*, 1988; Young *et al.*, 1990a). Figure 2 shows a predicted domain structure of human osteonectin (Young *et al.*, 1990). Domain (I), at the N-terminus from Ala1 to Pro54, shows an acidic region that is rich in glutamic and aspartic acids. Domain (II) is rich in cysteines (Cys55 to Cys138) and contains two potential N-glycosylation sites, at Asn71 and Asn99. Most of the cysteine residues (11 out of 15) are found in this region and probably form disulfide linkages within the molecule. This hypothesis is based on the observation that the reduced protein migrates slower in denaturing polyacrylamide gel electrophoresis. Do-

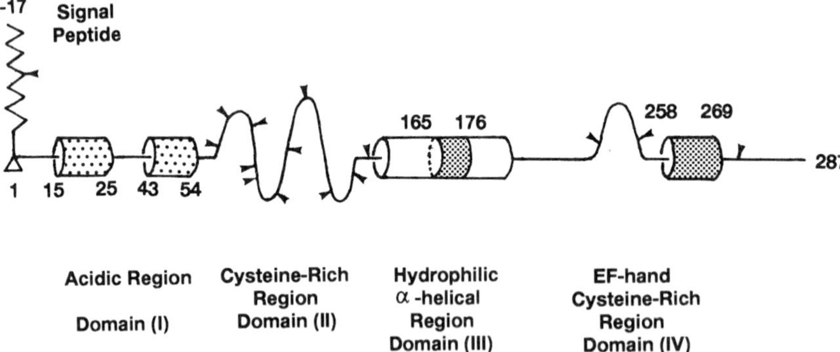

FIGURE 2 Predicted domain structure of human osteonectin. A protein model interpreted from amino acid structure analysis (Mason *et al.*, 1986a; Engel *et al.*, 1987; Swaroop *et al.*, 1988; Bolander *et al.*, 1988; Young *et al.*, 1990) was derived from complementary DNA sequencing. Shown are a signal peptide (17 amino acid residues) and domains (I) through (IV). Closed arrows (▶) indicate positions of cysteines and the open arrow (△) indicates N-terminus of the secreted osteonectin. Numbering is according to Bolander *et al.* (1988) and Young *et al.* (1990). α-Helical structure is predicted by computer analysis and is indicated by cylinders. Cylinders composed of amino acids 15–25 and 43–54 indicate internal homology and 165–176 and 258–269 indicate potential EF motifs.

main (III) is predominantly hydrophilic (Leu139 to Gly192), probably forming an α-helix. Domain (IV) contains three cysteines and two potential α-helical structures. The regions from Asp165 to Lys176 in domain (III) and from Asp258 to Glu269 in domain (IV) contain potential EF-hand calcium-binding loops (Bolander *et al.*, 1988). In Engel's model of SPARC structure, domain (I) was also predicted to have some Ca^{2+} binding activity presumably through the negative charges of the many acidic amino acids; however, an EF-hand motif was distinguished only in domain (IV) (Engel *et al.*, 1987).

Genomic analysis of both bovine (Young *et al.*, 1986) and human (Young *et al.*, 1990a) osteonectin and mouse (Mason *et al.*, 1986b) and human (Swaroop *et al.*, 1988) SPARC indicated that the osteonectin/SPARC/BM-40 gene is present in one copy per haploid genome. The gene was specifically localized to the central region of chromosome 11 in mouse (Mason *et al.*, 1986b) and to chromosome 5q31–33 in human (Swaroop *et al.*, 1988). Recently, a Southern analysis using DNA from vertebrate and nonvertebrate species indicated that the osteonectin gene appeared to be a single copy in all vertebrate genomes examined and that the comparative distances between species seemed to be consistent with the differences found between amino acid sequences (Ringuette *et al.*, 1991).

Northern analysis of tissues and cell lines from several species showed that the major transcript of osteonectin was 2.2 kilobases (kb) in mouse, bovine, and human (Mason *et al.*, 1986a; Young *et al.*, 1986, 1990a; Swaroop *et al.*, 1988; Nomura *et al.*, 1988; Lankat-Buttgereit *et al.*, 1988) with a small amount of an additional 3.0-kb transcript present in human tissues (Swaroop *et al.*, 1988; Lankat-Buttgereit *et al.*, 1988; Young *et al.*, 1990a). A 3.0-kb cDNA was described that contained a coding region identical to the 2.2-kb transcript but with utilization of a different downstream polyadenylation signal (Swaroop *et al.*, 1988). A relatively higher ratio of 3.0- to 2.2-kb message was observed among some tissues (muscle, Swaroop *et al.*, 1988; gingiva and periodontal ligament fibroblast from a patient, Young *et al.*, 1990a) as well as in normal keratinocytes and in the tumor cell line called MCF-7 (Howe *et al.*, 1990). In general, the message(s) of osteonectin was absent or very low in transformed cells and very variable in tumor cell lines (Mason *et al.*, 1986b; Swaroop *et al.*, 1988; Young *et al.*, 1986, 1990a; Howe *et al.*, 1990). The sizes of messenger RNA (mRNA) in *Xenopus laevis* (Damjanovski *et al.*, 1992) and chicken cartilage (Pacifici *et al.*, 1990) were 1.6 and 1.8 kb, respectively. Since no cDNA sequence containing potential alternatively spliced sequences has been found (Damjanovski *et al.*, 1992), it is likely that the shorter RNAs from *Xenopus* and chicken are the result of alternative utilization of different polyadenylation sites, as already described.

An analysis of the entire genomic structure (20–26.5 kb) of osteonectin revealed that in the mouse (McVey *et al.*, 1988), bovine (Findlay *et al.*, 1988), and human (Villarreal *et al.*, 1989), the osteonectin/SPARC/BM-40 genes were all encoded in 10 exons. Interestingly, the intron–exon junctions among these species were very similar. The sizes of exons were also relatively uniform (−130 basepairs [bp] long) except for exon 10, which was much larger (>1.0 kb). The first exon was found to span the 5′-noncoding region while exon 2 encoded the entire signal peptide sequence (17 amino acids) and the first two N-terminal amino acids of the mature protein. The most variable protein sequences among the species (see Fig. 1) were found in a single exon, 3 (McVey *et al.*, 1988; Findlay *et al.*, 1988). The acidic domain (I), Cys-rich (II), α-helix (III), and EF-hand motif (IV) in the protein are encoded in exons 3 and 4, exons 5 and 6, exons 7 and 8, and exons 9 and 10, respectively. Evolutionary conservation of the entire coding region of this gene and accumulation of mismatches in exon 3 during evolution may be strongly related to its function.

C. Osteonectin Binds to Ca^{2+}, Hydroxyapatite, Collagens, Thrombospondin, and Other Matrix Components

The association of osteonectin with mineral has been tested by adsorption of osteonectin to hydroxyapatite (HA) in 4 *M* guanidine HCl solution (Termine *et al.*, 1981a). Because osteonectin was also shown to stimulate the binding of both $^{45}Ca^{2+}$-labeled HA and $^{45}Ca^{2+}$ to insoluble collagen in a dose-dependent manner, it was speculated that osteonectin might play a role in the initiation of mineralization (Termine *et al.*, 1981b). Interestingly, 80% of BM-40 could be extracted from basement membrane using a physiological buffer containing 10 m*M* ethylenediaminetetraacetate, implicating that it, too, has Ca^{2+} binding activity (Mann *et al.*, 1987). A conformational change of BM-40 was observed in the presence of Ca^{2+} based on circular dichroism (CD) (Engel *et al.*, 1987). Romberg *et al.* (1985, 1986), on the other hand, did not detect a significant conformational change by CD; however, they found that bovine osteonectin strongly bound HA ($K_d = 8 \times 10^{-8}$ *M*) and that it inhibited HA formation more than five times as effectively as osteocalcin, suggesting that osteonectin might be an inhibitor rather than an initiator of mineralization. In addition to these studies, it should be noted that osteonectin is widely distributed in various nonmineralized tissues. These data together suggest that osteonectin may play a role as a modulator or organizer to assemble and/or stabilize the fibrillar network of collagens (as described later) and/or laminin–nidogen/entactin complexes (Paulsson *et al.*, 1987), perhaps in a Ca^{2+}-dependent manner.

Interactions between osteonectin and other extracellular matrix components also seem likely. Binding activity of bovine bone osteonectin to type I collagen was demonstrated in both a solid-phase (Termine *et al.*, 1981a, b) and an aqueous-phase study (Romberg *et al.*, 1985). Interestingly, complexes of procollagen and osteonectin were observed in the material immunoprecipitated from proteins biosynthesized by bone cells and from cell-free translation products of bone cell RNA using antibodies specific for osteonectin. This observation suggests that a complex formation between type I collagen and osteonectin may occur as early as when the proteins are translated and secreted (Young *et al.*, 1990b). On the other hand, a smaller fragment cleaved by a bone-derived metalloprotease was shown to have higher collagen-binding activity compared to an intact osteonectin (Tyree, 1989). It will be interesting to know when the interaction of these two proteins (presumably by complex formation) occurs and how it is regulated during the process of biosynthesis, secretion, and modification in the matrix.

An association of osteonectin with collagen was also observed in soft tissues. Sage *et al.* (1989) demonstrated that native SPARC preferentially bound to type III and type V collagens and thrombospondin. Further studies suggested that the Ca^{2+}-binding site of protein domain (IV) might contribute to the binding to collagens including type I (Lane and Sage, 1990). In BM-40 prepared from EHS tumor, specific calcium-dependent binding of domains (III) and (IV) to type IV collagen, but less to types I, III, V, and VI collagens, was observed (Mayer *et al.*, 1991). Interestingly, the bone (a high-mannose type) and platelet (a complex type) osteonectin had patterns of glycosylation that appeared to affect collagen binding activity (Kelm and Mann, 1991). Specifically, bone osteonectin bound to types I, III, and V collagens, and platelet osteonectin had no apparent affinity for them; these data suggest that post-translational modification of osteonectin may be controlled in a tissue-specific manner and, further, potentially associated with functions of osteonectin. Osteonectin was found in platelets (Stenner *et al.*, 1986) and shown to form a complex with thrombospondin, which is also a component of platelets (Clezardin *et al.*, 1988). Considering that thrombospondin is also found in the bone matrix and is produced by bone cells (Gehron Robey *et al.*, 1989), it is possible that osteonectin, collagen, and thrombospondin may form a network complex in bone matrix.

Indirect evidence for the co-regulation of collagen and osteonectin has been demonstrated. In bovine and rat endothelial cells undergoing angiogenesis *in vitro*, the transcription of both type I collagen and SPARC/osteonectin genes increased up to 12-fold compared to subconfluent cells (Iruela-Arispe *et al.*, 1991). Similar observations were found in the case of systemic sclerosis and localized scleroderma, where fibroblasts derived from patients overexpressed both collagen and osteonectin messages with a strong correlation (Vuorio *et al.*, 1991). Interestingly, while the release of thrombospondin and osteonectin from platelets by thrombin stimulation appeared to be co-regulated (Kelm and Mann, 1990), in angiogenesis of BAE cells, thrombospondin seemed to behave antagonistically to SPARC (Iruela-Arispe *et al.*, 1991). That is to say, thrombospondin message level decreased during angiogenesis, which was the reverse of that observed for collagen and SPARC. Furthermore, the presence of anti-thrombospondin antibodies increased vessel cord formation, indicating thrombospondin has an "anti-angiogenesis" function (Iruela-Arispe *et al.*, 1991). It is unknown whether the regulation of several matrix components, which possess potential binding activity to osteonectin, may be controlled in a coordinated tissue-specific manner.

The relationship between the structure and function of SPARC/osteonectin has been described in detail in the cultures of BAE cells using synthetic peptides and antibodies designed for the predicted functional domains (I) through (IV) (Sage *et al.*, 1989; Lane and Sage, 1990; Funk and Sage, 1991). Peptide 1.1 (acidic, potential Ca^{2+} binding) and peptide 4.2 (EF-hand, Ca^{2+} binding) were observed to efficiently inhibit the cell spreading in a dose-dependent manner (Lane and Sage, 1990). Peptide 4.2 was also shown to bind to collagens (types I, II, III, IV, V, and VIII) in a Ca^{2+}-dependent manner similar to that already described (Lane and Sage, 1990). Peptide 2.1 (Cys-rich) exhibited a dose-dependent inhibition of [^{3}H]thymidine uptake and delayed the G_0- to S-phase transition without anti-cell spreading (Funk and Sage, 1991). With regard to the study on cell spreading, it would be interesting in the future to test a peptide based on the bovine sequence instead of mouse, because domain (I) is the most variable in its amino acid sequence among species. In contrast to these studies, the overexpression of BM-40 by transfection of the full-length human BM-40 cDNA inserted in a eukaryotic vector was shown to have no effect on cell spreading or proliferation (Nischt *et al.*, 1991).

Quite surprisingly, two Ca^{2+}-binding glycoproteins called SC1 (Johnston *et al.*, 1990) and QR1 (Guermah *et al.*, 1991) (found in synaptic junctions in rat brain and in quail embryonic neuroretina, respectively) were shown to have a considerable homology to the EF-hand containing C-terminal sequence of osteonectin/SPARC/BM-40. The homology between the C-terminus of SC1 and mouse SPARC was 61% and that between SC1 and QR1 was 71%. Northern analysis of SC1 and mouse SPARC showed no cross-hybridization using high stringency conditions as well as differences in the two messages in both size and tissue abundance (Johnston *et al.*, 1990). Considering the evolutionary conservation of the osteonectin/SPARC/BM-40 gene, described earlier, and the appearance of its message in many tissues, SC1 and QR1 are probably encoded by one or two gene(s) that are distinct from osteonectin. It is still possible, however, that these proteins may be part of a superfamily related by their structure (EF-hand and binding activity to collagens) and function.

D. Osteonectin: Produced by Other Tissues and Predominantly Expressed in Bone

Studies on immunohistochemical localization of osteonectin in fetal bovine (Termine *et al.*, 1981b; Bianco *et al.*, 1988) and human bone (Jundt *et al.*, 1987; Bianco *et al.*, 1988) showed that osteonectin was localized to the cytoplasm of osteoblasts and osteoprogenitor cells as well as in young osteocytes and newly synthesized osteoid in bone matrix. Chondrocytes in the metaphyseal growth plate also stained positively using antibodies against osteonectin. Interestingly, in cartilage, osteonectin was found in the mineralizing zone but at a relatively very low level in resting, proliferating, and early hypertrophic zones, where osteonectin was mainly cell-associated (Pacifici *et al.*, 1990). In this latter study, osteonectin was demonstrated to be synthesized by both cells in nonmineralizing and mineralizing zones, indicating that osteonectin was produced by chondrocytes in each zone, but preferentially accumulating in the mineralizing zone. Considering the potential functions of osteonectin as a Ca^{2+}-binding protein, it is interesting to note that a considerable amount of osteonectin was found in the cytoplasm, particularly in the Golgi field and transporting or secretory vesicles (Romanowski *et al.*, 1990). *In situ* hybridization of developing human bones (Metsäranta *et al.*, 1989) and mouse chondyles (Strauss *et al.*, 1990) revealed that osteoblasts, cells in the periosteum, and hypertrophic chondrocytes expressed osteonectin at high levels. These studies were consistent with the observations in the mouse embryo by *in situ* hybridization using a SPARC/osteonectin complementing RNA probe (Holland *et al.*, 1987; Nomura *et al.*, 1988). In osteoblastlike cell culture, osteonectin was predominantly produced by cells derived from bovine (Whitson *et al.*, 1984), porcine (Otsuka *et al.*, 1984), and human (Gehron Robey and Termine, 1985) bones. Surprisingly, in rat bone and dentin, osteonectin was a minor component compared to other noncollagenous matrix proteins and a lower level of osteonectin was synthesized by rat calvarial cells than by fibroblasts derived from periodontal ligament (Zung *et al.*, 1986). Further study using rat calvarial cells showed that a larger amount of osteonectin was synthesized by more mature bone cell populations compared to more immature fibroblastic ones. In addition, the inducing effect of osteonectin by transforming growth factor β (TGF-β) was smaller in more mature cells (Wrana *et al.*, 1988). In rat (Nagata *et al.*, 1991a) and porcine calvaria (Domenicucci *et al.*, 1988), osteonectin was not found to be phosphorylated. In bovine bone (Termine *et al.*, 1981a) and mouse PYS cells (Engel *et al.*, 1987), however, there is evidence that osteonectin is phosphorylated. The differences among these studies reflect differences in both species and cell culture systems as well as the nature of an *in vitro* system compared to the situation *in vivo*.

Apparently, osteonectin is synthesized not only by cells from bone but also by the cells in other tissues including endothelial (Sage *et al.*, 1984), teratocarcinoma (Mason *et al.*, 1986a), and Reichert's membrane (Dziadek *et al.*, 1986). Distribution of osteonectin has been reported in several other nonbone sources including porcine periodontal ligament (Wasi *et al.*, 1984) and other dental tissues (Tung *et al.*, 1984), human platelets (Stenner *et al.*, 1986), mouse parietal endoderm, placenta, adult ovary, lung, and adrenal gland (Mason *et al.*, 1986b), and bovine tendon (Young *et al.*, 1986).

Although osteonectin is expressed by many tissues, it clearly has a precise temporal and spatial pattern during development. The distribution of SPARC/osteonectin in mouse development is summarized in Table I (Mason *et al.*, 1986b; Holland *et al.*, 1987; Nomura *et al.*, 1988). In brief, very high expression of SPARC/osteonectin was detected in the periosteal cells of membranous and endochondral bone, odontoblasts, and whisker follicles; relatively high expression was detected in ovary and testis in an adult; and very low expression in brain and liver. An increasing expression pattern of osteonectin in placenta was also detected from 14 to 18 days of gestation. Interestingly, abundant expression of osteonectin in mineralizing and remodeling tissues suggests that osteonectin may be closely related to tissue remodeling, cell migration and invasion, Ca^{2+} transport, and secretion.

E. Osteonectin Expression: Potentially Controlled by Multiple *cis*- and *trans*-Acting Factors

Compared to the numerous studies on the chemical and biological properties and localization of osteonectin/SPARC/BM-40, very few studies have been carried out on the regulation of this gene. In bone cell culture, osteonectin was shown to be synthesized and secreted within 40 min (Otsuka *et al.*, 1984). Expression of osteonectin during the differentiation of bovine bone cells has been demonstrated to be relatively high and stable from an early stage throughout differentiation (Ibaraki *et al.*, 1992b). The effects of hormones and growth factors on osteonectin expression have been also reported. Osteonectin message was revealed to decrease up to 50% by treatment with 1,25 dihydroxyvitamin D_3 in cultures of adult human bone cells (Gehron Robey *et al.*, 1986). A threefold increase in osteonectin expression was detected in ROS 17/2.8, a rat osteosarcoma cell line, when treated with TGF-β (Noda and Rodan, 1987; however, no effect was found with 1.0 ng/ml TGF-β in rat calvarial cells (Wrana *et al.*, 1988). In UMR 201 cells, which are nontransformed clonal cells derived from neonatal rat calvaria, osteonectin expression increased 1.5-fold by 10^{-6} *M* retinoic acid and 2-fold by 10^{-7} *M* dexamethasone (DEX) (Ng *et al.*, 1989) as well as 2.5-fold by 10^{-7} *M* insulinlike growth factor (Thiebaud *et al.*, 1990); however, no effects on expression were detected using 10^{-7} *M* growth hormone or 10^{-6} *M* insulin (Thiebaud *et al.*, 1990). As mentioned earlier, in undifferentiated F9 cells (Strickland and Mahdavi, 1978; Mason *et al.*, 1986b), retinoic acid plus dibutyryl cAMP is a strong inducer of osteonectin gene transcription. After differentiation, retinoic acid has little or no effect. Furthermore, transcription of osteonectin appears to be suppressed in several transformed cells and tissues as already described (Young *et al.*, 1986; Mason *et al.*, 1986b). It is unknown whether the differential responses to hormones and growth factors are tissue- and/or cell-specific, species-specific, or regulated by the transformed status of the cells tested.

Potential regulatory elements in SPARC/osteonectin gene were described in the mouse (McVey *et al.*, 1988; Nomura *et al.*, 1989) and bovine (Young *et al.*, 1989; Dominguez *et al.*, 1991) and showed that the 5′ end of the gene had several interesting features, which are listed as follows.

1. Neither a TATA nor a CCAAT were found upstream of the initiation site.
2. Multiple repeats of the sequence CCTG are found in exon 1 (human has four, mouse has seven, bovine has five; Swaroop *et al.*, 1988).
3. A 72-bp purine-rich stretch of direct repeats with the sequence motif GGGA and GGA ("GAGA box") as found from bases −55 to −126. The bovine promoter has a single long "GAGA box" and CCCT repeats further in the upstream, which, interestingly, are complementary to the GGGA sequence. In the mouse, the GA-rich region and CCCT repeats were considered as three separate "GAGA boxes" (Nomura *et al.*, 1989). In the bovine promoter, the GA-rich sequence is sensitive to the single-stranded nuclease called S1, indicating that it can assume a hinged or triplex conformation.
4. Approximately 500 bp of 5′ flanking sequence of the gene was necessary for the highest CAT (chloramphenicol transacetylase) activity.
5. DNA further upstream of the first 500 bp of the gene had an inhibitory effect on CAT activity and
6. the transcription factor activator protein 2 (AP-2) might be involved in the promoter activity.
7. A GC box for transcription factor SP1, and
8. cAMP-responsive element was also revealed in the bovine osteonectin promoter region (within the 500-bp 5′ flanking sequence of the gene) (Young *et al.*, 1989).

The relationship between the 5′ flanking sequence and the promoter activity is shown in Figure 3. Surprisingly, ONREX, which contains a portion of exon 1 (5′ noncoding region or DNA flanking the 5′ side of the

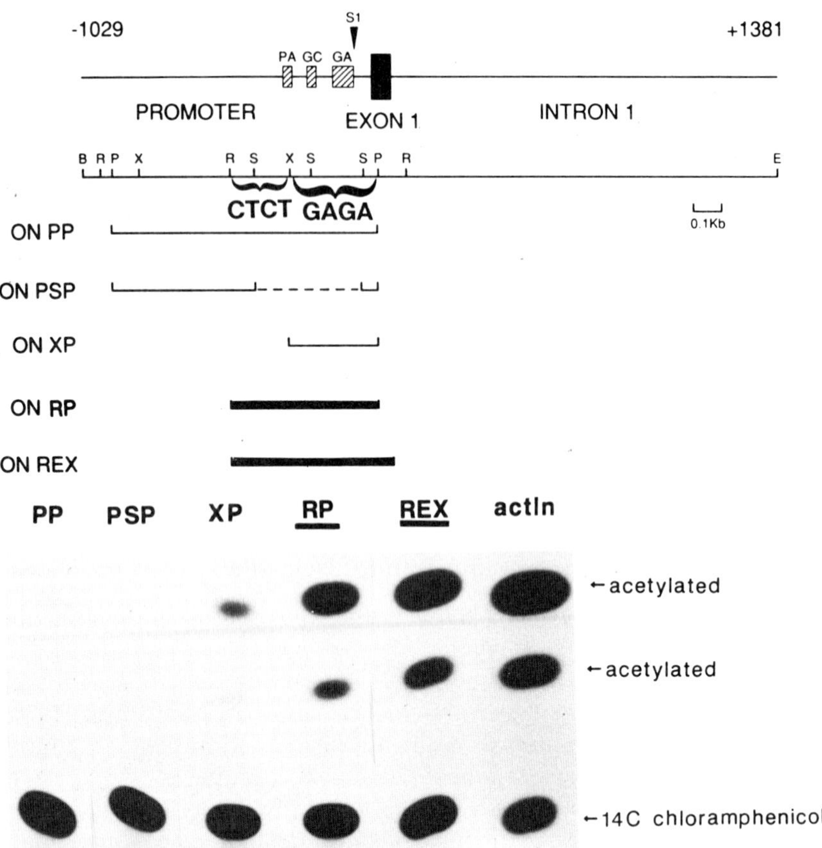

FIGURE 3 Promoter activity of the osteonectin gene. A genomic clone was isolated from a bovine library and a 2.0-kb fragment of DNA characterized by DNA sequencing. Several interesting features were revealed including an imperfect palindrome (PA), a GC-rich region (GC), and a region rich in purines (GA), which is sensitive to S1 endonuclease (S1) *in vitro* (Young *et al.*, 1989). A DNA fragment rich in purines extending from an *Xba*I site (X) to a *Pst*I site (P) is called "GAGA" while a pyrimidine-rich fragment located just upstream extends to an *Rsa*I site (R) is called "CTCT." Other restriction sites shown are *Bam*HI (B) and *Sst*I (S). To determine what role this DNA has in transcriptional control, it was ligated to the chloramphenicol acetyltransferase (CAT) gene and transfected into cultured bone cells. The name and corresponding part of the osteonectin promoter used in each transfection is indicated below the physical restriction map of the gene. The bottom of the figure shows the CAT activity of each construct measured by the conversion of [^{14}C]chloramphenicol to acetylated derivatives (shown with arrows). β-Actin promoter is shown in the last lane as a positive control.

translational start site) has greater activity than ONRP, which is a construct without exon 1. These data imply that exon 1 may, in part, regulate osteonectin gene activity. The promoter region contained within construct ONRP has been demonstrated to have potential cell-specific promoter activity (Dominguez *et al.*, 1991). As already described, the 5′ flanking sequence of the osteonectin gene has been shown to have positive (ONREX; −504/+40 [note: +1 represents the initiation site of transcription]) and negative regulatory elements (−1039/−505) (Dominguez *et al.*, 1991). ONREX can be digested by the restriction enzyme *Xba*I into two fragments designated "CTCT-rich" (−504/−287) and "GAGA" (−286/+40). These CTCT-rich and GAGA DNA fragments carry (*a*) CCCT repeats (which are complementary to the GGGA repeat in the cow, or "GAGA box 3," as termed in the mouse) and (*b*) a single long "GAGA box" (called GAGA boxes 1 and 2 in mouse). Several lines of evidence indicate that these two regions are likely to interact with each other, probably both by direct DNA–DNA interaction and through some transcription factors (Ibaraki *et al.*, 1992a).

To determine the nature of the proteins that bind to the GAGA box, nuclear proteins from bovine bone cells were cross-linked by ultraviolet light to a bromodeoxy uridine triphosphate (BrdUTP)-containing oligonucleotide that encoded most of the GAGA sequence. After DNase digestion and separation by sodium dodecyl sulfate (SDS) gel electrophoresis, a candidate transcription factor was detected that had a M_r 40,000 (Ibaraki *et al.*, 1992a). It will be interesting to determine the nature of this factor and, further, how it interacts with additional *trans*-acting factors in bone.

When dexamethasone (DEX) or retinoic acid (RA) as added to UMR 201 separately, a dose-dependent response was noted in CAT activity of the osteonectin promoter (Ng *et al.*, 1989). This response was attenuated when DEX and RA were added in combination. It was speculated, therefore, that the actions by either DEX or RA alone may act directly on transcription and that the attenuation by the combination of DEX and RA may involve competitive binding by more than two transcription factors to a single binding site rather than, for example, by synergistic cooperation. The regulation of the osteonectin gene during development and in response to some environmental changes is likely to be the accumulated result of a sequential cascade of interactions between *cis*- and *trans*-acting factors.

III. OSTEOPONTIN AND BONE SIALOPROTEIN

OPN and BSP are acidic, RGD-containing glycoproteins found in humans, rabbit, mouse, rat, ovine, bovine, porcine, canine, and chicken (Kiefer *et al.*, 1989; Young *et al.*, 1990c; Fisher *et al.*, 1983b, 1987, 1990; Kinne and Fisher, 1987; Miyazaki *et al.*, 1989; Oldberg *et al.*, 1986, 1988a; Prince *et al.*, 1987; Kerr *et al.*, 1991b; Franzen and Heinegard, 1985; Andrews *et al.*, 1967; Wrana *et al.*, 1989; Zhang *et al.*, 1990; Ullrich *et al.*, 1991; Moore *et al.*, 1991; Gotoh *et al.*, 1990a, b; Castagnola *et al.*, 1991).

TABLE II Growth Factor and Hormonal Regulation of Osteoportin and Bone Sialoprotein[a]

Factor	RNA	Protein
Transforming growth factor ß	↑ OPN MC3T3-E1[1]	↑ OPN MC3T3-E1[6]
	↑ OPN ROS 17/2.8[1–3]	↑ OPN ROS 17/2.8[2]
	– OPN ROS 17/2.8[4,5]	↓ OPN ROS 17/2.8[5]
	↑ OPN rat calvaria[2,4,5]	↑ OPN rat calvaria[2,4–6]
Platelet-derived growth factor	—	↑ OPN rat calvaria[5]
Epidermal growth factor	—	↑ OPN rat calvaria[5]
b Fibroblast growth factor	↑ OPN ROS 17/2.8[7]	—
	↑ OPN Swiss 3T3 cells[8]	—
Leukemia-inhibitory factor	↑ OPN MC3T3-E1[9]	—
Interleukin 1	↑ OPN MC3T3-E1[10]	↑ OPN MC3T3-E1[10]
	↑ OPN mouse calvaria [10]	—
Tumor necrosis factor	↑ OPN mouse calvaria[10]	—
	↑ OPN MC3T3-E1[10]	—
Retinoic acid	– OPN ROS 17–2.8[5]	↑ OPN ROS 17–2.8[5]
	—	↑ OPN rat calvaria [5]
Parathyroid hormone	↓ OPN ROS 17/2.8[3,5,11]	—
	↓ OPN rat calvaria[5]	↑ ↓ OPN rat calvaria[5]
Cyclic adenosine mono phosphate agonists	↓ OPN ROS 17–2.8[3,7]	—
Dexamethasone	↓ OPN ROS 17/2.8[3,12,13]	—
	↓ OPN rat calvaria[13]	—
	↑ OPN rat marrow stroma[14,15]	↑ OPN rat marrow stroma[15]
	↑ BSP ROS 17/2.8[13]	—
	↑ BSP rat calvaria[13]	—
1,25-Dihydroxyvitamin D_3	↑ OPN ROS 17/2.8[3,5,13,16]	↑ OPN ROS 17/2.8[5]
	↑ OPN MC3T3-E1[10]	—
	↑ OPN mouse calvaria [10]	—
	↑ OPN rat calvaria[5,13,17–19]	↑ OPN rat calvaria[5]
	↑ OPN rat marrow stroma[14]	↑ OPN mouse JB6 cells[6]
		—
	↓ BSP ROS 17/2.8[13]	—
	↓ BSP rat calvaria cells[13]	
Transformation	↑ OPN mouse epidermis[20,23]	↑ OPN mouse epidermis[20]
		↑ OPN mouse JB6 cells[20]
	↑ OPN mouse JB6 cells[22]	↑ OPN 3T3 cells[21]
	↑ OPN 3T3 cells[21]	↑ OPN rat kidney line[21]
	↑ OPN rat kidney line[21]	↑ OPN human osteosarcoma[24]
		↑ OPN human fibrosarcoma[24]
		↑ OPN human cervix[24]
		↑ OPN human colon[24]
		↑ OPN human ovary[24]
		↑ OPN human bladder[24]

(*continued*)

TABLE II (*Continued*)

Factor	RNA	Protein
		↑ OPN human macrophage[25] ↑ OPN hamster fibroblast[26,27] ↑ OPN rat fibroblast[26,27] ↑ OPN mouse fibroblast[26,27] ↑ OPN rabbit fibroblast[26,27] ↑ OPN rat liver epithelia[28] ↑ OPN mouse mammary tissue[28]

[a] Arrows indicate a relative increase(↑), relative decrease (↓), or no change (−) in gene expression. Superscript numbers indicate the references for each observation and are listed as follows: 1, Noda *et al.*, 1988; 2, Kubota *et al.*, 1989; 3, Noda and Rodan, 1989; 4, Wrana *et al.*, 1991; 5, Kasugai *et al.*, 1991a; 6, Ber *et al.*, 1991; 7, Rodan *et al.*, 1989; 8, Nomura *et al.*, 1988; 9, Noda *et al.*, 1990; 10, Jin *et al.*, 1990; 11, Noda *et al.*, 1989; 12, Yoon *et al.*, 1987; 13, Olberg *et al.*, 1989; 14, Leboy *et al.*, 1991; 15, Kasugai *et al.*, 1991b; 16, Prince *et al.*, 1987a; 17, Owen *et al.*, 1991; 18, Heath *et al.*, 1989; 19, Chang *et al.*, 1991; 20, Craig *et al.*, 1989; 21, Craig *et al.*, 1988; 22, Smith and Denhardt, 1987; 23, Craig *et al.*, 1990; 24, Senger and Perruzzi, 1985; 25, Senger *et al.*, 1988; 26, Senger *et al.*, 1979; 27, Senger *et al.*, 1980; 28, Senger *et al.*, 1983.

These proteins are regulated at multiple levels with respect to (1) post-transcriptional (OPN) and post-translational (OPN and BSP) modifications, (2) induction/enhancement by hormones, growth factors, and tumor promoters (Table II), and (3) tissue specificity (Table III).

A. Osteopontin Promoter: Potential Activator Protein-1 Regulation and an Unusual Vitamin D-Responsive Element

Little is known about the *cis*- and *trans*-acting regulatory elements that control these genes. The genomic organization and promoter sequence in the murine OPN gene were described by Miyazaki *et al.* (1990), with an additional noncoding exon (exon 1) detected by Craig and Denhardt (1991). Structural analysis of the promoter indicates that the gene, unlike osteonectin, contains TATA- and CCAAT-like consensus sequences. Transient transfections of the promoter region using JB6 epidermal cells (Craig and Denhardt, 1991) indicated that the promoter was functional and contained a positive transcription element between nucleotides −543 and −253 and a negative transcription element between nucleotides −777 and −543. Computer-generated comparisons between the

TABLE III Distribution of Osteopontin and Bone Sialoprotein in Nonskeletal Tissues or Cells

	RNA	Protein
Serum	—	BSP human[1]
	—	OPN human,[2] rat[2]
Plasma	—	BSP human[1]
	—	OPN human,[2] rat[2]
Platelets	—	BSP human[1]
Macrophages	—	OPN rat[2]
T lymphocytes	OPN mouse[3]	—
Adrenal gland	OPN mouse[4]	—
Kidney	OPN mouse,[4,5] rat [6,7]	OPN rat [14]
Corotid artery	OPN rat[7]	—
Aorta	OPN rat[7]	—
Ear	OPN mouse[4]	—
Stomach	OPN rat[7]	—
Heart	OPN rat[7]	—
Lung	OPN mouse[8]	—
Skin	OPN mouse,[8] cow[9]	—
Tendon	OPN cow[9]	—
Brain	OPN cow[9]	—
Fatty tissue	OPN mouse[8]	—
Uterus	OPN rat[7]	—
Placenta	OPN mouse,[4,8] human[10]	—
Decidua	OPN mouse,[4,8] human[10]	—
Trophoblast	BSP human[11]	BSP human[11]
Urine	—	OPN human[12]
Milk	—	OPN human[13]

Superscript numbers indicate the references used for each tissue and are listed as follows: 1, Chenu and Delmas, 1992; 2, Senger *et al.*, 1988; 3, Patarca *et al.*, 1989; 4, Nomura *et al.*, 1988; 5, Craig *et al.*, 1989; 6, Yoon *et al.*, 1987; 7, Giachelli *et al.*, 1991; 8, Craig and Denhardt, 1991; 9, Kerr *et al.*, 1991b; 10, Young *et al.*, 1990c; 11, Bianco *et al.*, 1991; 12, Shiraga *et al.*, 1992; 13, Senger *et al.*, 1989b; 14, Mark *et al.*, 1988b.

OPN promoter and known regulatory DNA elements indicated several potential activator protein (AP-1) binding sites. While these specific consensus sequences were not tested for transcriptional activity per se, addition of a phorbol ester (TPA) induced the activity of the reporter gene approximately two-fold and is consistent with the *in vivo* and *in vitro* induction of OPN message by this tumor promoter (Table II). The vitamin D metabolite, $1,25(OH)_2D_3$, has also been shown to stimulate OPN expression (Table II). A 24-bp sequence containing two direct repeats of 9 bp (AGGTTCACG) and located 761 bp upstream of the transcription start point was recently identified as a vitamin D response element in the murine OPN gene (Noda *et al.*, 1990). Analysis of the vitamin D response element by transient transfections in ROS 17/2.8 cells and by gel shifts indicated functionality of this regulatory element as well. Although estrogen and progesterone have been shown to induce OPN expression when applied to the skin of nonpregnant mice, DNA consensus sequences for these steroid response elements have not been identified as yet in the murine OPN promoter or upstream sequence (Craig and Denhardt, 1991).

B. Bone Sialoprotein Gene: An AC-Rich Structure in Intron 1 Composed Exclusively of Type 0 Intron–Exon Junctions

Characterization of the human BSP gene is currently in progress (Kerr *et al.*, 1991a). Of the 5′ upstream sequence from the transcriptional start site analyzed so far, BSP contains TATA and CCAAT consensus sequences in the reverse orientation (Kerr, unpublished data). Intron 1 contains several interesting features. Approximately 900 bp upstream from the 5′ side of exon 2 is a polypurine (C:T-rich) stretch that is also found in the OPN promoter (Miyazaki *et al.*, 1990). Interestingly, in the antisense orientation, this repetitive sequence is similar, but not identical, to the osteonectin "GAGA box," which can form intramolecular DNA triplexes (Bianchi *et al.*, 1990; Section II.E). Immediately 3′ to the polypurine repeat is an alternating purine : pyrimidine tract. In the sense orientation, this structure is composed of tandem repeats of A:C nucleotides that can adopt a left-handed helical configuration (Haniford and Pulleyblank, 1983; McLean and Wells, 1988) and has been implicated in stimulating homologous recombination events (Wahls *et al.*, 1990). Located in the upstream region of the rat prolactin gene, this purine–pyrimidine stretch has an inhibitory effect on transcription (Naylor and Clark, 1990), while at the 3′ end of genes, this sequence appears to be an important signal element for efficient cleavage and polyadenylation of mRNAs (Heath *et al.*, 1990).

Another interesting feature of the BSP gene is that the intron–exon junctions are type 0 (i.e., the codon remains intact). With this type of organization, any deletion or insertion event within any exon will still allow conservation of the open reading frame of downstream exons. The significance of the type 0 junctions and the unusual purine and purine : pyrimidine tracts is unknown at this time; however, these unusual features are intriguing given the fact that the expression of BSP is tightly regulated with respect to its tissue localization as well as its induction during specific stages of development, both *in vivo* and *in vitro*.

C. Distribution, Structure, and Potential Functions of Osteopontin and Bone Sialoprotein

In bone, both OPN and BSP can be detected in the osteoid; however, these proteins are more abundant at the mineralization front (Butler,

1989; Weinreb *et al.*, 1990; Alberius and Johnell, 1990; Langille and Solursh, 1990; Hultenby *et al.*, 1991; Chen *et al.*, 1991; Moore *et al.*, 1991; Bianco *et al.*, 1991; Mark *et al.*, 1987, 1988a). *In vitro* studies parallel this observation, showing the induction of these proteins with the more mature osteoblast phenotype and their association with a mineralized matrix. Fetal bovine (Ibaraki *et al.*, 1992b), rat (Owen *et al.*, 1990, 1991; Nagata *et al.*, 1991a; Chen *et al.*, 1991), and porcine (Nagata *et al.*, 1991b) bone cell cultures maximally express OPN and BSP at the onset of mineralization and are localized to the mineralized matrix; however, these proteins can be synthesized in the absence of a mineralizing matrix. In the presence of DEX, expression of these proteins is also observed in rat marrow stromal cell cultures (Leboy *et al.*, 1991; Kasugai *et al.*, 1991a; Nagata *et al.*, 1989), where they are secreted into the media.

Because these sialoproteins are localized in mineralizing matrix, both OPN and BSP may play roles in the mineralization process. BSP appears to be a likely candidate in this regard because its localization is restricted to skeletal tissue (Bianco *et al.*, 1991) and trophoblasts (Table III). It should be noted that in the latter system, focal points of mineralization have been detected. Because OPN is expressed in numerous tissues and fluids that normally are not calcified (Table III), it is possible that OPN has an inhibitory role in crystallization. This is supported, in part, by the observations of Shiraga *et al.* (1992) in which uropontin (human osteopontin from urine) inhibited calcium oxalate formation. Whether OPN can have a dual role in inhibiting crystal formation in the fluid phase (milk, serum, urine) and enhancing crystallization in the solid phase (i.e., in bone) is not clear. It is also not understood whether the overall composition of the extracellular matrix constituents, or the possible post-transcriptional and post-translational modifications observed in OPN regulate its function in the mineralization (or nonmineralization) process.

Originally isolated and characterized from bone, BSP was shown to contain approximately 50% protein with post-translational modifications including sialic acid, galactosamine, and glucosamine comprising the remainder of the molecule (Fisher *et al.*, 1983b; Franzen and Heinegard, 1985). In culture, osteoblasts obtained from newborn mouse calvaria were shown to synthesize sulfated BSP in the form of sulfated tyrosine residues (Ecarot-Charrier *et al.*, 1989). Midura *et al.* (1990) later reported the synthesis of a highly sulfated form of BSP in a rat osteosarcoma cell line, UMR 106-01. In this cell system, 50% of the sulfate was found as tyrosine sulfate while the remainder of the sulfate was associated with N- and O-linked oligosaccharides.

OPN (Prince, 1987) was demonstrated to contain several post-translational modifications including phosphorylation of approximately 1 threonine and 12 serine residues and glycosylation (1 *N*-glycoside and

5–6 *O*-glycosides). Recently, a post-transcriptional modification of human OPN involving the splicing of 14 amino acids (Young *et al.*, 1990c; Shiraga *et al.*, 1992) corresponding to exon 4 of the mouse OPN gene (Miyazaki *et al.*, 1990; exon 5 from Craig and Denhardt's [1991] structural analysis was described. Within this stretch of 14 amino acids is a potential site of serine phosphorylation or O-linked glycosylation. Since this modified protein has not been isolated and characterized by amino acid sequencing and analyzed for phosphorus and carbohydrate content, its biological relevance is difficult to determine. However, this modification in primary protein structure, as well as strictly post-translational modifications, must be of some significance because several observations of multiple isoforms have recently been reported.

The first indication of differential post-translational modifications was reported by Nemir *et al.* (1989) using normal rat kidney (NRK) cell line. Both nonphosphorylated (np69) and phosphorylated (pp69) forms of OPN were detected in conditioned media, but only pp69 was found to be associated with the cell surface. Furthermore, treatment of NRK cells with vanadyl sulfate (giving rise to an altered phenotype including anchorage-independent growth) resulted in an increase of pp69 at the cell surface, a decrease of secreted pp69, and an increase in secreted np69. Also noted from these studies was the co-precipitation of np69 with fibronectin. K. Singh *et al.* (1990) demonstrated that the relative amounts of secreted pp69 and np69 are dependent on the passage number of these cultures of NRK in which the pp69 is secreted in low and np69 is secreted in high number passages. Furthermore, these researchers noted that only np69 contains the N-linked glycoside (a consensus sequence for N-linked glycosylation is located immediately amino-terminal to the polyaspartic acid stretch). Inhibition of N-linked glycosylation or enzymatic removal of the N-linked glycoside inhibited the formation of an np69–fibronectin complex, whereas the formation of a complex of pp69 and cell surface-associated fibronectin remained unaltered. The only other report, to date, of a nonphosphorylated form of osteopontin is associated with a mouse epidermal cell line JB6. These cells are transformable by the tumor promoter, TPA, and transformation results in an induction of OPN RNA and subsequent synthesis of the phosphorylated form of the protein. In contrast to TPA, treatment of JB6 mouse epidermal cells with $1,25(OH)_2D_3$ has been shown to stimulate the synthesis of nonphosphorylated OPN (Chang and Prince, 1991). In this study, there was no apparent correlation between OPN expression and tumorigenicity.

Other forms of OPN have been detected. Kubota *et al.* (1989) showed two forms of OPN in conditioned media of rat bone cell populations, with the 55-kDa form (less phosphorylated) more predominant than the 44-kDa form (more phosphorylated) in these cultures. Using a rat calvarial cell population enriched in osteoblasts (RCIV), a rat osteosarcoma cell line (ROS 17/2.8), and a rat calvarial clonal line (RCAII), Kasugai *et al.* (1991b) showed that treatment with 1,25$(OH)_2D_3$ results in the equal induction of both forms of OPN in ROS 17/2.8, predominantly the 55-kDa form in RCIV, and only the 55-kDa form in RCAII. Kasugai *et al.* (1991a) also cultured rat bone marrow cells using mineralizing or nonmineralizing culture conditions. In this system, DEX induced the expression of OPN. Using nonmineralizing conditions, both the 55- and 44-kDa forms of OPN were isolated from the media, although the 55-kDa form was more predominant. Under mineralizing conditions, however, the more phosphorylated 44-kDa form of OPN appeared to be preferentially incorporated into the matrix (Kasugai *et al.*, 1991a; Nagata *et al.*, 1991a). If, *in vivo*, the nonphosphorylated form of OPN inhibits crystal nucleation, and 1,25$(OH)_2D_3$ induces the expression of this isoform, it would be interesting to speculate that this may be a potential mechanism by which vitamin D stimulates bone synthesis but inhibits the mineralization process *in vivo* (Hock *et al.*, 1986; Wronski *et al.*, 1986) and *in vitro* (Owen *et al.*, 1991). Sulfation also appears to be an important modification in that the sulfated form of OPN is restricted to the extracellular matrix under mineralizing conditions (Nagata *et al.*, 1989, 1991b). Whether the synthesis of the sulfated form of OPN is restricted to bone cells or is similarly modified in other tissues is unknown.

Both OPN and BSP enhance cell adhesion and cell spreading when cells are added to protein-coated plastic *in vitro* (Oldberg *et al.*, 1986, 1988a, b; Somerman *et al.*, 1987, 1988, 1989; Sommarin *et al.*, 1989; Sauk *et al.*, 1990). This is presumably through the RGD cell-attachment site binding to an integrin receptor (Ruoslahti and Pierschbacher, 1987), because addition of RGD-containing peptides in this *in vitro* assay greatly inhibits cell adhesion. While many proteins contain the RGD polypeptide, only a small percentage of these proteins serve as cell adhesion molecules. This property is presumably based on the conformation of the protein such that the RGD is exposed and able to assume a hydrophilic beta-turn (Cachau *et al.*, 1989). Aside from the cell adhesion studies, there is only one report in the literature on the localization of these sialoproteins to cell membranes. OPN appears to be adjacent to the clear zone of the plasma membrane of osteoclasts; an area in which vitronectin receptors are localized as well (Reinholt *et al.*, 1990). There is evidence indicating that the adhesive properties of both OPN and BSP initiate intracellular signaling in chicken osteoclasts (Miyauchi *et al.*, 1991). Exposure of chicken osteoclasts to BSP and OPN resulted in a decrease in cytosolic calcium (increased efflux) and an increase in resorptive activity. The signal transduction (i.e., calcium efflux) is thought to be mediated via the vitronectin receptor affecting a calmodulin-dependent Ca^{2+}-ATPase. Addition of RGD-containing OPN peptides resulted in an inhibition of both attachment of osteoclasts to bone particles and resorptive activity.

Whether this property of cell–matrix attachment is for the purpose of cell adhesion, disruption of focal adhesion points to initiate migration, or both is difficult to ascertain. At this time, the evidence in the literature is ambiguous. The fact that OPN is detected in proliferating populations of bone cells *in vitro* (Owen *et al.*, 1990; Stein *et al.*, 1990) may indicate the necessity for these cells to migrate through the matrix. In this case, OPN may serve either as an antiadhesion molecule to allow movement or, alternatively, as a tracking mechanism for the cells upon which to move. Several *in vivo* models would also support either hypothesis in that an increase in OPN appears to correlate with the proliferation and migratory activities of cells/tissues involved with the process of natural expansion. One example of this is decidua, which contains a population of granulated metrial gland cells that express elevated levels of OPN (Nomura *et al.*, 1988; Young *et al.*, 1990c). Based on the lytic potential of these cells, granulated metrial gland cells are postulated to play a role in the breakdown, invasion, and subsequent remodeling of uterine stroma and maternal blood vessels. An increased OPN expression has also been observed in arteries after an endothelial denuding injury (Giachelli *et al.*, 1991); this process initiates smooth muscle cell migration across the internal elastica to the intimal region where the cells proliferate and synthesize an extracellular matrix. Since milk contains OPN (Senger *et al.*, 1989b), it is reasonable to assume that expression of OPN occurs during the expansion (proliferation and concomitant matrix production) of breast tissue during pregnancy. This is supported by the observation that OPN is induced in lactating breast tissue of the mouse (Craig and Denhardt, 1991). Another scenario in which cells proliferate and migrate is in the carcinoma state. Utilizing a number of cell types, several investigators have noted a correlation of tumorigenicity and OPN expression (Craig *et al.*, 1988, 1989, 1990; Senger *et al.*, 1989a). Furthermore, OPN has been identified in plasma, and a significant increase in serum OPN is observed in patients with metastasizing carcinomas (Senger and Perruzi, 1985, 1988).

D. Osteopontin and Bone Sialoprotein in Bone Disease and Immune Function

The association of OPN and BSP with immunity or disease states has been recently recognized. An early T-lymphocyte activation 1 (Eta-1) protein was described by Patarca *et al.* (1989) to be synthesized by T cells in response to bacterial infection and may be associated with host resistance. This T-cell cytokine was cloned and determined to be OPN. Studies on the activity of Eta-1 indicated that this protein not only binds to macrophages (macrophages express approximately 10^4 Eta-1 receptors per cell with a K_d of approximately $5 \times 10^{-10} M$) but can stimulate macrophage infiltration to the site of Eta-1 administration *in vivo* (R. P. Singh *et al.*, 1990). Furthermore, overexpression of Eta-1 is associated with the onset of systemic autoimmune disease (Patarca *et al.*, 1990; Lampe *et al.*, 1991). BSP, on the other hand, has been associated with bone infections caused by *Staphylococcus aureus* in patients with osteomyelitis or septic arthritis (Ryden *et al.*, 1987, 1989). Binding studies indicated that there are approximately 10^3 BSP binding sites per bacterial cell and that this binding is restricted to the core protein at the amino-terminal half of the BSP molecule.

Few studies have described the relative changes in OPN and BSP in bone disease. In those described, OPN and BSP appear to be elevated. Osteoporotic bone, characterized as a loss of both protein and mineral, is thought to be a result of uncoupling of osteoblast and osteoclast activity at the time of menopause. Interestingly, bone from ovariectomized rats appears to have an elevated expression of OPN mRNA compared to ovariectomized rats given estrogen replacement (Turner *et al.*, 1990). Osteopetrotic bone, characterized by a thickening of bone and thought to be a result of malfunctional osteoclasts, also shows an overexpression of OPN (Marks *et al.*, 1989). In osteogenesis imperfecta, which is typified by fragile bone and considered a result of defects in collagen molecules, the bone contains higher levels of BSP (Vetter *et al.*, 1991b). What stimulates the overexpression of these proteins and how this overexpression impacts on bone integrity are unknown.

The observations presented here raise several questions. Which isoform(s) is expressed by the different cell types and under different physiological conditions *in vivo?* Are specific isoforms of OPN synthesized that do not promote crystallization? Can specific forms allow for cell attachment and spreading *in vivo,* and do these multiple forms bind to the integrin receptor with the same affinities and biological consequences? Or do the isoforms compete for binding, one form initiating adhesion and the other disrupting focal adhesion points? With respect to BSP, what are the factors (both *cis-* and *trans*-acting) that are involved with its regulation with respect to tissue specificity and time of induction? Furthermore, what role does BSP play in reproduction given that trophoblasts are the only nonskeletal site of synthesis of this sialoprotein? It is clear that while progress is being made in the understanding of the structure and function of these two sialoproteins, many questions remain unanswered at the nuclear, cellular, and extracellular levels, particularly *in vivo.*

IV. BIGLYCAN AND DECORIN

Biglycan (BGN, PG-I, DS/CS-PGI, PG-S1) and decorin (DCN, PG-II, PG-40, DS/CS-PGII, PG-S2) are two small proteoglycans that are pro-

TABLE IV Size and Location of the Major Noncollagenous Human Bone Matrix Genes

Gene	Chromosome[a]	Approximate size[b]
Osteopontin	4q13–21	5.4–8.2 kb (Young *et al.*, 1990)
Bone sialoprotein	4q28–31	>9.1 kb (Fisher *et al.*, 1990; Kerr *et al.*, 1991)
Decorin	12q21.3	>35 kb (McBride *et al.*, 1990; Vetter *et al.*, 1991)
Biglycan	Xq28–qter	~7 kb (Fisher *et al.*, 1991)
Osteonectin	5q31–q33	>20kb (Swaroop *et al.*, 1988)

[a]Locations of human genes are determined either by *in situ* hybridization of metaphase spreads of chromosomes or by Southern analysis of human–mouse or human–hamster hybrid cell lines.

[b]Sizes of human genes were estimated by analysis of genomic DNA digested with restriction enzymes followed by hybridization to radiolabeled fragments of complementary DNA.

duced and localized in bone and other tissues (Bianco *et al.*, 1990). Most experimental data are available for decorin and implicate it in the regulation of collagen formation (Vogel and Trotter, 1987). The function of biglycan is unknown. Both biglycan and decorin consist of a core protein of approximately 38,000 molecular weight (Fisher *et al.*, 1989; Krusius and Ruoslahti, 1986) and contain either one (decorin) or two (biglycan) glycosaminoglycan chains (Fisher, 1985). In bone, the glycosaminoglycan chains consist of chondroitin sulfate (Fisher *et al.*, 1983a), whereas in cartilage (Choi *et al.*, 1989), ligament (Hey *et al.*, 1990), sclera (Cöster and Fransson, 1981), kidney (Klein *et al.*, 1990; Thomas *et al.*, 1991), and skin (Choi *et al.*, 1989), they consist of dermatan sulfate. Although the length and L-iduronic acid content of the dermatan sulfate chains vary among different tissues, no differences in the molecular weight and amino acid sequence of the core proteins have been detected (Rauch *et al.*, 1986; Larjava *et al.*, 1988; Choi *et al.*, 1989). The primary structures of human biglycan and decorin have been deduced from cloned cDNAs (Fisher *et al.*, 1989; Krusius and Ruoslahti, 1986). At the amino acid level, there is 55% homology, suggesting that these two proteoglycans are a result of a gene duplication during evolution. The human biglycan gene has been mapped to Xq27–ter (Fisher *et al.*, 1991; McBride *et al.*, 1990) and the human decorin gene to 12q21.3 (Vetter *et al.*, 1993) (see Table IV).

A. Decorin and Biglycan: A Sequence Motif Rich in Leucine

The cDNA for decorin has been cloned from human embryonic fibroblasts (Krusius and Ruoslahti, 1986) and bovine bone (Day *et al.*, 1986, 1987). Although Northern blots showed the existence of two mRNA species (1.6 and 1.9 kb), all the human decorin cDNA clones that were sequenced had identical open reading frames (but two different sequences for the 3′ flanking region) (Krusius and Ruoslahti, 1986) and the translation resulted in a single protein (Sawhney *et al.*, 1991). The two mRNA transcripts are therefore thought to be the result of the heterogeneity of the 3′ flanking region. Human decorin cDNA codes for 359 amino acids including a hydrophobic signal prepeptide followed by a putative propeptide. The mature core protein consists of 329 amino acids corresponding to a molecular weight of 36,319. There are three Ser–Gly dipeptide sequences (possible attachment sites for glycosaminoglycan chains), but only the serine residue at position 4 is thought to be glycosylated (Chopra *et al.*, 1985; Roughley and White, 1989). In addition, three sites for possible N-glycosylation are present. All of these are probably occupied, because three N-linked glycans have been identified on the core protein (Scott and Dodd, 1990; Sawhney *et al.*, 1991).

The primary structure of biglycan has been deduced from cDNAs from human bone (Fisher *et al.*, 1989) and rat vascular smooth muscle (Dreher *et al.*, 1990) and from the amino acid sequencing of the biglycan core protein from bovine articular cartilage (Neame *et al.*, 1989). The amino acid sequences of the bovine, human, and rat biglycan are almost identical (97% homology) except for a hypervariable region between amino acids 7 and 23. Although decorin is less conserved (70%), a similar N-terminal hypervariable region has been identified in the human and bovine decorin core proteins (Day *et al.*, 1987). In human bone, the biglycan cDNA codes for a protein of 378 amino acids; like decorin, it starts with a hydrophobic signal prepeptide followed by a propeptide leaving a mature protein of 331 amino acids (M_r 37,983). There are two possible N-linked oligosaccharide attachment sites and four Ser–Gly sites for attachment of glycosaminoglycan chains. The serine residues at positions 5 and 10 are the most likely candidates for the two chondroitin sulfate–dermatan sulfate side chains (Fisher *et al.*, 1987; Roughley and White, 1989).

The hydrophobic signal prepeptides of biglycan and decorin correspond to a secretion signal common to many proteins, but the exact role of the putative propeptide is unknown. Recently, vascular smooth muscle cells from bovine and rat aortae have been shown *in vitro* to secrete biglycan with the propeptide still intact (Marcum and Thompson, 1991). However, further studies need to be conducted to determine the physiological relevance of these data.

Structurally, the decorin and biglycan core proteins are very similar (Fisher *et al.*, 1989). Both contain 10 (or 11, if one starts counting within the first cysteine cluster) leucine-rich domains flanked by two clusters of cysteine residues. The amino-terminal cluster contains four cysteine residues and the carboxy-terminal two. For bovine biglycan, disulfide bonds have been demonstrated between the first and fourth and between the fifth and sixth cysteines (Neame *et al.*, 1989). The leucine-rich repeat motif has been identified in numerous other proteins such as bovine osteoinductive factor (Smiley and Stuart, 1990), bovine fibromodulin, human leucine-rich α_2-glycoprotein, and the *Drosophila* proteins toll, slit, and chaoptin (Krantz *et al.*, 1991). Researchers have hypothesized that the repeats form an amphipathic secondary structure and are involved in cell–protein and protein–protein interactions (Krantz *et al.*, 1991).

B. Biglycan and Decorin Gene Structure

The human biglycan gene was recently isolated and characterized (Fisher *et al.*, 1991). The gene contains eight exons, the first being untranslated. The coding regions for the signal peptide, the propeptide, and the attachment sites for the two glycosaminoglycan chains are all contained within the second exon. There is no obvious correlation between the leucine-rich repeats and the intron–exon structure. The rat biglycan cDNA (Dreher *et al.*, 1990) and the mouse gene (Heegaard, unpublished results) contain a region of purine–pyrimidine dinucleotide repeats ($CT_{(n)}$,$CA_{(n)}$) in the untranslated 3′ end. Similar sequences have been identified in other eucaryotic genes (Naylor and Clark, 1990), including BSP (as previously described), and have been shown to favor the formation of Z-DNA, a DNA conformation that can influence gene transcription (Naylor and Clark, 1990). Whether it is important for the transcription of biglycan is not known.

The transcriptional regulation of the biglycan gene at the promoter level is currently under investigation. The promoter has no CCAAT or TATA boxes but contains the consensus sequences for *trans*-acting factors such as SP1, AP1, AP2, NF1, and NF-KB (Fisher *et al.*, 1991; Heegaard, unpublished results.) Transfection experiments have shown the biglycan promoter to be functional in human bone-forming cells, rat osteosarcoma cells (UMR 106) (Heegaard, unpublished results), and human skin fibroblasts (Heegaard *et al.*, 1991). The human decorin gene has recently been cloned and characterized (Vetter *et al.*, 1993; Danielson *et al.*, 1993). It spans more than 45 kb and contains 9 exons including two alternatively spliced leader exons in the untranslated 5′ region.

C. Expression of Biglycan and Decorin mRNA and Protein in Developing Human Tissues

The expression of biglycan and decorin in human fetal tissues has been investigated in detail by Bianco and co-workers (1990) using *in situ* hybridization and immunohistochemistry. In contrast to decorin, which is associated with all connective tissues rich in type I and type II collagen, biglycan is localized to a few specialized connective tissues and expressed in a range of specialized cells such as endothelial cells, skeletal myofibers, keratinocytes, and renal tubular epithelia.

In developing bone, the spatial and temporal distribution of the core proteins and mRNA also suggest different roles for the two proteins. The mRNA of biglycan and decorin are expressed in all cartilage and bone cells but with differences in specific regions of these tissues. For example, high levels of biglycan mRNA can be detected in articular regions, in forming growth plates, in perichondrium-derived mesenchymal cells inside vascular canals, and in preosteogenic cells of the subperiosteal bone. On the other hand, decorin mRNA can be found at high levels in a narrow rim of cartilage at peripheral subperichondral locations and in chondrocytes arranged around vascular canals ingrowing from the perichondrium to the epiphyseal cartilage. In contrast to biglycan, decorin is expressed uniformly throughout the periosteum.

The immunohistochemical localization of the core proteins in developing bone is equally divergent. Both biglycan and decorin are localized to cartilaginous and bony components but, again, with differences at specific sites. For instance, in the prospective articular cartilage, biglycan is primarily localized to the outer cap, whereas decorin is found deeper in the nonarticular resting cartilage. In the growth plate, biglycan is restricted to the territorial capsules of isogeneic groups of chondrocytes and decorin to the interterritorial matrix. These findings correspond to the distribution of decorin described for human adult tissue (Voss *et al.*, 1986).

D. Possible Functions of Biglycan and Decorin Include Cell–Matrix Interactions and Control of Cell Proliferation

Type I collagen is the major collagen species in bone. Decorin, but not biglycan, has been shown to bind to type I and type II collagen and to interfere with fibrillogenesis *in vitro* (Vogel *et al.*, 1984). When decorin was added to collagen in an *in vitro* fibrillogenesis assay, the rate of fibril formation was decreased and the width of the collagen fibrils diminished (Vogel and Trotter, 1987). The inhibitory effect of decorin is independent of the sugar moieties since the same effect was obtained using the core protein (Brown and Vogel, 1989). In one *in vitro* study, it was reported that decorin binds to type I collagen in a reversible fashion with a maximum binding at pH 7.4 and at physiological ionic strength and with an affinity constant of $3.3–8.2 \times 10^7 M^{-1}$ (Brown and Vogel, 1989). Using immunohistochemistry, the binding site(s) for decorin has been located to near the d and e bands in the D period of the type I collagen fibrils in nonmineralized tissue (Pringle and Dodd, 1990; Fleischmajer *et al.*, 1991). The interaction of decorin with collagen seems to be dependent on one or several of the disulfide bridges in decorin, because both the binding to collagen and the inhibition of fibrillogenesis are reduced by pretreatment of decorin with beta-mercaptoethanol (Scott *et al.*, 1986).

All the preceding mentioned experiments have been conducted using collagen from tendon, cartilage, or skin, and the relevance for bone tissue has yet to be elucidated. Experiments by Scott and Haigh (1985) showed a chondroitin sulphate proteoglycan localized to the collagen fibrils of bone but with a different staining pattern compared to other tissues. Fedarko *et al.* (1992) showed that there is a strong positive correlation of the synthesis of decorin and collagen in *in vitro* studies of human bone-forming cells from donors of different ages.

Decorin also interacts with fibronectin (Schmidt *et al.*, 1987), another component of the extracellular matrix. Electron micrographs of the surface of cultured human fibroblasts have shown decorin to be associated with fibronectin-containing fibrils. Furthermore, attachment of fibroblasts to fibronectin-coated surfaces is inhibited by the addition of decorin (Winnemöller *et al.*, 1991). This is thought to be a result of direct binding of decorin to fibronectin and not due to binding of decorin to a fibronectin receptor (Winnemöller *et al.*, 1991). The strongest binding of decorin to fibronectin was found at pH 5.0 and at low ionic strength but with binding still occurring at pH 7.4 and at physiological ionic strength (Schmidt *et al.*, 1987). Recently, the sequence Asn–Lys–Ile–Ser–Lys of the decorin core protein has been suggested to be involved in the binding (Schmidt *et al.*, 1991). Decorin binds both to the fragments of fibronectin containing the cell binding domains and to those containing the heparin binding domains (Winnemöller *et al.*, 1991). The interaction of decorin with fibronectin is similar whether the intact proteoglycan or the core protein is used, as long as care is taken to prevent denaturation of decorin (Winnemöller *et al.*, 1991). It is not clear whether biglycan can interact with fibronectin in a similar manner.

As previously mentioned, electron microscopy showed the association of decorin with fibrils on the fibroblast surface (Schmidt *et al.*, 1987). It also revealed clusters of decorin on the surface, supporting earlier work describing receptor-mediated uptake of proteoglycans (Truppe and Kresse, 1978; Glössl *et al.*, 1983). *In vitro* studies of monolayers of human skin fibroblasts showed that ${}^{35}SO_4{}^{2-}$-labeled decorin was endocytosed and degraded by the fibroblasts and that the uptake was probably mediated through the decorin core protein (Grössl *et al.*, 1983). The involvement of lysine residues was suggested. More recent studies have shown that within 4 hr approximately 50% of newly synthesized decorin molecules are endocytosed (Schmidt *et al.*, 1990). Addition of either decorin, monospecific anti-decorin antibodies, or collagen fibrils all reduce the uptake of decorin.

In an attempt to characterize a possible decorin receptor, endosomal preparations from human fibroblasts or MG-63 osteosarcoma cells were resolved by SDS–polyacrylamide gel electrophoresis (Hausser *et al.*, 1989). Immunoblots showed the binding of decorin to two proteins with molecular mass of 51 and 26 kDa (the 26 kDa protein might be a result of degradation). The putative receptor proteins co-precipitated with a decorin core protein–antibody complex and could be shown to bind to immobilized decorin. Detergent extracts from intact fibroblasts showed that the 51 kDa protein disappeared after trypsination of the cells and reappeared after a longer period of time (Hausser *et al.*, 1989). This is consistent with a location on the cell membrane. Subsequent studies showed that heparin and, to a lesser degree, dermatan sulfate compete with decorin for binding to the receptor(s) and for endocytosis (Hausser and Kresse, 1991). The biological significance of heparin and decorin sharing the same receptor is not clear. One could speculate that the receptor might provide a link in the regulation of cell proliferation, because both decorin and heparin bind growth factors (Lobb *et al.*, 1986).

Further evidence for a possible role of decorin in cell–matrix interactions was provided when Yamaguchi and Ruoslahti (1988) showed that expression of decorin in Chinese hamster ovary cells resulted in an inhibition of cell proliferation. When the Chinese hamster ovary cells were transfected with an expression vector containing the full-length human decorin cDNA, the decorin-expressing cells occupied three times more surface area than nonexpressing cells, had a slower growth rate, and became arrested in their growth at lower saturation densities. The effect was dose-dependent; the more decorin the cells produced the more surface area they occupied and the sooner they became quiescent. When decorin was purified from the media, a 25 kDa protein was co-purified. This was identified as TGB-β, a well-known modulator of cell proliferation and matrix deposition that is especially abundant in bone (Barnard *et al.*, 1990). Subsequent experiments showed that the decorin core protein (and maybe biglycan) binds to TGF-β *in vitro* with a dissociation constant of 1.5×10^{-9} *M* (Yamaguchi *et al.*, 1990). By binding to TGF-β, decorin neutralizes both the growth-stimulatory and growth-inhibitory activities of TGF-β *in vitro* (Yamaguchi *et al.*, 1990) and *in vivo* (Border *et al.*, 1992).

TGF-β affects the synthesis of both decorin and biglycan (Bassols and Massagué, 1988). In all the cell types tested, TGF-β increased the expression of biglycan but the effect on decorin depends on the cell type. In mesengial cells from rat kidney, TGF-β increases the expression of decorin (Border *et al.*, 1990), whereas in glomerular epithelial cells (Border and Ruoslahti, 1990), osteosarcoma MG-63 cells, human embryonic skin (Breuer *et al.*, 1990) and lung fibroblasts (Romaris *et al.*, 1991), and bovine fetal sclera fibroblasts (Westergren-Thorsson *et al.*, 1991) it has no effect. In adult human skin and gingival fibroblasts TGF-β reduced the production of both decorin mRNA and protein by 50–70% (Kähäri *et al.*, 1991). Both the effect on biglycan and on decorin could be blocked by the addition of either cycloheximide or actinomycin D (Bassols and Massagué, 1988; Romaris *et al.*, 1991), indicating the necessity of *de novo* protein synthesis and a regulation at the transcriptional level. TGF-β can also cause an increase in the length of the glycosaminoglycan chains of both biglycan and decorin (Kähäri *et al.*, 1991).

It has been suggested that decorin is the effector molecule in a negative feedback loop that regulates the activity of TGF-β (Ruoslahti and Yamaguchi, 1991). This was based on results describing an up-regulation of decorin synthesis by TGF-β. As mentioned, the effect of TGF-β on decorin synthesis has since been shown to vary among different cell types, and more information is needed to elucidate the regulatory relationship among TGF-β, decorin, and biglycan.

Very little is known about the function of biglycan. Retinoic acid suppresses its synthesis in bovine chondrocytes, *in vitro* (Pearson and Sasse, 1992) and it is positively regulated by TGF-β. Its expression in developing bone, predominantly at sites of forming growth plates and in preosteogenic cells, suggests a role in bone development and growth (Bianco *et al.*, 1990). In support of this theory are the results from an *in vitro* study of bone-forming cells from donors of different age. The expression of biglycan was found to be increased in connection with periods of rapid gain of bone mass (Fedarko *et al.*, 1992).

In addition, a recent study by Vetter *et al.* (1991a) describes reduced levels of biglycan in patients suffering from Turner's syndrome and increased levels of biglycan in patients with Klinefelter's disease. Turner patients are of short stature and Klinefelter patients are taller than the average population. These changes seem to be specific for biglycan, because the levels of both decorin and collagen were normal.

In summary, although the two small proteoglycans, biglycan and decorin, apparently are very similar in structure, they differ in distribution, regulation, and possible function. Most of the present information, although limited, points to a role for decorin in the regulation of matrix formation and cell proliferation and associates biglycan more generally with bone development and growth.

V. CONCLUSION

The functions of the bone matrix proteins are potentially multifaceted ranging from Ca^{2+} and hydroxyapatite binding, mineralization, bone cell attachment to tissue growth and differentiation. The genes encoding these proteins are also varied in size, structure, and location in the human genome (see Table IV). Although the genes have been mapped to precise regions of the genome, to date, none of the matrix genes described in this chapter link to, or are the primary cause of, any well-characterized inherited disease of bone. As molecular techniques become more sophisticated, however, it is likely that certain inherited diseases of bone (other than, perhaps, osteogenesis imperfecta) will be found to be caused by defects in one of these predominant noncollagenous matrix proteins. In addition to inherited disease, it is likely that certain acquired skeletal diseases potentially result from defects in the production or function of one or more of the predominant matrix proteins. By studying the function and genetic regulation of these genes in normal and diseased skeletal tissue, we will begin to get a better understanding of their role in the development and aging of bone tissue.

REFERENCES

Alberius, P., and Johnell, O. (1990). Immunohistochemical assessment of cranial *suture* development in rats. *J. Anat.* **173,** 61–68.

Andrews, A. T., De, B., Herring, G. M., and Kent, P. W. (1967). Some studies on the composition of bovine cortical-bone sialoprotein. *Biochem J* **104,** 705 715.

Barnard, J. A., Lyons, R. M., and Moses, H. L. (1990). The cell biology of transforming growth factor β. *Biochim. Biophys. Acta* **1032,** 79–87.

Bassols, A., and Massagué, J. (1988). Transforming growth factor β regulates the expression and structure of extracellular matrix chondroitin/dermatan sulfate proteoglycans. *J. Biol. Chem.* **263,** 3039–3045.

Ber, R., Kubota, T., Sodek, J., and Aubin, J. E. (1991). Effects of transforming growth factor-β on normal clonal bone cell populations. *Biochem. Cell Biol.* **69,** 132–140.

Bianchi, A., Wells, R. D., Heintz, N. H., and Caddle, M. S. (1990). Sequences near the origin of replication of the DHFR locus of Chinese hamster ovary cells adopt left-handed Z-DNA and triplex structures. *J. Biol. Chem.* **265,** 21789–21796.

Bianco, P., Silvestrini, G., Termine, J. D., and Bonucci, E. (1988). Immunohistochemical localization of osteonectin in developing human and calf bone using monoclonal antibody. *Calcif. Tissue Int.* **43,** 155–161.

Bianco, P., Fisher, L. W., Young, M. F., Termine, J. D., and Gehron Robey, P. (1990). Expression and localization of the two small proteoglycans biglycan and decorin in developing human skeletal and non-skeletal tissues. *J. Histochem. Cytochem.* **38,** 1549–1563.

Bianco, P., Fisher, L. W., Young, M. F., Termine, J. D., and Gehron Robey, P. (1991). Expression of bone sialoprotein (BSP) in developing human tissues. *Calcif. Tiss. Int.* **49,** 421–426.

Bolander, M. E., Young, M. F., Fisher, L. W., Yamada, Y., and Termine, J. D. (1988). Osteonectin cDNA sequence reveals potential binding regions for calcium and hydroxyapatite and shows homologies with both a basement membrane protein (SPARC) and a serine protease inhibitor (ovomucoid). *Proc. Natl. Acad. Sci. USA* **85,** 2919–2923.

Border, W. A., and Ruoslahti, E. (1990). Transforming growth factor-β1 induces extracellular matrix formation in glomerulonephritis. *Cell Diff. Dev.* **32,** 425–432.

Border, W. A., Okuda, S., Languino, L. R., and Ruoslahti, E. (1990). Transforming growth factor-β regulates production of proteoglycans by mesangial cells. *Kidney Int.* **37,** 689–695.

Border, W. A., Noble, N. A., Yamamoto, T., Harper, J. R., Yamaguchi, Y., Pierschbacher, M. D., and Ruoslahti, E. (1992). Natural inhibitor of transforming growth factor-β protects against scarring in experimental kidney disease. *Nature* **360,** 361–364.

Breuer, B., Schmidt, G., and Kresse, H. (1990). Non-uniform influence of transforming growth factor-β on the biosynthesis of different forms of small chondroitin sulphate/dermatan sulphate proteoglycan. *Biochem. J.* **269,** 551–554.

Brown, D. C., and Vogel, K. G. (1989). Characteristics of the *in vitro* interaction of a small proteoglycan (PG II) of bovine tendon with type I collagen. *Matrix* **9,** 468–478.

Butler, W. T. (1989). The nature and significance of osteopontin. *Connec. Tissue Res.* **23,** 123–126.

Cachau, R. E., Serpersu, E. H., Mildvan, A. S., August, T., and Amzel, L. M. (1989). Recognition in cell adhesion. A comparative study of the conformations of RGD-containing peptides by Monte Carlo and NMR methods. *J. Molec. Recog.* **2,** 179–186.

Castagnola, P., Bet, P., Quarto, R., Gennari, M., and Cancedda, R. (1991). cDNA cloning and gene expression of chick osteopontin. Expression of osteopontin mRNA in chondrocytes is enhanced by trypsin treatment of cells. *J. Biol. Chem.* **266,** 9944–9949.

Chang, P.-L., and Prince, C. W. (1991). 1α,25-Dihydroxyvitamin D_3 stimulates synthesis and secretion of nonphosphorylated osteopontin (secreted phosphoprotein 1) in mouse JB6 epidermal cells. *Cancer Res.* **51,** 2144–2150.

Chen, J., Shapiro, H. S., Wrana, S. R., Heersche, J. N. M., and Sodek, J. (1991). Localization of bone sialoprotein (BSP) expression to sites of mineralized tissue formation in fetal rat tissues by *in situ* hybridization. *Matrix* **11,** 133–143.

Chenu, C., and Delmas, P. D. (1992). Platelets contribute to circulating levels of bone sialoprotein in human. *J. Bone Miner. Res.* **7,** 47–54.

Choi, H. U., Johnson, T. L., Subhash, P., Tang, L.-H., and Rosenberg, L. (1989). Characterization of the dermatan sulfate proteoglycans, DS-PGI and DS-PGII, from bovine articular cartilage and skin isolated by octyl-sepharose chromatography. *J. Biol. Chem.* **264,** 2876–2884.

Chopra, R. K., Pearson, C. H., Pringle, G. A., Fackre, D. S., and Scott, P. G. (1985). Dermatan sulphate is located on serine-4 of bovine skin proteodermatan sulphate. *Biochem. J.* **232,** 277–279.

Clezardin, P., Malavel, L., Ehrensperger, A.-S., Delmas, P., Dechavanne, M., and McGregor, J. L. (1988). Complex formation of human thrombospondin with osteonectin. *Eur. J. Biochem.* **175,** 275–284.

Cöster, L., and Fransson, L.-A. (1981). Isolation and characterization of dermatan sulphate proteoglycans from bovine sclera. *Biochem. J.* **193,** 143–153.

Craig, A. M., and Denhardt, D. T. (1991). The murine gene encoding secreted phosphoprotein 1 (OPN, osteopontin): Promoter structure, activity, and induction in vivo by estrogen and progesterone. *Gene* **100,** 163–171.

Craig, A. M., Nemir, M., Mukherjee, B. B., Chambers, A. F., and Denhardt, D. T. (1988). Identification of the major phosphoprotein secreted by many rodent cell lines as 2ar/osteopontin: Enhanced expression in H-*RAS*-transformed 3T3 cells. *Biochem. Biophys. Res. Commun.* **157,** 166–173.

Craig, A. M., Smith, J. H., and Denhardt, D. T. (1989). Osteopontin, a transformation-associated cell adhesion phosphoprotein, is induced by 12-O-tetradecanoylphorbol-13-acetate in mouse epidermis. *J. Biol. Chem.* **264,** 9682–9689.

Craig, A. M., Bowden, G. T., Chambers, A. F., Spearman, M. A., Greenberg, A. H., Wright, J. A., McLeod, M., and Denhardt, D. T. (1990). Secreted phosphoprotein mRNA is induced during multistage carcinogenesis in mouse skin and correlates with the metastatic potential of murine fibroblasts. *Int. J. Cancer* **46,** 133–137.

Damjanovski, S., Liu, F., and Ringuette, M. (1992). Molecular analysis of *Xenopus laevis* SPARC (secreted protein, acidic, rich in cysteine): A highly conserved acidic calcium-binding extracellular-matrix protein. *Biochem. J.* **281,** 513–517.

Danielson, K. G., Fazzio, A., Cohen, I., Cannizzaro, L. A., Eichstetter, I., and Iozzo, R. V. (1993). The human decorin gene: Intron-exon organization, discovery of two alternative spliced exons in the first 5′ untranslated region, and mapping of the gene to 12q23. *Genomics* **15,** 146–160.

Day, A. A., Ramis, C. I., Fisher, L. W., Gehron Robey, P., Termine, J. D., and Young, M. F. (1986). Characterization of bone PG II cDNA and its relationship to PG II mRNA from other connective tissues. *Nucleic Acids Res.* **14,** 9861–9876.

Day, A. A., McQuillan, C. I., Termine, J. D., and Young, M. F. (1987). Molecular cloning and sequence analysis of the cDNA for small proteoglycan II of bovine bone. *Biochem. J.* **248,** 801–805.

Domenicucci, C., Wasi, S., and Sodek, J. (1988). Characterization of porcine osteonectin extracted from foetal calvariae. *Biochem. J.* **253,** 139–151.

Dominguez, P., Ibaraki, K., Gehron Robey, P., Hefferan, T. E., Termine, J. D., and Young, M. F. (1991). Expression of the osteonectin gene potentially controlled by multiple cis- and trans-acting factors in cultured bone cells. *J. Bone Miner. Res.* **6,** 1127–1136.

Dreher, K. L., Asundi, V., Matzura, D., and Cowan, K. (1990). Vascular smooth muscle biglycan represents a highly conserved proteoglycan within the arterial wall. *Eur. J. Cell. Biol.* **53,** 296–304.

Dziadek, M., Paulsson, M., Aumailley, M., and Timpl, R. (1986). Purification and tissue distribution of a small protein (BM-40) extracted from a basement membrane tumor. *Eur. J. Biochem.* **161,** 455–464.

Ecarot-Charrier, B., Bouchard, F., and Delloye, C. (1989). Bone sialoprotein II synthesized by cultured osteoblasts contains tyrosine sulfate. *J. Biol. Chem.* **264,** 20049–20053.

Engel, J., Taylor, W., Paulsson, M., Sage, H., and Hogan, B. (1987). Calcium binding domains and calcium-induced conformational transition of SPARC/BM-40/osteonectin, an extracellular glycoprotein expressed in mineralized and nonmineralized tissues. *Biochemistry* **26,** 6958–6965.

Fedarko, N. S., Vetter, U. K., Weinstein, S., and Gehron Robey, P. (1992). Age-related changes in hyaluronan, proteoglycan, collagen and osteonectin synthesis by human bone cells. *J. Cell. Physiol.* **151**(2), 215–227.

Findlay, D. M., Fisher, L. W., McQuillan, C. I., Termine, J. D., and Young, M. F. (1988). Isolation of the osteonectin gene evidence that a variable region of the osteonectin molecule is encoded within one exon. *Biochemistry* **27,** 1483–1489.

Fisher, L. W. (1985). The nature of the proteoglycans of bone. *In* "The Chemistry and Biology of Mineralized Tissues" (W. T. Butler, ed.), pp. 188–196. EBSCO Media, Birmingham, Alabama.

Fisher, L. W., Termine, J. D., Dejter Jr., S. W., Whitson, S. W., Yanagishita, M., Kimura, J. H., Hascall, V. C., Kleinman, H. K., Hassell, J. R., and Nilson, B. (1983a). Proteoglycans of developing bone. *J. Biol. Chem.* **258,** 6588–6594.

Fisher, L. W., Whitson, S. W., Avioli, L. V., and Termine, J. D. (1983b). Matrix sialoprotein of developing bone. *J. Biol. Chem.* **258,** 12723–12727.

Fisher, L. W., Hawkins, G. R., Tuross, N., and Termine, J. D. (1987). Purification and partial characterization of small proteoglycans I and II, bone sialoproteins I and II and osteonectin from the mineral compartment of developing human bone. *J. Biol. Chem.* **262,** 9702–9708.

Fisher, L. W., Termine, J. D., and Young, M. F. (1989). Deduced protein sequence of bone small proteoglycan I (biglycan) shows homology with proteoglycan II (decorin) and several non-connective tissue proteins in a variety of species. *J. Biol. Chem.* **264,** 4571–4576.

Fisher, L. W., McBride, O. W., Termine, J. D., and Young, M. F. (1990). Human bone sialoprotein. Deduced protein sequence and chromosomal localization. *J. Biol. Chem.* **265,** 2347–2351.

Fisher, L. W., Heegaard, A.-M., Vetter, U., Vogel, W., Just, W., Termine, J. D., and Young, M. F. (1991). Human biglycan gene. *J. Biol. Chem.* **266,** 14371–14377.

Fleischmajer, R., Fisher, L. W., MacDonald, E. D., Jacobs, L., Jr., Perlish, J. S., and Termine, J. D. (1991). Decorin interacts with fibrillar collagen of embryonic and adult human skin. *J. Structural Biol.* **106,** 82–90.

Franzen, A., and Heinegard, D. (1985). Isolation and characterization of two sialoproteins present only in bone calcified matrix. *Biochem. J.* **232,** 715–724.

Funk, S. E., and Sage, E. H. (1991). The Ca^{2+}-binding glycoprotein SPARC medulates cell cycle progression in bovine aortic endothelial cells. *Proc. Natl. Acad. Sci. USA* **88,** 2648–2652.

Gehron Robey, P., and Termine, J. D. (1985). Human bone cells *in vitro. Calcif. Tissue Int.* **37,** 453–460.

Gehron Robey, P., Findlay, D. M., Young, M. F., Bereford, J. N., Stubbs, J. N., Fisher, L. W., and Termine, J. D. (1986). Regulation of osteonectin production by 1,25-dihydroxyvitamin D_3 in human bone cells. *J. Bone Miner. Res.* **1,** 288.

Gehron Robey, P., Young, M. F., Fisher, L. W., and McClain, T. D. (1989). Thrombospondin is an osteoblast-derived component of mineralized extracellular matrix. *J. Cell Biol.* **108,** 719–727.

Giachelli, C., Bae, N., Lombardi, D., Majesky, M., and Schwartz, S. (1991). Molecular cloning and characterization of 2B7, a rat mRNA which distinguishes smooth muscle cell phenotypes *in vitro* and is identical to osteopontin (secreted phosphoprotein 1, 2ar). *Biochem. Biophys. Res. Commun.* **177,** 867–873.

Glössl, J., Schubert-Prinz, R., Gregory, J. D., Damle, S. P., Von Figura, K., and Kresse, H. (1983). Receptor-mediated endocytosis of proteoglycans by human fibroblasts involves recognition of the protein core. *Biochem. J.* **215,** 295–301.

Gotoh, Y., Gerstenfeld, L. C., and Glimcher, M. J. (1990a). Identification and characterization of the major chicken bone phosphoprotein. Analysis of its synthesis by cultured embryonic chick osteoblasts. *Eur. J. Biochem.* **187,** 49–58.

Gotoh, Y., Pierschbacher, M. D., Gresiak, J. J., Gerstenfeld, L., and Glimcher, M. J. (1990b). Comparison of two phosphoproteins in chicken bone and their similarities to the mammalian bone proteins, osteopontin and bone sialoprotein II. *Biochem. Biophys. Res. Commun.* **173,** 471–479.

Guermah, M., Crisanti, P., Laugier, D., Dezelee, P., Bidou, L., Pessac, B., and Calothy, G. (1991). Transcription of quail gene expressed in embryonic retinal cells is shut off sharply at hatching. *Proc. Natl. Acad. Sci. USA* **88,** 4503–4507.

Haniford, D. B., and Pulleyblank, D. E. (1983). Facile transition of poly [d (TG) d(CA)] into a left handed felix in physiological conditions. *Nature (London)* **302,** 632–634.

Hausser, H., and Kresse, H. (1991). Binding of heparin and of the small proteoglycan decorin to the same endocytosis receptor proteins leads to different metabolic consequences. *J. Cell Biol.* **114,** 45–52.

Hausser, H., Hoppe, W., Rauch, U., and Kresse, H. (1989). Endocytosis of a small dermatan sulphate proteoglycan. *Biochem. J.* **263,** 137–142.

Heath, C. V., Denome, R. M., and Cole, C. N. (1990). Spatial constraints on polyadenylation signal function. *J. Biol. Chem.* **265,** 9098–9104.

Heath, J. K., Rodan, S. B., Yoon, K., and Rodan, G. A. (1989). Rat calvarial cell lines immortalized with SV-40 large T antigen: Constitutive and retinoic acid-inducible expression of osteoblastic features. *Endocrinology* **124,** 3060–3068.

Heegaard, A.-M., Gehron Robey, P., Termine, J. D., and Young, M. F. (1991). Human biglycan promoter activity in human primary bone cells and fibroblasts. *J. Bone Miner. Res.* **6,** S159.

Hey, N. J., Handley, C. J., Ng, C. K., and Oakes, B. W. (1990). Characterization and synthesis of macromolecules by adult collateral ligament. *Biochim. Biophys. Acta* **1034,** 73–80.

Hock, J. M., Gunness-Hey, M., Poser, J., Olson, H., Bell, N. H., and Raisz, L. G. (1986). Stimulation of undermineralized matrix formation by 1,25-dihydroxyvitamin D_3 in long bones of rats. *Calcif. Tissue Int.* **38,** 79–86.

Holland, P. W. H., Harper, S. J., McVey, J. H., and Hogan B. L. M. (1987). In vivo expression of mRNA for the Ca^{++}-binding protein SPARC (osteonectin) revealed by in situ hybridization. *J. Cell Biol.* **105,** 473–482.

Howe, C. C., Kath, R., Manciatini, M. L., Herlyn, M., Mueller, S., and Cristofalo, V. (1990). Expression and structure of human SPARC transcripts: SPARC mRNA is expressed by human cells involved in extracellular matrix production and some of these cells show an unusual expression pattern. *Exp. Cell. Res.* **188,** 185–191.

Hultenby, K., Reinholt, F. P., Oldberg, Å, and Heinegård, D. (1991). Ultrastructural immunolocalization of osteopontin in metaphyseal and cortical bone. *Matrix,* **11,** 206–213.

Ibaraki, K., Gehron Robey, P., and Young, M. F. (1992a). The transcription of the osteonectin gene is potentially controlled by AP1, AP2, AP3, and a novel GAGA binding factor. *Connect. Tissue Res.* **27,** 130.

Ibaraki, K., Whitson, S. W., Termine, J. D., and Young, M. F. (1992b). Bone matrix mRNA expression in differentiating fetal bovine osteoblasts. *J. Bone Miner. Res.* **7,** 743–754.

Iruela-Arispe, M. L., Diglio, C. A., and Sage, E. H. (1991). Modulation of extracellular matrix proteins by endothelial cells undergoing angiogenesis in vitro. *Arteriosclero. Thrombo.* **11,** 805–815.

Jin, C. H., Miaura, C., Ishimi, Y., Hong, M. H., Sato, T., Abe, E., and Suda, T. (1990). Interleukin 1 regulates the expression of osteopontin mRNA by osteoblasts. *Mol. Cell. Endo.* **74,** 221–228.

Johnston, I. C., Paladino, T., Curd, J. W., and Brown, I. R. (1990). Molecular cloning of SC1: A putative brain extracellular matrix glycoprotein showing partial similarity of osteonectin/BM-40/SPARC. *Neuron* **2,** 165–176.

Jundt, G., Berghauser, K.-H., Termine, J. D., and Schulz, A. (1987). Osteonectin—A differentiation marker of bone cells. *Cell Tissue Res.* **248,** 409–415.

Kähäri, V.-M., Larjava, H., and Uitto, J. (1991). Differential regulation of extracellular matrix proteoglycan (PG) gene expression. *J. Biol. Chem.* **266,** 10608–10615.

Kasugai, S., Todescan, R., Jr., Nagata, T., Yao, K.-L., Butler, W. T., and Sodek, J. (1991a). Expression of bone matrix proteins associated with mineralized tissue formation by adult rat bone marrow cells *in vitro:* Inductive effects of dexamethasone on the osteoblastic phenotype. *J. Cell. Physiol.* **147,** 111–120.

Kasugai, S., Zhang, Q., Overall, C. M., Wrana, J. L., Butler, W. T., and Sodek, J. (1991b). Differential regulation of the 55 and 4 kDa forms of secreted phosphoprotein 1 (SPP-1, osteopontin) in normal and transformed rat bone cells by osteotropic hormones, growth factors and a tumor promoter. *Bone and Miner.* **13,** 235–250.

Kelm, R. J., Jr., and Mann, K. G. (1990). Human platelet osteonectin: Release, surface expression, and partial characterization. *Blood* **75,** 1105–1113.

Kelm, R. J., Jr., and Mann, K. G. (1991). The collagen binding specificity of bone and platelet osteonectin is related to differences in glycosylation. *J. Biol. Chem.* **266,** 9632–9639.

Kerr, J. M., Fisher, L. W., Termine, J. D., Wang, M. G., McBride, O. W., and Young, M. F. (1991a). Isolation and characterization of the human bone sialoprotein gene. *J. Dental Res.* **70,** 463.

Kerr, J. M., Fisher, L. W., Termine, J. D., and Young, M. F. (1991b). The cDNA cloning and RNA distribution of bovine osteopontin. *Gene* **108,** 237–243.

Kiefer, M. C., Bauer, D. M., and Barr, P. J. (1989). The cDNA and derived amino acid sequence for human osteopontin. *Nucleic Acids Res.* **17,** 3306.

Kinne, R. W., and Fisher, L. W. (1987). Keratan sulfate proteoglycan in rabbit compact bone is bone sialoprotein II. *J. Biol. Chem.* **262,** 10206–10211.

Klein, D. J., Brown, D. M., Kim, Y., and Oegema, T. R., Jr. (1990). Proteoglycans synthesized by human glomerular mesangial cells in culture. *J. Biol. Chem.* **265,** 9533–9543.

Krantz, D. D., Zidovetzki, R., Kagan, B. L., and Zipursky, S. L. (1991). Amphipathic β structure of a leucine-rich repeat peptide. *J. Biol. Chem.* **266,** 16801–16807.

Krusius, T., and Ruoslahti, E. (1986). Primary structure of an extracellular matrix proteoglycan core protein deduced from cloned cDNA. *Proc. Natl. Acad. Sci. USA* **83,** 7683–7687.

Kubota, T., Zhang, Q., Wrana, J. L., Ber, R., Aubin, J. E., Butler, W. T., and Sodek, J. (1989). Multiple forms of SPPI (secreted phosphoprotein, osteopontin) synthesized by normal and transformed rat bone cell populations: Regulation by TGF-B. *Biochem. Biophys. Res. Commun.* **162,** 1453–1459.

Lampe, M. A., Patarca, R., Iregui, M. V., and Cantor, H. (1991). Polyclonal B cell activation by the Eta-1 cytokine and the development of systemic autoimmune disease. *J. Immunol.* **147,** 2902–2906.

Lane, T. F., and Sage, H. (1990). Functional mapping of SPARC: peptides from two distinct $Ca^{(++)}$-binding sites modulates cell shape. *J. Cell Biol.* **111,** 3065–3076.

Langille, R. M., and Solursh, M. (1990). Formation of chondrous and osseous tissues in micromass cultures of rat frontonasal and mandibular ectomesenchyme. *Differentiation* **44,** 197–206.

Lankat-Buttgereit, B., Mann, K., Deutzmann, R., Timpl, R., and Krieg, T. (1988). Cloning and complete amino acid sequences of human and murine basement membrane protein BM-40 (SPARC, osteonectin). *FEBS Lett.* **236,** 352–356.

Larjava, H., Heino, J., Krusius, T., Vuorio, E., and Tammi, M. (1988). The small dermatan sulphate proteoglycans synthesized by fibroblasts derived from skin, synovium and gingiva show tissue-related heterogeneity. *Biochem. J.* **256,** 35–40.

Leboy, P. S., Beresford, J. N., Devlin, C., and Owen, M. E. (1991). Dexamethasone induction of osteoblast mRNAs in rat marrow stromal cell cultures. *J. Cell Physiol.* **146,** 370–378.

Lobb, R. R., Harper, J. W., and Fett, J. W. (1986). Purification of heparin-binding growth factors. *Analyt. Biochem.* **154,** 1–14.

Malaral, L., Fournier, B., and Delmas, P. (1987). Radioimmunoassay for mineralized tissues, and blood. *J. Bone Miner. Res.* **2,** 457–465.

Mann, K., Deutzmann, R., Paulsson, M., and Timpl, R. (1987). Solubilization of protein BM-40 from a basement membrane tumor with chelating agents and evidence for its identity with osteonectin and SPARC. *FEBS Lett.* **218,** 167–172.

Marcum, J. A., and Thompson, M. A. (1991). The amino-terminal region of a proteochondroitin core protein, secreted by aortic smooth muscle cells, shares sequence homology with the pre-propeptide region of the biglycan core protein from human bone. *Biochem. Biophys. Res. Commun.* **175,** 706–712.

Mark, M. P., Prince, C. W., Oosawa, T., Gay, S., Bronckers, A. L. J. J., and Butler, W. T. (1987). Immunohistochemical demonstration of a 44-kd phosphoprotein in developing rat bones. *J. Histochem. Cytochem.* **35,** 707–715.

Mark, M. P., Butler, W. T., Prince, C. W., Finkelman, R. D., and Ruch, J. V. (1988a). Developmental expression of 44-kDa bone phosphoprotein (osteopontin) and bone gamma-carboxy-glutamic acid (Gla)-containing protein (osteocalcin) in calcifying tissues of rat. *Differentiation* **37,** 123–126.

Mark, M. P., Prince, C. W., Gay, S., Austin, R. L., and Butler, W. T. (1988b). 44-kDal bone phosphoprotein (osteopontin) antigenicity at ectopic sites in newborn rats: kidney and nervous tissue. *Cell Tissue Res.* **251,** 23–30.

Marks, S. C., Jr., Mackowiak, S., Shaloub, V., Lian, J. B., and Stein, G. S. (1989). Proliferation and differentiation of osteoblasts in osteopetrotic rats: Modification in expression of genes encoding cell growth and extracellular matrix proteins. *Connect. Tiss. Res.* **21,** 107–116.

Mason, I. J., Taylor, A., Williams, J. G., Sage, H., and Hogan, B. L. M. (1986a). Evidence from molecular cloning that SPARC, a major product of mouse embryo parietal endoderm, is related to endothelial cell "culture shock" glycoprotein of Mr=43,000. *EMBO J.* **5,** 1465–1472.

Mason, I. J., Murphy, M., Munke, M., Franke, U., Eliott, R. W., and Hogan, B. L. M. (1986b). Developmental and transformation-sensitive expression of the Sparc gene on chromosome 11. *EMBO J.* **5,** 1831–1837.

Mayer, U., Aumailley, M., Mann, K., Timpl, R., and Engel, J. (1991). Calcium-dependent binding of basement membrane protein BM-40 (osteonectin, SPARC) to basement membrane collagen type IV. *Eur. J. Biochem.* **198,** 141–150.

McBride, O. W., Fisher, L. W., and Young, M. F. (1990). Localization of PGI (biglycan, BGN) and PGII (decorin, DCN, PG-40) genes on human chromosomes Xq13–qter and 12q, respectively. *Genomics* **6,** 219–225.

McLean, M. J., and Wells, R. D. (1988). The role of DNA sequence in the formation of Z-DNA versus cruciforms in plasmids. *J. Biol. Chem.* **263,** 7370–7377.

McVey, J. H., Nomura, S., Kelly, P., Mason, I. J., and Hogan, B. L. M. (1988). Characterization of the mouse SPARC/osteonectin gene. *J. Biol. Chem.* **263,** 11111–11116.

Metsäranta, M., Young, M. F., Sandberg, M., Termine, J. D., and Vuorio, E. (1989). Localization of osteonectin expression in human fetal skeletal tissues by in situ hybridization. *Calcif. Tissue Int.* **45,** 146–152.

Midura, R. J., McQuillan, D. J., Benham, K. J., Fisher, L. W., and Hascall, V. C. (1990). A rat osteogenic cell line (UMR 106-01) synthesizes a highly sulfated form of bone sialoprotein. *J. Biol. Chem.* **265,** 5285–5291.

Miyauchi, A., Alvarez, J., Greenfield, E. M., Teti, A., Grano, M., Colucci, S., Zambonin-Zallone, A., Patrick Ross, F., Teitelbaum, S. L., Cheresh, D., and Hruska, K. A. (1991). Recognition of osteopontin and related peptides by an a_vB_3 integrin stimulates immediate cell signals in osteoclasts. *J. Biol. Chem.* **266,** 20369–20374.

Miyazaki, Y., Setoguchi, M., Yoshida, S., Higuchi, Y., Akizuki, S., and Yamamoto, S. (1989). Nucleotide sequence of cDNA for mouse osteopontin-like protein. *Nucleic Acids Res.* **17,** 3298.

Miyazaki, Y., Setoguchi, M., Yoshida, S., Higuchi, Y., Akizuki, S., and Yamamoto, S. (1990). The mouse osteopontin gene: Expression in monocytic lineages and complete nucleotide sequence. *J. Biol. Chem.* **265,** 14432–14438.

Moore, M. A., Gotoh, Y., Rafidi, K., and Gerstenfeld, L. C. (1991). Characterization of a cDNA for chicken osteopontin: Expression during bone development, osteoblast differentiation, and tissue distribution. *Biochemistry,* **30,** 2501–2508.

Nagata, T., Todeacan, R., Goldberg, H. A., Zhang, Q., and Sodek, J. (1989). Sulphation of secreted phosphoprotein 1 (SPP1, osteopontin) is associated with mineralized tissue formation. *Biochem. Biophys, Res. Commun.* **165,** 234–240.

Nagata, T., Bellows, C. G., Kasugai, S., Butler, W. T., and Sodek, J. (1991a). Biosynthesis of bone proteins [SPP-1 (secreted phosphoprotein-1, osteopontin), BSP (bone sialoprotein) and SPARC (osteonectin)] in association with mineralized-tissue formation by fetal-rate clavarial cells in culture. *Biochem. J.* **274,** 513–520.

Nagata, T., Goldberg, H. A., Zhang, Q., Domenicucci, C., and Sodek, J. (1991b). Biosynthesis of bone proteins by fetal porcine calvariae *in vitro.* Rapid association of sulfated sialoproteins (secreted phosphoprotein-1 and bone sialoprotein) and chondroitin sulfate proteoglycan (CS-PGIII) with bone mineral. *Matrix* **11,** 86–100.

Naylor, L. H., and Clark, E. M. (1990). $d(TG)_n \cdot d(CA)_n$ sequences upstream of the rat prolactin gene form Z-DNA and inhibit gene transcription. *Nucleic Acids Res.* **18,** 1595–1601.

Neame, P. G., Choi, H. U., and Rosenberg, L. C. (1989). The primary structure of the core protein of the small Leucine-rich proteoglycan (PG I) from bovine articular cartilage. *J. Biol. Chem.* **264,** 8653–8661.

Nemir, M., DeVouge, M. W., and Mukherjee, B. B. (1989). Normal rat kidney cells secrete both phosphorylated and nonphosphorylated forms of osteopontin showing different physiological properties. *J. Biol. Chem.* **264,** 18202–18208.

Ng, K. W., Manji, S. S., Young, M. F., and Findlay, D. M. (1989). Opposing influences glucocorticoid and retinoic acid on transcriptional control in preosteoblasts. *Mol. Endo.* **3,** 2079–2085.

Nischt, R., Pottgiesser, J., Kreig, T., Mayer, U., Aumailley, M., and Timpl, R. (1991). Recombinant expression and properties of the human calcium-binding extracellular matrix protein BM-40. *Eur. J. Biochem.* **200,** 529–536.

Noda, M., and Rodan, G. (1987). Type β transforming growth factor (TGFβ) regulation of alkaline phosphatase expression and other phenotype-related mRNAs in osteoblastic rat osteosarcoma cells. *J. Cell. Physiol.* **133,** 426–437.

Noda, M., and Rodan, G. A. (1989). Transcriptional regulation of osteopontin production in rat osteoblast-like cells by parathyroid hormone. *J. Cell Biol.* **108,** 713–718.

Noda, M., Yoon, K., Prince, C. W., Butler, W. T., and Rodan, G. A. (1988). Transcriptional regulation of osteopontin production in rat osteosarcoma cells by type B transforming growth factor. *J. Biol. Chem.* **263,** 13916–13921.

Noda, M., Vogel, R. L., Craig, A. M., Prahl, J., DeLuca, H. F., and Denhardt, D. T. (1990). Identification of a DNA sequence responsible for binding of the 1,25-dihydroxyvitamin D_3 receptor and 1,25-dihydroxyvitamin D_3 enhancement of mouse secreted phosphoprotein 1 (Spp-1 or osteopontin) gene expression. *Proc. Natl. Acad. Sci. USA* **87,** 9995–9999.

Nomura, S., Wills, A. J., Edwards, D. R., Heath, J. K., and Hogan, B. L. M. (1988). Developmental expression of 2ar (osteopontin) and SPARC (osteonectin) RNA as revealed by in situ hybridization. *J. Cell Biol.* **106,** 441–450.

Nomura, S., Hashmi, S., McVey, J. H., Ham, J., Parker, M., and Hogan, B. L. M. (1989). Evidence for positive and negative regulatory elements in the 5′-flanking sequence of the mouse Sparc (osteonectin) gene. *J. Biol. Chem.* **264,** 12201–12207.

Oldberg, A., Franzen, A., and Heinegard, D. (1986). Cloning and sequence analysis of rat bone sialoprotein (osteopontin) cDNA reveals an Arg-Gly-Asp cell-binding sequence. *Proc. Natl. Acad. Sci. USA* **83,** 8819–8823.

Oldberg, A., Franzen, A., and Heinegard, D. (1988a). The primary structure of a cell-binding bone sialoprotein. *J. Biol. Chem.* **263,** 19430–19432.

Oldberg, A., Franzen, A., Heinegard, D., Pierschbacher, M., and Ruoslahti, E. (1988b). Identification of a bone sialoprotein receptor in osteosarcoma cells. *J. Biol. Chem.* **263,** 19433–19436.

Oldberg, A., Jirskog-Hed, B., Axelsson, S., and Heinegard, D. (1989). Regulation of bone sialoprotein mRNA by steroid hormones. *J. Cell Biol.* **109,** 3183–3186.

Otsuka, K., Yao, K.-L., Wasi, S., Tung, P. S., Aubin, J. E., Sodek, J., and Termine, J. D. (1984). Biosynthesis of osteonectin by fetal porcine calvarial cells in vitro. *J. Biol. Chem.* **259,** 9805–9812.

Owen, T. A., Aronow, M., Shalhoub, V., Barone, L. M., Wilming, L., Tassinari, M. S., Kennedy, M. B., Pockwinse, S., Lian, J. B., and Stein, G. S. (1990). Progressive development of the rat osteoblast phenotype *in vitro:* Reciprocal relationships in expression of genes associated with osteoblast proliferation and differentiation during formation of the bone extracellular matrix. *J. Cell. Physiol.* **143,** 420–430.

Owen, T. A., Aronow, M. S., Barone, L. M., Bettencourt, B., Stein, G. S., and Lian, J. B. (1991). Pleiotropic effects of vitamin D on osteoblast gene expression are related to the proliferative and differentiated state of the bone cell phenotype: Dependency upon basal levels of gene expression, duration of exposure, and bone matrix competency in normal rat osteoblast cultures. *Endocrinology* **128,** 1496–1504.

Pacifici, M., Oshima, O., Fisher, L. W., Young, M. F., Shapiro. I. M., and Leboy, P. S. (1990). Changes in osteonectin distribution and levels are associated with mineralization of the chicken tibial growth cartilage. *Calcif. Tissue Int.* **47,** 51–61.

Patarca, R., Freeman, G. J., Singh, P., Wei, F.-Y., Durfee, T., Blattner, F., Regnier, D. C., Kozak, C. A., Mock, B. A., Morse III, H. C., Jerrells, T. R., and Cantor, H. (1989). Structural and functional studies of the early T lymphocyte activation 1 (Eta-1) gene. Definition of a novel T cell-dependent response associated with genetic resistance to bacterial infection. *J. Exp. Med.* **170,** 145–161.

Patarca, R., Wei, F. Y., Singh, P., Morasso, M. I., and Cantor, H. (1990). Dysregulated expression of the T cell cytokine Eta-1 in $CD4^{-8-}$ lymphocytes during the development of murine autoimmune disease. *J. Exp. Med.* **172,** 1177–1183.

Paulsson, M., Aumailley, M., Deutzman, R., Timpl, R., Beck, K., and Engel, J. (1987). Laminin–nidogen complex. Extraction with chelating agents and structural characterization. *Eur. J. Biochem.* **166,** 11–19.

Pearson, D., and Sasse, J. (1992). Differential regulation of biglycan and decorin by retinoic acid in bovine chondrocytes. *J. Biol. Chem.* **267**(35), 25364–25370.

Prince, C. W. (1989). Secondary structure predictions for rat osteopontin. *Connect. Tissue Res.* **21,** 15–20.

Prince, C. W., and Butler, W. T. (1987). 1,25-dihydroxyvitamin D_3 regulates the biosynthesis of osteopontin, a bone derived cell attachment protein, a clonal osteoblast-like osteosarcoma cells. *Collagen Rel. Res.* **7,** 305–313.

Prince, C. W., Oosawa, T., Butler, W. T., Tomana, M., Bhown, A. S., Bhown, M., and Schrohenloher, R. E. (1987). Isolation, characterization, and biosynthesis of a phosphorylated glycoprotein from rat bone. *J. Biol. Chem.* **262,** 2900–2907.

Pringle, G. A., and Dodd, C. M. (1990). Immunoelectron microscopic localization of the core protein of decorin near the d and e bands of tendon collagen fibrils by use of monoclonal antibodies. *J. Histochem. Cytochem.* **38,** 1405–1411.

Rauch, U., Glössl, J., and Kresse, H. (1986). Comparison of small proteoglycans from skin fibroblasts and vascular smooth-muscle cells. *Biochem. J.* **238,** 465–474.

Reinholt, F. P., Hultenby, K., Oldberg, A., and Heinegard, D. (1990). Osteopontin–A possible anchor of osteoclasts to bone. *Proc. Natl. Acad. Sci. USA* **87,** 4473–4475.

Ringuette, M., Damjanovski, S., and Wheeler, D. (1991). Expression of SPARC/osteonectin in tissues of bony and cartilaginous vertebrates. *Biochem. Cell Biol.* **69,** 245–250.

Roden, S. B., Wesolowski, G., Yoon, K., and Rodan, G. A. (1989). Opposing effects of fibroblast growth factor and pertussis toxin on alkaline phosphatase, osteopontin, osteocalcin, and type I collagen mRNA levels in ROS 17/2.8 cells. *J. Biol. Chem.* **264,** 19934–19941.

Romanowski, R., Jundt, G., Termine, J. D., Mark, K., and Schulz, A. (1990). Immunoelectron microscopy of osteonectin and type I collagen in osteoblasts and bone matrix. *Calcif. Tissue Int.* **46,** 353–360.

Romaris, M., Heredia, A., Molist, A., and Bassols, A. (1991). Differential effect of transforming growth factor β on proteoglycan synthesis in human embryonic lung fibroblasts. *Biochim. Biophys. Acta* **1093,** 229–233.

Romberg, R. W., Werness, P. G., Lollar, P., Riggs, B. L., and Mann, K. G. (1985). Isolation and characterization of native adult osteonectin. *J. Biol. Chem.* **260,** 2728–2736.

Romberg, R. W., Werness, P. G., Riggs, B. L., and Mann, K. G. (1986). Inhibition of hydroxyapatite crystal growth by bone-specific and other calcium-binding protein. *Biochemistry* **25,** 1176–1180.

Roughley, P. J., and White, R. J. (1989). Dermatan sulphate proteoglycans of human articular cartilage. *Biochem. J.* **262,** 823–827.

Ruoslahti, E., and Pierschbacher, M. D. (1987). New perspectives in cell adhesion: RGD and integrins. *Science* **238,** 491–496.

Ruoslahti, E, and Yamaguchi, Y. (1991). Proteoglycans as modulators of growth factor activities. *Cell* **64,** 867–869.

Ryden, C., Maxe, I., Ljungh, A., Franzen, A., Heinegard, D., and Rubin, K. (1987). Selective binding of bone matrix sialoprotein to *Staphylococcus aureus* in osteomyelitis. *Lancet* **II,** 515.

Ryden, C., Yacoub, A. I., Maxe, I., Heinegard, D., Oldberg, A., Franzen, A., Ljungh, A., and Rubin, K. (1989). Specific binding of bone sialoprotein to *Staphylococcus aureus* isolated from patients with osteomyelitis. *Eur. J. Biochem.* **184,** 331–336.

Sage, H., Johnson, C., and Bornstein, P. (1984). Characterization of a novel serum albumin binding glycoprotein secreted by endothelial cells in culture. *J. Biol. Chem.* **259,** 3993–4007.

Sage, H., Tupper, J., and Bramson, R. (1986). Endothelial cell injury in vitro is associated with increased secretion of an Mr 43,000 glycoprotein ligand. *J. Cell Physiol.* **127,** 373–387.

Sage, H., Vernon, R. B., Funk, S. E., Everitt, E. A., and Angello, J. (1989). SPARC, a secreted protein associated with cellular proliferation, inhibits cell spreading in vitro and exhibits Ca^{+2}-dependent binding to the extracellular matrix. *J. Cell Biol.* **109,** 341–356.

Sauk, J. J., Van Kampen, C. L., Norris, K., Moehring, J., Foster, R. A., and Somerman, M. J. (1990). Persistent spreading of ligament cells on osteopontin/bone sialoprotein-I or collagen enhances tolerance to heat shock. *Exp. Cell Res.* **188,** 105–110.

Sawhney, R. S., Hering, T. M., and Sandell, L. J. (1991). Biosynthesis of small proteoglycan II (Decorin) by chondrocytes and evidence for a procore protein. *J. Biol. Chem.* **266,** 9231–9240.

Schmidt, G., Robenek, H., Harrach, B., Glössl, J., Nolte, V., Hörmann, H., Richter, H., and Kresse, H. (1987). Interaction of small dermatan sulfate proteoglycan from fibroblasts with fibronectin. *J. Cell Biol.* **104,** 1683–1691.

Schmidt, G., Hausser, H., and Kresse, H. (1990). Extracellular accumulation of small dermatan sulphate proteoglycan II by interference with the secretion-recapture pathway. *Biochem. J.* **266,** 591–595.

Schmidt, G., Hausser, H., and Kresse, H. (1991). Interaction of the small proteoglycan decorin with fibronectin. *Biochem. J.* **280,** 411–414.

Scott, P. G., and Dodd, C. M. (1990). Self-aggregation of bovine skin proteodermatan sulphate promoted by removal of the three N-linked oligosaccharides. *Connec. Tissue Res.* **24,** 225–236.

Scott, P. G., and Haigh, M. (1985). Proteoglycan-type I collagen fibril interactions in bone and non-calcifying connective tissues. *Biosci. Rep.* **5,** 71–81.

Scott, P. G., Winterbottom, N., Dodd, C. M., Edwards, E., and Pearson, C. H. (1986). A role for disulphide bridges in the protein core in the interaction of proteodermatan sulphate and collagen. *Biochem. Biophys. Res. Commun.* **138,** 1348–1354.

Senger, D. R., and Perruzzi, C. A. (1985). Secreted phosphoprotein markers for neoplastic transformation of human epithelial and fibroblastic cells. *Cancer Res.* **45,** 5818–5823.

Senger, D. R., Wirth, D. F., and Hynes, R. O. (1979). Transformed mammalian cells secrete specific proteins and phosphoproteins. *Cell* **16,** 885–893.

Senger, D. R., Wirth, D. F., and Hynes, R. O. (1980). Transformation-associated secreted phosphoproteins. *Nature (London)* **286,** 619–621.

Senger, D. R., Asch, B. B., Smith, B. D., Perruzzi, C. A., and Dvorak, H. F. (1983). A secreted phosphoprotein marker for neoplastic transformation of both epithelial and fibroblastic cells. *Nature (London)* **302,** 714–715.

Senger, D. R., Perruzzi, C. A., Gracey, C. F. Papadopoulos, A., and Tenen, D. G. (1988).

Secreted phosphoproteins associated with neoplastic transformation: Close homology with plasma proteins cleaved during blood coagulation. *Cancer Res.* **48,** 5770–5774.

Senger, D. R., Perruzzi, C. A., and Papadopoulos, A. (1989a). Elevated expression of secreted phosphoprotein 1 (osteopontin, 2ar) as a consequence of neoplastic transformation. *Anticancer Res.* **9,** 1291–1300.

Senger, D. R., Perruzzi, C. A., Papadopoulos, A., and Tenen, D. G. (1989b). Purification of human milk protein closely similar to tumor-secreted phosphoproteins and osteopontin. *Biochim. Biophys. Acta* **996,** 43–48.

Shiraga, H., Min, W., VanDusen, W. J., Clayman, M. D., Miner, D., Terrell, C. H., Sherbotie, J. R., Foreman, J. W., Przysiecki, C., Neilson, E. G., and Hoyer, J. R. (1992). Inhibition of calcium oxalate crystal growth *in vitro* by uropontin: Another member of the aspartic acid-rich protein superfamily. *Proc. Natl. Acad. Sci. USA* **89,** 426–430.

Singh, K., DeVouge, M. W., and Mukherjee, B. B. (1990). Physiological properties and differential glycosylation of phosphorylated and non-phosphorylated forms of osteopontin secreted by normal rat kidney cells. *J. Biol. Chem.* **265,** 18696–18701.

Singh, R. P., Patarca, R., Schwartz, J., Singh, P., and Cantor, H. (1990). Definition of a specific interaction between the early T lymphocyte activation 1 (Eta-1) protein and murine macrophages in vitro and its effect upon macrophages in vivo. *J. Exp. Med.* **171,** 1931–1942.

Smiley, B. L., and Stuart, K. (1990). Leucine-rich sequence in osteoinductive factor. *J. Bone Miner. Res.* **5,** 1189–1190.

Smith, J. H., and Denhardt, D. T. (1987). Molecular cloning of a tumor promoter-inducible mRNA found in JB6 mouse epidermal cells: Induction is stable at high, but not at low, cell densities. *J. Cell Biochem.* **34,** 13–22.

Somerman, M. J., Prince, C. W., Sauk, J. J., Foster, R. A., and Butler, W. T. (1987). Mechanism of fibroblast attachment to bone extracellular matrix: Role of a 44 kilodalton protein. *J. Bone Miner. Res.* **2,** 259–265.

Somerman, M. J., Fisher, L. W., Foster, R. A., and Sauk, J. J. (1988). Human bone sialoprotein I and II enhance fibroblast attachment *in vitro. Calcif. Tissue Int.* **43,** 50–53.

Somerman, M. J., Prince, C. W., Butler, W. T., Foster, R. A., Moehring, J. M., and Sauk, J. J. (1989). Cell attachment activity of the 44 kilodalton bone phosphoprotein is not restricted to bone cells. *Matrix* **9,** 49–54.

Sommarin, Y., Larsson, T., and Heinegard, D. (1989). Chondrocyte–matrix interactions. Attachment to proteins isolated from cartilage. *Exp. Cell Res.* **184,** 181–192.

Stein, G. S., Lian, J. B., and Owen, T. A. (1990). Relationship of cell growth to the regulation of tissue-specific gene expression during osteoblast differentiation. *FASEB J.* **4,** 3111–3123.

Stenner, D. D., Tracy, R. P., Riggs, B. L., and Mann, K. G. (1986). Human platelets contain and secrete osteonectin, a major protein of mineralized bone. *Proc. Natl. Acad. Sci. USA* **83,** 6892–6896.

Strauss, P. G., Closs, E. I., Schmidt, J., and Erfle, V. (1990). Gene expression during osteogenic differentiation in mandibular condyles in vitro. *J. Cell Biol.* **110,** 1369–1378.

Strickland, S., and Mahdavi, V. (1978). The induction of differentiation in teratocarcinoma stem cells by retinoic acid. *Cell* **15,** 393–403.

Swaroop, A., Hogan, B. L. M., and Franke, U. (1988). Molecular analysis of the cDNA for human SPARC/osteonectin/BM-40: Sequence, expression and localization of the gene to chromosome 5q31–q33. *Genomics* **2,** 37–47.

Termine, J. D., Belcourt, A. B., Conn, K. M., and Kleinman, H. K. (1981a). Mineral and collagen-binding proteins of fetal calf bone. *J. Biol. Chem.* **256,** 10403–10408.

Termine, J. D., Kleinman, H. K., Whitson, S. W., Conn, K. M., McGarvey, M. L., and Martin, G. R. (1981b). Osteonectin, a bone-specific protein linking mineral to collagen. *Cell* **26,** 99–105.

Thiebaud, D., Ng, K. W., Findlay, D. M., Harker, M., and Martin, T. J. (1990). Insulinlike growth factor 1 regulates mRNA levels of osteonectin and pro-α1(I)-collagen in clonal preosteoclastic calvarial cells. *J. Bone Miner. Res.* **5,** 761–767.

Thomas, G. J., Mason, R. M., and Davies, M. (1991). Characterization of proteoglycans synthesized by human adult glomerular mesangial cells in culture. *Biochem. J.* **277,** 81–88.

Truppe, W., and Kresse, H. (1978). Uptake of proteoglycans and sulfated glycosaminoglycans by cultured skin fibroblasts. *Eur. J. Biochem.* **85,** 351–356.

Tung, P. S., Domenicucci, C., Wasi, S., and Sodek, J. (1984). Specific immunohistochemical localization of osteonectin and collagen type I and II in fetal and adult porcine dental tissues. *J. Histochem. Cytochem.* **55,** 531–540.

Turner, R. T., Colvard, D. S., and Spelsberg, T. C. (1990). Estrogen inhibition of periosteal bone formation in rat long bones: Down-regulation of gene expression for bone matrix proteins. *Endocrinology* **127,** 1346–1351.

Tyree, B. (1989). The partial degradation of osteonectin by a bone-derived metalloprotease enhances binding to type I collagen. *J. Bone Miner. Res.* **4,** 877–883.

Ullrich, O., Mann, K., Haase, W., and Koch-Brandt, C. (1991). Biosynthesis and secretion of an osteopontin-related 20-kDa polypeptide in the Madin-Darby canine kidney cell line. *J. Biol. Chem.* **266,** 3518–3525.

Vetter, U., Fedarko, N. S., Young, M. F., Termine, J. D., Just, W., Vogel, W., and Gehron Robey, P. (1991a). Biglycan synthesis in fibroblasts of patients with Turner's syndrome and other X chromosome abnormalities. *J. Bone Miner. Res.* **6,** S133.

Vetter, U., Fisher, L. W., Mintz, K. P., Kopp, J. B., Tuross, N., Termine, J. D., and Gehron Robey, P. (1991b). Osteogenesis Imperfecta: Changes in noncollagenous proteins in bone. *J. Bone Miner. Res.* **6,** 501–505.

Vetter, U., Vogel, W., Just, W., Young, M. F., and Fisher, L. W. (1993). Human decorin gene: Intron–exon junctions and chromosomal localization. *Genomics* **15,** 161–168.

Villarreal, X. C., Mann, K. G., and Long, G. L. (1989). Structure of human osteonectin based upon analysis of cDNA and genomic sequences. *Biochemistry* **28,** 6483–6491.

Vogel, K. G., and Trotter, J. A. (1987). The effect of proteoglycans on the morphology of collagen fibrils formed *in vitro. Collagen Rel. Res.* **7,** 105–114.

Vogel, K. G., Paulsson, M., and Heinegård, D. (1984). Specific inhibition of type I and type II collagen fibrillogenesis by the small proteoglycan of tendon. *Biochem. J.* **223,** 587–597.

Voss, B., Glössl, J., Cully, Z., and Kresse, H. (1986). Immunocytochemical investigation on the distribution of small chondroitin sulfate-dermatan sulfate proteoglycan in the human. *J. Histochem. Cytochem.* **34,** 1013–1019.

Vuorio, T., Kähäri, V. M., Black, C., and Vuorio, E. (1991). Expression of oteonectin, decorin, and transforming growth factor beta 1 genes in fibroblasts cultured from patients with systemic sclerosis and morphia. *J. Reumato.* **18,** 247–251.

Wahls, W. P., Wallace, L. J., and Moore, P. D. (1990). The Z-DNA motif d$(TG)_{30}$ promotes reception of information during gene conversion events while stimulating homologous recombination in human cells in culture. *Mol. Cell. Biol.* **10,** 785–793.

Wasi, S., Otsuka, K., Yao, K.-L., Tung, P. S., Aubin, J. E., Sodek, J., and Termine, J. D. (1984). An osteonectin like protein in porcine periodontal ligament and its synthesis by periodontal ligament fibroblast. *Can. J. Biochem. Cell Biol.* **62,** 470–478.

Weinreb, M., Shinar, D., and Rodan, G. A. (1990). Different pattern of alkaline phosphatase, osteopontin, and osteocalcin expression in developing rat bone visualized by *in situ* hybridization. *J. Bone Miner. Res.* **5,** 831–842.

Westergren-Thorsson, G., Antonsson, P., Malmström, A., Heinegård, D., and Oldberg, Å. (1991). The synthesis of a family of structurally related proteoglycans in fibroblasts is differently regulated by TGF-β. *Matrix* **11,** 177–183.

Whitson, S. W., Harrison, W., Dunlap, M. K., Bowers, D. E., Jr., Fisher, L. W., Gehron

Robey, P., and Termine, J. D. (1984). Fetal bovine bone cells synthesize bone-specific matrix proteins. *J. Cell Biol.* **99,** 607–614.

Winnermöller, M., Schmidt, G., and Kresse, H. (1991). Influence of decorin on fibroblast adhesion to fibronectin. *Eur. J. Cell Biol.* **54,** 10–17.

Wrana, J. L., Maeno, M., Hawrylyshyn, B., Yao, K.-L., Domenicucci, C., and Sodek, J. (1988). Differential effects of transforming growth factor-β on the synthesis of extracellular matrix proteins by normal fetal rat calvarial bone cell populations. *J. Cell Biol.* **106,** 915–924.

Wrana, J. L., Zhang, Q., and Sodek, J. (1989). Full length cDNA sequence of porcine secreted phosphoprotein-1 (SPP-1, osteopontin). *Nucleic Acids Res.* **17,** 10119.

Wronski, T. J., Halloran, B. P., Bikle, D. D., Globus, R. K., and Morey-Holton, E. R. (1986). Chronic administration of 1,25-dihydroxyvitamin D_3: Increased bone but impaired mineralization. *Endocrinology* **119,** 2580–2585.

Yamaguchi, Y., and Ruoslahti, E. (1988). Expression of human proteoglycan in Chinese hamster ovary cells inhibits cell proliferation. *Nature (London)* **336,** 244–246.

Yamaguchi, Y., Mann, D. M., and Ruoslahti, E. (1990). Negative regulation of transforming growth factor-β by the proteoglycan decorin. *Nature (London)* **346,** 281–284.

Yoon, K., Buenaga, R., and Rodan, G. (1987). Tissue specificity and developmental expression of rat osteopontin. *Biochem. Biophys. Res. Commun.* **148,** 1129–1136.

Young, M. F., Bolander, M. E., Day, A. A., Ramis, C. I., Robey, P. G., Yamada, Y., and Termine, J. D. (1986). Osteonectin mRNA: Distribution in normal and transformed cells. *Nucleic Acids Res.* **14,** 4483–4497.

Young, M. F., Findlay, D. M., Dominguez, P., Burbelo, P. D., MaQuilla, C., Kopp, J. B., Robey, P. G., and Termine, J. D. (1989). Osteonectin promoter DNA sequence analysis and S1 endonuclease site potentially associated with transcriptional control in bone cells. *J. Biol. Chem.* **264,** 450–456.

Young, M. F., Day, A. A., Dominguez, P., McQuillan, C. I., Fisher, L. W., and Termine, J. D. (1990a). Structure and expression of osteonectin mRNA in human tissue. *Connec. Tissue Res.* **24,** 17–28.

Young, M. F., Day, A. A., Gehron Robey, P., and Termine, J. D. (1990b). Interaction of osteonectin and type I collagen in bone cells. *In* "Structure, Molecular Biology, and Pathology of Collagen" (R. Fleischmajor, B. R. Olsen, and K. Kuhn, eds.), Vol. 580, pp. 526–528. *Ann. N.Y. Acad. Sci.*

Young, M. F., Kerr, J. M., Termine, J. D., Wewer, U. M., Ge Wang, M., McBride, O. W., and Fisher, L. W. (1990c). cDNA cloning, mRNA distribution and heterogeneity, chromosomal location, and RFLP analysis of human osteopontin (OPN). *Genomics* **7,** 491–502.

Zhang, Q., Domenicucci, C., Goldberg, H. A., Wrana, J. L., and Sodek, J. (1990). Characterization of fetal porcine bone sialoproteins, secreted phosphoprotein I (sppI, osteopontin), bone sialoprotein, and a 23-kDa glycoprotein. Demonstration that the 23-kDa glycoprotein is derived from the carboxyl terminus of sppI. *J. Biol. Chem.* **265,** 7583–7589.

Zung, P., Domenicucci, C., Wasi, S., Kuwata, F., and Sodek, J. (1986). Osteonectin is a minor component of mineralized connective tissues in rat. *Biochem. Cell Biol.* **64,** 356–362.

7

THE OSTEOCALCIN GENE AS A MOLECULAR MODEL FOR TISSUE-SPECIFIC EXPRESSION AND 1,25-DIHYDROXYVITAMIN D_3 REGULATION

J. Wesley Pike, Teruki Sone, Keiichi Ozono, Robert A. Kesterson, and Sandra A. Kerner

Cellular and Molecular Biology of Bone

I. INTRODUCTION

Osteoblasts are derived from undifferentiated mesenchymal progenitor cells, cells that are capable of differentiating into chondrocytes and muscle cells as well as adipocytes (Rodan and Martin, 1981; Owen, 1985). Features of mature osteoblasts include the capacity to express alkaline phosphatase, bone Gla protein or osteocalcin [OC] and collagen, and to express intracellular and membrane receptors that mediate the actions of retinoic acid, 1,25-dihydroxyvitamin D_3 (1,25$(OH)_2D_3$), glucocorticoids, parathyroid hormone, and a variety of growth factors such as transforming growth factor β and bone morphogenic protein-2 (Epstein, 1989; Hauschka *et al.*, 1975; Price *et al.*, 1976; Malone *et al.*, 1982). Despite considerable knowledge of osteoblast function, however, little is known of the transcriptional events that regulate osteoblast development and maturation or of the factors that mediate these events. Experiments that provide molecular detail about the nature of osteoblastic gene expression are thus warranted.

OC is one of the most abundant noncollagenous proteins found in adult bone. It is a highly conserved γ-carboxyglutamic acid-containing protein that is believed to be produced exclusively by osteoblasts (Hauschka *et al.*, 1975; Price *et al.*, 1976). The function of OC remains unknown, although its appearance is associated with mineral deposition (Malone *et al.*, 1982), and its capacity to bind hydroxyapatite (Gerstenfeld *et al.*, 1986) and to act as a chemoattractant for bone resorbing cells (Glowacki and Lian, 1989) suggest a role for the protein as an index of bone turnover (Hauschka *et al.*, 1975). Indeed, serum levels of OC are widely considered as clinically diagnostic for osteoblast function and active bone remodeling (Price and Nishimoto, 1980; Lian and Gundberg, 1988; Price *et al.*, 1980; Gundberg *et al.*, 1983; Delmas *et al.*, 1973). Despite the lack of clarity regarding OC function in bone, it is clear that this protein is highly regulated by a variety of hormones that act upon the osteoblast (Price and Baukol, 1980; Lian *et al.*, 1985, 1987). Early work identified 1,25$(OH)_2D_3$ as a major upregulator of this protein, and critical investigation suggested that control was exerted at the level of the gene (Pan and Price, 1984). Indeed, as will be discussed in this chapter, the OC gene has become a leading model for elucidation of the molecular pathway by which the vitamin D hormone acts to induce transcription.

II. OSTEOCALCIN GENE TRANSCRIPTION UNIT

Probes derived from the amino acid sequence of OC enabled recovery of human genomic sequence comprising the gene for human OC (Celeste *et al.*, 1986). Determination of the structural organization of this gene revealed four exons interspersed by three introns that together spanned approximately a kilobase of DNA. Exon 1 contained an in-frame ATG initiator codon as well as codons that direct synthesis of a highly hydrophobic signal sequence responsible for transmembrane transfer. As anticipated, exon 4 contained the TGA termination codon, 3′ noncoding sequences, and a canonical AATAAA polyadenylation sequence. The subsequent cloning and structural determination of the rat gene for OC indicated an overall organization highly conserved with respect to the human gene (Lian *et al.*, 1989). Considerable homology within the 5′ regulatory region suggested a functional importance for these sequences.

III. FUNCTIONAL ACTIVITY OF THE OSTEOCALCIN PROMOTER

Visual inspection of the 5′ portions of both the human and rat genes for OC suggested the presence of putative TATA box sequences as well as consensus sequences for a variety of important transcription factors (Lian *et al.*, 1989; Celeste *et al.*, 1986). These observations, together with previous studies demonstrating that the OC gene was highly regulated transcriptionally by hormones such as retinoic acid, $1,25(OH)_2D_3$, and glucocorticoids (Price and Baukol, 1980; Nishimoto *et al.*, 1987; Jowell *et al.*, 1987), prompted us and others to examine the molecular basis for promoter activity. Determination of activity was made at the level of transcriptional control by using chimeric genes containing the OC promoter and upstream 5′ flanking regions fused to the chloramphenicol acetyltransferase structural gene and an SV40 polyadenylation signal. Plasmids were introduced by DNA-mediated gene transfer into cultured cells of osteoblastic lineage such as ROS 17/2.8, UMR 106, and MC3T3-E1 and basal as well as hormone-inducible activities of the reporter examined.

A. Basal Activity of the Osteocalcin Promoter *in Vitro*

An OC chimeric gene containing over a kilobase of upstream human promoter sequence exhibited strong basal activity when introduced into ROS 17/2.8 cells (McDonnell *et al.*, 1989). Importantly, these osteosarcoma-derived cells express high levels of endogenous OC (Price and Baukol, 1980). In contrast, subsequent examination of this fusion gene in NIH3T3 and CV-1 fibroblasts as well as in the preosteoblastic cell line MC3T3-E1, none of which produce endogenous OC, revealed an extremely low basal level of activity (Pike, 1990; Uchida *et al.*, 1991). Studies with the rat OC promoter led to similar results; high levels of basal activity in ROS 17/2.8 cells and low levels of activity in OC-nonproducing cells such as UMR 106 (Yoon *et al.*, 1988). These observations suggest that a

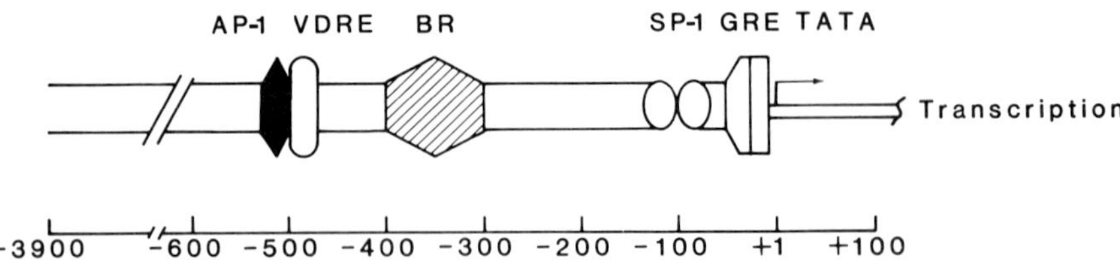

FIGURE 1 Functional regions of the human osteocalcin gene promoter. The start site of transcription is +1 from the TATA box. AP-1 (activator protein-1) binding site; BR, basal repressor region; GRE, glococorticoid response element; SP-1, - GC box region; TATA, TATA box; VDRE, vitamin D response element.

factor(s) important for the endogenous expression of OC may be acting similarly upon the exogenously added foreign chimeric genes and that this factor(s) together with the DNA sequence elements with which it interacts might be identified. Moreover, it is possible that this activator(s) represents a major determinant of osteoblast-specific expression of OC.

Critical regions that contributed to basal expression of the human OC promoter in ROS 17/2.8 cells were identified by utilizing a series of promoter sequences that contained the TATA box and increasing upstream sequence (Kerner *et al.*, 1989). The results of these experiments, as illustrated in Figure 1, suggest that important regions lie immediately proximal to the region of initiation as well as more than 400 basepairs (bp) upstream. Studies with the rat OC promoter similarly have identified a potentially functional region near the TATA sequence that appears as a CAAT box (Markose *et al.*, 1990). Certain sequence homologies exist within this region between the rat and human genes. Nevertheless, the CAAT box in the rat gene appears as a CAAT sequence, whereas in the human promoter this sequence is replaced by CAAAT and bounded on both sides by GC boxes. Indeed, selective mutagenesis of each of these three sites followed by transfection into ROS 17/2.8 cells suggests that only the two GC boxes in the human promoter contribute to basal activity (K. Ozono and J. W. Pike, unpublished data). Distal basal elements include an activator protein-1 (AP-1) site within the human promoter at −500 (Ozono *et al.*, 1990) and an apparent unmapped element within the rat promoter approximately a kilobase upstream (Markose *et al.*, 1990; Terpening *et al.*, 1991). These elements clearly contribute to the capacity of certain steroid hormones to induce both promoters (see later). Despite these studies, the contribution of these elements to tissue-specific expression of OC remain unknown.

B. Bone-Specific Expression of a Reporter Gene in Transgenic Mice

To determine whether the OC promoter was capable of directing bone-specific expression *in vivo*, we introduced a gene chimera containing the

OC promoter and approximately 4 kilobases of upstream DNA fused to CAT into the germ line of mice (Kesterson *et al.*, 1993). Several mouse lines were generated, and a subset of those containing integrated DNA expressed CAT. While each mouse line expressed a unique level of CAT, the expression of the enzyme activity was predominantly found in bone-associated tissues that included extracted calvaria and femur. Activity was also observed in bone marrow and brain. Immunohistochemical evaluation of femur sections with an antibody to CAT further revealed the presence of peroxidase stain not only in osteoblastlike cells lining the bone surface but also in chondrocytes. These studies suggest that indeed the OC promoter is capable of directing bone-specific expression of its linked gene, although unexpected sites of expression are clearly evident. Further mouse strains are currently being generated to determine whether the sites of expression of the reporter outside bone or in chondrocytes are bona fide, are due to the incomplete nature of the DNA that has been introduced, or are due to the random nature of the site of integration.

C. Hormonal Responsiveness of the Osteocalcin Promoter

A DNA construction carrying approximately 1300 bp of the 5′ flanking region of the human gene fused to CAT was introduced transiently into ROS 17/2.8 cells, and its activity in the presence of a variety of hormones was evaluated (McDonnell *et al.*, 1989; Morrison *et al.*, 1989). As observed in Figure 2, enhanced activity was identified when the cells were treated with $1,25(OH)_2D_3$ and retinoic acid, and suppression of the gene was observed when glucocorticoids were utilized (McDonnell *et al.*, 1989; Pike, 1990; Kerner *et al.*, 1989; Morrison *et al.*, 1989). Each result was consistent with the effects of these hormones on endogenous gene expression (Price and Baukol, 1980; Nishimoto *et al.*, 1987; Jowell *et al.*, 1987). Indeed, similar results with $1,25(OH)_2D_3$ were observed with the rat OC promoter (Terpening *et al.*, 1991; Demay *et al.*, 1990). Thus, it is clear that $1,25(OH)_2D_3$ as well as retinoic acid alters the level of expression of OC directly through an action at the level of the gene's promoter.

That OC was regulated at the level of its promoter was also evaluated *in vivo* utilizing the transgenic mouse strains previously described. Several mouse strains homozygous for the osteocalcin promoter transgene were treated with strontium chloride to suppress circulating levels of $1,25(OH)_2D_3$, and a subset of these "inhibited" mice were then further treated by injection with 25 ng of exogenous $1,25(OH)_2D_3$ per day for 3 days to restore circulating levels of the hormone. Calvaria, femur, and brain tissues were removed and evaluated for CAT activity, and these activities were contrasted with those derived from dietary normal animals. The results suggest that, indeed, removal of the hormone from circulation led to a significant suppression of tissue transgene activity,

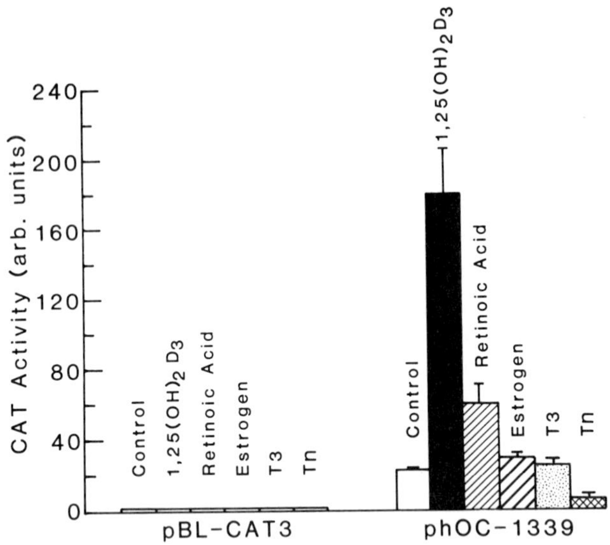

FIGURE 2 Transcriptional response of the human osteocalcin promoter to various hormones. A plasmid composed of the human osteocalcin promoter (−1339 to +10, phOC-1339) fused to chloramphenicol acetyltransferase (CAT) was introduced into the ROS 17/2.8 cell line by transfection methods and examined for response to 1,25-dihydroxyvitamin D_3 (1,25$(OH)_2D_3$), retinoic acid, estrogen, thyroid hormone (T3), and triamcinalone (Tn). CAT activity (arbitrary [arb.] units) was determined 48 hr following treatment. Error bars are indicated (SEM).

and restoration of the hormone resulted in recovery and in several cases increased activity of the reporter transgene. These *in vivo* studies confirm the *in vitro* observation that the OC promoter contains elements that are solely responsible for mediating the induction conferred by 1,25$(OH)_2D_3$.

D. Mapping the Vitamin D- and Retinoic Acid-Responsive Elements

The precise location of the vitamin D-responsive element (VDRE) within the OC gene promoter was mapped functionally using unidirectional and/or internal deletion analysis. As previously, the individual constructions were introduced into ROS 17/2.8 cells by transfection, and the activity of the construction assessed in the presence and absence of the vitamin D hormone. Our studies revealed that the region that conferred vitamin D response was located in the human gene between −512 and −485 relative to the start site of transcription (Kerner *et al.*, 1989; Ozono *et al.*, 1990). Deletion of this sequence in the human promoter led to loss of vitamin D response. Perhaps more definitively, transfer of these sequences in either orientation to heterologous viral promoters not normally inducible by vitamin D, such as those of the mouse mammary tumor virus LTR or the herpes simplex virus thymidine kinase gene, conferred a novel new responsiveness to the hormone (Pike, 1990; Kerner *et al.*, 1989). Interestingly, retinoic acid responsiveness also mapped to the same DNA sequence (Pike, 1990; Schule *et al.*, 1990), suggesting that the action of both hormones was mediated through the same *cis*-acting element. Similar studies using the rat gene promoter were carried out, leading to the identification of a structurally similar element lying between −458 and −433 (Demay *et al.*, 1990; Terpening *et al.*, 1991). These results suggest that indeed DNA sequences identified within the OC gene promoter are capable of directing the response to vitamin D and that these sequences retain properties typical of hormone-responsive enhancers (Beato, 1989).

E. Structural Organization of the Vitamin D-Responsive Elements

The human OC VDRE is composed of two direct but imperfect repeats separated by three nucleotide pairs, as documented in Fig. 3 (Ozono *et al.*, 1990). Studies in the rat OC gene (Terpening *et al.*, 1991) as well as more recent examination of the mouse osteopontin gene (Noda *et al.*, 1990) have identified cognate VDREs that are structurally identical to those found in the human gene. The consensus sequences for the two hexanucleotide half-sites of these three VDREs are G G G/T T G/C A and G G G/T T G/T/A A. These direct repeats are clearly reminiscent of the structural motif and sequence of elements that mediate both thyroid hormone and retinoic acid response in other genes (Glass *et al.*, 1988; Umesono *et al.*, 1988; de The *et al.*, 1990). Thus, it is not surprising, perhaps, that retinoic acid acts on the OC gene through this element. However, the observation that thyroid hormone does not stimulate the activity of this promoter (Ozono and Pike, unpublished data) also indicates the importance of promoter context in the hormonal response mechanism. Recent systematic evaluation of these elements for determinants that confer hormonal specificity for vitamin D, retinoic acid, thyroid hormone, and other hormones suggests that the spacing between the hexanucleotide halfsites is a critical factor, although it is clearly not the only specifier (Umesono *et al.*, 1991). Thus, the "3-4-5" rule predicts that a maximal vitamin D hormone response is registered using two directly repeated A G G T C A hexanucleotides when three nucleotides are located in between. Similarly, maximal thyroid hormone and retinoic acid response is obtained when the two repeats are interspersed with 4 and 5 bp, respectively. A final prediction of this rule is that perhaps additional unknown hormones may prefer repeated halfsites spaced by 1 or 2, and possibly 6 or more, base separations. Despite these findings, it is clear that the specific sequence of the repeated hexanucleotides is also influential in determining hormonal specificity.

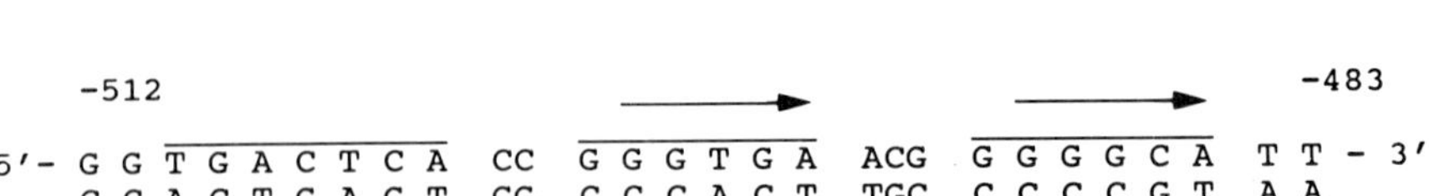

FIGURE 3 Structural organization of the human osteocalcin vitamin D-responsive element (VDRE). The activator protein 1 (AP-1) and VDRE sites are indicated over the nucleotide sequence. Arrows indicate the two direct repeats within the VDRE.

F. The Activator Protein-1 Site within the Human Osteocalcin Promoter Synergizes with the Vitamin D-Responsive Element

Immediately juxtaposed to the VDRE in the human OC is a unique consensus DNA sequence for the AP-1 protooncogene family (see Fig. 3). This site is crucial for vitamin D hormone activation, although the mechanism by which this contribution to hormonal response is manifested remains undefined. The structure of this responsive region indicates that it is composite in nature (Diamond *et al.*, 1990). That this site represents a functional component of OC expression is derived from several lines of evidence. First, administration of phorbol esters such as TPA to cultured cells containing either transfected or integrated OC promoters enhances reporter function independent of vitamin D activation (Ozono *et al.*, 1991). This observation suggests that the role of the AP-1 site is to enhance OC response. Consistent with this interpretation is the observation that internal point mutagenesis of the AP-1 site, at nucleotide bases known to compromise the ability of c-*fos*/c-*jun* heterodimers to interact, results in a down-regulation of OC promoter activity. Indeed, comparison of both basal and 1,25$(OH)_2D_3$-inducible levels of the mutant and intact promoters indicate the presence of a functional synergism when both *cis*-acting sites and their transactivating proteins are present.

The mechanism by which this synergism occurs remains to be clarified. However, *in vitro* DNA binding studies have revealed that AP-1 protein complexes such as Fos/Jun or osteoblastic AP-1 complexes of unknown identity and the vitamin D receptor (VDR) (see later for VDR DNA binding studies) can simultaneously bind to this dual element (Ozono *et al.*, 1991). Thus, incubation of both the VDR and extracts containing AP-1 to the AP-1/VDRE DNA under conditions of probe excess reveal protein–DNA complexes following bandshift analysis comprised of VDR alone, AP-1 alone, and a highly retarded complex of both VDR and AP-1. Thus, synergism is likely to arise through complex protein–protein interactions at the site of DNA binding, although alternative possibilities must also be considered. Importantly, while synergism is apparent in the current studies, inhibition of $1,25(OH)_2D_3$ response might also arise in the context of the osteoblast if an inhibitor of the Fos/Jun heterodimer were to be expressed at an appropriate time during this bone cell's development. Interestingly, in the rat OC gene promoter, an upstream element that remains to be identified appears to represent a functional homolog of the human AP-1 sequence (Yoon *et al.*, 1988; Terpening *et al.*, 1991).

IV. STEROID RECEPTOR SUPERFAMILY

The actions of steroid, thyroid, and retinoid hormones are known to be mediated by specific intracellular receptors. Indeed, the recent molecular cloning of the receptors for most if not all of the currently known hormonal ligands has revealed each to be part of an enlarging family of gene products (Evans, 1988; O'Malley, 1990), all of which function in a fundamentally similar manner.

A. Structural Aspects of the Steroid Receptor Family

A salient feature of each of these receptors is a highly homologous central or amino-terminal DNA binding core (Evans, 1988; O'Malley, 1990). The region is highly basic in nature, contains nine positionally conserved cysteine residues, and is structurally comprised of two finger loops. Eight of the nine cysteine sulfur residues are believed to tetrahedrally coordinate two zinc atoms and are essential for the proper folding of each of the proteins into an active conformation. This hypothesis has been confirmed through extended X-ray adsorption fine-structure analysis of the recombinantly expressed glucocorticoid receptor DNA binding domain fragment (Freedman *et al.*, 1988). More recently, the three-dimensional solution structures for this region of the glucocorticoid (Hard *et al.*, 1990) and estrogen (Schwabe *et al.*, 1990) receptors have been determined by nuclear magnetic resonance spectroscopy. These experiments suggest that the molecular organization of this region is similar between the two receptors and is comprised of two important α-helical stretches located on the carboxyl-terminal sides of each finger loop. A model taken from the molecular organization of the estrogen receptor structure (Schwabe *et al.*, 1990) is observed for the DNA binding domain of the VDR in Figure 4. The high degree of homology within this region among all the steroid receptors predicts that the overall three-dimensional organization may be very similar.

The carboxyl-terminal portion of the protein is less highly conserved among the receptors, containing several localized regions of moderate

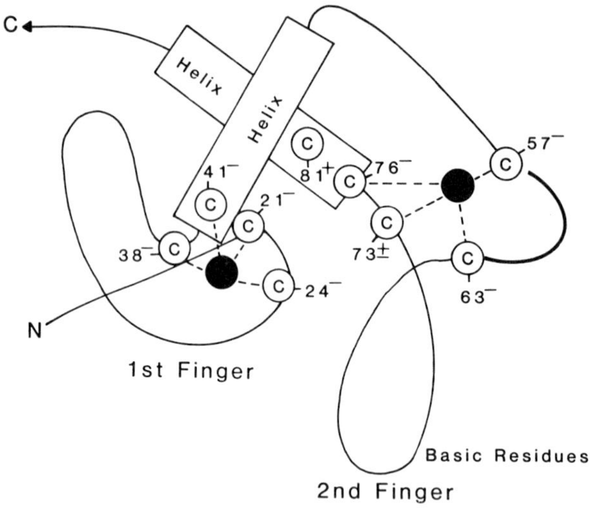

FIGURE 4 Model of the vitamin D receptor (VDR) DNA binding domain derived from the nuclear magnetic resonance structure of the estrogen receptor defined in Schwabe *et al.* (1990). C and N represent the carboxy- and amino-terminal orientation of the sequence. The nine positionally conserved cysteines (C) are indicated. The position of the two zinc atoms is indicated by a solid circle. Numbers indicate the amino acid residue within the VDR protein. The superscript + or − indicates the transcriptional activity of a mutant form of the VDR bearing a cysteine to serine mutation at that site (see Sone *et al.*, 1991a). The two helical structures are indicated with a rectangle.

conservation and predominant regions of little homology (Evans, 1988; O'Malley, 1990). While this region is known to confer capacity for high-affinity ligand binding to the receptor gene family, internal deletion analysis has not pinpointed significantly the location of the hydrophobic hormone binding pocket. Thus, a broad region of more than several hundred amino acids appears to be required presently for this binding function. This region also contributes an important dimerization function to the steroid receptor family (Kumar and Chambon, 1988; Fawell *et al.*, 1990). The interaction of receptors on their cognate *cis*-acting DNA elements is conferred through the formation of either homo- or heterodimers (see later). Receptor dimers exhibit a considerable increase in affinity for appropriate DNA binding sites over that of their monomeric forms, largely due to the stability achieved by the duplex on corresponding specific DNAs. The dimerization domain within the receptors appears to be bipartite, again spanning a majority of the carboxyl terminus and co-localizing to two of the conserved regions within this portion of the receptors. Additional functional activities within this complex region have also been mapped. They include such activities as nuclear transfer signals (Picard and Yamamoto, 1988) and transactivation domains (Evans, 1988; O'Malley, 1990). As the functions of dimerization, nuclear transfer, and transactivation are all modulated by ligand binding *in vivo*, it is not surprising from a structural point of view that these functional regions might be interconnected.

B. Cloning and Structure of the Vitamin D Receptor

The development of antibodies directed toward the VDR, together with the introduction of a new lambda-phage cloning vector, provided the technical means whereby the receptor's rate structural gene was recovered from a complementary DNA (cDNA) library. Partial cDNAs for the chicken VDR were initially identified and confirmed (McDonnell *et al.*, 1987), whereupon cDNAs for human (Baker *et al.*, 1988) and rat (Burmester *et al.*, 1988) also were obtained. Importantly, the deduced primary amino acid sequences of these cDNA clones revealed the receptor to be a member of the steroid receptor superfamily, which at the time of cloning included the glucocorticoid, progesterone, and estrogen receptors. Despite this, definitive identity of the cloned VDR was obtained through recombinant expression of the cDNA in a receptor-free mammalian cell background (Baker *et al.*, 1988). The expressed protein was of the appropriate size and immunoreactivity and bound $1,25(OH)_2D_3$ with normal receptor affinity and specificity. Moreover, while homologous regions characteristic of the intracellular receptor family were present within the VDR, deletion fragments of the protein containing these putative functional regions were demonstrated to retain the anticipated functional activity (McDonnell *et al.*, 1989). A current model for active functional domains of the human VDR is illustrated in Figure 5.

V. ROLE OF THE VITAMIN D RECEPTOR IN OSTEOCALCIN GENE ACTIVATION

A. The Vitamin D Receptor Is Essential to 1,25-Dihydroxyvitamin D_3 Activation of Osteocalcin

Strong activation of the OC promoter by $1,25(OH)_2D_3$ was obtainable in cells such as ROS 17/2.8 and MC3T3-E1 that naturally express the VDR. In contrast, no hormonal response was evident when the OC promoter was transfected into cells that did not contain the VDR. While these data supported the contention that the VDR mediated the observed hormonal response, no direct evidence of this fact could be concluded from the studies. Evidence for the involvement of the VDR in OC transcription was obtained, therefore, by introducing an expression vector containing the receptor cDNA into VDR-negative cells (McDonnell *et al.*, 1989). Under these conditions, $1,25(OH)_2D_3$ response was promptly restored. This restoration depended on receptor concentration and required syn-

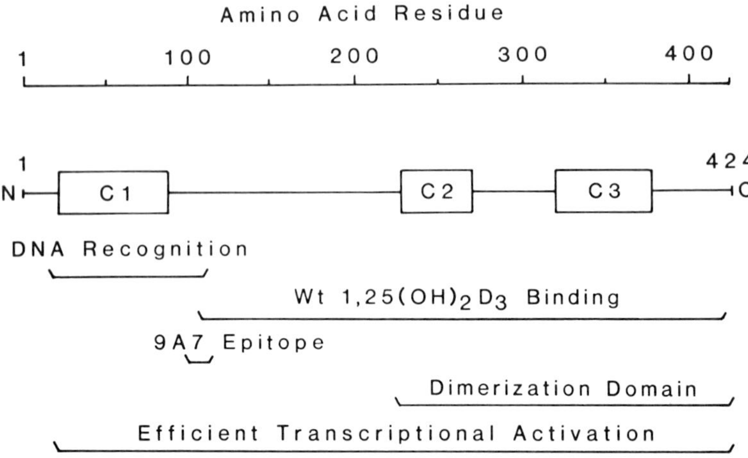

FIGURE 5 The functional domain structure of the human vitamin D receptor (VDR). A linear model of the human VDR is shown below an amino acid residue scale. Three regions that exhibit conservation with the steroid receptor superfamily are designated C1, C2, and C3. C1 comprises the putative DNA binding zinc finger domain, whereas C2 and C3 lie within the steroid binding domain. Regions that confer functional activity are indicated below the linear diagram. $1,25(OH)_2D_3$, 1,25-dihydroxyvitamin D_3.

thesis of a normal receptor; receptors that were not full-length, did not bind $1,25(OH)_2D_3$, or contained mutations within the DNA binding domain were inactive (McDonnell *et al.*, 1989; Sone *et al.*, 1989, 1991a). These studies provide the most definitive evidence to date that the VDR is an essential component that acts in *trans* to mediate $1,25(OH)_2D_3$ action.

In addition to establishing the relevance of the VDR in transactivation, the development of the preceding assay also permitted an evaluation of receptor structural requirements necessary for functional activation of OC transcription. Deletion and point mutagenesis revealed that the DNA binding domain was essential but not sufficient for transactivation (McDonnell *et al.*, 1989; Sone *et al.*, 1991a). Also revealed was the critical nature of conserved amino acids in maintaining the structural integrity of the DNA binding domain. Thus, mutagenesis of basic amino acid residues or eight of the nine conserved cysteine residues led to loss of proper folding, DNA binding capability, and capacity to activate transcription (Sone *et al.*, 1991a). Likewise, the ligand binding domain was essential to transactivation capability, suggesting that a transactivation domain was located in the VDR within that region. In contrast, deletion of the 21 amino acids lying amino-terminal to the DNA binding domain did not compromise any of the apparent functions of the receptor (Sone *et al.*, 1991a).

B. Characteristics of Binding of the Vitamin D Receptor to Vitamin D-Responsive Element DNA *in Vitro*

Direct interaction of mammalian cell-derived VDR with VDRE DNA *in vitro* has been demonstrated (Ozono *et al.*, 1990; Terpening *et al.*, 1991; Demay *et al.*, 1990; Noda *et al.*, 1990; Sone *et al.*, 1991a). Indeed, the VDR exhibits sufficiently high affinity for these elements that protein–DNA complexes can be identified following resolution by bandshift assays as well as through VDRE affinity chromatography. Determination of the affinity of this interaction by incubating receptor with increasing amounts of labeled DNA probe has revealed an equilibrium dissociation constant of approximately 0.2 n*M* (Sone *et al.*, 1991a). This affinity measurement is consistent with that of other steroid receptor–DNA interactions. Interestingly, the *in vitro* interaction of the VDR with VDRE DNA is modulated by the presence of hormone (Sone *et al.*, 1991a; Liao *et al.*, 1990; Sone *et al.*, 1990). Thus, while incubation of unoccupied receptor with VDRE DNA led to a poor yield of VDR–DNA complexes following bandshift analysis, addition of $1,25(OH)_2D_3$ to the incubation prior to assay dramatically increased receptor DNA binding. Therefore, the capacity of $1,25(OH)_2D_3$ to activate the VDR and increase transcriptional response *in vivo* is mimicked by this specific action *in vitro.*

In addition to hormone, recent experiments have suggested that high-affinity binding of the VDR to DNA requires a heterologous protein factor (Sone *et al.*, 1990, 1991a,b; Liao *et al.*, 1990). We observed that when the VDR was produced by *in vitro* translation from synthetic transcripts, or prepared by recombinant expression in either bacteria or yeast, the protein would not bind DNA with high affinity. Indeed, chromatographic studies using concatemerized VDRE DNA linked to an inert support revealed that even in the presence of $1,25(OH)_2D_3$, these sources of VDR bound with weak affinity to specific DNA as noncooperative monomers (Sone *et al.*, 1991a). Single VDRE halfsites and the VDR DNA binding domain alone were the only requirements for this weak interaction. In contrast, the addition of mammalian cell extraction to recombinantly produced VDR strongly enhanced the latter's capacity to bind VDRE DNA, and a fully intact response element was essential for this high-affinity interaction. A summary of the DNA binding properties of the VDR is documented in Table I. Additional investigation revealed that the interaction between the VDR and this trypsin- and temperature-sensitive protein factor designated nuclear accessory factor (NAF) did not require the presence of DNA. Thus, immobilized VDR could be utilized to recover NAF from a soluble protein mixture, supporting the idea that VDR–NAF heterodimeric complexes form in solution. These experiments collectively suggest that the binding of the VDR to appropriate DNA elements requires the presence of a protein

TABLE I Characteristics of Human Vitamin D Receptor (VDR) Binding to Specific Vitamin D-Responsive Element (VDRE) DNA

Weak interaction	Strong interaction
Low affinity	High affinity
VDRE halfsite required	Functional VDRE required
Noncooperative binding	Cooperative binding
Monomeric VDR	VDR–accessory factor heteromer

partner and that formation of the heteromer, while likely influenced by DNA, does not require its presence. Recent experiments with retinoic acid and thyroid hormone receptors have led to similar conclusions (Glass *et al.*, 1990; Burnside *et al.*, 1990). If these heterodimeric complexes for thyroid hormone, retinoic acid, and vitamin D prove to be units that function *in vivo*, they will contrast clearly with those for the sex steroids and the glucocorticoids, where the functional receptor units are homodimeric in nature (Kumar and Chambon, 1988; Fawell *et al.*, 1990). These observations may reflect the differing structural nature of the responsive elements; the former are composed of direct repeats, whereas the latter are palindromic in nature.

Two additional observations about the interaction of the VDR and NAF with DNA have been identified. First, studies using VDRs containing large deletions of the DNA or ligand binding domains have revealed that dimerization occurs through the steroid binding region of the molecule and does not appear to require the DNA binding core (Sone *et al.*, 1991a). Thus, the carboxy-terminal portion of the receptor is sufficient for interaction with NAF. More detailed studies have in fact localized the dimerization region of the VDR to the amino acid segments that remain conserved within the receptor family. Second, and perhaps most important, $1,25(OH)_2D_3$ has been demonstrated to enhance the formation of VDR–NAF heterodimers in solution (Sone *et al.*, 1991a). Thus, incubation of yeast-produced VDR with $1,25(OH)_2D_3$, followed by incubation with increasing concentrations of protein extract containing NAF results in an enhancement of salt-extractable NAF when compared to an identical experiment carried out in the absence of the vitamin D hormone. Determination of the equilibrium dissociation constant for both conditions revealed an increase in affinity of the VDR for NAF of 10-fold when $1,25(OH)_2D_3$ was present. These results suggest that at least one role of the vitamin D hormone may be to alter the equilibrium between VDR and NAF monomers and the VDR–NAF heterodimer such that an increase in DNA binding becomes evident. While formation of the heteromer, DNA binding, and transactivation may be synonymous, this hypothesis remains to be proven.

VI. PROPERTIES OF NUCLEAR ACCESSORY FACTOR AND OTHER STEROID RECEPTOR ACCESSORY FACTORS

Several protein factors that contribute to the specific DNA binding capacity of vitamin D, thyroid hormone, and retinoic acid receptors (RARs) have been characterized recently (Liao *et al.*, 1990; Glass *et al.*, 1990; Burnside *et al.*, 1990). It is important to note, however, that although these proteins clearly function to promote cooperative association of the respective receptor with its cognate response element, evidence linking these macromolecules to function *in vivo* remains inconclusive. The forthcoming identity of these protein partners will aid in experiments designed to test functional relevance. Moreover, the likely possibility that at least a subset of these accessory factors may belong to the steroid receptor gene superfamily suggests an added complexity—that their function may also be modified by a hormonal ligand.

A. Distribution of Nuclear Accessory Factor

We have determined the distribution of NAF in tissues and in cultured cells by employing a complementation bandshift assay wherein extracts that contain NAF or NAF-like components facilitate yeast-derived VDR binding (Sone *et al.*, 1991a,b; Liao *et al.*, 1990). As observed in Table II, NAF is detectable in both mouse liver and kidney extracts. This experiment indicates that NAF is not co-expressed with the VDR, as the latter is not produced in liver. Thus, it is likely that NAF retains an inherent function of its own. As seen in Table II, NAF is also expressed widely in cultured cells, including fibroblasts, osteoblasts, hepatocytes, lymphocytes, and cancer cells. Again, while many of these cells synthesize endogenous VDR, several of them do not. To date, the only cell types that do not express NAF activity are bacteria and yeast. These studies

TABLE II Tissue and Cellular Distribution of Nuclear Accessory Factor

Mouse tissues
Liver
Kidney
Cultured cells
Kidney fibroblasts (CV-1, COS-1)
Cervical carcinoma (HeLa)
Breast cancer (T47D)
Hepatoma (HepG2)
Osteosarcoma (ROS 17/2.8)
Calvaria (MC3T3-E1)
Fibroblasts (human primary)
Lymphoblasts (human, Epstein Barr virus transformed)

TABLE III Vitamin D Receptor Interaction with Natural Vitamin D-Responsive Elements Requires Nuclear Accessory Factor

Element	Sequence			NAF Required
	H	S	H	
Human osteocalcin	5′-GGGTGA	ACG	GGGGCA-3′	+
Rat osteocalcin	5′-GGGTGA	ATG	AGGACA-3′	+
Mouse osteopontin	5′-GGTTCA	CAG	GGTTCA-3′	+
Mouse retinoic acid receptor ß	5′-GGTTCA	CCGAA	AGTTCA-3′	+

indicate that NAF is widely distributed and not co-expressed together with the VDR. Thus, it is likely that in addition to the potential role it plays in $1,25(OH)_2D_3$ activation, NAF protein(s) retains a function independent of this activation.

B. Properties of Nuclear Accessory Factor

In addition to distribution, we have investigated other physical and functional features of NAF. We examined the requirement for NAF in VDR binding to a variety of VDREs (Sone *et al.*, 1991b). Natural VDREs derived from the human and rat OC genes and the osteopontin gene were investigated. As documented in Table III, VDR binding to each of these DNA sequences required NAF. Thus, it is clear that heterodimer formation is a general feature of VDR DNA binding and not a specific function of the element evaluated. The molecular mass of NAF was determined both by cross-linking VDR and NAF and by assessing the activity of renatured protein fractions derived from sodium dodecyl sulfate–polyacrylamide gel electrophoresis (Sone *et al.*, 1991a,b). In both experiments, NAF behaves as a 53- to 55-kDa protein. Finally, chromatographic analysis has revealed that NAF binds to a series of ion exchange supports and immobilized ligands that include diethylaminoethyl- and phosphocelluloses, calf thymus DNA, and VDRE DNA. Like the VDR, NAF exhibits a weak affinity for immobilized VDRE DNA when applied alone but strong affinity when it is combined with the VDR. These behavioral attributes of NAF support the idea that NAF may be a member of the steroid receptor gene superfamily.

C. Retinoid X Receptors Display Nuclear Accessory Factor Activity

Three highly related but separate receptors for all-*trans*-retinoic acid have been cloned and identified as members of the steroid receptor superfamily of genes (O'Malley, 1990). More recently, a class of receptors that exhibit low homology to the RARs in both the DNA and ligand binding domains have been identified (Mangelsdorf *et al.*, 1990). This class of proteins, termed retinoid X receptors (RXRs), comprises α, β, and γ forms (Mangelsdorf *et al.*, 1992) and binds and is activated specifically by the retinoic acid metabolite 9-*cis* retinoic acid (Heyman *et al.*, 1992). Indeed, 9-*cis* retinoic acid may be the first natural small molecule hormone to be identified since the discovery of $1,25(OH)_2D_3$. Interestingly, the RXR family of proteins appears to form heterodimeric complexes with the thyroid hormone receptors, RARs, and VDR (Yu *et al.*, 1991; Leid *et al.*, 1992; Zhang *et al.*, 1992; Kliewer *et al.*, 1992). Indeed, the gene for an accessory protein that heterodimerized with both the RAR and the thyroid hormone receptor was cloned by two independent groups by separate methodologies and shown to be RXR-β (Yu *et al.*, 1991; Leid *et al.*, 1992). As observed in Figure 6, recombinantly expressed RXR-α binds to the VDR to form VDR heterodimers. In contrast, addition of similar extracts of RAR-α, has no effect on VDR binding to the VDRE. These experiments as well as those previously reported (Yu *et al.*, 1991; Leid *et al.*, 1992; Zhang *et al.*, 1992; Kliewer *et al.*, 1992) suggest that members of the RXR family contain unique properties with regard to their capacity to form complexes with other members of the steroid receptor family. At present, however, little functional evidence exists to suggest that an alteration in the cellular complement of RXR, achieved through transfection of an RXR expression vector, is effective in modifying cellular response to the vitamin D hormone. Moreover, it is also unclear at present whether NAF is identical to RXR. Thus, additional experiments will be required to determine whether the two proteins are identical as well as to evaluate the *in vivo* significance of these *in vitro* observations.

VII. SUMMARY

The studies outlined in this chapter describe recent advances in our understanding of the molecular mechanisms by which the osteoblast-specific bone protein OC is regulated at the genomic level. The gene is controlled at the basal level by a series of *cis*-acting elements that function together with their cognate transactivators at appropriate times during osteoblast maturation. Transgenic studies suggest the presence of tissue-specific activators, although these have not yet been identified. Finally, the gene is modulated by a series of steroid hormones that include $1,25(OH)_2D_3$. The action of this hormone is mediated by the VDR. Interestingly, however, *in vitro* studies have suggested an additional complexity in this regulation. Thus, it appears that the VDR requires a non-VDR protein for cooperative interaction with specific DNA. Current evidence suggests that the dimerization partner may be another member of the steroid receptor superfamily, namely the RXR subfamily. This increase in complexity may well extend the diversity of response

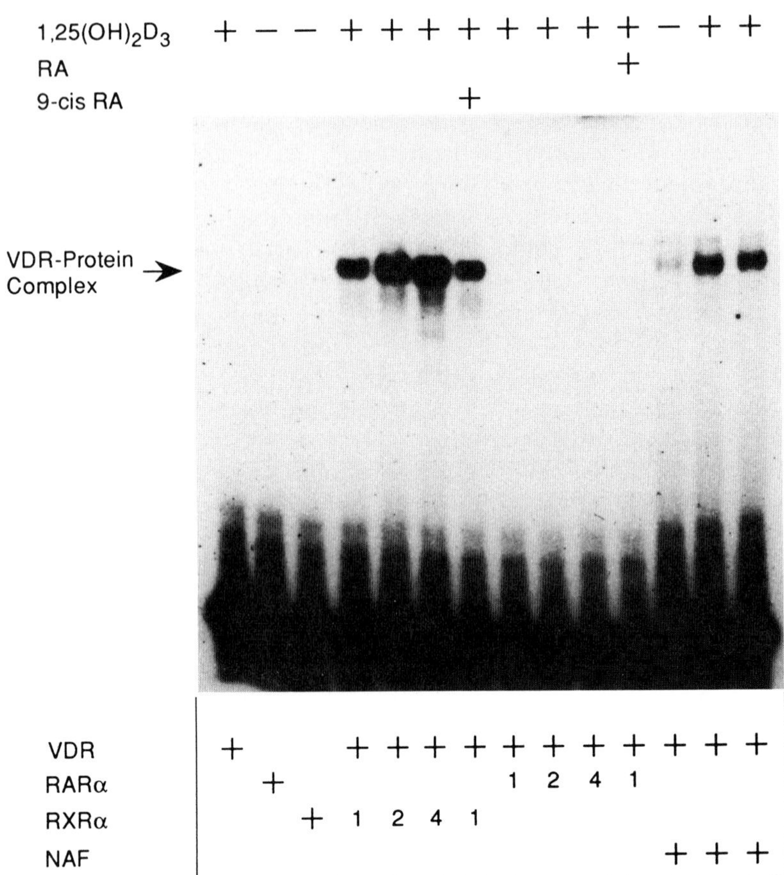

FIGURE 6 Nuclear accessory factor (NAF) and retinoid X receptor α (RXR-α) form complexes with the vitamin D receptor (VDR) on vitamin D responsive element (VDRE) DNA. VDRE probe was mixed with VDR (lane 1); RAR-α (lane 2); RXR-α (lane 3); VDR + 1 μg (lane 4), 2 μg (lane 5), or 4 μg (lane 6) of RXR-α; VDR + 1 μg RXR-α + 9-*cis* retinoic acid (lane 7); VDR + 1 μg (lane 8), 2 μg (lane 9), or 4 μg (lane 10) of RAR-α; VDR + 1 μg RAR-α + retinoic acid (lane 11); VDR + 1 μg NAF (lane 12); VDR + 1 μg NAF + 1,25-dihydroxyvitamin D_3 ($1,25(OH)_2D_3$) (lanes 13 and 14). Lanes 1 and 4–11 contained $1,25(OH)_2D_3$. Bandshift analysis was carried out as in Sonne *et al.* (1991a). Microgram additions represent total protein from baculovirus extracts containing the respective receptor. RA, retinoic acid.

generated by the vitamin D hormone and its receptor protein, particularly if ligands for the non-VDR protein partners are found to play a role in functional response.

REFERENCES

Baker, A. R., McDonnell, D. P., Hughes, M. R., Crisp, T. M., Mangelsdorf, D. J., Haussler, M. R., Pike, J. W., Shine, J., and O'Malley, B. W. (1988). Molecular cloning and expression of human vitamin D_3 receptor complementary DNA. *Proc. Natl. Acad. Sci. USA* **85,** 3294–3298.

Beato, M. (1989). Gene regulation by steroid hormones *Cell* **5,** 335–344.

Burmester, J. K., Maeda, N., and DeLuca, H. F. (1988). Isolation and expression of rat 1,25-dihydroxyvitamin D receptor cDNA. *Proc. Natl. Acad. Sci. USA* **85,** 1005–1009.

Burnside, J., Darling, D. S., and Chin, W. W. (1990). A nuclear factor that enhances binding of thyroid hormone receptors to a thyroid hormone response element. *J. Biol. Chem.* **265,** 2500–2504.

Celeste, A. J., Rosen, V., Buecker, J. L., Kriz, Wang, E. A., and Wozney, J. M. (1986). Isolation of the human gene for bone gla protein utilizing mouse and rat cDNA clones. *EMBO J.* **5,** 1885–1890.

Delmas, P. D., Stenner, D., Wahmer, H. W., Mann, K. G., and Riggs, B. I. (1973). Assessment of bone turnover in post menopausal osteoporosis by measurement of serum bone gla-protein. *J. Clin. Invest.* **71,** 1316–1323.

Demay, M. B., Gerardi, J. M., DeLuca, H. F., and Kronenberg, H. M. (1990). DNA sequences in the rat osteocalcin gene that bind the 1,25-dihydroxyvitamin D_3 receptor and confer responsiveness to 1,25-dihydroxyvitamin D_3. *Proc. Natl. Acad. Sci. USA* **87,** 369–373.

de The, H., del Mar Vivanco-Ruiz, N. M., Tiollais, P., Stunnenberg, H., and Dejean, A. (1990). Identification of a retinoic acid responsive element in the retinoic acid receptor beta gene. *Nature (London)* **343,** 177–180.

Diamond, M. I., Miner, J. N., Yoshinaga, S. K., and Yamamoto, K. R. (1990). Transcription factor interactions: Selectors of positive or negative regulation from a single DNA element. *Science* **249,** 1266–1272.

Epstein, S. (1989). Bone-derived proteins. *Trends Endocrinol. Metab.* **1,** 9–14.

Evans, R. M. (1988). The steroid and thyroid hormone receptor superfamily. *Science* **240,** 889–895.

Fawell, W. E., Lees, J. A., White, R., and Parker, M. G. (1990). Characterization and colocalization of steroid binding and dimerization activities in the mouse estrogen receptor. *Cell* **60,** 953–962.

Freedman, L. P., Luisi, B. F., Korszun, Z. R., Basavappa, R., Sigler, P. B., and Yamamoto, K. R. (1988). The function and structure of the metal coordination sites within the glucocorticoid receptor DNA binding domain. *Nature (London)* **334,** 543–546.

Gerstenfeld, L. C., Chipman, S. D., Glowacki, J., and Lian, J. B. (1986). Expression of differentiate function by mineralizing cultures of chicken osteoblasts. *Dev. Biol.* **122,** 49–60.

Glass, C. K., Franco, R., Weinberger, C., Albert, V. R., Evans, R. M., and Rosenfeld, M. G. (1988). A c-erb A binding site in rat growth hormone gene mediates trans-activation by thyroid hormone. *Nature (London)* **329,** 738–741.

Glass, C. K., Devary, O. V., and Rosenfeld, M. G. (1990). Multiple cell type-specific proteins differentially regulate target sequence recognition by the alpha retinoic acid receptor. *Cell* **63,** 729–736.

Glowacki, J., and Lian, J. B. (1989). Impaired recruitment of osteoclast progenitors by osteocalcin-deficient bone implants. *Cell Diff.* **21,** 247–254.

Gundberg, C. M., Lian, J. B., Gallop, P. M., and Steinberg, J. J. (1983). Urinary γ-carboxyglutamic acid and serum osteocalcin as bone markers: Studies in osteoporosis and Paget's disease. *J. Clin. Endocrinol. Metab.* **57,** 1221–1225.

Hard, T., Kellenbach, E., Boelens, R., Maler, B., Dahlman, K., Freedman, L. P., Carlstedt-Duke, J., Yamamoto, K. R., Gustafsson, J.-A., and Kaptein, R. (1990). Solution structure of the glucocorticoid receptor DNA-binding domain. *Science* **249,** 157–160.

Hauschka, P. V., Lian, J. B., and Gallop, P. M. (1975). Direct identification of the calcium binding amino acid γ-carboxyglutamate in mineralized tissues. *Proc. Natl. Acad. Sci. USA* **72,** 3925–3929.

Heyman, R. A., Mangelsdorf, D. J., Dyck, J. A., Stein, R. B., Eichele, G., Evans, R. M., and Thaller, C. (1992). 9-*Cis* retinoic acid is a high affinity ligand for the retinoid X receptor. *Cell* **68,** 397–406.

Jowell, P. S., Epstein, S., Fallon, M. D., Reinhardt, T. A., and Ismail, F. (1987). 1,25-Dihydroxyvitamin D_3 modulates glucocorticoid-induced alteration in serum bone gla protein and bone histomorphometry. *Endocrinology* **120,** 531–536.

Kerner, S. A., Scott, R. A., and Pike, J. W. (1989). Sequence elements in the human osteocalcin gene confer basal activation and inducible response to hormonal vitamin D_3. *Proc. Natl. Acad. Sci. USA* **86,** 4455–4459.

Kesterson, R. A., Stanley, L., Finegold, M. J., DeMayo, F., and Pike, J. W. (1993). The human osteocalcin promoter directs bone-specific vitamin D-regulatable expression in transgenic mice. *Mol. Endo.* **7,** 462–467.

Kliewer, S. A., Umesono, K., Mangelsdorf, D. J., and Evans, R. M. (1992). Retinoid X receptor interacts with nuclear receptors in retinoic acid, thyroid hormone and vitamin D_3 signaling. *Nature (London)* **355,** 446–449.

Kumar, V., and Chambon, P. (1988). The estrogen receptor binds tightly to its responsive element as a ligand-induced homodimer. *Cell* **55,** 145–156.

Leid, M., Kastner, P., Lyons, R., Nakshatro, H., Saunders, M., Zacharewski, T., Chen, J.-Y., Staub, A., Garnier, J.-M., Mader, S., and Chambon, P. (1992). Purification, cloning, and RXR identity of the Hela cell factor with which RAR or TR heterodimerizes to bind target sequences efficiently. *Cell* **68,** 377–395.

Lian, J. B., and Gundberg, C. M. (1988). Osteocalcin: Biochemical considerations and clinical application. *Clin. Orthop. Relat. Res.* **226,** 267–291.

Lian, J. B., Coutts, M. N. C., and Canalis, E. (1985). Studies of hormonal regulation of osteocalcin synthesis in cultured fetal rat calvariae. *J. Biol. Chem.* **260,** 8706–8710.

Lian, J. B., Carnes, D. L., and Glimcher, M. (1987). Bone and serum concentrations of osteocalcin as a function of 1,25-dihydroxyvitamin D_3 circulating levels in bone disorders in rats. *Endocrinology* **120,** 2123–2130.

Lian, J. B., Stewart, C., Puchacz, E., Mackowiak, S., Shalhoub, V., Collart, D., Sambetti, G., and Stein, G. (1989). Structure of the rat osteocalcin gene and regulation of vitamin D-dependent expression. *Proc. Natl. Acad. Sci. USA* **86,** 1143–1147.

Liao, J., Ozono, K., Sone, T., McDonnell, D. P., and Pike, J. W. (1990). Vitamin D receptor interaction with specific DNA requires a nuclear protein and 1,25-dihydroxyvitamin D_3. *Proc. Natl. Acad. Sci. USA* **87,** 9751–9755.

Malone, J. D., Teitelbaum, S. L., Griffin, G. L., Senior, R. M., and Kahn, A. J. (1982). Recruitment of osteoclast precursors by purified bone matrix constituents. *J. Cell. Biol.* **92,** 227–230.

Mangelsdorf, D. J., Ong, E. S., Dyck, J. A., and Evans, R. M. (1990). Nuclear receptor that identifies a novel retinoic acid response pathway. *Nature (London)* **345,** 224–229.

Mangelsdorf, D. J., Borgmeyer, U., Heyman, R. A., Zhou, J. Y., Ong, E. S., Oro, A. E., Kakizuka, A., and Evans, R. M. (1992). Characterization of three RXR genes that mediate the action of 9-cis retinoic acid. *Genes Dev.* **6,** 329–344.

Markose, E. R., Stein, J. L., Stein, G. S., and Lian, J. B. (1990). Vitamin D-mediated modifications in protein–DNA interactions at two promoter elements of the osteocalcin gene. *Proc. Natl. Acad. Sci. USA* **87,** 1701–1705.

McDonnell, D. P., Mangelsdorf, D. J., Pike, J. W., Haussler, M. R., and O'Malley, B. W. (1987). Molecular cloning of complementary DNA encoding the avian receptor for vitamin D. *Science* **235,** 1214–1217.

McDonnell, D. P., Scott, R. A., Kerner, S. A., O'Malley, B. W., and Pike, J. W. (1989). Functional domains of the human vitamin D_3 receptor regulate osteocalcin gene expression. *Mol. Endo.* **3,** 635–644.

Morrison, N. A., Shine, J., Fragonas, J.-C., Verkest, V., McMenemy, L., and Eisman, J. A. (1989). 1,25-Dihydroxyvitamin D-responsive element and glucocorticoid repression in the osteocalcin gene. *Science* **246,** 1158–1161.

Nishimoto, S. K., Salka, C., and Nimni, M. E. (1987). Retinoic acid and glucocorticoids enhance the effect of 1,25-dihydroxyvitamin D_3 on bone γ-carboxyglutamic acid protein synthesis by rat osteocarcoma cells. *J. Bone. Miner. Res.* **2,** 571–577.

Noda, M., Vogel, R. L., Craig, A. M., Prahl, J., DeLuca, H. F., and Denhardt, D. (1990). Identification of a DNA sequence responsible for binding of the 1,25-dihydroxyvitamin D_3 receptor and 1,25-dihydroxyvitamin D_3 enhancement of mouse secreted phosphoprotein 1 (Supp-1 or osteopontin) gene expression. *Proc. Natl. Acad. Sci. USA* **87,** 9995–9999.

O'Malley, B. W. (1990). The steroid receptor superfamily: More excitement predicted for the future. *Mol. Endo.* **4,** 363–369.

Owen, M. (1985). Lineage of osteogenic cells and their relationship to the stromal system. *In* "Bone and Mineral Research" (W. A. Peck, ed.), Vol. 3, pp. 1–24. Elsevier, Amsterdam.

Ozono, K., Liao, J., Kerner, S. A., Scott, R. A., and Pike, J. W. (1990). The vitamin D responsive element in the human osteocalcin gene: Association with a nuclear proto-oncogene enhancer. *J. Biol. Chem.* **265,** 21881–21888.

Ozono, K., Sone, T., and Pike, J. W. (1991). Functional interactions between vitamin D receptors and the AP-1 proto-oncogene family. *J. Bone Miner. Res.* **6**(Suppl. 1), 677 (abstract).

Pan, L. C., and Price, P. A. (1984). The effect of transcriptional inhibitors on the bone γ-carboxylic acid protein response to 1,25-dihydroxyvitamin D in osteosarcoma cells. *J. Biol. Chem.* **259,** 5844–5847.

Picard, D., and Yamamoto, K. R. (1988). Two signals mediate hormone-dependent translocation of the glucocorticoid receptor. *EMBO J.* **6,** 3333–3340.

Pike, J. W. (1990). *Cis* and *trans*-regulation of osteocalcin gene expression by vitamin D and other hormones. *In* "Calcium Regulation and Bone Metabolism" (D. V. Cohn, F. H. Glorieux, and T. J. Martin, eds.), Vol. 10, pp. 127–136. Excerpta Medica, Amsterdam.

Price, P. A., and Baukol, S. A. (1980). 1,25-Dihydroxyvitamin D_3 increases the synthesis of the vitamin D-dependent bone protein by osteosarcoma cells, *J. Biol. Chem.* **255,** 11660–11663.

Price, P. A., and Nishimoto, S. K. (1980). Radioimmunoassay for the vitamin D-dependent protein of bone and its discovery in plasma. *Proc. Natl. Acad. Sci. USA* **77,** 2234–2238.

Price, P. A., Otsuka, A. S., Posner, J. W., Kristaponis, J., and Raman, N. (1976). Characterization of a γ-carboxyglutamic acid-containing protein from bone. *Proc. Natl. Acad. Sci. USA* **73,** 1447–1451.

Price, P. A., Parthemore, J. G., and Deftos, L. F. (1980). New biochemical marker for bone metabolism: Measurement by radioimmunoassay of bone gla-protein in the plasma of normal subjects and patients with bone disease. *J. Clin. Invest.* **66,** 878–883.

Rodan, G. A., and Martin, T. J. (1981). Role of osteoblasts in hormonal control of bone resorption. *Calcif. Tissue Int.* **33,** 349–351.

Schule, R., Umesono, K., Mangelsdorf, D. J., Bolado, J., Pike, J. W., and Evans, R. M. (1990). Jun-Fos and receptors for vitamins A and D recognize a common response element in the human osteocalcin gene. *Cell* **61,** 497–504.

Schwabe, J. W. R., Neuhaus, D., and Rhodes, D. (1990). Solution structure of the DNA-binding domain of the oestrogen receptor. *Nature (London)* **348,** 458–461.

Sone, T., Scott, R. A., Hughes, M. R., Malloy, P. J., Feldman, D., O'Malley, B. W., and Pike, J. W. (1989). Mutant vitamin D receptors which confer hereditary resistance to 1,25-dihydroxyvitamin D_3 in humans are transcriptionally inactive in vivo. *J. Biol. Chem.* **264,** 20230–20234.

Sone, T., McDonnell, D. P., O'Malley, B. W., and Pike, J. W. (1990). Expression of human vitamin D receptor in *Saccharomyces cerevisiae:* Purification, properties, and generation of polyclonal antibodies. *J. Biol. Chem.* **265,** 21997–22003.

Sone, T., Kerner, S. A., and Pike, J. W. (1991a). Vitamin D receptor interaction with specific

DNA: Association as a 1,25-dihydroxyvitamin D_3-modulated heterodimer. *J. Biol. Chem.* **266,** 23296–23305.

Sone, T., Ozono, K., and Pike, J. W. (1991b). A 55 kilodalton accessory factor facilitates vitamin D receptor DNA binding. *Mol. Endo.* **5,** 1578–1586.

Terpening, C. M., Haussler, C. A., Jurutka, P. W., Galligan, M. A., Komm, B. S., and Haussler, M. R. (1991). The vitamin D-responsive element in the rat bone gla protein gene is an imperfect direct repeat that cooperates with other cis-elements in 1,25-dihydroxyvitamin D_3-mediated transcriptional activation. *Mol. Endo.* **5,** 373–385.

Uchida, M., Ozono, K., and Pike, J. W. (1991). 24R,25-dihydroxyvitamin D_3 activates osteocalcin gene transcription through the vitamin D receptor. *In* "Vitamin D, Gene Regulation, Structure–Function Analysis and Clinical Application" (A. W. Norman, R. Bouillon, and M. Thomasset, eds.), pp. 304–305. de Gruyter, Berlin.

Umesono, K., Giguere, V., Glass, C. K., Rosenfeld, M. G., and Evans, R. M. (1988). Retinoic acid and thyroid hormone induce gene expression through a common responsive element. *Nature (London)* **336,** 262–265.

Umesono, K., Murikami, K. K., Thompson, C. C., and Evans, R. M. (1991). Direct repeats as selective response elements for the thyroid hormone, retinoic acid, and vitamin D_3 receptors. *Cell* **65,** 1255–1266.

Yoon, D., Turledge, S. J. C., Buenaga, R. F., and Rodan, G. A. (1988). Characterization of the rat osteocalcin gene: Stimulation of promoter activity by 1,25-dihydroxyvitamin D_3. *Biochemistry* **27,** 8521–8526.

Yu, V., Delsert, C., Andersen, B., Holloway, J. M., Devary, O. V., Naar, A. M., Kim, S. Y., Boutin, J.-M., Glass, C. K., and Rosenfeld, M. G. (1991). RXRβ: A coregulator that enhances binding of retinoic acid, thyroid hormone, and vitamin D receptors to their cognate response elements. *Cell* **67,** 1251–1266.

Zhang, X.-K., Hoffmann, B., Tran, P. B.-V., Graupner, G., and Pfahl, M. (1992). Retinoid X receptor is an auxiliary protein for thyroid hormone and retinoic acid receptors. *Nature (London)* **355,** 441–446.

8
MOLECULAR MECHANISMS OF ESTROGEN AND THYROID HORMONE ACTION

CHRISTOPHER K. GLASS and MYLES A. BROWN

Cellular and Molecular Biology of Bone

I. OVERVIEW OF ESTROGEN AND THYROID HORMONE ACTION

Bone is an important target tissue of both estrogens and thyroid hormones. During puberty, estrogens act to cause closure of the epiphyseal plates in females, terminating axial growth of the long bones. In the adult female, estrogens play an important role in maintaining bone mass, as evidenced by accelerated rates of bone resorption in the postmenopausal state. Thyroid hormones also influence bone development and metabolism. Congenital hypothyroidism results in several developmental abnormalities, including decreased axial growth of the long bones. In adults, thyroid hormone excess causes accelerated bone loss and represents an important cause of osteoporosis.

Although the mechanisms by which estrogens and thyroid hormones influence specific aspects of bone metabolism are poorly understood, significant progress has been made in defining the mechanisms responsible for estrogen and thyroid hormone action at the molecular level. Both estrogens and thyroid hormones exert their principal effects on patterns of gene expression by binding to specific receptors that are members of a superfamily of ligand-dependent transcription factors. In addition to estrogen and thyroid hormone receptors, members of this superfamily include receptors for all of the other known steroid hormones (i.e., glucocorticoids, mineralocorticoids, androgens, and progestins) as well as receptors for metabolites of vitamin D and vitamin A (for general reviews covering the nuclear receptor superfamily, see Evans, 1988; Beato, 1989).

The focus of this chapter is to summarize current thinking regarding the mechanisms of action of estrogen and thyroid hormones at the molecular level. Rather than providing a comprehensive review of what has become a very large field, our approach will be to illustrate major findings and concepts that have arisen from the study of highly developed model systems. We then consider the relevance of these findings to potential mechanisms by which estrogens and thyroid hormones regulate bone metabolism.

General schemes for the mechanisms of estrogen and thyroid hormone action are illustrated in Figure 1. In the absence of estrogen, estrogen receptors are complexed to a high molecular weight multiprotein complex containing heat shock protein 90 (hsp 90) and perhaps other heat shock proteins (Joab *et al.*, 1984; Pratt *et al.*, 1988). Following diffusion into the cell, estrogen binds to estrogen receptors, resulting in their dissociation from the heat shock protein-containing complex. Estrogen receptors subsequently dimerize and bind to specific DNA sequences termed estrogen response elements, which are generally located in the vicinity of the promoter sequences of target genes. Following DNA bind-

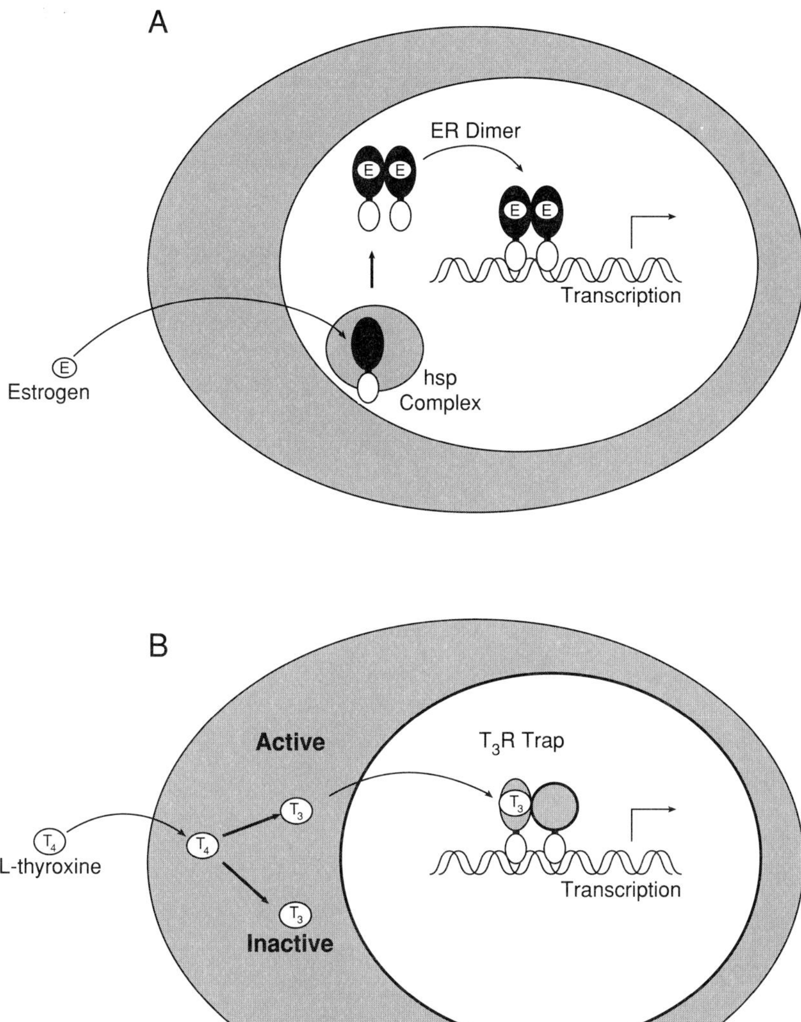

FIGURE 1 Models for mechanisms of action of estrogens and thyroid hormones. (A) Mechanisms of estrogen action. The estrogen receptor is prototypic of nuclear receptors that bind steroid ligands, as illustrated in the upper half of the figure. Estrogen diffuses into the cell, where it interacts with the estrogen receptor as part of a multiprotein complex containing heat shock protein 90 (hsp complex). Following estrogen binding, the receptor dimerizes and interacts with estrogen response elements present in target genes, resulting in increased rates of transcription. (B) Mechanisms of thyroid hormone action. L-Thyroxine (T_4), the major circulating form of thyroid hormone, enters the cell by diffusion. It is then converted to the more active metabolite, 3,3′,5′-triiodothyronine (T_3), or the inactive metabolite reverse T_3 (rT_3) by the actions of two distinct deiodinases. T_3 binds to chromatin-associated T_3 receptors and converts the receptor into a form that is capable of activation transcription. T_3 receptors appear to be primarily bound to T_3 response elements as heterodimeric complexes that have been termed co-regulator T_3 receptor auxiliary proteins (TRAPs).

ing, estrogen receptors act to increase target gene transcription, resulting in increased messenger RNA (mRNA) and protein levels.

Although general aspects of thyroid hormone action are similar to the mechanisms of action of estrogens, there are several distinct differences with respect to the localization of thyroid hormone receptors in the absence of ligand and mechanisms by which they interact with DNA (Fig. 1B). In contrast to estrogen receptors, thyroid hormone receptors are bound to their cognate response element sequences in target genes whether or not thyroid hormone is present (Spindler *et al.*, 1975, Damm *et al.*, 1989). Rather than binding to their cognate response element sequences as homodimers, thyroid hormone receptors may preferentially interact with these elements as heterodimers (discussed in detail later). In addition, many cells have the ability to metabolize thyroid hormone. Following entry into the cell, the major circulating form of thyroid hormone, L-thyroxine (T_4), can be metabolized to yield either a biologically more active form, 3,3′,5′-triiodothyroinine (T_3), or the inactive metabolite, reverse T_3. This capability potentially enables cells to employ an additional level of regulation in their response to a particular level of circulating thyroid hormone.

II. CHARACTERIZATION OF ESTROGEN- AND THYROID HORMONE-RESPONSIVE GENES

Estrogens and thyroid hormones influence gene expression at multiple levels. In addition to regulation of transcription, each of these hormones has been demonstrated to be capable of regulating mRNA stability, protein translation, and protein stability (see, e.g., regulation of vitellogenin mRNA stability by estrogens in Shapiro *et al.* [1989] and regulation of rat growth hormone mRNA stability by thyroid hormones in Diamond and Goodman [1985]). Direct participation of estrogen and thyroid hormone receptors in these processes, however, has primarily been demonstrated at the level of regulation of gene transcription. It is possible that many or all of the posttranscriptional actions of estrogens and thyroid hormones are indirect effects that are mediated by the products of genes that are transcriptionally regulated by these hormones.

Several criteria have been used to define immediate target genes of steroid and thyroid hormones. As a rule, mRNA and protein levels corresponding to immediate target genes are observed to increase rapidly following hormone treatment. This increase in mRNA is sensitive to inhibitors of transcription by RNA polymerase II, such as α-amanitin, but is generally insensitive to inhibitors of protein synthesis, such as cycloheximide. These observations reflect the ability of preformed estrogen or thyroid hormone receptors to regulate transcription of direct target genes in response to ligand without the synthesis of additional protein factors. Specific measurements of changes in rates of gene transcription in response to hormone treatment are accomplished by performing nuclear runoff assays (see, e.g., Diamond and Goodman, 1985), which demonstrate rapid increases in new transcripts.

Based on these criteria, several genes that are transcriptionally regulated by estrogens or thyroid hormones have been identified. Genes that are positively regulated by estrogens that have been intensively studied as models for estrogen-dependent transcription include vitellogenin A_2 (e.g., Jost *et al.*, 1984; Klein-Hitpass *et al.*, 1986), prolactin (e.g., Waterman *et al.*, 1988; Maurer and Notides, 1987), ovalbumin (e.g., Lai *et al.*, 1983; Tora *et al.*, 1988), and oxytocin (e.g., Richard and Zingg, 1990). Genes that are transcriptionally up-regulated by thyroid hormones include rat growth hormone (e.g., Yaffe and Samuels, 1984; Nyborg *et al.*, 1984), α-myosin heavy chain (e.g., Izumo *et al.*, 1986), malic enzyme (Song *et al.*, 1988; Petty *et al.*, 1990), liver spot 14 (Jump, 1989), and the chicken lysozyme (Baniahmad *et al.*, 1990) genes.

In addition to their stimulatory actions, estrogens and thyroid hormones are also capable of inhibiting gene expression. For example, estrogen inhibits leutenizing hormone gene transcription during the late luteal phase of the menstrual cycle. Well-characterized model systems for study of negative regulation of transcription by estrogen receptors have not been extensively developed, however. Genes that are negatively regulated by thyroid hormones that have served as important model systems in examining potential mechanisms of transcriptional inhibition include the α and β subunits of thyroid-stimulating hormone (TSH) genes (e.g., Shupnik *et al.*, 1985) and the epidermal growth factor (EGF) receptor gene (Hudson *et al.*, 1990; Kesevan *et al.*, 1991). Promoter sequences corresponding to these genes confer appropriate negative regulation to reporter genes in response to thyroid hormone administration in transient transfection studies.

III. IDENTIFICATION AND CHARACTERIZATION OF HORMONE RESPONSE ELEMENTS

Direct regulation of gene transcription by estrogen and thyroid hormones requires binding of their cognate receptors to response element sequences in target genes. Two criteria are generally applied to the identification of response element sequences. First, putative response element sequences must be functionally defined as sequences that are essential for hormone-dependent transcriptional responses in the presence of physiologic concentrations of receptor and hormone. To identify hormone response elements, genomic sequences corresponding to the promoter and enhancer regions of hormonally responsive genes are linked to reporter genes that encode easily assayable products that

Vit	A G G T C A C T G T G A C C T
apo VLDL (1)	A G G T C A G A C T G A C C T
apo VLDL (2)	G G G G C T C A G T G A C C C
Oxytocin	A G G T G A C C T T G A C C
Calbindin	A G G T C A G G G T G A T C T
pS2	A G G T C A C G G T G G C C A
c-fos	C C G G C A G C G T G A C C C
	→ ←
Consensus ERE	G G T C A N N N T G A C C
Consensus GRE	G A A C A N N N T G T T C

FIGURE 2 Estrogen response elements. Minimal sequences that confer estrogen responsiveness are illustrated for the following genes: Vit, *Xenopus* vitellogenin 2A ERE; apo VLDL(1) and apo VLDL(2), two ERE sequences from the chicken apo very low-density lipoprotein II gene; oxytocin, human oxytocin ERE; calbindin, rat calbindin D-9K ERE; pS2, human pS2 ERE; and c-*fos*, human c-*fos* ERE.

are not expressed in mammalian cells, such as chloramphenicol acyltransferase (CAT) or firefly luciferase. When transfected into cells that contain the appropriate receptor, these fusion genes will exhibit hormone-dependent activation of transcription if response element sequences are contained within the promoter or enhancer elements (see, e.g., Klein-Hitpass *et al.*, 1986). Subsequent mutational analysis then permits identification of discreet sequences, usually of 16–20 basepairs (bp), that are required for hormonal activation and are capable of transferring this regulation to otherwise unresponsive promoters. Once functional response elements are identified, the second criteria to be established is that the appropriate receptors bind to these sequences with high affinity and sequence specificity. Mutations that decrease receptor binding *in vitro* should reduce ligand-dependent transcription when promoters containing these mutations are assayed in transfected cells.

Several estrogen response elements (EREs) have been identified that satisfy these criteria, some of which are illustrated in Figure 2. These elements are generally located within 200 bp of the transcriptional start site but can also be located kilobases away. For example, the estrogen response element of the rat prolactin gene resides approximately 1.5 kilobases (kb) upstream of the transcriptional start site (Waterman *et al.*, 1988). In general, the sequences that confer transcriptional responses to estrogen exhibit palindromic or near palindromic characteristics. The first estrogen response element to be characterized was that present in the *Xenopus* vitellogenin A_2 gene (Klein-Hitpass *et al.*, 1986). In this case, the minimal sequence conferring hormonal responsiveness exhibited perfect dyad symmetry (Fig. 2) and probably represents the most effective estrogen response element yet identified. Subsequent characterization of other genes that are transcriptionally regulated by estrogen has led to the identification and characterization of additional estrogen response elements. These include EREs present in the prolactin gene (Waterman *et al.*, 1988), the apo VLDL II gene (Wijnholds *et al.*, 1988), the pS2 gene (Berry *et al.*, 1989), the oxytocin gene (Richard and Zingg, 1990), the c-*fos* gene (Weisz and Rosales, 1990), and the calbindin D-9K gene (Darwish *et al.*, 1991). Although considerable variation in the precise sequences of these elements exists, each exhibits a near-palindromic motif that is related to the sequence of the vitellogenin A_2 estrogen response element. Comparison of these elements yields a palindromic consensus sequence: AGGTCA NNN TGACCT.

An interesting exception to this consensus is the *cis*-active element conferring estrogen responsiveness present in the ovalbumin promoter. Analysis of the sequences required for estrogen induction of this gene revealed a single GGTCA halfsite with no corresponding palindromic counterpart (Tora *et al.*, 1988). Subsequent analysis of the mechanisms of estrogen-dependent activation indicated that a direct interaction of the estrogen receptor with the promoter was not required, because mutant estrogen receptors lacking a functional DNA-binding domain were also capable of stimulating transcription (Gaub *et al.*, 1990). In these studies, estrogen receptors appeared to exert their stimulatory effects through c-*jun* and c-*fos*, components of the AP-1 class of transcriptional activators. AP-1 proteins were found to bind to the GGTCA element, and mutations that abolished their binding also abolished estrogen regulation in transient transfection experiments.

The sequences of estrogen response elements that are directly involved in estrogen receptor binding also show significant similarity to glucocorticoid response elements (Strahle *et al.*, 1987). As shown in Figure 2, the consensus GRE sequence differs at only 4 of 12 bp from that of the consensus ERE. Predictably, interchanging these basepairs permits conversion of the consensus GRE to a functional ERE and vice versa (Martinez *et al.*, 1987; Klock *et al.*, 1987). The similarity between glucocorticoid and estrogen response elements presumably reflects evolution of hormone response elements parallel to the evolution of ligand-dependent nuclear receptors.

Response element sequences mediating transcriptional responses to thyroid hormones have been characterized in several cellular genes. These include the rat growth hormone gene (Koenig *et al.*, 1987; Glass *et al.*, 1987), α-myosin heavy chain (Izumo and Mahdavi, 1988), malic enzyme (Petty *et al.*, 1990), hepatic spot 14 (Zilz *et al.*, 1990), and the chicken lysozyme gene (Baniahmad *et al.*, 1990). In addition, a thyroid hormone response element has been identified within transcriptional regulatory elements of the Maloney murine leukemia virus (Sap *et al.*, 1989).

The first thyroid hormone response element to be extensively characterized was that present in the rat growth hormone gene (Fig. 3). Mutational analysis of this response element revealed the presence of three functionally important motifs that are denoted by A, B, and C in Figure 3 (e.g., Glass *et al.*, 1988; Brent *et al.*, 1989). The first two of these motifs can be considered to be arranged as direct repeats. The third motif is inverted with respect to the second and therefore exhibits partial palindromic characteristics. Each of these motifs represent variations of the consensus estrogen response element halfsite, AGGTCA. Although all three motifs are necessary for maximal induction by thyroid hormones, partial thyroid hormone responsiveness is observed with just the direct repeat of motifs A and B or the imperfect palindrome of motifs B and C. Point mutations of the rat growth hormone TRE in motifs B and C that resulted in a perfect palindrome consisting of the ERE AGGTCA halfsites, but lacking the 3-bp spacer (i.e., AGGTCA TGACCT), resulted in an idealized element, termed TRE-pal, that conferred a more efficient transcriptional response to T_3 than the wild-type sequence (Glass *et al.*, 1988). These observations suggested that the estrogen and T_3 receptors were capable of recognizing highly related recognition motifs. However, as additional T_3 response elements became further characterized, it became evident that palindromic arrangements of core recognition motifs were not required for a transcriptional response to T_3. Although most positively regulated response elements appear to contain two or more degenerate copies of the core recognition motif, these putative core sequences are arranged not only as palindromes (e.g., growth hormone elements B and C) but also as direct repeats (e.g., growth hormone elements A and B and the α-myosin heavy-chain T_3 response element) or inverted palindromes (e.g., the chicken lysozyme TRE).

This unusual feature of thyroid hormone response elements has also been observed in the recognition sequences mediating the transcriptional effects of other nuclear receptors that do not bind steroid ligands, including the retinoic acid receptor and the vitamin D receptor. For example, the retinoic acid receptor is capable of activating transcription from the TRE-pal element (Umesono *et al.*, 1988) as well as from elements that contain direct repeats of the same AGGTCA core recognition motif spaced by 5 bp (Umesono *et al.*, 1991).

These observations led to the hypothesis that variations in the spacing and/or orientation of core binding motifs may constitute a "code" that plays a role in determining selective or overlapping patterns of control by related members of the nuclear receptor superfamily. Recent

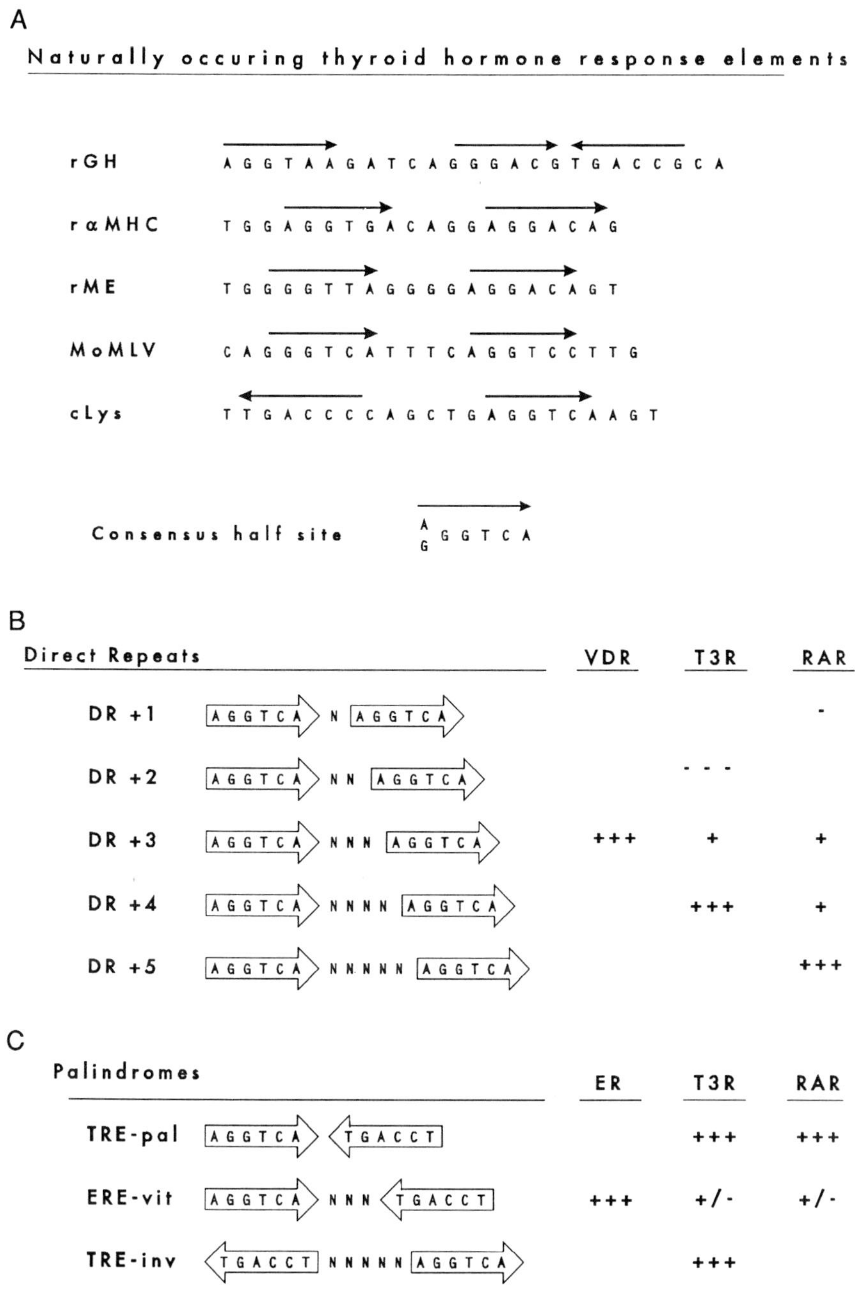

A

Naturally occuring thyroid hormone response elements

rGH	A G G T A A G A T C A G G G A C G T G A C C G C A
rαMHC	T G G A G G T G A C A G G A G G A C A G
rME	T G G G G T T A G G G G A G G A C A G T
MoMLV	C A G G G T C A T T T C A G G T C C T T G
cLys	T T G A C C C C A G C T G A G G T C A A G T

Consensus half site $^{A}_{G}$ G G T C A

B

Direct Repeats		VDR	T3R	RAR
DR +1	AGGTCA N AGGTCA			-
DR +2	AGGTCA NN AGGTCA		- - -	
DR +3	AGGTCA NNN AGGTCA	+++	+	+
DR +4	AGGTCA NNNN AGGTCA		+++	+
DR +5	AGGTCA NNNNN AGGTCA			+++

C

Palindromes		ER	T3R	RAR
TRE-pal	AGGTCA TGACCT		+++	+++
ERE-vit	AGGTCA NNN TGACCT	+++	+/-	+/-
TRE-inv	TGACCT NNNNN AGGTCA		+++	

FIGURE 3 Thyroid hormone response elements. (A) Naturally occurring thyroid hormone response elements. rGH, rat growth hormone TRE, rαMHC, rat α-myosin heavy-chain TRE; rME, rat malic enzyme TRE; MoMLV, Moloney murine luekemia virus TRE; cLys, chicken lysozyme TRE. Arrows denote regions with similarity to estrogen response element halfsites. (B) Hormonal responsiveness conferred by direct repeats of the core recognition sequence AGGTCA at different spacings. Positive transcriptional effects are denoted by + to +++. Inhibitory effects are denoted by − to −−−. (C) Hormonal responsiveness conferred by palindromic arrangements of the AGGTCA recognition motif.

evidence has indicated that selective actions of thyroid hormone, vitamin D, and retinoic acid receptors can be observed as the consequence of differences in spacing between direct repeats of the core recognition motif (Umesono *et al.*, 1991; Fig. 3B). Using promoters containing two directly repeated copies of the idealized core motif AGGTCA, separation of these motifs by 5 bp resulted in preferential regulation of promoter activity by retinoic acid receptors. At a spacing of 4 bp, which is comparable to the spacing of putative halfsites in the spacing observed in the α-myosin heavy-chain element, thyroid hormone receptors exerted relatively specific transcriptional effects. Elements containing the core motifs spaced by 3 bp were preferentially activated by the vitamin D receptor (Umesono *et al.*, 1991).

Spacing of the core binding motif also played an important role in determining receptor specificity when arranged in a palindromic configuration (Glass *et al.*, 1988; Näär *et al.*, 1991). As previously described, the unspaced configuration (TRE-pal) was capable of mediating transactivation by both thyroid hormone and retinoic acid receptors but was ineffective as an estrogen response element. Conversely, T_3 and retinoic acid receptors failed to function effectively as transactivators on elements containing the palindromic arrangement spaced by 3 bp (i.e., as in the vitellogenin ERE). Intriguingly, both the thyroid hormone and retinoic acid receptors were capable of binding to ERE sequences with high affinity and antagonizing estrogen-dependent transcription from promoters containing these elements (Glass *et al.*, 1988; Holloway *et al.*, 1990; Lipkin *et al.*, 1992). An inverted arrangement of the palindrome, corresponding to the core recognition motif organization found in the chicken lysozyme TRE, was found to confer selective activation to the T_3 receptor, but not the vitamin D, retinoic acid, or estrogen receptors (Näär *et al.*, 1991).

These observations indicate that two copies of a single recognition motif, AGGTCA, can result in selective or overlapping transcription responses to estrogen, thyroid hormone, retinoic acid, and vitamin D in a manner that depends on their relative spacing and orientation. The extent to which these observations define a molecular "code" that is utilized by these receptors to recognize target genes *in vivo* remains to be established. Naturally occurring response elements vary not only in the spacing and orientation of recognition motifs, but also in their precise sequence composition. Because the thyroid hormone, retinoic acid, and vitamin D receptors appear to preferentially recognize their cognate response elements as heterodimers, associations with different heterodimeric partners could, at least in principle, result in different preferences for specific arrangements of core binding motif or their sequence composition.

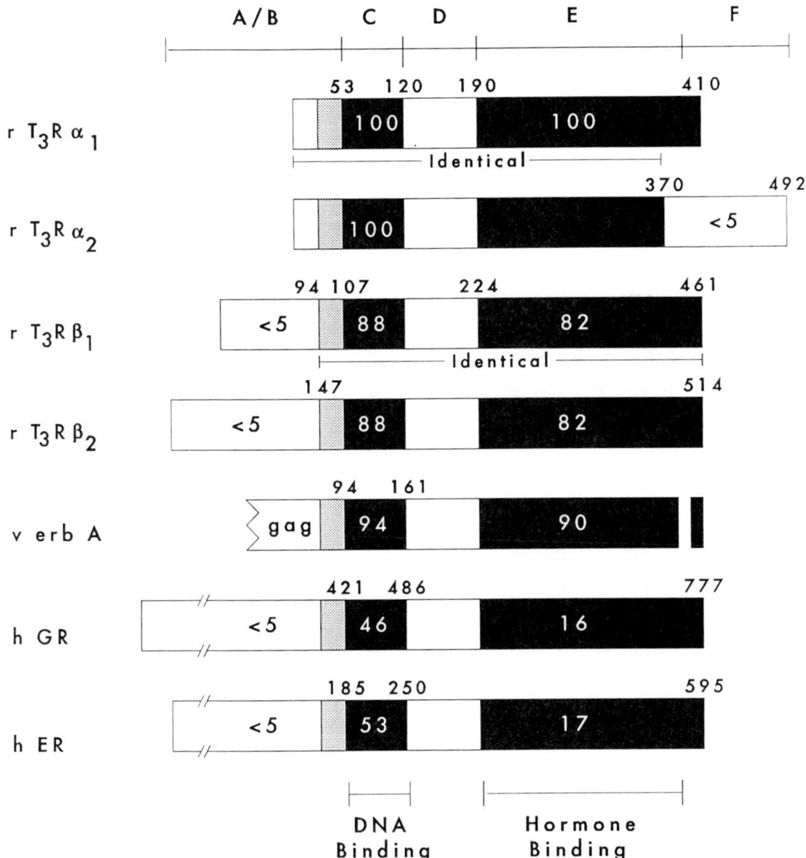

FIGURE 4 Thyroid hormone and estrogen receptor complementary DNAs. Schematic representation of the primary sequences of thyroid hormone receptor isoforms, v erb A, the estrogen receptor, and the glucocorticoid receptor. Numbers within clear or shaded regions represent amino acid identity with respect to the rat α_1 T_3 receptor. Numbers above each region represent the amino acids forming the junctions between domains or the carboxy terminus.

IV. ESTROGEN AND THYROID HORMONE RECEPTOR GENES

The cloning of glucocorticoid receptor (Hollenberg *et al.*, 1985) and estrogen receptor (Green *et al.*, 1986b) complementary DNAs (cDNAs) permitted the first structure–function analysis of the mechanisms of action of eukaryotic transcription factors. Based on primary amino acid sequence and the properties of mutant receptors, the predicted amino acid sequences of nuclear receptors have been divided into six domains, labeled A through F (Fig. 4). Comparison of the primary amino acid sequences of the glucocorticoid and estrogen receptors demonstrates two regions of significant homology. The most highly conserved region (the C domain) consists of 66 amino acids containing nine invariant cysteine residues that function as the DNA-binding domain (Fig. 5). The E domain, which is less highly conserved, encompasses approximately 220 residues that participate in ligand binding, receptor dimerization, and transactivation. The structural and functional properties of these two domains are discussed in further detail below.

Analysis of estrogen receptor mRNA levels in various tissues indicates a pattern of expression that is consistent with the results obtained from biochemical studies based on hormone binding. Relatively high levels of mRNA are observed in classically defined target tissues, such as uterus, pituitary, and breast, but detectable levels are also found in a variety of other tissues. Although initially difficult to demonstrate, estrogen receptors have now been identified in bone and bone-derived cells by several laboratories. Estrogen binding and receptor mRNA were first described in primary human osteoblastlike cells derived from explants of trabecular bone (Eriksen *et al.*, 1988) and osteoblastlike osteosarcoma cells (Komm *et al.*, 1988). Estrogen receptor mRNA levels have also been demonstrated to be increased in callus during fracture healing in the rat (Boden *et al.*, 1989). More recently, estrogen receptors have been identified by immunochemical techniques in osteoclasts of membranous bone (Pensler *et al.*, 1990). Analysis of highly purified populations of primary avian osteoclasts also revealed the existence of saturable and specific estrogen binding (5600 receptors/nucleus), immunoassayable estrogen receptor, and estrogen receptor mRNA (Oursler *et al.*, 1991). The latter two studies are of particular significance because the effects of estrogen treatment on maintenance of bone mass following menopause appear to be primarily due to decreased bone resorption by osteoclasts. The demonstration of estrogen receptors in these cells, in concert with direct biological effects of estrogens on cultured osteoclasts and osteoblasts, support the idea that at least some of the effects of estrogen on patterns of gene expression in bone are mediated by a classical receptor-dependent mechanism.

Comparison of the predicted protein sequences of the glucocorticoid and estrogen receptors with previously cloned genes indicated a close relationship with the v-erb-A oncogene in the DNA binding domain (Weinberger *et al.*, 1985; Green *et al.*, 1986b). This observation suggested that the cellular counterpart of v-erbA encoded a ligand-dependent transcription factor. Using v-erbA sequences as probes, a number of related cDNAs were obtained. Analysis of the proteins encoded by these clones revealed that some, but not all, encoded high-affinity receptors for T_3 (Sap *et al.*, 1986; Weinberger *et al.*, 1986; Thompsen *et al.*, 1987; Benbrook and Pfahl, 1987; Koenig *et al.*, 1988; Lazar *et al.*, 1988). In mammals,

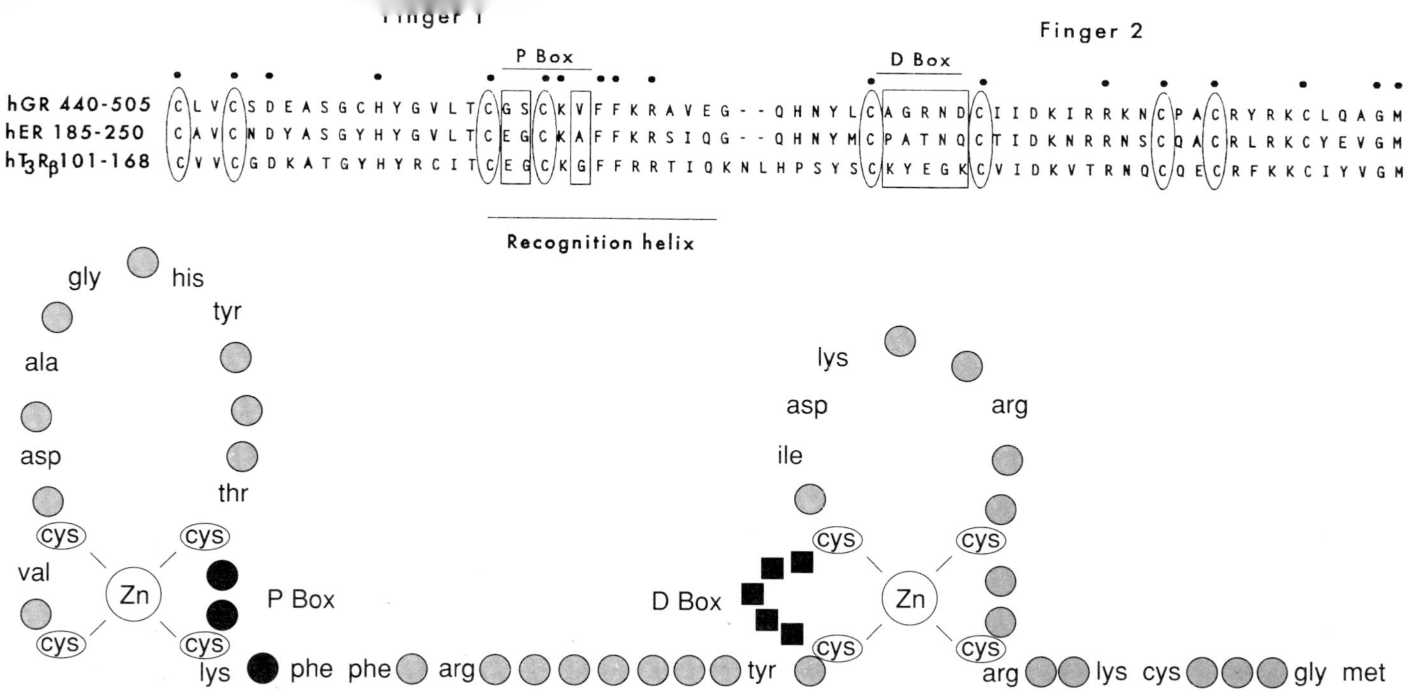

FIGURE 5 Characteristics of the DNA-binding domains of the estrogen, thyroid hormone, and glucocorticoid receptor DNA-binding domains. In the upper panel, the primary amino acid sequences of the human estrogen receptor, thyroid hormone receptor, and glucocorticoid receptor are compared. Conserved cysteine residues that coordinate zinc metal ions, are enclosed by ellipses. Amino acid residues that are highly conserved among all members of the nuclear receptor superfamily are indicated by closed circles. Amino acids that are involved in contacting specific basepairs of the glucocorticoid response element in the case of the glucocorticoid receptor are enclosed by boxes in the first finger and are designated the P-box. These residues form the beginning of an α-helical structure, designated the recognition helix, that resides in the major groove of the glucocorticoid response sequence. Amino acid residues that form part of a dimerization interface within the DNA-binding domain, and that are sensitive to spacing between halfsites of response element sequences are enclosed by a box in the second finger, termed the D-box. The organization of finger structures is illustrated in the lower panel. P-box and D-box amino acids are represented as filled circles and boxes, respectively. Shaded circles represent nonconserved residues between the three receptors.

thyroid hormone receptors are encoded by two separate but closely related genes, labeled α and β. The α forms of the T_3 receptor are most closely related to the v-erb-A oncogene, and the human gene encoding this form of the receptor has been localized to chromosome 17. Transcripts encoded by the α gene are alternatively spliced in the hormone-binding domain, generating two isoforms, labeled α_1 and α_2. The α_1 isoform binds T_3 with high affinity, while the α_2 isoform does not (Izumo and Mahdavi, 1988; Lazar *et al.*, 1988; Mitsuhashi *et al.*, 1988; Miyajima *et al.*, 1989; Fig. 4). Both the α_1 and α_2 isoforms of the T_3 receptor are widely expressed, with particularly high mRNA levels found in the brain (Thompson *et al.*, 1987; Bradley *et al.*, 1989).

The β thyroid hormone receptor is highly related to the α_1 isoform in the DNA- and ligand-binding domains but diverges considerably in the amino-terminus (Fig. 4). The β form of the T_3 receptor is expressed in most tissues that exhibit thyroid hormone responsiveness, including liver and pituitary. Intriguingly, in the pituitary, alternative promoter usage gives rise to a second isoform, β_2, that is negatively regulated by thyroid hormone (Hodin *et al.*, 1989).

The biological requirement for multiple forms of the thyroid hormone receptor is not understood. Presumably, each form plays a specific developmental or homeostatic role. Studies of the transcriptional properties of the α_1 and β thyroid hormone receptors have not as yet revealed clear functional differences, however. Both forms exert similar effects on model promoters in transient transfection studies. In addition to intrinsic differences in transcriptional properties, it is also possible that the existence of two separate genes facilitates different patterns of regulation in different cell types. It is likely that experimental approaches that exploit more physiologic settings will be needed to gain an understanding of the specific roles of the two thyroid hormone receptor genes. In this regard, it is interesting to note that the clinical syndrome of thyroid hormone resistance is tightly linked to the β thyroid hormone receptor locus (Usala *et al.*, 1988), and studies of thyroid hormone receptors from these patients have usually led to the identification of mutations in the ligand-binding domain that reduce the binding affinity for thyroid hormone.

The functional role of the α_2 variant is also poorly understood, despite the fact that the mRNA encoding this protein is expressed in several tissue types, including brain (e.g., Bradley *et al.*, 1989). Consistent with its failure to bind thyroid hormone, the α_2 variant also fails to confer a transcriptional response to T_3 when expressed in otherwise receptor-deficient cells. However, when coexpressed with the α_1 or β isoforms, the α_2 protein has been shown to inhibit their ability to mediate T_3-dependent transcription (Koenig *et al.*, 1989). This has led to the suggestion that the α_2 isoform modulates the activities of the α_1 and β thyroid hormone receptors.

Relatively little information is available on the expression of thyroid hormone receptors in bone. Evidence for a direct effect of thyroid hormones on bone turnover has been obtained using cultured neonatal mouse calvaria (Klaushofer *et al.*, 1989). Thyroid hormones have also been demonstrated to stimulate alkaline phosphatase activity in osteoblastlike (MC3T3) cells (Kasono *et al.*, 1988). Ligand-binding studies with radiolabeled T_3 have indicated the presence of specific and saturable binding sites in bone (Krieger *et al.*, 1988). However, mRNA analysis of T_3-receptor isoforms using specific cDNA probes has not been reported.

V. DNA-BINDING PROPERTIES OF THE ESTROGEN RECEPTOR

A detailed understanding of the mechanisms by which ligand binding regulates the subcellular localization, DNA binding, and transcriptional activation functions of the estrogen receptor has only been partially developed. Although classic experiments employing hypotonic lysis of cells leads to the appearance of estrogen receptor in the cytosol (Jensen *et al.*, 1968), more recent experiments employing immunohistochemical staining with antibodies against estrogen receptor show the receptor to be predominantly nuclear, even in the absence of hormone (Greene *et al.*, 1980). However, the nature of the interaction of estrogen receptor with the nucleus does change after hormone binding, as evidenced by the increased ionic strength required to extract the receptor from the nucleus when bound to estradiol.

Several members of the nuclear receptor superfamily, including estrogen receptor, are known to be associated *in vivo* with hsp90. The domain required for either glucocorticoid receptor or estrogen receptor to bind hsp90 includes the ligand-binding domain (Chambraud *et al.*, 1990). The function of this interaction remains controversial. One model holds that hsp90 acts to keep the unliganded receptor in an inactive state (Cadepond *et al.*, 1991). However, when this model was tested directly in a *Saccharomyces cerevisiae* strain mutant for the yeast hsp90 homolog, hsp82, the results were not consistent with hsp90 being a repressive subunit (Picard *et al.*, 1990). In fact, the results of these experiments suggested that hsp90 was required to allow receptors to be activated by hormone. The exact function of hsp90 remains to be elucidated, but given the well-established finding that hsp90 dissociates from the estrogen receptor in the presence of estrogen, its function must be upstream of those factors that serve to transduce the hormone-dependent transactivation signal.

The estrogen receptor binds the estrogen response element as a homodimer (Kumar and Chambon, 1988). This is evidenced by the ability of mixed heterodimers between the receptor and truncated receptors to bind the ERE *in vitro* and the very high degree of cooperativity in binding demonstrated on the palindromic ERE. In contrast to the thyroid hormone receptor, the affinity of the estrogen receptor monomer for DNA is low. Under conditions in which the estrogen receptor binds tightly to the palindromic ERE, binding to a sequence in which one of the two AGGTCA "halfsites" has been mutated is not detectable (Brown and Sharp, 1990). Also in contrast to the thyroid hormone receptor attempts to demonstrate heterodimer formation between estrogen receptor and other members of the nuclear receptor superfamily have failed.

The core DNA-binding domain of the estrogen receptor extends from cysteine-185 to methionine-250 (Fig. 5). Substitution of the homologous domain from the glucocorticoid receptor is able to convert the DNA-binding specificity of estrogen receptor to that of glucocorticoid receptor (Green and Chambon, 1987). More subtle mutations of this domain have demonstrated that three amino acids (Glu-203, Gly-204, Ala-207) at the base of the first zinc finger (P box; Fig. 5) are important for the ability of estrogen receptor to distinguish EREs from the closely related sequence of the GRE (Mader *et al.*, 1989; Umesono and Evans, 1989).

The three-dimensional solution structures of both the estrogen (Schwabe *et al.*, 1990) and glucocorticoid (Härd *et al.*, 1990) receptor DNA-binding domains have been determined using nuclear magnetic resonance techniques. In addition, the crystallographic structure of the glucocorticoid receptor DNA-binding domain bound to two different GRE's has been solved (Luisi *et al.*, 1991). The structures confirm the prediction that the two sets of four cysteine residues characteristic of the nuclear receptor DNA-binding domain each form a tetrahedral coordination complex containing a zinc metal ion. These so-called zinc fingers are distinct from the zinc-finger motif found in a second, unrelated class of DNA-binding proteins represented by transcription factor IIIA (TFIIIA). Whereas the TFIIIA-type zinc fingers form independent domains, the two zinc fingers of the glucocorticoid and estrogen receptor DNA-binding domains form an integrated globular structure. This structure resembles the classic helix–turn–helix motif found in other classes of sequence-specific DNA-binding proteins in that the structure of a recognition helix that makes specific basepair contacts with DNA is stabilized by a second helix running at right angles to the first. The recognition helix contains the P-box amino acids involved in distinguishing between estrogen and glucocorticoid response elements, while the stabilization helix is provided by the second finger of the DNA-binding domain (Fig. 5). The glucocorticoid receptor DNA-binding domain binds DNA as a dimer even though it is a monomer in solution. The ability of each subunit to make specific contacts with a GRE halfsite depends on the correct spacing between the halfsites. The structure formed on a GRE containing a four-nucleotide spacer rather than the normal 3 bp showed specific binding of the glucocorticoid DNA-binding domain dimer to only one of the halfsites. Thus, the interface between the two DNA-binding domains determines the ability of the dimer to interact appropriately with halfsites of specific spacing.

Although *in vivo* studies suggest that hormone binding plays an important role in mediating DNA binding of the estrogen receptor, it does not appear to be essential when DNA-binding assays are performed *in vitro*. Estrogen receptor expressed *in vitro* or in recombinant systems is able to bind the ERE even in the absence of hormone. The cause of this discrepancy is not well understood. Given the strong preference of estrogen receptor to bind as a homodimer, the role that ligand plays in stabilizing dimerization is significant. Both estradiol and partial agonists such as tamoxifen stabilize dimerization and subsequent DNA binding (Kumar and Chambon, 1988). Pure antiestrogens such as ICI 164,384 prevent dimerization and subsequent DNA binding (Parker, 1990).

VI. DNA-BINDING PROPERTIES OF THE THYROID HORMONE RECEPTOR

Unlike the estrogen receptor and other steroid hormone receptors, the thyroid hormone receptor is localized to chromatin in the presence or absence of its regulatory ligand (Spindler *et al.*, 1975; Damm *et al.*, 1989). Analysis of T_3 receptor–DNA complexes obtained following micrococcal DNase treatment of chromatin resulted in a nucleoprotein complex of 130,000 Da that contained the T_3 receptor and approximately 30 bp of DNA (Perlman *et al.*, 1982). Subtracting the molecular weight of the associated DNA, these results implied that the receptor bound as a homodimer or was associated with an additional protein of similar size.

In vitro studies of the DNA-binding properties of partially purified T_3 receptors derived from cultured cells or tissue sources have demonstrated high-affinity, sequence-specific binding to several functional thyroid hormone response elements (TREs), including those present in the rat growth hormone (e.g., Glass *et al.*, 1987), α-myosin heavy chain (e.g., Izumo and Mahdavi, 1988), and the TSH α- and β-subunit genes (e.g., Darling *et al.*, 1989). These studies have generally demonstrated a requirement for two or more copies of a sequence related to the AGGTCA core recognition motif for high-affinity binding, although binding of monomeric T_3 receptor to a single copy of the core motif has been reported (Lazar *et al.*, 1991). In the case of the palindromic TRE, purified T_3 receptors have been shown to bind to the palindromic T_3 response element as a homodimer (e.g., Holloway *et al.*, 1990; Lazar *et al.*, 1991), consistent with the observations for steroid hormone receptors. However, in the presence of nuclear proteins derived from liver, pituitary tumor cells, or other cell types, the T_3 receptor appears to preferentially interact with DNA as a heterodimer (Murray and Towle, 1989; Burnside *et al.*, 1990; Beebe *et al.*, 1991; Darling *et al.*, 1991). As defined biochemically, these additional proteins, which have been termed T_3 receptor auxiliary proteins (TRAPs), range from 44 to 65 kD, and appear to exhibit different patterns of expression in different cell types. The ability of the T_3 receptor to interact with these proteins depends on sequences within its ligand-binding domain that also appear to serve as a dimerization interface (O'Donnel *et al.*, 1991).

Although the identities of the factors that interact with the T_3 receptor remain to be firmly established, the T_3 receptor has been demonstrated to be capable of forming heterodimers with two additional members of the nuclear receptor superfamily, the retinoic acid receptor (RAR) (Glass *et al.*, 1989; Forman *et al.*, 1989) and the retinoid X receptor (RXR) (Yu *et al.*, 1991). These observations suggest that the T_3 receptor might be capable of forming interactions with several members of the nuclear receptor superfamily. Analysis of the sequence requirements for heterodimer formation between the T_3 receptor and the RAR indicated that an extensive region of the carboxy-terminus of each receptor was necessary. Point mutations and small deletions or insertions further indicated that these sequences included a region with homology to a portion of the estrogen receptor C-terminus required for the formation of homodimers (Fawell *et al.*, 1990). These observations suggested that the formation of heterodimers between the T_3 and RARs might be similar to the formation of homodimers between estrogen receptors.

Although the T_3 receptor/RAR heterodimer may serve as a useful model for studies of receptor interactions, several lines of evidence indicate that RARs do not represent the major activities that interact with the T_3 receptor in most cell types. First, the molecular weight profiles cannot be accounted for by RARs. Second, relatively high levels of these factors are expressed in cells that do not express significant numbers of RARs. Third, the DNA-binding properties of T_3 receptor/RAR heterodimers were quite distinct from those of T_3 receptor heterodimers derived from tissue sources.

More recently, the T_3 receptor has been demonstrated to be capable of interacting with members of the RXR subclass of nuclear receptors. The initial member of this subclass to be identified was H2RIIBP (also referred to as RXR-β) (Hamada *et al.*, 1989), a protein that binds to a *cis*-active element present in the promoter of the major histocompatability class I gene. Subsequently, a highly related nuclear receptor (referred to as RXR-α) was identified and demonstrated to activate transcription in response to high concentrations of retinoic acid (Mangelsdorf *et al.*, 1990). Because this protein was not observed to bind retinoic acid directly, it was hypothesized to respond to the binding of a metabolite produced *in vivo*.

Both the RXR-α and RXR-β interact with the thyroid hormone receptor to enhance its binding to specific T_3 response elements (Yu *et al.*, 1991). Furthermore, co-expression of the RXR-β and the T_3 receptor results in an increase in T_3 receptor-dependent transcription from promoters containing these response elements. These properties suggest that RXR isoforms may represent at least a subset of the TRAP activities that have been characterized biochemically. Development of isoform-specific antibodies to RXR will be necessary to establish this point definitively. It is likely, however, that RXR-α and RXR-β do not represent all forms of TRAP, because the higher molecular weight forms (65 kDa) are too large to be encoded by either of these isoforms.

Intriguingly, RXR isoforms also interact efficiently with the RAR, acting to specifically increase its binding to retinoic acid response elements and enhancing its transcription activity on these elements (Yu *et al.*, 1991). These observations suggest the possibility of complex interactions between thyroid hormone, RAR, and RXR isoforms that may underly cell type-specific patterns of hormone responsiveness for some target genes of retinoic acid and thyroid hormones.

VII. MECHANISMS RESPONSIBLE FOR TRANSCRIPTIONAL ACTIVATION BY ESTROGEN AND THYROID HORMONE RECEPTORS

The basic mechanisms by which nuclear receptors, including the estrogen and thyroid hormone receptors, influence rates of transcription once bound to target genes remain poorly understood. Investigation of this question has primarily utilized three methodologies: functional analysis of receptor mutants, biochemical analysis of receptor function by *in vitro* transcription assays, and, more recently, searching for downstream components of the transcriptional machinery through the utilization of yeast genetics (Metzger *et al.*, 1988).

As has been the case for other classes of transcription factors, discreet functional domains have been identified in the estrogen receptor that are required for transcriptional activation independent of the DNA-binding domain. The definition of these domains required for gene regulation comes largely from transient co-transfection assays of the receptor mutants and reporter plasmids (Green *et al.*, 1986a; Kumar *et al.*, 1986; Waterman *et al.*, 1988; Thompson and Evans, 1989; Holloway *et al.*, 1990). The estrogen receptor contains two transactivation domains: a constitutive domain in the A/B region termed TAF-1 and a ligand-dependent transactivation domain within the E region termed TAF-2 (Kumar *et al.*, 1986, 1987; Webster *et al.*, 1988; Parker *et al.*, 1989; Tora *et al.*, 1989; Fig. 4). Intriguingly, the activity of TAF-1 is cell type- and promoter context-dependent (Berry *et al.*, 1989), such that it exerts a strong effect on transcription on target promoters in some cell types but has little or no activity in other cell types or on other promoters. In these latter cases, the primary determinant of transcriptional activation is the hormone-inducible TAF-2 domain. In contrast to the estrogen receptor, the thyroid hormone receptor has a relatively short A/B region and the activation function resides primarily in domain E. The thyroid hormone receptor is capable of exerting cell type-specific effects on patterns of gene expression (Izumo *et al.*, 1986), possibly by forming heterodimers with cell type-specific proteins as described earlier.

In the case of the estrogen receptor, three classes of ligand exist: pure agonists such as estradiol, mixed agonist–antagonists such as tamoxifen, and pure antagonists such as ICI 164,384. Insights into how these ligands differ in activity has recently been provided by studies of their effects on estrogen receptor activity in yeast, in which tamoxifen functioned equivalently to estradiol as an agonist (Berry *et al.*, 1989). The conclusion drawn from these studies was that the TAF-2 domain does not function in yeast and that the transactivation activity observed was due to the action of TAF-1. In concert with the observed cell-specific differences in the activity of TAF-1 in mammalian cells, these findings suggest the following model to explain the mixed agonist–antagonist activity of tamoxifen. Tamoxifen, which is able to stabilize ER dimerization and subsequent DNA binding, fails to activate TAF-2 but allows the constitutive activity of TAF-1 to function in permissive cell types. The variability of cellular responses to TAF-1 explains why tamoxifen is only an agonist in certain tissues. Estradiol is capable of carrying out yet a second important step in ER activation and by an unknown mechanism activate the function of TAF-2. This activation of TAF-2 is manifest in DNA–protein complexes as an increase in mobility in bandshift assays. Pure antiestrogens such as ICI 164,384 inhibit the first step of ER activation by inhibiting dimerization and subsequent DNA binding (Parker, 1990) and therefore inhibit the activity of both TAF-1 and TAF-2.

It has recently been possible to demonstrate estrogen receptor-dependent transcription *in vitro.* Studies employing HeLa nuclear extracts (Elliston *et al.*, 1990) demonstrated that stimulation of *in vitro* transcription by ER was ERE- but not estrogen-dependent. In addition, the authors demonstrated that the receptor functioned in this system to facilitate the formation of a stable preinitiation complex. In contrast to these studies, *Xenopus* liver nuclear extracts retain hormone-dependent activation (Corthesy *et al.*, 1988, 1990). A protein fraction responsible for the hormone-dependent function has also been identified (Corthesy *et al.*, 1991). Despite these significant achievements, the specific cellular proteins required for transducing the hormone-dependent transcriptional signal are as yet unknown (Meyer *et al.*, 1989). In addition, *in vitro* transcription systems have not as yet been capable of permitting the estrogen receptor to act at relatively long distances from the site of transcriptional initiation as it does *in vivo.*

The potential role that chromatin may play in modulating the ability of steroid receptors to activate transcription is an area of very active research. The most intensively studied systems involve the role that nucleosome phasing on the MMTV promoter plays in GR-stimulated transcription (for a review, see Beato *et al.*, 1989). In addition, very recent studies of the function of both GR and ER expressed in yeast have suggested that genes first identified as regulators of yeast mating type, Swi-1,2,3 and Sin-1,2, may play a role in the function of the glucocorticoid and estrogen receptors (K. Yamamoto, personal communication). These genes encode regulators of chromatin structure or chromatin components themselves. The exact relationship between the animal cell homologs of these genes and the *in vivo* function of steroid receptors remains to be established.

VIII. IMPLICATIONS FOR THE STUDY OF THE MECHANISMS OF ACTION OF ESTROGEN AND THYROID HORMONES ON BONE DEVELOPMENT AND HOMEOSTASIS

An understanding of the mechanisms by which estrogen and thyroid hormones regulate bone metabolism will require the identification of pertinent target genes that directly or indirectly mediate their actions. To date, this has proven to be a difficult task, in part because it has not been clear whether the effects of estrogen or thyroid hormones are intrinsic to bone or reflect the regulation of humoral factors that are produced at distant sites. The recent demonstration of estrogen and thyroid hormone effects on cultured bone and bone-derived cells, and the demonstration of receptors in these cells, indicates that at least some of the actions of these hormones are direct.

Two general approaches are likely to be useful in providing further information on biologically relevant target genes of estrogen and thyroid hormones in bone. One approach would be to evaluate the levels of expression of specific genes that serve important roles in bone development, structure, or homeostasis under conditions of hormone deficiency and excess. For example, this list might include genes encoding bone matrix proteins, growth factors, and enzymes involved in bone resorption (e.g., Sato *et al.*, 1987). A directed approach of this type is biased by the availability of the reagents needed to detect specific gene products, such as antibodies or cDNA clones. However, an increasing array of cDNA clones is continuously becoming available that encode proteins likely to be involved in bone biology. Furthermore, an advantage of this type of approach is that it can be performed at multiple levels, beginning with the *in vivo* setting and progressing through more highly defined systems such as organ cultures or primary cultures of bone cell subclasses. For example, the levels of stromal cell production of interleukin-6, a potent trophic factor for osteoclasts, have been found to be markedly increased by ovarectomy and reduced by estrogen replacement (Jilka *et al.*, 1992).

A second approach to the identification of target genes would be to utilize cDNA cloning strategies designed to detect genes that are differentially expressed in hormone-deficient versus hormone-excess states. Several strategies have been developed that permit the production of cDNA libraries that are highly enriched for clones corresponding to transcripts expressed in one source of tissue or cells but not another (see, e.g., Palazzolo and Meyerowitz, 1987). Although technically difficult and labor-intensive, this approach provides the possibility of identifying previously unrecognized genes that are regulated by the hormone of interest.

The study of the mechanisms by which estrogens and thyroid hormones influence the development and maintenance of bone is likely to provide important insights into mechanisms of osteoporosis and other bone disorders. By virtue of the fact that estrogen and thyroid hormone receptors are regulated by small molecular weight ligands, they represent important targets for pharmacologic intervention. As in virtually every other field of biology and medicine, the application of increasingly powerful molecular and cellular approaches promises to yield not only an improved understanding of the biology and pathology of bone but also more powerful therapeutic interventions.

ACKNOWLEDGMENTS

CKG is a Lucille P. Markey Fellow for Biomedical Science and is supported, in part, by the Lucille P. Markey Foundation. MB is supported by an NIH clinical investigator award (KO8-CA-01363).

REFERENCES

Adler, S., Waterman M. L., He, X., and Rosenfeld, M. G. (1988). Steroid receptor-mediated inhibition of rat prolactin gene expression does not require the receptor DNA-binding domain. *Cell* **52,** 685–695.

Baniahmad, A., Steiner, C., Kühne, A. C., and Renkawitz, R. (1990). Modular structure of a chicken lysozyme silencer: Involvement of an unusual thyroid hormone receptor binding site. *Cell* **61,** 505–514.

Beato, M. (1989). Gene regulation by steroid hormones. *Cell* **56,** 335–344.

Beato, M., Chalepakis, G., Schauer, M., and Slater, E. P. (1989). DNA regulatory elements for steroid hormones. *J. Steroid Biochem.* **32,** 737–747.

Beebe, J. S., Darling, D. S., and Chin, W. W. (1991). 3,5,3′-Triiodothyronine receptor auxiliary protein (TRAP) enhances receptor binding by interactions within the thyroid hormone response element. *Mol. Endo.* **5,** 85–93.

Benbrook, D., and Pfahl, M. (1987). A novel thyroid hormone receptor encoded by a cDNA clone from a human testes library. *Science* **238,** 788–791.

Berry, M., Nunez, A.-M., and Chambon, P. (1989) Estrogen-responsive element of the human pS2 gene is an imperfectly palindromic sequence. *Proc. Natl. Acad. Sci. USA* **86,** 1218–1222.

Boden, S. D., Joyce, M. E., Oliver, B., Heydermann, A., and Bolander, M. E. (1989). Receptor mRNA expression in callus during fracture healing in the rat. *Calcif. Tissue Int.* **45,** 324–325.

Bradley, D. J., Young III, W. S., and Weinberger, C. (1989). Differential expression of alpha and beta thyroid hormone receptor genes in rat brain and pituitary. *Proc. Natl. Acad. Sci. USA* **86,** 7250–7254.

Brent, G. A., Harney, J. W., Chen, Y., Warne, R. L., Moore, D. D., and Larsen, P. R. (1989). Mutations of the rat growth hormone promoter which increase and decrease response to thyroid hormone define a consensus thyroid hormone response element. *Mol. Endo.* **3,** 1996–2004.

Brown, M., and Sharp, P. A. (1990). Human estrogen receptor forms multiple protein–DNA complexes. *J. Biol. Chem.* **265,** 11238–11243.

Burch, J. B. E., and Weintraub, H. (1983). Temporal order of chromatin structural changes associated with activation of the chicken vitellogenin gene. *Cell* **33,** 65–76.

Burnside, J., Darling, D. S., and Chin, W. W. (1990). A nuclear factor that enhances binding of thyroid hormone receptors to thyroid hormone response elements. *J. Biol. Chem.* **265,** 2500–2504.

Cadepond, F., Schweizer, G. G., Segard, M. I., Jibard, N., Hollenberg, S. M., Giguere, V., Evans, R. M., and Baulieu, E. E. (1991). Heat shock protein 90 as a critical factor in maintaining glucocorticosteroid receptor in a nonfunctional state. *J. Biol. Chem.* **266,** 5834–5841.

Chambraud B., Berry, M., Redeuilh, G., Chambon, P., and Baulieu, E. E. (1990). Several regions of human estrogen receptor are involved in the formation of receptor-heat shock protein 90 complexes. *J. Biol. Chem.* **265,** 20686–20691.

Corthesy, B., Hipskind, R., Theulaz, I., and Wahli, W. (1988). Estrogen-dependent in vitro transcription from the vitellogenin promoter in liver nuclear extracts. *Science* **239,** 1137–1139.

Corthesy, B., Claret, F. X., and Wahli, W. (1990). Estrogen receptor level determines sex-specific in vitro transcription from the *Xenopus* vitellogenin promoter. *Proc. Natl. Acad. Sci. USA* **87,** 7878–7882.

Corthesy, B., Corthesy, T. I., Cardinaux, J. R., and Wahli, W. (1991). A liver protein fraction regulating hormone-dependent in vitro transcription from the vitellogenin genes induces their expression in *Xenopus* oocytes. *Mol. Endo.* **5,** 159–169.

Damm, K., Thompson, C. C., and Evans, R. M. (1989). Protein encoded by v-erbA functions as a thyroid-hormone receptor antagonist. *Nature (London)* **339,** 593–597.

Darling, D. S., Burnside, J., and Chin, W. W. (1989). Binding of thyroid hormone receptors to the rat thyrotropin-beta gene. *Mol. Endo.* **3,** 1359–1368.

Darling, D. S., Beebe, J. S., Burnside, J., Winslow, E. R., and Chin, W. W. (1991). 3,5,3′-Triiodothyronine (T_3) receptor-auxiliary protein (TRAP) binds DNA and forms heterodimers with the T_3 receptor. *Mol. Endo.* **5,** 73–84.

Darwish, H., Krisinger J., Furlow, J. D., Smith, C., Murdoch, F. E., and DeLuca, H. F. (1991). An estrogen-responsive element mediates the transcriptional regulation of calbindin D-9K gene in rat uterus. *J. Biol. Chem.* **266,** 511–558.

Diamond, D. J., and Goodman, H. M. (1985). Regulation of growth hormone messenger RNA synthesis by dexamethasone and triiodothyronine: Transcriptional rate and mRNA stability changes in pituitary tumor cells. *J. Mol. Biol.* **181,** 41–62.

Elliston, J. F., Fawell, S. E., Klein-Hitpass, L., Tsai, S. Y., Tsai, M. J., Parker, M. G., and

O'Malley, B. W. (1990). Mechanism of estrogen receptor-dependent transcription in a cell-free system. *Mol. Cell. Biol.* **10,** 6607–6612.

Eriksen, E. F., Colvard, D. S., Berg, N. J., Graham, M. L., Mann, K. G., Spelsberg, T. C., and Riggs, B. C. (1988). Evidence of estrogen receptors in normal osteoblast-like cells. *Science* **241,** 84–86.

Evans, R. M. (1988). Steroid and thyroid hormone receptors as transcriptional regulators of development and physiology. *Science* **240,** 889–895.

Fawell, S. E., Lees, J. A., White, R., and Parker, M. G. (1990). Characterization and colocalization of steroid binding and dimerization activities in the mouse estrogen receptor. *Cell* **60,** 953–962.

Forman, B. M., Yang, C., Au, M., Casanova, J., Ghysdael, J., and Samuels, H. H. (1989). A domain containing leucine-zipper-like motifs mediate novel in vivo interactions between the thyroid hormone and retinoic acid receptors. *Mol. Endo.* **3,** 1610–1626.

Gaub, M. P., Bellard, M., Scheuer, I., Chambon, P., and Sassone-Corsi, P. (1990). Activation of the ovalbumin gene by the estrogen receptor involves the fos-jun complex. *Cell* **63,** 1267–1276.

Glass, C. K., Franco, R., Weinberger, C., Albert, V., Evans, R. M., and Rosenfeld, M. G. (1987). A c-erb-A binding site in the rat growth hormone gene mediates transactivation by thyroid hormone. *Nature (London)* **329,** 738–741.

Glass, C. K., Holloway, J. M., Devary, O. V., and Rosenfeld, M. G. (1988). The thyroid hormone receptor binds with opposite transcriptional effects to a common sequence motif in thyroid hormone and estrogen response elements. *Cell* **54,** 313–323.

Glass, C. K., Lipkin, S. M., Devary, O. V., and Rosenfeld, M. G. (1989). Positive and negative regulation of gene transcription by a retinoic acid-thyroid hormone receptor heterodimer. *Cell* **59,** 697–708.

Green, S., and Chambon, P. (1987). Oestradiol induction of a glucocorticoid-responsive gene by a chimeric receptor. *Nature (London)* **325,** 75–78.

Green, S., Kumar, V., Krust, A., Walter, P., and Chambon, P. (1986a). Structural and functional domains of the estrogen receptor. *Cold Spring Harb. Symp. Quant. Biol.* **2,** 751–758.

Green, S., Walter, P., Kumar, V., Krust, A., Bornet, J. M., Argos, P., and Chambon, P. (1986b). Human estrogen receptor cDNA: Sequence, expression and homology to v erbA. *Nature (London)* **320,** 134–139.

Greene, G. L., Fitch, F. W., and Jensen, E. V. (1980). Monoclonal antibodies to estrophilin: Probes for the study of estrogen receptors. *Proc. Natl. Acad. Sci. USA* **77,** 157–161.

Hamada, K., Gleason, S. L., Levi, B.-Z., Hirschfeld, S., Appella, E., and Ozato, K. (1989). H-2RIIBP, a member of the nuclear hormone receptor superfamily that binds to both the regulatory element of major histocompatibility class I genes and the estrogen response element. *Proc. Natl. Acad. Sci. USA* **86,** 8289–8293.

Härd, T., Kellenbach, E., Boelens, R., Maler, B. A., Dahlman, K., Freedman, L. P., Carlstedt-Duke, J., Yamamoto, K. R., Gustafsson, J.-Å., and Kaptein, R. (1990). Solution structure of the glucocorticoid receptor DNA-binding domain. *Science* **249,** 157–160.

Hodin, R. A., Lazar, M. A., Wintman, B. I., Darling, D. S., Koenig, R. J., Larsen, P. R., Moore, D. D., and Chin, W. W. (1989). Identification of thyroid hormone receptor that is pituitary specific. *Science* **244,** 76–79.

Hollenberg, S. M., Weinberger, C., Ong, E. S., Cerelli, G., Oro, A., Lebo, R., Thompson E. B., Rosenfeld, M. G., and Evans, R. M. (1985). Primary structure and expression of a functional human glucocorticoid receptor. *Nature (London)* **318,** 635–641.

Holloway, J. M., Glass, C. K., Adler, S., Nelson, C., and Rosenfeld, M. G. (1990). The C-terminal interaction domain of the T_3 receptor confers the avility of the DNA site to dictate positive or negative transcriptional activity. *Proc. Natl. Acad. Sci. USA* **87,** 8160–8164.

Hudson, L. G., Santon, J. B., Glass, C. K., and Gill, G. N. (1990). Ligand-activated thyroid hormone and retinoic acid receptors inhibit growth factor receptor promoter expression. *Cell* **62,** 1165–1175.

Izumo, S. B., and Mahdavi, V. (1988). Thyroid hormone receptor alpha isoforms generated by alternative splicing differentially activate myosin heavy chain gene transcription. *Nature (London)* **334,** 539–542.

Izumo, S. B., Nadal-Ginard, B., and Mahdavi, V. (1986). All members of the MHC multigene family respond to thyroid hormone in a highly tissue-specific manner. *Science* **231,** 597–600.

Jensen, E. V., Suzuki, T., Kawashima, T., Stumpf, W. E., Jungblutt, P. W., and DeSombre, E. R. (1968). A two-step mechanism for the interaction of estradiol with rat uterus. *Proc. Natl. Acad. Sci. USA* **59,** 632–638.

Jilka, R. L., Hangoc, G., Girasole, G., Passeri, G., Williams, D. C., Abrams, J. S., Boyce, B., Broxmeyer, H., and Manolagas, S. C. (1992). Increased osteoclast development after estrogen loss: Mediation by interleukin-6. *Science* **257,** 88–91.

Joab, I., Radanyi, C., Renoir, M., Buchou, T., Catelli, M. G., Binaut, N., Mester, J., and Baulieu, E. S. (1984). Common non-hormone binding component in non-transformed chick oviduct receptor of four steroid hormones. *Nature (London)* **308,** 850–853.

Jost, J. P., Seldran, M., and Geiser, M. (1984). Preferential binding of estrogen-receptor complex to a region containing the estrogen-dependent hypomethylation site preceding the chicken vitellogenin A2 gene. *Proc. Natl. Acad. Sci. USA* **81,** 8429–8433.

Jump, D. B. (1989). Rapid induction of rat liver S14 gene transcription by thyroid hormone. *J. Biol. Chem.* **264,** 4698–4703.

Kasono, K., Sato, K., Han, D. C., Fujii, Y., Tsushima, T., and Shizume, K. (1988). Stimulation of alkaline phosphatase activity by thyroid hormone in mouse osteoblast-like cells (MC3T3-E1): A possible mechanism of hyperalkaline phospatasia in hyperthyroidism. *Bone Miner.* **4,** 355–363.

Kerner, S. A., Scott, R. A., and Pike, J. W. (1989). Sequence elements in the human osteocalcin gene confer basal activation and inducible response to hormonal vitamin D_3. *Proc. Natl. Acad. Sci. USA* **86,** 4455–4459.

Kesavan, P., Mukhopadhayay, S., Murphy, S., Rengaraju, M., Lazar, M. A., and Das, M. (1991). Thyroid hormone decreases the expression of epidermal growth factor receptor. *J. Biol. Chem.* **266,** 10282–10286.

Klaushofer, K., Hoffman, O., Gleispach, H., Leis, H. J., Czerwenka, E., Koller, K., and Peterlik, M. (1989). Bone-resorbing activity of thyroid hormones is related to prostaglandin production in cultured nional mouse calvaria. *Bone Miner.* **4,** 305–312.

Klein-Hitpass, L., Schorpp, M., Wagner, U., and Ryffel, G. U. (1986). An estrogen-responsive element derived from the 5′ flanking region of the *Xenopus* vitellogenin A2 gene functions in transfected human cells. *Cell* **46,** 1053–1061.

Klock, G., Strahle, U., and Schutz, G. (1987). Oestrogen and glucocorticoid responsive elements are closely related but distinct. *Nature (London)* **329,** 734–736.

Koenig, R. J., Brent, G. A., Warner, R. L., Larsen, P. R., and Moore, D. D. (1987). Thyroid hormone receptor binds to a site in the rat growth hormone promoter required for induction by thyroid hormone. *Proc. Natl. Acad. Sci. USA* **84,** 5670–5674.

Koenig, R. J., Warner, R. L., Brent, G. A., Harney, J. W., Larsen, P. R., and Moore, D. D. (1988). Isolation of a cDNA clone encoding a biologically active thyroid hormone receptor. *Proc. Natl. Acad. Sci. USA* **85,** 5031–5035.

Koenig, R., Lazar, M. A., Hodin, R. A., Brent, G. A., Larsen, P. R., Chin, W. W., and Moore, D. D. (1989). Inhibition of thyroid hormone action by a non-hormone binding c-erbA protein generated by alternative mRNA splicing. *Nature (London)* **332,** 659–663.

Komm, B. S., Terpening, C. M., Benz, D. J., Graeme, K. A., Gallegos, A., Korc, M., Greene, G. L., O'Malley, B. W., and Haussler, M. R. (1988). Estrogen binding, receptor mRNA and biological response in osteoblast-like osteosarcoma cells. *Science* **241,** 81–84.

Krieger, N. S., Stappenbeck, T. S., and Stern, P. H. (1988). Characterization of specific thyroid hormone receptors in bone. *J. Bone Miner.* **3,** 473–478.

Kumar, V. J., and Chambon, P. (1988). The estrogen receptor binds tightly to its responsive element as a ligand-induced homodimer. *Cell* **55,** 145–156.

Kumar, V., Green, S., Stack, G., Berry, M., Jin, J., and Chambon, P. (1987). Functional domains of the human estrogen receptor. *Cell* **51,** 941–951.

Kumar, V., Green, S., Staub, A., and Chambon, P. (1986) Localization of the estradiol binding and putative DNA binding domains of the human estrogen receptor. *EMBO J.* **5,** 2231–2236.

Lai, E. C., Riser, M. E., and O'Malley, B. W. (1983). Regulated expression of the chicken ovalbumin gene in a human estrogen-responsive cell line. *J. Biol. Chem.* **258,** 12693–12701.

Lazar, M. A., Hodin, R. A., Darling, D. S., and Chin, W. W. (1988). Identification of a rat c-erbA related protein which binds deoxyribonucleic acid but does not bind thyroid hormone. *Mol. Endo.* **2,** 893–901.

Lazar, M. A., Berrodin, T. J., and Harding, H. P. (1991). Differential DNA binding by monomeric, homodimeric, and potentially hetermomeric forms of the thyroid hormone receptor. *Mol. Cell. Biol.* **11,** 5005–5015.

Lipkin, S., Nelson, C. A., Glass, C. K., and Rosenfeld, M. G. (1992). A negative retinoic acid response element in the rat oxytocin promote restricts transcriptional stimulation by heterologous transactivation domains. *Proc. Natl. Acad. Sci. USA* **89,** 1209–1213.

Luisi, B. F., Xu, W. X., Otwinowski, Z., Freedman, L. P., Yamamoto, K. R., and Sigler, P. B. (1991). Crystallographic analysis of the interaction of the glucocorticoid receptor with DNA. *Nature (London)* **352,** 497–505.

Mader, S., Kumar, V., de Verneuil, H., and Chambon, P. (1989). Three amino acids of the estrogen receptor are essential to its ability to distinguish an estrogen from a glucocorticoid-response element. *Nature (London)* **338,** 271–274.

Mangelsdorf, D. J., Ong, E. S., Dyck, J. A., and Evans, R. M. (1990). Nuclear receptor that identifies a novel retinoic acid response pathway. *Nature (London)* **345,** 224–229.

Martinez, E., Givel, F., and Wahli, W. (1987). The estrogen responsive element as an inducible enhancer: DNA sequence requirements and conversion to a glucocorticoid response element. *EMBO J.* **6,** 3719–3727.

Maurer, R.-A., and Notides, A. C. (1987). Identification of an estrogen-responsive element from the 5'-flanking region of the rat prolactin gene. *Mol. Cell. Biol.* **7,** 4247–4254.

Metzger, D., White, J. H., and Chambon, P. (1988). The human estrogen receptor functions in yeast. *Nature (London)* **334,** 31–36.

Meyer, M. E., Gronemeyer, H., Turcotte, B., Bocquel, M. T., Tasset, D., and Chambon, P. (1989). Steroid hormone receptors compete for factors that mediate their enhancer function. *Cell* **57,** 433–442.

Miglisccio, A., DiDomenico, M., Green, S., deFalco, A., Kajtaniak, E. L., Blasi, F., Chambon, P., and Auricchio, F. (1989). Phosphorylation on tyrosine of in vitro synthesized human estrogen receptor activates its hormone binding. *Mol. Endo.* **3,** 1061–1069.

Mitsuhashi, T., Tennyson, G. E., and Nikodem, V. M. (1988). Alternative splicing generates messages encoding rat c-erbA proteins that do not bind thyroid hormone. *Proc. Natl. Acad. Sci. USA* **85,** 5804–5808.

Miyajima, N., Horiuchi, R., Shibuya, Y., Fukushige, S.-I., Matsubara, K.-I., Toyoshima, K., and Yamamoto, T. (1989). Two erbA homologs encoding proteins with different T_3 binding capacities are transcribed from opposite DNA strands of the same genetic locus. *Cell* **57,** 31–39.

Murray, M. B., and Towle, H. C. (1989). Identification of nuclear factors that enhance binding of the thyroid hormone receptor to a thyroid hormone response element. *Mol. Endo.* **3,** 1434–1442.

Näär, A. M., Boutin, J.-M., Lipkin, S. M., Yu, V. C., Holloway, J. M., Glass, C. K., and Rosenfeld, M. G. (1991). The orientation and spacing of core DNA-binding motifs dictate selective transcriptional responses to three nuclear receptors. *Cell* **65,** 1267–1279.

Nyborg, J. K., Nguyen, A. P., and Spindler, S. R. (1984). Relationship between thyroid and glucocorticoid hormone receptor occupancy, growth hormone gene transcription, and mRNA accumulation. *J. Biol. Chem.* **259,** 12377–12382.

Nyborg, K., and Spindler, S. (1986). Alternations in chromatin structure accompany thyroid hormone induction of growth hormone gene transcription. *J. Biol. Chem.* **261,** 5685–5688.

O'Donnell, A. L., Rosen, E. D., Darling, D. S., and Koenig, R. J. (1991). Thyroid hormone receptor mutations that interfere with transcriptional activation also interfere with receptor interaction with a nuclear protein. *Mol. Endo.* **5,** 94–99.

Oursler, M. J., Osdoby, P., Pyfferoen, J., Riggs, B. L., and Spelsberg, T. C. (1991). Avian osteoclasts as estrogen target cells. *Proc. Natl. Acad. Sci. USA* **88,** 6613–6617.

Palazzolo, M. J., and Meyerowitz, E. M. (1987). A family of lambda phage cDNA cloning vectors, λSWAJ, allowing the amplification of RNA sequences. *Gene* **52,** 197–206.

Parker, M. G. (1990). Inhibition of estrogen receptor-DNA binding by the "pure" antiestrogen ICI 164,384 appears to be mediated by impaired receptor dimerization. *Proc. Natl. Acad. Sci. USA* **87,** 6883–6887.

Parker, M. G., Lehman, J. A. J., and Martin, D. E. (1989). Identification of constitutive and steroid-dependent transactivation domains in the mouse oestrogen receptor. *J. Steroid Biochem.* **34,** 33–39.

Pensler, J. M., Radosevich, J. A., Higbee, R., and Langman, C. B. (1990). Osteoclasts isolated from membranous bone in children exhibit nuclear estrogen and progesterone receptors. *J. Bone Miner.* **5,** 797–802.

Perlman, A. J., Stanley, F., and Samuels, H. H. (1982). Thyroid hormone nuclear receptor. *J. Biol. Chem.* **257,** 930–938.

Petty, K. S., Desvergne, B., Mitsuhashi, T., and Nikodem, V. M. (1990). Identification of a thyroid hormone response element in the malic enzyme gene. *J. Biol. Chem.* **265,** 7395–7400.

Picard, D., Khursheed, B., Garabedian, M. J., Fortin, M. G., Lindquist, S., and Yamamoto, K. R. (1990). Reduced levels of hsp90 compromise steroid receptor action in vivo. *Nature (London)* **348,** 166–168.

Pratt, W., Jolly, D. J., Pratt, D. V., Hollenberg, S. M., Giguere, V., Cadepon, F. M., Schweizer-Groyer, G., Cartelli, M. G., Evans, R. M., and Baulieu, E. E. (1988). A region in the steroid-binding domain determines formation of the non-DNA binding glucocorticoid receptor complex. *J. Biol. Chem.* **263,** 267–273.

Rhodes, D., and Schwabe, J. W. (1991). Structural biology. Complex behavior. *Nature (London)* **352,** 478–479.

Richard, S., and Zingg, H. H. (1990). The human oxytocin gene promoter is regulated by estrogens. *J. Biol. Chem.* **265,** 6098–6103.

Sap, J., Munoz, A., Schmitt, J., Stunnenberg, H., and Vennstrom, B. (1989). Repression of transcription mediated at a thyroid hormone response element by the v-erb-A oncogene product. *Nature (London)* **340,** 242–244.

Sap, J., Munoz, A., Damm, K., Goldberg, Y., Ghysdael, J., Leutz, A., Beug, H., and Vennstrom, B. (1986). The c-erb-A protein is a high affinity receptor for thyroid hormone. *Nature (London)* **324,** 635–640.

Sato, K., Han, D. C., Fujii, Y., Tsushima, T., and Shizume, K. (1987). Thyroid hormone stimulates alkaline phosphatase activity in cultured rat osteoblastic cells (ROS 17/2.8) through 3,5,3′-triiodo-L-thyronine nuclear receptors. **120,** 1873–1881.

Schüle, R., Muller, M., Kaltschmidt C., and Renkawitz, R. (1988). Many transcription factors interact synergistically with steroid receptors. *Science* **242,** 1418–1420.

Schwabe, J. W., Neuhaus, D., and Rhodes, D. (1990). Solution structure of the DNA-binding domain of the oestrogen receptor. *Nature (London)* **348,** 458–461.

Shapiro, D. J., Barton, M. C., McKearin, D. M., Chang, T. C., Lew, D., Blume, J., Nielsen, D. A., and Gould, L. (1989). Estrogen regulation gene transcription and mRNA stability. *Rec. Prog. Hormone Res.* **45,** 29–58.

Shupnik, M. A., Chin, W. W., Ross, D. S., Habener, J. F., and Ridgway, E. C. (1985). Transcriptional regulation of the thyrotropin subunit genes by thyroid hormone. *J. Biol. Chem.* **260,** 2900–2903.

Song, M.-K. H., Dozin, B., Grieco, D., Rall, J. E., and Nikodem, V. M. (1988). Transcriptional activation and stabilization of malic enzyme mRNA precursor by thyroid hormone. *J. Biol. Chem.* **263,** 17970–17974.

Spindler, B. J., MacLeod, K. M., Ring, J., and Baxter, J. D. (1975). Thyroid hormone receptors: Binding characteristics and lack of hormonal dependency for nuclear localization. *J. Biol. Chem.* **250,** 4113–4119.

Strahle, U., Klock, G., and Schutz, G. (1987). A DNA sequence of 15 base pairs is sufficient to mediate both glucocorticoid and progesterone induction of gene expression. *Proc. Natl. Acad. Sci. USA* **84,** 7871-7875.

Thompson, C. C., and Evans, R. M. (1989). Trans-activation by thyroid hormone receptors: Functional parallels with steroid hormone receptors. *Proc. Natl. Sci. USA* **86,** 3494–3498.

Thompson, C. C., Weinberger, C., Lebo, R., and Evans, R. M. (1987). Identification of a novel thyroid hormone receptor expressed in the mammalian central nervous system. *Science* **237,** 1610–1614.

Tora, L., Gaub, M.-P., Mader, S., Dierich, A., Bellard, M., and Chambon, P. (1988). Cell-specific activity of a GGTCA half-palindromic oestrogen-responsive element in the chicken ovalbumin gene promoter. *EMBO J.* **7,** 3771–3778.

Tora, L., White., Brou, C., Tasset, D., Webster, N., Scheer, E., and Chambon, P. (1989). The human estrogen receptor has two independent nonacidic transcriptional activation functions. *Cell* **59,** 477–487.

Umesono, K., and Evans, R. M. (1989). Determinants of target gene specificity for steroid/thyroid hormone receptors. *Cell* **57,** 1139–1146.

Umesono, K., Giguere, V., Glass, C. K., Rosenfeld, M. G., and Evans, R. M. (1988). Retinoic acid and thyroid hormone induce gene expression through a common response element. *Nature (London)* **336,** 262–265.

Umesono, K., Murakami, K. K., Thompson, C. C., and Evans, R. M. (1991). Direct repeats as selective response elements for the thyroid hormone, retinoic acid, and vitamin D_3 receptors. *Cell* **65,** 1255–1266.

Usala, S. J., Bale, A. E., Gesundheit, N., Weinberger, C., Lash, R. W., Wondisford, F. E., McBride, O. W., and Weintraub, B. D. (1988). Tight linkage between the syndrome of generalized thyroid hormone resistance and the human c-erbA beta gene. *Mol. Endo.* **2,** 1217–1220.

Waterman, M. L., Adler, S., Nelson, C., Greene, G. L., Evans, R. M., and Rosenfeld, M. G. (1988). A single domain of the estrogen receptor confers deoxyribonucleic acid binding and transcriptional activation of the rat prolactin gene. *Mol. Endo.* **2,** 14–21.

Webster, N. J., Green, S., Jin, J. R., and Chambon, P. (1988). The hormone-binding domains of the estrogen and glucocorticoid receptors contain an inducible transcription activation function. *Cell* **54,** 199–207.

Weinberger, C., Hollenberg, S. M., Rosenfeld, M. G., and Evans, R. M. (1985). Domain structure of human glucocorticoid receptor and its relationship to the v-erb-A oncogene product. *Nature (London)* **318,** 670–672.

Weinberger, C., Thompson, C., Ong, E. S., Lebo, R., Gruol, D. J., and Evans, R. M. (1986). The *c-erb-A* gene encodes a thyroid hormone receptor. *Nature (London)* **324,** 641–646.

Weisz, A., and Rosales, R. (1990). Identification of an estrogen response element upstream of the human c-fos gene that binds the estrogen receptor and the AP-1 transcription factor. *Nucleic Acids Res.* **18,** 5097–5106.

Wijnholds, J., Philipsen, J. N. J., and Ab, G. (1988). Tissue-specific and steroid-dependent interaction of transcription factors with the oestrogen-inducible apoVLDL II promoter in vivo. *EMBO J.* **7,** 2757–2763.

Yaffe, B. M., and Samuels, H. H. (1984). Hormonal regulation of the growth hormone gene. *J. Biol. Chem.* **259,** 6285–6291.

Yu, V. C., Delsert, C., Anderson, B., Holloway, J. M., Devary, O. V., Näär, A. M., Kim, S. Y., Boutin, J.-M., Glass, C. K., and Rosenfeld, M. G. (1991). RXRβ: A co-regulator that enhances binding of retinoic acid, thyroid hormone, and vitamin D receptors to their cognate response elements. *Cell* **67,** 1251–1266.

Zilz, N. D., Murray, M. B., and Towle, H. C. (1990). Identification of multiple thyroid hormone response elements located far upstream from the qrat S14 promoter. *J. Biol. Chem.* **265,** 8136–8143.

9

RECENT ADVANCES IN THE BIOLOGY OF RETINOIDS

GREGOR EICHELE, CHRISTINA THALLER, and SUSAN M. SMITH

Cellular and Molecular Biology of Bone

I. INTRODUCTION

Retinoids are signaling molecules involved in a broad range of biological processes including epithelial differentiation, morphogenesis, and vision. Most physiological effects of retinoids are meditated by specific nuclear receptors which are ligand-dependent transcription factors that regulate the expression of specific target genes (Fig. 1). The main exception to this receptor-dependent mechanism of action occurs is vision, in which the 11-*cis* isomer of retinal is converted in a light-dependent reaction to all-*trans*-retinal (for a review, see Rando, 1990). During a subsequently occurring light-independent process, the all-*trans* isomer is recycled to 11-*cis*-retinal. Thus, vision does not involve a receptor but depends on a series of reactions, the most critical of which is *cis–trans* isomerization. As discussed later, *cis–trans* isomerization may also be physiologically important for retinoid action outside the visual system (see Section III).

The importance of retinoids in differentiation and maintenance of epithelial tissues was discovered by Wolbach and Howe (1925). Likewise, the requirement of retinoids for normal growth and development has been known for a considerable time and has been used to assay the relative potencies of various natural and synthetic retinoids. In contrast, evidence indicating that retinoids are involved in morphogenesis is more recent, although studies exploring the teratogenic effects of retinoids go back to the 1950s and 1960s (Cohlan, 1953; Kochhar, 1967). To respond to retinoids, vertebrates utilize six different retinoid receptors that are classified into two families. The first family consists of three genetically distinct receptors, all with high affinity for all-*trans* retinoic acid. They are referred to as retinoic acid receptors α, β, and γ (RAR-α, RAR-β, and RAR-γ, respectively) (see Section II for details about retinoid receptors). The second family of retinoid receptors is known as the retinoid X receptors (RXRs): RXR-α, RXR-β, and RXR-γ. RXRs can bind to and be activated by 9-*cis*-retinoic acid (see Section III). While RXRs do not bind all-*trans*-retinoic acid, RARs bind both isomers. That 9-*cis*-retinoic acid is a ligand for RXRs emphasizes that isomers other than the all-*trans* form play a role in nonvisual functions of retinoids.

Retinoid receptors are operative in the nucleus (for a review, see Evans, 1988; Beato, 1989). They bind to specific sites in the regulatory region of target genes. These DNA-binding sites are known as retinoid-responsive elements (Fig. 1). Upon ligand binding, the receptor is activated (i.e., becomes capable of enhancing [or repressing] target gene transcription). The mode of action outlined in Fig. 1 is also the basis for a convenient method to assay receptor activity (e.g., Giguère and Evans, 1990; Pfahl *et al.*, 1990). In a typical receptor assay, two DNA constructs are co-transfected into cultured cells. One construct is an expression

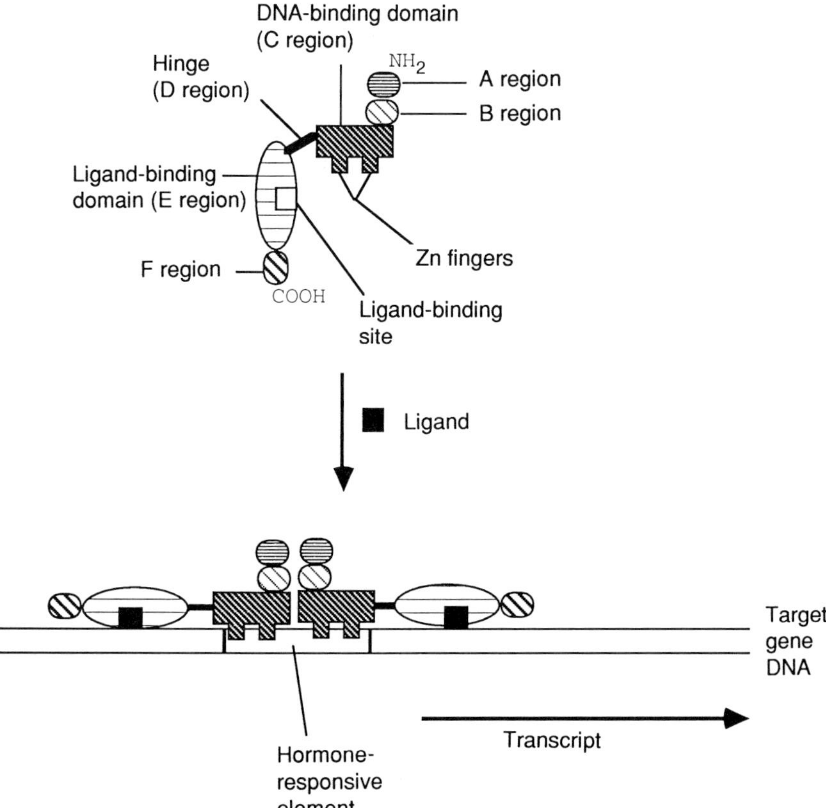

FIGURE 1 Mechanism of action and domain structure of nuclear hormone receptors. Receptors have a multidomain structure. Ligand binds to a site in the ligand-binding domain. Ligand binding may induce conformational changes allowing the receptor to become functional. Some nuclear receptors such as the glucocorticoid receptor require ligand binding to enter the nucleus and interact with DNA. Other receptors are normally bound to DNA and require ligand binding for activation. In the case shown here, two receptor molecules are bound on the same side of the hormone-responsive element (HRE) and are arranged with dyad symmetry reflecting the dyad symmetry of the HRE. However, HREs vary and the geometry of binding might be different from that shown in the scheme. Binding of receptor results in activation (repression) of the target gene, because the receptor proteins interact with other components of the transcriptional machinery presumably through the A region.

vector containing the retinoid receptor complementary DNA (cDNA), and the other is a plasmid containing a reporter gene (e.g., luciferase) downstream of a retinoid-responsive element. Upon treating transfected cells with retinoid, the reporter gene is induced. The main limitation of this assay is that it involves living cells that are capable of converting the added retinoid to another metabolite that may or may not be a receptor ligand. For example, 9-*cis*-retinoic acid, a ligand of RXR is generated *in situ* from all-*trans*-retinoic acid (see Section III). Another potential complication is that such reporter cells may also express other transcription factors that can interact with, for example, RARs. Indeed, several groups have recently reported that RXRs, which are expressed in many cell types, can form heterodimers with RARs (see later).

The structure of retinoids is relatively simple (Fig. 2A). A six-membered ionylidene ring is linked to a polyene chain containing four conjugated double bonds. The ring as well as the polyene chain can carry various functional groups (e.g., at positions 4 and 15, respectively). In addition, each of the double bonds can assume either a *cis* or a *trans* conformation. Because retinoids contain four double bonds, a total of 16 stereoisomers can arise. It should be pointed out, though, that some of these isomers are energetically unstable. Not all retinoids bind to RARs or RXRs. For example, all-*trans*-retinol (C_{15} is CH_2OH) activates neither RARs nor RXRs but serves as a physiological precursor for all-*trans*-retinoic acid (Dowling and Wald, 1960; Thaller and Eichele, 1988). In contrast, 3,4-didehydroretinoic acid (Fig. 4B), which has an additional double bound between C_3 and C_4, is an efficient activator of RARs (Mangelsdorf *et al.*, 1992).

In addition to establishing synthetic routes to generate naturally occurring retinoids (for a review, see Mayer and Isler, 1971; Frickel, 1984), chemists have synthesized hundreds of retinoid analogs (reviewed by

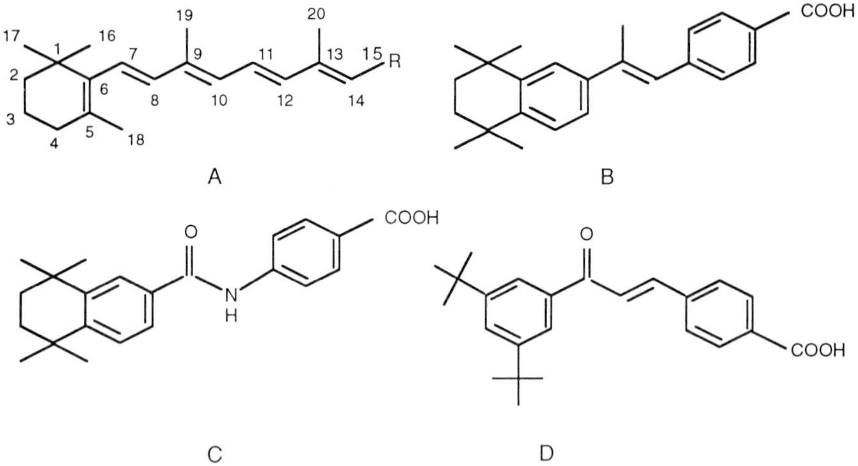

FIGURE 2 Structure of natural (A) and synthetic retinoids (B–D). (A) In the case of retinol R=CH_2 OH, in retinal R=CHO, and in retinoic acid R=COOH. In the case of 4-oxo-retinoic acid, C_4 carries an oxygen atom (=O). (B) TTNPB (*p*-[(*E*)-2-(5,6,7,8-tetrahydro-5,5,8,8-tetramethyl-2-naphthyl)propenyl]benzoic acid). (C) Am 580 (*p*-[(*E*)-2-(5,6,7,8-tetrahydro-5,5,8,8-tetramethyl-2-naphtalene-carboxyamido)]benzoic acid). (D) (*E*)-4-[3-(3,5-Di-*tert*-butylphenyl)-3-oxo-1-propenyl]benzoic acid).

Dawson and Hobbs, 1990; Klaus, 1990; Sani and Hill, 1990). The main reason for these efforts is to improve the therapeutic usefulness of retinoids while simultaneously reducing side effects such as embryo toxicity. Moreover, it is expected that synthetic retinoids can be made that are selective for a particular receptor type and, more generally speaking, for a specific therapeutic application. Typical structures of synthetic retinoids are shown in Fig. 2B–D. Similar to natural retinoids, synthetic analogs consist of a lipophilic moiety at one end of the molecule, a polar group at the opposite end, and a largely hydrophobic spacer connecting these two parts. In most analogs, the spacer is sterically restrained. This results in molecules in which the two end moieties are in a well-defined relative spatial orientation.

Taken together, despite the fact that McCollum and Davis discovered retinoids 80 years ago, the mechanism of action of these broadly acting substances is not yet fully understood. However, it is now clear that, except for vision, retinoids operate at the level of gene expression through specific nuclear receptors. One of the challenges is to identify these genes and elucidate their physiological function. Tools are now at hand to tackle these issues in a productive way. In what follows, we discuss some of the recent advances in this field with a particular emphasis on retinoid receptors, new retinoids, and the role of retinoids in embryonic development. A comprehensive review of various aspects of retinoid action can be found in Sporn *et al.* (1984) and in a volume edited by Saurat (1991). Research on the transport and storage of retinoids has been reviewed by Blomhoff *et al.* (1990). For methods in retinoid research, the reader may consult two recent volumes of *Methods in Enzymology* edited by Packer (1990).

II. RETINOID RECEPTORS

Several laboratories almost simultaneously discovered that retinoic acid operates by binding to specific nuclear receptors that are transcription factors regulating target gene expression (Fig. 1). Retinoid receptors show striking structural and functional homologies to steroid hormone receptors and belong to the gene superfamily of nuclear hormone receptors (for a review, see Chambon *et al.*, 1991). Therefore, we can consider the action of retinoic acid in terms of a signal transduction process: Ligand (the signal) interacts with receptor (the sensor), which then converts the input into a physiological response in the form of gene expression (Fig. 1). Obviously, retinoic acid action is far more complicated than this simple model would predict. The effects of retinoids vary greatly with dose, the cell type, the retinoid used, the duration of treatment, etc. What mechanisms would allow a relatively simple small molecule compound to have such diverse effects?

1. There are two families of retinoid receptors (the RARs and the RXRs), and each family has three members (RAR-α, RAR-β, and RAR-γ; RXR-α, RXR-β, and RXR-γ). Each of the RARs and RXRs can give rise to so-called isoforms that differ in the amino-terminal region. Moreover, RARs and RXRs are capable of forming heterodimers with each other, and RXRs form heterodimers with other members of the nuclear receptor superfamily (Yu *et al.*, 1991; Kliewer *et al.*, 1992a; Leid *et al.*, 1992; Zhang *et al.*, 1992a).
2. Expression studies demonstrate that the various types of receptors as well as their isoforms are differentially expressed.
3. Retinoid-responsive elements to which receptors bind exhibit sequence diversity (e.g., Umesono *et al.*, 1991).
4. There are several retinoids capable of binding to RARs and/or RXRs (see Section III).
5. Cells differ in their levels of cellular retinoic acid-binding protein (CRABP), which is required for most retinoid-dependent processes. CRABP binds retinoic acid with high affinity, and its concentration in the cell might determine the concentration of free retinoic acid available to the RARs and RXRs (for a discussion, see Maden *et al.*, 1989; S. M. Smith *et al.*, 1989; Boylan and Gudas, 1991).

This list makes it easy to explain, in general terms, why retinoids have such pleiotropic effects: The combination of several receptors, ligands, and responsive elements could, in principle, account for a very large number of distinct physiological processes. It is important to realize that different ligand–receptor–response element combinations might result in the *same* physiological output. Such a redundancy would make a process less susceptible to disturbances (e.g., mutations in receptors, fluctuations in ligand concentrations). Unfortunately, redundant mechanisms are also more difficult to elucidate, because simple cause–effect relationships are obscured by the alternate pathway(s).

RARs and RXRs have been discovered in all vertebrate species examined to date. In mammals (mouse and/or human), RAR-α, RAR-β, and RAR-γ (Giguère *et al.*, 1987; Petkovich *et al.*, 1987; Benbrook *et al.*, 1988; Brand *et al.*, 1988; Krust *et al.*, 1989; Zelent *et al.*, 1989) and RXR-α, RXR-β, and RXR-γ (Hamada *et al.*, 1989; Mangelsdorf *et al.*, 1990, 1992; Yu *et al.*, 1991; Leid *et al.*, 1992) have been identified. In newts, there is an RAR-δ, which is related to mammalian RAR-γ (Ragsdale *et al.*, 1989). In chick, RAR-β (Noji *et al.*, 1991; Rowe *et al.*, 1991b; Smith and Eichele, 1991) and RXR-γ (Rowe *et al.*, 1991a) have so far been identified. Ellinger-Ziegelbauer and Dreyer (1991) and Blumberg *et al.* (1992) have found RAR-α, RAR-γ, RXR-α, and RXR-γ in *Xenopus*. All retinoid receptors share the structural features characteristic of the members of the nuclear receptor superfamily. Based on comparison of sequence of RARs and

RXRs with each other and with other members of the superfamily, retinoid receptors are organized into several distinct regions (Figs. 1 and 3). The most conserved domain is the DNA-binding region with two zinc-cysteine fingers that interact with the DNA of the response element (Luisi *et al.*, 1991). Linked to the C domain by a "hinge" (D region) is the ligand-binding domain. The hinge region is somewhat less conserved (60–70%), but the ligand-binding domains between receptors within the same family are ≤85% identical. In RARs but not in RXRs, the C-terminus has greatly diverged. Likewise, the N-termini (A and B regions) of RARs and RXRs are divergent (see later). Between species, N- and C-termini of a particular family member are highly conserved. For example, RAR-β sequence identity in the N- and C-termini of chick and human are 85 and 75%, respectively (Smith and Eichele, 1991). The conservation between species of the N- and C-termini of a particular family member indicates that these regions are functionally important. In contrast, divergence of the N- and C-termini within a receptor family indicates that each receptor gene encodes a protein with some unique property.

As if six receptors were not sufficient, nature has created a further level of diversity. It is now well established that most retinoid receptor genes give rise to multiple variants (referred to as isoforms) that differ in their N-termini (A region). Isoforms have been best characterized for RARs in mouse (Giguère *et al.*, 1990; Kastner *et al.*, 1990; Leroy *et al.*, 1991; Zelent *et al.*, 1991; for a review, see Chambon *et al.*, 1991). By combining cDNA library screens and anchored polymerase chain reaction analysis, seven RAR-α, three RAR-β, and seven RAR-γ isoforms were identified in mouse. Isoforms arise by alternative splicing and/or differential promoter usage. The region of divergence between RAR isoforms begins upstream of the 5′ end of the B region. In other words, different isoforms have distinct A regions. Downstream of the divergence site, all isoforms of particular receptor subtypes seem to have identical sequences. Although less well studied, Northern blots and sequence comparisons demonstrate that RXR subtypes also give rise to isoforms (e.g., Leid *et al.*, 1992; Mangelsdorf *et al.*, 1992).

It would seem logical that the structural diversity of retinoid receptors would result in functional differences between receptor isoforms. This evidence is slow in coming but is being investigated using several different approaches. One obvious strategy is to determine the spatiotemporal distributions of the various messenger receptor RNAs (mRNAs). A differential expression would support the idea of a distinct physiological role. *In situ* hybridization analysis of embryos indeed reveals that mRNAs distribution for each receptor is quite different. A detailed description of retinoid receptor expression patterns is beyond the scope of this chapter, and the interested reader should consult the primary literature (Dollé *et al.*, 1989, 1990; Osumi-Yamashita *et al.*, 1990;

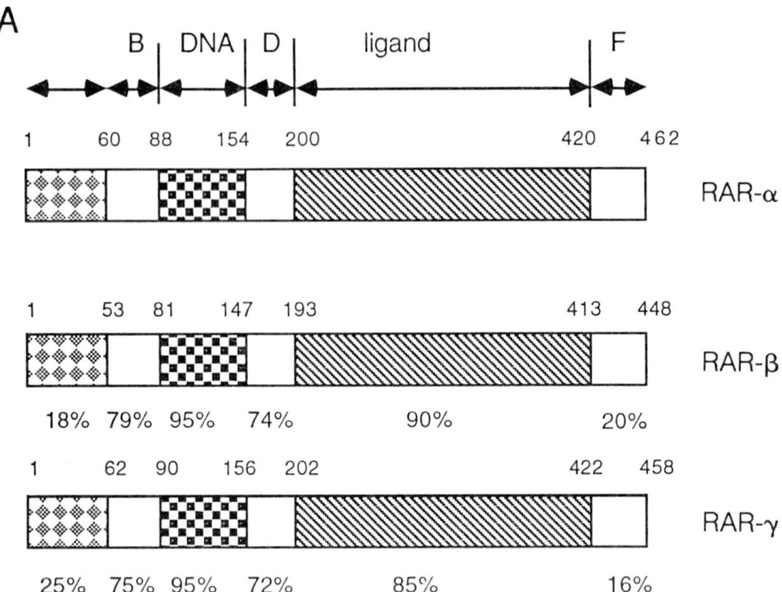

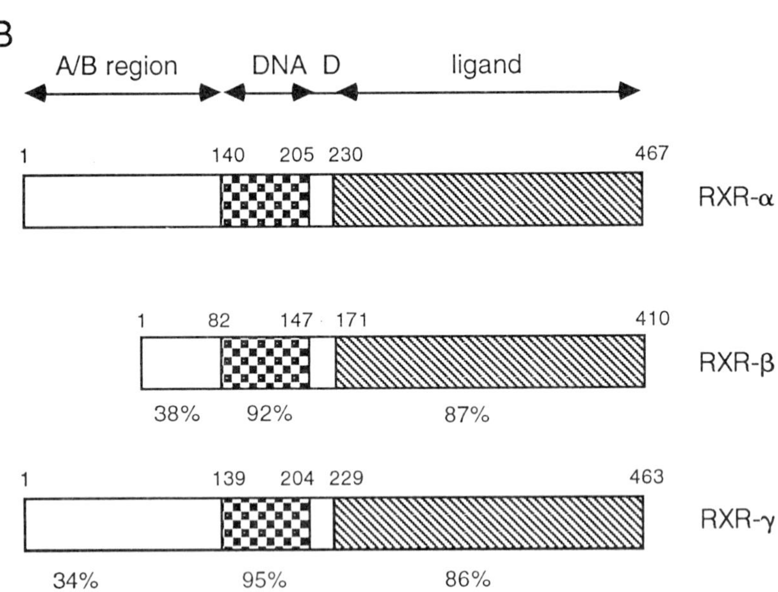

FIGURE 3 Amino acid sequence comparisons between murine retinoic acid receptors (RARs) (A) and retinoid X receptors (RXRs) (B). Percentage identities are shown with respect to RAR-α and RXR-α, respectively. In the case of RARs, the sequences shown are those of (m)RAR-α1, mRAR-β2, and mRAR-γ1 isoforms. Sequence numbering and alignments are as described in Chambon *et al.* (1991) (RARs) and Mangelsdorf *et al.* (1992) (RXRs).

Ruberte *et al.*, 1990, 1991; Ellinger-Ziegelbauer and Dreyer, 1991; Mendelson *et al.*, 1991; Noji *et al.*, 1989, 1991; Rowe *et al.*, 1991b; Smith and Eichele, 1991). In mouse embryos, for example, RAR-α is expressed in many tissues, sometimes at low levels (Dollé *et al.*, 1990). RAR-β and RAR-γ each have transcript distributions that are overlapping but nonidentical. In the early developing limb, RAR-α and RAR-γ are expressed throughout the limb bud mesenchyme, while RAR-β transcripts are found in most proximal mesenchyme. Later in limb development, when cartilage begins to form, RAR-α is still uniformly expressed but RAR-γ is found in the precartilagenous blastemata, suggesting some role of this receptor in bone formation. In contrast, RAR-β is excluded from regions of cartilage condensation. Regions in the limb rudiment undergoing programmed cell death tend to express RAR-β. Smith and Eichele (1991) provided evidence indicating that in chick RAR-β isoforms display somewhat different expression patterns: RAR-β2 mRNA is restricted to a subset of those cells that express RAR-β1.

Expression patterns of RXRs in embryos have recently been examined by Rowe *et al.* (1991a), Blumberg *et al.* (1992), and Mangelsdorf *et al.* (1992). RNase protection assays of RNA isolated from *Xenopus* embryos show that RXR-α and RXR-γ are expressed during oogenesis and in early cleavage stages but rapidly disappear prior to gastrulation (Blumberg *et al.*, 1992). This temporal expression pattern is indicative for RXR-α and RXR-γ being maternal transcripts. Later in development (tailbud stage), the RXR-α gene resumes expression and transcript levels gradually increase. Northern blots of mouse embryos reveal RXR expression at least from Day 10.5 onward (Mangelsdorf *et al.*, 1992). *In situ* hybridization studies of Day 16.5 embryos show that RXR-α is expressed mainly in the epithelia of intestine, the liver, and in skin. There is little expression in the nervous system and in the skeletal elements. RXR-β transcripts are detected in almost all tissues including the central nervous system. RXR-γ mRNA is much more restricted. High levels of expression are seen in the *corpus striatum,* in the pituitary, and in the neck and the skeletal muscles. In earlier embryos, RXR-γ signal is also seen in somites and in the ventral horn region of the spinal cord. Rowe *et al.* (1991a) report high RXR-γ expression in liver and in tissues of the developing peripheral nervous system of chick.

In summary, the various RAR and RXR transcripts are often expressed in distinct and distinguishable patterns. This lends support to the idea that the corresponding proteins have a specific biological function. Expression domains, however, often overlap, raising the possibility that two or perhaps even more receptor variants could have the *same* function (redundancy). To determine which of the two mechanisms applies to a particular region of expression, it will be necessary to knock out transcripts selectively using a gene targeting approach.

III. NEW RETINOIDS

As already pointed out, tissues contain numerous retinoids. Most of them appear not to interact directly with RARs or RXRs but either serve as precursors or are catabolites of all-*trans*-retinoic acid. Retinol, for example, a key precursor of all-*trans*-retinoic acid is incapable of directly activating RARs. Recent work has resulted in the identification of new naturally occurring retinoids that activate RARs and/or RXRs as effectively as all-*trans*-retinoic acid. Of particular relevance is the finding that 9-*cis*-retinoic acid is a high-affinity ligand of the RXR family (Heymann *et al.*, 1992; Levin *et al.*, 1992). When RXR-β, the first member of the RXR subfamily, was discovered, it was noted that this receptor is readily activated by all-*trans*-retinoic acid (Mangelsdorf *et al.*, 1990). However, it seemed that the all-*trans* isomer has a very low affinity for RXR. One possible explanation for this finding is that RXR has no ligand on its own but, rather, is activated through heterodimer interaction with, for example, one of the RARs. An alternative explanation assumes that all-*trans*-retinoic acid added to the cells is converted to a metabolite that binds to RXR. If so, it should be possible to isolate metabolites from cells treated with all-*trans*-retinoic acid and then demonstrate that one of these metabolites is capable of binding to RXR. Two research groups have independently identified 9-*cis*-retinoic acid (Fig. 4A) as a physiological ligand of RXR. In the following, we summarize the approaches used that led to this discovery.

Heymann *et al.* (1992) incubated either CV-1 kidney cells or *Drosophila* S2 Schneider cells with 1 μM all-*trans*-retinoic acid. After 24 hr, cells were pelleted and retinoids were extracted with organic solvent. The extract was fractionated over a reversed phase high-performance liquid chromatography (HPLC) column, and individual fractions were assayed for the ability to activate a RXR-α-dependent reporter plasmid. This resulted in the identification of a distinct activity peak different from that of all-*trans*-retinoic acid. Subsequent mass spectroscopic analyses, chemical derivatization reactions, and chromatography with authentic standards led to the identification of 9-*cis*-retinoic acid as the active agent. Importantly, binding studies revealed that 9-*cis*-retinoic acid binds to RXR-α with high affinity ($K_d \sim 12$ nM).

An alternative approach was chosen by Levin *et al.* (1992), who overexpressed RXR-α in COS-1 cells and cultured these cells in the presence of radioactive all-*trans*-retinoic acid. If cells were to generate a ligand for RXR-α, then this ligand would bind to the receptor. Upon extraction of nucleosol fractions with organic solvent, the "trapped" ligand would be released from RXR and appear as a peak in a subsequent HPLC analysis. Using this strategy, Levin *et al.* (1992) were able to recover a ligand that they subsequently identified as 9-*cis*-retinoic acid. Al-

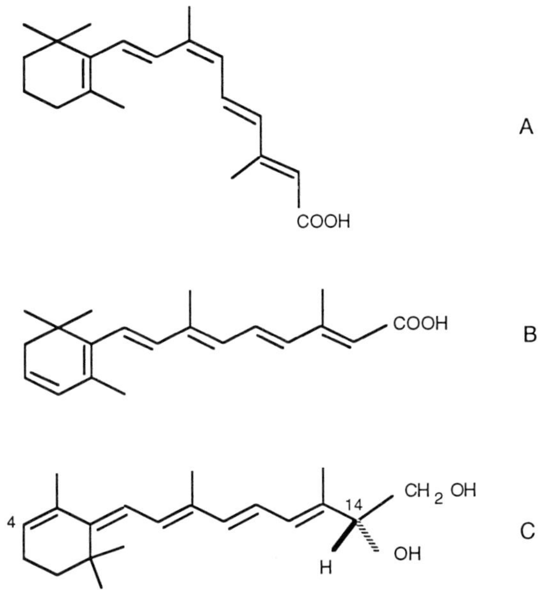

A

COOH

COOH

B

CH 2 OH

14

4

C

OH

H

FIGURE 4 Structures of newly identified natural retinoids. (A) 9-*cis*-retinoic acid. (B) 3,4-Didehydro-all-*trans*-retinoic acid. (C) 14-Hydroxy-4,14-*retro*-retinol. Note that the latter compound has a chiral C_{14}.

though the initial 9-*cis*-retinoic acid activation experiments were carried out with RXR-α, it is now clear that 9-*cis*-retinoic acid also activates RXR-β and RXR-γ (Mangelsdorf *et al.*, 1992). Several laboratories have identified so-called orphan receptors, that is, nuclear receptors whose ligands are yet unknown but based on DNA sequence belong to the steroid–thyroid–retinoid–vitamin D receptor superfamily. Orphan receptors are found throughout the animal kingdom and presumably have important regulatory functions. It is likely that the strategies successful in finding a ligand for RXR can be adapted to searching for ligands of orphan receptors.

As already pointed out, all-*trans*-retinoic acid does not bind to RXRs, and activation of this receptor required conversion to 9-*cis*-retinoic acid. In contrast, 9-*cis*-retinoic acid activates both RXRs and RARs. One possible explanation is that 9-*cis*-retinoic acid is isomerized to all-*trans*-retinoic acid during the incubation of cells with ligand. Alternatively, 9-*cis*-retinoic acid may directly bind to RARs. Thus, *in vitro* 9-*cis*-retinoic acid is presumably a bifunctional ligand capable of activating two distinct retinoid receptor pathways. To what extent this capability is operative in a physiological context is an important question. To address this issue, it will be necessary to synthesize retinoids selective for RARs or RXRs.

Such synthetic ligands should, in principle, allow one to distinguish between a RAR- or RXR-mediated process. A complication is that because RARs and RXRs can form heterodimers some of the retinoid-dependent processes may require both receptors.

An obvious question is whether function of RXR homodimers and of heterodimers between RXR and other receptors depends on the presence of 9-*cis*-retinoic acid. Two recent reports provide some insights into this problem. Zhang *et al.* (1992b) reported that RXR-α itself binds poorly to a palindromic thyroid hormone response element (TRE_p is an artificial response element), but addition of either RAR or thyroid hormone receptor results in binding. This is consistent with the findings of others that RXR heterodimers display enhanced binding to response elements and that RXR can function as an auxiliary receptor even in the absence of 9-*cis*-retinoic acid (Yu *et al.*, 1991; Kliewer *et al.*, 1992a; Leid *et al.*, 1992; Zhang *et al.*, 1992a). Interestingly, however, RXR-α will also effectively bind to TRE_p if 9-*cis*-retinoic acid is present. Similarly, effective RXR-α binding to the retinoid-responsive element of cellular retinol binding protein II (identified by Mangelsdorf *et al.*, 1991) requires 9-*cis*-retinoic acid. Moreover, this isomer will induce RXR homodimer formation in solution, even in the absence of an appropriate response element. Kliewer *et al.* (1992b) explored how 9-*cis*-retinoic acid affects the interaction between RXR-α and peroxisome proliferator-activated receptor (PPAR), a receptor involved in peroxisome proliferation and, thus, in lipid metabolism (for a discussion on PPAR, see Issemann and Green, 1990). It was found that in cells expressing RXR-α–PPAR heterodimers, an appropriate reporter gene carrying a PPAR-binding site, is activated either by 9-*cis*-retinoic acid or clofibric acid (a compound that activates PPAR; see Issemann and Green, 1990; Dreyer *et al.*, 1992). This finding and those by Zhang *et al.* (1992b) suggest that 9-*cis*-retinoic acid is required for transcriptional activations that are regulated either by RXR homodimers or, as in the case of PPAR, by RXR heterodimers. In conclusion, RXRs seem to have a dual mechanism of action, and they can operate in a 9-*cis*-retinoic acid-dependent manner but also as auxiliary receptors that do not require ligand.

In addition to 9-*cis*-retinoic acid, all-*trans*-3,4-didehydroretinoic acid is also a biologically active retinoid. 3,4-Didehydro retinoids that have an additional double bond in the ionylidene ring between carbons 3 and 4 (Fig. 4B). Although 3,4-didehydroretinol, the precursor compound, has long been known to occur, for example, in fish liver, it was not until recently that 3,4-didehydroretinoic acid was identified as a physiological substance present in chick embryos and chick limb buds (Thaller and Eichele, 1990). In the chick digit pattern duplication assay (see later), 3,4-didehydroretinoic acid and all-*trans*-retinoic acid are equipotent. Embryonic tissues from chick are capable of forming didehydroretinol from retinol (Wagner *et al.*, 1990), indicating that retinol is the precursor of didehydroretinoids (Thaller and Eichele, 1990).

At present, it is not clear whether didehydroretinoids are unique to birds. Human serum and mouse embryos (G. Eichele and C. Thaller, unpublished data) do not contain detectable levels of didehydroretinoids. However, Vahlquist (1980) provided evidence for the presence of didehydroretinol in human skin. It may be that in mammals didehydroretinoids are found only in certain tissues where they play the role of a local mediator.

Buck *et al.* (1991) recently found that retinol is essential for the growth of lymphocytes in culture. Unlike for most other retinoid-dependent processes, retinoic acid apparently cannot substitute for retinol in this system. This finding suggests that retinol or, more likely, a metabolite of retinol other than retinoic acid has growth-promoting activity in lymphocytes. To identify the responsible agent(s), lymphoblastoid cells (or HeLa cells) were treated with retinol bound to retinol-binding protein. The resulting metabolites were extracted and fractionated by HPLC. This allowed the isolation of one of the major retinol metabolites in amounts sufficient for chemical identification. Using a combination of spectroscopic techniques, this metabolite was identified as 14-hydroxy-4,14-*retro*-retinol (Fig. 4C). This retinoid is not only found in lymphoblastoid cells but also in fibroblasts, B lymphocytes, and even in cells from *Drosophila*. 14-Hydroxy-4,14-*retro*-retinol is quite abundant, accounting for 0.5–9% of total cellular retinol. Taken together, these findings suggest that 14-hydroxy-4,14-*retro*-retinol and possibly also other *retro*-retinoids are general intracellular mediators whose physiological function has been conserved over a long evolutionary period. It will be important to determine the mechanism of action of *retro*-retinoids (i.e., to find out whether there are specific receptors and how these compounds feed into the mechanisms that control cell growth).

In summary, modern analytical methods have made it possible to search for new retinoids. Remarkably, the newly discovered retinoids are not, as one might have expected, a minor variation of an already known theme but exhibit novel and largely unexpected physiological functions.

IV. RETINOIDS IN DEVELOPMENT

One of the reasons for the broad interest in retinoids relates to their presumed role in vertebrate morphogenesis. In the following, we briefly review some recent research on the mechanism of action of retinoids in development.

To generate a complex organism from a fertilized egg, two types of processes must take place. First, the egg must divide many times and give rise to a large number of different cell types that make up an organism (e.g., neurons, chondrocytes, muscle cells, epidermis, blood cells). The establishment of cell type diversity is known as *histogenesis.* In an organism, cells are not randomly intermingled but, rather, form highly organized tissues and structures, characteristic for a particular species. For example, long bones and cartilage in limbs generate specific patterns, the digits in a hand are arranged in a particular sequence, and individual nerves of the nervous system are connected in a particular way. The establishment of ordered biological structures is referred to as *pattern formation.* It seems that retinoids are involved in histogenesis as well as pattern formation. The role of retinoids in pattern formation has been investigated most extensively in the establishment of the anteroposterior axis of vertebrate limbs and in the formation of the primary body axis. Other developmental processes involving retinoids are urodele limb regeneration (see, e.g., Maden, 1982; Mohanty-Hejmadi *et al.*, 1992; Monkemeyer *et al.*, 1992) and eye–visual system development (Manns and Fritzsch, 1991, 1992; McCaffery *et al.*, 1992).

A. Retinoids and the Anteroposterior Pattern in Vertebrate Limbs

Vertebrate limbs develop from small buds protruding from the embryonic flank. Limb buds consist of an ectodermal hull filled with mesenchymal cells (Fig. 5A). In the course of a few days, the mesenchyme develops into the intricate pattern of a vertebrate extremity. Limbs have three axes. The proximodistal axis extends between the hand plate and the shoulder, the anteroposterior axis extends between the thumb and the little finger, and the dorsoventral axis is perpendicular to the two other axes. In the case of a chick wing bud, the mesenchyme begins to differentiate terminally around Day 4.5 of development, and within a week a wing forms that consists of humerus, radius, and ulna and three distinguishable digits. Digit 2, the smallest digit, is located anteriormost, followed by digit 3, and digit 4 is situated at the posterior wing margin. The normal pattern along the anteroposterior limb axis can be changed by applying retinoic acid from a bead implanted at the anterior limb bud margin (Fig. 5B) (Tickle *et al.*, 1982, 1985; Summerbell, 1983). Retinoic acid turns a normal 2-3-4 digit pattern into a 4-3-2-2-3-4 mirror-symmetrical duplication. The concentration of applied retinoic acid required in the limb bud tissue to induce duplication is in the range 20–50

FIGURE 5 (A) Chick wing development. lb, leg bud; wb, wing bud. (B–D) Experimental manipulations that affect the wing pattern. A crude fate map is superimposed onto the wing bud. The presumptive regions shown are humerus (h); radius (r); ulna (u); and digits 2, 3, and 4 (2, 3, and 4, respectively). Anteroposterior and proximodistal axes are indicated. RA, retinoic acid; ZPA, zone of polarizing activity.

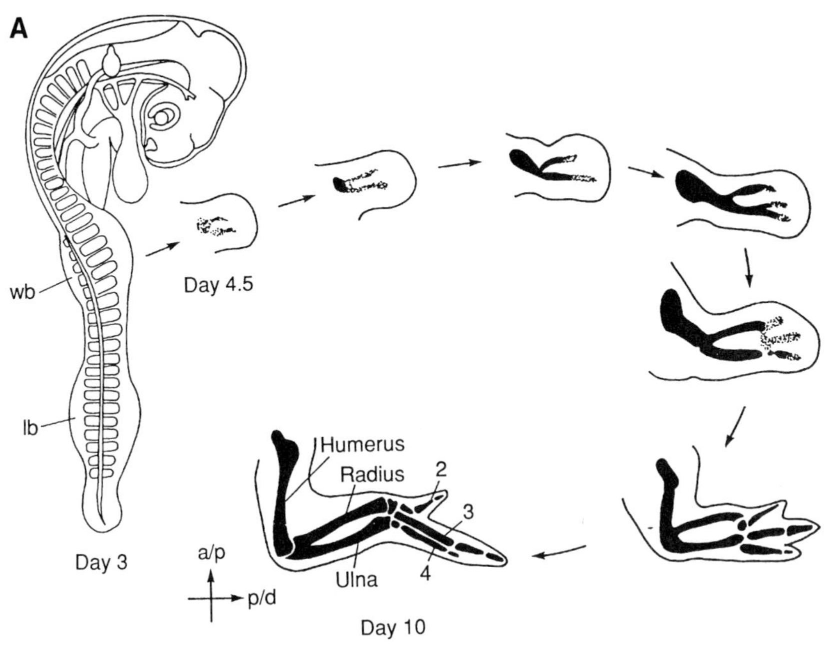
A
wb
lb
Day 3
Day 4.5
Humerus
Radius
2
3
4
Ulna
a/p
p/d
Day 10

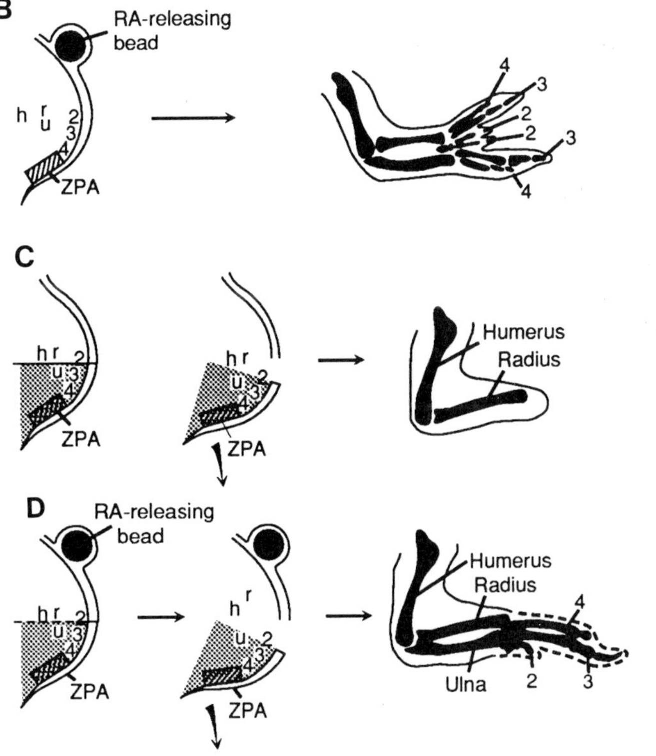
B
RA-releasing
bead
h
r
u
2
3
4
ZPA
4
3
2
2
3
4
C
h r
u
2
3
4
ZPA
ZPA
Humerus
Radius
D
RA-releasing
bead
h r
u
2
3
4
ZPA
ZPA
Humerus
Radius
Ulna
4
2
3

n*M* (Thaller and Eichele, 1987). This is close to the concentration of endogenous retinoic acid (Thaller and Eichele, 1987, 1990). Duplications also develop after grafting to the same anterior position a piece of posterior limb bud tissue that is known as the zone of polarizing activity (ZPA) (Saunders and Gasseling, 1968; Tickle *et al.*, 1975). It has been suggested that the ZPA releases a morphogen that specifies the limb pattern along the anteroposterior axis (Wolpert, 1969). Because retinoic acid mimics the ZPA, it is possible that retinoic acid is the hypothetical ZPA morphogen (Tickle *et al.*, 1982). A wing lacking digits (Fig. 5C) develops after removing the posterior limb bud half, which includes the ZPA and the regions destined to form the ulna and the digits (Hinchliffe and Gumpel-Pinot, 1981). However, either application of retinoic acid (Eichele, 1989) or grafting a ZPA (Saunders and Gasseling, 1968) will restore in the residual anterior half limb bud an essentially complete wing with humerus, forearm, and digits 2, 3, and 4 (Fig. 5D). Because the inducing agent comes from the anterior limb bud margin and not as in the normal limb from the posteriorly located ZPA, the rescued wing reveals a reverse digit configuration (4-3-2, not 2-3-4). Thus, retinoic acid "transforms" anterior cells into posterior cells capable of forming digits.

Experiments in which natural and synthetic retinoids were applied to chick wing buds demonstrate that induction of digits is a dose- and time-dependent process (Summerbell, 1986; Eichele *et al.*, 1985; Tickle *et al.*, 1985). Low doses of, for example, retinoic acid induce a pattern with an additional digit 2 (2-2-3-4 pattern), intermediate doses create a pattern with extra digits 2 and 3 (3-2-2-3-4 or 3-2-3-4 patterns), and higher doses create patterns such as 4-3-2-2-3-4 or 4-3-2-3-4, that is, patterns with additional digits 2, 3, and 4.

One obvious question is whether applied retinoic acid immediately affects limb development or induction of additional digits requires prolonged exposure to retinoic acid. An answer to this question comes from studies in which a retinoic acid-releasing bead was implanted and subsequently removed after different times (Eichele *et al.*, 1985). These investigations revealed that treatment with retinoic acid needs to persist for at least 10 hr in order to create an additional digit 2, about 14 hr to induce an extra digit 3, and approximately 18 hr to induce an additional digit 4. One plausible explanation for this delay is that retinoic acid released from the bead first induces a new ZPA in the surrounding host tissue. In support of this, Wanek *et al.* (1991) reported that tissue next to the bead, when transplanted into another host wing bud, will induce a digit pattern duplication in the new host. Excluding the trivial explanation that retinoic acid was transferred with the graft, this experiment argues in favor of a mechanism in which applied retinoic acid first generates a new ZPA. If that were the case, one would expect that a ZPA graft would induce digits more rapidly than a retinoic acid bead. A difficulty in first grafting and subsequently removing ZPA is that grafted ZPA tissue will tightly integrate into the host limb bud. To avoid this problem, Smith (1980) implanted ZPA irradiated with X rays. Irradiation does not abolish the digit-inducing capacity of ZPA, but the graft no longer integrates into the host limb bud and, thus, can later be removed. These studies established that ZPA shows a kinetics of digit induction similar to that of retinoic acid-releasing beads; therefore, it may not be necessary that retinoic acid first induces a new ZPA (unless, of course, one assumes that the irradiated graft first induces a new ZPA in the host as well). It goes without saying that this conclusion does not invalidate the experimental findings by Wanek *et al.* (1991). A unique property of the most posterior tissue is its ability to act as a ZPA. Because retinoic acid converts anterior tissue into posterior tissue (see Fig. 5D), it is not entirely unexpected that a new ZPA could arise. In conclusion, the evidence for direct action of retinoic acid in wing bud pattern formation remains somewhat circumstantial. One strategy to resolve the issue is to identify retinoic acid target genes in the limb bud. Knowing where these genes are expressed and what cellular function they have would help to argue more clearly for or against a direct mode of action.

B. Retinoids and the Formation of the Primary Body Axis of Vertebrates

One of the hallmarks of the external morphology of vertebrates is the distinct anteroposterior (head-to-tail) axis. Likewise, internal organs such as the nervous system and the digestive tract also display pronounced anteroposterior specializations. Although slightly less obvious than in insects, vertebrates have a metameric body plan (i.e., an organization in which a structural unit is present in multiple copies along the anteroposterior axis). Examples of metamerism are the vertebrae that constitute the vertebral column or the rhombomeres, transient segments found in the developing hindbrain of all vertebrate embryos (for a review, see Keynes and Lumsden, 1990). Anteroposterior polarity is seen almost from the inception of embryonic development, at the time when the primitive streak is formed (Bellairs, 1986). A fundamental question is how this initial polarity is established and how it develops into the distinct anteroposterior pattern of, for example, the axial skeleton or the nervous system. Having discussed the evidence for an involvement of retinoic acid in the formation of the anteroposterior limb axis, we now examine the role of retinoids in patterning of the anteroposterior body axis.

One of the important advances in modern developmental biology was the discovery of genes that contain a homeobox (for a review, see, e.g., Gehring, 1987). The homeobox is a 180-basepair DNA motif encoding a 60-amino acid protein domain that folds into a helix-turn-helix motif capable of sequence-specific DNA binding (Quian *et al.*, 1989). There are now well over a hundred genes identified that contain a homeobox (for a review, see Scott *et al.*, 1989). Of relevance for the subsequent discussion is a particular class of homeobox genes known as Hox genes, whose homeobox is closely related to the archetypal homeobox of the Antennapedia gene of *Drosophila.* In insects and vertebrates, this highly conserved family of transcription factors plays a key role in determining the position of structures along the anteroposterior body axis (for a review, see McGinnis and Krumlauf, 1992). Hox genes are arranged in four clusters (loci), and between clusters there are equivalent genes (so-called paralog groups) that are highly conserved (Fig. 6).

Hox genes are expressed in vertebrate embryos, and expression seems to follow the co-linearity rule, first observed in *Drosophila* (Lewis, 1978). This rule states that along the anteroposterior body axis, 3′ genes in each of the Hox clusters (Fig. 6, right side) are expressed in more anterior regions of the embryo, while 5′ genes are expressed more posteriorly. For example, the anterior boundary of the domain of Hox 2.9 expression is in the front part (presumptive hindbrain) of the embryo (see Fig. 8A and C), whereas the anterior boundary of expression of a 5′ gene is in the posterior region of the embryo. It is important to note, though, that Hox expression domains often overlap.

Expression of Hox genes can be influenced by retinoic acid. Initial evidence for this came from studies in which mouse embryonal carcinoma cells were induced to differentiate by treatment with retinoic acid. Levels of various homeobox mRNAs were found to rise during the course of this differentiation process (Simeone *et al.*, 1990, 1991; Papa-

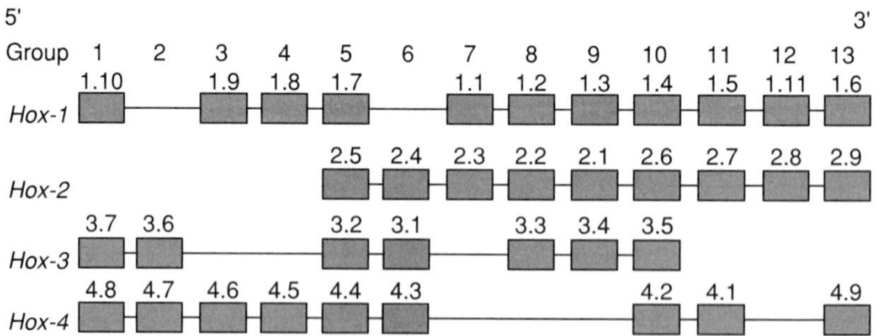

FIGURE 6 Schematic representation of Hox gene clusters of mouse that are located on four different chromosomes. In other higher vertebrates, the relative position of Hox genes is identical to that shown here for mouse. Hox genes are subdivided into 13 groups of paralogs. Within a group, sequences are strongly conserved, suggesting that the four clusters derive from an ancestral locus. In embryos, 3′ genes are expressed earlier than 5′ genes and, at least *in vitro*, 3′ genes but not 5′ are inducible by retinoic acid. The boundary between retinoic acid-responsive genes and nonresponsive genes is at paralog group 5.

lopulu *et al.*, 1991). While a number of homeobox genes respond to treatment with retinoic acid, it has been found that the rate of response varies considerably and correlates with the localization of a particular gene within the cluster. A systematic study of 38 Hox genes in two human embryo carcinoma cell lines (NT2/D1 and Tera-2/clone 13) revealed the following rules (see Simeone *et al.*, 1991): In all four loci, 3′ genes are sequentially activated by retinoic acid in a 3′ to 5′ order (i.e., co-linear with their relative position within the cluster). (Figure 6 shows the four Hox clusters in mouse. Hox clusters in humans and mouse are organized identically, but the nomenclature differs. So as not to confuse the reader, we use mouse nomenclature, although the cell lines studied by Simeone *et al.* [1991] are, of course, human cell lines.) With the exception of genes in the Hox 2 and 4 clusters, 5′ genes are not or are minimally expressed in the EC cells, whether or not retinoic acid is present. The boundary between responsive and nonresponsive genes is approximately at the level of Hox 1.7 and its paralogs (Hox 2.5, 3.2, and 4.4; see Fig. 6). All members of the Hox 2 cluster respond positively to retinoic acid, because in this locus there are no genes 5′ to the response boundary gene (Hox 2.5). Experiments using the sensitive RNase protection assay show that in NT2/D1 cells 5′ members of the Hox 4 cluster (Hox 4.5, 4.6, 4.7, and 4.8) are expressed at low levels in uninduced embryonal carcinoma cells and are down-regulated by retinoic acid. In contrast, in Tera-2/clone 13 cells, expression of Hox 4.5, 4.6, 4.7, and 4.8 is not influenced by retinoic acid. This discrepancy between cell lines, however, vanishes at lower concentrations of retinoic acid. Taken together, these findings suggest that retinoic acid regulates expression of Hox genes in a complex and carefully poised fashion.

These observations made in embryonal carcinoma cells raise the question of whether retinoic acid affects the expression of Hox genes in embryos. This is indeed the case. Several investigators have examined the expression pattern of Hox genes in embryos treated with retinoic acid (e.g., Sive *et al.*, 1990; Morriss-Kay *et al.*, 1991; Ruiz i Altaba and Jessell, 1991a,b; Sharpe, 1991; Sive and Cheng, 1991; Sundin and Eichele, 1992). In the following, we discuss the effect of exogenously provided retinoic acid on (1) the axial organization of the hindbrain and (2) the pattern of vertebrae along the anteroposterior body axis. Since Hox genes are involved in axis formation, one would expect retinoic acid to alter these axial structures.

The chick embryo is well suited to the study of hindbrain morphogenesis because a considerable body of data is available on this subject in this organism (for a review, see Keynes and Lumsden, 1990). Moreover, chick embryos can be removed from the egg and cultured in the presence of retinoic acid. Retinoic acid treatment of primitive streak chick embryos (Hamilton–Hamburger Stage 4) shows few immediate effects but results in pronounced morphological changes as development proceeds to later stages (for details, see Sundin and Eichele, 1992). Embryos are smaller, there is less mesodermal tissue, and the heart is not properly developed (Osmond *et al.*, 1991). In addition, there are considerable changes in the organization of the central nervous system (Fig. 7). The anteriormost structures such as forebrain and eye primordia are somewhat diminished in size, but most striking is the absence of a normal midbrain and anterior hindbrain. Instead of these structures, there is only a short portion of neural tube (Fig. 7D, region between dashed lines). In normal embryos of this stage, several distinct segmental units, known as rhombomeres, are usually visible as a series of constrictions along the axis of the hindbrain (Fig. 7B). In the retinoic acid-treated embryos, these constrictions are either absent or poorly defined. One interpretation of these findings is that upon retinoic acid treatment the entire midbrain and the anterior half of the hindbrain (region between dashed lines in Fig. 7B) have been transformed into the small segment of neural tube seen between the two dashed lines in Fig. 7D. Thus, retinoic acid evokes major structural reorganizations of the chick hindbrain. It should be added that similar findings have been reported for mouse (Moriss-Kay *et al.*, 1991; Conlon and Rossant, 1992).

Given the role of Hox genes in establishing the anteroposterior body axis, the question arises of whether retinoic acid treatment has any effect on the expression of Hox genes. This question has been examined in mouse (Moriss-Kay *et al.*, 1991; Conlon and Rossant, 1992) and chick (Sundin and Eichele, 1992). Particularly interesting are those Hox genes that are normally expressed when the morphologically affected structures are specified. A well-characterized case of an early Hox gene is Hox 2.9, referred to as Ghox 2.9 in the chick. As already pointed out, Stage 4 embryos treated with retinoic acid for 4 hr develop essentially normally for quite some time; however, when these embryos are stained with an antibody to Ghox 2.9 protein, one observes a number of marked changes (compare Figs. 8A and 8B). There is significant ectopic expression of Ghox 2.9 protein in anterior regions (Fig. 8B) where the gene is normally not active (Fig. 8A). Hensen's node and midline structures extending up to the prechordal plate show a striking increase in Ghox 2.9 antibody labeling (Fig. 8B), while there is no induction in the prechordal plate or its overlying ectoderm. When retinoic acid-treated embryos are withdrawn from retinoic acid and allowed to reach the two-somite stage (Stage 7), they continue to develop with a normal timetable of notochord elongation, node regression, and somite formation. The ectoderm and mesoderm of treated embryos are slightly thinner, but key landmarks appear in their normal location and display normal spacing relative to each other. Nevertheless, the pattern of Ghox 2.9 expression remains markedly different from that of untreated embryos. In the nor-

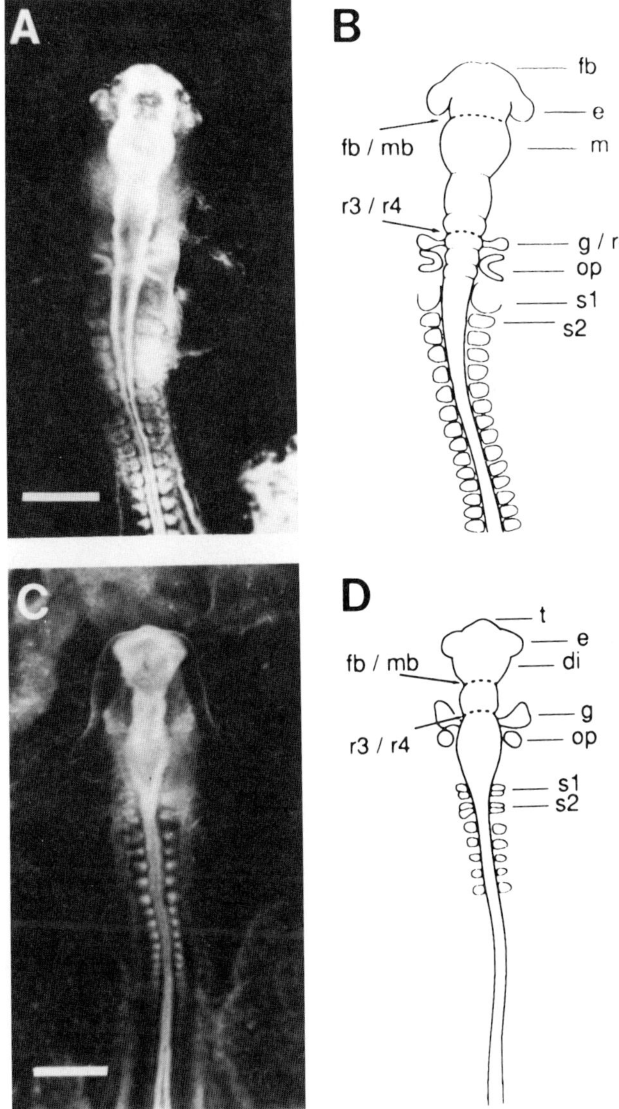

FIGURE 7 Effect of retinoic acid treatment on the morphology of Hamburger-Hamilton Stage 12 embryos. (A) Control embryo grown in culture from Stage 4 to Stage 12. (B) Camera lucida drawing corresponding to (A). Dashed lines indicate the forebrain–midbrain boundary (fb/mb) and the anterior boundary of rhombomere 4 (r3/r4) respectively. (C) Embryo treated at Hamburger–Hamilton Stage 4 with 6 μM retinoic acid for 4 hr. (D) Camera lucida drawing corresponding to (C). Dashed lines are as in (B). Bars, 400 μm. di, diencephalon; e, eye; fb, forebrain; g/r4 (B) and g (D), primordium of acousticofacial ganglion derived from rhombomere 4 neural crest; m, midbrain; op, otic placode; s1, first somite; s2, second somite; t, telencephalon.

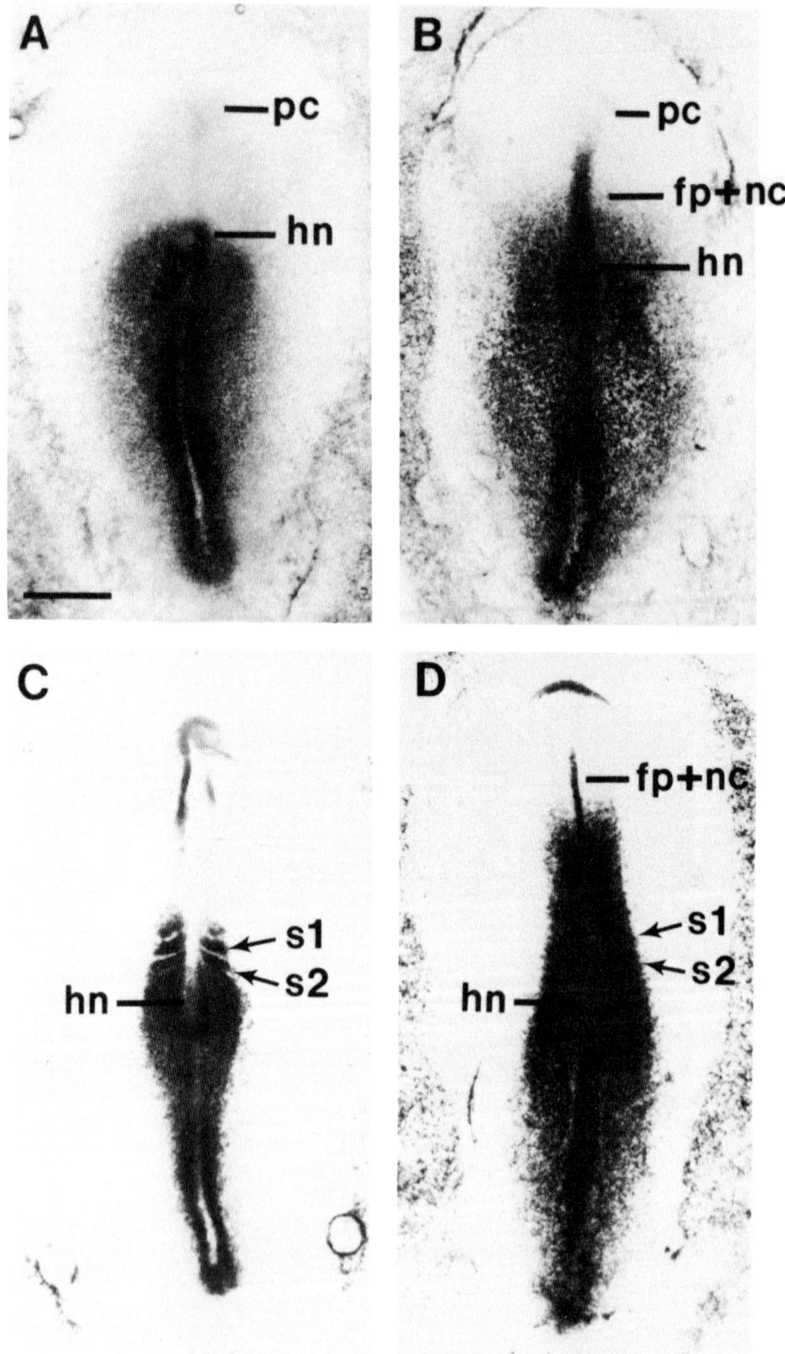
A
—pc
—hn
B
—pc
—fp+nc
—hn
C
s1
s2
hn
D
—fp+nc
s1
s2
hn

mal embryo, Ghox 2.9 expression terminates a short distance anterior to the first somite (Fig. 8C). In contrast, Fig. 8D makes it very clear that in a treated embryo the Ghox 2.9 domain extends far anterior to somite 1. These data show that retinoic acid treatment results in an ectopic expression of Ghox 2.9 in more anterior regions. In other words, upon retinoic acid treatment, cells in anterior hindbrain and midbrain express Ghox 2.9, a gene characteristic for posterior hindbrain. Therefore, the morphological changes shown in Fig. 7D could be the result of a retinoic acid induced anterior to posterior homeotic transformation.

In two elegant studies, Kessel and Gruss (1991) and Kessel (1992) examined the effect of retinoic acid on the morphology of the axial skeleton of mice. Similar to segments of insects, the vertebral column is highly invariant within a strain of mice and, thus, suitable for examining mechanisms of pattern formation. When pregnant mice were treated with retinoic acid (either all-*trans* or 13-*cis* isomer), the pattern of the vertebral column changed in a systematic way. Treatment at Day 6.5 of gestation had no effect on the vertebral pattern. In contrast, treatment around Day 7–7.5 (and also later; see Kessel, 1992) resulted in a series of structural changes including transformations of vertebrae. A striking transformation was noted in the seventh cervical vertebra (C7), which in normal embryos has no ribs. Following retinoic acid treatment around Day 7.3, C7 frequently developed a partial or a complete set of ribs that is typical for T1, the first thoracic vertebra, which lies immediately posterior to C7. This structural change is an example of a homeotic transformation in which an anterior structure acquires a posterior character (Bateson, 1894). Treatment of embryos around Day 8 and later often led to anterior transformations. For example, normally thoracic vertebra T7 had a rib, and thoracic vertebra T8 did not. Upon retinoic acid treatment, half of the mice had a rib on T8. Therefore, the more posterior T8 had acquired the character of the more anterior T7.

Remarkably, these structural changes were accompanied by reprogramming of expression patterns of Hox genes. For example, the anterior boundary of Hox 1.1 was normally in prevertebrae 10/11 (T3/4). Upon retinoic acid treatment, this boundary moved anteriorly to prever-

FIGURE 8 Effect of a 4-hr retinoic acid treatment of Stage 4 chick embryos on the expression of the homeodomain protein Ghox 2.9 (B and D). (A and C). Normal embryos. (B and D) Embryos treated with retinoic acid *in vitro*. The embryo in (B) was stained with anti-Ghox 2.9 antibody after 4 hr of retinoic acid exposure, whereas that in (D) was left to develop in the absence of retinoic acid to Stage 7. Dark areas represent cells expressing Ghox 2.9 protein. Note the distinct "spike" of expression anterior to Hensen's node in (B). Upon further development, this zone further enlarges (D). In control embryos of equivalent age (A and B), the anterior boundary of Ghox 2.9 expression is clearly more posterior than in treated embryos. Bar, 400 μm. fp + nc, floor plate plus notochord; hn, Hensen's node; pc, prechordal plate; s1, first somite; s2, second somite.

tebra 9 (T2). The anterior boundary of Hox 3.1 was normally at prevertebra 12 (T5), and in an exposed embryo this boundary was at prevertebra (T4). Thus, in both cases the expression domain shifted anteriorly by one segment. This induced expression of a posterior gene in a more anterior region was in agreement with the morphological posterior transformation observed in the thoracic region (see earlier). Moreover, these findings were consistent with a model proposed by Lewis (1978), which postulated that a posterior transformation results from a gain-of-function event, which in this case is an ectopic expression of a normally more posterior gene in an anterior region.

Based on studies of expression patterns of Hox genes in normal embryos and in embryos in which Hox expression is altered either by retinoic acid or by transgenic techniques (Kessel *et al.*, 1990), and based on morphological transformations resulting from these interferences, Kessel and Gruss (1991) formulated the *Hox code model.* In essence, this model posits that the identity of a vertebral segment is specified by a particular combination of actively expressed Hox genes. If the Hox code of a prevertebra is altered by ectopic expression of another Hox gene normally not expressed in that region, the Hox code model predicts a homeotic transformation. As already summarized, this seems to be what happens. It goes without saying that more research needs to be done to further substantiate the Hox code model. Apart from additional gain-of-function experiments, good tests are provided by loss-of-function mutations created by gene targeting (Chiasaka and Capecchi, 1991; Lufkin *et al.*, 1991; Le Mouellic *et al.*, 1992). Taken together, these findings led Kessel and Gruss (1991) to the following conclusion concerning the specification of the body axis:

> a retinoic acid signal generated from the midline embryonic structures [e.g., Hensen's node, primitive streak] during gastrulation is received by ingressing cells, which respond with the sequential activation of more and more Hox genes, leading to overlapping, nonidentical expression domains of Hox genes along the anteroposterior axis. The activation of a more 5′ Hox gene defines a more posterior axial level. The combination of functionally active Hox genes, the Hox code, specifies the identity of a body region, for example, a vertebral segment. . . . Hox codes can be altered by exposure to retinoic acid during the specification phase, leading to homeotic transformations. (p. 101)

It goes without saying that a similar line of reasoning could be used to rationalize the function of retinoids in the establishment of hindbrain segments, as discussed earlier.

One of the key questions of the Kessel–Gruss model of body axis formation is which midline structure generates retinoic acid. Recent studies by Hogan *et al.* (1992) suggested that Hensen's node and, to a lesser extent, the primitive streak are sources of retinoic acid. One strategy to search for retinoic acid-producing tissues in embryos is based on the observation that locally applied retinoic acid generates digit pattern duplications in the chick wing bud (see Section IV.A). A simple-minded idea is that tissues capable of inducing duplications may also synthesize retinoic acid. Hornbruch and Wolpert (1986) found that chick Hensen's node (a thickening at the anterior end of the primitive streak; see Fig. 8A) induces digit duplications when grafted to chick wing buds. Hogan *et al.* (1992) found that node from mouse embryos also induces duplications. In 6.5-day-old embryos, the node had relatively little inducing capacity; however, node from 7- to 7.5-day-old embryos were very efficient. Control grafts with tissue taken from very anterior regions of an embryo did not induce pattern duplications. These findings suggested that the node but not anterior tissue of Day 7.5 embryos might be a source of retinoic acid. Incidentally, this is the period when embryos are most responsive to applications of exogenous retinoic acid (see earlier and Kessel and Gruss, 1991). Because the node is very small, it is not possible to directly measure its retinoic acid content. Instead, nodes were dissected from 10–15 embryos and incubated in radioactive retinol, the precursor of retinoic acid. Subsequent HPLC revealed that node tissue effectively produces retinoic acid from retinol (4 pg/μg DNA). In contrast, the biologically ineffective anterior tissue produced an order of magnitude less retinoic acid (0.3 pg/μg DNA). Primitive streak tissue (i.e., tissue located just behind the node) generated intermediate amounts of retinoic acid (~1 pg/μg DNA). This intermediate level is consistent with a moderate duplicating activity of streak tissue in the wing bud assay (Hogan *et al.*, 1992).

Based on these findings, Hogan *et al.* (1992) proposed a model describing how retinoic acid released from the node affects axial patterning. Figure 9 shows a dorsal view of the node. According to this model, the node consists of two types of cells, those migrating into it from the primitive streak (bold arrows) and those that are born in it (Stern, 1979; Selleck and Stern, 1991; Lawson and Pedersen, 1992; Schoenwolf *et al.*, 1992). Mitotic activity studies (Stern, 1979; Lawson and Pedersen, 1992) and dye labeling experiments in the chick (Selleck and Stern, 1991) raise the possibility that the node contains a blastemalike proliferative zone. The model posits that retinoic acid is steadily produced in the node and spreads by diffusion into the surrounding region (Fig. 9, small arrows). Within the node, the product of retinoic acid concentration and time increases with time so that cells that leave the node early will have been exposed to less retinoic acid for a shorter time than cells exiting later. Note that some cells leave the node in an elongated array that gives rise to midline structures (notochord, floor plate, and midline endoderm) and other cells leave laterally to contribute to the medial half of somites (Selleck and Stern, 1991; Schoenwolf *et al.*, 1992). One requirement of this model is that cells need to remember how much retinoic acid they

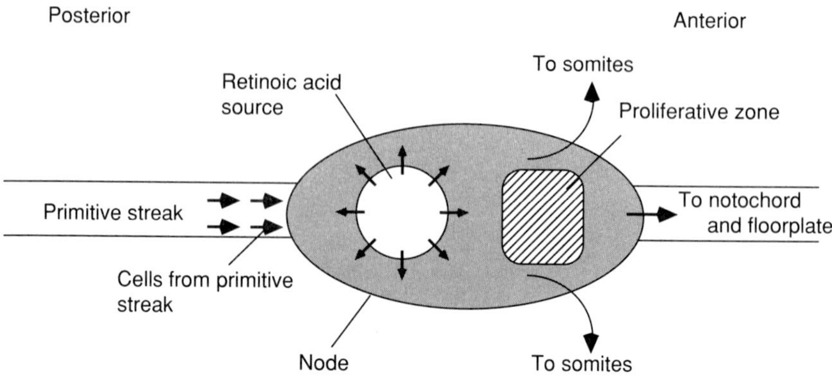

FIGURE 9 Hypothetical mechanism of action of retinoic acid in early embryogenesis. A group of cells in Hensen's node (large oval) produces and releases retinoic acid. Cells surrounding the source are exposed to retinoic acid and begin to express Hox genes. Importantly, at that stage of development, node cells express retinoic acid receptor β (S. M. Smith and Eichele, 1991). Of particular relevance to this model is the proliferative zone. These cells as well as their descendents give rise to midline structures (notochord, floorplate), which are subsequently involved in other patterning processes. Note that cells from the node also contribute to the somites that will later form vertebrae. Cells leaving the node early will form anterior somites; cells that exit later will form more posterior somites. The model suggests that these cells are programmed by retinoic acid to express certain Hox genes. Early in development, these cells are exposed to a low level of retinoic acid for a short time and turn on only 3′ Hox genes. Thus, anterior somites developing from them should express predominantly 3′ Hox genes. This is indeed the case.

(and their ancestors) have been exposed to. This cellular memory could involve chromosomal "imprinting" of Hox genes in cells leaving the node (Gaunt and Singh, 1990). This model is consistent, for example, with the fact that in cultured embryonal carcinoma cells, the expression of 3′ Hox genes is induced by retinoic acid more rapidly and at a lower retinoic acid concentration than more 5′ genes (Simeone *et al.*, 1990, 1991; Papalopulu *et al.*, 1991). In terms of the model, a short exposure of cells to retinoic acid in the node or early primitive streak would first activate transcription of the most 3′ genes and successively longer exposures would turn on more 5′ genes. Although the data base is still somewhat limited, this is exactly what is observed in the embryo (Murphy and Hill, 1991; McGinnis and Krumlauf, 1992). It should be emphasized that the model proposed by Hogan *et al.* (1992) is not meant to imply that retinoic acid from the node is the sole signal required for anteroposterior axial patterning, but that it could be part of a network of interacting factors, also including polypeptide signaling molecules, which have also been proposed to act as morphogens during this process (see Jessell and Melton, 1992).

V. CONCLUSION

In summary, exogenously provided retinoids have profound effects on the development of skeletal structures, the hindbrain, and the limb. Data summarized in Section IV suggest that retinoids have a role in *normal* development. It appears that Hox genes are key mediators of retinoid function in embryogenesis. It is a bit surprising that so far no conclusive data have been reported that would show direct binding of retinoid receptors to regulatory regions of Hox genes. In part, this lack of information may be due to the complex organization of Hox loci. Alternatively, to interact directly with Hox genes retinoid receptors may require other yet unknown auxiliary factors. It is worth pointing out that retinoic acid also influences the expression of other homeobox genes that do not belong to the Hox family. Examples are the *goosecoid* (Cho *et al.*, 1991) and *Xlim-1* (Taira *et al.*, 1992), two genes probably involved in patterning mesoderm during gastrulation.

REFERENCES

Bateson, W. (1894). "Materials for the Study of Variation Treated with Special Regard to Discontinuity of species." Macmillan, London.

Beato, M. (1989). Gene regulation by steroid hormones. *Cell* **56,** 335–344.

Bellairs, R. (1986). The primitive streak. *Anat. Embryol.* **174,** 1–14.

Benbrook, D., Lernhardt, E., and Pfahl, M. (1988). A new retinoic acid receptor identified from a hepatocellular carcinoma. *Nature (London)* **333,** 669–672.

Blomhoff, R., Green, M. H., Berg, T., and Norum, K. R. (1990). Transport and storage of vitamin A. *Science* **250,** 399–403.

Blumberg, B., Mangelsdorf, D. J., Dyck, J. A., Bittner, D. A., Evans, R. M., and De Robertis, E. (1992). Multiple retinoid-responsive receptors in a single cell: Families of RXRs and RARs in the *Xenopus* egg. *Proc. Natl. Acad. Sci. USA* **89,** 2321–2325.

Boylan, J. F., and Gudas, L. J. (1991). Overexpression of the cellular retinoic acid binding protein-I (CRABP-I) results in a reduction of differentiation-specific gene expression in F9 teratocarcinoma cells. *J. Cell. Biol.* **112,** 965–979.

Brand, N., Petkovich, M., Krust, A., Chambon, P., de The, H., Marchio, A., Tiollais, P., and Dejean, A. (1988). Identification of a second human retinoic acid receptor. *Nature (London)* **332,** 850–853.

Buck, J., Derguini, F., Levi, E., Nakanishi, K., and Hämmerling, U. (1991). Intracellular signaling by 14-hydroxy-4,14-*retro*-retinol. *Science* **254,** 1654–1656.

Chambon, P., Zelent, A., Petkovich, M., Mendelsohn, C., Leroy, P., Krust, A., Kastner, P., and Brand, N. (1991). The family of retinoic acid nuclear receptors. *In* "Retinoids: 10 Years On" (J.-H. Saurat, ed.). (pp. 10–27) Karger, Basel.

Chisaka, O., and Capecchi, M. R. (1991). Regionally restricted developmental defects resulting from targeted disruption of the mouse homeobox gene *hox*-1.5. *Nature (London)* **350,** 473–479.

Cho, K. W. Y., Blumberg, B., Steinbeisser, H., and De Robertis, E. M. (1991). Molecular nature of Spemann's organizer: The role of the *Xenopus* homeobox gene *goosecoid*. *Cell* **67,** 1111–1120.

Cohlan, S. Q. (1953). Excessive intake of vitamin A as a cause of congenital anomalies in the rat. *Science* **117,** 535–536.

Conlon, R. A., and Rossant, J. (1992). Exogenous retinoic acid rapidly induces anterior expression of murine Hox-2 genes *in vivo*. *Development* **116,** 357–368.

Dawson, M. I., and Hobbs, P. D. (1990). Synthetic retinoic acid analogs: Handling and characterization. *Methods Enzymol.* **189,** 15–43.

Dollé, P., Ruberte, E., Kastner, P., Petkovich, M., Stoner, C. M., Gudas, L. J., and Chambon, P. (1989). Differential expression of genes encoding α, β and γ retinoic acid receptors and CRABP in the developing limbs of the mouse. *Nature (London)* **342,** 702–705.

Dollé, P., Ruberte, E., Leroy, P., Morriss-Kay, G., and Chambon, P. (1990). Retinoic acid receptors and cellular retinoid binding proteins. I. A systematic study of their differential pattern of transcription during mouse organogenes is. *Development* **110,** 1133–1151.

Dowling, J. E., and Wald, G. (1960). The biological function of vitamin A acid. *Proc. Natl. Acad. Sci. USA* **46,** 587–608.

Dreyer, C., Krey, G., Keller, H., Givel, F., Helftenbein, G., and Wahli, W. (1992). Control of the peroxisomal β-oxidation pathway by a novel family of nuclear hormone receptors. *Cell* **68,** 879–887.

Eichele, G. (1989). Retinoic acid induces a pattern of digits in anterior half wing buds that lack the zone of polarizing activity. *Development* **107,** 863–868.

Eichele, G., Tickle, C., and Alberts, B. M. (1985). Studies on the mechanism of retinoid-induced pattern duplications in the early chick limb bud: Temporal and spatial aspects. *J. Cell Biol.* **101,** 1913–1920.

Ellinger-Ziegelbauer, H., and Dreyer, C. (1991). A retinoic acid receptor expressed in the early development of *Xenopus laevis*. *Genes Dev.* **5,** 94–104.

Evans, R. M. (1988). The steroid and thyroid hormone receptor superfamily. *Science* **240,** 889–895.

Frickel, F. (1984). Chemistry and physical properties of retinoids. *In* "The Retinoids" (M. B. Sporn, A. B. Roberts, and D. S. Goodman, eds.), pp. 7–145. Academic Press, Orlando, Florida.

Gaunt, S. J., and Singh, P. B. (1990). Homeogene expression patterns and chromosomal imprinting. *Trends Genet.* **6,** 208–212.

Gehring, W. J. (1987). Homeoboxes in the study of development. *Science* **236,** 1245–2521.

Giguère, V., and Evans, R. (1990). Identification of receptors for retinoids as members of the steroid and thyroid hormone superfamily. *Methods Enzymol.* **189,** 223–232.

Giguère, V., Ong, E. S., Segui, P., and Evans, R. M. (1987). Identification of a receptor for the morphogen retinoic acid. *Nature (London)* **330,** 624–629.

Giguère, V., Shago, M., Zirngibl, R., Tate, P., Rossant, J., and Varmuza, S. (1990). Identification of a novel isoform of the retinoic acid receptor γ expressed in the mouse embryo. *Mol. Cell. Biol.* **10,** 2335.

Hamada, L., Gleason, S. L., Levi, B.-Z., Hirschfeld, S., Appella, E., and Ozato, K. (1989). H-2RIIBP, a member of the nuclear hormone receptor superfamily that binds to both the regulatory element of major histocompatibility class I genes and the estrogen response element. *Proc. Natl. Acad. Sci. USA* **86,** 8289–8293.

Heyman, R. A., Mangelsdorf, D. J., Dyck, J. A., Stein, R., Eichele, G., Evans, R. M., and Thaller, C. (1992). 9-*cis* retinoic acid is a high affinity ligand for the retinoid X receptor. *Cell* **68,** 397–406.

Hinchliffe, J. R., and Gumpel-Pinot, M. (1980). Control of maintenance and anteroposterior skeletal differentiation of the anterior mesenchyme of the chick wing bud by its posterior margin (the ZPA). *J. Embryol. Exp. Morph.* **62,** 63–82.

Hogan, B. L., Thaller, C., and Eichele, G. (1992). Evidence that Hensen's node is a site of retinoic acid synthesis. *Nature (London)* **359,** 237–241.

Hornbruch, A., and Wolpert, L. (1986). Positional signaling by Hensen's node, when grafted to the chick limb bud. *J. Embryol. Exp. Morph.* **94,** 257–265.

Issemann, I., and Green, S. (1990). Activation of a member of the steroid hormone receptor superfamily by peroxisome proliferators. *Nature (London)* **347,** 645–650.

Jessell, T. M., and Melton, D. A. (1992). Diffusible factors in vertebrate embryonic induction. *Cell* **68,** 257–270.

Kastner, P., Krust, A., Mendelsohn, C., Garnier, J. M., Leroy, P., Staub, A., and Chambon, P. (1990). Murine isoforms of retinoic acid receptor-g with specific patterns of expression. *Proc. Natl. Acad. Sci. USA* **87,** 2700.

Kessel, M. (1992). Respecification of vertebral identities by retinoic acid. *Development* **115,** 487–501.

Kessel, M., and Gruss, P. (1991). Homeotic transformations of murine vertebrae and concomitant alteration of *Hox* codes induced by retinoic acid. *Cell* **67,** 89–104.

Kessel, M., Ballig, R., and Gruss, P. (1990). Variations of cervical vertebrae after expression of a Hox-1.1 transgene in mice. *Cell* **61,** 301–308.

Keynes, R., and Lumsden, A. (1990). Segmentation and the origin of regional diversity in the vertebrate central nervous system. *Neuron* **4,** 1–9.

Klaus, M. (1990). Structure characteristics of natural and synthetic retinoids. *Methods Enzymol.* **189,** 3–14.

Kliewer, S. A., Umesono, K., Mangelsdorf, D. J., and Evans, R. M. (1992a). The retinoid X receptor interacts directly with nuclear receptors involved in retinoic acid, thyroid hormone and vitamin D_3 signaling. *Nature (London)* **355,** 446–449.

Kliewer, S. A., Umesono, K., Noonen, D. J., Heyman, R. A., and Evans, R. M. (1992b). Convergence of 9-*cis*-retinoic acid and peroxisome proliferator signalling pathways trough heterodimer formation of their receptors. *Nature (London)* **358,** 771–774.

Kochhar, D. M. (1967). Teratogenic activity of retinoic acid. *Acta Path. Microbiol. Scand.* **70,** 393–404.

Krust, A., Kastner, P., Petkovich, M., Zelent, A., and Chambon, P. (1989). A third human retinoic acid receptor, hRAR-γ. *Proc. Natl. Acad. Sci. USA* **86,** 5310–5314.

Lawson, K. A., and Pedersen, R. A. (1992). Clonal analysis of cell fate during gastrulation and early neurulation in the mouse. *In* "Post Implantation Development in the Mouse" (D. J. Chadwick and J. Marsh, eds.), pp. 3–26. Ciba Foundation Symposium 165. Wiley, Chichester.

Leid, M., Kastner, P., Lyons, R., Nakshatri, H., Saunders, M., Zacharewski, T., Chen, J.-Y., Staub, A., Garnier, J.-M., Mader, S., and Chambon, P. (1992). Purification, cloning and RXR identity of the HeLa cell factor with which RAR or TR heterodimerizes to bind target sequences efficiently. *Cell* **68,** 377–396.

LeMouellic, H., Lallemand, Y., and Brûlet, P. (1992). Homeosis in the mouse induced by a null mutation in the *Hox-3.1* gene. *Cell* **69,** 251–264.

Leroy, P., Krust, A., Zelent, A., Mendelsohn, C., Garnier, J. M., Kastner, P., Dierich, A., and Chambon, P. (1991). Multiple isoforms of the mouse retinoic acid receptor alpha are generated by alternative splicing and differential induction by retinoic acid. *EMBO J.* **10,** 59.

Levin, A. A., Sturzenbecker, L. J., Kazmer, S., Bosakowski, T., Huselton, C., Allenby, G., Speck, J., Kratzeisen, C., Rosenberger, M., Lovey, A., and Grippo, J. F. (1992). 9-*cis*-retinoic acid stereoisomer binds and activates the nuclear receptor RXRα. *Nature (London)* **355,** 359–361.

Lewis, E. B. (1978). A gene complex controling segmentation in *Drosophila. Nature (London)* **276,** 565–570.

Lufkin, T., Dierich, A., LeMeur, M., Mark, M., and Chambon, P. (1991). Disruption of the Hox-1.6 homeobox gene results in defects in a region corresponding to its rostral domain of expression. *Cell* **66,** 1105–1119.

Luisi, B. F., Xu, W. X., Otwinowski, Z., Freedman, L. P., Yamamoto, K. R., and Sigler, P. B. (1991). Crystallographic analysis of the interaction of the glucocorticoid receptor with DNA. *Nature (London)* **352,** 497–505.

Maden, M. (1982). Vitamin A and pattern formation in the regenerating limb. *Nature (London)* **295,** 672–675.

Maden, M., Ong, D. E., Summerbell, D., and Chytil, F. (1989). The role of retinoid-binding proteins in the generation of pattern in the developing limb, the regenerating limb and the nervous system. *Development* **107**(Suppl.), 109–119.

Mangelsdorf, D. J., Ong, E. S., Dyck, J. A., and Evans, R. M. (1990). Nuclear receptor that identifies a novel retinoic acid response pathway. *Nature (London)* **345,** 224–229.

Mangelsdorf, D., Umesono, K., Kliewer, S., Borgmeyer, U., Ong, E., and Evans, R. (1991). A direct repeat in the cellular retinol-binding protein type II gene confers differential regulation by RXR and RAR. *Cell* **66,** 555–561.

Mangelsdorf, D. J., Borgmeyer, U., Heyman, R. A., Zhou, J. Y., Ong, E. S., Oro, A. E., Kakizuka, A., and Evans, R. M. (1992). Characterization of three RXR genes that mediate the action of 9-*cis* retinoic acid. *Genes Dev.* **6,** 329–344.

Manns, M., and Fritzsch, B. (1991). The eye in the brain: Retinoic acid effects morphogenesis of the eye and pathway selection of axons but not the differentiation of the retina in *Xenopus laevis. Neurosci. Lett.* **127,** 150–154.

Manns, M., and Fritzsch, B. (1992). Retinoic acid affects the organization of reticulospinal neurons in developing *Xenopus. Neurosci. Lett.* **139,** 253–256.

Mayer, H., and Isler, O. (1971). Total syntheses. *In* "Carotenoids" (O. Isler, ed.), pp. 325–575. Birkhäuser, Basel.

McCafferey, P., Lee, M.-O., Wagner, M. A., Sladek, N. E., and Dräger, U. C. (1992). Asymetrical retinoic acid synthesis in the dorsoventral axis of the retina. *Development* **115,** 371–382.

McGinnis, W., and Krumlauf, R. (1992). Homeobox genes and axial patterning. *Cell* **68,** 283–302.

Mendelsohn, C., Ruberte, E., LeMeur, M., Morriss-Kay, G., and Chambon, P. (1991). Developmental analysis of the retinoic acid-induced RAR-β2 promoter in transgenic animals. *Development* **113,** 723–734.

Mohanty-Hejmadi, P., Dutta, S. K., and Mahapatra, P. (1992). Limbs generated at the site of tail amputation in marbeled balloon frog after vitamin A treatment. *Nature (London)* **355,** 352–353.

Monkemeyer, J., Ludolph, D. C., Cameron, J. A., and Stocum, D. L. (1992). Retinoic acid-induced change in anteroposterior positional identity in regenerating axolotl limbs is dose-dependent. *Dev. Dynam.* **193,** 286–294.

Morriss-Kay, G. M., Murphy, P., Hill, R. E., and Davidson, D. R. (1991). Effects of retinoic acid excess on expression of Hox 2.9 and Krox-20 and on morphological segmentation in the hindbrain of mouse embryos. *EMBO J.* **19,** 2985–2995.

Murphy, P., and Hill, R. E. (1991). Expression of the mouse *labial*-like homeobox-containing genes, Hox 2.9 and Hox 1.6, during segmentation of the hindbrain. *Development* **111,** 61–74.

Noji, S., Yamaai, T., Koyama, E., Nohno, T., and Taniguchi, S. (1989). Spatial and temporal expression pattern of retinoic acid receptor genes during mouse bone development. *FEBS Lett.* **264,** 93–96.

Noji, S., Nohno, T., Koyama, E., Muto, K., Ohyama, K., Yoshinobu, A., Tamura, K., Ohsugi, K., Ide, H., Taniguchi, S., and Saito, T. (1991). Retinoic acid induces polarizing activity but is unlikely to be a morphogen in the chick limb bud. *Nature (London)* **350,** 83–86.

Osmond, M. K., Butler, A. J., Voon, F. C. T., and Bellairs, R. (1991). The effect of retinoic acid on heart formation in the early chick embryo. *Development* **113,** 1405–1417.

Osumi-Yamashita, N., Noji, S., Nohno, T., Koyama, E., Doi, H., Eto, K., and Taniguchi, S. (1990). Expression of retinoic acid receptor genes in neural crest-derived cells during mouse facial development. *FEBS Lett.* **264,** 71–74.

Packer, L. (ed.) (1990). Retinoids. In Methods in Enzymology, Vols. 189 and 190. Academic Press, San Diego, California.

Papalopulu, N., Lovell-Badge, R., and Krumlauf, R. (1991). The expression of murine Hox-2 genes is dependent on the differentiation pathway and displays a colinear sensitivity to retinoic acid in of cells and *Xenopus* embryos. *Nucleic Acids Res.* **19,** 5497–5506.

Petkovich, M., Brand, N. J., Krust, A., and Chambon, P. (1987). A human retinoic acid receptor which belongs to the family of nuclear receptors. *Nature (London)* **330,** 444–450.

Pfahl, M., Tzukerman, M., Zhang, X., Lehmann, J., Hermann, T., Wills, K. N., and Graupner, G. (1990). Nuclear retinoic acid receptors: Cloning, analysis, and function. *Methods Enzymol.* **189,** 256–270.

Quian, Y. Q., Billeter, M., Otting, G., Müller, M., Gehring, W. J., and Wüthrich, K. (1989). The structure of the *Antennapedia* homeodomain determined by NMR spectroscopy in solution: comparison with prokaryotic repressors. *Cell* **59,** 573–580.

Ragsdale, C. W., Petkovich, M., Gates, P. B., Chambon, P., and Brockes, J. P. (1989). Identification of a novel retinoic acid receptor in regenerative tissues of the newt. *Nature (London)* **341,** 654–657.

Rando, R. R. (1990). The chemistry of vitamin A and vision. *Angew. Chem. Int. Ed. Engl.* **29,** 461–480.

Rowe, A., Eager, N. S. C., and Brickell, P. M. (1991a). A member of the RXR nuclear receptor family is expressed in the neural-crest-derived cells of the developing peripheral nervous system. *Development* **111,** 771–778.

Rowe, A., Richman, J. M., and Brickell, P. M. (1991b). Retinoic acid treatment alters the distribution of retinoic acid receptor-β transcripts in the embryonic chick face. *Development* **111,** 1007–1016.

Ruberte, E., Dolle, Krust, A., Zelent, A., Morriss-Kay, G., and Chambon, P. (1990). Specific spatial and temporal distribution of retinoic acid receptor gamma transcripts during mouse embryogenesis. *Development* **108,** 213–222.

Ruberte, E. Dollé, P., Chambon, P., and Morriss-Kay, G. (1991). Retinoic acid receptors and cellular retinoid binding proteins. II. Their differential pattern of transcription during early morphogenesis in mouse embryos. *Development* **111,** 45–60.

Ruiz i Altaba, A., and Jessell, T. M. (1991a). Retinoic acid modifies mesodermal patterning in early *Xenopus* embryos. *Genes Dev.* **5,** 175–187.

Ruiz i Altaba, A., and Jessell, T. M. (1991b). Retinoic acid modifies the pattern of cell differentiation in the central nervous system of neurula stage *Xenopus* embryos. *Development* **112,** 945–958.

Sani, B. P., and Hill, D. L. (1990). Characteristics of synthetic retinoids. *Methods Enzymol.* **189,** 43–50.

Saunders, J. W., and Gasseling, M. T. (1968). Ectodermal–mesenchymal interactions in the origin of wing symmetry. *In* "Epithelial–Mesenchymal Interactions" (R. Fleischmajer, R. E. Billingham, eds.), pp. 78–97. Williams and Wilkins, Baltimore, Maryland.

Saurat, J.-H. (1991). "Retinoids: 10 Years On." Karger, Basel.

Schoenwolf, G. C., Garcia-Martinez, V., and Dias, M. S. (1992). Mesoderm movement and fate during avian gastrulation and neurulation. *Dev. Dynamics* **193,** 235–248.

Scott, M. P., Tamkun, J. W., and Hartzell, G. W. (1989). The structure and function of the homeodomain. *Biochim. Biophys. Acta* **989,** 25–48.

Selleck, M. A. J., and Stern, C. (1991). Fate mapping and cell lineage analysis of Hensen's node in the chick embryo. *Development* **112,** 615–626.

Sharpe, C. R. (1991). Retinoic acid can mimic endogenous signals involved in transformation of the *Xenopus* nervous system. *Neuron* **7,** 239–247.

Simeone, A., Acampora, D., Arcioni, L., Andrew, P. W., Boncinelli, E., and Mavilio, F.

(1990). Sequential activation of HOX2 homeobox genes by retinoic acid in human embryonal carcinoma cells. *Nature (London)* **346,** 763–766.

Simeone, A., Acampora, D., Nigro, V., Faiella, A., D'Esposito, M., Stornaiuolo, A., Mavilio, F., and Boncinelli, E. (1991). Differential regulation by retinoic acid of the homeobox genes of the four HOX loci in human embryonal carcinoma cells. *Mech. Dev.* **33,** 215–228.

Sive, H. L., and Cheng, P. F. (1991). Retinoic acid perturbs the expression of Xhox.lab genes and alters mesodermal determination in *Xenopus laevis. Genes Dev.* **5,** 1321–1332.

Sive, H. L., Draper, B. W., Harland, R. L., and Weintraub, H. (1990). Identification of a retinoic acid-sensitive period during primary axis formation in *Xenopus laevis. Genes Dev.* **4,** 932–942.

Smith, J. C. (1980). The time required for positional signalling in the chick wing bud. *J. Embryol. Exp. Morph.* **60,** 321–328.

Smith, S. M., and Eichele, G. (1991). Temporal and regional differences in the expression pattern of distinct retinoic acid receptor-β transcripts in the chick embryo. *Development* **111,** 245–252.

Smith, S. M., Pang, K., Sundin, O., Wedden, S. E., Thaller, C., and Eichele, G. (1989). Molecular approaches to vertebrate limb morphogenesis. *Development* **107**(Suppl.), 121–131.

Sporn, M. B., Roberts, A. B., and Goodman, D. S. (eds.). (1984). "The Retinoids." Academic Press, Orlando, Florida.

Stern, C. D. (1979). A re-examination of mitotic activity in the early chick embryo. *Anat. Embryol.* **156,** 319–329.

Summerbell, D. (1983). The effect of local application of retinoic acid to the anterior margin of the developing chick limb. *J. Embryol. Exp. Morph.* **78,** 269–289.

Sundin, O., and Eichele, G. (1992). An early marker of axial pattern in the chick embryo and its respecification by retinoic acid. *Development* **114,** 841–852.

Taira, M., Jamrich, M., Good, P. J., and Dawid, I. B. (1992). The LIM domain containing homeo box gene *Xlim-1* is expressed specifically in the organizer region of *Xenopus* gastrula embryos. *Genes Dev.* **6,** 356–366.

Thaller, C., and Eichele, G. (1987). Identification and spatial distribution of retinoids in the developing chick limb bud. *Nature (London)* **327,** 625–628.

Thaller, C., and Eichele, G. (1988). Characterization of retinoid metabolism in the developing chick limb bud. *Development* **103,** 473–483.

Thaller, C., and Eichele, G. (1990). Isolation of 3,4-didehydroretinoic acid, a novel morphogenetic signal in the chick wing bud. *Nature (London)* **345,** 815–819.

Tickle, C., Summerbell, D., and Wolpert, L. (1975). Positional signalling and specification of digits in chick limb morphogenesis. *Nature (London)* **254,** 199–202.

Tickle, C., Alberts, B. M., Wolpert, L., and Lee, J. (1982). Local application of retinoic acid to the limb bud mimics the action of the polarizing region. *Nature (London)* **296,** 564–565.

Tickle, C., Lee, J., and Eichele, G. (1985). A quantitative analysis of the effect of all-*trans*-retinoic acid on the pattern of chick wing development. *Dev. Biol.* **109,** 82–95.

Umesono, K., Murakami, K., Thompson, C., and Evans, R. (1991). Direct repeats as selective response elements for the thyroid hormone, retinoic acid and vitamin D_3 receptors. *Cell* **65,** 1255–1266.

Vahlquist, A. (1980). The identification of dehydroretinol (vitamin A_2) in human skin. *Experientia* **36,** 317–318.

Wagner, M., Thaller, C., Jessell, T., and Eichele, G. (1990). Polarizing activity and retinoid synthesis in the floor plate of the neural tube. *Nature (London)* **345,** 819–822.

Wanek, N., Gardiner, D. M., Muneoka, K., and Bryant, S. V. (1991). Conversion by retinoic acid of anterior cells into ZPA cells in the chick wing bud. *Nature (London)* **350,** 81–83.

Wolbach, S. B., and Howe, P. R. (1925). Tissue changes following deprivation of fat soluble A vitamin. *J. Exp. Med.* **42,** 753–777.

Wolpert, L. (1969). Positional information and the spatial pattern of cellular differentiation. *J. Theor. Biol.* **25,** 1–47.

Yu, V. C., Delsert, C., Anderson, B., Holloway, J. M., Devary, O., Näär, A. M., Kim, S. Y., Boutin, J.-M., Glass, C. K., and Rosenfeld, M. G. (1991). RXRβ: A coregulator that enhances binding of retinoic acid, thyroid hormone, and vitamin D receptors to their cognate response elements. *Cell* **67,** 1251–1266.

Zelent, A., Krust, A., Petkovich, M., Kastner, P., and Chambon, P. (1989). Cloning of murine α and β retinoic acid receptors and a novel receptor γ predominantly expressed in skin. *Nature (London)* **339,** 714–717.

Zelent, A., Mendelsohn, C., Kastner, P., Garnier, J. M., Ruffenach, F., Leroy, P., and Chambon, P. (1991). Differentially expressed isoforms of the mouse retinoic acid receptor beta are generated by usage of two promoters and alternative splicing. *EMBO J.* **10,** 71.

Zhang, X., Hoffmann, B., Tran, P., Graupner, G., and Pfahl, M. (1992a). Retinoid X receptor is an auxillary protein for thyroid hormone and retinoic acid receptors. *Nature (London)* **355,** 441–446.

Zhang, X., Lehmann, J., Hoffmann, B., Dawson, M. I., Cameron, J., Graupner, G., Hermann, T., Tran, P., and Pfahl, M. (1992b). Homodimer formation of retinoid X receptor induced by 9-*cis* retinoic acid. *Nature (London)* **358,** 587–591.

10

PARATHYROID HORMONE BIOSYNTHESIS AND ACTION: Molecular Analysis Of the Parathyroid Hormone Gene and Parathyroid Hormone/Parathyroid Hormone-Related Peptide Receptor

MARIE DEMAY, HARALD JÜPPNER, ABDUL-BADI ABOU-SAMRA, GINO SEGRE, and HENRY KRONENBERG

Cellular and Molecular Biology of Bone

I. INTRODUCTION

A biochemical understanding of the mechanisms of calcium homeostasis requires detailed characterization of the molecules involved. Over the past 15 years, recombinant DNA technology has allowed the isolation and the experimental manipulation of the parathyroid hormone (PTH) gene and its derived messenger RNA (mRNA) and encoded protein. More recently, the cloning of DNA encoding the PTH receptor has made possible the detailed characterization of the mechanisms of PTH action. This chapter summarizes current understanding of the regulation of PTH biosynthesis and describes the early lessons learned from the isolation of complementary DNA (cDNA) encoding PTH receptors.

II. PARATHYROID HORMONE GENE STRUCTURE

The human PTH gene is a single-copy gene that extends over 4200 basepairs (bp) of DNA (Vasicek *et al.*, 1983) on the short arm of chromosome 11 (Naylor *et al.*, 1983). The gene contains three exons, the first of which is noncoding. The second exon contains the initiator ATG and encodes all of the signal or "pre" sequence and most of the short "pro" peptide. The third exon encodes the remainder of the "pro" sequence, the residues corresponding to the mature hormone, as well as the 3′ noncoding region. The intron–exon structure is preserved across species to the bovine (Weaver *et al.*, 1984) and rat (Heinrich *et al.*, 1984) genes.

The human PTH gene encodes two transcripts of 822 and 793 bp (Hendy *et al.*, 1981). Primer extension analysis has demonstrated that two functional TATA boxes exist in the human gene, resulting in these two mRNAs that differ in length by 29 bp. Similarly, the bovine gene is thought to have two functional promoters, resulting in transcripts of 701 and 672 bp (Weaver *et al.*, 1982). The rat cDNA is thought to contain a single functional promoter, which directs transcription of an mRNA transcript of 712 bp (Heinrich *et al.*, 1984). The chicken mRNA also has a single functional promoter (Khosla *et al.*, 1988); the gene has a long 3′ untranslated region and, as a consequence, a long transcript of 2.3 kilobases (kb) (Russell and Sherwood, 1989). Extensive nucleic acid homology exists among the different species, notably in the coding region, where there is 90% homology between the bovine and human mRNAs and 78% between the rat and human mRNAs (Heinrich *et al.*, 1984). The chicken cDNA is the only nonmammalian PTH sequence cloned to date (Khosla *et al.*, 1988; Russell and Sherwood, 1989). Although there is 68% identity between the chicken and human cDNAs in the bases corresponding to amino acids −31 to +32, marked differences from mammalian PTH, including deletions, exist in the sequences coding for amino acids 32–37 in the chicken cDNA. Between the residues coding for amino acids 37–53 in the chicken sequence (45–61 in mammalian PTH), there is once again marked homology with the human sequence (78%). Divergence among the species is observed in the form of an insertion in the chicken sequence coding for amino acids 54–75. The bases coding for amino acids 76–88 in the chicken are once again homologous (74%) to those encoding the carboxy-terminal region of the human peptide. The evolutionary conservation of large portions of the PTH sequence distal to residue 32 remains unexplained, because the function of this region remains obscure.

III. REGULATION OF PARATHYROID HORMONE GENE EXPRESSION

A. Role of Calcium

PTH, like 1,25-dihydroxyvitamin D_3 (1,25$(OH)_2D_3$), plays a major role in calcium homeostasis. Extracellular calcium modulates several steps in the biosynthesis of PTH, including transcription (Naveh-Many *et al.*, 1989; Naveh-Many and Silver, 1990a; Russell *et al.*, 1983; Yamamoto *et al.*, 1989), intracellular processing and degradation of the protein (Chu *et al.*, 1973; Habener *et al.*, 1975; Mayer *et al.*, 1979), and hormone secretion (Habener *et al.*, 1975).

With the establishment of PTH immunoassays, the acute regulation of PTH secretion by calcium was studied. Provocative and suppressive tests to investigate the relationship among hypocalcemia, hypercalcemia, and PTH levels were performed *in vivo* in cows and goats (Sherwood *et al.*, 1966). These studies demonstrated that there was a reciprocal relationship between plasma calcium and PTH immunoreactivity in the blood. The secretory changes observed in these acute studies led to investigations of the effects of calcium on PTH biosynthesis. In studies of rats fed a hypocalcemic diet, Chu *et al.* (1973) examined compensatory changes in the parathyroid gland. After 13 days on a 0.02% calcium diet, although parathyroid gland size and weight appeared unchanged, the content of PTH and the rate of PTH synthesis doubled relative to that of rats on a high calcium (2%) diet. In more acute *in vitro* studies (Habener *et al.*, 1975), synthesis of pro-PTH and its conversion to PTH were independent of extracellular calcium concentration. In contrast, increased intracellular degradation of hormone (up to 50%) was noted when parathyroid slices were incubated in high calcium (2.5 m*M*) conditions. In conditions of low calcium, less than 10% of PTH was degraded intracellularly. Effects of calcium on intracellular PTH cleavage have also been noted *in vivo*. Secretion of carboxy-terminal fragments, thought to be biologically inactive, is increased with hypercalcemia. Mayer *et al.* (1979) monitored the effects of calcium concentration on the relative proportion of amino- and carboxy-terminal immunoreactivity in calf parathyroid venous effluent. The carboxy- to amino-terminal ratio of 1.3, observed in hypocalcemia, increased to a value of 3 during hypercalcemia. This was accompanied by a 15-fold decrease in amino-terminal PTH immunoreactivity between hypocalcemia and hypercalcemia. Thus, early studies suggested that calcium regulated PTH biosynthesis, intracellular degradation, and secretion by a variety of mechanisms.

Increases in PTH mRNA levels in response to low extracellular calcium have been observed both *in vitro* (Russell *et al.*, 1983) and *in vivo* (Naveh-Many *et al.*, 1989, 1990a; Yamamoto *et al.*, 1989). *In vivo* experiments (Naveh-Many *et al.*, 1989; Naveh-Many and Silver, 1990a; Yamamoto *et al.*, 1989) have demonstrated that, in intact rats, hypocalcemia can increase PTH mRNA levels two- to seven-fold. However, in dispersed bovine parathyroid cells, although a reversible decrease in PTH mRNA was observed after 16–72 hr exposure to 2.5 m*M* calcium, low levels (0.5 m*M*) of extracellular calcium failed to increase PTH mRNA; this result suggests that, in this culture system, the basal transcription rate was already at its maximal level (Russell *et al.*, 1983). The discrepancy between the *in vitro* and *in vivo* observations may be explained by a variety of mechanisms. The digestion of bovine parathyroid glands, required for the generation of primary cultures, may disrupt cell surface calcium signaling mechanisms and thereby impair the normal physiological response of these cells to hypocalcemia. Artifacts seem less likely in models using intact animals. Hypocalcemia was induced by phosphate therapy (Naveh-Many *et al.*, 1989), ethyleneglycol tetra-acetic acid or calcitonin (Yamamoto *et al.*, 1989), or prolonged (3-week) dietary deficiency (Naveh-Many and Silver, 1990a). This variety of stimuli all increased PTH mRNA levels. It is unlikely that such diverse stimuli result in increased PTH levels by a mechanism other than hypocalcemia, although, in these *in vivo* experiments, it is not possible to establish that hypocalcemia acts directly on the parathyroid cell. Identification of the signal transduction mechanisms by which hypocalcemia modulates PTH gene transcription will depend on identification and characterization of the parathyroid cell calcium sensors and signaling pathways.

Studies aimed at isolating the DNA sequences responsible for mediating the effects of extracellular calcium on PTH gene transcription have been undertaken. Culturing of various nonparathyroid cell lines in high calcium conditions (3.0 m*M*) resulted in increased binding of nuclear proteins to silencer sequences in the human PTH gene. These DNA sequences could confer calcium suppressibility to a neutral promoter in transfection experiments. Thus, upstream sequences (at −2.5 and −3.5 kb) may play a role in the transcriptional down-regulation of the PTH gene by high extracellular calcium (Okazaki *et al.*, 1991a). Whether the relevant binding proteins are also present and active in parathyroid cells remains to be established.

B. Role of 1,25-Dihydroxyvitamin D_3

In vivo and *in vitro,* $1,25(OH)_2D_3$ has been shown to suppress PTH levels, independently of extracellular calcium concentration. *In vivo* studies performed in rats (Chertow *et al.*, 1975) have demonstrated a rapid decrease in serum PTH levels in response to $1,25(OH)_2D_3$. The doses of $1,25(OH)_2D_3$ used did not produce a measurable increase in serum calcium levels. Delmez *et al.* (1989) studied the effects of intravenous $1,25(OH)_2D_3$ in uremic patients with secondary hyperparathyroidism. Two weeks of treatment with intravenous $1,25(OH)_2D_3$ significantly decreased PTH levels without an apparent increase in calcium concentrations. When these patients were dialyzed against low calcium dialysate (1 meq/liter), the PTH response to this hypocalcemic stimulus was attenuated relative to the response seen before the administration of $1,25(OH)_2D_3$. Furthermore, calcium infusion caused a more marked suppression of PTH levels.

$1,25(OH)_2D_3$ binding has been found to be suppressed in parathyroid glands obtained from uremic patients (Korkor, 1987). Naveh-Many *et al.* (1990) showed that $1,25(OH)_2D_3$ can increase $1,25(OH)_2D_3$ receptor mRNA levels in the parathyroid glands and duodenum of intact rats. This suggests that the apparent decrease in $1,25(OH)_2D_3$ receptor levels observed in parathyroid glands from uremic patients may reflect $1,25(OH)_2D_3$ deficiency. Alternatively, uremia may alter $1,25(OH)_2D_3$ receptor levels by other, as yet unclarified mechanisms.

$1,25(OH)_2D_3$ may also decrease PTH levels by suppressing parathyroid cell growth. In dispersed bovine parathyroid cells, 10^{-7} to 10^{-9} M $1,25(OH)_2D_3$ suppresses cell proliferation (Ishimi *et al.*, 1990). No decreased proliferation was seen with exposure of these cells to high calcium (2.5 mM).

Down-regulation of PTH gene transcription by $1,25(OH)_2D_3$ has been studied extensively both *in vitro* and *in vivo*. Levels of PTH mRNA in dispersed bovine parathyroid cells have been shown to be repressed by doses as low as 10^{-11} M $1,25(OH)_2D_3$ (Silver *et al.*, 1985). Using doses of 10^{-7} M, suppression is seen as early as 6 hr posttreatment (Russell *et al.*, 1986). Nuclear runoff studies suggest that this is a direct transcriptional effect. Other analogs, notably $24,25(OH)_2D_3$ and $25(OH)_2D_3$, were not as effective at suppressing PTH mRNA levels in the bovine parathyroid cells (Silver *et al.*, 1985). *In vivo* experiments performed following intraperitoneal injection of 100 pmol of $1,25(OH)_2D_3$ in rats have demonstrated a decrease in PTH mRNA levels 48 hr postinjection, with no concomitant increase in serum calcium (Silver *et al.*, 1986). As in the *in vitro* studies $1,25(OH)_2D_3$ was more potent than either $24,25(OH)_2D_3$ or $25(OH)_2D_3$. In contrast to the rapid inhibition of PTH secretion by calcium (within 2 hr), effects of $1,25(OH)_2D_3$ were best seen 48 hr post treatment. The decrease in PTH secretion paralleled a decrease in PTH mRNA levels, suggesting that the effect of $1,25(OH)_2D_3$ was on transcription or on mRNA stability. Nuclear runoff experiments directly demonstrated suppression of gene transcription by $1,25(OH)_2D_3$.

Investigations directed at isolating the sequences responsible for down-regulation of PTH transcription by $1,25(OH)_2D_3$ were undertaken by Okazaki *et al.* (1988). Stable GH4 cell lines were established, which expressed a human PTH 5′ regulatory region–neomycin resistance fusion gene. A 60% decrease in reporter gene mRNA levels was observed in response to $1,25(OH)_2D_3$. This effect was mediated by 684 bp of the 5′ flanking region of the human PTH gene. Attempts to further characterize the region in the human PTH gene that mediated $1,25(OH)_2D_3$ repression focused on $1,25(OH)_2D_3$ receptor binding assays. Sequences in the bovine PTH gene between −485 and −100 were shown to bind a protein that co-migrated with the $1,25(OH)_2D_3$ receptor on Southwestern blots (Farrow *et al.*, 1990). Using the assay developed for studying the binding of the $1,25(OH)_2D_3$ receptor to the $1,25(OH)_2D_3$ response element of the rat osteocalcin gene (Demay *et al.*, 1990), sequences in the human PTH gene were identified that could compete with this up-regulatory response element for binding to the $1,25(OH)_2D_3$ receptor (Demay *et al.*, 1991). These sequences, located approximately 125 bp upstream from the start of exon 1 have been shown, in gel retardation assays, to bind the $1,25(OH)_2D_3$ receptor directly. Furthermore, when placed upstream to a heterologous viral promoter, they are able to mediate transcriptional down-regulation in response to $1,25(OH)_2D_3$, after transfection into GH4 cell lines. These sequences do not fit the criteria for up-regulatory $1,25(OH)_2D_3$ response elements (Demay *et al.*, 1992). They do, however, contain a single motif that is identical to that repeated in the mouse osteopontin $1,25(OH)_2D_3$ response element (Noda *et al.*, 1990). The interplay of this motif with the $1,25(OH)_2D_3$ receptor and with factors that bind to adjacent DNA sequences needs further investigation in order to clarify the mechanism of transcriptional repression of the PTH gene by $1,25(OH)_2D_3$. Parathyroid cell cultures will be needed to understand the relative importance and contributions of these DNA sequences in the $1,25(OH)_2D_3$ regulation of the PTH gene *in vivo*.

Other effects of $1,25(OH)_2D_3$ may be mediated by nongenomic actions. It has been demonstrated that, in isolated bovine parathyroid glands, $1,25(OH)_2D_3$ can increase intracellular calcium concentration within 2 min (Sugimoto *et al.*, 1988). Therefore, $1,25(OH)_2D_3$ may also affect PTH transcription by modulating intracellular calcium concentrations. Clarifying the relative contributions of the genomic and non-genomic actions of $1,25(OH)_2D_3$ on PTH gene transcription will require the availability of a parathyroid cell line that expresses PTH in a $1,25(OH)_2D_3$-responsive fashion.

C. Role of Other Factors

Studies of the tissue-specific expression of the PTH gene has led to the identification of silencer elements in the 5′ flanking region of the human PTH gene. Okazaki *et al.* (1991b) identified sequences that suppress the transcriptional activation of a reporter gene in BHK, CV1, and 3T3 cells. Two elements (110 and 113 bp) can silence a previously active human PTH–chloramphenicol acetyltransferase fusion gene in BHK cells. Nuclear factors from these cells have been shown to bind specifically to these parathyroid sequences. The absence of a cell line that normally expresses PTH has hampered further clarification of the biological significance of these sequences.

The human PTH gene has also been shown to contain a functional cyclic adenosine monophosphate (cAMP) response element. A sequence differing by one base from a classical cAMP response element is located at −81 from the start of exon 1 (Rupp *et al.*, 1990). Similar sequences are also present in the rat and bovine PTH gene. The sequences flanking the cAMP consensus sequence in the bovine gene are thought to attenuate its responsiveness (Deutsch *et al.*, 1988).

The effects of glucocorticoids on PTH levels have been studied *in vitro*. Peraldi *et al.* (1990) observed an increase in PTH release after dexamethasone treatment of human parathyroid cells from secondarily hyperplastic glands. This effect was blunted by RU 486, a glucocorticoid antagonist. $1,25(OH)_2D_3$ inhibited both the immunoreactive PTH and the PTH mRNA increase, as did α-amanitin, an inhibitor of RNA synthesis. Karmali *et al.* (1989) demonstrated that 24-hr treatment with cortisol caused a 20% decrease in $1,25(OH)_2D_3$ receptors in bovine parathyroid cells. This result suggests that glucocorticoids may increase PTH levels, in part, by preventing $1,25(OH)_2D_3$-mediated PTH repression. Other investigators (Sugimoto *et al.*, 1989) found a similar increase in PTH levels with dexamethasone, but no change in $1,25(OH)_2D_3$ receptors was observed after 48 hr of dexamethasone treatment.

D. Summary

Homeostatic stability mandates that the parathyroid gland be regulated by calcium and by $1,25(OH)_2D_3$, because PTH tightly regulates the blood levels of both calcium and $1,25(OH)_2D_3$. Calcium acts acutely to modulate the secretion of PTH; both calcium and $1,25(OH)_2D_3$ modulate more chronic activities of the parathyroid cell. Calcium acts on the parathyroid cell at several distinct levels. The major stimulus for PTH production is hypocalcemia. Decreases in extracellular calcium are associated with increased PTH mRNA levels and increased PTH biosynthesis. In contrast, hypercalcemia does not have a major effect on mRNA levels. Experimental models do suggest that hypercalcemia increases intracellular degradation of PTH and increases the fraction of PTH secreted as carboxy-terminal fragments. By affecting metabolism of the peptide, therefore, increased levels of extracellular calcium result in decreased secretion of bioactive PTH. The secretory activity of the parathyroid cell is closely regulated by the ambient level of extracellular calcium. The precise mechanisms whereby levels of extracellular calcium modulate PTH synthesis and secretion will require the characterization of the parathyroid cell calcium sensor.

1,25$(OH)_2D_3$ plays an important role in PTH regulation. It has been shown to decrease parathyroid cell proliferation *in vivo* and *in vitro*. 1,25$(OH)_2D_3$ also decreases parathyroid hormone gene transcription by binding to its nuclear receptor, which then interacts with upstream regulatory elements in the PTH gene. Rapid effects of 1,25$(OH)_2D_3$ on intracellular concentration of calcium have been observed and may play a role as well.

IV. PARATHYROID HORMONE/PARATHYROID HORMONE-RELATED PEPTIDE RECEPTOR

A. Isolation and Structure of cDNAs: Encoding the Parathyroid Hormone/Parathyroid Hormone-Related Peptide Receptor

PTH binds to specific, G protein-coupled receptors in bone and kidney, which have been extensively characterized over the past years. The parathyroid hormone-related peptide (PTHrP) is structurally related to PTH. PTHrP was initially isolated from several carcinomas (Suva *et al.*, 1987; Mangin *et al.*, 1988; Strewler *et al.*, 1987) and is responsible for the humoral hypercalcemia associated with many malignancies (Orloff *et al.*, 1989b). However, PTHrP immunoreactivity and PTHrP mRNA transcripts have been described in a large variety of adult and fetal organs (Henderson *et al.*, 1990; Burton *et al.*, 1990; McCauley *et al.*, 1992; Thiede and Rodan, 1988; Thiede *et al.*, 1990, 1991; Nickols *et al.*, 1990; Hongo *et al.*, 1991), and the peptide circulates in at least some normal subjects (Henderson *et al.*, 1990; Burtis *et al.*, 1990; Budayr *et al.*, 1989). The peptide may regulate calcium metabolism in certain settings (Abbas *et al.*, 1989; Budayr *et al.*, 1990) and may act as a paracrine factor. Only 8 of the 13 amino-terminal residues of PTH and PTHrP are identical, and the remainder of their sequences share no sequence homology and are different in length. Despite this very limited homology, recent radioreceptor studies and analysis of ligand–receptor complexes after affinity cross-linking with amino-terminal analogs of either PTH or PTHrP suggest that both ligands bind to the same receptors in various target tissues (Shigeno *et al.*, 1988c; Orloff *et al.*, 1989a; Nissenson *et al.*, 1988; Jüppner *et al.*, 1988). PTH/PTHrP receptors have been shown to be 60,000- to 80,000-Da glycoproteins with N-linked oligosaccharides that are sensitive to reducing agents and enzymatic degradation (Karpf *et al.*, 1987, 1991; Shigeno *et al.*, 1988a,b).

Complementary DNA clones that encode common PTH/PTHrP receptors from cDNA libraries prepared from opossum kidney (OK) (Jüppner *et al.*, 1991b) and rat osteosarcoma (ROS 17/2.8) (Abou-Samra *et al.*, 1992a) cells have recently been isolated. Both receptors were cloned independently by detecting expression in mammalian COS-7 cells using photoemulsion autoradiography (Gearing *et al.*, 1989). Complementary DNAs were prepared from poly(A)$^+$-selected RNA of both cell lines (Fig. 1). After second-strand synthesis, ligation of BstXI adaptors, and size selection (>1600 bp), cDNA libraries were constructed in the plas-

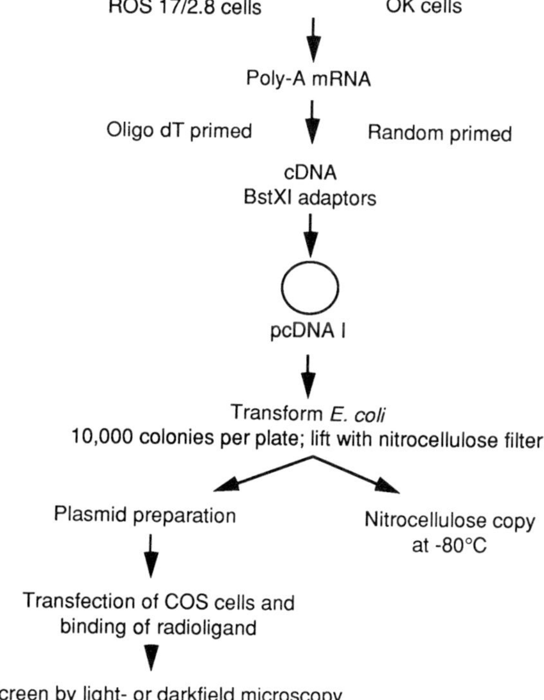

FIGURE 1 Construction of cDNA libraries from ROS 17/2.8 and OK cells and the initial screening to isolate the parathyroid hormone/parathyroid hormone-related peptide receptor. mRNA, pcDNAI, plasmid receptor for expression cloning.

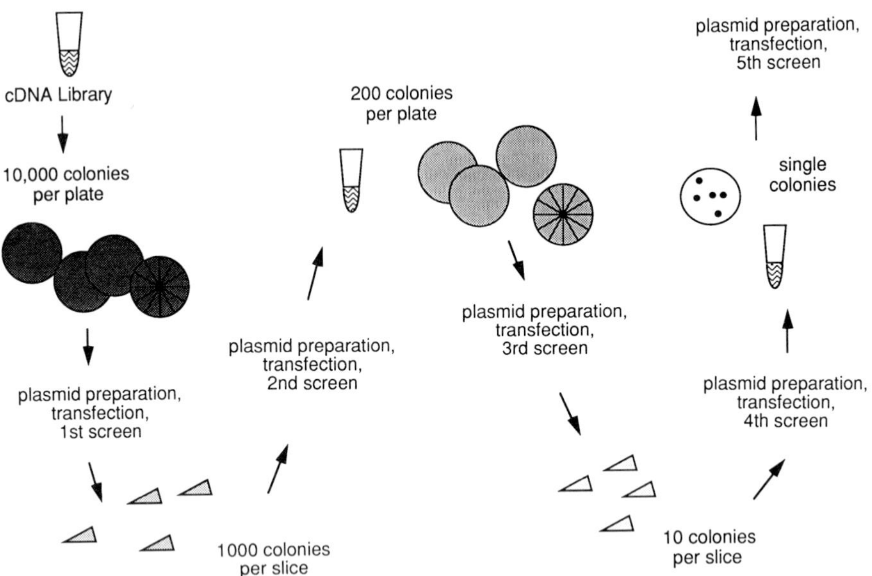

FIGURE 2 Strategy to isolate the parathyroid hormone/parathyroid hormone related peptide receptor from cDNA pools each representing about 10,000 individual bacterial colonies.

mid vector pcDNAI (Seed, 1987). *Escherichia coli* (MC 1061/P3) were then transformed with aliquots of each of the two libraries, and plasmid cDNA pools from ≈10,000 independent colonies were transfected by the diethylaminoethyl–dextran method (Seed and Aruffo, 1987) into COS-7 cells that were grown in "slideflasks." After 48 hr, the cells were incubated with ^{125}I-labeled [Tyr^{36}]PTHrP(1-36)amide (PTHrP) for 2 hr at room temperature, rinsed with binding buffer, and fixed with 1.25% glutaraldehyde in phosphate-buffered saline. The slides were then dipped into photographic emulsion for autoradiography. After 2–4 days exposure at 4°C, the slides were developed and examined by light microscopy. The screening of plasmid pools representing a total of ≈200,000 colonies of each cDNA library revealed three independent positive DNA pools; two from the OK cell library and one from the ROS 17/2.8 cell library. The bacterial pools responsible for generating the positive autoradiographic signals were then sequentially subdivided and screened until independent clones, OK-O and OK-H, and R15B were isolated from the two cDNA libraries (Fig. 2).

The nucleotide sequences of OK-O and R15B revealed open-reading frames that encode a 585- and a 591-amino acid protein, respectively. OK-H confirmed the nucleotide sequence of OK-O with one exception, the presence of 5 instead of 6 Gs (nucleotides 1629–1634) as a result of a cloning artifact. The amino acid sequences encoded by OK-O and R15B are 78% identical, indicating remarkable conservation between the renal receptor from a marsupial and the bone receptor from a placental mammal (Fig. 3). As expected, the recently cloned human kidney PTH/PTHrP receptor (Schipani *et al.*, 1993) has an even higher amino acid identity of 91% with the rat bone receptor; the homology with the OK receptor is only 80%. Analysis by the method of Klein *et al.* (1985) predicts that all three receptors have 10 hydrophobic domains, 7 of which are thought to be membrane-spanning. The hydrophobic amino-terminal region A is likely to represent a signal peptide, whereas the second and third hydrophobic regions B and C are predicted to be located in the amino-terminal, extracellular portion and the first extracellular loop, respectively. Four conserved, potential N-linked glycosylation sites are located in the extracellular region adjacent to the first transmembrane domain. The amino-terminal extensions of the PTH/PTHrP receptors are, therefore, longer than those present in the receptors for opsins, amines, and acetylcholine (Lefkowitz and Caron, 1988) and about 200 amino acids shorter than those found in receptors that bind the glycopeptide hormones (McFarland *et al.*, 1989; Parmentier *et al.*, 1989; Loosfelt *et al.*, 1989; Sprengel *et al.*, 1990).

All three PTH/PTHrP receptors contain eight conserved cysteine residues, six in the putative amino-terminal, extracellular region (with two additional cysteines in the putative signal sequence) and one in each of the first two extracellular loops. Loss of ligand-binding capacity by treatment of PTH/PTHrP receptors with reducing agents suggests that sulfhydryl bonds confer a receptor conformation that is favorable for ligand binding (Karpf *et al.*, 1991). Potential sites for phosphorylation are present in the third intracellular loop and the carboxy-terminal tail.

Data base searches for nucleic acid or protein sequences that are similar to the PTH/PTHrP receptors revealed that, out of 120 reported G protein-linked receptors, only the calcitonin (Lin *et al.*, 1991) and the secretin (Ishihara *et al.*, 1991) receptor share significant identity of 32 and 42%, respectively. In addition to their overall structural similarity with seven transmembrane domains, at least seven of eight putative extracellular cysteines and at least two of four potential N-glycosylation sites are conserved in all four receptors (Fig. 3).

B. Function of Cloned Parathyroid Hormone/Parathyroid Hormone-Related Peptide Receptors

The peptide-binding characteristics of the cloned receptors were determined in transiently transfected COS-7 cells, using both ^{125}I-labeled [Nle^{8,18},Tyr^{34}]bPTH(1-34)amide (PTH) and ^{125}I-labeled PTHrP as ligands. COS-7 cells expressing OK-O, OK-H, or R15B bound both PTH

```
OKOR 1     ··MGAPRISHSLALLLCCSVLSSVYALVDADDVITKEEQIILLRNAQ··A
R15B 1     ··MGAARIAPSLALLLCCPVLSSAYALVDADDVFTKEEQIFLLHRAQ··A
SECR 1     ·MLSTMRPRLSLLLL·······················RLLLLTKAAHTV
CTR  1     MRFTLTRWCLTLFIFLNRPLPVLPDSADGAHTPTLEPEPFLYILGKQRML

OKOR 47    QCEQRLKEVLR·VPELAESAKDW··MSRSAKTKKEKPAEKLYPQAEESRE
R15B 47    QCDKLLKEVLHTAANIMESDKGWTPASTSGKPRKEKASGKFYPESKENKD
SECR 27    GVPPRLCDVRRVL··LEERAHCLQQLSKEKKG············ALGPET
CTR  51    EAQHRCYDRMQKLPPYQG································

                      +        +             +
OKOR 94    VSDRSRLQDGFCLPEWDNIVCWPAGVPGKVVAVPCPDYFYDFNHK·GRAY
R15B 97    VPTGSRRRGRPCLPEWDNIVCWPLGAPGEVVAVPCPDYIYDFNHK·GHAY
SECR 63    ASG········CEGLWDNMSCWPSSAPARTVEVQCPKFLLMLSNKNGSLF
CTR  69    ·······EGLYCNRTWDGWSCWDDTPAGVLAEQYCPDYFPDFD·AAEKVT

                                                  I
                                            ——————————————
             +            #    #   +
OKOR 143   RRCDSNGSWELVPGNNRTWANYSECVKFLTN·ETREREVFDRLGMIYTVG
R15B 146   RRCDRNGSWEVVPGHNRTWANYSECLKFMTN·ETREREVFDRLGMIYTVG
SECR 105   RNCTQDGWSETFP·····RPDLACGVNINNSFNERRHAYLLKLKVMYTVG
CTR  111   KYCGEDGDWYRHPESNISWSNYTMCNAFTPDKLQNAYILY····YLAIVG

                                                 I I
           ———————————————————           ————————————————————
OKOR 192   YSISLGSLTVAVLILGYFRRLHCTRNYIHMHLFVSFMLRAVSIFIKDAVL
R15B 195   YSMSLASLTVAVLILAYFRRLHCTRNYIHMHMFLSFMLRAASIFVKDAVL
SECR 150   YSSSLAMLLVALSILCSFRRLHCTRNYIHMHLFVSFILRALSNFIKDAVL
CTR  157   HSLSILTLLISLGIFMFLRSISCQRVTLHKNMFLTYVLNSIIIIVHLVVI

           ———————                             +
OKOR 242   YSGVSTDEIERITEEELRAFTE···PPPADKAGFVGCRVAVTVFLYFLTT
R15B 245   YSGFTLDEAERLTEEELHIIAQVPPPPAAAAVGYAGCRVAVTFFLYFLAT
SECR 200   FSSDDVTYCD·····················AHKVGCKLVMIFFQYCIMA
CTR  207   ···········VPNGEL············VKRDPPICKVLHFFHQYMMSC
```

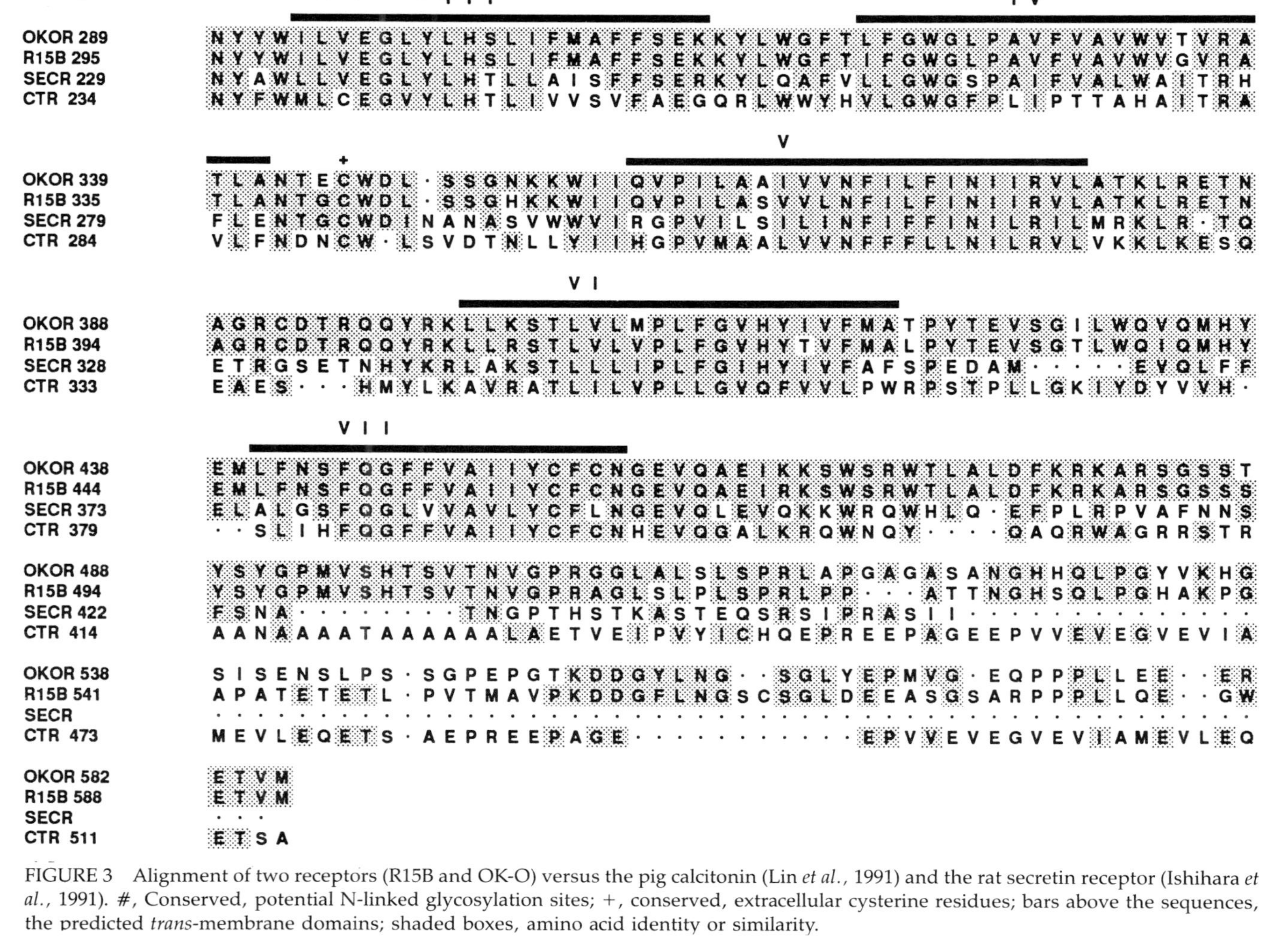

FIGURE 3 Alignment of two receptors (R15B and OK-O) versus the pig calcitonin (Lin *et al.*, 1991) and the rat secretin receptor (Ishihara *et al.*, 1991). #, Conserved, potential N-linked glycosylation sites; +, conserved, extracellular cysterine residues; bars above the sequences, the predicted *trans*-membrane domains; shaded boxes, amino acid identity or similarity.

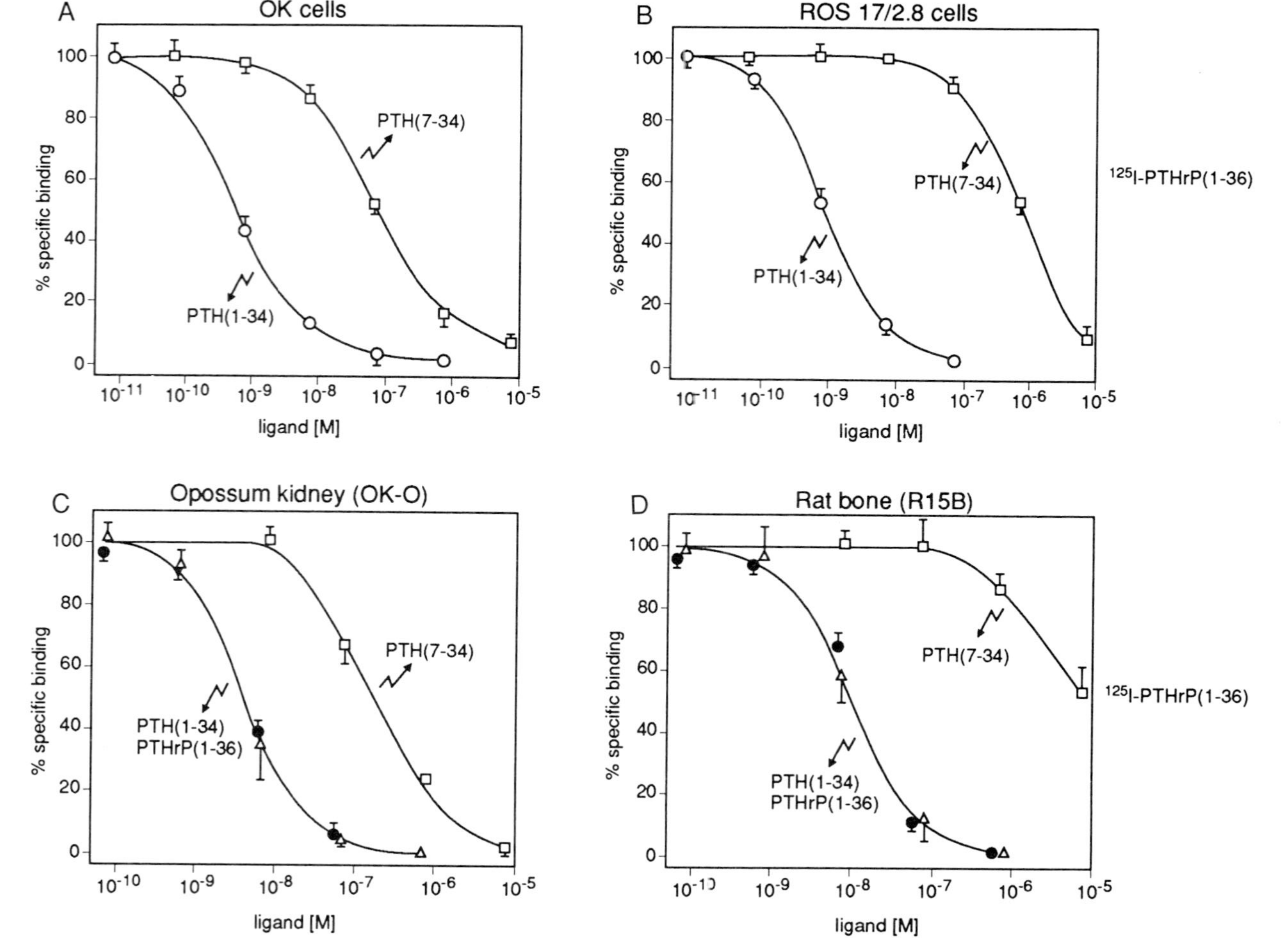

FIGURE 4 Radioreceptor studies using ^{125}I-labeled [Tyr^{36}]PTHrP (1-36)amide: Native OK cells (A), native ROS 17/2.8 cells (B), COS-7 cells transiently expressing OK-O (C), or R15B (D). Inhibition of radioligand binding by [$Nle^{8,18}$,Tyr^{34}]bPTH(1-34)amide,

and PTHrP equivalently, when either peptide was used as radioligand. Their apparent K_d values were 4 n*M* for COS-7 cells transfected with OK-O and OK-H compared to 0.5 n*M* in native OK cells. COS-7 cells transfected with R15B revealed an apparent K_d of 10 n*M* compared to 0.9 n*M* in native ROS 17/2.8 cells (Fig. 4). However, when stably transfected into either AtT-20 mouse pituitary cells or LLC-PK_1 cells at a receptor copy number of 10,000–100,000/cell (a receptor density approximating that of the native cells), the apparent K_d value of all three cloned receptors were about 0.5 n*M* (Bringhurst *et al.*, 1993; Abou-Samra *et al.*, 1993). Thus, the lower affinity of these receptors in transiently transfected COS-7 cells appears to be related to their high copy number (more than 10^6/COS-7 cell). The failure of many of these overexpressed receptors to couple efficiently to G proteins may well explain the lower affinity of the receptors for ligands (Gilman, 1987). PTH analogs, (3-34)amide and (7-34)amide, bind with progressively lower affinities in COS-7 cells transiently transfected with all three clones. PTH(7-34)amide binds less well to ROS 17/2.8 cells than to OK cells. In transfection experiments, this truncated ligand also binds less well to the receptor cloned from ROS 17/2.8 cells than to the receptor cloned from OK cells (Jüppner *et al.*, 1991a; Miyauchi *et al.*, 1990; Yamaguchi *et al.*, 1987). PTH and PTHrP stimulated cAMP accumulation about 10-fold in COS-7 cells transfected transiently with OK-O, OK-H, or R15B, yet more than 200-fold when stably expressed in either LLC-PK_1, CHO, or AtT_{20} cells (Bringhurst *et al.*, 1993; Abou-Samra *et al.*, 1993). The ED_{50}s for cAMP generation in all transfected cells were similar to those obtained with these ligands in the native cell lines (0.4–1 n*M*). Northern blot analysis of total RNA and poly(A)$^+$ RNA from OK cells, ROS 17/2.8 cells, and rat kidney, using cDNA probes that contain the entire coding region of either OK-O or R15B, revealed one major species of approximately 2.5 kb.

Previous studies in bone- and kidney-derived cells suggested that PTH and PTHrP not only activate adenylate cyclase but also activate additional second-messenger systems. Recent experiments with the cloned receptors show that these receptors can generate multiple signals (Abou-Samra *et al.*, 1992b). Fura 2-loaded COS-7 cells expressing OK-O and R15B gave a prompt, dose-dependent increase in intracellular free calcium when treated with either PTH or PTHrP. COS-7 cells transfected with both receptor clones showed an increase in their concentrations of inositol-trisphosphate when treated with either PTH or PTHrP. Interestingly, COS-7 cells transfected with OK-H showed little or no increase in intracellular free calcium and no increase in inositol phosphate metabolites in response to either PTH or PTHrP. These data indicate that activated renal and bone PTH/PTHrP receptors can stimulate two effectors, adenylate cyclase and phospholipase C. Additionally, while most of the carboxy-terminal tail of the PTH/PTHrP receptors appears to be unimportant for ligand binding or coupling to Gs, it seems to play an important role in coupling to the transducer molecule responsible for activating phospholipase C (Abou-Samra *et al.*, 1992b).

C. Summary

PTH/PTHrP receptors from marsupial kidney and from kidney and bone of two mammalian species, human and rat, are highly homologous. Our data establish that PTH and PTHrP bind with equal affinity to similar, perhaps identical, receptors in kidney and bone. PTH/PTHrP receptors stimulate adenylate cyclase with the same efficacy when activated by either ligand and are capable of activating at least two effector molecules, adenylate cyclase and phospholipase C. PTH/PTHrP receptors have significant homology only with the calcitonin and the secretin receptor and little or no similarities with other G protein-linked receptors. It thus appears likely that these three receptor types represent the first members of a novel family of G protein-linked receptors.

REFERENCES

Abbas, S. K., Pickard, D. W., Rodda, C. P., Heath, J. A., Hammonds, R. G., Wood, W. I., Caple, I. W., Martin, T. J., and Care, A. D. (1989). Stimulation of ovine placental calcium transport by purified natural and recombinant parathyroid hormone-related protein (PTHrP) preparations. *J. Exp. Physiol.* **74,** 549–552.

Abou-Samra, A. B., Jüppner, H., Khalifa, A., Karga, H., Kong, X. F., Schiffer-Alberts, D., Xie, L. Y., and Segre, G. V. (1993). Parathyroid hormone (PTH) stimulates adrenocorticotropin release in AtT-20 cells stably expressing a common receptor for PTH and PTH-related peptide. *Endocrinology* **132,** 801–805.

Abou-Samra, A. B., Jüppner, H., Force, T., Freeman, M. W., Kong, X. F., Schipani, E., Urena, P., Richards, J., Bonventre, J. V., Potts, J. T., Jr., Kronenberg, H. M., and Segre, G. V. (1992a). Expression cloning of a common receptor for parathyroid hormone and parathyroid-related peptide receptor from rat osteoblast-like cells: A single receptor stimulates intracellular accumulation of both cAMP and inosital trisphosphates and increases intracellular free calcium. *Proc. Natl. Acad. Sci. USA* **89,** 2732–2736.

Abou-Samra, A. B., Jüppner, H., Force, T., Karga, H., Iida-Klein, A., Xie, L. X., Bonventre, J. V., Kronenberg, H. K., and Segre, G. V. (1992b). The PTH/PTHrP receptor activates adenylate cyclase and phospholipase C through two distinct molecular domains, *74th Annual Meeting of the Endocrine Society,* abstract #43. San Antonio, TX.

Bringhurst, F. R., Jüppner, H., Guo, J., Urena, J. T., Potts, Jr., Kronenberg, H. M., Abou-Samra, A.-B., and Segre, G. V. (1993). Cloned, stably expressed parathyroid hormone PTH/PTH-related peptide receptors activate multiple messenger signals and biologic responses in LLC-PK_1 kidney cells. *Endocrinology* **132,** 2090–2098.

Budayr, A. A., Nissenson, R. A., Klein, R. F., Pun, K. K., Clark, O. H., Diep, D., Arnand, C. D., and Strewler, G. F. (1989). Increased serum levels of a parathyroid hormone-like protein in malignancy associated hypercalcemia. *Ann. Intern. Med.* **111,** 807–812.

Budayr, A. A., Halloran, B. P., King, J. C., Diep, D., Nissenson, R. A., and Strewler, G. J. (1990). High levels of a PTH-like protein in milk. *Proc. Natl. Acad. Sci. USA* **86,** 7183–7185.

Burtis, W. J., Brady, T. G., Orloff, J. J., Ersbak, J. B., Warrell, R. P., Jr., Olson, B. R., Wu, T. L., Mitnick, M. E., Broadus, A. E., and Stewart, A. F. (1990). Immunochemical characterization of circulating parathyroid hormone-related protein in patients with humoral hypercalcemia of cancer. *N. Engl. J. Med.* **322,** 1106–1112.

Burton, P. B. J., Moniz, C., Quirke, P., Tzannatos, C., Pickles, A., Dixit, M., Triffit, J. T., Jüppner, H., Segre, G. V., and Knight, D. E. (1990). Parathyroid hormone-related peptide in the uro-genital tract. *Mol. Cell Endocrinol.* **69,** R13–R17.

Chertow, B. S., Baylink, D. J., Wergedal, J. E., Su, M. H. H., and Norman, A. W. (1975). Decrease in serum immunoreactive parathyroid hormone in rats and in parathyroid hormone secretion *in vitro* by 1,25-dihydroxycholecalciferol. *J. Clin. Invest.* **56,** 668–678.

Chu, L. L. H., MacGregor, R. R., Anast, C. S., Hamilton, J. W., and Cohn, D. V. (1973). Studies on the biosynthesis of rat parathyroid hormone and proparathyroid hormone: Adaptation of the parathyroid gland to dietary restriction of calcium. *Endocrinology* **93,** 915–924.

Delmez, J. A., Tindira, C., Grooms, P., Dusso, A., Windus, D. W., and Slatopolsky, E. (1989). Parathyroid hormone suppression by intravenous 1,25-dihydroxyvitamin D. A role for increased sensitivity to calcium. *J. Clin. Invest.* **83,** 1349–1355.

Demay, M. B., Gerardi, J. M., DeLuca, H. F., and Kronenberg, H. M. (1990). DNA sequences in the rat osteocalcin gene that bind the 1,25-dihydroxyvitamin D_3 receptor and confer responsiveness to 1,25-dihydroxyvitamin D_3. *Proc. Natl. Acad. Sci. USA* **87,** 369–373.

Demay, M. B., DeLuca, H. F., and Kronenberg, H. M. (1991). Identification of sequences in the human parathyroid hormone gene that bind the 1,25-dihydroxyvitamin D_3 receptor. *J. Bone Miner. Res.* **6** (Suppl. 1), abstract #616.

Demay, M. B., Kiernan, M. S., DeLuca, H. F., and Kronenberg, H. M. (1992). Characterization of 1,25-dihydroxyvitamin D_3 receptor interactions with target sequences in the rat osteocalcin gene. *Mol. Endocrinol.* **6,** 557–562.

Deutsch, P. J., Hoeffler, J. P., Jameson, J. L., Lin, J. C., and Habener, J. F. (1988). Structural determinants for transcriptional activation by cAMP-responsive DNA elements. *J. Biol. Chem.* **263,** 18466–18472.

Farrow, S. M., Hawa, N. S., Karmali, R., Hewison, M., Walters, J. C., and O'Riordan, J. L. H. (1990). Binding of the receptor for 1,25-dihydroxyvitamin D_3 to the 5'-flanking region of the bovine parathyroid hormone gene. *J. Endocrinol.* **126,** 355–359.

Gearing, D. P., King, J. A., Gough, N. M., and Nicola, N. A. (1989). Expression cloning of a receptor for human granulocyte–macrophage colony-stimulating factor. *Embo. J.* **8,** 3667–3676.

Gilman, A. G. (1987). G proteins: Transducers of receptor-generated signals. *Annu. Rev. Biochem.* **56,** 615–649.

Habener, J. F., Kemper, B., and Potts, J. T., Jr. (1975). Calcium-dependent intracellular degradation of parathyroid hormone: A possible mechanism for the regulation of hormone stores. *Endocrinology* **97,** 431–441.

Heinrich, G., Kronenberg, H. M., Potts, J. T., Jr., and Habener, J. F. (1984). Gene encoding parathyroid hormone: Nucleotide sequence of the rat gene and deduced amino acid sequence of rat preproparathyroid hormone. *J. Biol. Chem.* **259,** 3320–3329.

Henderson, J. E., Shustik, C., Kremer, R., Rabbani, S. A., Hendy, G. N., and Goltzman, D. (1990). Circulating concentrations of parathyroid hormone-like peptide in malignancy and in hyperparathyroidism. *J. Bone Miner. Res.* **5,** 105–113.

Hendy, G. N., Kronenberg, H. M., Potts, J. T., Jr., and Rich, A. (1981). Nucleotide sequence of cloned cDNAs encoding human preproparathyroid hormone. *Proc. Natl. Acad. Sci. USA* **78,** 7365–7369.

Hongo, T., Kupfer, J., Enomoto, H., Sharifi, B., Giannela-Neto, D., Forrester, J. S., Singer,

F. R., Goltzman, D., Hendy, G. N., Pirola, C., Fagin, J. A., and Clemens, T. L. (1991). Abundant expression of parathyroid hormone-related protein in primary rat aortic smooth muscle cells accompanies serum-induced proliferation. *J. Clin. Invest.* **88,** 1841–1847.

Ishihara, T., Nakamura, S., Kaziro, Y., Takahashi, T., Takahashi, K., and Nagata, S. (1991). Molecular cloning and expression of a cDNA encoding the secretin receptor. *Embo. J.* **10,** 1635–1641.

Ishimi, Y., Russell, J., and Sherwood, L. M. (1990). Regulation by calcium and 1,25-$(OH)_2D_3$ of cell proliferation and function of bovine parathyroid cells in culture. *J. Bone Miner. Res.* **5,** 755–760.

Jüppner, H., Abou-Samra, A. B., Uneno, S., Gu, W. X., Potts, J. T., Jr., and Segre, G. V. (1988). The parathyroid hormone-like peptide associated with humoral hypercalcemia of malignancy and parathyroid hormone bind to the same receptor on the plasma membrane of ROS 17/2.8 cells. *J. Biol. Chem.* **263,** 8557–8560.

Jüppner, H., Abou-Samra, A. B., Freeman, M. W., Kong, X., Richards, J., Schipani, E., Hock, J., Potts, J. T., Jr., Kronenberg, H. M., and Segre, G. V. (1991a). Characterization of cDNAs encoding bone and renal parathyroid hormone (PTH)/PTH-related peptide (PTHrP) receptors, *73rd Annual Meeting of the Endocrine Society,* Washington, D.C., abstract #884.

Jüppner, H., Abou-Samra, A.-B., Mason, F., Kong, X. F., Schipani, E., Richards, J., Kolakowski, L. F., Jr., Kronenberg, H. M., and Segre, G. V. (1991b). A G protein-linked receptor for parathyroid hormone and parathyroid hormone-related peptide. *Science* **254,** 1024–1026.

Karmali, R., Farrow, S., Hewison, M., Barker, S., and O'Riordan, J. L. H. (1989). Effect of 1,25-dihydroxyvitamin D_3 and cortisol on bovine and human parathyroid cells. *J. Endocrinol.* **123,** 137–142.

Karpf, D. B., Arnaud, C. D., King, K., Bambino, T., Winer, J., Nyiredy, K., and Nissenson, R. A. (1987). The canine renal parathyroid hormone receptor is a glycoprotein: Characterization and partial purification. *Biochemistry* **26,** 7825–7833.

Karpf, D. B., Bambino, T., Alford, G., and Nissenson, R. A. (1991). Features of the renal parathyroid hormone–parathyroid hormone-related protein receptor derived from structural studies of receptor fragments. *J. Bone Miner. Res.* **6,** 173–182.

Khosla, S., Demay, M., Pines, M., Hurwitz, S., Potts, J. T., Jr., and Kronenberg, H. M. (1988). Nucleotide sequence of cloned cDNAs encoding chicken preproparathyroid hormone. *J. Bone Miner. Res.* **3,** 689–698.

Klein, P., Kanehisa, M., and DeLis, C. (1985). The deduction and classification of membrane-spanning proteins. *Biochim. Biophys. Acta* **815,** 468–476.

Korkor, A. B. (1987). Reduced binding of [^{3}H]1,25-dihydroxyvitamin D_3 in the parathyroid glands of patients with renal failure. *N. Engl. J. Med.* **316,** 1573–1577.

Lefkowitz, R. J., and Caron, M. G. (1988). Adrenergic receptors. *J. Biol. Chem.* **263,** 4993–4996.

Lin, H. Y., Harris, T. L., Flanery, M. S., Aruffo, A., Kaji, E. H., Gorn, A., Kolakowski, L. F., Jr., Lodish, H. F., and Goldring, S. R. (1991). Expression cloning of an adenylase cyclase-coupled calcitonin receptor. *Science* **254,** 1022–1024.

Loosfelt, H., Misrahi, M., Atger, M., Salesse, R., Vu-Hai-Luu-Thi-M.-T., Jolivet, A., Guiochon-Mantel, A., Sar, S., Jallal, B., Garnier, J., and Milgrom, E. (1989). Cloning and sequencing of porcine LH-hCG receptor cDNA: Variants lacking transmembrane domain. *Science* **245,** 525–528.

Mangin, M., Webb, A. C., Dreyer, B. E., Posillico, J. T., Ikeda, K., Weir, E. C., Stewart, A. F., Bander, N. H., Milstone, L., Barton, D. E., Francke, U., and Broadus, A. E. (1988). Identification of a cDNA encoding a parathyroid hormone-like peptide from a human tumor associated with humoral hypercalcemia of malignancy. *Proc. Natl. Acad. Sci. USA* **85,** 597–601.

Mayer, G. P., Keaton, J. A., Hurst, J. G., and Habener, J. F. (1979). Effects of plasma calcium concentration on the relative proportion of hormone and carboxyl fragments in parathyroid venous blood. *Endocrinology* **104,** 1778–1784.

McCauley, L. K., Rosol, T. J., Merryman, J. E., and Capen, C. C. (1992). Parathyroid hormone-related protein binding to human T-cell lymphotropic virus type 1-infected lymphocytes. *Endocrinology* **130,** 300–306.

McFarland, K. C., Sprengel, R., Phillips, H. S., Köhler, M., Rosemblit, N., Nikolics, K., Segaloff, D. L., and Seeburg, P. H. (1989). Lutropin-choriogonadotropin receptor: An unusual member of the G protein-coupled receptor family. *Science* **245,** 494–499.

Miyauchi, A., Dobre, V., Rickmeyer, M., Cole, J., Forte, L., and Hruska, K. A. (1990). Stimulation of transient elevations in cytosolic Ca2+ is related to inhibition of Pi transport in OK cells. *Am. J. Physiol.* **259,** F485–F493.

Naveh-Many, T., and Silver, J. (1990). Regulation of parathyroid hormone gene expression by hypocalcemia, hypercalcemia, and vitamin D in the rat. *J. Clin. Invest.* **86,** 1313–1319.

Naveh-Many, T., Friedlaender, M. M., Mayer, H., and Silver, J. (1989). Calcium regulates parathyroid hormone messenger ribonucleic acid (mRNA), but not calcitonin mRNA *in vivo* in the rat. Dominant role of 1,25-dihydroxyvitamin D. *Endocrinology* **125,** 275–280.

Naveh-Many, T., Marx, R., Keshet, E., Pike, J. W., and Silver, J. (1990). Regulation of 1,25-dihydroxyvitamin D_3 receptor gene expression by 1,25-dihydroxyvitamin D_3 in the parathyroid *in vivo*. *J. Clin. Invest.* **86,** 1968–1975.

Naylor, S. L., Sakaguchi, A. Y., Szoka, P., Hendy, G. N., Kronenberg, H. M., Rich, A., and Shows, T. B. (1983). Human parathyroid hormone gene (PTH) is on short arm of chromosome 11. *Somatic Cell Genet.* **9,** 609–616.

Nickols, G. A., Nickols, M. A., and Helwig, J.-J. (1990). Binding of parathyroid hormone and parathyroid hormone-related protein to vascular smooth muscle of rabbit renal microvessels. *Endocrinology* **126,** 721–727.

Nissenson, R. A., Diep, D., and Strewler, G. J. (1988). Synthetic peptides comprising the amino-terminal sequence of a parathyroid hormone-like protein from human malignancies. *J. Biol. Chem.* **263,** 12866–12871.

Noda, M., Vogel, R. L., Craig, A. M., Prahl, J., DeLuca, H. F., and Denhardt, D. T. (1990). Identification of a DNA sequence responsible for binding of the 1,25-dihydroxyvitamin D_3 receptor and 1,25-dihydroxyvitamin D_3 enhancement of mouse secreted phosphoprotein 1 (Spp-1 or osteopontin) gene expression. *Proc. Natl. Acad. Sci. USA* **87,** 9995–9999.

Okazaki, T., Igarashi, T., and Kronenberg, H. M. (1988). 5'-Flanking region of the parathyroid hormone gene mediates negative regulation by 1,25-$(OH)_2$ vitamin D_3. *J. Biol. Chem.* **263,** 2203–2208.

Okazaki, T., Igarashi, T., and Ogata, E. (1991a). The conserved mechanism of negative gene regulation by extracellular calcium. *J. Bone Miner. Res.* **6**(Suppl. 1), abstract #390.

Okazaki, T., Zajac, J. D., Igarashi, T., Ogata, E., and Kronenberg, H. M. (1991b). Negative regulatory elements in the human parathyroid hormone gene. *J. Biol. Chem.* **266,** 21903–21910.

Orloff, J. J., Wu, T. L., Heath, H. W., Brady, T. G., Brines, M. L., and Stewart, A. F. (1989a). Characterization of canine renal receptors for the parathyroid hormone-like protein associated with humoral hypercalcemia of malignancy. *J. Biol. Chem.* **264,** 6097–6103.

Orloff, J. J., Wu, T. L., and Steward, A. F. (1989b). Parathyroid hormone-like proteins: Biochemical responses and receptor interactions. *Endocr. Rev.* **10,** 476–495.

Parmentier, M., Libert, F., Maenhaut, C., Lefort, A., Gerard, C., Perret, J., Van Sande, J., Dumont, J. E., and Vassart, G. (1989). Molecular cloning of the thyrotropin receptor. *Science* **246,** 1620–1622.

Peraldi, M.-N., Rondeau, E., Jousset, V., El M'Selmi, A., Lacave, R., Delarue, F., Garel,

J.-M., and Sraer, J.-D. (1990). Dexamethasone increases preproparathyroid hormone messenger RNA in human hyperplastic parathyroid cells *in vitro*. *Eur. J. Clin. Invest.* **20,** 392–397.

Rupp, E., Mayer, H., and Wingender, E. (1990). The promoter of the human parathyroid hormone gene contains a functional cyclic AMP-response element. *Nucleic Acids Res.* **18,** 5677–5683.

Russell, J., and Sherwood, L.-M. (1989). Nucleotide sequence of the DNA complementary to avian (chicken) preproparathyroid hormone mRNA and the deduced sequence of the hormone precursor. *Mol. Endocrinol.* **3,** 325–331.

Russell, J., Lettieri, D., and Sherwood, L. M. (1983). Direct regulation by calcium of cytoplasmic messenger ribonucleic acid coding for pre-proparathyroid hormone in isolated bovine parathyroid cells. *J. Clin. Invest.* **72,** 1851–1855.

Russell, J., Lettieri, D., and Sherwood, L. M. (1986). Suppression by $1,25(OH)_2D_3$ of transcription of the pre-proparathyroid hormone gene. *Endocrinology* **119,** 2864–2866.

Schipani, E., Karga, H., Karaplis, A. C., Potts, J. T., Jr., Kronenberg, H. M., Segre, G. V., Abou-Samra, A.-B., and Jüppner, H. (1993). Identical complementary deoxyribonucleic acids encode a human renal and bone parathyroid hormone (PTH)/PTH-related peptide receptor. *Endocrinology,* **132,** 2157–2165.

Seed, B. (1987). An LFA-3 cDNA encodes a phospholipid-linked membrane protein homologous to its receptor CD2. *Nature (London)* **329,** 840–842.

Seed, B., and Aruffo, A. (1987). Molecular cloning of the CD2 antigen, the T-cell erythrocyte receptor, by a rapid immunoselection procedure. *Proc. Natl. Acad. Sci. USA* **84,** 3365–3369.

Sherwood, L. M., Potts, J. T., Jr., Care, A. D., Mayer, G. P., and Aurbach, G. D. (1966). Evaluation by radioimmunoassay of factors controlling the secretion of parathyroid hormone: Intravenous infusions of calcium and ethylenediamine tetraacetic acid in the cow and goat. *Nature (London)* **209,** 52–55.

Shigeno, C., Hiraki, Y., Westerberg, D. P., Potts, J. T., Jr., and Segre, G. V. (1988a). Photoaffinity labeling of parathyroid hormone receptors in clonal rat osteosarcoma cells. *J. Biol. Chem.* **263,** 3864–3871.

Shigeno, C., Hiraki, Y., Westerberg, D. P., Potts, J. T., Jr., and Segre, G. V. (1988b). Parathyroid hormone receptors are plasma membrane glycoproteins with asparagine-linked oligosaccharides. *J. Biol. Chem.* **263,** 3872–3878.

Shigeno, C., Yamamoto, I., Kitamura, N., Noda, T., Lee, K., Sone, T., Shiomi, K., Ohtaka, A., Fujii, N., Yajima, H., and Konishi, J. (1988c). Interaction of human parathyroid hormone-related peptide with parathyroid hormone receptors in clonal rat osteosarcoma cells. *J. Biol. Chem.* **263,** 18369–18377.

Silver, J., Russell, J., and Sherwood, L. M. (1985). Regulation by vitamin D metabolites of messenger ribonucleic acid for preproparathyroid hormone in isolated bovine parathyroid cells. *Proc. Natl. Acad. Sci. USA* **82,** 4270–4273.

Silver, J., Naveh-Many, T., Mayer, H., Schmelzer, H. J., and Popovtzer, M. M. (1986). Regulation by vitamin D metabolites of parathyroid hormone gene transcription *in vivo* in the rat. *J. Clin. Invest.* **78,** 1296–1301.

Sprengel, R., Braun, T., Nikolics, K., Segaloff, D. L., and Seeburg, P. H. (1990). The testicular receptor for follicle stimulating hormone: Structure and functional expression of clones cDNA. *Mol. Endocrinol.* **4,** 525–530.

Strewler, G. J., Stern, P. H., Jacobs, J. W., Eveloff, J., Klein, R. F., Leung, S. C., Rosenblatt, M., and Nissenson, R. A. (1987). Parathyroid hormone-like protein from human renal carcinoma cells. Structural and functional homology with parathyroid hormone. *J. Clin. Invest.* **80,** 1803–1807.

Sugimoto, T., Ritter, C., Ried, I., Morrissey, J., and Slatopolsky, E. (1988). Effect of 1,25-dihydroxyvitamin D_3 on cytosolic calcium in dispersed parathyroid cells. *Kidney Int.* **33,** 850–854.

Sugimoto, T., Brown, A. J., Ritter, C., Morrissey, J., Slatopolsky, E., and Martin, K. J. (1989). Combined effects of dexamethasone and 1,25-dihydroxyvitamin D_3 on parathyroid hormone secretion in cultured bovine parathyroid cells. *Endocrinology* **125,** 638–641.

Suva, L. G., Winslow, G. A., Wettenhall, R. E. H., Hammonds, R. G., Moseley, J. M., Diefenbach-Jagger, H., Rodda, C. P., Kemp, B. E., Rodriguez, H., Chen, E. Y., Hudson, P. J., Martin, T. J., and Wood, W. I. (1987). A parathyroid hormone-related protein implicated in malignant hypercalcemia: Cloning and expression. *Science* **237,** 893–896.

Thiede, M. A., and Rodan, G. A. (1988). Expression of a calcium-mobilizing parathyroid hormone-like peptide in lactating mammary tissue. *Science* **242,** 278–280.

Thiede, M. A., Daifotis, A. G., Weir, E. C., Brines, M. L., Burtis, W. J., Ikeda, K., Dreyer, B. E., Garfield, R. E., and Broadus, A. E. (1990). Intrauterine occupancy controls expression of the parathyroid hormone-related peptide gene in preterm rat myometrium. *Proc. Natl. Acad. Sci. USA* **87,** 6969–6973.

Thiede, M. A., Harm, S. C., McKee, R. L., Grasser, W. A., Duong, L. T., and Leach, R. M., Jr. (1991). Expression of the parathyroid hormone-related protein gene in the avian oviduct: Potential role as a local modulator of vascular smooth muscle tension and shell gland motility during the egg-laying cycle. *Endocrinology* **129,** 1958–1966.

Vasicek, T. J., McDevitt, B. E., Freeman, M. W., Fennick, B. J., Hendy, G. N., Potts, J. T., Jr., Rich, A., and Kronenberg, H. M. (1983). Nucleotide sequence of the human parathyroid hormone gene. *Proc. Natl. Acad. Sci. USA* **80,** 2127–2131.

Weaver, C. A., Gordon, D. F., and Kemper, B. (1982). Nucleotide sequence of bovine parathyroid hormone messenger RNA. *Mol. Cell Endocrinol.* **28,** 411.

Weaver, C. A., Gordon, D. F., Kissil, M. S., Mead, D. A., and Kemper, B. (1984). Isolation and complete nucleotide sequence of the gene for bovine parathyroid hormone. *Gene* **28,** 319–329.

Yamaguchi, D. T., Hahn, T. J., Iida-Klein, A., Kleeman, C. R., and Muallem, S. (1987). Parathyroid hormone-activated calcium channels in an osteoblast-like clonal osteosarcoma cell line. *J. Biol. Chem.* **262,** 7711–7718.

Yamamoto, M., Igarashi, T., Muramatsu, M., Fukagawa, M., Motokura, T., and Ogata, E. (1989). Hypocalcemia increases and hypercalcemia decreases the steady-state level of parathyroid hormone messenger RNA in the rat. *J. Clin. Invest.* **83,** 1053–1056.

11

MOLECULAR MECHANISMS OF CALCITONIN GENE TRANSCRIPTION AND POST-TRANSCRIPTIONAL RNA PROCESSING

SARA PELEG, GILBERT J. COTE, and ROBERT F. GAGEL

Cellular and Molecular Biology of Bone

I. INTRODUCTION

The peptide hormone calcitonin (CT) was discovered by Copp *et al.* (1962), who identified a humoral factor with potent hypocalcemic action in salmon. The 32-amino acid peptide was later purified from porcine thyroid gland and sequenced (Potts and Aurbach, 1976). In *in vitro* translation studies, the cloning of a complementary DNA encoding the rat CT precursor (Jacobs *et al.*, 1981) and the cloning of the chromosomal gene led to the realization that additional peptides are encoded by the CT gene (Rosenfeld *et al.*, 1984). The most important is the calcitonin gene-related peptide (CGRP). It is produced by a unique series of RNA processing events resulting in a messenger RNA (mRNA) species different from that of the CT mRNA (Fig. 1). Other peptides, produced by proteolytic cleavage of the precursor peptide for either CT or CGRP include amino-terminal proCT, carboxy-terminal proCT or katacalcin, amino-terminal proCGRP, and carboxy-terminal proCGRP (Fig. 1) (Jacobs *et al.*, 1981; Rosenfeld *et al.*, 1984).

In addition to the first calcitonin gene (CALC I in human or CT-α-CGRP in rat), a second gene has been identified (Amara *et al.*, 1985; Steenbergh *et al.*, 1985a). This gene (CALC II in human or β-CGRP in rat) resembles the first gene, but only a CGRP-like mRNA and a polypeptide are derived from its expression. A third gene (a pseudogene, CALC III in human), similar to the first and second genes, does not produce a functional protein product (Hoppener *et al.*, 1988).

The goal of this chapter is to discuss new information of importance regarding molecular mechanisms of transcription and RNA processing of the CALC I/CT-α-CGRP gene. To simplify the nomenclature, the human CALC I and the rat CT-α-CGRP genes will be referred to as the CT gene.

II. THE CELLULAR DISTRIBUTION OF CALCITONIN AND CALCITONIN GENE-RELATED PEPTIDE

The primary site of CT production in mammals is the parafollicular cell or C cell of the thyroid gland (Munson, 1976). This cell type originates in the neural crest and migrates, during embryonic life, to the fourth pharyngeal pouch. A mass of the C-cell precursors, collectively known as the ultimobranchial body, fuses with the thyroid gland as it migrates caudally during neck morphogenesis (Le Douarin, 1982). These cells account for greater than 90% of CT production. However, specific neuroendocrine cells in the thymus, lung, bronchial mucosa (Kulchitsky cells), prostate, gastrointestinal tract, adrenal medulla, and pituitary also transcribe the CT gene and process the transcript to CT mRNA and

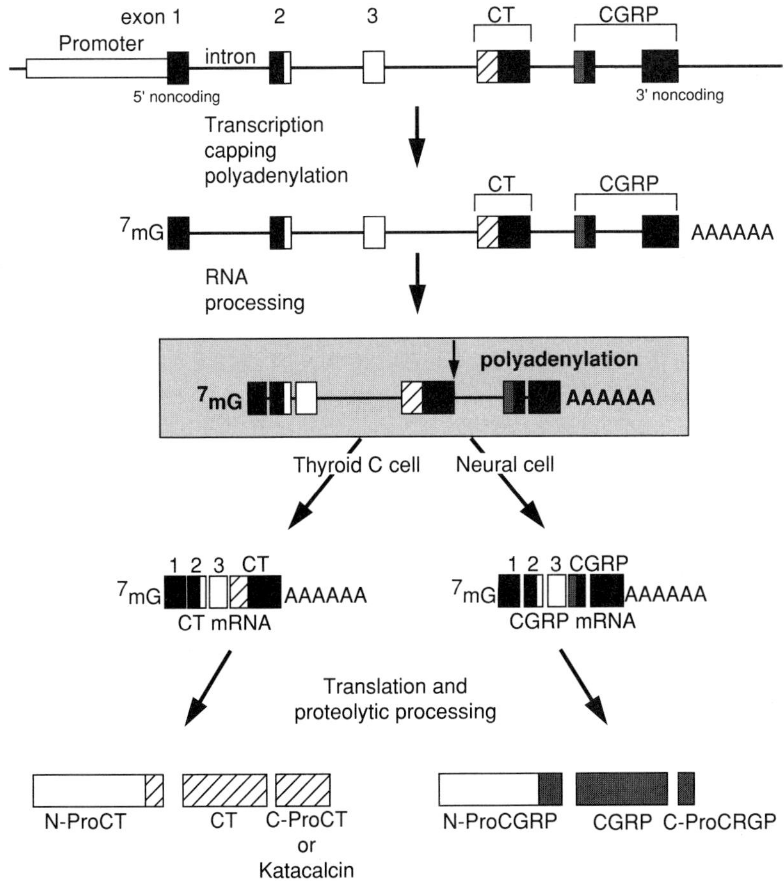

FIGURE 1 Schematic presentation of the structural calcitonin/calcitonin gene-related peptide (CT/CGRP) gene and alternative RNA processing pathways. The gene contains six exons and five introns. The hatched boxes represent transcribed, noncoding sequences. In C cells, CT messenger RNA (mRNA) is generated by splicing of exons 1–4. In neuronal cells, CGRP mRNA is produced by splicing of exons 1–3 to exons 5 and 6. Translation of each mRNA produces a polypeptide precursor, which is further processed by cleavage of a signal peptide, NH_2-terminal peptide, and COOH-terminal peptide. The principal bioactive peptides are CT (32 amino acids) and CGRP (37 amino acids).

protein (Grauer *et al.*, 1990). It is thought, though not proven, that these cells also originate from the neural crest.

Production of CGRP (from CALC I/α-CGRP or CALC II/β-CGRP genes) is restricted to certain areas of the central and peripheral nervous systems. The highest concentration of CGRP is in the trigeminal ganglia and dorsal root ganglia and in nerve fibers in the dorsal horn of the spinal cord (Amara *et al.*, 1985).

The CT gene is also transcribed in tumors containing neuroendocrine-type cells. These tumors process the primary transcript to both CT and CGRP mRNA. Medullary thyroid carcinoma (MTC), a tumor derived from the C cells, almost always expresses the CT gene. Other neuroendocrine-type malignancies including small cell carcinoma of the lung (SCLC) and tumors of the gastrointestinal tract, adrenal medulla, pancreatic islets, breast, and prostate express the CT gene with varying frequency (Grauer *et al.*, 1990). To date, there is no report describing transcription of the CT gene in tumors derived from the central or peripheral nervous systems.

III. TRANSCRIPTIONAL REGULATION OF THE CALCITONIN GENE

It is important to view transcriptional regulation of the CT gene within the context of general transcriptional regulatory mechanisms. Transcription of protein-encoding genes by eukaryotic RNA polymerase II is regulated by interaction of *trans*-acting factors with specific DNA sequences (*cis* elements). These sequences are most commonly located in a region of the DNA upstream of the transcription start site, defined as the promoter (Lewin, 1985). One of these elements, the TATA box, is located approximately 25 basepairs (bp) upstream of the transcription start site. The function of the TATA box is to direct RNA polymerase II-initiated transcription. Another element, defined by the sequence CCAAT, is found in the promoter of most eukaryotic genes and is located immediately upstream of the TATA box.

The cell-restricted transcription of a gene is mediated by one or a combination of *cis* elements that function to enhance or repress transcription. These elements are located in regions flanking the transcription unit or within the transcription unit itself. The function of enhancer elements, unlike that of the TATA box and the CCAAT element, is independent of distance or orientation with respect to the transcription start site.

A. Identification of *cis* Elements

The biological assay commonly applied for the identification of *cis* elements of transcription requires the construction of an artificial gene. The promoter and other genomic sequences of the target gene are fused to a promoterless reporter gene. The fusion gene is then transfected into cultured cells (normal or transformed) that express the gene of interest but not the reporter gene (Lewin, 1985). The transcriptional activity induced by the sequences of the target gene is reflected by the synthesis of the reporter mRNA and protein (Fig. 2). Important *cis* elements are identified by modification of the DNA sequence through deletion or

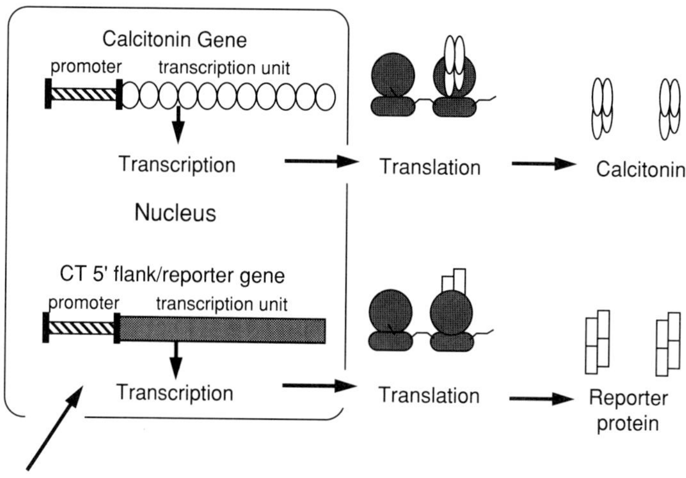

FIGURE 2 Schematic presentation of transient transfection assays. Transcription of a target gene is driven by promoter and enhancer elements, located at the 5′ flanking DNA of the gene. To examine transcriptional activation by enhancer elements, the promoter and 5′ flanking DNA of the target gene are isolated and subcloned upstream of a promoterless reporter gene. The new fusion gene is introduced into a cell type expressing the gene of interest and promoter/enhancer activities are determined by measurement of reporter gene messenger mRNA and protein. CT, calcitonin.

targeted mutation. A deletion or mutation that impairs transcription of the fusion gene indicates the likely location of a regulatory sequence. The possibility exists that the function of various *cis* elements in cultured cells may not represent accurately their function *in vivo*. The creation of transgenic animal models permits the study of transgene expression in the animal instead of cultured cells (Lewin, 1985). Moreover, the function of a *cis* element may be compromised by its separation from additional regulatory sequences located elsewhere on the chromosomal DNA. The introduction of selective mutations into the genome by homologous recombination techniques permits the study of functional elements of transcription in the context of the intact chromosomal DNA (Capecchi, 1989).

In this chapter, we describe the identification of *cis* elements of transcription of the CT gene by DNA transfer into cultured cells. The occasional difficulties in the interpretation of the results in this system are discussed.

1. Characterization of a Neuroendocrine-Specific Enhancer of Calcitonin Gene Transcription

The neuronal/neuroendocrine-restricted expression of the rat and human CT genes suggests that it is modulated by cell-specific *cis* elements of transcription. To identify and characterize *cis* elements of transcription, the rat and human chromosomal CT genes have been cloned from genomic libraries (Amara *et al.*, 1984; Steenbergh *et al.*, 1985b). Fragments from these clones have been subcloned and sequenced to identify the location of the transcription start site with respect to the isolated genomic fragment. Sequence information from the human CT gene showed a typical RNA polymerase II promoter: a TATA box at −30 to −25 (position of sequence upstream of the transcription start site is indicated by a minus symbol) and overlapping GGGCGG and CCAAT sequences between −75 and −65 (Cote *et al.*, 1990a). A 1.5 to 2.0-kilobase (kb) genomic fragment that includes the proximal promoter sequences and extends upstream of the transcription start site has been subcloned and fused in an upright orientation to promoterless reporter genes. These constructs or deletion mutants were transfected into a human MTC cell line that continually expresses the CT gene. The transcriptional activity of the transgene was monitored by measurement of reporter gene products (Peleg *et al.*, 1989, 1990; de Bustros *et al.*, 1990).

These experiments led to the identification of a 140-bp transcriptional enhancer region (Fig. 3), 1 kb upstream of the transcription start site for the human CT gene (Peleg *et al.*, 1990). The sequence fulfilled the definitions of an enhancer by virtue of retaining function at various distances and orientations with respect to the transcription start site. The human enhancer functions constitutively and contains at least two important regions: a 30-bp sequence between −1060 to −1030 (the upstream element [USE]) and a second element between −1030 and −920 (the downstream element [DSE]). Removal of the USE reduced basal transcriptional activity of the fusion gene by three- to fourfold, but there was significant residual activity from the DSE. The USE could not function on its own to enhance transcription but required the cooperation of the DSE. The DSE could function on its own to enhance transcription because attachment of this sequence to the proximal promoter of the CT gene or to a heterologous promoter enhanced transcription of the reporter gene by 30- to 100-fold. The full-length enhancer or the DSE alone functioned in CT producing MTC and SCLC cells, but not in cell lines that do not express the CT gene: HeLa (cervical carcinoma), CV1 (kidney), HIT (hamster insulinoma), and 3T3 (fibroblasts) (Table I). CT-negative MTC and SCLC cell lines did not utilize the cell-specific enhancer, thereby further confirming its specificity (Table I). Further mapping of the DSE by deletions and point mutations showed that the cell-specific component is located between −948 and −920 of the human CT gene (Fig. 3). We have defined this cell-specific enhancer as the *neuroendocrine-specific enhancer.*

Studies of the rat CT gene by Fredman *et al.* (1990) revealed the presence of a similar enhancer. Although the rat cell-specific enhancer

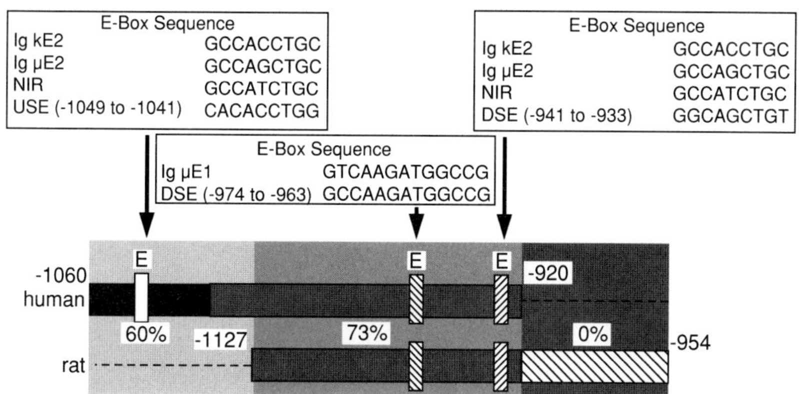

FIGURE 3 Schematic presentation of the cell-specific enhancer of the human CT gene and comparison with the rat enhancer. The fragment of the human CT gene located between 1060 and 920 upstream of the transcription start site has been mapped by deletion analyses and transient transfection assays. Two functional elements have been identified: an upstream element (USE, lightest shading) and a downstream element (DSE, medium shading). The functional sequences of these elements are E motifs (CANNTG) that resemble *cis* elements of the immunoglobulin light-chain enhancer (κE2), the heavy-chain enhancer (μE1, μE2), and the insulin NIR box. A comparison of human and rat sequence shows considerable homology in the regions containing the E motifs but no homology in a downstream region (darkest shading) thought to be involved in constitutive transcription of the rat gene. The boxes above show the specific sequence for the E motifs involved in cell-specific expression of the human CT gene and a comparison with several other known E motifs.

has not been mapped in detail, a comparison of the human and the rat sequence has revealed striking homology, particularly in the region corresponding to the human DSE (Fig. 3). In addition to sequence homology, the similarity extends to cell specificity: The rat enhancer is functional in MTC cells but not in nonendocrine CT-negative cells (Fredman *et al.*, 1990). One exception, however, was the finding that the rat enhancer was functional in a CT- and CGRP-negative neuronal cell line. These results suggest that other factors in addition to the cell-specific transcription factor may be required for transcription of the CT gene in neuronal cells. Unfortunately, a neuronal cell line that transcribes the CT gene is not available, making it difficult to determine which enhancer sequences are crucial for CT gene transcription in these cell types.

The enhancer described is not functional in the pluripotent teratocarcinoma F9 cells, although these cells transcribe the CT gene (Evain *et al.*, 1984). We found that F9 cells utilize an enhancer of transcription located between −250 and −129 of the human CT gene, instead of the neuroendocrine-specific enhancer. The important enhancer located between −250 and −129 will be described later in this chapter. This se-

TABLE I Cell Specificity of the CT/CGRP Enhancer of Basal Transcription

Cell type	Enhancer activity (fold stimulation)[a]
Medullary thyroid carcinoma	
TT	16
MZ-CRC	5
CA-77	35
rMTC 6-23	24
RO-D81[b]	2
Small cell lung carcinoma	
NCI-H209	7.0
NCI-H446[b]	0.8
Nonendocrine	
HeLa	0.2
C6	0.1
HIT[c]	0.9
CV1	0.5
3T3	1.0
F9	0.5

[a] Cells were transfected with a "promoter-only" construct (CT genomic fragment from −129 to +90 attached in an upright orientation to the promoterless growth hormone gene [reporter]) or with a "promoter + enhancer construct (CT fragment from −1333 to −734 attached in an upright orientation to the "promoter only" construct). Enhancer activity is expressed as fold stimulation of growth hormone production by the enhancer + promoter construct compared to growth hormone (GH) production by the "promoter-only" construct.

[b] CT-negative neural crest-derived neuroendocrine tumors.

[c] CT-negative endodermal-derived neuroendocrine tumor.

quence also enhances transcription of a transgene in CT-negative HeLa cells. Therefore, the function of this particular enhancer may be crucial for the transcription of the gene in pluripotent embryonic stem cells but is not sufficient for transactivation of the gene in HeLa cells.

To summarize, the neuroendocrine-specific enhancer sequences located between −1060 and −920 of the human CT gene are probably responsible for its expression in MTC and SCLC cell lines. There are, however, a number of fundamental but unanswered questions. These include the role of the neuroendocrine-specific enhancer in regulation of CT gene transcription in normal C cells and its role in normal or transformed neuronal cells. Generation of transgenic animal models in which the CT enhancer attached to a reporter is inserted into the germ line of mice may provide answers to these questions.

2. *trans*-Acting Factors of the Neuroendocrine-Specific Enhancer

Close inspection of the nucleic acid sequence of the human neuroendocrine-specific enhancer revealed that the functional elements defined by the transient transfection assays are E motifs (CANNTG). These motifs share considerable homology with *cis* elements of the immunoglobulin light- and heavy-chain enhancers as well as the insulin NIR box (Fig. 3); (Peleg *et al.*, 1990). Experiments were devised to examine the binding pattern of nuclear proteins derived from CT-producing cells to the E boxes of the CT enhancer and to define their relationship to nuclear factors that interact with similar E boxes.

The binding pattern of nuclear proteins from human CT-producing cells to DNA fragments containing the USE of the enhancer was examined by an electrophoretic mobility shift assay. In this assay system, the nuclear extract was combined with the radiolabeled USE, and the DNA–protein complexes formed were separated by nondenaturing polyacrylamide gel electrophoresis. Specificity of binding of TT-cell proteins was determined by addition of cold USE to the reaction mixture prior to the addition of the radiolabeled probe. Protein(s) in one complex interacted specifically with the USE (Fig. 4; Peleg *et al.*, 1990). Further characterization by competition with several oligonucleotides whose sequence was derived from related E motifs revealed a binding specificity for proteins in this complex to the immunoglobulin μE2 motif but not to μE1 or κE2 sequences and only to a limited extent to the insulin NIR box sequence (Fig. 4).

Examination of TT-cell nuclear protein binding to the labeled downstream cell-specific E motif revealed that several complexes were formed. Competition experiments with oligonucleotides indicated that only one complex included proteins that bound specifically to the E element. The proteins in this complex did not bind to the μE1, μE2, or κE2 sequences. Therefore, the protein(s) that bind to the neuroendocrine-specific component of the enhancer appear to be different from the protein(s) that bind to the upstream component of the enhancer (Fig. 5).

3. Is the Neuroendocrine-Specific Transcription Factor a Member of the Helix–Loop–Helix Group of Proteins?

The finding that E boxes modulate the cell-specific transcriptional activity of the CT enhancer provides a clue to the nature of the *trans* activa-

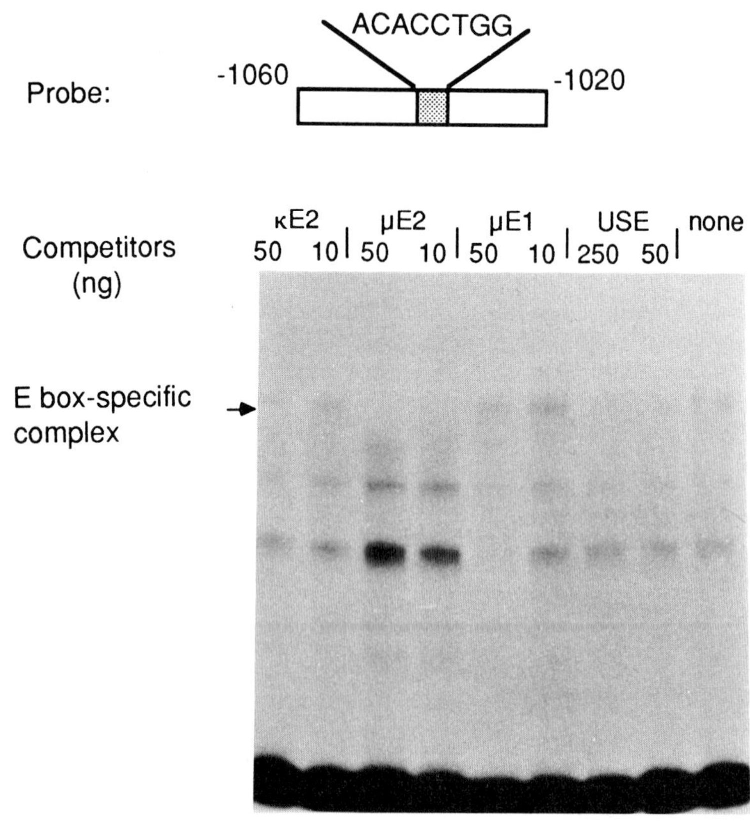

FIGURE 4 Analysis of protein interaction with the upstream element (USE) of the human CT enhancer. A crude nuclear extract from TT cells (10 μg/reaction) was incubated with ^{32}P-labeled USE. Incubations were at 22°C for 30 min ± competitors. Samples were then loaded onto a 4% PAGE (acrylamide : *bis*-acrylamide, 19 : 1), and electrophoretic separation of the protein–DNA complexes was performed at 4°C. The autoradiogram of the gel shows the DNA–protein complexes formed. Specificity of binding was determined by competition with unlabeled double-stranded oligonucleotides. The type of oligonucleotide competitor and concentration (ng/reaction) is indicated above each lane (sequences of competitors are shown in Fig. 3). The probe is illustrated above the autoradiogram.

tor(s) of these enhancer elements. Cell-specific *trans*-activation of gene expression via an E motif is a common feature of several proteins of the helix–loop–helix (H-L-H) family. The best characterized members of this family are the products of the *Drosophila* achaete-scute gene complex and the mammalian E2A gene (E12 and E47) as well as the *Drosophila* daughterless, mammalian MyoD, and c-*myc* proteins (Murre *et al.*, 1989). Based on this similarity, we hypothesize that the *trans*-acting factor of the cell-specific enhancer of the CT gene is also a member of the H-L-H family of proteins.

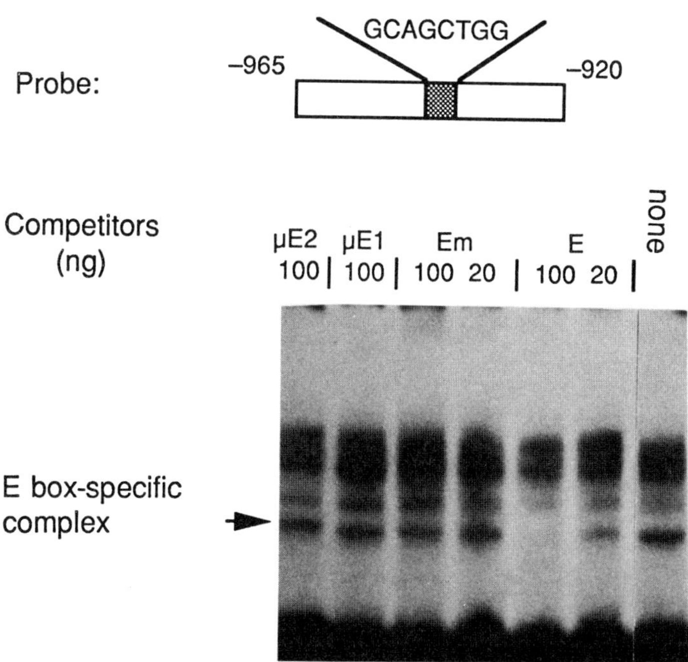

FIGURE 5 Analysis of protein interaction with the downstream element (DSE) of the human CT enhancer. Binding assays to the ^{32}P-labeled DSE probe illustrated above the autoradiogram were performed as described in Fig. 4. Competition studies were carried out with several oligonucleotides including the DSE probe used for binding (E), an oligonucleotide in which the E motif has been mutated from wild type (CAGCTG) to CTTCCG (Em) or oligonucleotides containing the μE1 and μE2 sequences.

In addition to *trans*-activation of gene expression, H-L-H proteins take part in regulation of cellular differentiation. For example, the H-L-H proteins encoded by the *Drosophila* achaete–scute gene complex regulate development by the peripheral nervous system (Cabrera *et al.*, 1987). Two mammalian homologs of the *Drosophila* achaete-scute have been recently cloned from a sympathoadrenal progenitor cell line (MASH-1 and MASH-2). MASH-1 is transiently transcribed in the brain during embryonic development (Johnson *et al.*, 1990). Its gene product is barely detectable in normal adult adrenal medulla but is expressed in pheochromocytoma and MTC, both transformed neural crest-derived endocrine cells. Treatment of chromaffin cells (normal or transformed) with nerve growth factor causes a shift in cell morphology and function toward the neuronal phenotype (away from an endocrine phenotype) and is associated with enhanced expression of the MASH-1 gene (Johnson *et al.*, 1990). These results are consistent with the hypothesis that MASH-1 is involved in neuronal differentiation of the neural crest. It is tempting to speculate that another H-L-H protein, perhaps closely related to

MASH, is responsible for the endocrine differentiation of neural crest-derived C cells and for the cell-restricted transcription of the CT gene.

B. Transcriptional Activation of the Calcitonin Gene by cAMP and Phorbol Esters

The treatment of human and rat MTC cells with cyclic adenosine monophosphate (cAMP) analogs or phorbol esters causes a rapid increase in the steady-state level of CT and CGRP mRNA (de Bustros *et al.*, 1986). Nuclear runoff transcription assays have confirmed that the changes in the levels of these mRNA species are due to an enhanced rate of transcription rather than modification in stability. Combined treatment with these two agents had an additive rather than a synergistic effect on transcription, suggesting that the two drugs activated transcription via separate pathways.

Cyclic AMP and phorbol ester activate various transcription factors. Cyclic AMP induces protein kinase A (PKA)-mediated phosphorylation of the cAMP response element binding (CREB) proteins (Hoeffler *et al.*, 1988). Phosphorylation is essential for the CREB proteins to function as enhancers of transcription. PKA-mediated phosphorylation also augments transcriptional activity of activator protein 2 (AP 2) (Imagawa *et al.*, 1987). An indirect PKA-mediated phosphorylation event can also activate progesterone receptor (Denner *et al.*, 1990). Phorbol esters or their cellular homolog, diacylglycerol, induce another phosphorylation pathway mediated by protein kinase C (PKC). Transcription factors activated by this mechanism are the activator protein 1 (AP 1) complex (*jun* and *fos* proteins) (Chiu *et al.*, 1987, 1988) and AP 2 (Imagawa *et al.*, 1987).

A homology search for cAMP and phorbol ester response elements on the CT 5′ flanking DNA revealed several potential binding sites for AP 1, AP 2, and CREB (Fig. 6; Cote *et al.*, 1990a). Treatment of human MTC cells, transfected with the human CT 5′ flanking DNA attached to a reporter gene, with either cAMP or phorbol esters, enhanced the basal production of reporter protein by severalfold (Cote *et al.*, 1990a). Detailed mapping of the CT 5′ flanking DNA by deletion mutation by De Bustros *et al.* (1989) located a cAMP-induced enhancer between −250 and −129. The cAMP-induced enhancer is not cell-specific in its function: HeLa and F9 cells utilize this sequence as a constitutive enhancer, the activity of which can be further augmented by cAMP treatment (Peleg, Monla, and Gagel, unpublished). Close inspection of this enhancer revealed two CREB-like binding sites and one AP 2-like binding site (Fig. 6).

Additional studies are needed to characterize this region further and to determine whether these elements function independently or in con-

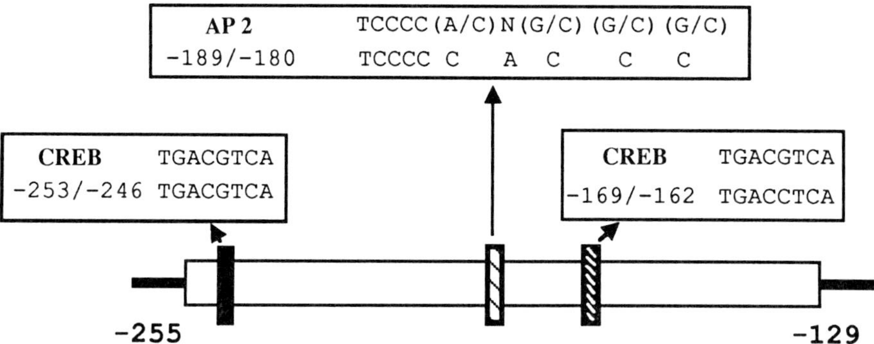

FIGURE 6 Schematic presentation of the cyclic adenosine monophosphate (cAMP)-induced enhancer of the calcitonin (CT)/calcitonin gene-related peptide gene. This figure shows potential *cis* elements within the region of the gene shown, by transient transfection assays, to mediate cAMP enhancement of CT gene transcription. The specific location of potential binding sites for cAMP response element binding (CREB) and activator protein II (AP II) is indicated by boxes that contain a comparison of the CT sequences and the consensus sequence for CREB and AP II.

cert. One targeted mutation that destroyed a CREB site within this region had no effect on the overall response of the cAMP-induced enhancer (Peleg, unpublished results). Which of the remaining elements is functional by individually mutating each element remains to be seen. The nature of the *trans*-acting factors that interact with these elements has not been elucidated, and their role in the *trans*-activation of CT transcription in the normal C cell is not known.

Preliminary evidence suggests the possibility of an interaction between the cell-specific enhancer (−1060 to −920) and the cAMP-induced enhancer (−250 to −129). Deletion mutation experiments (Peleg, unpublished; de Bustros *et al.*, 1990) show independent functioning of each of these enhancers. The combined activity of the two enhancers, however, is synergistic rather than additive, suggesting the possibility of positive cooperation. It seems that transcription of the CT gene is the result of a complex interaction between neuroendocrine-specific E motifs located between −1060 and −920 and the cAMP-induced enhancer elements between −250 and −129. Additional investigation will be required to define these relationships.

C. Regulation of the Calcitonin Gene Transcription by Cell Cycle Events

Two lines of evidence indicate a relationship between cell growth and transcriptional rate of the CT gene. First, studies by Nelkin *et al.* (1989) in human MTC cells demonstrate an inverse relationship between cell growth and transcription of the CT gene. Steady-state levels of CT and CGRP mRNA are relatively low during the logarithmic growth phase but increase as the cells approach confluence. Second, several pharmacologic agents that inhibit cell growth, including cAMP analogs, phorbols esters, and sodium butyrate, enhance the transcriptional rate of the CT gene (de Bustros *et al.*, 1986; Nakagawa *et al.*, 1988). In addition, stable expression of the activated oncogene, *v*-Harvey-*ras* (*v*-Ha-*ras*) in MTC and SCLC cell lines reduces the rate of cell growth and is associated with an increased transcription of the CT gene (Mabry *et al.*, 1989; Nakagawa *et al.*, 1987).

The enhancement of transcription by cAMP and phorbol esters is rapid and precedes the growth inhibitory effect. It is reasonable to assume that enhancement of transcription by these agents depends on PKA- or PKC-mediated phosphorylation of *trans*-acting factors. In contrast, the effects of sodium butyrate and *v*-HA-*ras* are delayed and probably indirect. Enhanced transcription of the CT gene is observed only after 48 hr of treatment with sodium butyrate (Nakagawa *et al.*, 1988) and 8 days after transfection of MTC cells with *v*-HA-*ras* (Nakagawa *et al.*, 1987).

The indirect mechanisms of *v*-Ha-*ras* action may involve a change in the availability of transcription factors. Nelkin *et al.* (1990) showed that stable transfection of *v*-Ha-*ras* into MTC cells was associated with an increase in *jun* expression. Similar results, indicating increased expression and function of *jun*, due to stable expression of *v*-Ha-*ras*, have been described in other cell culture systems (Binetruy *et al.*, 1991). A functional *jun*-mediated response element (AP 1 site) has not been detected so far on the CT 5′ flanking DNA. However, because *jun* can form a transcriptionally active heterodimer with CREB (Benbrook and Jones, 1990), the activity of the cAMP response elements described in the previous section might be affected by changes in functional *jun* protein.

D. Down-Regulation of Transcription of the Calcitonin Gene By Vitamin D_3

In the previous sections, we described positive regulation of CT gene transcription. The positive mechanisms involve two principal enhancers of transcription: a constitutive neuroendocrine-specific enhancer and a cAMP-induced enhancer. The negative regulation of CT gene transcription by vitamin D appears to occur through a complicated mechanism involving both constitutive transcription and a cAMP-induced transcription.

Constitutive transcription of the CT/CGRP gene is down-regulated in the normal thyroidal C cell by the biologically active metabolite of vitamin D_3, 1,25-dihydroxyvitamin D_3 ($1,25(OH)_2D_3$) (Naveh-Many and

Silver, 1988). The *in vivo* effect of vitamin D is rapid: The steady-state level of CT mRNA decreases to 6% of control within 6 hr. Nuclear runoff transcription assays confirm that the inhibitory effect is on nascent CT mRNA synthesis and not due to destabilization of the RNA. The effect is specific: $24,25(OH)_2D_3$, an inactive metabolite of vitamin D_3, has little effect on transcription. Moreover, steady-state levels of mRNA encoding somatostatin, another peptide hormone produced by the normal C cell, are not affected by this treatment.

The rapid response of normal C cells to vitamin D treatment suggests the possibility of a direct effect of vitamin D on CT gene transcription, perhaps by interaction of the vitamin D receptor with a *cis* element of transcription conferring a negative effect (a negative vitamin D response element). To further investigate the molecular mechanism of vitamin D action on CT gene transcription we have utilized transformed C-cell line (TT cells). Treatment with $1,25(OH)_2D_3$ but not $24,25(OH)_2D_3$ resulted in a dose-related decrease in steady-state levels of CT and CGRP mRNA (Cote *et al.*, 1987). Steady-state CT and CGRP mRNA levels were reduced to 50% of control at $1,25(OH)_2D_3$ concentrations of 10^{-8} to 10^{-7} *M* within 48 hr after initiation of sterol treatment. However, inhibition was not complete even after 96 hr of incubation with the sterol. The inhibitory effect of vitamin D in MTC cells is not specific to CT gene transcription; the sterol also inhibits somatostatin (Rogers *et al.*, 1989) and parathyroid hormone-related peptide gene transcription (Ikeda *et al.*, 1989). The effect of vitamin D did not appear to be related to changes in growth rate of the cells (Cote *et al.*, 1987).

To identify *cis* elements involved in vitamin D action, 1.5 kb of CT 5′ flanking DNA fused to a promoterless growth hormone gene was transfected into TT cells (Fig. 7). Treatment of the transfected cells for 5 days with $1.25(OH)_2D_3$ (10^{-7} *M*) induced less than 20% inhibition of the constitutive transcription of the transgene (Fig. 7). Treatment, however, with the same concentration of $1,25(OH)_2D_3$ was sufficient to completely abolish the cAMP-induced transcriptional response of the fusion gene. There was no effect of $24,25(OH)_2D_3$ on either basal or cAMP-induced transcription. The specificity of the effect for the CT 5′ flanking DNA was confirmed by demonstrating that transcription of the Rous sarcoma virus long terminal repeat attached to chloramphenicol acetyltransferase gene was not altered by cAMP or vitamin D treatment when cotransfected into TT cells along with the CT/growth hormone fusion gene (Fig. 7). The conclusion drawn from these experiments was that sequence within the 1.5 kb of CT 5′ flanking DNA probably confers the negative response to vitamin D.

It is intriguing but not surprising that $1,25(OH)_2D_3$ inhibited cAMP-mediated transcription but not basal transcription of the CT–reporter fusion gene in MTC cells. In previous sections of this chapter, it was

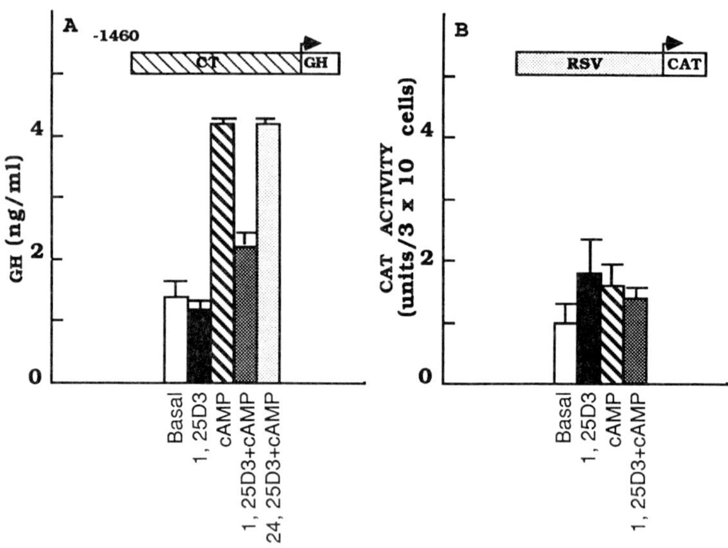

FIGURE 7 Vitamin D inhibits cyclic adenosine monophosphate (cAMP) enhancement of calcitonin gene transcription but has little effect on basal enhancement. TT cells were cotransfected with a plasmid that contains the 1460-basepairs of the CT/calcitonin gene-related peptide flanking DNA fused to the growth hormone gene (CT-GH) and a plasmid containing the Rous sarcoma virus long terminal repeat attached to the chloramphenicol acetyl transferase gene (RSV-CAT). Transcriptional activity by the transgenes was examined in untreated (BASAL), vitamin D-treated ($1{,}25D_3$), cAMP-treated (cAMP), vitamin D- and cAMP-treated ($1{,}25D_3$ + cAMP), and $24{,}25D_3$- and cAMP-treated (cAMP + $24{,}25\ D_3$) cells. Expression of CT-GH was monitored by radioimmunoassay of GH secreted into the medium. Expression of RSV-CAT was monitored by assay of CAT activity in cell extracts.

shown that basal and cAMP-induced transcription of the CT/growth hormone fusion gene are mediated by two widely separated enhancers of transcription. The vitamin D receptor may bind a sequence adjacent to or overlapping the cAMP-induced enhancer and interfere with the DNA binding or function of the cAMP-induced *trans*-activators without affecting the function of the cell-specific enhancer. Other potential explanations include a vitamin D receptor interference with the phosphorylation of CREB-like factor or an inactivation by direct interaction between the vitamin D receptor and the CREB-like transcription factor.

The finding that $1{,}25(OH)_2D_3$ inhibited only cAMP-mediated transcription with little or no effect on the constitutive cell-specific transcription of the transgene in MTC cells is not consistent with observations in the intact animal where $1{,}25(OH)_2D_3$ treatment reduces the constitutive transcription of the CT gene in normal C cells (Naveh-Many and Silver, 1988). Although these differences have not been completely reconciled, there are several possible explanations. First, the cAMP-induced enhancer might be more active in normal C cells, whereas the cell-specific enhancer is predominant in transformed C cells. Second, the basal transcription in normal C cells may be derived from constitutive activity of the cAMP-induced enhancer, although this enhancer is silent in MTC cells in the absence of added cAMP. Finally, the effects of vitamin D in transformed cells may be completely unrelated to those observed in normal cells, a possibility that seems unlikely but which will be excluded only by additional studies.

E. A Model for Transcription Regulation of the Calcitonin/Calcitonin Gene-Related Peptide Gene

The studies reviewed identified two regions of 5′ flanking DNA involved in transcription of the CT gene (Fig. 8). A region located approximately 1 kb upstream of the start of transcription contains several E motifs (CANNTG) that function together to drive cell-restricted, constitutive transcription in transformed neuroendocrine cells. It seems likely that a unique neuroendocrine-specific *trans*-acting factor of the H-L-H protein family is responsible for the cell-specific transcription of the gene. The direct correlation between CT gene transcription and the function of the neuroendocrine-specific enhancer in CT-positive and CT-negative tumor cells strongly suggests that this enhancer regulates the transcription of the endogenous gene in transformed neuroendocrine cells. Whether this enhancer is active in the normal C cells and in neuronal cells that express the CT gene is not clear at present, in part because of the absence of suitable model systems.

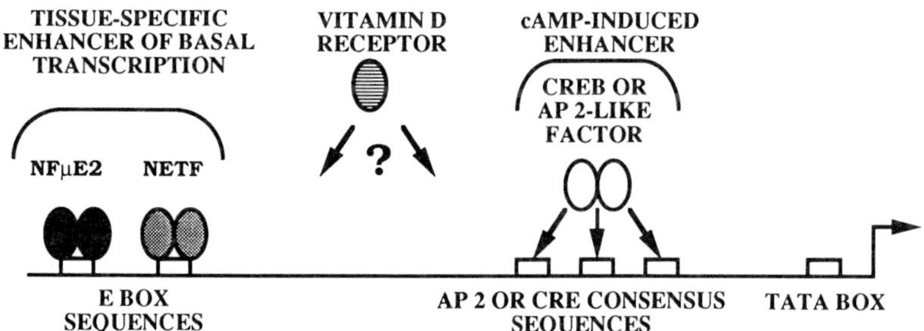

FIGURE 8 A model for transcriptional regulation of the calcitonin (CT) gene. Two principal enhancers of transcription have been identified within the 1460 nucleotides upstream of the transcription start site of the CT gene. A constitutive neuroendocrine-specific enhancer at approximately −1000 and a cyclic adenosine monophosphate (cAMP)-induced enhancer at −250. The *cis* elements of the cell-specific enhancer are E motifs, a likely binding site for *trans*-acting factors of the helix-loop-helix family. The *cis* elements of the cAMP-induced enhancer are similar to activator protein II (AP II) and cAMP response element (CREB) binding protein binding sites. Vitamin D appears to inhibit exclusively the cAMP-induced transcription, but its precise response element has not been identified.

The second enhancer is located approximately 250 bp upstream of the start of transcription. In MTC cells, this enhancer is silent unless activated by addition of cAMP analogs. However, some evidence suggests that this enhancer functions constitutively in the normal C cells. The relative importance of each enhancer region in normal cells may be defined by generating transgenic lines of mice that carry CT–reporter fusion transgenes including in these enhancer elements.

The relationship between the cell-specific and the cAMP-induced transcription factors is also unclear at present. Several lines of evidence suggest a significant synergism of cAMP-induced transcription by the neuroendocrine-specific *trans*-acting factors. To demonstrate interaction between proteins it will be necessary to purify and characterize the neuroendocrine-specific as well as the cAMP-induced *trans*-acting factors. A combination of studies in transgenic animals and *in vitro* should lead to a broader understanding of this fascinating transcription unit.

IV. POST-TRANSCRIPTIONAL PROCESSING OF THE CALCITONIN GENE PRIMARY TRANSCRIPT

The study of transcriptional and post-transcriptional aspects of CT production have been greatly facilitated through the use of MTC models. Serially passaged rat MTC cell lines produce and secrete high levels of CT (Roos *et al.*, 1979); however, some of these tumor lines lose the ability to produce CT when serially passaged. Studies of these tumors revealed no change in transcription rate of the CT gene and the presence of a CT-like precursor RNA. Analysis of this new mRNA species determined that it shared sequence homology (exons 1–3) with the CT mRNA and contained unique 3′ sequence that coded for the production of a previously unidentified peptide, CGRP (Fig. 1). This led to the proposal of an alternative RNA processing pathway for the CT gene primary transcript (Rosenfeld *et al.*, 1981). Cloning and sequence analysis of the CT gene as well as tissue expression and nuclear runoff studies provided convincing evidence for the alternative RNA processing pathway depicted in Fig. 1 (Amara *et al.*, 1982, 1984; Rosenfeld *et al.*, 1982, 1983). Two mRNA species are produced through a pathway involving alternative RNA processing of a single precursor RNA. Initial processing steps involve RNA capping and polyadenylation of the terminal exon 6. Subsequent processing steps are cell-specific; however, it is believed that both RNA processing pathways proceed through a common intermediate (Fig. 1) (Bovenberg *et al.*, 1986; Rosenfeld *et al.*, 1981, 1982). In thyroid C cells, processing continues by polyadenylation at the end of exon 4 and splicing of exon 3 to exon 4. This choice results in a mRNA encoding for the production of CT. In neuronal cells, an alternative RNA processing pathway is observed in which exon 3 is spliced to exon 5. This final mRNA encodes for the production of CGRP.

The genomic structure and RNA processing of the primary transcript of the human CT gene was subsequently shown to be identical to that found in the rat gene (Bovenberg *et al.*, 1986; Broad *et al.*, 1989; Edbrooke *et al.*, 1985; Jonas *et al.*, 1985; Nelkin *et al.*, 1984; Steenbergh *et al.*, 1985b). A similar processing choice also occurs for the chicken CT gene (Minvielle *et al.*, 1987). The remainder of this chapter focuses on alternative RNA processing of the rat and human CT genes.

A. Models for Studying RNA Processing of the Calcitonin Gene

In mammalian species, there is tissue-specific processing of the CT pre-mRNA. C cells produce 95–99% CT (the remainder being CGRP), and neuronal cells produce ~99% CGRP with little CT-specific processing (Amara *et al.*, 1982; Rosenfeld *et al.*, 1983, 1984). In contrast, the transformed C cell (a MTC cell) frequently processes the one precursor RNA to produce equivalent amounts of CT and CGRP (Cote and Gagel, 1986).

1. RNA Processing Decisions in the Malignant Thyroid C Cell

The co-expression of CT and CGRP in MTC cell lines provides a model for examining the unregulated processing decision. Several groups have examined factors affecting the CT/CGRP splicing ratios in these cell lines. Treatment of a human MTC cell line (TT cells) with the glucocorticoid dexamethasone resulted in an increased CT/CGRP mRNA ratio (Cote and Gagel, 1986). A similar increase is also observed in the rat CA-77 cell line (Russo *et al.*, 1988) but not in the rat 44-2C cell line (Zeytin *et al.*, 1987). Additional studies on the TT cell line revealed that treatment with sodium butyrate, high-density culturing, or transformation with *v*-Ha-*ras* oncogene also led to an increased CT/CGRP ratio (Nakagawa *et al.*, 1987, 1988; Nelkin *et al.*, 1989). Each of these experimental treatments inhibits cell growth and produces a more endocrine phenotype. This has led to the proposal that the state of differentiation may play a role in this RNA processing choice. This is likely to be an oversimplification. Treatment of these cells with phorbol esters or cAMP analogs (agents that slow cell growth) does not affect processing of the CT pre-mRNA (de Bustros *et al.*, 1986, 1990), suggesting that other factors must be involved in addition to cell growth.

2. *In Vivo* Models of Calcitonin Gene Alternative Splicing

The ideal model for studying alternative processing of the CT gene is a cell type that produces either CT or CGRP. MTC cells are a poor model for studying specific RNA processing choices because of their co-production of CT and CGRP. This problem was overcome by the development of tissue culture models. The CT gene was transfected into cells that make RNA processing decisions analogous to those observed *in vivo.* A variety of cell lines were stably transfected with the rat CT gene constructs driven by the immunoglobulin enhancer. When the intact rat CT gene was introduced into the lymphocytic A20 cell, it processed the CT pre-mRNA in a CT-specific fashion, whereas in PC12 adrenal medullary carcinoma or F9 teratocarcinoma cells, CGRP mRNA was the predominant product (Leff *et al.*, 1987). Two additional cell lines, HeLa and 293 (transformed human kidney) have since been shown to process RNA in a CT-specific pattern (Adema *et al.*, 1990; Cote *et al.*, 1991; Emeson *et al.*, 1989).

A second approach was the creation of transgenic mice expressing the CT gene via a heterologous mouse metallothionein-I transcription promoter (Crenshaw *et al.*, 1987). In these animals, CT-specific processing of the RNA precursor was observed in most tissues. CGRP-specific mRNA expression was restricted to neuronal cells. Based on these observations, they proposed that neuronal cells contain a unique RNA processing environment that is required for CGRP-specific RNA processing. Because CT production could occur in a number of tissues, they reasoned that the RNA processing environment of the thyroid C cell was not unique and that in the absence of the neuron-specific factors CT production is the default RNA processing pathway.

3. *In Vitro* Models of Calcitonin Gene Alternative Splicing

In vivo models of CT gene alternative RNA processing are well suited for the identification of RNA *cis*-regulatory sequences but have limited usefulness for the identification of regulatory factors. These limitations led to the development of *in vitro* systems capable of duplicating the CT alternative RNA processing decision. These systems rely on nuclear extracts prepared from the same cell lines used for transfection experiments. RNA processing is followed by using radiolabeled RNA precursors. The advantages of *in vitro* systems are that they allow detailed analysis of reaction kinetics and components, thereby providing a mechanism for identifying important *trans*-acting factors. Disadvantages include inefficient RNA processing and the possibility of artifacts due to an inability to reproduce the *in vivo* system completely.

B. Regulatory Elements of Calcitonin/Calcitonin Gene-Related Peptide pre-mRNA Splicing

Several groups have utilized the various model systems to identify important regulatory elements involved in the CT/CGRP processing system. While a precise regulatory mechanism is still unclear, several important observations have been made. The following sections summarize results of multiple laboratories, utilizing both *in vitro* and *in vivo* systems.

1. *cis*-Acting Elements of Calcitonin Gene Alternative Splicing

There are three major sites of RNA processing activity: the 5′ splice site, 3′ splice site, and polyadenylation site. The sequences of these sites demarcate exon–intron boundaries. For the CT pre-mRNA, the key processing decision is the inclusion or exclusion of exon 4; therefore, work has focused on known exon 4 regulatory elements and the sequences surrounding these sites. These sequence elements are illustrated in Figure 9. The rat and human RNA sequences at these positions have been compared to the vertebrate consensus.

Identification of important *cis*-regulatory sequence has been performed by three routes: deletion, substitution, or mutation. This methodology has been applied to both *in vivo* transfection systems and *in vitro* splicing assays. The combined observations of several investigators using rat and human gene sequences are summarized in Figure 10 and Table II. It is clear from these studies that the sequences comprising the 3′ splice site region of exon 4 play a major role in CT/CGRP RNA processing. Deletion of this region (rat alteration 3 and human alteration 4) results in CGRP-specific splicing in CT-producing systems both *in vivo* and *in vitro* (Cote *et al.*, 1990b, 1992; Emeson *et al.*, 1989; Leff *et al.*, 1987). Substitution or mutation of this region (rat alteration 6 and human alter-

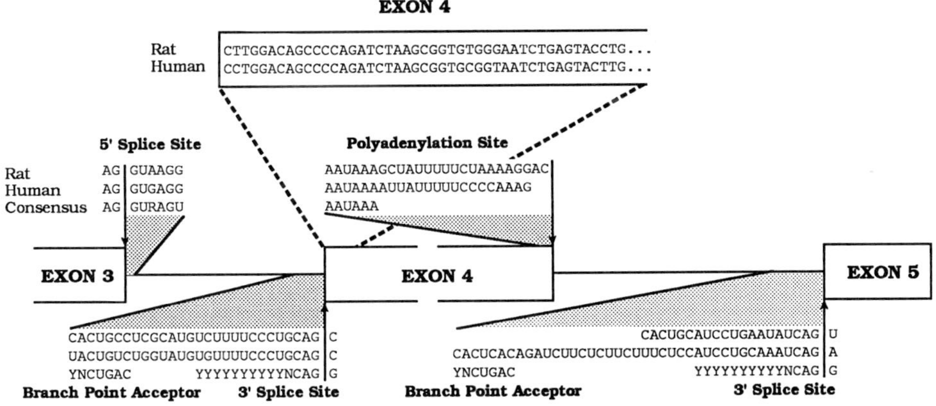

FIGURE 9 The calcitonin (CT) gene RNA processing signals deviate from consensus. The CT exon 4 and flanking exons are schematically represented. Relevant sequence for the rat and human genes is compared to the mammalian 5′ splice site, branch point acceptor, 3′ splice site, and polyadenylation consensus sequences. The first 45 nucleotides of exon 4 shown may also play a role in RNA processing (see text).

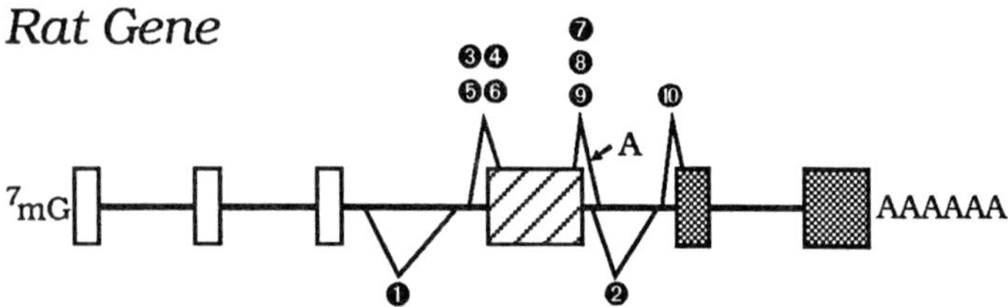

❶ - Intron 3 deletion: -587 to -207
❷ - Intron 4 deletion: -710 to -120
❸ - Deletions of the 3' splice site/branch point region: -63 to +13
❹ - Region: -58 to -38
❺ - Regions: -58 to -27 and -58 to -17
❻ - Substitution of -58 to +11 with analogous exon 3 sequence
❼ - *Bgl* 2 deletion of exon 4 polyadenylation signals
❽ - Substitution of polyadenylation sequence with analogous exon 6 sequence
❾ - Mutation of polyadenylation signals
❿ - Deletion of 3' splice site/branch point region: -187 to +18

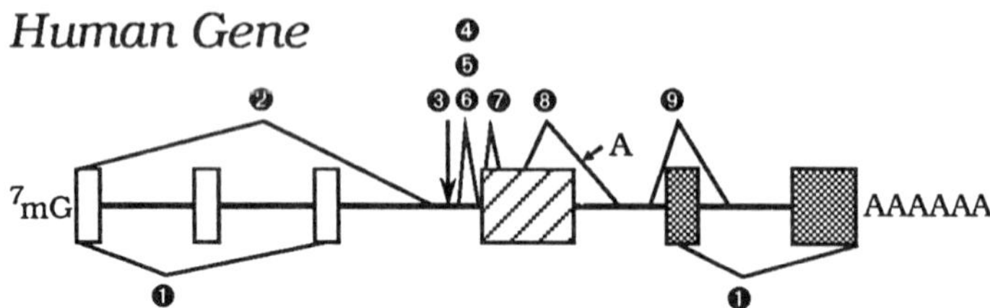

❶ - Simultaneous substitution of sequence through +53 of exon 3 with viral cap sequence and from +139 of exon 5 with human ß globin polyadenylation signal sequence
❷ - Substitution of sequence through -151 of intron 3 with viral cap and 5' splice site sequence
❸ - Mutation of the branch point acceptor: U>A
❹ - Deletion of 3' splice site region: -5 to -2
❺ - Insertion of polypyrimidine sequence
❻ - Substitution of -58 to -1 with analogous adenovirus exon 2 sequence
❼ - Mutation of exon 4: +2 to +29
❽ - Deletion of exon 4 polyadenylation signals
❾ - Substitution of exon 5 and flanking splice site signals with analogous adenovirus exon 3 sequence

FIGURE 10 Schematic representation localizing important *cis*-regulatory elements controlling calcitonin (CT) alternative RNA processing. The sequence of both the rat and human CT genes has been altered in numerous experiments to identify important *cis*-regulatory elements. The general location of these changes is illustrated within schematics of each pre-messenger mRNA. Detailed information regarding the alteration at specific regions is shown below each diagram. Sequence is identified relative to the intron–exon boundary, with the last intronic nucleotide −1 and the first exonic nucleotide +1. The RNA processing phenotypes resulting from the sequence alterations depicted are discussed in Table II.

TABLE II Effect of Sequence Alterations on CT/CGRP RNA Processing

	CT processing model		CGRP processing model		
	In vivo	*In vitro*	*In vivo*	*In vitro*	References[a]
Rat gene alteration					
1	CT[b]	NT[b]	CGRP[b]	NT[b]	1[a]
2	CT	NT	CGRP	NT	1
3	CGRP	NT	CGRP	NT	1
4	CT	NT	CGRP	NT	1
5	CT	NT	16–28% CT	NT	1
6	CT	NT	CT	NT	1
7	NP	NT	NP	NT	2
8	CT	NT	CGRP	NT	1
9	NP	NT	CGRP	NT	2
10	CT	NT	NP	NT	1
Human gene alteration					
1	CT	CGRP	NT	CGRP	3, 4, 6
2	CT	NT	CGRP	CGRP	7–9
3	CT	CT	CT	CGRP	5, 6, 9, 10
4	CGRP	CGRP	CGRP	NT	6
5	CGRP	CGRP	NT	NT	3
6	CT	CT	CT	CT	11
7	CGRP	CGRP	NT	NT	6
8	NT	NP	NT	CGRP	4
9	NT	CGRP	NT	CGRP	4

[a]*References*: 1, Emeson *et al.* (1989); 2, Leff *et al.* (1987); 3, Bovenberg *et al.* (1988); 4, Bovenberg *et al.* (1989); 5, Adema *et al.* (1988); 6, Adema *et al.* (1990); 7, Cote *et al.* (1990b); 8, Cote *et al.* (1991); 9, Cote *et al.* (1992); 10, Adema *et al.* (1991); 11, Cote *et al.* (unpublished observation).

[b]CGRP, calcitonin gene-related peptide; CT, calcitonin; NP, no processing; NT, not tested.

ations 3 and 6) results in CT-specific splicing in CGRP-producing systems (Adema and Baas, 1991; Adema *et al.*, 1990; Emeson *et al.*, 1989). A key element appears to be a noncanonical branch point acceptor (U in humans, C in rats) first identified by Adema *et al.* (1988). Recent experiments by Adema *et al.* (1991) have shown that a simple U to A mutation created in an essentially intact CT gene results in significant CT mRNA production in transfected F9 cells. There is, therefore, an absolute requirement for this suboptimal branch acceptor to maintain regulated CGRP expression in neuronal cells. The role of other sequences within the 3′ splice site region is unclear. Emeson *et al.* (1989) suggested that sequences upstream of the polypyrimidine tract serve to inhibit CT production in neuronal cells (rat alteration 5); however, their experiments introduce a potential cryptic branch point acceptor while deleting the predicted naturally occurring site, thereby raising questions about the validity of the observation (note that the CT-specific branch point acceptor has not yet been mapped for the rat gene). The sequence content of the pyrimidine tract itself may also play an important role. Adema *et al.* (1990) observed that increasing the polypyrimidine tract length by substitution of −5 to −2 (from the start of exon 4) with −17 to −2 of human β-globin exon 2 resulted in CGRP-specific splicing in transfected 293 cells (human alteration 5). Specific mutagenesis of the polypyrimidine sequence has not yet been performed to determine the significance of this observation.

In addition to the CT-specific 3′ splice site, the adjacent CT exon 4 sequence appears to play an important role in the regulation of splice site selection. Deletion or mutation of the sequence contained within the first 45 nucleotides of exon 4 results in CGRP production both *in vitro* and *in vivo*. CT-producing model systems (Cote *et al.*, 1990b, 1992). These sequences are unrelated to the constitutive splice sites involved in RNA processing and suggest this exonic element as a likely candidate for regulation of the "alternative" processing decision. The sequences within this region show no remarkable features other than a GCGGT direct repeat, observed only in the human gene (Fig. 9). The proximity of these sequences to the 3′ splice site and branch point suggest that these sequences might encompass a large *cis*-regulatory element spanning both regions.

The nature of the alternative processing decision suggests that exon 4 polyadenylation could also be a regulatory site. Replacement of exon 4 polyadenylation sequences with exon 6 sequences does not alter regulation of alternative processing in transfected cells (Emeson *et al.*, 1989), making it unlikely that regulation of polyadenylation is an important regulatory feature. Mutation of the exon 4 hexanucleotide consensus (rat alteration 9) blocks polyadenylation and inhibits splicing of exon 3 to exon 4 in transfected A20 cells (CT-producing cells) (Leff *et al.*, 1987). While inhibition of splicing by deletion of the exon 4 splice site was sufficient to allow CGRP production in the A20 cell line, inhibition of polyadenylation is not. This provides further support for the importance of *cis*-regulatory elements in the 3′ splice site region.

Finally, it is apparent from Fig. 9 that the 3′ splice sequences preceding exon 5 deviate from consensus. The significance of this deviation is unclear. *In vitro* experiments suggest that the exon 5 splice site acceptor is kinetically "weak" compared to constitutive splice sites. Replacement of exon 5 and the flanking splice site with an exon containing constitutive splice sites leads to enhanced skip splicing *in vitro* (Cote *et al.*, 1990b). Therefore, there may be a role for the 3′ splice sites preceding both exon 4 and exon 5.

2. *trans*-Acting Factors of Calcitonin Gene Alternative Splicing

The use of *in vivo* systems has resulted in the identification of several important *cis*-regulatory elements, but *trans*-acting factors that recognize these sequences remain to be elucidated. Our *in vitro* processing system does not splice in a CT-specific fashion, but recognition of the CGRP-specific splice site occurs in an extract-specific manner that mimics the *in vivo* processing pathway (Cote *et al.*, 1991). This suggests that splicing activity may not be reproducible *in vitro*, but exon recognition may. We have used cell-specific nuclear extracts (HeLa and F9 cells) to look for factors that recognize the *cis* elements discussed earlier. Thus far, we have failed to observe sequence-specific binding of a specific factor in CGRP-producing nuclear extracts; however, an unexpected finding has been the observation of an exon 4 binding protein from CT-producing cell nuclear extract (Cote *et al.*, 1992). This protein recognizes a CT sequence contained within the first 40 nucleotides of exon 4 but does not bind mutated sequence capable of introducing skip splicing *in vivo* and *in vitro*. Furthermore, binding of this protein is inhibited by the addition of CGRP-producing nuclear extract. These observations suggest an intriguing model for regulation, presented in the following section.

C. Proposed Mechanism for Calcitonin/Calcitonin Gene-Related Peptide RNA Processing

The removal of intronic sequence between adjacent exons occurs in two steps. Exon sequence is first defined by recognition processing signals and then splicing proceeds by removal of sequence between defined exons (Niwa *et al.*, 1990; Robberson *et al.*, 1990). The key factors involved in this process are large ribonuclear protein particles referred to as snRNPs (Green, 1989; Mattaj, 1990; Mowry and Steitz, 1988; Swanson, 1990). Several classes of snRNPs are involved in RNA splicing and polyadenylation. In addition, a variety of factors facilitate in this process; only a few of these factors have been positively identified. Any hypothesis proposed for regulation of CT alternative RNA processing not only must include the unique aspects of the CT pre-mRNA but also must superimpose a rather elaborate constitutive splicing process.

The key event in the processing of the CT pre-mRNA is the inclusion or exclusion of exon 4. We hypothesize that recognition of exon 4 and its inclusion in the final mRNA is an *unfavored event* that occurs only with the aid of one or more facilitating factors (Fig. 11). Furthermore, we believe that these factor(s) interact with sequence located in the proximal portion of exon 4. In the absence of facilitated recognition, CGRP production is the default pathway.

A number of observations support such a mechanism. The first is

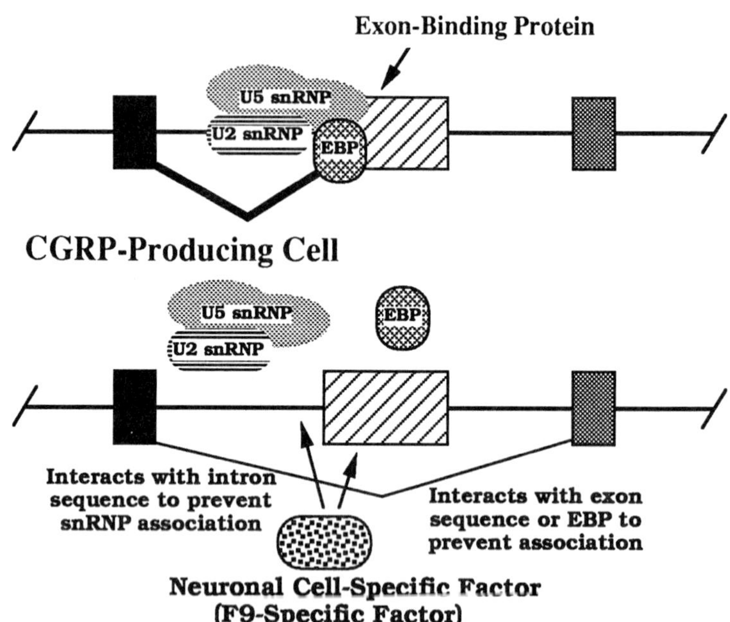

FIGURE 11 A model for regulation of cell-specific RNA processing. The decision to include or exclude exon 4 in the final messenger mRNA determines production of calcitonin (CT) or calcitonin gene-related peptide (CGRP). In this model, recognition of exon 4 is facilitated by an exon-binding protein (EBP) that recruits snRNPs to the splice site. Recognition and RNA splicing of exon 4 results in CT production. In CGRP-producing cells, a neuronal cell factor prevents recognition of the exon 4 splice site. The default pathway becomes RNA splicing to exon 5.

the presence of 3′ splice site sequences preceding exon 4 that deviate significantly from consensus sequences (Fig. 9). There is a relatively short polypyrimidine tract, and the branch point sequence contains a uridine, instead of the usual adenosine (Adema *et al.*, 1988, 1990). Therefore, it is possible that the snRNPs that normally bind these sequences during constitutive splicing would either fail to recognize exon 4 or do so only very slowly. Experimental evidence supports this idea. An intermediate of CT/CGRP RNA processing accumulates for both endogeneous and transfected gene transcripts (Fig. 1). In addition, splicing to exon 4 is dismal in *in vitro* systems, regardless of the cell type from which nuclear extract is derived. Finally, mutation of the branch acceptor to the canonical adenosine results in markedly enhanced CT-specific splicing *in vitro* (Adema *et al.*, 1988, 1990; Cote *et al.*, 1991) and *in vivo* (Adema and Baas, 1991; Adema *et al.*, 1990).

A second observation that supports this hypothesis is the identification of a *cis*-acting exonic element adjacent to the 3′ splice site. A specific RNA-binding factor interacting with sequences in exon 4 could function to assist recognition of the 3′ splice site by snRNPs. A similar mechanism has been proposed for the *Drosophila doublesex* gene (Hedley and Maniatis, 1991; Hoshijima *et al.*, 1991; Mattox and Baker, 1991). In this six-exon gene, inclusion or exclusion of exon 4 is the key regulatory choice. Exon 4 inclusion in *doublesex* proceeds by a pathway that involves facilitated recognition of a "weak" 3′ splice site (Hedley and Maniatis, 1991; Hoshijima *et al.*, 1991; Mattox and Baker, 1991). This facilitated recognition requires association of the RNA-binding protein, *transformer-2,* with exon 4 sequences. Mutation of the *transformer-2* binding sequences in exon 4 results in an inability to exclude exon 4 in the final mRNA.

We have recently identified a protein that associates with the exon 4 sequences required for CT-specific splicing (Cote *et al.*, 1992). This factor, like *transformer-2,* may function to facilitate recognition of the CT-specific 3′ splice site and prevent skip splicing in CT-producing cells (Fig. 11). The production of CGRP would require that recognition of exon 4 be suppressed. This could be accomplished by preventing the association of either the exon-binding protein or the snRNPs. *In vivo* and *in vitro* results suggest that a dominant factor is present in CGRP-producing cell types (Cote *et al.*, 1991; Crenshaw *et al.*, 1987). Nuclear extract prepared from CGRP-producing cells inhibits association of the exon 4 binding protein *in vitro* (Cote *et al.*, 1992). These results suggest that CGRP production occurs by disruption of the facilitated recognition of exon 4 (Fig. 11).

The model presented in Figure 11 allows several predictions. First, mutation of the deviant splice site sequences to consensus should uncouple the alternative processing pathway. A perfect consensus 3′ splice site at exon 4 would remove the requirement for a facilitatory factor. In fact, mutations of exon 4 and exon 5 splicing signals uncouple the splicing pathway. Second, in the absence of regulatory factors, CGRP production would be the predominant pathway. Deletion or mutation of the exon 4 binding site results in CGRP-specific splicing in HeLa cells. Deletion of exon 5 splicing signals does not produce CT in F9 cells. Lability of the exon 4 binding protein may also explain the observation that CGRP-specific splicing is the preferred pathway in many *in vitro* systems. Finally, the widespread tissue distribution of CT-specific splicing in the transgenic mouse model suggests that the exon 4 binding protein is not unique to the thyroid C cell. When one considers the increasing number of alternatively processed RNAs and their tissue distribution, this binding protein may play a generalized role. Only additional studies will be able to determine the relevance of our proposed regulatory pathway.

ACKNOWLEDGMENTS

Supported by U.S. Public Health Service Grants DK38146 to R.F.G. and RR-05425 to S.P. and Veterans Affairs (Merit and Clinical Investigator) Grants to R.F.G. and G.J.C.

REFERENCES

Adema, G. J., and Baas, P. D. (1991). Deregulation of alternative processing of calcitonin/CGRP-1 pre-mRNA by a single point mutation. *Biochem. Biophys. Res. Commun.* **178,** 985–992.

Adema, G. J., Bovenberg, R. A., Jansz, H. S., and Baas, P. D. (1988). Unusual branch point selection involved in splicing of the alternatively processed Calcitonin/CGRP-I premRNA. *Nucleic Acids Res.* **16,** 9513–9526.

Adema, G. J., van Hulst, K. L., and Baas, P. D. (1990). Uridine branch acceptor is a cis-acting element involved in regulation of the alternative processing of calcitonin/CGRP-1 pre-mRNA. *Nucleic Acids Res.* **18,** 5365–5373.

Amara, S. G., Jonas, V., Rosenfeld, M. G., Ong, E. S., and Evans, R. M. (1982). Alternative RNA processing in the calcitonin gene expression generates mRNA encoding different polypeptide products. *Nature (London)* **298,** 240–244.

Amara, S. G., Evans, R. M., and Rosenfeld, M. G. (1984). Calcitonin/calcitonin gene-related peptide transcription unit: Tissue-specific expression involves selective use of alternative polyadenylation sites. *Mol. Cell. Biol.* **4,** 2151–2160.

Amara, S. G., Arriza, J. L., Leff, S. E., Swanson, L. W., Evans, R. M., and Rosenfeld, M. G. (1985). Expression in brain of a messenger RNA encoding a novel neuropeptide homologous to calcitonin gene-related peptide. *Science* **229,** 1094–1097.

Benbrook, D. M., and Jones, N. C. (1990). Heterodimer formation between CREB and JUN proteins. *Oncogene* **5,** 295–302.

Binetruy, B., Smeal, T., and Karin, M. (1991). Ha-Ras augments c-Jun activity and stimulates phosphorylation of its activation domain. *Nature (London)* **351,** 122–127.

Bovenberg, R. A., van de Meerendonk, W. P. M., Baas, P. D., Steenbergh, P. H., Lips, C. J., and Jansz, H. S. (1986). Model for alternative RNA processing in human calcitonin gene expression. *Nucleic Acids Res.* **14,** 8785–8803.

Broad, P. M., Symes, A. J., Thakker, R. V., and Craig, R. K. (1989). Structure and methylation of the human calcitonin/alpha-CGRP gene. *Nucleic Acids Res.* **17,** 6999–7011.

Cabrera, C. V., Martinez-Arias, A., and Bate, M. (1987). Expression of three members of the Achaete Scute gene complex correlates with neuroblast segregation in *Drosphilia. Cell* **50,** 425–433.

Capecchi, M. R. (1989). Altering the genome by homologous recombination. *Science* **244,** 1288–1292.

Chiu, R., Imagawa, M., Imbra, R. J., Bochoven, J. R., and Karin, M. (1987). Multiple cis- and trans-acting elements mediate transcriptional response to phorbol esters. *Nature (London)* **329,** 648–651.

Chiu, R., Boyle, W. J., Meek, J., Smeal, T., Hunter, T., and Karin, M. (1988). The c-fos protein interacts with c-jun/AP-1 to stimulate transcription of AP-1 responsive genes. *Cell* **54,** 541–552.

Copp, D. H., Cameron, E. C., Cheney, B. A., Davidson, A. G. F., and Henze, K. G. (1962). Evidence for calcitonin: A new hormone which lowers blood calcium. *Endocrinology* **70,** 638–647.

Cote, G. J., and Gagel, R. F. (1986). Dexamethasone differentially affects the levels of calcitonin and calcitonin gene-related peptide mRNA's expressed in a human medullary thyroid carcinoma cell line. *J. Biol. Chem.* **261,** 15524–15528.

Cote, G. J., Rogers, D. G., Huang, E. S., and Gagel, R. F. (1987). The effect of 1,25-

dihydroxyvitamin D3 treatment on calcitonin and calcitonin gene-related peptide mRNA levels in cultured human thyroid C-cells. *Biochem. Biophys. Res. Commun.* **149,** 239–243.

Cote, G. J., Abruzzese, R. V., Lips, C. J., and Gagel, R. F. (1990a). Transfection of calcitonin gene regulatory elements into a cell culture model of the C cell. *J. Bone Miner. Res.* **5,** 165–171.

Cote, G. J., Nguyen, I. N., Berget, S. M., and Gagel, R. F. (1990b). Calcitonin exon sequences influences alternative RNA processing. *Mol. Endocrinol.* **4,** 1744–1749.

Cote, G. J., Nguyen, I. N., Berget, S. M., and Gagel, R. F. (1991). Validation of an in vitro RNA processing system for CT/CGRP precursor mRNA. *Nucleic Acids Res.* **19,** 3601–3606.

Cote, G. J., Stolow, D. T., Peleg, S., Berget, S. M., and Gagel, R. F. (1992). Identification of exon sequences and an RNA-binding protein involved in alternative RNA splicing of calcitonin/CGRP. *Nucleic Acids Res.* **20,** 2361–2366.

Crenshaw, E. B., Russo, A. F., Swanson, L. W., and Rosenfeld, M. G. (1987). Neuron-specific alternative RNA processing in transgenic mice expressing a metallothionein-calcitonin fusion gene. *Cell* **49,** 389–398.

de Bustros, A., Baylin, S. B., Levine, M. A., and Nelkin, B. D. (1986). Cyclic AMP and phorbol esters separately induce growth inhibition, calcitonin secretion, and calcitonin gene transcription in cultured human medullary thyroid carcinoma. *J. Biol. Chem.* **261,** 8036–8041.

de Bustros, A., Lee, R. Y., Compton, D., Tsong, T. Y., Baylin, S. B., and Nelkin, B. D. (1990). Differential utilization of calcitonin gene regulatory DNA sequences in cultured lines of medullary thyroid carcinoma and small-cell lung carcinoma. *Mol. Cell. Biol.* **10,** 1773–1778.

Denner, L. A., Weigel, N. L., Maxwell, B. L., Schrader, W. T., and O'Malley, B. W. (1990). Regulation of progesterone receptor-mediated transcription by phosphorylation. *Science* **250,** 1740–1743.

Edbrooke, M. R., Parker, D., McVey, J. H., Sorenson, G. D., Pettengill, O. S., and Craig, R. K. (1985). Expression of the human calcitonin/CGRP gene in lung and thyroid carcinoma. *EMBO J.* **4,** 715–724.

Emeson, R. B., Hedjran, F., Yeakley, J. M., Guise, J. W., and Rosenfeld, M. G. (1989). Alternative production of calcitonin and CGRP mRNA is regulated at the calcitonin-specific splice acceptor. *Nature (London)* **341,** 76–80.

Evain, B. D., Binet, E., Donnadieu, M., Laurent, P., and Anderson, W. B. (1984). Production of immunoreactive calcitonin and parathyroid hormone by embryonal carcinoma cells: Alteration with retinoic acid-induced differentiation. *Dev. Biol.* **104,** 406–412.

Fredman, L. S., Leff, S. E., Klein, E. S., Crenshaw III, E. B., Yeakley, J., and Rosenfeld, M. G. (1990). A tissue specific enhancer in the rat calcitonin/CGRP gene is active in both neural and endocrine cell types. *Mol. Endocrinol.* **4,** 497–504.

Grauer, A., Raue, F., and Gagel, R. F. (1990). Changing concepts in the management of hereditary and sporadic medullary thyroid carcinoma. *Endocrinol. Metab. Clin. North Am.* **19,** 613–35.

Green, M. R. (1989). Pre-mRNA processing and mRNA nuclear export. *Curr. Opin. Cell Biol.* **1,** 519–525.

Hedley, M. L., and Maniatis, T. (1991). Sex-specific splicing and polyadenylation of dsx pre-mRNA requires a sequence that binds specifically to tra-2 protein in vitro. *Cell* **65,** 579–586.

Hoeffler, J. P., Meyer, T. E., Yun, Y., Jameson, J. L., and Habener, J. F. (1988). Cyclic AMP-responsive DNA-binding protein: Structure based on a cloned placental cDNA. *Science* **242,** 1430–1433.

Hoppener, J. W., Steenbergh, P. H., Zandberg, J., Adema, G. J., Geurts, v. K. A. H., Lips, C. J., and Jansz, H. S. (1988). A third human CALC (pseudo)gene on chromosome 11. *FEBS Lett.* **233,** 57–63.

Hoshijima, K., Inoue, K., Higuchi, I., Sakamoto, H., and Shimura, Y. (1991). Control of doublesex alternative splicing by transformer and transformer-2 in *Drosophila*. *Science* **252,** 833–836.

Ikeda, K., Lu, C., Weir, E. C., Mangin, M., and Broadus, A. E. (1989). Transcriptional regulation of the parathyroid hormone-related peptide gene by glucocorticoids and vitamin D in a human C-cell line. *J. Biol. Chem.* **264,** 15743–15746.

Imagawa, M., Chiu, R., and Karin, M. (1987). Transcription factor AP-2 mediates induction by two different signal transduction pathways; protein kinase C and cAMP. *Cell* **51,** 251–260.

Jacobs, J. W., Goodman, R. H., Chin, W. W., Dee, P. C., Habener, J. F., Bell, N. H., and Potts, J. T., Jr. (1981). Calcitonin messenger RNA encodes multiple polypeptides in a single precursor. *Science* **213,** 457–459.

Johnson, J. E., Birren, S. J., and Anderson, D. J. (1990). Two rat homologues of *Drosophila* achaete-scute specifically expressed in neuronal precursors. *Nature (London)* **346,** 858–861.

Jonas, V., Lin, C. R., Kawashima, E., Semon, D., Swanson, L. W., Mermod, J. J., Evans, R. M., and Rosenfeld, M. G. (1985). Alternative RNA processing events in human calcitonin/calcitonin gene-related peptide gene expression. *Proc. Natl. Acad. Sci. USA* **82,** 1994–1998.

Le Douarin, N. (1982). "The Neural Crest." Cambridge University Press, Cambridge.

Leff, S. E., Evans, R. M., and Rosenfeld, M. G. (1987). Splice commitment dictates neuron-specific alternative RNA processing in calcitonin/CGRP gene expression. *Cell* **48,** 517–24.

Lewin, B. (1985). "Genes," 2nd ed. Wiley and Sons, New York.

Mabry, M., Nakagawa, T., Baylin, S., Pettengill, O., Sorenson, G., and Nelkin, B. (1989). Insertion of the v-Ha-ras oncogene induces differentiation of calcitonin-producing human small cell lung cancer. *J. Clin. Invest.* **84,** 194–199.

Mattaj, I. W. (1990). Splicing stories and poly(A) tales: An update on RNA processing and transport. *Curr. Opin. Cell. Biol.* **2,** 528–538.

Mattox, W., and Baker, B. S. (1991). Autoregulation of the splicing of transcripts from the transformer-2 gene of *Drosophila*. *Genes Dev.* **5,** 786–796.

Minvielle, S., Cressent, M., Delehaye, M. C., Segond, N., Milhaud, G., Jullienne, A., Moukhtar, M. S., and Lasmoles, F. (1987). Sequence and expression of the chicken calcitonin gene. *FEBS Lett.* **223,** 63–68.

Mowry, K. L., and Steitz, J. A. (1988). snRNP mediators of 3′ end processing: Functional fossils? *Trends Biochem. Sci.* **13,** 447–451.

Munson, P. L. (1976). Physiology and pharmacology of thyrocalcitonins. *In* "Handbook of Physiology" (G. D. Aurbach, eds.), Vol. VII, pp. 443–465. American Physiological Society, Washington, D.C.

Murre, C., McCaw, P. S., and Baltimore, D. A. (1989). A new DNA binding and dimerization motif in immunoglobulin enhancer binding, daughterless, Myo D, and myc proteins. *Cell* **56,** 777–783.

Nakagawa, T., Mabry, M., de Bustros, A., Ihle, J. N., Nelkin, B. D., and Baylin, S. B. (1987). Introduction of v-Ha-ras oncogene induces differentiation of cultured human medullary thyroid carcinoma cells. *Proc. Natl. Acad. Sci. USA* **84,** 5923–5927.

Nakagawa, T., Nelkin, B. D., de Bustros, A., and Baylin, S. B. (1988). Transcriptional and posttranscriptional modulation of calcitonin gene expression by sodium n-butyrate in cultured human medullary thyroid carcinoma. *Cancer Res.* **48,** 2096–2100.

Naveh-Many, T., and Silver, J. (1988). Regulation of calcitonin gene transcription by vitamin D metabolites in vivo in the rat. *J. Clin. Invest.* **81,** 270–273.

Nelkin, B. D., Rosenfeld, K. I., de Bustros, A., Leong, S. S., Roos, B. A., and Baylin, S. B. (1984). Structure and expression of a gene encoding human calcitonin and calcitonin gene related peptide. *Biochem. Biophys. Res. Commun.* **123,** 648–655.

Nelkin, B. D., Chen, K. Y., de Bustros, A., Roos, B. A., and Baylin, S. B. (1989). Changes in calcitonin gene RNA processing during growth of a human medullary thyroid carcinoma cell line. *Cancer Res.* **49,** 6949–6952.

Nelkin, B. D., Borges, M., Mabry, M., and Baylin, S. B. (1990). Transcription factor levels in medullary thyroid carcinoma cells differentiated by Harvey ras oncogene: c-jun is increased. *Biochem. Biophys. Res. Commun.* **170,** 140–146.

Niwa, M., Rose, S. D., and Berget, S. M. (1990). In vitro polyadenylation is stimulated by the presence of an upstream intron. *Genes Dev.* **4,** 1552–1559.

Peleg, S., Cote, G. J., Abruzzese, R. V., and Gagel, R. F. (1989). Transcriptional regulation of the human calcitonin gene: a progress report. *Henry Ford Hosp. Med. J.* **37,** 194–197.

Peleg, S., Abruzzese, R. V., Cote, G. J., and Gagel, R. F. (1990). Transcription of the human calcitonin gene is mediated by a C cell-specific enhancer containing E-box-like elements. *Mol. Endocrinol.* **4,** 1750–1757.

Potts, J. T., and Aurbach, G. D. (1976). Chemistry of the calcitonins. *In* "Handbook of Physiology" (G. D. Aurbach, eds.), Vol. VII, pp. 423–430. American Physiological Society, Washington, D.C.

Robberson, B. L., Cote, G. J., and Berget, S. M. (1990). Exon definition may facilitate splice site selection in multiexon RNAs. *Mol. Cell. Biol.* **10,** 84–94.

Rogers, D. G., Cote, G. J., Huang, E. S., and Gagel, R. F. (1989). 1,25-Dihydroxyvitamin D3 silences 3′,5′-cyclic adenosine monophosphate enhancement of somatostatin gene transcription in human thyroid C cells. *Mol. Endocrinol.* **3,** 547–551.

Roos, B. A., Yoon, M. J., Frelinger, A. L., Pensky, A. E., Birnbaum, R. S., and Lambert, P. W. (1979). Tumor growth and calcitonin during serial transplantation of rat medullary thyroid carcinoma. *Endocrinology* **105,** 27–32.

Rosenfeld, M. G., Amara, S. G., Roos, B. A., Ong, E. S., and Evans, R. M. (1981). Altered expression of the calcitonin gene associated with RNA polymorphism. *Nature (London)* **290,** 53–65.

Rosenfeld, M. G., Lin, C. R., Amara, S. G., Stolarsky, L. S., Ong, E. S., and Evans, R. M. (1982). Calcitonin mRNA polymorphism: Peptide switching associated with alternative RNA splicing events. *Proc. Natl. Acad. Sci. USA* **79,** 1717–1721.

Rosenfeld, M. G., Mermod, J. J., Amara, S. G., Swanson, P. E., Sawchenko, P. E., Rivier, J., Vale, W., and Evans, R. M. (1983). Production of a novel neuropeptide encoded by the calcitonin gene via tissue-specific RNA processing. *Nature (London)* **304,** 129–135.

Rosenfeld, M. G., Amara, S. G., and Evans, R. M. (1984). Alternative RNA processing: Determining neuronal phenotype. *Science* **225,** 1315–1320.

Russo, A. F., Nelson, C., Roos, B. A., and Rosenfeld, M. G. (1988). Differential regulation of the coexpressed calcitonin/alpha-CGRP and beta-CGRP neuroendocrine genes. *J. Biol. Chem.* **263,** 5–8.

Steenbergh, P. H., Hoppener, J. W., Zandberg, J., Lips, C. J., and Jansz, H. S. (1985a). A second human calcitonin/CGRP gene. *FEBS Lett.* **183,** 403–407.

Steenbergh, P. H ., Hoppener, J. W. M., Zandberg, J., Van de Ven, W. J. M., Jansz, H. S., and Lips, C. J. M. (1985b). Structure of the human calcitonin gene and its transcripts in medullary thyroid carcinoma. *In* "Calcitonin" (A. Pecile, ed.), pp. 23–31. Elsevier, Amsterdam.

Swanson, M. S. (1990). Heterogeneous nuclear ribonucleoprotein complexes. *Mol. Biol. Rep.* **14,** 79–82.

Zeytin, F. N., Rusk, S., and Leff, S. E. (1987). Calcium, dexamethasone, and the antiglucocorticoid RU-486 differentially regulate neuropeptide synthesis in a rat C cell line. *Endocrinology* **121,** 361–370.

12

CYTOKINES IN BONE: Local Translators in Cell-To-Cell Communications

TOSHIYUKI YONEDA

Cellular and Molecular Biology of Bone

I. INTRODUCTION

The skeleton is a unique organ that includes two physically distinctive structures. The outer part is composed of a rigid and hard connective tissue of mineralized matrix, and the inner part is composed of a richly cellular connective tissue, which is called bone marrow. The bone marrow is the major hematopoietic tissue that provides the microenvironments which influence the growth and differentiation of hematopoietic cells and nonhematopoietic cells such as stromal cells. It contains diverse hematopoietic stem cells that can differentiate into each of the blood cell types (Golde, 1991). The hematopoietic stem cells are also known to give rise to osteoclasts, which play the principal role in bone resorption (Mundy and Roodman, 1987). Marrow stromal cells have been shown to differentiate into osteoblasts, which subsequently form bone (Owen and Friedenstein, 1988). Thus, there is an intimate and complex relationship between cells in bone marrow and surrounding bone, particularly at the endosteal surfaces of trabecular bone. Although much is unknown about how this relationship is maintained, accumulating evidence suggests that a variety of cytokines which are produced in bone marrow mediate the communications between marrow cells and bone cells by organizing complex cytokine networks. Thus, the cytokines function as translators in the mixed population of cells that reside in bone. In this chapter, the roles of cytokines as local mediators between cells in the bone microenvironment are described.

II. OSTEOTROPIC CYTOKINES AND GROWTH FACTORS IN BONE

A list of osteotropic cytokines and growth factors that play a role in bone function and bone remodeling are described in Tables I and II. Most of these cytokines and growth factors were originally found in nonskeletal tissues or cells. They have important regulatory roles in many essential organ processes such as embryogenesis, angiogenesis, tumorigenesis, wound healing, and immune responses. Thus, interleukin 1 (IL-1), tumor necrosis factor (TNF), IL-6, and transforming growth factor β (TGF-β) all have multiple effects that are not related to bone. However, there may be other cytokines that have specific, or at least relatively specific, effects on bone such as bone morphogenetic proteins (BMPs). With improvements in techniques for detecting bone-specific factors, it is highly likely that additional factors with specific effects on bone cells will be identified.

TABLE I Osteotropic Cytokines

Cytokine	Cellular sources[a]	Effects on	
		Osteoblast	Osteoclast
IL-1[b]	Macrophages, stromal cells, endothelial cells, osteoblasts	P↑ or ↓, D↑ or ↓	↑
IL-4	T cells	? ↑	↓
IL-6	Macrophages, T cells, stromal cells, endothelial cells, osteoblasts, osteoclasts	No effects or P↑, D↓	↑
GM-CSF	T cells, macrophages, stromal cells, endothelial cells, osteoblasts	P↑, D↓	↑
M-CSF	Macrophages, stromal cells, endothelial cells, osteoblasts	?	↑
IL-3	T cells, osteoblasts	?	↑
G-CSF	Macrophages, stromal cells, endothelial cells, osteoblasts	?	↑
TNF	Macrophages, stromal cells, endothelial cells, T cells, osteoblasts	↓	↑
LT	T cells	↓	↑
IFN-γ	T cells	P↓, D↑	↓
LIF (DIF)	stromal cells, osteoblasts	↑	↑
OPF	?	?	↑

[a] Only cells present in or adjacent to bone are listed

[b] Abbreviations: D, differentiation; DIF, differentiation-inducing factor; G-, GM-, and M-CSF, granulocyte, granulocyte–macrophage, and macrophage colony-stimulating factor, respectively; IL, interleukin; LIF, leukemia inhibitory factor; P, proliferation; TNF, tumor necrosis factor; IFN-γ, interferon-γ; OPF, osteoclast poietic factor.

III. CELLULAR COMPONENTS IN BONE MICROENVIRONMENT

A. Cellular Source of Osteotropic Cytokines and Growth Factors

The local factors that modulate bone cell function and bone remodeling are referred to as osteotropic cytokines or growth factors. The immune cells of the bone marrow are the likely major producers of osteotropic cytokines. For example, granulocyte–macrophage colony-stimulating factor (GM-CSF), M-CSF, IL-1, TNF, lymphotoxin, and IL-6 are well-known products of the monocyte–macrophage series (Groopman *et al.*, 1989), and GM-CSF, IL-3, and interferon-γ (IFN-γ) are of T lymphocytes (Paul, 1989).

However, immune cells are not the only cellular sources of osteotropic cytokines. Stromal cells (fibroblasts), endothelial cells, and osteoblasts produce GM-CSF, M-CSF, IL-1, IL-6, TNF, and differentiation-inducing factor (DIF) or leukemia inhibitory factor (LIF) (Mundy, 1990).

TABLE II Growth Factors Produced in Bone [a]

Growth factor	Effects on: Osteoblast	Effects on: Osteoclast
EGF[b]	P↑, D↓	↑
aFGF	↑	?
bFGF	↑	?
IGF-I	↑	?
IGF-II	↑	?
PDGF	↑	↑
TGFα	↓	↑
TGFβ	↑	↓
BMPs	↑	↓

[a] Details of each growth factor are not described in the text. Refer to the reviews of Canalis *et al.* (1988) and Mundy (1990).

[b] Abbreviations: P, proliferation. D, differentiation; EGF, Epidermal growth factor; aFGF, acidic fibroblast growth factor; bFGF, basic fibroblast growth factor; IGF-I, Insulinlike growth factor I; IGF-II, Insulinlike growth factor II; PDGF, Platelet-derived growth factor; TGFα, Transforming growth factor α; TGFβ, Transforming growth factor β; BMPs, Bone morphogenetic proteins.

As shown in Table I, osteoblasts may be sources of essentially all of the bone resorbing cytokines. This may be relevant because the presence of osteoblasts is required for osteoclasts to resorb bones. However, parathyroid hormone (PTH), 1,25 dihydroxyvitamin D_3, and calcitonin show no effect on the production of IL-1 and TNF by osteoblasts, although lipopolysaccharide (LPS) increases the production of these cytokines (Gowen *et al.*, 1990; Keeting *et al.*, 1991). Several groups have reported that PTH stimulates IL-6 production by osteoblastic cells (Feyen *et al.*, 1989; Löwik *et al.*, 1989; Li *et al.*, 1991), but other groups have reported it does not (Littlewood *et al.*, 1991b; Linkhart *et al.*, 1991). It is, therefore, still unclear whether production of these cytokines by osteoblasts is regulated in a physiologically important manner.

In certain disease conditions such as Paget's disease and giant cell tumors, osteoclasts have recently been shown to secrete IL-6, which is a stimulator of osteoclast formation from hemopoietic stem cells (Kurihara *et al.*, 1990; Ohsaki *et al.*, 1991; Roodman *et al.*, 1992). This suggests that osteoclasts as well as osteoblasts are secretory cells and IL-6 is an autocrine growth and differentiation factor for osteoclasts.

Current data make it unlikely that early hematopoietic stem cells are a source of cytokines.

B. Target Cells for Osteotropic Cytokines and Growth Factors

Progenitor cells present in the bone marrow are major targets of the cytokines and growth factors. In particular, hemopoietic precursors of osteoclasts are strongly influenced by these cytokines during differentiation into mature osteoclasts (Mundy and Roodman, 1987). Mature osteoclasts are also targets of cytokines such as TNF and IL-1 in the presence of intermediary cells (Thomson *et al.*, 1986, 1987).

Proliferation and differentiation of osteoblasts, which are sources of many cytokines, are modulated by the cytokines through autocrine and/or paracrine mechanisms. The effects may be stimulatory or inhibitory depending on their maturational stages.

It has recently been demonstrated that yet-unidentified osteoprogenitor cells in the marrow nonadherent cell populations become highly proliferative in response to cytokines such as TGF-β and eventually differentiate into osteoblasts (Long *et al.*, 1990).

Marrow stromal fibroblasts, which are known to have osteogenic potential (Owen *et al.*, 1987), have been induced to differentiate into osteoblasts and form mineralized nodules *in vivo* in the presence of decalcified bone matrix extract, which presumably contains BMP-like molecules (Harada *et al.*, 1988).

Thus, it is likely that most of the cells that reside in bone and bone marrow can be either producers or targets of osteotropic cytokines and growth factors.

IV. MEDIATION OF CELL-TO-CELL COMMUNICATION BY OSTEOTROPIC CYTOKINES

The osteotropic cytokines bind to specific high-affinity receptors to exert their actions. Until recently, all of these cytokines have been thought of as soluble molecules and the receptors on their target cells as integral membrane-bound proteins. In these circumstances, soluble cytokines are likely to elicit their actions through conventional autocrine or paracrine mechanisms (Fig. 1). However, evidence accumulated in the past few years indicates that several of the cytokines can exert their effects in a membrane- or matrix-bound form and, conversely, that several of the receptors can exist as soluble molecules with ligand-binding activity (Fernandez-Botran, 1991). The roles of these soluble cytokine receptors are not known. They may function as cytokine inhibitors or as carrier proteins. The lists of these examples are described in Tables III and IV.

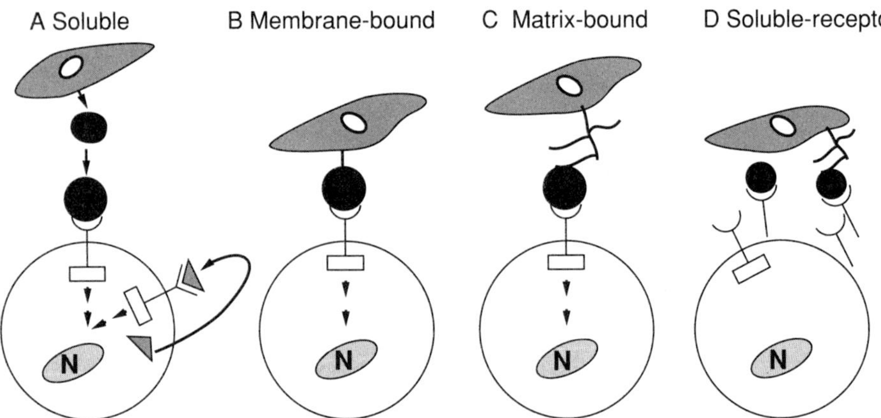

FIGURE 1 Mediation of cell-to-cell communication by cytokines. (A) Autocrine/paracrine mediation. Soluble cytokines (closed circle and triangle) released by accessory cells (top) or stem cells (bottom) bind to receptors on stem cells and exert their actions. (B) Membrane-bound cytokine mediation. Cytokines produced by accessory cells are not secreted and bound to accessory cell membrane. Membrane-bound cytokines transduce signals through receptors on stem cells, which are in close contact with accessory cells. (C) Matrix-bound cytokines mediation. Cytokines produced by accessory cells are secreted and trapped in extracellular matrix. Matrix-bound cytokines transduce signals through receptors on stem cells, which are in close contact with accessory cells. (D) Soluble-receptor binding to membrane- or matrix-bound cytokines. Receptors are released from the membrane of stem cells and react with membrane- or matrix-bound cytokines produced by accessory cells. Soluble receptors may function as inhibitors or carriers. N, nucleus.

Thus, there are at least four types of mediation of cell-to-cell communication by osteotropic cytokines and growth factors, as shown in Figure 1. Much attention has been paid in particular to matrix-bound cytokines in the last few years. The process by which target cells respond to matrix-anchored cytokines and growth factors is termed juxtacrine (Nathan and Sporn, 1991). It is now believed that juxtacrine regulation is a widespread phenomenon in the control of a variety of cell functions. Among matrices that anchor cytokines and growth factors, proteoglycans have been extensively studied (Ruoslahti and Yamaguchi, 1991). Basic fibroblast growth factor interacts with a proteoglycan called syndecan and heparin; TGF-β with betaglycan, which is a receptor protein, and decorin, which neutralizes TFG-β activity; and platelet factor 4 with serglycin and heparin. GM-CSF and IL-3 also bind to heparin. There are several examples in which a juxtacrine mechanism may play a role in bone. Cellulose ester membranes that were coated with matrices (glycoproteins, proteoglycans, glycosaminoglycans) secreted by marrow stromal cells have been shown to induce bone formation when implanted subcu-

TABLE III Osteotropic Cytokines and Growth Factors Bound to Cell Surface

Membrane-bound	
M-CSF[a]	Stein *et al.*, 1990
GM-CSF	Gough *et al.*, 1985
IL-1	Kurt-Jones *et al.*, 1985
TGFα	Wong *et al.*, 1989
TNF	Kriegler *et al.*, 1988
Matrix-bound	
GM-CSF	Gordon *et al.*, 1987
IL-3	Roberts *et al.*, 1988
IFN-γ	Gordon, 1991
LIF (DIF)	Rathjen *et al.*, 1990
bFGF	Yayon *et al.*, 1991
TGFβ	Yamaguchi *et al.*, 1990

[a] Abbreviations: M-CSF, macrophage colony-stimulating factor; GM-CSF, granuocyte–macrophage colony-stimulating factor; IL-1, interleukin 1; TGFα, transforming growth factor α; TNF, tumor necrosis factor; GM-CSF, granulocyte–macrophage colony-stimulating factor; IL-3, interleukin 3; IFNγ, interferon-γ; LIF (DIF), leukemia inhibitory factor (differentiation-inducing factor); bFGF, fibroblast growth factor; TGFβ, transforming growth factor β.

TABLE IV Soluble Receptors for Osteotropic Cytokines and Growth Factors

IL-1[a]	Giri *et al.*, 1990
IL-4	Mosley *et al.*, 1989
IL-6	Novick *et al.*, 1989
IFN-γ	Novick *et al.*, 1989
TNF	Nophar *et al.*, 1990
M-CSF	Downing *et al.*, 1989
EGF	Weber *et al.*, 1984
IGF-II	MacDonald *et al.*, 1989

[a] Abbreviations: IL-1, interleukin 1; IL-4, interleukin 4; IL-6, interleukin 6; IFN-γ, interferon-γ; TNF, tumor necrosis factor; M-CSF, macrophage colony-stimulating factor; EGF, epidermal growth factor; IGF-II, insulinlike growth factor II.

taneously in animals (Knospe *et al.*, 1989). This may be due to accumulation of host-derived cytokines in the matrices, which in turn promote the differentiation of osteoprogenitor cells. During the processes of osteoclast formation, direct contact of accessory cells (either osteoblasts or stromal cells) with hematopoietic stem cells may be necessary for osteoclast differentiation (Udagawa *et al.*, 1989). The requirement of a similar relationship between osteoclast progenitor cells and bone marrow stromal cells for osteoclast formation in multiple myeloma has also been reported (Caligaris-Cappis *et al.*, 1991). IL-1 bound to macrophage cell membrane causes bone resorption in organ cultures of neonatal mouse calvariae (Nishihara *et al.*, 1989). NaCl extracts or collagenase digests of osteoblasts stimulated bone resorption by disaggregated neonatal rat osteoclasts in the presence of heparin but not in the absence of heparin (Fuller *et al.*, 1991). Production of this activity was stimulated by 1,25D_3. Cytokines bound to accessory cell membranes or matrix may act as transducers through direct contact to the receptors on plasma membranes of osteoclast precursor cells. Whether a juxtacrine process or other cell-to-cell interactions such as communication by gap junctions (Stagg and Fletcher, 1990) or adherence junctions (Tsukita *et al.*, 1991) are involved in these situations remains to be determined.

V. ACTIONS OF CYTOKINES AND GROWTH FACTORS ON BONE

A. Interleukin 1

1. Effects on Bone Formation

IL-1 shows complex and apparently paradoxical effects on bone formation. Accumulating evidence indicates important roles of IL-1 in proliferation and differentiation of osteoblasts. Newborn mouse calvariae (Lorenzo *et al.*, 1990) and human osteoblastlike cells (Keeting *et al.*, 1991) produce IL-1α and IL-1β. Transformed (Rodan *et al.*, 1990a) and non-transformed osteoblastic cells (Shen *et al.*, 1990) possess IL-1 receptors with a molecular size of 70–80 kDa. Thus, it appears that IL-1 is an autocrine regulatory cytokine for osteoblasts.

Several groups have reported that IL-1 stimulates the proliferation of osteoblasts (Gowen *et al.*, 1985; Evans *et al.*, 1990), whereas other groups have proposed that IL-1 is inhibitory on osteoblast proliferation (Stashenko *et al.*, 1987b; Dedhar, 1989). Effects of IL-1 on markers of differentiation in osteoblasts such as alkaline phosphatase activity, osteocalcin, and type I collagen synthesis are either inhibitory (Stashenko *et al.*, 1987a; Evans *et al.*, 1990), stimulatory (Dedhar, 1989), or biphasic (Canalis, 1986). Thus, the effects of IL-1 on overall bone-forming capacities of osteoblasts are not yet clear from these *in vitro* findings. Recently,

Boyce *et al.* (1989) carefully studied the long-term effects of IL-1 *in vivo* and found that subcutaneous injections of IL-1 for 3 days caused increased bone formation subsequent to osteoclastic bone resorption after 3–4 weeks, indicating profound long-term local effects of IL-1 on bone turnover. These effects of IL-1 may be mediated through prostaglandins, since indomethacine inhibited these processes. Whatever the mechanisms are, the findings suggest that IL-1 could be a direct or an indirect inducer of bone formation. However, more extensive studies are required to elucidate the precise roles of IL-1 in bone formation.

2. Effects on Bone Resorption

a. Bone-resorbing activity of interleukin 1α and 1β in vitro

It is widely accepted that IL-1α and IL-1β are powerful bone-resorbing cytokines. In 1983, Gowen *et al.* (1983) first showed that native human purified IL-1 stimulated bone resorption. Following this report, porcine IL-1 (catabolin) (Heath *et al.*, 1985) and recombinant human IL-1α and IL-1β (Gowen and Mundy, 1986) at concentrations as low as 10^{-11} *M* were also found to increase bone resorption.

Almost at the same time, IL-1 was identified in bone resorbing activity in phytohemagglutinin-activated leukocyte culture supernatants, which was then loosely called osteoclast-activating factor, by N-terminal amino acid sequence (Dewhirst *et al.*, 1985). Consistent with this finding, Lorenzo *et al.* (1987a) showed that neutralizing antibodies to IL-1β inhibited bone-resorbing activity in leukocyte culture supernatants.

Stimulation of bone resorption by IL-1 is most likely due to promotion of proliferation and differentiation of hematopoietic progenitor cells of osteoclasts. IL-1 increases the formation of osteoclastlike cells by promoting proliferation of committed precursors and subsequent differentiation in human marrow cultures (Pfeilschifter *et al.*, 1989). Furthermore, IL-1 also activates mature osteoclasts to resorb bone (Thomson *et al.*, 1986). This effect of IL-1 is at least in part indirect and mediated through other cells such as bone lining cells. The precise molecular mechanisms by which IL-1 activates mature osteoclasts to resorb bone are unknown.

b. Bone-resorbing activity of interleukin 1 in vivo

IL-1 is a potent stimulator of osteoclastic bone resorption not only *in vitro* but also *in vivo*. Three-day infusions of IL-1 in normal mice caused increased osteoclastic bone resorption that was accompanied with hypercalcemia (Sabatini *et al.*, 1988). Daily subcutaneous injection of IL-1 for 3 days over the murine calvaria caused hypercalcemia and increased bone resorption that lasted for 7–10 days (Boyce *et al.*, 1989a). In these experiments, bone resorption was evaluated by histologic means. König *et al.* (1988) reported that subcutaneous administration of IL-1α stimulates bone resorption measured by [^{3}H]tetracycline excretion in urine from [^{3}H]tetracycline-prelabeled animals. It appears that IL-1 induces bone resorption at local sites, which in turn systemically induces hypercalcemia *in vivo*.

3. Cooperation between Interleukin 1 and Other Cytokines and Systemic Hormones in Bone Resorption

IL-1 was found to be synergistic with TNF and lymphotoxin in bone resorption in organ cultures of fetal rat long bones (Stashenko *et al.*, 1987a). Sabatini *et al.* (1987) and Lorenzo *et al.* (1988) demonstrated that the effects of IL-1 on bone resorption were enhanced by TNF and TGF-α. Synergy between IL-1 and PTH in stimulation of bone resorption (Dewhirst *et al.*, 1987) and PGE_2 secretion by osteoblasts was also demonstrated (Tatakis *et al.*, 1988). More recently, we have found that IL-1α and IL-1β produce synergistic effects on bone when given together with IL-6 (Black *et al.*, 1990). Similar observations have been described by another group (Ishimi *et al.*, 1990). *In vivo*, Sato *et al.* (1989) found that IL-1α and parathyroid hormone-related peptide (PTHrP) produced synergistic effects on plasma calcium in rats. Because IL-1 is produced by osteoblasts, as described earlier, is always present in the bone microenvironment, and exerts its actions in the presence of circulating hormones such as PTH and PTHrP or other locally produced cytokines, these interactions are clearly important for determining the effects of IL-1 on bone resorption in both physiological and pathological situations.

4. Interleukin 1 Receptors and Its Antagonist in Bone Resorption

At least, two IL-1 receptors (IL-1Rs) are identified. Murine (Sims *et al.*, 1988) and human (Sims *et al.*, 1989) IL-1R with a molecular size of 80K were cloned. This IL-1R is present on fibroblasts and T cells and binds to IL-1α and IL-1β with equal affinity (Bird and Saklatvala, 1986). Murine IL-1R has 69% homology with human IL-1R at amino acid level and both IL-1Rs belong to members of the immunoglobulin gene superfamily (Sims *et al.*, 1988). A 60-kDa receptor is expressed on B cells and macrophages and binds to Il-1α and IL-1β with different affinities (Benjamin and Dower, 1990). Recently, soluble forms of IL-1R with a molecular size of 60 kDa (Symons and Duff, 1990) and 35–45 kDa (Giri *et al.*, 1990), which are presumably proteolytically cleaved forms of IL-1R on the plasma membrane, have been reported.

The effects of IL-1 on osteoclastic bone resorption appear to be mediated through the 80-kDa receptor, because antibodies to the 80-kDa receptor block the capacity of IL-1 to stimulate osteoclastic bone resorption (Garrett *et al.*, 1990).

Recently, a cytokine related to the IL-1 family but that has no IL-1 effects has been identified (Carter *et al.*, 1990; Hannum *et al.*, 1990; Eisenberg *et al.*, 1990; Arend *et al.*, 1989). This cytokine interferes with some of the biological effects of IL-1 by binding to one of the IL-1 receptors. This cytokine, which is a naturally occurring endogenous IL-1 receptor antagonist, competes with IL-1 for binding to the 80-kDa receptor. The protein has been purified and the gene molecularly cloned and expressed (Carter *et al.*, 1990; Hannum *et al.*, 1990; Eisenberg *et al.*, 1990). It has 152 amino acids and a molecular weight of 17,000. It has 26% homology with IL-1β and 19% with IL-1α. The gene has very similar organization to the IL-1 genes, and this peptide is most likely to be part of the IL-1 family. It is a secreted protein with a well-defined signal sequence. It is produced by monocytes, but it is clearly under separate regulatory control from that of IL-1 (Arend *et al.*, 1989). It has similar affinity for the 80-kDa receptor as IL-1. We have examined the effects of this IL-1R antagonist on bone resorption *in vitro* and on blood ionized calcium *in vivo*. It inhibits the effects of IL-1 to increase blood-ionized calcium (Garrett *et al.*, 1990). It has identical effects on bone *in vitro* to those of antibodies to the 80-kDa IL-1R (Garrett *et al.*, 1990). This indicates that the effects of IL-1 on bone resorption are mediated through the 80-kDa receptor. IL-1R antagonists block the effects of both IL-1α and IL-1β on bone resorption in organ cultures (Garrett *et al.*, 1990; Seckinger *et al.*, 1990). It partially inhibits TNF and lymphotoxin but has no effects on bone resorption, which is stimulated by PTH or $1,25D_3$. The naturally occurring IL-1R antagonist is shown to have no agonist effects (Dinarello, 1991). It should, therefore, provide us new experimental and clinical approaches in the study of the diverse diseases caused by IL-1. However, it must be given in relatively large doses to produce effects on bone, so whether this will be a useful therapeutic agent for skeletal diseases remains to be determined (Garrett *et al.*, 1990).

5. Roles of Interleukin in Diseases Associated with Abnormal Bone and Calcium Metabolism

a. Malignant diseases

i. Solid tumors. IL-1 has been implicated in the bone resorption and hypercalcemia that occurs in malignant diseases. Production of IL-1α in several human squamous carcinomas that are associated (Sato *et al.*, 1989) or unassociated (Fried *et al.*, 1989) with hypercalcemia has been demonstrated. In some cases, IL-1α synergistically work with other tumor products such as PTHrP (Sato *et al.*, 1989) and IL-6 (Kawano *et al.*, (1989b).

ii. Myeloma. Bone destruction and hypercalcemia seen in myeloma have been ascribed to IL-1β (Cozzolino *et al.*, 1989; Yamamoto *et al.*, 1989). In these studies, freshly isolated marrow cells from patients with myeloma were studied and found to produce IL-1β and TNF but no lymphotoxin (Lichtenstein *et al.*, 1989; Kawano *et al.*, 1989a). However, in our studies, elevated lymphotoxin but no IL-1β production was observed in human myeloma cell lines established in culture (Garrett *et al.*, 1987). It is possible that IL-1 in freshly isolated myeloma cultures might be derived from the contaminating nonneoplastic mononuclear or stromal cells (Caligaris-Cappis *et al.*, 1991).

iii. Adult T-Cell Leukemia. IL-1α has been implicated in the hypercalcemia that occurs in adult T-cell leukemia (Tanaka *et al.*, 1990). Production of IL-1α by cultured adult T-cell leukemia cells is shown to be modulated by calcium concentrations in the culture medium (Tanaka *et al.*, 1990). There might be a positive feedback loop between calcium and IL-1α production in adult T-cell leukemia.

b. Chronic inflammation

Roles of IL-1 in chronic inflammatory diseases have also been described. In rheumatoid arthritis, increased production of IL-1β by infiltrating immune cells in synovial fluids has been linked to destruction of bone and cartilage in the joints (Harris, 1990).

Similarly, in periodontal disease, accumulation of IL-1 in inflamed gingiva is likely to account for resorption of adjacent alveolar bone (Matsuki *et al.*, 1991).

In cystic lesions of the jaw, production of IL-1 by stromal fibroblasts of the cyst capsule may be responsible for the fibroblast proliferation and bone resorption that results in cyst expansion (Meghji *et al.*, 1989).

c. Osteoporosis

Increased constitutive production of IL-1 by peripheral blood mononuclear cells in some patients with osteoporosis compared to that in corresponding control subjects has been linked to high bone turnover osteoporosis (Pacifici *et al.*, 1987, 1989). Increased IL-1 production may be due to stimulation of peripheral blood mononuclear cells by bone matrix constituents released from excessively resorbing bone (Pacifici *et al.*, 1991). These investigators have also shown that treatment with estrogen decreased IL-1 production by mononuclear cells in postmenopausal patients, which is relevant to clinical situations. Estrogen also inhibits the production of IL-6, which is induced by IL-1 in human bone cells (Girasole *et al.*, 1990). Thus, it seems likely that decreased levels of estrogen at postmenopause may lead to increased production of IL-1, IL-6, and other cytokines by mononuclear and bone cells, which in turn could lead to increased bone resorption, and increased production of these cytokines could be suppressed by treatment with estrogens.

B. Interleukin 4

IL-4 is a 20-kDa glycoprotein that is produced by activated T cells and mast cells (Howard *et al.*, 1982). Recently, accumulating evidence indicated that IL-4 influences bone and calcium metabolism. Watanabe *et al.* (1990) found that IL-4 inhibits bone resorption that is stimulated by PTH, PTHrP, 1,25D_3, IL-1α, and IL-1β in organ cultures of mouse fetal long bones, and Nagasaki *et al.* (1991) reported that IL-4 blocks PTHrP-induced hypercalcemia *in vivo*. Because IL-4 is shown to decrease the production of IL-1, IL-6, TNFα, and prostaglandin E_2 by activated monocytes (Hart *et al.*, 1989; Essner *et al.*, 1989; Gibbons *et al.*, 1991) that are all potent stimulators of osteoclastic bone resorption, these effects of IL-4 are likely to be due to inhibition of osteoclast formation induced by these cytokines (Shioi *et al.*, 1991). However, IL-4 is also a promoter of mature monocyte proliferation and differentiation and, thus, may stimulate mature osteoclasts. Because of these biphasic actions, evaluation of the therapeutic potential of IL-4 for treatment of hypercalcemia and metabolic bone diseases associated with increased osteoclastic bone resorption remains problematic.

The stimulatory effect of IL-4 on bone formation has recently been reported (Veno *et al.*, 1992).

C. Interleukin 6

IL-6, which was originally identified as a B-cell differentiation factor, has turned out to have multiple regulatory roles on the immune system, hematopoiesis, inflammation, and tumorigenesis (Akira *et al.*, 1990). In many circumstances, its functions overlap those of other cytokines such as IL-1, TNF, and G-CSF.

1. Effects of Bone Formation

In bone, IL-6 is produced by osteoblasts (Table I; Feyen *et al.*, 1989; Löwik *et al.*, 1989; Ishimi *et al.*, 1990; Li *et al.*, 1991; Linkhart *et al.*, 1991), chondrocytes (Guerne *et al.*, 1990, and stromal cells (Caligaris-Cappis *et al.*, 1991; Gimble *et al.*, 1991). Human osteoblasts constitutively express IL-6 receptor messenger RNA (mRNA) (Littlewood *et al.*, 1991a). Nonetheless, surprisingly these human osteoblasts did not respond to exogenously added IL-6 (Littlewood *et al.*, 1991b). This may be due to saturation of the receptors by high levels of endogenously produced IL-6. In contrast, it has been demonstrated that IL-6 has direct effects on osteoblastic osteosarcoma UMR 106 cells and stimulates their proliferation and suppresses their collagen synthesis (Fang and Hahn, 1991). There are IL-6 transgenic mice (Suematsu *et al.*, 1989). However, the histologic appearance of the bones in these transgenic mice is unknown. Bones of animals to which IL-6 is administered have not been specifically studied for bone formation. IL-6 may exert its action on bone formation in concert with other cytokines and growth factors. At present, the effects of IL-6 on bone formation remain to be elucidated.

2. Effects on Bone Resorption

a. Bone resorbing activity of interleukin 6 in vitro

As its effects on bone formation are complex and paradoxical, IL-6 effects on bone resorption are also conflicting. Earlier studies have shown that IL-6 produced by osteoblasts in response to PTH, PTHrP, IL-1α, IL-1β, or TNF-α stimulates bone resorption in organ cultures of fetal murine metacarpal bones (Löwik *et al.*, 1989) and of fetal murine calvariae (Ishimi *et al.*, 1990). In contrast, more recent studies have reported that IL-6 has no stimulatory effects on bone resorption in organ cultures of neonatal mouse calvariae (Al-Humidan *et al.*, 1991). As raised by Bataille and Klein (1991), this conflict may be due to differences in maturity of cells in the osteoclast lineage present in each bone organ culture system employed. In support of this notion, Kurihara *et al.* (1990) demonstrated that IL-6 stimulates formation of multinucleated cells with osteoclast phenotype in human long-term marrow cultures, indicating a capacity of IL-6 to promote early progenitors of the osteoclast. Thus, IL-6 may be active in fetal metacarpal bones and calvariae because these bones contain more precursors than the neonatal mouse calvariae or fetal rat long bones.

b. Bone resorbing activity of interleukin 6 in vivo

Single or multiple bolus injections of IL-6 fail to cause osteoclastic bone resorption or changes in plasma calcium, although they increase leukocyte and platelet counts (Hill *et al.*, 1990). Bones and plasma calcium levels in IL-6 transgenic mice have not apparently been altered (Suematsu *et al.*, 1989). However, we have used Chinese hamster ovarian cells (CHO cells) that were transfected with the complementary DNA (cDNA) for murine IL-6 (Black *et al.*, 1991a). These cells stably express IL-6 and can be carried as tumors in nude mice. The mice bearing this tumor developed hypercalcemia and increased osteoclastic bone resorption. It has been found that anti-IL-6 monoclonal antibody treatment decreases serum calcium levels in a patient with plasma cell leukemia (Klein *et al.*, 1991). These results support that IL-6 is a stimulator of bone resorption *in vivo*.

3. Roles of Interleukin 6 in Diseases Associated with Abnormal Bone and Calcium Metabolism

a. Malignant diseases

i. Solid tumors. Elevated serum levels of IL-6 were found in a patient with pheochromocytoma who manifested humoral hypercalcemia of malignancy (Fukumoto *et al.*, 1991). The tumor produced both IL-6 and PTHrP, and removal of the tumor resulted in a decrease in serum IL-6 levels to undetectable levels. In this situation, there may be a synergistic interaction between IL-6 and PTHrP in causing hypercalcemia.

We have recently studied a human squamous cancer of the maxilla named MH-85 that was associated with leukocytosis, hypercalcemia, and cachexia (Yoneda *et al.*, 1989). MH-85 cells established in culture secreted G-CSF, GM-CSF, and M-CSF (Yoneda *et al.*, 1991a). In addition to CSFs, MH-85 expressed IL-6 mRNA and released IL-6 into the culture medium (Yoneda *et al.*, 1993). MH-85 tumor-bearing nude mice developed hypercalcemia and showed elevated plasma IL-6 levels. Treatment of these MH-85-bearing mice with hypercalcemia with monoclonal antibodies to human IL-6 decreased blood Ca^{2+} levels. Monoclonal antibodies to mouse IL-6 failed to reduce blood Ca^{2+}. These findings indicate that IL-6 produced by human tumor MH-85 is responsible at least in part for the hypercalcemia in tumor-bearing animals.

More recently, Ohsaki *et al.* (1991) demonstrated that bone resorption seen in giant cell tumors of bone is attributable to IL-6 secreted by multinucleated giant cells in the tumor that express many of osteoclast phenotype. They also found IL-6 mRNA expression in these cells and in normal osteoclasts. Therefore, IL-6 may be an autocrine regulatory factor for osteoclasts as well as for osteoblasts.

ii. Hematologic malignancies. The roles of IL-6 in myeloma have been studied in some detail. Initially, IL-6 was thought to be an autocrine growth factor for freshly isolated myeloma cells (Kawano *et al.*, 1988). However, the heterogenous cell population in primary myeloma cell culture does not allow definitive conclusions on whether this is really the case. Recently, several bodies of evidence have revealed that IL-6 is a paracrine rather than autocrine growth factor for myeloma cells that is produced by stromal cells (Zhang *et al.*, 1989; Klein *et al.*, 1989; Bataille *et al.*, 1989a; Caligaris-Cappis *et al.*, 1991). However, IL-6 is still proposed to be an autocrine growth factor in the human myeloma cell line U266 (Schwab *et al.*, 1991). Recently, it has been reported that autocrine production of IL-6 in several myeloma cell lines is stimulated by IFN-α (Jourdan *et al.*, 1991). IL-6 probably plays a role in recruitment of osteoblasts and osteoclasts, which may be an important step in the pathophysiology of multiple myeloma (Bataille *et al.*, 1991). Under these circumstances, IL-6 may be a stimulator of osteoclastic bone resorption in its own right and may also enhance the bone resorbing effects of other cytokines such as IL-1 and TNF.

As two of the rare cases, elevated secretion of IL-6 in KI-1-positive large cell anaplastic lymphoma (Agematsu *et al.*, 1991) and Castleman's disease (Yabuhara and Komiyama, 1990), which are associated with extensive bone destruction, has been described.

b. Paget's disease of bone

Paget's disease of bone is characterized by imbalanced bone remodeling that results from abnormal osteoclast activity (Krane and Simon, 1987). The osteoclasts in pagetic bone are much larger in size and have greater numbers of nuclei than do the osteoclasts in uninvolved bone. Long-term cultures of bone marrow cells from patients with Paget's disease produce 10- to 20-fold greater numbers of cells with osteoclast characteristics than do normal marrow cultures. Osteoclasts formed in pagetic marrow cultures resemble osteoclasts in pagetic bone. They are much larger in size and have more nuclei per cell (Kukita *et al.*, 1990). Interestingly enough, conditioned media from pagetic marrow cultures stimulate osteoclast formation in normal marrow cultures. Roodman *et al.* (1992) found that this stimulatory effect of pagetic marrow culture conditioned media is, at least in part, ascribed to IL-6 that is secreted in an autocrine/paracrine fashion. IL-6 is identified by *in situ* hybridization and by immunohistochemistry in both osteoclasts and stromal cells. IL-6 therefore may play a pathophysiologic role in the abnormality in osteoclast function that occurs in Paget's disease. Whether this is an epiphenomenon occurring as a consequence of viral disease or an important proximal event in the pathophysiology of Paget's disease remains to be determined.

c. Chronic inflammation

One of the most characteristic histologic features of chronic inflammatory diseases is an infiltration of activated B cells and plasma cells at local sites. Therefore, accumulation of cytokines that influence proliferation and differentiation of these infiltrating cells at local lesions is predictable.

i. Rheumatoid arthritis. Like IL-1, IL-6 has also been implicated in the pathophysiology of rheumatoid arthritis (Al-Balaghi *et al.*, 1984). Human synoviocytes constitutively secrete IL-6. IL-6 production by synoviocytes is stimulated by IL-1 and TNF-α (Guerne *et al.*, 1989). Elevated IL-6 activity is detected in the synovial fluids from patients with a variety of inflammatory arthropathies and rheumatoid arthritis and in cultures of freshly isolated mononuclear cells in synovial fluids (Hirano *et al.* 1988). Cytokines including IL-1, IL-6, and TNF may play in harmony a central role in causing inflammatory joint destruction through osteoclastic bone resorption.

ii. Periodontal disease. Gingival mononuclear cells isolated from inflamed tissues of patients with periodontal disease constitutively produce IL-6 (Kono *et al.*, 1991). However, peripheral blood mononuclear cells from the same patients secrete IL-6 only when they are activated. Again, other cytokines such as IL-1, TNF-α, and lymphotoxin are involved and cooperate with IL-6 in the processes of alveolar bone destruction and periodontal tissue damage (Williams, 1990).

D. Colony-Stimulating Factors

1. Effects on Bone Formation

Osteoblasts are a rich local source of all CSFs. Murine osteoblasts show constitutive production of M-CSF, which is increased upon stimulation with LPS or $1,25D_3$ (Elford *et al.*, 1987). They also secrete G-CSF, GM-CSF, and IL-3 in response to LPS (Felix *et al.*, 1988), PTH (Weir *et al.*, 1989), or TNF (Felix *et al.*, 1991). CSFs produced by osteoblasts are likely to regulate their own proliferation and phenotypic expression by an autocrine fashion. Evans *et al.* (1990) showed that recombinant human GM-CSF promotes human osteoblast proliferation but decreases $1,25D_3$-induced elevation of osteocalcin synthesis and alkaline phosphatase activity. GM-CSF showed no effects on PGE_2 production and plasminogen activator activity. They suggest that GM-CSF may stimulate immature osteoblast to proliferate and inhibit mature osteoblast to further differentiate. Because GM-CSF induces the secretion of TNF-α and IL-1 in a variety of cells (Glasson, 1991), its actions on bone cells may be mediated by other cytokines.

In op/op murine osteopetrosis, M-CSF production by stromal cells, osteoblasts, and immune cells is defective, which results in impaired function of osteoclasts. Recently, Takahashi *et al.* (1991b) demonstrated that M-CSF restores the capacity of osteoblasts of op/op mouse to release an unknown factor that in turn confers bone resorbing activity on op/op mouse osteoclasts.

2. Effects on Bone Resorption

a. Bone resorbing activity of colony-stimulating factor in vitro

Because osteoclasts arise from hematopoietic stem cells and CSFs are growth and differentiation-regulating factors for hematopoietic stem cells, it is not surprising that CSFs exert their effects on osteoclasts formation. However, earlier studies produced conflicting results. Several groups using human (MacDonald *et al.*, 1986) or monkey (Povolony *et al.*, 1990) bone marrow cells have reported that GM-CSF and M-CSF increase osteoclast formation, whereas other groups using mouse bone marrow cells (Van der Wijngaert *et al.*, 1987; Shinar *et al.*, 1990; Hattersley and Chambers, 1990) have shown that GM-CSF and M-CSF decrease osteoclast formation. Moreover, it was also shown that GM-CSF and M-CSF inhibit bone resorption by rat mature osteoclasts (Hattersley *et al.*, 1988). However, more recent studies have clearly shown that actions of GM-CSF and M-CSF depend on proliferative stages of hematopoietic cells that are used in experiments (Lorenzo *et al.*, 1987b). Bone marrow cells cultured in the presence of 1,25D_3 together with GM-CSF or M-CSF produced decreased numbers of osteoclast compared to the cells cultured with 1,25D_3 alone, indicating inhibitory action of GM-CSF and G-CSF (Takahashi *et al.*, 1991b). In contrast, when bone marrow cells that are initially induced to proliferate by either GM-CSF or M-CSF are subsequently treated with 1,25D_3, they generate increased numbers of osteoclast compared to cells that are not treated with GM-CSF or M-CSF prior to 1,25D_3 treatment. Consistent with this finding, stimulation by M-CSF of osteoclastic bone resorption in metatarsals in which immature cells are abundant is diminished by irradiation, which prohibits cell division (Corboz *et al.*, 1992). These results suggest that cellular proliferation may be a prerequisite for subsequent cellular differentiation induced by CSFs.

b. Bone resorbing activity of colony-stimulating factors in vivo

Recombinant human G-CSF and GM-CSF are now administered to patients with various hematologic diseases and with solid tumors who are under intensive anticancer therapy (Groopman *et al.*, 1989). No abnormalities of skeletal tissues have been reported to date except for mild to moderate bone pain in CSF-treated patients. Further clinical studies may clarify the effects of CSFs on bone and calcium metabolism in humans.

In an experimental animal model, long-term administration of both G-CSF and GM-CSF increased osteoclastic bone resorption at endosteal surfaces, which eventually caused reduction of bone thickness (Lee *et al.*, 1991). These animals did not develop hypercalcemia. The results suggest that CSFs may have direct and/or indirect actions on hematopoietic osteoclast precursors *in vivo*.

3. Roles of Colony-Stimulating Factors in Diseases Associated with Abnormal Bone and Calcium Metabolism

a. Osteopetrosis

Osteopetrosis, an inherited metabolic bone disease in which osteoclast function is impaired, has provided numerous insights into osteoclast biology (Marks, 1989). Especially in the last few years, studies on one variant of murine osteopetrosis op/op mouse have demonstrated convincing evidence the M-CSF plays a key role in the formation of the osteoclast in the hematopoietic microenvironment. Osteopetrotic op/op mice have a deficiency of M-CSF due to a defect in the coding region of M-CSF (Yoshida *et al.*, 1990). Administration of M-CSF to these animals cures the abnormalities in bones (Felix *et al.*, 1990; Kodama *et al.*, 1991). In this circumstance, M-CSF most likely acts on accessory cells (stromal cells or osteoblasts) that do not support hematopoietic cell differentiation into osteoclasts due to a loss of capacity to produce M-CSF (Takahashi *et al.*, 1991a). However, M-CSF alone in the absence of accessory cells fails to support osteoclast formation, suggesting that M-CSF induces the production of other unknown cytokines in the accessory cells, which in turn modulate osteoclast progenitors. These experiments show that M-CSF is necessary for osteoclast formation and marrow cavity development in physiological conditions.

b. Tumor-associated leukocytosis

For the last two decades or so, it has been reported that a number of human and animal tumors induce leukocytosis (granulocytosis) by releasing CSFs (Sato, *et al.*, 1986; Kondo *et al.*, 1983). Interestingly and perhaps importantly, reports of this association have accumulated in Japan and have involved squamous cell carcinomas of the head and neck (Yoneda *et al.*, 1991e). Many of these tumors cause not only leukocytosis but also hypercalcemia associated with increased osteoclastic bone resorption in nude mice. Occurrence of both leukocytosis and hypercalcemia are not uncommon in patients with cancers. We have shown that the paraneoplastic leukocytosis–hypercalcemia syndrome occurs more frequently together than would be expected by chance occurrence (Yoneda *et al.*, 1991e). We retrospectively studied 225 patients with oral cancers and found 10 patients had hypercalcemia, 11 had leukocytosis, and 4 had both syndromes. The occurrence of the two paraneoplastic syndromes in the same patient was far greater than could have been expected by chance.

We have studied a tumor isolated from one of these patients. This tumor MH-85 produces G-CSF, GM-CSF, M-CSF, and IL-6 in culture and induces leukocytosis, hypercalcemia, and cachexia in nude mice (Yoneda *et al.*, 1991a,c). These paraneoplastic syndromes are abrogated by splenectomy as well as by excision of the tumor. This suggests that not only tumor but also immune cells in the spleen play a contributory role in the pathophysiology of these paraneoplastic syndromes.

In another tumor model we have studied, GM-CSF was identified as a tumor product that was responsible for oversecretion of TNF-α by host immune cells. Cachexia and hypercalcemia developed in this tumor A375-bearing nude mice may be due to excessive production of TNF-α (Sabatini *et al.*, 1990b).

Findings obtained from these two tumor models suggest that CSFs could elicit their actions on osteoclast through direct and indirect mechanisms involving immune cells.

c. *Chronic inflammation*

Increased levels of GM-CSF and M-CSF were detected in synovial fluids and in the supernatants of cultured synovial tissues from patients with rheumatoid arthritis (Xu *et al.*, 1989). These authors have proposed that there may be a positive loop between IL-1, which is produced by macrophages in response to prolonged local inflammation, and GM-CSF, which is produced by fibroblasts. These interactions may progress to destruction of local tissues in rheumatoid arthritis.

E. Tumor Necrosis Factor and Lymphotoxin

1. Effects on Bone Formation

The first report describing the effects of TNF and lymphotoxin on bone formation have shown that they inhibit collagen synthesis and stimulate cell growth in organ cultures of fetal rat calvariae (Bertolini *et al.*, 1986). Similar results are found in osteosarcoma cells (Smith *et al.*, 1987; Tatakis and Dziak, 1989) and fetal rat osteoblasts (Centrella *et al.*, 1988). It has been shown that TNF down-regulates PTH receptors (Schneider *et al.*, 1991) and impairs cyclic adenosine monophosphate response to PTH in UMR 106 osteosarcoma cells (Gutierrez *et al.*, 1987). It has also been shown to inhibit $1,25D_3$-induced bone Gla protein synthesis in rat osteosarcoma cells (Nanes *et al.*, 1991). It appears, thus, that TNF promotes proliferation but suppresses phenotypic expression in osteoblasts. The local source of TNF in bone is the osteoblast as well as the macrophage. Production of TNF by these cells may be enhanced by IL-1 and GM-CSF (Gowen *et al.*, 1990). Systemic osteotropic hormones such as $1,25D_3$,

PTH, and calcitonin have no effects on TNF secretion from osteoblasts. Therefore, TNF seems to act on osteoblasts as a local paracrine/autocrine regulator and its production may be regulated locally but not systemically.

Lymphotoxin has identical effects on osteoblast to TNF. Lymphotoxin may inhibit rat osteoblastic cell proliferation via a prostaglandin-mediated pathway (Tatakis and Dziak, 1989).

2. Effects on Bone Resorption

a. Bone resorbing activity of tumor necrosis factor and lymphotoxin in vitro

Bone resorbing activity of TNF and lymphotoxin was first described in organ cultures of fetal rat long bones (Bertolini *et al.*, 1986) and neonatal mouse calvariae (Tashjian *et al.*, 1987). Bone resorption of mouse calvariae by TNF appears to be mediated through prostaglandin synthesis. Prostaglandin E_2 and M-CSF are secreted by osteoblasts upon stimulation with TNF (Sato *et al.*, 1987). Osteoblasts mediate TNF-stimulated osteoclastic bone resorption (Thomson *et al.*, 1987). Thus, TNF effects on bone resorption seem to be mediated by both cell–cell and cytokine–cytokine interactions.

Stimulation of bone resorption by TNF is probably due to increased osteoclast formation. TNF as well as IL-1 have been shown to increase osteoclastlike multinucleated cell formation in human marrow cultures (Pfeilschifter *et al.*, 1989). In the same culture system, Roodman *et al.* (1987) found that TNF elevates IL-1 levels in the culture supernatants and IL-1 antibodies partially neutralize TNF effects. Furthermore, Garrett *et al.* (1990) showed that 80-kDa IL-R antagonist also in part inhibits TNF-caused bone resorption. More recently, we found that IL-6 antibodies can inhibit the effects of TNF on bone (Black *et al.*, 1991b). Synergism among IL-1, TNF, and lymphotoxin in bone resorption has been described (Stashenko *et al.*, 1987a). All of these data indicate that there are complex interactions between TNF and other cytokines during bone resorption.

b. Bone resorbing activity of tumor necrosis factor and lymphotoxin in vivo

Tashjian *et al.* (1986) demonstrated the first evidence that repeated injections of TNF cause hypercalcemia *in vivo*. Following this, König *et al.* (1988) reported that intravenous or subcutaneous administration of recombinant human or murine TNF stimulates bone resorption in a dose-dependent manner using a new quantitative technique to measure bone resorption that they developed. Lymphotoxin has also been shown to increase blood calcium levels (Garrett *et al.*, 1987). Our group has provided further evidence that TNF has bone resorbing action *in vivo* (Johnson *et al.*, 1989). We studied CHO cells that were transfected with the human TNF gene. When these cells were inoculated into nude mice, they caused a marked increase in osteoclastic bone resorption and hypercalcemia, whereas CHO cells that were transfected with the empty vector did not. More recently, we found that neutralizing antibodies to IL-6 decrease blood calcium levels that are elevated by TNF (unpublished observation).

3. Roles of Tumor Necrosis Factor and Lymphotoxin in Diseases Associated with Abnormal Bone and Calcium Metabolism

a. Malignant diseases

i. Solid tumors. Oversecretion of TNF may account for the pathophysiology of paraneoplastic syndromes such as cachexia, hypercalcemia, and leukocytosis in some tumor models. We have studied a human squamous carcinoma of the maxilla that was associated with hypercalcemia, cachexia, and leukocytosis (Yoneda *et al.*, 1989). This tumor MH-85 caused the same three paraneoplastic syndromes in nude mice (Yoneda *et al.*, 1991c). Because of the development of severe cachexia, we measured plasma levels of TNF/cachectin by radioimmunoassay using polyclonal antibodies specific to mouse TNF and found that plasma mouse TNF levels increased fourfold in tumor-bearing animals compared to non-tumor-bearing animals (Yoneda *et al.*, 1991a). Hypercalcemia, leukocytosis, and cachexia were reversed by administration of polyclonal antibodies to mouse TNF but not by polyclonal antibodies to human TNF. The tumor-bearing mice also showed a marked splenomegaly, and removal of the enlarged spleens resulted in a reversal of these paraneoplastic syndromes. These results suggested that TNF overproduced by host mouse immune cells rather than human MH-85 tumor cells was responsible for the pathophysiology of the paraneoplastic syndromes. In fact, MH-85 cells in culture did not release TNF and did not express TNF mRNA, whereas host immune cells including peritoneal, spleen, and peripheral blood macrophages secreted increased levels of TNF in the presence of MH-85 culture supernatants. The studies using antibodies to various cytokines that are known to induce TNF production suggested that the activity in MH-85 culture supernatants may be a unique molecule. This activity has an apparent molecular weight of 40K.

A similar mechanism is likely to play a role in several other tumor models such as the Leydig cell tumor (Sabatini *et al.*, 1990b) and A375 melanoma. In the latter model, GM-CSF was identified to be responsible for induction of TNF overexpression by host immune cells (Sabatini *et al.*, 1990a).

ii. Myeloma. Lymphotoxin has been implicated in myeloma. Myeloma cells in culture released lymphotoxin and expressed lymphotoxin mRNA (Garrett *et al.*, 1987; Bataille *et al.*, 1989b). Antibodies to lymphotoxin partially blocked bone resorption caused by myeloma culture supernatants. Obviously, this is not the only mechanism by which osteoclasts are stimulated in myeloma. Production of other cytokines such as IL-1β (Kawano *et al.*, 1989a) and IL-6 (Kawano *et al.*, 1988; Zhang *et al.*, 1989) may also be important.

b. Chronic inflammation

TNF as well as IL-1 and IL-6 may play a role in the localized bone destruction that occurs in chronic inflammatory diseases such as rheumatoid arthritis and periodontal disease. TNF increases the production of collagenase and prostaglandins in synovial cells (Dayer *et al.*, 1985). TNF stimulates glycosaminoglycan destruction and inhibits glycosaminoglycan synthesis in cartilage (Saklatvala, 1986). More recently, transgenic mice carrying 3′-modified human TNF transgenes have been found to develop chronic inflammatory polyarthritis (Keffer *et al.*, 1991). In this model, a monoclonal antibody to human TNF completely prevents the occurrence of polyarthritis. This study convincingly indicates the importance of TNF in the pathogenesis of arthritis.

c. Osteoporosis

Ralston *et al.* (1990) showed that TNF secretion by peripheral blood mononuclear cells in postmenopausal osteoporotic women is inhibited by 17-β-estradiol. 17-β-estradiol showed no effect on TNF secretion by the mononuclear cells in men or premenopausal women. Testosterone had no effect on TNF secretion. They proposed that this might be part of the mechanism by which 17-β-estradiol prevents osteoporosis in postmenopausal women.

F. Leukemia Inhibitory Factor (Differentiation-Inducing Factor)

LIF is a recently characterized cytokine that is produced by Krebs II ascites tumor cells. LIF is a single-chain glycoprotein of 58 kDa, and its murine cDNA clone (Gearing *et al.*, 1987) and human genomic clone (Gough *et al.*, 1988) were isolated. LIF is characterized by its capacity to induce differentiation and suppress proliferation of mouse leukemia M1 cells. Because of this capacity, it appears that LIF is identical to a cytokine called DIF, which was purified from mouse L929 cells and Ehrlich's ascites tumor and which stimulated M1 cell differentiation (Tomida *et al.*, 1984).

1. Effects on Bone Formation

Mouse osteoblastic cell line MC3T3-E1 constitutively produce DIF (Abe *et al.*, 1988), and several osteoblastic cells such as UMR 106-01 (Allan *et al.*, 1990) and immortalized rat calvarial cell RCT-1 (Rodan *et al.*, 1990b) have been shown to possess high-affinity receptors to LIF. Thus, LIF seems to be an autocrine/paracrine regulator of osteoblast functions. However, its effect on osteoblast proliferations is controversial. Reid *et al.* (1990) and Lowe *et al.* (1991) reported that LIF stimulates proliferation of mouse calvarial cells and UMR 106 osteoblastic osteosarcoma cells, respectively. In contrast, Noda *et al.* (1990) showed that LIF inhibits MC3T3-E1 mouse osteoblast proliferation.

On osteoblast phenotypic expression, LIF alone decreased mRNA levels of alkaline phosphatase, type I collagen (Noda *et al.*, 1990), and plasminogen activator activity (Allan *et al.*, 1990) and increased osteopontin mRNA levels (Noda *et al.*, 1990). However, LIF enhanced retinoic acid induction of alkaline phosphatase (Rodan *et al.*, 1990b; Allan *et al.*, 1990). Therefore, overall effects of LIF on osteoblasts are not clear and remain to be determined.

Recently, using a transgenic mice model, interesting actions of LIF have been demonstrated (Metcalf and Gearing, 1989). Mice carrying the murine hematopoietic cell line FDC-P1 that were transfected stably with LIF cDNA and produced high levels of LIF showed excess bone formation, suggesting that LIF may play a role in osteoblast function. However, the bone changes were not described in detail, and precise evaluation of LIF effects on bone formation needs further studies.

2. Effects on Bone Resorption

Abe *et al.* (1986) first reported that DIF that was produced by mitogen-activated spleen cells stimulated bone resorption in organ cultures of neonatal mouse calvariae. They also found that DIF was constitutively produced by mouse osteoblast MC3T3-E1 and stimulated osteoclastlike cell formation in mouse marrow cultures (Abe *et al.*, 1988). More recently, recombinant mouse and human LIF have been shown to stimulate bone resorption in organ cultures of neonatal mouse calvariae (Reid *et al.*, 1990).

In LIF cDNA transgenic mice, increased bone resorption was occasionally noted (Metcalf and Gearing, 1989). Bone changes in animals that have been administered LIF have not been reported to date.

G. Interferon-γ

1. Effects on Bone Formation

IFN-γ has been shown to inhibit both proliferation and differentiation of osteoblasts (Smith *et al.*, 1987). It inhibits collagen synthesis in organ cultures of fetal rat calvariae (Smith *et al.*, 1987). This effect is additive with those of TNF.

2. Effects on Bone Resorption

a. In vitro

IFN-γ is a multifunctional cytokine that shares many of the biological activities with TNF and IL-1. However, IFN-γ antagonizes the effects of IL-1 and TNF on bone resorption (Gowen *et al.*, 1986; Gowen and Mundy, 1986; Klaushofer *et al.*, 1989). Its inhibitory action is more effective on bone resorption induced by IL-1 and TNF than on bone resorption induced by systemic hormones such as PTH and $1{,}25D_3$ (Gowen *et al.*, 1986). An ultrastructural study has demonstrated that the effects of IFN-γ on osteoclast morphology is comparable to the calcitonin (Klaushofer *et al.*, 1989). Inhibition of bone resorption by IFN-γ seems to be due to inhibition of osteoclast formation rather than activation (Takahashi *et al.*, 1986), since IFN-γ shows no discernible effects on mature osteoclasts *in vitro* and *in vivo* (Klaushofer *et al.*, 1989). IFN-γ may inhibit proliferation or differentiation of hematopoietic progenitors of osteoclasts.

It has been proposed that IFN-γ inhibits bone resorption by suppressing prostaglandin synthesis (Hoffmann *et al.*, 1987). However, other groups did not find IFN-γ's inhibitory effects on prostaglandin synthesis (Gowen *et al.*, 1986). Involvement of prostaglandins in IFN-γ actions remains to be clarified.

b. In vivo

Sato *et al.* (1992) recently reported that IFN-γ reverses hypercalcemia associated with a human squamous carcinoma in nude mice. They also found that bone marrow cells from IFN-γ treated animals produce fewer osteoclasts than do bone marrow cells from untreated animals. This raises the possibility that IFN-γ may be a potential therapeutic agent for hypercalcemia associated with increased osteoclastic bone resorption.

H. Osteoclastpoietic Factor

Since effects of all osteotropic cytokines described earlier were originally found in tissues other than bone and subsequently tested in bone, their effects are diverse and not specific or selective for bone. In an attempt to search for a cytokine that specifically influences bone resorption, we reasoned that there may be a growth and differentiation-regulating factor for osteoclasts akin to the CSFs, because CSFs specifically control proliferation and differentiation of the hematopoietic stem cells within which osteoclast precursors reside. In search for a source of such a factor, we found that a human squamous cancer of the maxilla was a suitable source. This tumor, MH-85, was associated with leukocytosis, hypercalcemia, and cachexia (Yoneda *et al.*, 1989). In nude mice, MH-85 induced the same three paraneoplastic syndromes and increased osteoclastic bone resorption in bone that was far distant from the site of tumor inoculation. When the tumor was surgically removed, osteoclastic bone resorption was markedly diminished. These results led us to hypothesize that MH-85 tumor produced a circulating factor that caused increased osteoclastic bone resorption. We found that MH-85 produced G-CSF, GM-CSF, and IL-6. In addition to these cytokines, MH-85 secreted a factor that stimulated osteoclastlike cell formation in human and mouse marrow cultures (Yoneda *et al.*, 1991d) and in cultures of human leukemia cell line HL-60 (Yoneda *et al.*, 1991b). We then attempted to purify and identify this factor using a mouse marrow culture assay. This factor, which we tentatively named osteoclastpoietic factor (OPF), was purified and analyzed for its amino acid sequence. The result demonstrated that OPF had a unique amino acid sequence and was a novel polypeptide. Whether OPF is a specific osteotropic cytokine that is produced in bone microenvironment and plays an important role in normal bone turnover and whether it is implicated in metabolic bone diseases associated with abnormal osteoclast function remain to be elucidated.

VI. CONCLUSION

There are so many types of cells and diverse cytokines in bone that it seems almost impossible to clearly elucidate their roles and to understand how they interact and collaborate in normal and pathologic bone turnover. Conventional techniques do not appear to be able to produce sufficient information to answer these important but complicated questions. However, with the help of recent dramatic progress in the development of modern techniques in cell and molecular biology, it is now possible for us to learn more about bone physiology from different points of view. For instance, targeted disruption of a specific gene by homologous recombination (Cappecchi, 1989) appears to be one of the powerful tools that will give us opportunities to study the role of a specific gene in the regulation of a variety of cell functions. It was recently reported that knock-out of the c-*src* protooncogene results in osteopetrosis in transgenic mice (Soriano *et al.*, 1991). Since osteopetrosis is characterized by dysfunction of osteoclasts, c-*src* expression may be crucial for normal osteoclast function. We have been analyzing this animal model by using histologic and tissue culture techniques. Continuous attempts to incorporate new techniques in combination with conventional techniques are always required. As long as we maintain this attitude, we should be able to gather the information on the role of osteotropic cytokines in bone cell function that will advance bone research.

REFERENCES

Abe, E., Tanaka, H., Ishimi, Y., Chisato, M., Hayashi, T., Nagasawa, H., Tomida, M., Yamaguchi, Y., Hozumi, M., and Suda, T. (1986). Differentiation-inducing factor purified from conditioned medium of mitogen-treated spleen cell cultures stimulates bone resorption. *Proc. Natl. Acad. Sci.* **83,** 5958–5962.

Abe, E., Ishimi, Y., Takahashi, N., Akatsu, T., Ozawa, H., Yamana, H., Yoshiki, S., and Suda, T. (1988). A differentiation-inducing factor produced by the osteoblastic cell line MC3T3-E1 stimulates bone resorption by promoting osteoclast formation. *J. Bone Miner. Res.* **3,** 635–645.

Agematsu, K., Takeuchi, S., Ichikawa, M., Yabuhara, A., Uehara, Y., Kawai, H., Miyagawa, Y., Ishii, K., Katsuyama, T., Nakahata, T., and Komiyama, A. (1991). Spontaneous production of interleukin-6 by Kl-1-positive large-cell anaplastic lymphoma with extensive bone destruction. *Blood* **77,** 2299–2303.

Akira, S., Hirano, T., Taga, T., and Kishimoto, T. (1990). Biology of multifunctional cytokines: IL-6 and related molecules (IL-1 and TNF). *FASEB J.* **4,** 2860–2867.

Al-Balaghi, S., Hakan, S., and Moller, E. (1984). B cell differentiation factor in synovial fluids of patients with rheumatoid arthritis. *Immunol. Rev.* **78,** 7–23.

Al-Humidan, A., Ralston, S. H., Hughes, D. E., Chapman, K., Aarden, L., Russell R. G. G., and Gowen, M. (1991). Interleukin-6 does not stimulate bone resorption in neonatal mouse calvariae. *J. Bone Miner. Res.* **6,** 3–7.

Allan, E. H., Hilton, D. J., Brown, M. A., Evely, R. S., Yumita, S., Metcalf, D., Gough, N. M., Ng, K. W., Nicola, N. A., and Martin, T. J. (1990). Osteoclasts display receptors for and responses to leukemia-inhibitory factor. *J. Cell. Physiol.* **145,** 110–119.

Arend, W. P., Joslin, F. G., Thompson, R. C., and Hannum, C. H. (1989). An IL-1 inhibitor from human monocytes. Production and characterization of biologic properties. *J. Immunol.* **15,** 1851–1858.

Bataille, R., and Klein, B. (1991). The bone resorbing activity of IL-6. *J. Bone Miner. Res.* **6,** 1143–1144.

Bataille, R., Jourdan, M., Zhang, X. G., and Klein, B. (1989a). Serum levels of interleukin-6, a potent myeloma cell growth factor, as a reflect of disease severity in plasma cell dyscrasias. *J. Clin. Invest.* **84,** 2008–2011.

Bataille, R., Klein, B., Jourdan, M., Rossi, J. F., and Durie, B. G. M. (1989b). Spontaneous secretion of tumor necrosis factor-beta by human myeloma cell lines. *Cancer* **63,** 877–880.

Bataille, R., Chappard, D., Marcelli, C., Dessauw, P., Baldet, P., Sany, J., and Alexandre, C. (1991). Recruitment of new osteoblasts and osteoclasts is the earliest critical event in the pathogenesis of human multiple myeloma. *J. Clin. Invest.* **88,** 62–66.

Benjamin, D., and Dower, S. K. (1990). Human B cells express two types of interleukin-1 receptors. *Blood* **75,** 2017–2023.

Bertolini, D. R., Nedwin, G. E., Bringman, T. S., and Mundy, G. R. (1986). Stimulation of bone resorption and inhibition of bone formation *in vitro* by human tumour necrosis factor. *Nature (London)* **319,** 516–518.

Bird, T. A., and Saklatvala, J. (1986). Identification of a common class of high affinity receptors for both types of porcine interleukin on consecutive tissue cells. *Nature (London)* **324,** 263–265.

Black, K., Mundy, G. R., and Garrett, I. R. (1990). Interleukin-6 causes hypercalcemia *in vivo,* and enhances the bone resorbing potency of interleukin-1 and tumor necrosis factor by two orders of magnitude *in vitro. J. Bone Miner. Res.* **5**(Suppl. 2), abstract #787.

Black, K., Garrett, I. R., and Mundy, G. R. (1991a). Chinese hamster ovarian cells transfected with the murine interleukin-6 gene cause hypercalcemia as well as cachexia, leukocytosis and thrombocytosis in tumor-bearing nude mice. *Endocrinology* **128,** 2657–2659.

Black, K., Yoneda, T., Garrett, I. R., and Mundy, G. R. (1991b). Antibodies to interleukin-6 inhibit effects of bone resorbing factors both *in vivo* and *in vitro*. *J. Bone Miner. Res.* **6**(Suppl. 1), S287 (abstract).

Boyce, B. F., Aufdemorte, T. B., Garrett, I. R., Yates, A. J. P., and Mundy, G. R. (1989a). Effects of interleukin-1 on bone turnover in normal mice. *Endocrinology* **125,** 1142–1150.

Caligaris-Cappis, F., Bergui, L., Gregoretti, M. G., Gaidano, G., Gaboli, M., Schena, M., Zambonin-Zallone, A., and Marchisio, P. C. (1991). Role of bone marrow stromal cells in the growth of human multiple myeloma. *Blood* **77,** 2688–2693.

Canalis, E. (1986). Interleukin-1 has independent effects on deoxyribonucleic acid and collagen synthesis in cultures of rat calvariae. *Endocrinology* **118,** 74–81.

Canalis, E., McCarthy, T., and Centrella, M. (1988). Growth factors and the regulation of bone remodeling. *J. Clin. Invest.* **81,** 277–281.

Cappecchi, M. R. (1989). Altering the genome by homologous recombination. *Science* **244,** 1288–1292.

Carter, D. B., Deibel, M. R., Jr., Dunn, C. J., Tomich, C. S. C., Laborde, A. L., Slightom, J. L., Berger, A. E., Bienkowski, M. J., Sun, F. F., McEwan, R. N., Harris, P. K. W., Yem, A. W., Waszak, G. A. Ghosay, J. G., Sieu, L. C., Hardee, M. M., Zurcher-Neely, H. A., Reardon, I. M., Heinrikson, R. L., Truesdell, S. E., Shelly, J. A. Eessalu, T. E., Taylor, B. M., and Tracey, D. E. (1990). Purification, cloning, expression and biological characterization of an interleukin-1 receptor antagonist protein. *Nature (London)* **344,** 633–638.

Centrella, M., McCarthy, T. L., and Canalis, E. (1988). Tumor necrosis factor-α inhibits collagen synthesis and alkaline phosphatase activity independently of its effects on deoxyribonucleic acid synthesis in osteoblast-enriched bone cell cultures. *Endocrinology* **123,** 1442–1448.

Corboz, N. A., Cecchini, M. G., Felix, R., Fleisch, H., Pluijm, G., and Löwik, C. W. G. M. (1992). Effect of macrophage colony-stimulating factor on *in vitro* osteoclast generation and bone resorption. *Endocrinology* **130,** 437–442.

Cozzolino, F., Torcia, M., Aldinucci, D., Rubartelli, A., Miliani, A., Shaw, A. R., Lansdorp, P. M., and Diguglielmo, R. (1989). Production of interleukin-1 by bone marrow myeloma cells. *Blood* **74,** 387–390.

Dayer, J. M., Beutler, B., and Cerami, A. (1985). Cachectin/tumor necrosis factor stimulates collagenase and prostaglandin E_2 production by human synovial cells and dermal fibroblasts. *J. Exp. Med.* **162,** 2163–2168.

Dedhar, S. (1989). Regulation of expression of the cell adhesion receptors, integrins, by recombinant human interleukin-1β in human osteosarcoma cells: Inhibition of cell proliferation and stimulation of alkaline phosphatase activity. *J. Cell. Physiol.* **138,** 291–299.

Dewhirst, F. E., Stashenko, P. P., Mole, J. E., and Tsurumachi, T. (1985). Purification and partial sequence of human osteoclast-activating factor: Identity with interleukin-1 beta. *J. Immunol.* **135,** 2562–2568.

Dewhirst, F. E., Ago, J. M., Peros, W. J., and Stashenko, P. (1987). Synergism between parathyroid hormone and interleukin-1 in stimulating bone resorption in organ culture. *J. Bone Miner. Res.* **2,** 127–134.

Dinarello, C. A. (1991). Interleukin-1 and interleukin-1 antagonism. *Blood* **77,** 1627–1652.

Downing, J. R., Roussel, M. F., and Sherr, C. J. (1989). Protein kinase C 'transmodulates' the colony stimulating factor receptor by activating a membrane-associated protease that specifically releases the receptor ligand-binding domain from the cell. *Clin. Res.* **37,** 544A.

Eisenberg, S. P., Evans, R. J., Arend, W. P., Verderber, E., Brewer, M. T., Hannum, C. H., and Thompson, R. C. (1990). Primary structure and functional expression from complementary DNA of a human interleukin-1 receptor antagonist. *Nature (London)* **343,** 341–346.

Elford, P. R., Felix, R., Cecchini, M., Trechsel, V., and Fleisch, H. (1987). Murine osteoblast-like cells and the osteogenic cell MC3T3-E1 release a macrophage clony-stimulating activity in culture. *Calcif. Tissue Int.* **41,** 151–156.

Essner, R., Rhodes, K., McBride, W. H., Morton, D. L., and Economau, J. S. (1989). IL-4 down-regulates IL-1 and TNF gene expression in human monocytes. *J. Immunol.* **142,** 3857–3861.

Evans, D. B., Bunning, R. A. D., and Russell, R. G. G. (1990). The effects of recombinant human interleukin-1β on cellular proliferation and the production of prostaglandin E_2, plasminogen activator, osteocalcin and alkaline phosphatase by osteoblast-like cells derived from human bone. *Biochem. Biophys. Res. Commun.* **155,** 208–216.

Fang, M. A., and Hahn, T. J. (1991). Effects of interleukin-6 on cellular function in UMR-106-01 osteoblast-like cells. *J. Bone Miner. Res.* **6,** 133–139.

Felix, R., Elford, P. R., Stoerckle, C., Cecchini, M., Wetterwald, A., Trechsel, V., Fleisch, H., and Stadler, B. M. (1988). Production of hemopoietic growth factors by bone tissue and bone cells in culture. *J. Bone Miner. Res.* **3,** 27–36.

Felix, R., Cecchini, M. G., and Fleisch, H. (1990). Macrophage colony stimulating factor restores *in vivo* bone resorption in the OP/OP osteopetrotic mouse. *Endocrinology* **127,** 2592–2594.

Felix, R., Cecchini, M. G., Hofstetter, W., Guenther, H. L., and Fleisch, H. (1991). Production of granulocyte–macrophage (GM-CSF) and granulocyte colony-stimulating factor (G-CSF) by rat clonal osteoblastic cell population CRP 10/30 and the immortalized cell line IRC 10/30-myc 1 stimulated by tumor necrosis factor α. *Endocrinology* **126,** 661–667.

Fernandez-Botran, R. (1991). Soluble cytokine receptors: Their role in immunoregulation. *FASEB J.* **5,** 2567–2574.

Feyen, J. H. M., Elford, P., Dipadova, F. E., and Trechsel, U. (1989). Interleukin-6 is produced by bone and modulated by parathyroid hormone. *J. Bone Miner. Res.* **4,** 633–638.

Fried, R. M., Voelkel, E. F., Rice, R. H., Levine, L., Gaffney, E. V., and Tashjian, A. H. (1989). Two squamous cell carcinomas not associated with humoral hypercalcemia produce a potent bone resorption-stimulating factor which is interleukin-1 alpha. *Endocrinology* **125,** 742–751.

Fukumoto, S., Matsumoto, T., Harada, S., Fujisaki, J., Kawano, M., and Ogata, E. (1991). Pheochromocytoma with pyrexia and marked inflammatory signs: A paraneoplastic syndrome with possible relation to interleukin-6 production. *J. Clin. Endocrinol. Metab.* **73,** 877–881.

Fuller, K., Gallagher, A. C., and Chambers, T. J. (1991). Osteoclast resorption-stimulating activity is associated with the osteoblast cell surface and/or the extracellular matrix. *Biochem. Biophys. Res. Commun.* **181,** 67–73.

Garrett, I. R., Durie, B. G. M., Nedwin, G. E., Gillespie, A., Bringman, T., Sabatini, M., Bertolini, D. R., and Mundy, G. R. (1987). Production of the bone resorbing cytokine lymphotoxin by cultured human myeloma cells. *N. Engl. J. Med.* **317,** 526–532.

Garrett, I. R., Black, K. S., and Mundy, G. R. (1990). Interactions between interleukin-6 and interleukin-1 in osteoclastic bone resorption in neonatal mouse calvariae. *Calcif. Tissue Int.* **46**(Suppl. 2), abstract #140.

Gasson, J. C. (1991). Molecular physiology of granulocyte–macrophage colony-stimulating factor. *Blood* **77,** 1131–1145.

Gearing, D. P., Gough, N. M., King, J. A., Hilton, D. J., Nicola, N. A., Simpson, R. J., Nice, E. C., Kelso, A., and Metcalf, D. (1987). Molecular cloning and expression of cDNA encoding a murine myeloid leukaemia inhibitory factor (LIF). *EMBO J.* **13,** 3995–4002.

Gibbons, G., Martinez, O., Matli, M., Heinzel, F., Bernstein, M., and Warren, R. (1991). Recombinant IL-4 inhibits IL-6 synthesis by adherent peripheral blood cells *in vitro*. *Lymph. Res.* **9,** 283–293.

Gimble, J. M., Hudson, J., Henthorn, J., Hua, X., and Burstein, S. A. (1991). Regulation of interleukin-6 expression in murine bone marrow stromal cells. *Exp. Hematol.* **19,** 1055–1060.

Girasole, G., Sakagami, Y., Hustmyer, F. G., Yu, X. P., Derrigs, H. G., Boswell, S., Peacock, M., Boder, G., and Manolagas, S. C. (1990). 17-β estradiol inhibits cytokine induced IL-6 production by bone marrow stromal cells and osteoblasts. *J. Bone Miner. Res.* **5**(Suppl. 2), abstract #795.

Giri, J. G., Newton, R. C., and Howk, R. (1990). Identification of soluble interleukin-1 binding protein in cell-free supernatants. Evidence for soluble interleukin-1 receptor. *J. Biol. Chem.* **265,** 17416–17419.

Golde, D. W. (1991). The stem cell. *Sci. Am.* **265,** 86–93.

Gordon, M. Y. (1991). Hemopoietic growth factors and receptors: Bound and free. *Cancer Cells* **3,** 127–133.

Gordon, M. Y., Riley, G. P., Watt, S. M., and Greaves, M. F. (1987). Compartmentalization of a haematopoietic growth factor (GM-CSF) by glycosaminoglycans in the bone marrow microenvironment. *Nature (London)* **326,** 403–405.

Gough, N. M., Metcalf, D., Gough, J., Grail, D., and Dunn, A. R. (1985). Structure and expression of the mRNA for murine granulocyte–macrophage colony-stimulating factor. *EMBO J.* **4,** 645–653.

Gough, N. M., Gearing, D. P., King, J. A., Willson, T. A., Hilton, D. J., Nicola, N. A., and Metcalf, D. (1988). Molecular cloning and expression of the human homologue of the murine gene encoding myeloid leukemia-inhibitory factor. *Proc. Natl. Acad. Sci. USA* **85,** 2623–2627.

Gowen, M., and Mundy, G. R. (1986). Actions of recombinant interleukin-1, interleukin-2 and interferon gamma on bone resorption *in vitro*. *J. Immunol.* **136,** 2478–2482.

Gowen, M., Meikle, M. C., and Reynolds, J. J. (1983). Stimulation of bone resorption *in vitro* by a non-prostanoid factor released by human monocytes in culture. *Biochim. Biophys. Acta* **762,** 471–474.

Gowen, M., Wood, D. D., and Russell, R. G. (1985). Stimulation of the proliferation of human bone cells *in vitro* by human monocyte products with interleukin-1 activity. *J. Clin. Invest.* **75,** 1223–1229.

Gowen, M., Nedwin, G., and Mundy, G. R. (1986). Preferential inhibition of cytokine stimulated bone resorption by recombinant interferon gamma. *J. Bone Miner. Res.* **1,** 469–474.

Gowen, M., Chapman, K., Littlewood, A., Hughes, D., Evans, D., and Russell, G. (1990). Production of tumor necrosis factor by human osteoblasts is modulated by other cytokines, but not by osteotropic hormones. *Endocrinology* **126,** 1250–1255.

Groopman, J. E., Molina, J. M., and Scadden, D. T. (1989). Hematopoietic growth factors. Biology and clinical applications. *N. Engl. J. Med.* **321,** 1449–1459.

Guerne, P. A., Zuraw, B. L., Vaughan, J. H., Carson, D. A., and Lotz, M. (1989). Synovium as a source of interleukin-6 *in vitro*. Contribution to local and systemic manifestations of arthritis. *J. Clin. Invest.* **83,** 585–592.

Guerne, P. A., Carson, D. A., and Lotz, M. (1990). IL-6 production by human articular chondrocytes. Modulation of its synthesis, growth factors, and hormones *in vitro*. *J. Immunol.* **144,** 499–505.

Gutierrez, G. E., Mundy, G. R., Derynck, R., Hewlett, K. L., and Katz, M. S. (1987). Inhibition of parathyroid hormone-responsive adenylate cyclase in clonal osteoblast-like cells by transforming growth factor alpha and epidermal growth factor. *J. Biol. Chem.* **262,** 15845–15850.

Gutierrez, G. E., Mundy, G. R., Manning, D. R., Hewlett, E. L., and Katz, M. S. (1990). Transforming growth factor β enhances parathyroid hormone stimulation of adenylate cyclase in clonal osteoblast-like cells. *J. Cell. Physiol.* **144,** 438–447.

Hannum, C. H., Wilcox, C. J., Arend, W. P., Joslin, F. G., Dripps, D. J., Heimdal, P. L., Armes, L. G., Sommer, A., Eisenberg, S. P., and Thompson, R. C. (1990). Interleukin-1 receptor antagonist activity of a human interleukin-1 inhibitor. *Nature (London)* **343,** 336–340.

Harada, K., Oida, S., and Sasaki, S. (1988). Chondrogenesis and osteogenesis of bone marrow-derived cells by bone-inductive factor. *Bone* **9,** 177–183.

Harris, E. D. (1990). Rheumatoid arthritis. Pathophysiology and implications for therapy. *N. Engl. J. Med.* **322,** 1277–1289.

Hart, P. H., Vitti, G. F., Burgess, D. R., Whitty, G. A., Picolli, D. S., and Hamilton, J. A. (1989). Potential antiinflammatory effects of interleukin-4: Suppression of human monocyte tumor necrosis factor α, interleukin-1, and prostaglandin E_2. *Proc. Natl. Acad. Sci. USA* **85,** 3803–3807.

Hattersley, G., and Chambers, T. J. (1990). Effects of interleukin-3 and of granulocyte–macrophage colony stimulating factors on osteoclast differentiation from mouse hemopoietic tissue. *J. Cell. Physiol.* **142,** 201–209.

Hattersley, G., Dorey, E., Horton, M. A., and Chambers, T. J. (1988). Human macrophage colony-stimulating factor inhibits bone resorption by osteoclasts disaggregated from rat bone. *J. Cell. Physiol.* **137,** 199–203.

Heath, J. K., Saklatvala, J., Meikle, M. C., Atkinson, S. J., and Reynolds, J. J. (1985). Pig interleukin-1 (catabolin) is a potent stimulator of bone resorption *in vitro*. *Calcif. Tissue Int.* **37,** 95–97.

Hill, R. J., Warren, M. K., and Levine, J. (1990). Stimulation of thrombopoiesis in mice by human recombinant interleukin-6. *J. Clin. Invest.* **85,** 1242–1247.

Hirano, T., Matsuda, T., Turner, M., Miyasaka, N., Buchan, G., Tang, B., Sato, K., Shimizu, M., Maini, R., Feldmann, M., and Kishimoto, T. (1988). Excessive production of interleukin-6/B cell stimulatory factor-2 in rheumatoid arthritis. *Eur. J. Immunol.* **18,** 1797–1801.

Hoffmann, O., Klaushofer, K., Gleispach, H., Leis, H. J., Luger, T., Koller, K., and Peterlik, M. (1987). Gamma interferon inhibits basal and interleukin-1 induced prostaglandin production and bone resorption in neonatal mouse calvaria. *Biochem. Biophys. Res. Commun.* **143,** 38–43.

Howard, M., John, F., Hilfiker, M., Johnson, B., Takatsu, K., Hamaoka, T., and Paul, W. E. (1982). Identification of a T cell-derived B cell growth factor distinct from interleukin-2. *J. Exp. Med.* **155,** 914–923.

Ishimi, Y., Miyaura, C., Jin, C. H., Akatsu, T., Abe, T., Nakamura, Y., Yamaguchi, A., Yoshiki, S., Matsuda, T., Hirano, T., Kishimoto, T., and Suda, T. (1990). IL-6 is produced by osteoblasts and induces bone resorption. *J. Immunol.* **145,** 3297–3303.

Johnson, R. A., Boyce, B. F., Mundy, G. R., and Roodman, G. D. (1989). Tumors producing human TNF induce hypercalcemia and osteoclastic bone resorption in nude mice. *Endocrinology* **124,** 1424–1427.

Jourdan, M., Zhang, X. G., Portier, M., Boiron, J.-M., Bataille, R., and Klein, B. (1991). IFN-α induces autocrine production of IL-6 in myeloma cell lines. *J. Immunol.* **147,** 4402–4407.

Kawano, M., Hirano, T., Matsuda, T., Taga, T., Horii, Y., Iwato, K., Asaoku, H., Tang, B., Tanabe, O., Tanaka, H., Kuramoto, A., and Kishimoto, T. (1988). Autocrine generation and requirement of BSF-2/IL-6 for human multiple myelomas. *Nature (London)* **332,** 83–85.

Kawano, M., Yamamoto, I., Iwato, K., Tanaka, H., Asaoku, H., Tanabe, O., Ishikawa, H., Nobuyoshi, M., Ohmoto, Y., Hirai, Y., and Kuramoto, A. (1989a). Interleukin-1 beta rather than lymphotoxin as the major bone resorbing activity in human multiple myeloma. *Blood* **73,** 1646–1649.

Kawano, M., Tanaka, H., Ishikawa, H., Nobuyoshi, M., Iwato, K., Asaoku, H., Tanabe, O., and Kuramoto, A. (1989b). Interleukin-1 accelerates autocrine growth of myeloma cells through interleukin-6 in human myeloma. *Blood* **73,** 2145–2148.

Keeting, P. E., Rifas, L., Harris, S. A., Colvard, D. S., Spelsberg, T. C., Peck, W. A., and Riggs, B. L. (1991). Evidence for interleukin-1β production by cultured normal human osteoblast-like cells. *J. Bone Miner. Res.* **6,** 827–833.

Keffer, J., Probert, L., Cazlaris, H., Georgeopoulos, S., Kaslaris, E., Kioussis, D., and Kollias, G. (1991). Transgenic mice expressing human tumour necrosis factor: A predictive genetic model of arthritis. *EMBO J.* **10,** 4025–4031.

Klaushofer, K., Hörandser, H., Hoffmann, O., Czerwenka, E., König, U., Koller, K., and Peterlik, M. (1989). Interferon γ and calcitonin induce differential changes in cellular kinetics and morphology of osteoclasts in cultured neonatal mouse calvaria. *J. Bone Miner. Res.* **4,** 585–606.

Klein, B., Zhang, X. G., Jourdan, M., Content, J. Houssiau, F., Aarden, L., Piechaczyk, M., and Bataille, R. (1989). Paracrine rather than autocrine regulation of myeloma-cell growth and differentiation by interleukin-6. *Blood* **73,** 517–526.

Klein, B., Wijdenese, J., Zhang, X. G., Jourdan, M., Boiron, J. M., Brochier, J., Liautard, J., Merlin, M., Clement, C., Moral-Fournier, B., Lu, Z. Y., Mannoni, P., Sany, J., and Bataille, R. (1991). Murine anti-interleukin-6 monoclonal antibody therapy for a patient with plasma cell leukemia. *Blood* **78,** 1198–1204.

Knospe, W. H., Husseini, S. G., and Fried, W. (1989). Hematopoiesis on cellulose ester membranes. XI. Induction of new bone and a hematopoietic microenvironment by matrix factors secreted by marrow stromal cells. *Blood* **74,** 66–70.

Kodama, H., Yamasaki, A., Nose, M., Niida, S., Ohgame, Y., Abe, M., Kumegawa, M., and Suda, T. (1991). Congenital osteoclast deficiency in osteopetrotic (op/op) mice is cured by injections of macrophage colony-stimulating factor. *J. Exp. Med.* **173,** 269–272.

Kondo, Y., Sato, K., Ohkawa, H., Ueyama, Y., Okabe, T., Sato, N., Asano, S., Mori, M., Ohsawa N., and Kosaka, K. (1983). Association of hypercalcemia with tumors producing colony-stimulating factor(s). *Cancer Res.* **43,** 2368–2374.

König, A., Muhlbauer, R. C., and Fleisch, H. (1988). Tumor necrosis factor α and interleukin-1 stimulate bone resorption *in vivo* as measured by urinary [^{3}H] tetracyclin excretion from prelabeled mice. *J. Bone Miner. Res.* **3,** 621–627.

Kono, Y., Beagley, K. W., Fujihashi, K., McGhee, J. R., Taga, T., Hirano, T., Kishimoto, T., and Kiyono, H. (1991). Cytokine regulation of localized inflammation. Induction of activated B cells and IL-6-mediated polyclonal IgG and IgA synthesis in inflamed human gingiva. *J. Immunol.* **146,** 1812–1821.

Krane, S. M., and Simon, L. S. (1987). Metabolic consequences of bone turnover in Paget's disease of bone. *Clin. Orthop.* **217,** 26–36.

Kriegler, M., Perez, C., De Fay, K., Albert, I., and Lu, S. D. (1988). A novel form of TNF/cachectin is a cell surface cytotoxic transmembrane protein: Ramifications for the complex physiology of TNF. *Cell* **53,** 45–53.

Kukita, A., Chenu, C., McManus, L. M., Mundy, G. R., and Roodman, G. D. (1990). Atypical multinucleated cells form in long-term marrow cultures from patients with Paget's disease. *J. Clin. Invest.* **85,** 1280–1286.

Kurihara, N., Bertolini, D., Suda, T., Akiyama, Y., and Roodman, G. D. (1990). IL-6 stimulates osteoclast-like multinucleated cell formation in long term human marrow cultures by inducing IL-1 release. *J. Immunol.* **144,** 4226–4230.

Kurt-Jones, E. A., Beller, D. I., Mizel, S. B., and Unanue, E. R. (1985). Identification of a membrane-associated interleukin-1 in macrophages. *Proc. Natl. Acad. Sci. USA* **83,** 1204–1208.

Lee, M. Y., Fukumaga, R., Lee, T. J., Lottsfeld, J. L., and Nagata, S. (1991). Bone modulation in sustained hematopoietic stimulation in mice. *Blood* **77,** 2135–2141.

Li, N., Ouchi, Y., Okamato, Y., Masuyama, A., Kanaki, M., Futami, A., Hossi, T., Nakamura, T., and Orimo, H. (1991). Effect of parathyroid hormone on release of interleukin-1 and interleukin-6 from cultured mouse osteoblastic cells. *Biochem. Biophys. Res. Commun.* **179,** 236–242.

Lichtenstein, A., Berenson, J., Norman, D., Chang, M. P., and Carlile, A. (1989). Production of cytokines by bone marrow cells obtained from patients with multiple myeloma. *Blood* **74,** 1266–1273.

Linkhart, T. A., Linkhart, S. G., MacCharles, D. C., Long, D. L., and Strong, D. D. (1991). Interleukin-6 messenger RNA expression and interleukin-6 protein secretion in cells isolated from normal human bone: Regulation by interleukin-1. *J. Bone Miner. Res.* **6,** 1285–1294.

Littlewood, A. J., Aarden, L. A., Evans, D. B., Russell, R. G. G., and Gowen, M. (1991a). Human osteoblastlike cells do not respond to interleukin-6. *J. Bone Miner. Res.* **6,** 141–148.

Littlewood, A. J., Russell, J., Harvey, G. R., Hughes, D. E., Russell, R. G. G., and Gowen, M. (1991b). The modulation of the expression of IL-6 and its receptor in human osteoblasts *in vitro. Endocrinology* **129,** 1513–1520.

Long, M. W., Williams, J. L., and Mann, K. G. (1990). Expression of human bone-related proteins in the hematopoietic microenvironment. *J. Clin. Invest.* **86,** 1387–1395.

Lorenzo, J. A., Sousa, S. L., Alander, C., Raisz, L. G., and Dinarello, C. A. (1987a). Comparison of the bone resorbing activity in the supernatants from phytohemagglutinin stimulated human peripheral blood mononuclear cells with that of cytokines through the use of an antiserum to interleukin-1. *Endocrinology* **121,** 1164–1170.

Lorenzo, J. A., Sousa, S. L., Fonseca, J. M., Hock, J. M., and Medlock, E. S. (1987b). Colony-stimulating factors regulate the development of multinucleated osteoclasts from recently replicated cells *in vitro. J. Clin. Invest.* **80,** 160–164.

Lorenzo, J. A., Sandra, L. S., and Centrella, M. (1988). Interleukin-1 in combination with transforming growth factor-α produces enhanced bone resorption *in vitro. Endocrinology* **123,** 2194–2200.

Lorenzo, J. A., Sousa, S. L., Brink-Webb, S. E., and Korn, J. H. (1990). Production of both interleukin-1α and β by newborn mouse calvarial cultures. *J. Bone Miner. Res.* **5,** 77–83.

Lowe, C., Cornish, J., Callon, K., Martin, T. J., and Reid, J. R. (1991). Regulation of osteoblast proliferation by leukemia inhibitory factor. *J. Bone Miner. Res.* **6,** 1277–1283.

Löwik, C. W. G. M., Vanderpluijm, G., Bloys, H., Hoekman, K., Bijvoet, O. L. M., Aarden, L. A., and Papapoulos, S. E. (1989). Parathyroid hormone (PTH) and PTH-like protein (Plp) stimulate interleukin-6 production by osteogenic cells—A possible role of interleukin-6 in osteoclastogenesis. *Bioc. Biop. Res.* **162,** 1546–1552.

MacDonald, B. R., Mundy, G. R., Clark, S., Wang, E. A., Kuehl, T. J., Stanley, E. R., and Roodman, G. D. (1986). Effects of human recombinant CSF-GM and highly purified CSF-1 on the formation of multinucleated cells with osteoclast characteristics in long term bone marrow cultures. *J. Bone Miner. Res.* **1,** 227–233.

MacDonald, R. G., Tepper, M. A., Clairmon, K. B., Perregaux, S. B., and Czech, M. P. (1989). Serum form of the rat insulin-like growth factor-II/mannose 6-phosphate receptor is truncated in the carboxyl-terminal domain. *J. Biol. Chem.* **264,** 3256–3261.

Marks, S. C., Jr. (1989). Osteoclast biology: Lessons from mammalian mutations. *Am. J. Med. Genet.* **34,** 43–54.

Matsuki, Y., Yamamoto, T., and Hara, K. (1991). Interleukin-1 mRNA-expressing macrophages in human chronically inflamed gingival tissue. *Am. J. Pathol.* **138,** 1299–1305.

Meghji, S., Harvey, W., and Harris, M. (1989). Interleukin-1-like activity in cystic lesions of the jaw. *Br. J. Oral Maxilla. Surg.* **27,** 1–11.

Metcalf, D., and Gearing, D. P. (1989). Fatal syndrome in mice engrafted with cells producing high levels of the leukemia inhibitory factor. *Proc. Natl. Acad. Sci. USA* **86,** 5948–5952.

Mosley, B., Beckmann, M. P., March, C. J., Idzera, R. L., Gimpel, S. D., Vanden Bos, T., Friend, D., Alpert, A., Anderson, D., Jackson, J., Wignall, J. M., Smith, C., Gallis, B., Sims, J. E. Urdal, D., Widner, M. B., Cosman, D., and Park, L. S. (1989). The murine interleukin-4 receptor: Molecular cloning and characterization of secreted and membrane bound forms. *Cell* **59,** 335–348.

Mundy, G. R. (1990). Immune system and bone remodeling. Trends Endocrinol. Metab. **1,** 307–311.

Mundy, G. R., and Roodman, G. D. (1987). Osteoclast ontogeny and function. *In* "Bone and Mineral Research" (W. A. Peck, ed.), Vol. 5, pp. 209–280. Elsevier, Amsterdam.

Nagasaki, K., Yamaguchi, K., Watanabe, K., Eto, S., and Abe, K. (1991). Interleukin-4 blocks parathyroid hormone-related protein-induced hypercalcemia *in vivo. Biochem. Biophys. Res. Commun.* **178,** 694–698.

Nanes, M. S., Rubin, J., Titus, L., Hendy, G. N., and Catherwood, B. (1991). Tumor necrosis factor-α inhibits 1,25-dihydroxyvitamin D_3-stimulated bone Gla protein synthesis in rat osteosarcoma cells (ROS 17/2.8) by a pretranslational mechanism. *Endocrinology* **128,** 2577–2582.

Nathan, C., and Sporn, M. (1991). Cytokines in context. *J. Cell Biol.* **113,** 981–986.

Nishihara, T., Ishihara, Y., Noguchi, and Koga, T. (1989). Membrane IL-1 induces bone resorption in organ culture. *J. Immunol.* **143,** 1881–1886.

Noda, M., Vogel, R., Hasson, D. M., and Rodan, G. A. (1990). Leukemia inhibitory factor suppresses proliferation, alkaline phosphatase activity and typel collagen messenger ribonucleic acid level and enhances osteopontin mRNA level in murine osteoblast-like (MC3T3E1) cells. *Endocrinology* **127,** 185–190.

Nophar, Y., Kemper, O., Brokbusch, C., Engelmann, H., Zwang, R. Adarka, D., Holtmann, H., and Wallach, D. (1990). Soluble forms of tumor necrosis factor receptors (TNF-Rs). The cDNA for the type-*T* TNF-R, cloned using amino acid sequence data of its soluble form, encodes both the cell surface and a soluble form of the receptor. *EMBO J.* **9,** 3269–3278.

Novick, D., Engelmann, H., Wallach, D., and Rubinstein, M. (1989). Soluble cytokine receptors are present in normal human urine. *J. Exp. Med.* **170,** 1409–1414.

Ohsaki, Y., Scarcez, T., Williams, R., and Roodman, G. (1991). Interleukin-6 is a potential autocrine factor responsible for the intensive bone resorption by multinucleated giant cells in giant cell tumors of bone. *J. Bone Miner. Res.* **6**(Suppl. 1): S265 (abstract).

Owen, M., and Friedenstein, A. J. (1988). Stromal stem cells: Marrow-derived osteogenic precursors in cell and molecular biology of vertebrate hard tissues. *CIBA Found. Symp.* **136,** 42–60.

Owen, M. E., Cave, J., and Joyner, C. J. (1987). Clonal analysis *in vitro* of osteogenic differentiation of marrow stromal CFU-F. *J. Cell Sci.* **87,** 731–738.

Pacifici, R., Rifas, L., Teitelbaum, S., Slatopolsky, E., McCracken, R., Bergfeld, M., Lee, W., Avioli, L. V., and Peck, W. A. (1987). Spontaneous release of interleukin-1 from human blood monocytes reflects bone formation in idiopathic osteoporosis. *Proc. Natl. Acad. Sci. USA* **84,** 4616–4620.

Pacifici, R., Rifas, L., McCracken, R., Vered, I., McMurty, C., Avioli, L., and Peck, W. A. (1989). Ovarian steroid treatment blocks a postmenopausal increase in blood monocyte interleukin-1 release. *Proc. Natl. Acad. Sci. USA* **86,** 2398–2402.

Pacifici, R., Carao, A., Santoro, S. A., Rifas, L., Jeffrey, J. J., Malone, J. D., McCrachen, R., and Avioli, L. V. (1991). Bone matrix constituents stimulate interleukin-1 release from human blood mononuclear cells. *J. Clin. Invest.* **87,** 221–228.

Paul, W. E. (1989). Pleiotropy and redundancy: T cell-derived lympokines in the immune response. *Cell* **57,** 521–524.

Pfeilschifter, J., Mundy, G. R., and Roodman, G. D. (1989). Interleukin-1 and tumor necrosis factor stimulate the formation of human osteoclast-like cells *in vitro*. *J. Bone Miner. Res.* **4,** 113–118.

Povolny, B., Lee, M., and Hall, S. (1990). Modulation of tartrate-resistant acid phosphatase expression by calcitriol in CSF-induced macrophage colonies. *Exp. Hematol.* **18,** 283–288.

Ralston, S. H., Russell, R. G. G., and Gowen, M. (1990). Estrogen inhibits release of tumor necrosis factor from peripheral blood mononuclear cells in postmenopausal women. *J. Bone Miner. Res.* **5,** 983–988.

Rathjen, P. D., Toth, S., Willis, A., Heath, J. K., and Smith, A. G. (1990). Differentiation inhibiting activity is produced in matrix-associated and diffusible forms that are generated by alternate promoter usage. *Cell* **62,** 1105–1114.

Reid, L. R., Lowe, C., Cornish, J., Skinner S. J. M., Hilton, D. J., Willson, T. A., Gearing, D. P., and Martin, T. J. (1990). Leukemia inhibitor factor—A novel bone-active cytokine. *Endocrinology* **126,** 1416–1420.

Roberts, R., Gallagher, J., Spooncer, E., Allen, T. D., Bloomfield, F., and Dexter, T. M. (1988). Heparan sulfate bound growth factors: A mechanism for stromal cell mediated haemopoiesis. *Nature (London)* **332,** 376–378.

Rodan, S. B., Wesolowski, G., Chin, J., Limjuco, G. A., Schmidt, J. A., and Rodan, G. A. (1990a). IL-1 binds to high affinity receptors on human osteosarcoma cells and potentiates prostaglandin E_2 stimulation of cAMP production. *J. Immunol.* **145,** 1231–1237.

Rodan, S. B., Wesolowski, G., Hilton, D. J., Nicola, N. A., and Rodan, G. A. (1990b). Leukemia inhibitory factor binds with high affinity to pre-osteoblastic RCT-1 cells and potentiates the retinoic acid induction of alkaline phosphatase. *Endocrinology* **127,** 1602–1608.

Roodman, G. D., Takahashi, N., Bird, A., and Mundy, G. R. (1987). Tumor necrosis factor α (TNF) stimulates formation of osteoclast-like cell (OCL) in long term human marrow cultures by stimulating production of interleukin-1 (IL-1). *Clin. Res.* **35,** 515A.

Roodman, G. D., Kurihara, N., Ohsaki, Y., Kukita, A., Hosking, D., Demulder, A., Smith, J. F., and Singer F. R. (1992). Interleukin-6. A potential autocrine/paracrine factor in Paget's disease of bone. *J. Clin. Invest.* **89,** 46–52.

Ruoslahti, E., and Yamaguchi, Y. (1991). Proteoglycans as modulators of growth factor activities. *Cell* **64,** 87–89.

Sabatini, M., Garrett, I. R., and Mundy, G. R. (1987). TNF potentiates the effects of interleukin-1 on bone resorption *in vitro*. *J. Bone Miner. Res.* **2,** abstract #34.

Sabatini, M., Boyce, B., Aufdemorte, T., Bonewald, L., and Mundy, G. R. (1988). Infusions of recombinant human interleukin-1 alpha and beta cause hypercalcemia in normal mice. *Proc. Natl. Acad. Sci. USA* **85,** 5235–5239.

Sabatini, M., Chavez, J., Mundy, G. R., and Bonewald, L. F. (1990a). Stimulation of tumor necrosis factor release from monocytic cells by the A375 human melanoma via granulocyte-macrophage colony stimulating factor. *Cancer Res.* **50,** 2673–2678.

Sabatini, M., Yates, A. J., Garrett, R., Chavez, J., Dunn, J., Bonewald, L., and Mundy, G. R. (1990b). Increased production of tumor necrosis factor by normal immune cells in a model of the humoral hypercalcemia of malignancy. *Lab. Invest.* **63,** 676–681.

Saklatvala, J. (1986). Tumor necrosis factor α stimulates resorption and inhibits synthesis of proteoglycan in cartilage. *Nature (London)* **322,** 547–549.

Sato, K., Mimura, H., Han, D. C., Kariuchi, T., Ueyama, Y., Ohkawa, H., Okabe, T., Kondo, Y., Ohsawa, N., Tsushima, T., and Shizume, K. (1986). Production of bone-resorbing activity and colony-stimulating activity *in vivo* and *in vitro* by a human squamous cell carcinoma associated with hypercalcemia and leukocytosis. *J. Clin. Invest.* **78,** 145–154.

Sato, K., Kasono, K., Fujii, Y., Kawakami, M., Tsushima, T., and Shizume, K. (1987). Tumor necrosis factor type α (cachectin) stimulates mouse osteoblast-like cells

(MC3T3-E1) to produce macrophage–colony stimulating activity and prostaglandin E_2. *Biochem. Biophys. Res. Commun.* **145,** 323–329.

Sato, K., Fujii, Y., Kasono, K., Ozawa, M., Imamura, H., Kanaji, Y., Kurosawa, H., Tsushima, T., and Shizume, K. (1989). Parathyroid hormone-related protein and interleukin-1α synergistically stimulate bone resorption *in vitro* and increase the serum calcium concentration in mice *in vivo*. *Endocrinology* **124,** 2172–2178.

Sato, K., Satoh, T., Shizume, K., Yamakawa, Y., Ono, Y., Demura, H., Akatsu, T., Takahashi, N., and Suda, T. (1992). Prolonged decrease of serum calcium concentration by murine γ-interferon in hypercalcemic, human tumor (EC-GI)-bearing nude mice. *Cancer Res.* **52,** 444–449.

Schneider, H.-G., Allan, E. H., Moseley, J. M., Martin, T. J., and Findlay, D. M. (1991). Specific down-regulation of parathyroid hormone (PTH) receptors and responses to PTH by tumor necrosis factor α and retinoic acid in UMR-106 osteoblast-like osteosarcoma cells. *Biochem. J.* **280,** 451–457.

Schwab, G., Siegall, C. B., Aarden, L. A., Neckers, L. M., and Nordan, R. P. (1991). Characterization of an interleukin-6-mediated autocrine growth loop in the human multiple myeloma cell line, U266. *Blood* **77,** 587–593.

Seckinger, P., Klein-Nulend, J., Alander, C., Thompson, R. C., Dayer, J. M., and Raisz, L. G. (1990). Natural and recombinant human IL-1 receptor antagonists block the effects of IL-1 on bone resorption and prostaglandin production. *J. Immunol.* **145,** 4181–4184.

Shen, V., Cheng, S. L., Kohler, N. G., and Peck, W. A. (1990). Characterization and hormonal modulation of IL-1 binding in neonatal mouse osteoblast-like cells. *J. Bone Miner. Res.* **5,** 507–515.

Shinar, D. M., Sato, M., and Rodan, G. A. (1990). The effects of hemopoietic growth factors on the generation of osteoclast-like cells in mouse bone marrow cultures. *Endocrinology* **126,** 1728–1735.

Shioi, A., Teitelbaum, S. L., Ross, F. P., Welgres, H. G., Suzuki, H., Ohara, J., and Lacey, D. L. (1991). Interleukin-4 inhibits murine osteoclast formation *in vitro*. *J. Cell. Biochem.* **47,** 272–277.

Sims, J. E., March, C. J., Cosman, D., Widmer, M. B., MacDonald, H. R., McMohan, C. J., Grubin, C. E., Wignall, J. M., Jackson, J. L., Call, S. M., Friend, D., Alpert, A. R. Gills, S., Urdal, D. L., and Dower, S. K. (1988). cDNA expression cloning of the IL-1 receptor of the immunoglobulin superfamily. *Science* **241,** 585–589.

Sims, J. E., Acres, R. B., Gubin, C. E., McMahan, C. J., Wignall, J. M., March, C. J., and Dower, S. K. (1989). Cloning the interleukin-1 receptor from human T cells. *Proc. Natl. Acad. Sci. USA* **86,** 8946–8950.

Smith, D., Gowen, M., and Mundy, G. R. (1987). Effects of interferon gamma and other cytokines on collagen synthesis in fetal rat bone cultures. *Endocrinology* **120,** 2494–2499.

Soriano, P., Montgomery, C., Geske, R., and Bradley, A. (1991). Targeted disruption of the c-src proto-oncogene leads to osteopetrosis in mice. *Cell* **64,** 693–702.

Stagg, R. B., and Fletcher, W. H. (1990). The hormone-induced regulation of contact-dependent cell-cell communication by phosphorylation. *Endocr. Rev.* **11,** 302–325.

Stashenko, P., Dewhirst, F. E., Peros, W. J., Kent, R. L., and Ago, J. M. (1987a). Synergistic interactions between interleukin-1, tumor necrosis factor, and lymphotoxin in bone resorption. *J. Immunol.* **138,** 1464–1468.

Stashenko, P., Dewhirst, F. E., Rooney, M. L., Desjardins, L. A., and Hceley, J. D. (1987b). Interleukin-1β is a potent inhibitor of bone formation *in vitro*. *J. Bone Miner. Res.* **2,** 559–565.

Stein, J., Borzillo, G. V., and Rettenmier, C. W. (1990). Direct stimulation of cells expressing receptors for macrophage colony stimulating factor (CSF-1) by a plasma membrane-bound precursor of human CSF-1. *Blood* **76,** 1308–1314.

Suematsu, S., Matsuda, T., Aozasa, K., Akira, S., Nakano, N., Ohno, S., Miyazaki, J., Yamamura, K., Hirano, T., and Kishimoto, T. (1989). IgG, plasmacytosis in interleukin-6 transgenic mice. *Proc. Natl. Acad. Sci. USA* **86,** 7547–7551.

Symons, J. A., and Duff, G. W. (1990). A soluble form of the interleukin-1 receptor produced by a human B cell line. *FEBS Lett.* **272,** 133–136.

Takahashi, N., Mundy, G. R., and Roodman, G. D. (1986). Recombinant human gamma interferon inhibits formation of human osteoclast like cells. *J. Immunol.* **137,** 3541–3549.

Takahashi, N., Udagawa, N., Akatsu, T., Tanaka, H., Isogai, Y., and Suda, T. (1991a). Deficiency of osteoclasts in osteopetrotic mice is due to a defect in the local microenvironment provided by osteoblastic cell. *Endocrinology* **128,** 1792–1796.

Takahashi, N., Udagawa, N., Akatsu, T., Tanaka, H., Shionome, M., and Suda, T. (1991b). Role of colony-stimulating factors in osteoclast development. *J. Bone Miner. Res.* **6,** 977–985.

Tanaka, Y., Yamashita, V., Watanabe, K., Nagata, K., Mori, N., Oda, S., and Eto, S. (1990). Calcium dependency of the production of interleukin-1 and the expression of interleukin-1 receptors of human adult T cell leukemia cells *in vitro*. *Cancer Res.* **50,** 4344–4348.

Tashjian, A. H., Voelkel, E. F., Lloyd, W., Derynck, R., Winkler, M. E., and Levine, L. (1986). Actions of growth factors on plasma calcium. *J. Clin. Invest.* **78,** 1405–1409.

Tashjian, A. H., Voelkel, E. F., Lazzaro, M., Goad, D., Bosma, T., and Levine, L. (1987). Tumor necrosis factor α (cachectin) stimulates bone resorption in mouse calvaria via a prostaglandin-mediated mechanism. *Endocrinology* **120,** 2029–2036.

Tatakis, D. N., Schneeberger, G., and Dziak, R. (1988). Recombinant interleukin-1 stimulates prostaglandin E_2 production by osteoblastic cells: Synergy with parathyroid hormone. *Calcif. Tissue Int.* **42,** 358–362.

Tatakis, D. N., and Dziak, R. (1989). Recombinant human lymphotoxin effects on osteoblastic cells. *Biochem. Biophys. Res. Commun.* **162,** 435–440.

Thomson, B. M., Saklatvala, J., and Chambers, T. J. (1986). Osteoblasts mediate interleukin-1 stimulation of bone resorption by rat osteoclasts. *J. Exp. Med.* **164,** 104–112.

Thomson, B. M., Mundy, G. R., and Chambers, T. J. (1987). Tumor necrosis factors alpha and beta induce osteoblastic cells to stimulate osteoclastic bone resorption. *J. Immunol.* **138,** 775–779.

Tomida, M., Yamamoto-Yamaguchi, Y., and Hozumi, M. (1984). Purification of a factor inducing differentiation of mouse myeloid leukemic M1 cells from conditioned medium of mouse fibroblast L929 cells. *J. Biol. Chem.* **259,** 10978–10982.

Tsukita, S., Oishi, K., Akiyama, T., Yamanashi, Y., Yamamoto, T., and Tsukita, S. (1991). Specific proto-oncogenic tyrosine kinases of src family are enriched in cell-to-cell adherens junctions where the level of tyrosine phosphorylation is elevated. *J. Cell Biol.* **113,** 867–879.

Udagawa, N., Takahashi, N., Akatsu, T., Sasaki, T., Yamaguchi, A., Kodama, H., Martin, T. J., and Suda, T. (1989). The bone marrow-derived stromal cell lines MC3T3-G2/PA6 and ST2 support osteoclast-like cell differentiation in co-cultures with mouse spleen cells. *Endocrinology* **125,** 1805–1813.

Van der Wijngaert, F. P., Tas, M. C., van der Meer, J. W. M., and Burger, E. H. (1987). Growth of osteoclast precursor-like cells from whole mouse bone marrow: Inhibitory effect of CSF-1. *Bone Miner.* **3,** 97–110.

Vilcek, J., and Lee, T. H. (1991). Tumor necrosis factor. New insights into the molecular mechanisms of its multiple actions. *J. Biol. Chem.* **266,** 7313–7316.

Watanabe, K., Tanaka, Y., Morimoto, I., Yahata, K., Zeki, K., Fujihara, T., Yamashita, V., and Eto, S. (1990). Interleukin-4 as a potent inhibitor of bone resorption. *Biochem. Biophys. Res. Commun.* **172,** 1035–1041.

Weber, W., Gill, G., and Spiess, J. (1984). Production of an epidermal growth factor receptor-related protein. *Science* **224,** 294–298.

Weir, E. C., Insogna, K. L., and Horowitz, M. C. (1989). Osteoblast-like cells secrete granulocyte–macrophage colony-stimulating factor in response to parathyroid hormone and lipopolysaccharide. *Endocrinology* **124,** 899–904.

Williams, R. C. (1990). Periodontal disease. *N. Engl. J. Med.* **322,** 373–382.

Wong, S. T., Winchell, L. F., McCune, B. K., Earp, H. S. Teixido, J., Massague, J., Herman, B., and Lee, D. C. (1989). The TGFα precursor expressed on the cell surface binds to the EGF receptor on adjacent cells, leading to signal transduction. *Cell* **56,** 495–506.

Xu, W. D., Firestein, G. S., Taetle, R., Kaushansky, K., and Zvaifler, N. J. (1989). Cytokines in chronic inflammatory arthritis. *J. Clin. Invest.* **83,** 876–882.

Yabuhara, A., and Komiyama, A. (1990). Castleman's disease and interleukin-6. *Leuk. Lymphoma* **2,** 369–373.

Yamaguchi, Y., Mann, D. M., and Ruoslahti, E. (1990). Negative regulation of transforming growth factor β by the proteoglycan decorin. *Nature (London)* **346,** 281–284.

Yamamoto, I., Kawano, M., Sone, T., Iwato, K., Tanaka, H., Ishikawa, H., Kitamura, N., Lee, K., Shigeno, K., Konishi, J., Asashu, H., Tanabe, O., Nobuyoshi, M., Ohmoto, Y., Hirai, Y., Higuchi, M., Ohsawa, T., and Kuramoto, A. (1989). Production of interleukin-1β, a potent bone resorbing cytokine, by cultured human myeloma cells. *Cancer Res.* **49,** 4242–4246.

Yayon, A. Klagsburn, M., Esko, J. D., Leder, P., and Ornitz, D. M. (1991). Cell surface, heparin-like molecules are required for binding of basic fibroblast growth factor to its high affinity receptor. *Cell* **64,** 841–848.

Yoneda, T., Nishikawa, N., Nishimura, R., Kato, I., and Sakuda, M. (1989). Three cases of oral squamous cancer associated with leukocytosis, hypercalcemia or both. *Oral Surg.* **68,** 604–611.

Yoneda, T., Alsina, M. M., Chavez, J. B., Bonewald, L., Nishimura, R., and Mundy, G. R. (1991a). Evidence that splenic cytokines play a pathogenetic role in the paraneoplastic syndromes of cachexia, hypercalcemia and leukocytosis in a human tumor in nude mice. *J. Clin. Invest.* **87,** 977–985.

Yoneda, T., Alsina, M. A., Garcia, J. L., and Mundy, G. R. (1991b). Differentiation of HL-60 cells into cells with the osteoclast phenotype. *Endocrinology* **129,** 683–689.

Yoneda, Y., Aufdemorte, T. B., Nishimura, R., Nishikawa, N., Sakuda, M., Alsina, M. M., Chavez, J. B., and Mundy, G. R. (1991c). Occurrence of hypercalcemia and leukocytosis with cachexia in a human squamous cell carcinoma of the maxilla in athymic nude mice. A novel experimental model of three concomitant paraneoplastic syndromes. *J. Clin. Oncol.* **9,** 468–477.

Yoneda, T., Kato, I., Bonewald, L. F., Chisoku, H., Burgess, W. H., and Mundy, G. R. (1991d). A novel osteoclastpoietic peptide: Purification and characterization. *J. Bone Miner. Res.* **6**(Suppl. 1), S197 (abstract).

Yoneda, T., Nishimura, R., Kato, I., Ohmae, M., Takita, M., and Sakuda, M. (1991e). Frequency of the hypercalcemia–leukocytosis syndrome in oral malignancies. *Cancer* **68,** 617–622.

Yoneda, T., Nakai, M., Moriyama, K., Scott, L., and Mundy, G. R. (1993). Neutralizing antibodies to human interleukin-6 reverse hypercalcemia associated with a human squamous carcinoma. *Cancer Res.***53,** 737–740.

Yoshida, H., Hayashi, S. I., Kunisada, T., Ogawa, M., Nishikawa, S., Okamura, H., Sudo, T., Shultz, L. D., and Nishikawa, S. I. (1990). The murine mutation osteopetrosis is in the coding region of the macrophage colony stimulating factor gene. *Nature (London)* **345,** 442–444.

Zhang, X. G., Klein, B., and Bataille, R. (1989). Interleukin-6 is a potent myeloma-cell growth factor in patients with aggressive multiple myeloma. *Blood* **74,** 11–13.

13

SIGNAL TRANSDUCTION IN OSTEOBLASTS AND OSTEOCLASTS

KEITH A. HRUSKA, FELICE ROLNICK, RANDALL L. DUNCAN, MEETHA MEDHORA, and KENSUKE YAMAKAWA

Cellular and Molecular Biology of Bone

I. INTRODUCTION

Signal transduction in the major bone cells, osteoblasts and osteoclasts, is a broad topic. Osteoblasts are somewhat mysterious cells in that their ontogeny and differentiation are still being described. For purposes of this chapter, an osteoblast will refer to cells that can be isolated from tissue sources that express a specific phenotype. This phenotype is the early expression of type 1 collagen production and alkaline phosphatase followed by secretion of specific noncollagenous bone matrix proteins such as osteocalcin, osteonectin, osteopontin, and others. Responsiveness to parathyroid hormone and the ability of the cells to calcify the extracellular matrix, which may be stimulated by ascorbic acid and β-glycerol phosphate, are additional properties of the osteoblast phenotype. Surprisingly, the identification of these cells *in vivo,* and their relationship to such cells as the lining cells, the osteocytes, and precursor cells off of the bone surface, has not been clearly established. Osteoblasts have proven to be extremely pleiotropic cells possibly related to their ontogeny. Because they share common ancestry with adipose tissue, muscle cells, and fibroblasts, it is not be surprising that they exhibit many responses of these cells to specific substances. As a result, the topic of signal transduction in the osteoblast is a vast one. The list of substances that activate the cells and the mechanisms by which activation produces specific biological effects are exceedingly complex and incompletely described. In this chapter, the general mechanisms of signal transduction are discussed along with the general list of specific substances and their mechanisms of action. A specific example of cell activation and the pathways of signal transduction from parathyroid hormone are discussed for osteoblasts.

Osteoclasts are also enigmatic cells. Their ontogeny has recently been somewhat elucidated. One of the fascinating features of osteoclast development is the loss of many receptors that are expressed in progenitor cells and that are present in osteoblasts. Several substances that regulate bone resorption do so despite the absence of receptors for these substances on the differentiated osteoclast. Thus, in the osteoclast, signal transduction by paracrine substances and cell-to-cell communication are important mechanisms of regulating cell function. In addition, the osteoclast has several unique mechanisms of signal transduction, which are described in this chapter. Also, a novel mechanism of signal transduction from occupancy of an integrin by matrix proteins of specific integrins has recently been described in the author's laboratory, and some of these preliminary studies are described.

II. SUBSTANCES WITH EFFECTS IN BONE CELLS

Because of the diverse nature of bone cell ontogeny, the list of substances that affects osteoblasts and osteoclasts is prodiguous. The list may be too large to enumerate usefully in a chapter such as this. The list in Table I is by no means inclusive, but it sets the stage for description of the mechanisms of signal generation used by several classes of substances. Classical bone physiology has considered the actions of calcitropic hormones and their effects on bone remodeling to play a central role in skeletal homeostasis. Although this basic tenant still has substance, our current understanding of bone physiology is much more complex. Nevertheless, systemic hormones play key roles in skeletal remodeling.

A. Hormones

The hormones that affect skeletal remodeling can be divided into two general groups: the peptide hormone class and the steroid hormone class.

1. Peptide Hormones

The major peptide hormones that effect osteoblast skeletal remodeling are parathyroid hormone (PTH), calcitonin, calcitonin gene-related peptide, and growth hormone. Other circulating peptides such as thrombin that are not considered to be calcitropic hormones also have dramatic effects on osteoblast function *in vitro*. Peptide hormones generally activate their target cells through binding to surface receptor protein with membrane-spanning domains. The receptors, in turn, couple to intracellular effectors through their cytoplasmic domains, which generally interact through guanosine triphosphate (GTP)-binding proteins. The effectors include adenylate cyclase, phospholipases, and ion channels.

2. Steroid Hormones

The list of steroid hormones with major actions on skeletal remodeling is large. Generally, the mechanism of steroid hormone action is thought to occur through receptors that are transiently in the cytoplasm but that have the capability of translocating the steroid–receptor complex into the nucleus. Here, the receptor complex binds to DNA along with accessory proteins serving as regulatory factors in gene transcription. Recent

TABLE I Substances That Have Effects in Bone

	Site of action			Net effect	
Substance[a]	Osteoblast	Osteoclast	Other	Formation	Resorption
Peptides					
Parathyroid	+	?	+	+	+
Calcitonin					−
Calcitonin Gene-Related Peptide					
Steroid					
Vitamin D	+	+			+
glucocorticoid	+			−	+
Estrogen	+	+		+	−
Testosterone	+			+	−
Progesterone					
Thyroid					+
Retinoic acid	+		+		
Cytokines					
IL-1	+				
	+				
TNFα	+				+
TNFβ					+
IL-3		+			+
IL-4		+			+
IL-6		+			+
IL-8					
γ-Interferon					
Leukemia inhibitory factor					
Growth factors					
Macrophage colony-stimulating factor		+			
Granulocyte–macrophage colony-stimulating factor					
IGF-I	+				+
IGF-II	+				+
Platelet-derived growth factor	+				+
Epidermal growth factor	+				+
Paracrines/autocrines					
Transforming growth factor β	+	+		+	+
Bone morphogenetic proteins 1–7	+	+			
Matrix proteins		+	+		
aFGF	+				
bFGF	+				

(continued)

TABLE I Substances That Have Effects in Bone

	Site of action			Net effect	
Substance[a]	Osteoblast	Osteoclast	Other	Formation	Resorption
Prostaglandins	+				
Prostaglandins leukotrienes		+			
Endothelin		+		+	+
EDRF		+			−
O_2 radicals		+			+
H^+		+			+
Ca^{2+}		+			−
Na^+					
Cell adhesion molecules					
Integrins					
Cadherins					

[a]FGF, fibroblast growth factor; IGF, insulinlike growth factor; IL, interleukin; TNF, tumor necrosis factor.

attention to nongenomic mechanisms of steroid hormone action, however, has demonstrated important effects of these factors independent of regulating gene transcription. The mechanism of these nongenomic actions of steroid hormones remains uncertain.

B. Cytokines

The list of locally produced cytokines that have major actions on skeletal remodeling is also large. Many of these substances are produced by hematopoietic cells resident in the bone marrow and affect trabecular osteoclasts and osteoblasts in a paracrine fashion. Recent attention to the role of interleukin 1, tumor necrosis factor α, and granulocyte–macrophage colony-stimulating factor (GM-CSF) in the stimulation of bone resorption following oophorectomy (Pacifici *et al.*, 1991a) and in osteoporosis (Pacifici *et al.*, 1991b) has been impressive and recently confirmed (Girasde, 1992). In addition, several of the interleukins serve as specific growth factors during osteoclast ontogeny, including interleukin 3, interleukin 6, whereas other interleukins serve as inhibitory factors related to osteoclast production such as interleukin 4.

C. Growth Factors

Numerous growth factors play key roles in the differentiation in production of osteoblast and osteoclast. Many of their physiologic roles in skeletal homeostasis remain to be clearly elucidated; however, their mechanisms of signal transduction have recently been described, and bone cells exhibit these pathways of signal generation.

D. Paracrine/Autocrine Factors

The group of paracrine/autocrine factors is also extremely large and disparate in the type of substance included, ranging from specific peptide growth factors present in the bone matrix in inactive forms, to TGF-β, fibroblast growth factor, and others. In addition, other matrix proteins serve as specific recognition sites for bone cell adhesion molecules and have direct action on regulating cell physiology through the process of integrin occupancy. A recently described novel pathway of signal transduction from a noncollagenous bone matrix protein, osteopontin with an Arg–Gly–Asp (RGD) sequence has been described in osteoclast regulation and is discussed herein. In addition, arachidonic acid metabolites such as the prostaglandins luekotrienes and cytochrome P450 products also have bone cell actions. Products of endothelial cells such as endothelin, endothelium-derived relaxing factor, and oxygen radicals have also been shown to play important roles in bone remodeling. Finally, individual ions appear to play unique roles in the regulation of bone cell function. Protons, calcium, and sodium appear to have extracellular receptors that recognize these ions and control function independent of their transport processes.

E. Cell Adhesion Molecules

A wide variety of cell adhesion molecules are expressed in bone cells. These include integrins and cadherins. Their roles in skeletal physiology are just becoming elucidated, and unique mechanisms of signal transduction from these substances are now being described. These substances also appear to represent a mechanism for cell-to-cell communication within cells of the same family and also among cells of different origins. For instance, the recently described mechanism of osteoblast–osteoclast cell adhesion between VLA 4 and v-CAM may explain the basis for the necessary step of stromal cell contact in osteoclast differentiation (Suda *et al.*, 1992).

III. MECHANISMS OF SIGNAL GENERATION

The myriad of bone active substances, some listed in Table I, generate signals in bone cells through multiple mechanisms. In general, these mechanisms involve binding of a substance to a recognition molecule (i.e., a receptor). In turn, the receptor is associated in the ligand-bound state with effector molecules that, when activated by the ligand–receptor complex, are capable of generating cell signals.

A. Receptors

The receptors for the substances listed in Table I fall into several types. There are three known classes of cell surface receptor proteins: G protein-linked, catalytic, and channel-linked (Berridge, 1985; Kahn, 1976; Levitski, 1984; Rees *et al.*, 1982; Snyder, 1985). Cell surface receptor proteins are defined by the signal transduction mechanism used. The GTP-binding protein-linked receptors indirectly activate or inactivate a separate plasma membrane-bound enzyme or ion channel. The interaction between the receptor and the enzyme or ion channel is mediated by a third protein, a GTP-binding regulatory protein (or G protein). The G protein-linked receptors usually activate a chain of events that alters the concentration of one or more small intracellular signaling molecules, often referred to as intracellular messengers. These intracellular messengers act, in turn, to alter the behavior of yet other target proteins in the cell. The G protein-linked receptors fit into several classes. They are generally proteins with seven membrane-spanning domains with large cytoplasmic loops and carboxyl-terminal tails. The β-adrenergic receptor is perhaps the best described (Benovic *et al.*, 1990). Recently, two receptors for calcitropic hormones have been cloned that appear to form a new subfamily of the class of seven membrane-spanning domain receptors. These receptors are the PTH receptor and the calcitonin receptor (Juppner *et al.*, 1991; Lin *et al.*, 1991).

B. Signal-Generating Complexes

Association of ligands with their receptors generates complexes capable of producing cell signals. The signal-generating complex may consist only of the receptor and the bound ligand in the case of steroid hormones. These ligand–receptor complexes are then capable of translocation into the nucleus, where they bind to DNA along with an accessory protein forming dimers that activate DNA transcription. The translocation of the receptor–ligand complex is affected by phosphorylation, which in some instances results in activation of the hormone receptor complex (Orti *et al.*, 1992). For other ligands that bind to receptors with seven transmembrane-spanning domains, which classically associate with G proteins, the signal-generating complex is much more complicated. In this instance, a hormone, receptor, G protein, and effector element together form the signal-generating complex (Fig. 1). Changes in the receptor associated with ligand binding increases the association of the receptor with the trimeric forms of G protein. The α subunit of the

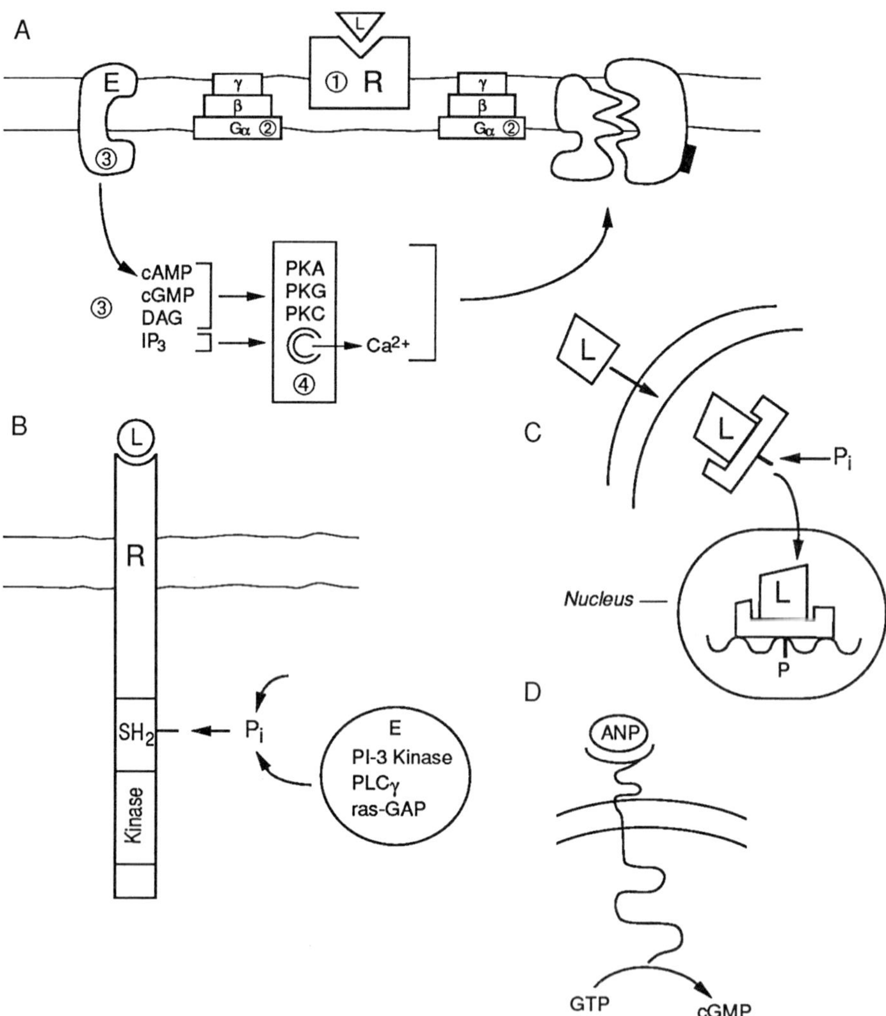

FIGURE 1 Ligand (L) binding to receptors (R) leads to generation of multi-unit complexes referred to here as signal generating complexes. Several examples are diagrammatically described. (A) Hormones that are ligands for receptors with seven membrane spanning domains (1) classically associate with trimeric GTP-binding proteins, (2) associated with numerous effector elements, and (3) forming a signal generating complex producing multiple second messengers (as shown in the figure). Second messengers activate a series of enzymes and release of calcium from intracellular stores as diagrammed (4). The kinases and calcium produce direct biological effects and/or participate in phosphorylation cascades leading to biological effects. (B) Growth factor receptors (R) are usually single transmembrane spanning proteins with intrinsic protein tyrosine kinase activity. The receptors contain src-2 (SH2) and src-3 (SH3) domains, which produce association with other signal generating enzymes including the src family of tyrosine kinases (E), PI-3 kinase, PLCγ, and ras-GAP. These enzymes associate with the tyrosine phosphorylated receptor through the

G proteins then is stimulated to bind GTP and associate and activate various effectors. In some instances, the effector may represent an ion channel that directly couples to α subunits of trimeric G proteins (Fig. 1). A third general type of signal-generating complex is represented by a class of growth factor receptors that are tyrosine kinases. Upon ligand association, these receptors activate the receptor tyrosine kinase and become autophosphorylated on src homology 2 (SH2) domains. The phosphorylation of the SH2 domains then produces association of multiple effectors with the activated receptor (Fig. 1).

C. Effector Elements

Several effector elements with prominent actions in bone cells are listed in Table II. There are two general categories for discussion. The first is the G protein-linked effectors, regulated by a heteotrimeric class of GTP-binding proteins; the second are the effectors, regulated through tyrosine phosphorylation.

1. G Protein-Linked Effectors

a. Adenylate cyclase

The plasma membrane-bound enzyme, adenylate cyclase, when activated produces the ubiquitous intracellular messenger cyclic adenosine monophosphate (cAMP). Cyclic AMP is rapidly and continuously synthesized and destroyed. Destruction occurs by one or more cAMP phosphodiesterases, which hydrolyze catrial natriuretic peptide (ANP) to AMP. Receptor proteins, which activate adenylate cyclase, usually do so by a stimulatory G protein (G_s) (Gautier *et al.*, 1989). Individuals who are genetically deficient in G_s have decreased responses to many hormones, and this includes the action of PTH. Reconstitution of cAMP production by insertion of epinephrine receptors, G_s, and adenylate cyclase molecules into phospholipid vesicles indicates that no other proteins are required for activation of the adenylate cyclase effector. The adenylate cyclase molecule is activated by the receptor hormone complex through binding of GTP to G_s. G_s keeps the adenylate cyclase active as long as

SH2 domain. (C) Steroid hormones bind to receptors that are transiently in the cytosol. On binding of the steroid hormone, the ligand receptor complex translocates to the nucleus where it associates with other protein factors serving as *trans*-activating factors for specific *cis* elements on multiple genes effecting gene transcription (D) Atrial natriuretic peptide is an example wherein the receptor is also the effector element, an enzyme particulate guanylate cyclase, producing the second messenger, cGMP.

TABLE II Effectors

G protein-linked effectors	Tyrosine kinase-linked effectors
Adenylate cyclase	Phospholipase C
Guanylate cyclase	Phosphatidylinositol 3-kinases
Phospholipases A	ras-guanosine triphosphatase-activating protein
Phospholipases C	src family tyrosine kinases
Phospholipases D	Tyrosine phosphatases
Ion channels	

GTP is intact. Hydrolysis of GTP to guanosine diphosphate by the G_s protein terminates activation of the cyclase. The adenylate cyclase effector protein can also be inhibited by coupling to ligand-occupied inhibitory receptors and inhibitory G proteins. The α_2-adrenergic receptors coupled to the inhibitory G protein (G_i) inhibit adenylate cyclase activity. G_i α and the $\beta\gamma$ subunits are believed to contribute to the inhibition of adenylate cyclase activity.

Bacterial toxins have been shown to have specific actions on G proteins that have assisted greatly in the elucidation of their biological roles. Cholera toxin is an enzyme that catalyzes the transfer of adenosine diphosphate ribose from intracellular nicotinamide adenine dinucleotide to the α subunit G_s. The ribosylation alters G_s so that it can no longer hydrolyze bound GTP. This produces an indefinitely active state of the G protein and results in prolonged elevations in cAMP levels. Pertussis toxin, made by the bacterium that causes whooping cough, produces the same effect by ADP ribosylating $G_{i\alpha}$. In this case, however, the G_i complex is prevented from interacting with receptors and therefore fails to inhibit adenylate cyclase in response to receptor activation.

b. Guanylate cyclase

The discovery that atrial natriuretic peptide (ANP) activates particulate guanylate cyclase leading to the production of the second-messenger cyclic guanosine monophosphate (cGMP) has renewed interest in guanylate cyclase in recent years. Besides atrial natiuretic peptide, brain naturetic peptide, *Escherichia coli* toxin, and nitrous oxide serve to increase cGMP levels in target tissues. In bone cells, nitrous oxide has been shown to have a major inhibitory action on osteoclast function (MacIntyre *et al.*, 1991). This action was reported to be independent of guanylate cyclase activity, although confirmation of the latter point is required.

In target tissues of peptide substances capable of activating guanylate cyclase, the functions correlate with the distribution of particulate guanylate cyclase rather than that of the soluble form of the enzyme. Recently, the isolation sequencing and expression of a complete complementary DNA (cDNA) clone coding for the membrane guanylate cyclase of rat brain clearly showed that an ANP receptor domain is present in the enzyme. The ANP receptor–guanylate cyclase molecule is a transmembrane protein that contains an extracellular ANP-binding domain and an intracellular guanylate cyclase catalytic domain (Chinkers *et al.*, 1989). Another type of guanylate cyclase receptor cloned recently appears to be more specific for brain natiuretic peptide than ANP. However, because of the high concentrations needed to stimulate guanylate cyclase activity, it is possible that other natural endogenous ligands of this receptor exist. Radiation inactivation studies have indicated that particulate guanylyl cyclase is a multidomain protein with separate domains for ANP binding and cGMP synthesizing activity. There is an additional functional domain on particulate guanylate cyclase with high homology to protein kinases. This domain appears to function as regulatory element of the enzyme (Potier *et al.*, 1991).

The soluble form of guanylate cyclase has been reported to exist as a heterodimer; it appears to contain heme as a prosthetic group, and it is activated by nitroprusside, nitric oxide, and reactive free radicals. Soluble guanylate cyclase is a heterodimer of 82- and 70-kDa proteins (Waldman *et al.*, 1991), and its activation by endothelium-derived vasodilators suggest that it may have a role in the angiogenesis associated with bone modeling and remodeling.

c. Phospholipases

Phospholipases are a family of enzymes responsible for phospholipid hydrolysis. They are designated by letter, depending wherein the phospholipid molecule hydrolytic cleavage is stimulated by the enzyme. Phospholipase A_2 is responsible for removing the fatty acid from the second position of the glycerol backbone of the target phospholipid. This phospholipid is usually arachidonic acid. Phospholipase A_2 is a major source of arachidonate release leading to eicosanoid production. In contrast, phospholipase C acts to cleave at the phosphoric acid residue coupling the glycerol backbone to the polar head group of phospholipids. A specific group of phospholipase C enzymes, phosphatidylinositol-specific phospholipase C, is responsible for hydrolysis of phosphoinositides into diacylglycerol and inositol phosphates. Phosphatidylinositol-specific phospholipase C are enzymes that have been identified to couple with transmembrane-spanning receptors and G proteins. Thus, they are activated by a host of signal-transducing molecules active in bone cells. Phospholipase D is an enzyme that cleaves at the head group of phospholipids producing phosphatidic acid and the free polar head group. Phosphatidic acid is an important signal molecule. In addition, glycosyl phosphatidylinositol (GPI)-specific phospholipase D degrades the GPI anchor of alkaline phosphatase. This anchor-degrading activity is abundant in mammalian plasma and serum. Although the physiologic function of this enzyme remains to be determined, it is proposed to play a role in the regulation of cell surface expression of GPI-anchor proteins.

d. Ion channels

Because certain ion channels couple directly to receptor proteins, they must be considered as effector elements of activated membrane receptors. In addition, G proteins may directly activate ion channels, indicating that the latter are G-protein effectors (Brown 1991). The first pathway for which a membrane-delimited G-protein activation was deduced was that involving the muscarinic M_2 atrial receptor, the G protein called G_k, and the specific atrial potassium channel gated by this protein (Brown and Birbaumer, 1990).

2. Tyrosine Kinase-Linked Effectors

In recent years, several steps involved in signal transduction pathways mediated by receptors with intrinsic tyrosine kinase activity have been elucidated. Early responses to ligand occupancy of these receptors include the clustering and internalization of the receptors, activation of the intrinsic tyrosine kinase activity, autophosphorylation of the cytoplasmic domain of the receptor, phosphorylation of exogenous substrates on tyrosine residues, generation of ion fluxes, stimulation of P_i phosphatidylinositol turnover, and induction of the protooncogenes c-*myc* and c-*fos* (Bjorge *et al.*, 1990). The induction of phosphatidylinositol turnover is produced by the association of an isoform of phospholipase C, phospholipase C_γ, to the receptor through its homology two domains, SH2. Likewise, phosphatidylinositol is phosphorylated in the 3 position by the association of phosphatidylinositol 3 (OH) kinases with receptors through its SH2 domain. Also stimulated to associate to the receptor through its SH2 domain is a GTP-activating protein, which binds to the *ras* oncogene product forming the *ras*–gap complex. In this setting, the small molecular weight G-protein *ras* is provided with guanosine triphosphate hydrolytic capabilities and is thus activated.

There exists an additional mechanism activating enzymes with SH2 domains when the receptor protein is not an intrinsic tyrosine kinase. This is the association of a large family of cytosolic tyrosine kinases, which are myristolated and associated with the inner leaflet of the plasma membrane, with ligand–receptor complexes. This family of tyrosine kinases, the src family, possesses SH2 domains and is capable of forming signal generating complexes similar to the epidermal growth factor (EGF) and plate-derived growth factor (PDGF) receptors, which are intrinsic tyrosine kinases. This mechanism of activation also appears to operate for the family of tyrosine phosphatases, the prototype of which is CD45, the leukocyte common antigen.

a. Phospholipase Cγ

Cloning of the various isoforms of phospholipase C has revealed that three members of the family have two conserved regions considered to be catalytic domains for phospholipase C activity in common. Phospholipase Cα has a totally different amino acid sequence showing similarity to the dioxin of *E. coli*. Phospholipase Cγ contains homologous regions related to the NH_2 terminal regulatory domains of oncogenes of the src family. Two isoforms of phospholipase Cγ, $PLC\gamma_1$ and $PLC\gamma_2$ have been a cloned (Ryu *et al.*, 1987; Takenawa and Nagai, 1981). The distribution of phospholipase C isoforms in bone are unknown, but both forms of phospholipase Cγ are ubiquitously expressed.

b. Phosphatidylinositol 3-OH-kinase

A new category of phosphoinositides phosphorylated at the 3 position of the inositol ring have recently stimulated significant interest. Phosphoinositide 3-OH-kinase (type 1) is associated with ligand occupied PDGF and EGF receptors (Zhang *et al.*, 1992; Auger *et al.*, 1989; Bjorge *et al.*, 1990; Whitman *et al.*, 1988; Stephens *et al.*, 1989). Mutant PDGF receptors competent to activate PLCγ, but unable to bind and activate phosphatidylinositol 3-OH-kinase, do not exert mitogenic effects in fibroblasts (Coughlin *et al.*, 1989). This implies an important signaling function for 3-phosphorylated phosphoinositides. Phosphoinositide 3-OH-kinase from several organs has an apparent size of 190 kDa, determined by gel filtration, and is a heterodimer consisting of 85- and 110-kDa subunits (Carpenter *et al.*, 1990; Shibasaki *et al.*, 1991; Morgan *et al.*, 1990). The P85 subunit is phosphorylated on serine, threonine, and tyrosine after stimulation (Kaplan *et al.*, 1987; Escobedo *et al.*, 1991; Courtneidge and Heber, 1987). It contains one SH3 and two SH2 regions homologous to the nonkinase regions of PP60 c-*src* (Otsu *et al.*, 1991) which appear to mediate the specific association of the phosphoinositide 3-OH-kinase with tyrosine protein kinases of both receptor and nonreceptor classes. In contrast, the P110 protein is considered to be the catalytic component of the 3-kinase (Otsu *et al.*, 1991). We have recently demonstrated activation of PI3 kinase by matrix proteins, and this enzyme appears to be an important regulator of osteoclast function.

c. ras-guanosine triphosphotase-activating protein

The ras protein is a GTP-binding protein that acts as a transducer mediating the signals of growth or differentiation in many types of cells (Barbacid, 1987; Kaziro *et al.*, 1991). In fibroblasts, accumulation of active ras-GTP complexes was observed in response to EGF or PDGF (Satoh *et al.*, 1990a,b). ras also accumulates in response to other ligands that bind to tyrosine kinase receptors. The oncogene products of the src family, which are tyrosine kinases, also induce the increase of ras-GTP (Satoh *et al.*, 1990b; Gibbs *et al.*, 1990). GTPase-activating protein (GAP) is rapidly phosphorylated on tyrosine residues when cells are stimulated by EGF, PDGF, or oncogenes encoding tyrosine kinases (Morla *et al.*, 1988; Ellis *et al.*, 1990; Kaplan *et al.*, 1990). Tyrosine phosphorylation of GAP reduces GTPase-stimulating activity and causes the accumulation of active ras-GTP. GAP forms complexes in a ligand-dependent manner with EGF and PDGF receptors, which include phosphatidylinositol-OH-3 kinase, phospholipase Cγ, and src family tyrosine kinases. This complex triggers signal-transducing events (Ullrich and Schlessinger, 1990). In the case of interleukin 3 and GM-CSF, two receptors that are important in bone and activate ras-GAP complexes and ras-GTP levels, the receptors are not tyrosine kinases, and the tyrosine kinase associated with these receptors has not been described. (Satoh *et al.*, 1992).

d. src family of tyrosine kinases

The src family of non-receptor cytosolic protein tyrosine kinases includes eight closely related representatives whose proteins, when tested, are localized to the inner face of the cell membrane by amino-terminal myristylation (Cooper, 1990). One isoform of $p59^{fyn}$, found primarily in lymphocytes, has been shown to regulate T-cell receptor signaling (Cooke *et al.*, 1991). The src family of tyrosine kinases is extremely important in bone cell physiology. Recent knock-out experiments using an anti-sense strategy have demonstrated that transgenic mice devoid of src develop a metabolic bone disease similar to osteopetrosis (Soriano *et al.*, 1991). The nonreceptor protein tyrosine kinases are capable of associating with multiple receptors following ligand occupancy (Eiseman and Bolen, 1992). This process thus enables multiple receptors that are not intrinsic tyrosine kinases to activate signaling complexes associated with activation of the src family and production of the effector complexes through src homology-binding domains (SH2 and SH3 domains).

e. Tyrosine phosphatases

Protein tyrosine phosphatases have an increasingly appreciated and important role in signal transduction. A prototype for a transmembrane protein tyrosine phosphatase is CD45. CD45 is a structurally heteogenous family of isoforms distributed in cells of the hematopoietic system. The structure of CD45 indicates that it has a single transmembrane-spanning protein with an extracellular NH_2-terminal domain rich in O-linked sugars. It has a large, highly conserved, cytoplasmic domain that possesses protein tyrosine phosphatase activity. Thus, CD45 is a prototype of a novel class of receptors that play an active role in the regulation of cell growth. The ligand for CD45 has not been discovered yet. However, in lymphocytes, the adhesion molecule, CD22, interacts with T cells by binding to the smallest isoform of CD45. The protein phosphatase family is a large group of proteins that include transmembrane proteins of the single transmembrane-spanning type and cytosolic proteins with carboxyl-terminal regions that are important in determining their intracellular localization in regulation of their enzymic activity. The intracellular protein tyrosine phosphatases are associated with the particulate fraction of cell homogenates. They have hydrophobic carboxyl termini that may serve as membrane anchors. One intracellular protein tyrosine phosphatase, PTP1C, is characterized by the presence of SH2 domains. Another, PTPH1, is characterized by a talin-related domain suggesting that it may play an important role in focal adhesions and regulation of the actin cytoskeleton. Tyrosine phosphatases are important in the regulation of the cell cycle and cell transformation. They do not block function simply by dephosphorylation of proteins. They can synergize with kinases to produce specific functions. CD45 specifically activates the src family of kinases through dephosphorylation of the tyrosine residue in their regulatory domain. The protein tyrosine phosphatase PTPH1 exhibits homology to erzin and is associated with the cytoskeleton.

D. Signals

A list of the substances produced by the process of signal-generating complexes activating effector molecules is provided in Table III. Many of these intracellular signal substances are known to have major effects in the process of bone remodeling. Others have been shown to play signifi-

TABLE III Signals

Cyclic adenosine monophosphate	Cl^-
Cyclic guanosine monophosphate	Phosphatidic acid
Inositol phosphates	Lysolipids
Diacylglycerols	Prostaglandins
glycoshingolipids	Leukotrienes
Ca^{2+}	Lipoxins
H^+	Phosphate
Na^+	

cant roles in isolated cells, especially osteosarcoma cells, but their action remains to be determined *in vivo.* Also, many of these signals participate in cell function through a complex array and cascade of events. This clouds our ability to interpret the biologic role of these signals. Thus, in this chapter an attempt will not be made to describe the biologic effects of each of these signals, because much is still required before such a task could be successfully performed. Rather, specific examples of signal transduction in bone cells will be provided, and the roles of individual signals will be discussed in this context.

IV. SPECIFIC EXAMPLES OF SIGNAL TRANSDUCTION IN OSTEOBLASTS: PARATHYROID HORMONE/PARATHYROID HORMONE-RELATED PEPTIDES

PTH regulates calcium and phosphorous homeostasis by binding to specific G protein-coupled receptors in bone and kidney (Rosenblatt *et al.*, 1989). Parathyroid hormone-related peptide (PTHrP) which shares 8 of the 13 amino-terminal residues of PTH, binds to the same 80-kDa receptor glycoprotein (Orloff *et al.*, 1989; Jüppner *et al.*, 1988; Shigeno *et al.*, 1988; Karpf *et al.*, 1987, 1991). An important issue yet to be resolved is the mechanism by which non-homologous domains of PTH and PTHrP activate the same receptor or whether additional specific receptors will be discovered.

A. The Parathyroid Hormone/Parathyroid Hormone-Related Peptide Receptor

The receptor for PTH/PTHrP has recently been cloned (Juppner *et al.*, 1991) from a cDNA library prepared from opposum kidney cells using an expression cloning strategy. The cloned PTH receptor bound PTH 1-34 and PTHrP 1-36 equivalently. Nucleotide sequencing revealed an open reading frame encoding a 585-amino acid protein that showed no similar sequences in nucleic acid or protein data bases. The receptor protein is predicted to have seven membrane-spanning domains similar to other G protein-coupled receptors (Fig. 1). The subsequent cloning of the calcitonin receptor (Lin *et al.*, 1991) and the secretin receptor indicates conservation of glycosalation sites, and extracellular cysteines, suggesting that these receptors form a subfamily of receptors sharing functional features and distinguishing them from the other G protein-linked receptors.

B. Parathyroid Hormone/Parathyroid Hormone-Related Peptide Signal-Generating Complexes

The cloning of the PTH receptor should clarify the nature of the signal-generating complex associated with the PTH receptor. At the present time, the complex is known to contain several G proteins and multiple effector elements. The G proteins associated with the PTH receptor include the classically described G_s for adenylate cyclase. In addition, the cloning of the PTH receptor has confirmed work from this laboratory and others (Dulay and Hruska, 1990) indicating that PTH couples through a G protein, probably G_q to phospholipase C. However, the nature of the G protein responsible for parathyroid activation of phospholipase C activity remains to be determined. Studies also suggest that the small molecular weight GTP-binding protein *rho* may be associated with PTH function (Reshkin and Murer, 1992). Whether or not this protein associates with the signal-generating complex remains to be determined. Thus, as suggested in Fig. 1 (top), one possibility regarding the diversity of signals generated through a single receptor for both PTH and PTHrP is variable activation of multiple GTP binding proteins.

C. Effectors

The effector elements associated with the signal-generating complex of the PTH receptor include adenylate cyclase, phospholipase C, and ion channels. There exists additional data that suggests phospholipase A_2 and D are also activated by PTH. PTH receptor coupling through the G_s and activation of adenylate cyclase is a classic pathway of PTH-based signal transduction. However, it is clear that in the osteoblast PTH also activates phospholipase C (Civitelli *et al.*, 1988; Abou-Samra *et al.*, 1989; Farnndale *et al.*, 1988; Suzuki *et al.*, 1989).

We have recently shown that PTH modulates the action of stretch-activated cation channels in the osteoblastic osteogenic carcoma cell line UMR 106, (Duncan *et al.*, 1992), an effect independent of cAMP generation. The mechanism of this ion channel activation remains to be determined. However, it is unlikely that this channel associates directly with the receptor, and, thus, it must be excluded as a potential effector component of the PTH signal-generating complex. Other ion channels have also been shown to be activated by PTH (Chesnoy-Marchais, 1989; Edelman *et al.*, 1986; Ferrier and Ward, 1986; Ferrier *et al.*, 1988). The PTH effects on these ion channels appear to be mediated through the actions of cAMP and calcium. However, neither cAMP nor calcium mimic the actions of PTH on the stretch-activated cation channel of the UMR 106 cell (Duncan *et al.*, 1992).

D. Signals

The signals associated with the PTH receptor include cAMP, calcium (Reid *et al.*, 1987; van Leeuwen *et al.*, 1988; Donahue *et al.*, 1988;

Yamaguchi, 1987; Bidwell *et al.*, 1991), inositol phosphate (Civitelli *et al.*, 1988; Farndale *et al.*, 1988), and diacylglycerol (Civitelli *et al.*, 1988; Abou-Samra *et al.*, 1989).

Cyclic AMP is the most important signal generated by the PTH signal-generating complex. By stimulating protein kinase A-mediated protein phosphorylation, it directly regulates numerous protein functions in its target cells. However, although detection of cAMP by sensitive radioimmunoassay methods has been accomplished, a discrepancy remains between the physiologic levels of circulating PTH and the ability to determine cAMP production. PTH circulates at the 10^{-11}–$10^{-12}M$ levels, whereas stimulation of cAMP generation, at best, can be accomplished at $10^{-10}M$ doses. Utilizing the effect of cAMP to dissociate the regulatory subunit from protein kinase A, one can measure the saturation of protein kinase a catalytic activity with the regulatory subunit. This indirect measure of cAMP generation is more sensitive than cAMP assays and affords greater correlation between protein kinase A activity and the biological effects of PTH. However, many of the biologic effects of PTHs have not been carefully correlated with activation of protein kinase A, and, thus, doubt remains regarding the role of cAMP in some of PTH biologic effects. Cyclic AMP besides stimulating protein kinase A also serves as a gene transcription factor through the cAMP response element and cAMP response element-binding proteins (Habener, 1990). Many of the long-term actions of the PTH are regulated through cAMP-dependent regulation of gene transcription.

Observations from the laboratory of Herrmann-Erlee *et al.* (1983) demonstrated a failure to correlate PTH-stimulated bone resorption and cAMP production. These studies were supported by the further observation that PTH fragments, shortened at the amino-terminus and unable to stimulate cAMP production, were still capable of stimulating bone resorption. Subsequently, studies from multiple laboratories, including our own, have indicated direct effects of the PTH signal-generating complex on increasing cytosolic calcium. The mechanisms by which PTH increases calcium fluxes in target cells have only been partially elucidated. First, the activation of phospholipase C produces a release of calcium from intracellular stores through the actions of inositol 1,4,5-trisphosphate, serving to increase open time of calcium channels in the endoplasmic reticulum or closely associated organelles (Hruska *et al.*, 1987; Reid *et al.*, 1988). In addition, PTH stimulates calcium entry through calcium channels of the plasma membrane. These calcium channels appear to be of two types: voltage-operated calcium channels of the L type and receptor-operated calcium channels (Bidwell *et al.*, 1991; Yamaguchi *et al.*, 1987; Reid *et al.*, 1988). In PTH target cells which do not exhibit voltage-operated calcium channels, PTH stimulates a calcium entry through a putitive receptor-operated calcium channel, which has yet to be clearly described. However, in cells that do express the receptor-operated calcium channel, convincing studies apparently indicate that PTH serves to increase the open probability of these channels (Yamaguchi *et al.*, 1987; Reid *et al.*, 1988). The direct action of an increase in cytosolic calcium on osteoblast function remains to be clearly elucidated. Many of the effects of cAMP appear to be enhanced by the change in cytosolic calcium, possibly through an amplification of calcium calmodulin-dependent kinase activities.

PTH and PTHrP increase inositol trisphosphate production upon binding to the PTH receptor of osteoblasts (Civitelli *et al.*, 1989; Farndale *et al.*, 1988). The increase in inositol trisphosphate appears to occur through activation of phospholipase C. However, the isoform of phospholipase C affected by PTH has not been determined, nor has the mechanism of phospholipase C activation clearly been determined. One possibility is that the PTH signal-generating complex includes association with G_q and a phospholipase C isoform; however, this remains to be determined. Another possibility would be that the direct actions of PTH on ion channels could produce an activation of phospholipase C through changes in either sodium concentration or calcium concentration. This would explain recently described differences between thrombin, a substance known to activate a plasma membrane phospholipase C, and PTH in osteoblast-like cells. The biologic effects of inositol trisphosphate, besides contributing to the changes in cytosolic calcium stimulated by PTH, are unclear.

Associated with the stimulation of phospholipase C activity by PTH, diacylglycerol has also been shown to be produced, leading to protein kinase C translocation to the plasma membrane (Abou-Samra, 1989). The activation of protein kinase C activity by a PTH suggests a multitude of actions that have largely yet to be clearly demonstrated for the hormone. This continues to be a puzzling issue related to PTH-based signal transduction. One area of special concern is whether or not at some stage in osteoblast development PTH is a growth factor through its actions on phospholipase C.

E. Biological Effects

The biologic effects of PTH and PTHrP on the osteoblast are prodigous. They include many actions that have as yet to be described. This topic is beyond the scope of this chapter. One interesting aspect of the topic of biologic effects is the mechanism of action of the nonamino terminal regions to these molecules, which have recently been shown to stimulate placental Ca^{2+} transport and inhibits osteoblast function (Fenton *et al.*, 1991; Care *et al.*, 1990).

V. SPECIFIC EXAMPLES OF SIGNAL TRANSDUCTION IN OSTEOCLASTS

The cellular basis of bone remodeling is not completely understood. The osteoclast, the multinucleated cell involved in bone resorption, is a complex unit that develops a specialized apparatus for dissolving the bone matrix (King and Holtrop, 1975; Holtrop and King, 1977). Using cell culture systems, several advances have recently been made indicating the molecular events involved in osteoclast bone resorbing activity. For bone resorption to be initiated, the osteoclast polarizes (Baron *et al.*, 1985) and directly attaches to the bone surface by a specialized area termed the clear zone (Holtrop and King, 1977), in which the contact with the substrate is established by specific adhesion structures called podosomes (Marchisio *et al.*, 1984, 1987; Zambonin-Zallone *et al.*, 1988). Morphologically, podosomes appear as short membrane protrusions with a core of microfilaments linked to the plasma membrane by talin and vinculin (Marchisio *et al.*, 1984, 1987). Recent data suggest that podosomes play a pivotal role in substrate recognition by osteoclasts because a specific β_3 integrin of the RGD superfamily of matrix receptors is expressed on their cell membrane surface (Davies *et al.*, 1989; Zambonin-Zallone *et al.*, 1989). Substrate recognition is a necessary early step in the initiation of bone resorption, and it may induce phenotypic differences in cellular responses as the osteoclast changes from a motile cell seeking bone substrate to an actively resorbing cell.

The organization of the podosome-containing clear zone allows tight sealing of the resorbing compartment between the osteoclast plasma membrane and the bone surface. The acidification of this extracellular microenvironment (Baron *et al.*, 1985; Blair *et al.*, 1989) produces hydroxyapatite solubilization. Lysosomal enzymes, secreted into this space by a mannose 6-phosphate receptor-driven mechanism (Baron *et al.*, 1988; Blair *et al.*, 1988) and activated by the acid pH, digest the organic components of the bone matrix (Blair *et al.*, 1986). Tight sealing of the compartment is needed to maintain the pH of 5 and the Ca^{2+} concentrations of up to 40 m*M* (Silver *et al.*, 1988).

While the mechanisms of osteoclast regulation are incompletely understood, it is clear that the osteoclast is a unique cell in that its plasma membrane is devoid of many receptors that activate osteoblast and regulate bone remodeling. For example, PTH, interleukin 1, prostaglandin E_2, $1,25(OH)_2D_3$, and several other hormones known to stimulate bone resorption do so despite the absence of receptors in the osteoclast. Thus, it appears that paracrine factors will have a special importance on the regulation of osteoclast function.

A. Role of Specific Paracrine Substances

An example of a paracrine substance stimulated by a systemic hormone known to function in stimulation of bone resorption is the release of

GM-CSF from murine osteoblasts by PTH (Horowitz *et al.*, 1989). GM-CSF does not directly induce bone resorption (Lorenzo *et al.*, 1988) but, rather, induces the increased formation of osteoclasts from marrow precursors. This suggests that these cytokines augment resorption by increasing the number of osteoclasts available for activation. The exact role of this action of PTH to increase osteoblast GM-CSF production in the bone resorption stimulated by PTH remains to be determined, but this is an important example of potential mechanisms of indirect osteoclast regulation functioning in bone remodeling.

Another paracrine substance stimulated by PTH has recently been shown to directly activate osteoclasts function is interleukin-6 (Girasole *et al.*, 1989). This important observation indicates that one factor released by osteoblast and other cells in the bone micro-environment, IL-6, can account for some of the paracrine stimulation of osteoclast function induced by PTH and other factors, especially interleukin-1. Whether the combined actions of GM-CSF and interleukin-6 account for the osteoclast stimulation induced by PTH and other factors remains to be determined.

B. Hydrogen Ion

Because metabolic acidosis is known to stimulate bone resorption *in vitro* and *in vivo*, we have analyzed the effects of extracellular protons on the regulation of osteoclast function. We have found that exposure of the osteoclast to metabolic acids produces a fall in intracellular pH and cytosolic calcium. The reductions in both of these ions participates in rearrangement of the microfilament cytoskeleton with a rapid increase in the expression of podosomes (Fig. 2). The increase in podosome formation is followed shortly by a very significant stimulation in bone resorption (Teti *et al.*, 1989). The mechanism of the reduction in cytosolic calcium appeared to be an activation of the plasma membrane residing Ca^{2+}-ATPase. Because the Ca^{2+}-ATPase is electrogenic and functions as a Ca^{2+}/H^{+} exchanger, it is possible that the effect of intracellular protons was directly on the Ca^{2+}-ATPase at an internal modifier site. The direct role of intracellular calcium in the regulation of podosome formation may represent the function of cytoskeletal-associated proteins such as gelsolin or profilin. These proteins regulate actin filament polymerization and severing through regulation of phosphatidylinositol bisphosphate levels bound to the proteins in a Ca^{2+}-dependent complex (Bryan and Coluccio, 1985; Chaponnier *et al.*, 1986).

C. Calcium

Because of the role of Ca^{2+} in the control of podosome formation, we have analyzed the mechanisms of Ca^{2+} entry in the osteoclast. We have

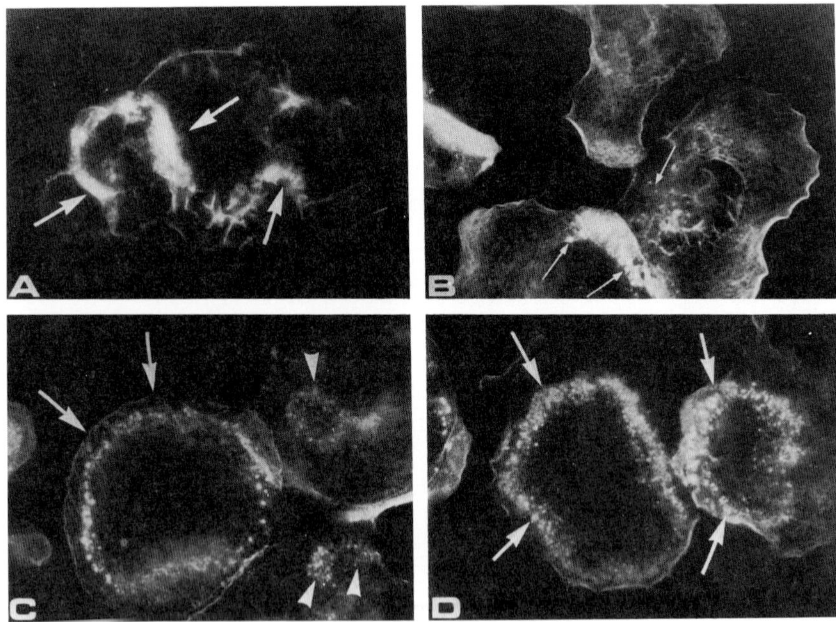

FIGURE 2 Fluorescence microscopy of osteoclast microfilaments detected by rhodamine phalloidin (R-PHD). (A) Osteoclasts with F actin distributed in membrane ruffles (arrows). Such cells make up 66% of 2-d control avian osteoclast cultures. (B) Osteoclasts with F actin distributed in a fine network and containing a small number of podosomes (arrows). Such cells make up 43% of 2-d control cultures. (C and D) Examples of osteoclasts with well organized podosomes (Na butyrate treated for 90 min). In (C), an osteoclast with a peripheral ring of podosomes (arrows) is surrounded by osteoclasts in which podosomes are scanty and organized in small clusters (arrowheads). In (D), a well-organized clear zone containing several layers of podosomes is visible in two osteoclasts (arrows).

demonstrated that increasing extracellular Ca^{2+} produces a remarkable increase in cytosolic Ca^{2+}, which derives mainly from Ca^{2+} release from intracellular stores (Miyauchi *et al.*, 1990; Malgaroli *et al.*, 1989; Zaidi *et al.*, 1988). Furthermore, the increase in intracellular Ca^{2+} produced by changes in extracellular Ca^{2+} are associated with rapid reorganization of the actin cytoskeleton and disruption of podosome expression. This is associated with a remarkable reduction in bone resorptive activity (Fig. 3) (Miyauchi *et al.*, 1990).

Recently, we have shown that the increase in extracellular Ca^{2+} activates an osteoclast plasma membrane-associated phospholipase C, suggesting that the osteoclast possesses a Ca^{2+} sensor protein. Furthermore, this Ca^{2+} sensor appears to represent a G protein-linked receptor because the Ca^{2+}-induced activation of phospholipase C is markedly increased by AlF_4^- and fluoride. This is analogous to the function of a Ca^{2+} sensor protein on the parathyroid chief cell, where changes in

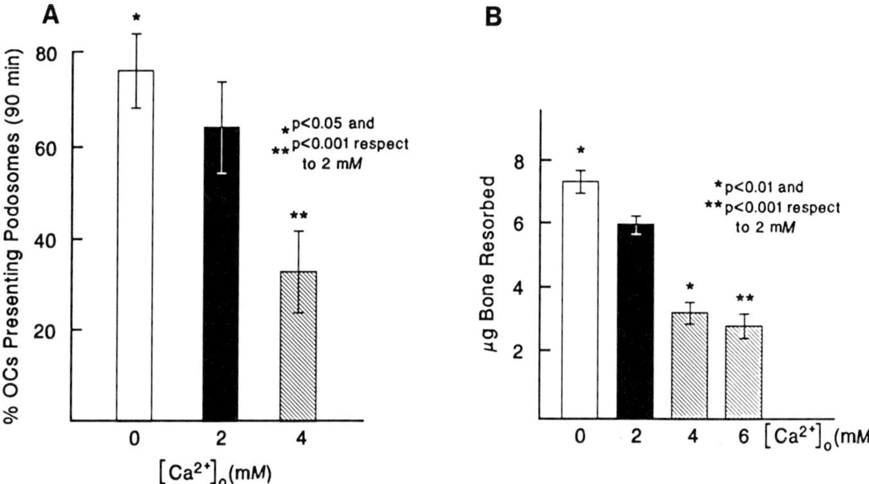

FIGURE 3 Effects of extracellular Ca^{2+} on podosome expression and bone resorption (A) increasing $[Ca^{2+}]_0$ from nominally absent to 4 m*M* resulted in dose-dependent inhibition of podosome expression in osteoclasts. 4 m*M* extracellular Ca^{2+} (final concentration) induced 42% inhibition of osteoclast presenting podosomes with respect to the cultures treated with 2 m*M* Ca^{2+}. Data are ± SE of three experiments performed in triplicate. (B) Dose-dependent inhibition of bone resorption in osteoclasts treated with increasing doses of extracellular calcium. 4 m*M* extracellular calcium (final concentration) reduced osteoclast resorbing activity by 50% of the value obtained at 2 m*M*. Data are mean ± SE of at least three experiments performed in triplicate.

extracellular calcium regulate PTH secretion (Brown, 1991). We have also shown that osteoclast precursors exhibit voltage-operated Ca^{2+} channels in their plasma membrane that rapidly disappear upon binding of the osteoclast to bone matrix.

These data suggest that the osteoclast regulates Ca^{2+} concentrations in the resorption space by changes in the adhesion of the cell to the bone and the sealing of the resorption space. As resorption space Ca^{2+} concentrations increase, the resorption products stimulate the osteoclast to decrease podosome expression and osteoclast bone adhesion. This results in incompetency of the resorption space and release of its contents to the interstitial bone fluid. This is an energy-conserving mechanism of returning the bone resorption products to the interstitial and eventually plasma fluid. It avoids the energy-expensive transcellular transport of ions and resorption products through the osteoclast cell.

D. Osteopontin/$\alpha_v\beta_3$ Integrin Signaling

The mechanism of osteoclast attachment to bone has not been clearly determined, but the $\alpha_v\beta_3$ integrin is thought to be a key mechanism of attachment to matrix proteins (Horton, 1988; Davies *et al.*, 1989;

Zambonin-Zallone *et al.*, 1989). The bone matrix proteins recognized by the vitronectin receptor ($\alpha_v\beta_3$) of the podosome have recently been identified (Fig. 4). Reinholdt *et al.*, (1990) suggested that osteopontin, a protein with tight binding hydroxyapatite, is one protein recognized by the vitronectin receptor in bone. Osteopontin contains a functional RGD cell-binding sequence by cDNA cloning and sequencing (Olgberg *et al.*, 1986, 1988). Osteopontin is an osteoblast product whose synthesis is genomically regulated by 1,25-dihydroxycholecalciferol (Yoon *et al.*, 1987; Heath *et al.*, 1989; Butler, 1989). We have recently shown that osteopontin plays a key role in anchoring the osteoclast to the bone surface, and that another candidate protein for a function similar to that proposed for osteopontin is bone sialoprotein, another RDG-containing bone matrix protein. (Miyauchi *et al.*, 1991; Ross *et al.*, 1993).

Recently, we demonstrated that recognition of osteopontin peptides from the osteopontin and bone sialoprotein sequence stimulate immediate reductions in osteoclast cytosolic Ca^{2+}. The changes in cytosolic Ca^{2+} required the RGD sequence and were blocked by a monoclonal antibody to the $\alpha_v\beta_3$ integrin (Fig. 4). The decrease in cytosolic Ca^{2+} stimulated by osteopontin and related peptides appeared to be due to activation of a plasma membrane Ca^{2+}-ATPase. The mechanism of signal transduction from the occupied integrin to activation of the Ca^{2+}-ATPase is not clear. Recent studies demonstrating the critical role of cytosolic protein tyrosine kinases (c-src) in osteoclast function raises the possibility that this protein, through association with the occupied integrin, could serve to regulate osteoclast function. We have preliminary evidence indicating that the src protein is, in fact, associated with the $\alpha_v\beta_3$ integrin and serves to regulate osteoclast function through effector elements with SH2 to domains, as already discussed.

E. Calcitonin

The recent cloning of the calcitonin receptor (Lin *et al.*, 1991) has clarified the mechanisms of signal transduction related to the ability of this hormone to inhibit osteoclast function. The calcitonin receptor is closely related to the PTH receptor, discussed earlier. Although the PTH/PTHrP receptor is more than a 100 amino acids longer than the calcitonin receptor, overall there is 32% identity and 56% similarity between the sequences of the two receptors. Both receptors activate adenylate cyclase (Lin *et al.*, 1991; Juppner *et al.*, 1991). The calcitonin receptor is thought to couple to G_s and an additional signaling pathway has been reported through a pertussis toxin-sensitive G_i protein in isolated osteoclasts and in LLC-PK-1 cells (Zaidi *et al.*, 1988; Chakerborty *et al.*, 1991). Zaidi (1990) demonstrated that calcitonin, besides increasing cAMP production, also stimulates an increase in osteoclast cytosolic Ca^{2+}, release of inositol

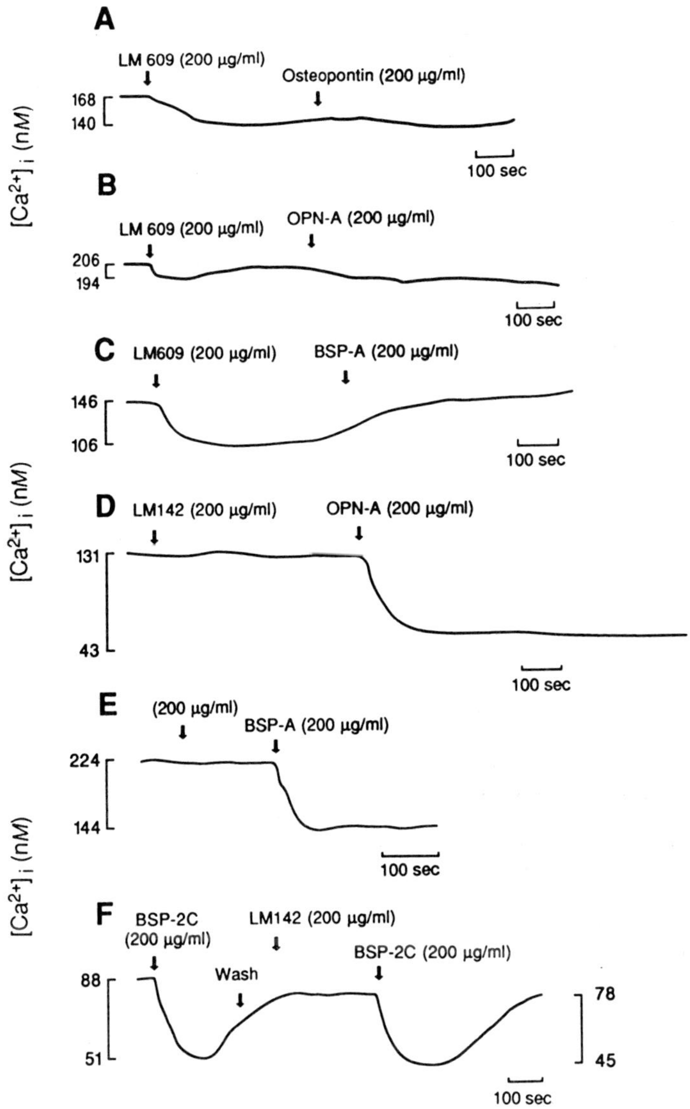

FIGURE 4 Effect of the monoclonal antibody, LM609, on changes in cytosolic calcium induced by osteopontin and bone sialoprotein peptides. LM609 (antivitronectin receptor) recognizes the osteoclast $\alpha_v\beta_3$ integrin and completely inhibited the effects of intact osteopontin (A), OPN-A (B), and BSP-A (C). However, the monoclonal antibody, LM142 (D and F) had no effect on changes in cytosolic calcium stimulated by OPN-A or BSP-2C. A peptide, CD4β, from an irrelevant IgG also failed to affect the changes in cytosolic calcium produced by osteopontin (not shown) and bone sialoprotein peptides (E). The representative tracings demonstrate results observed with each antibody at least five times. These studies demonstrate that bone matrix proteins are ligands for the $\alpha_v\beta_3$ integrin and that integrin occupancy generates immediate cell signals.

trisphosphate, and production of diacylglycerol. These mechanisms of signal transduction by the calcitonin receptor signal-generating complex appear to control the biologic effects of calcitonin in the osteoclast. Specifically, osteoclast retraction and inhibition of bone resorption appear to result from an increase in cAMP, $[Ca^{2+}]_i$, and protein kinase C activity. Recent studies by Teti *et al.* (1990) confirmed that protein kinase C activity is inhibitory to osteoclast function. This raises a paradigm that mechanisms of signal transduction such as cAMP and phospholipase C activity are inhibitory in the osteoclast. It raises specific issues about the mechanisms of signal transduction related to osteoclast stimulation. A key role of the src tyrosine kinases in this are suggested by our studies on the stimulatory effects of matrix proteins and by recent preliminary studies from Kato *et al.* (1991); Boyce *et al.*, (1992).

REFERENCES

Abou-Samra, A.-B., Jüppner, H., Westerberg, D., Potts, Jr., J. T., Westerberg, D. (1989). Parathyroid hormone causes translocation of protein kinase-C from cytosol to membranes in rat osteosarcoma cells. *Endocrinology* **124,** 1107–1113.

Auger, K. R., Serunian, L. A., Soltoff, S. P., Cantley, L. C. (1989) PDGF-dependent tyrosine phosphorylation stimulates production of novel polyphosphoinositides in intact cells. *Cell* **57,** 167–175.

Barbacid, M. (1987). *ras* genes. *Annu. Rev. Biochem.* **56,** 779–827.

Baron, R., Neff, L., Louvard, D., Courtoy, P. J. (1985). Cell-mediated extracellular acidification and bone resorption: Evidence for a low pH in resorbing lacunae and localization of a 100-KD lysosomal membrane protein on the osteoclast ruffled border. *J. Cell. Biol.* **101,** 2210–2222.

Baron, R., Neff, L., Brown, W., Courtoy, P. J., Louvard, D., Farquhar, M. G. (1988). Polarized secretion of lysosomal enzymes: Co-distribution of cation-independent mannose-6-phosphate receptors and lysosomal enzymes along the osteoclast exocytic pathway. *J. Cell. Biol.* **106,** 1863–1872.

Benovic, J. L., Onorato, J. J., Caron, M. G., Lefkowitz, R. J. (1990). Regulation of G protein-coupled receptors by agonist-dependent phosphorylation. *Soc. Gen. Physiol. Ser.* **45,** 87–103.

Berridge, M. (1985). The molecular basis of communication within the cell. *Sci. Am.* **253**(4) 142–152.

Bidwell, J. P., Carter, W. B., Fryer, M. J., Heath, III, H. (1991). Parathyroid hormone (PTH)-induced intracellar Ca^{2+} signaling in naive and PTH-desensitized osteoblast-like cells (ROS 17/2.8): Pharmacological characterization and evidence for synchronous oscillation of intracellular Ca^{2+}. *Endocrinology* **129,** 2993–3000.

Bjorge, J. D., Chan, T.-O., Antczak, M., Kung, H.-J., Fujita, D. J. (1990). Activated type I phosphatidylinositol kinase is associated with the epidermal growth factor (EGF) receptor following EGF stimulation. *Proc. Natl. Acad. Sci.* **87,** 3816–3820.

Blair, H. C., Kahn, A. J., Crouch, E. C., Jeffrey, J. J., Teitelbaum, S. L. (1986). Isolated osteoclasts resorb the organic and inorganic components of bone. *J. Cell. Biol.* **102,** 1164–1172.

Blair, H. C., Teitelbuam, S. L., Ghiselli, R., Gluck, S. L. (1989). Osteoclastic bone resorption by a polarized vacuolar proton pump. *Science* **245,** 855–857.

Blair, H. C., Teitelbaum, S. L., Schimke, P. A., Konsek, J. D., Koziol, C. M., Schlesinger,

P. H. (1988). Receptor-mediated uptake of a mannose-6-phosphate bearing glycoprotein by isolated chicken osteoclasts. *J. Cell. Physiol.* **137,** 476–482.

Boyce, B.F., Yoneda, T., Lowe, C., Soriano, P., and Mundy, G.F. (1992). Requirement of $PP60^{c-src}$ expression for steoclasts to form ruffled boarders and resorb bone in mice. *J. Clin. Invest.* **90,** 1622–1627.

Brown, A. M. and Birnbaumer. (1988). Direct G protein gating of ion channels. *Am. J. Physiol.* **254,** H401–H410.

Brown, A. M. (1991). Ion channels as G protein effectors. *NIPS,* **6,** 158–160.

Bryan, J., Coluccio, L. M. (1985). Kinetic analysis of F-actin depolymerization in the presence of platelet gelsolin and gelsolin-actin complexes. *J. Cell Biol.* **101,** 1236–1244.

Butler, W. T. (1989). The nature and significance of osteopontin. *Connect. Tissue Res.* **23,** 123–136.

Carpenter, C. L., Duckworth, B. C., Auger, K. R., Cohen, B., Schaffhausen, B. S., Cantley, L. C. (1990). Purification and characterization of phosphoinositide 3-kinase from rat liver. *J. Biol. Chem.* **265,** 19704–19711.

Chakraborty, M., Chatterjee, D., Kellokunysu, S., Rasmussen, H., and Boron, R. (1991). Cell cycle-dependent coupling of the calcitonin receptor to different G proteins. *Science* **251,** 1078–1082.

Chaponnier, C., Yanmey, P. A., Yin, H. L. (1986). The actin filament-severing domain of plasma gelsolin. *J. Cell Biol.* **103,** 1473–1481.

Chesnoy-Marchais, D., and Fritsch, J. (1989). Chloride current activated by cyclic AMP and parathyroid hormone in rat osteoblasts. *Pflugers Arch.* **415,** 104–114.

Chinkers, M., Garbers, D.L., Chang, M.S., Lowe, D.G., Chin, H.M., Goeddle, D.V., and Schultz, S. (1989). A membrane form of guanylate cyclase in an atrial natriuretic peptide receptor. *Nature* **338,** 78–83.

Civitelli, R., Martin, T. J., Fausto, A., Gunsten, S. L., Hruska, K. A., Avioli, L. V. (1989). Parathyroid hormone-related peptide transiently increases cytosolic calcium in osteoblast-like cells. Comparison with parathyroid hormone. *Endocrinology* **125,** 1204–1210.

Civitelli, R., Reid, I. R., Westbrook, S., Avioli, L. V., Hruska, K. A. (1988). PTH elevates inositol polyphosphates and diacylglycerol in a rat osteoblast-like cell line. *Am. J. Physiol.* **255,** E660–E667.

Cooke, M. P., Abraham, K. M., Forbush, K. A., Perlmutter, R. M. (1991). Regulation of T cell receptor signaling by src family protein-tyrosine kinase ($p59^{fyn}$). *Cell* **65,** 281–291.

Cooper, J. A. (1990). The Src family of protein-tyrosine kinases. *In* "Peptides and Protein Phosphorylation" (B. E. Kemp, ed.) pp. 85–113. CRC Press. Boca Raton, FL.

Coughlin, S. R., Escobedo, J. A., and Williams, L. T. (1989). Role of phosphatidylinositol kinase in PDGF receptor signal transduction. *Science* **243,** 1191–1194.

Courtneidge, S. A., and Heber, A. (1987). An 81 kd protein complexed with middle T antigen and PP60C-src: A possible phosphatidylinositol kinase. *Cell* **50,** 1031–1037.

Davies, J., Warwick, J., Totty, N., Philip, R., Helfrich, M., Horton, M. (1989). The osteoclast functional antigen, implicated in the regulation of bone resorption, is biochemically related to the vitronectin receptor. *J. Cell. Biol.* **109,** 1817–1826.

Donahue, H. J., Fryer, M. J., Eriksen, E. F., Hunter, H. III. (1988). Differential effects of parathyroid hormone and its analogues on cytosolic calcium ion and cAMP levels in cultured rat osteoblast-like cells. *J. Biol. Chem.* **263,** 13522–13527.

Duncan, R. L., Hruska, K. A., Misler, S. (1992). Parathyroid hormone activation of stretch-activated cation channels in osteosarcoma cells (UMR-106.01). *Fed. European Biochem. Soc.* **307,** 219–223.

Edelman, A., Fritsch, J., and Balsan, S. (1986). Short-term effects of PTH on cultured rat osteoblasts: Changes in membrane potential. *Am. J. Physiol* **251,** C483–C490.

Eiseman, E., and Bolen, J. B. (1992). Engagement of the high-affinity IgE receptor activates src protein-related tyrosine kinases. *Nature* **355,** 78–80.

Ellis, C., Moran, M., McCormick, F., Pawson, T. (1990). Phosphorylation of GAP and GAP-associated proteins by transforming and mitogenic tyrosine kinases. *Nature* **343,** 377–381.

Escobedo, J. A., Kaplan, D. R., Kavanaugh, W. M., Turck, C. W., William, L. T. (1991). A phosphatidylinositol-3 kinase binds to platelet-derived growth factor receptors through specific receptor sequence containing phosphotyrosine. *Mol. Cell Biol.* **11,** 1125–1132.

Farndale, R. W., Sandy, J. R., Atkinson, S. J., Pennington, S. R., Meghji, S., Meikle, M. C. (1988). Parathyroid hormone and prostaglandin E_2 stimulate both inositol phosphates and cyclic AMP accumulation in mouse osteoblast cultures. *Biochem. J.* **252,** 263–268.

Ferrier, J., and Ward, A. (1986). Electrophysiological differences between bone cell clones: Membrane potential responses to parathyroid hormone and correlation with the cAMP response. *J. Cell. Physiol.* **126,** 237–242.

Ferrier, J., Ward-Kesthely, A., Heersche, J. N. M., Aubin, J. E. (1988). Membrane potential changes, cAMP stimulation and contraction in osteoblast-like UMR 106 cells in response to calcitonin and parathyroid hormone. *Bone and Mineral* **4,** 133–145.

Gautier, J., Matsukawa, T., Nurse, P., Maller, J. (1989). Dephosphorylation and activation of *Xenopus* $p34^{cdc2}$ protein kinase during the cell cycle. *Nature (London)* **339,** 626–629.

Gibbs, J. B., Marshall, M. S., Scolnick, E. M., Dixon, R. A. F., Vogel, U. S. (1990). Modulation of guanine nucleotides bound to Ras in NIH3T3 cells by oncogenes, growth factors, and the GTPase activating protein (GAP). *J. Biol. Chem.* **265,** 20437–20442.

Girasole, G., Jilka, R. L., Passeri, G., Boswell, S., Boder, G., Williams, D. C., Manolagas S. C. (1992). 17β-estradiol inhibits interleukin-6 production by bone marrow-derived stromal cells and osteoblasts *in vitro.* A potential mechanism for the antiosteoportic effect of estrogens. *J. Clin. Invest.* **89,** 883–891.

Habener, J. (1990). Cyclic AMP response element binding proteins: a cornucopia of transcription factors. *Mol. Endocrinol.* **4,** 1087–1094.

Heath, J. K., Rodan, S. B., Yoon, K., and Rodan, G. A. (1989). SV-40 large-T immortalization of embryonic bone cells: establishment of osteoblastic clonal cell lines. *Connect. Tissue Res.* **20,** 15–21.

Herrmann-Erlee, M. P. M., Nijweide, P. J., van der Meer, J. M., Ooms, M. A. (1983). Action of bPTH and bPTH fragments on embryonic bone in vitro: dissociation of the cyclic AMP bone resorbing response. *Calcif. Tissue Int.* **35,** 70–77.

Holtrop, M. E., and King, G. J. (1977). The ultrastructure of osteoclast and its functional implications. *Clin. Orthop. Rel. Res.* **123,** 177–196.

Horowitz, M. C., Coleman, D. L., Flood, P. M., Kupper, T. S., Jilka, R. L. (1989). Parathyroid hormone and lipopolysaccharide induce murine osteoblast-like cells to secrete a cytokine indistinguishable from granulocyte-macrophage colony-stimulating factor. *J. Clin. Invest.* **83,** 149–157.

Horton, M. A. (1988). *ISI Atlas Sci. Immunol.* **1,** 35–43.

Hruska, K. A., Moskowitz, D., Esbrit, P., Civitelli, R., Westbrook, S., Huskey, M. (1987). Stimulation of inositol trisphosphate and diacylglycerol production in renal tubular cells by parathyroid hormone. *J. Clin. Invest.* **79,** 230–239.

Jüppner, H., Abou-Samra, A.-B., Freeman, M., Kong, S. F., Schipani, E., Richards, J., Kolakowski, Jr., L. F., Hock, J., Potts, Jr., J. T., Kronenberg, H. M., Segre, G. V. (1991). A G protein-linked receptor for parathyroid hormone and parathyroid hormone-related peptide. *Science* **254,** 1024–1026.

Jüppner, H., Abou-Samra, A. B., Uneno, S., Gu, W.-X., Potts, J. T., Jr., Segre, G. V. (1988). The parathyroid hormone-like peptide associated with humoral hypercalcemia of malignancy and parathyroid hormone bind to the same receptor on the plasma membrane of ROS 17/2.8 cells. *J. Biol. Chem.* **263,** 8557.

Kahn, C. R. (1976). Membrane receptors for hormones and neurotransmitters. *J. Cell Biol.* **70,** 261–286.

Kaplan, D. R., Morrison, D. K., Wong, G., McCormick, F., and Williams, L. T. (1990). PDGF β receptor stimulates tyrosine phosphorylation of GAP and association of GAP with a signaling complex. *Cell* **61,** 125–133.

Kaplan, D. R., Whitman, M., Schaffhausen, B., Pallas, D. C., White, M., Cantley, L., Roberts, T. M. (1987). Common elements in growth factor stimulation and oncogenic transformation: 85 kd phosphoprotein and phosphatidylinositol kinase activity. *Cell* **50,** 1021–1029.

Karpf, D. B., Arnaud, C. D., King, K., Bambino, T., Winer, J., Nyiredy, K., Nissenson, R. A. (1987). The canine renal parathyroid hormone receptor is a glycoprotein: characterization and partial purification. *Biochemistry* **26,** 7825.

Karpf, D. B., Bambino, T., Alford, G., Nissenson, R. A. (1991). Features of the renal parathyroid hormone-parathyroid hormone-related protein receptor derived from structural studies of receptor fragments. *J. Bone Miner. Res.* **6,** 173–182.

Kato, I., Yoneda, T., Izbicka, E., Lee, C., Gutierrez, G., Boyce, B., Soriano, P., Mundy, G. R. (1991). Osteoclasts deficient in c-src tyrosine kinase fail to resorb bone *in vitro* and *in vivo* (abstract). *J. Bone Miner. Res.* **6,** S197.

Kaziro, Y., Itoh, H., Kozasa, T., Nakafuku, M., Satoh, T. (1991). Structure and function of signal-transducing GTP-binding proteins. *Annu. Rev. Biochem.* **60,** 349–400.

King, G. J., and Holtrop, M. E. (1975). Actin-like filaments in bone cells of cultured mouse calvaria as demonstrated by binding to heavy meromyosin. *J. Cell Biol.* **66,** 445–451.

Levitki, A. (1984). *In* "Receptors: A Quantitative Approach". Benjamin-Cummings. Menlo Park, CA.

Lin, H. Y., Harris, T. L., Flannery, M. S., Aruffo, A., Kaji, E. H., Gorn, A., Kolakowski, Jr., L. F., Lodish, H. F., Goldring, S. R. (1991). Expression cloning of anadenylate cyclase-coupled calcitonin receptor. *Science* **254,** 1022–1024.

Lorenzo, J. A., Sousa, S. L., and Centrella, M. (1988). Interleukin-1 in combination with transforming growth factor-α produces enhanced bone resorption *in vitro. Endocrinology* **123,** 2194–2200.

Low, M. G., and Saltiel, A. R. Structural and functional rates of glycosyl-phosphatidylinositol in membranes. (1988). *Science* **239,** 268–275.

MacIntyre, I., Zaidi, M., Towhidual Alam, A. S. M., Datta, H. K., Moonga, B. S., Lidbury, P. S., Hecker, M., Vane, J. R. (1991). Osteoclastic inhibition: An action of nitric oxide not mediated by cyclic GMP. *Proc. Natl. Acad. Sci. USA* **88,** 2936–2940.

Malgaroli, A., Meldolesi, J., Zambone-Zallone, A., Teti, A. (1989). Control of cytosolic free calcium in rat and chicken osteoclasts. The role of extracellular calcium and calcitonin. *J. Biol. Chem.* **264,** 14342–14349.

Marchisio, P. C., Cirillo, D., Naldini, L., Primavera, M. V., Teti, A., Zambonin-Zallone, A. (1984). Cell-substratum interaction of cultured avian osteoclasts is mediated by specific adhesion structures. *J. Cell Biol.* **99,** 1696–1705.

Marchisio, P. C., Cirillo, D., Teti, A., Zambonin-Zallone, A., Tarone, G. (1987). Rous sarcoma virus-transformed fibroblasts and cells of monocytic origin display a peculiar dot-like organization of cytoskeletal proteins involved in microfilament-membrane interactions. *Exp. Cell Res.* **169,** 202–214.

Miyauchi, A., Alvarez, J., Greenfiled, E., Teti, A., Zambonin-Zallone, A., Ross, F. P., Teitelbaum, S. L., Cheresh, D., Hruska, K. (1991). Matrix protein binding to the osteoclast adhesion integrin ($\alpha_v\beta_3$) mediates a reduction in $[Ca^{2+}]_i$. *J. Bone and Min. Res.* **6,** S96.

Miyauchi, A., Hruska, K. A., Greenfield, E. M., Barattolo, R., Colucci, S., Zambonin-Zallone, A., Teitelbaum, S., Teti, A. (1990). Osteoclast cytosolic calcium, regulated by voltage operated calcium channels and extracellular calcium, controls podosome assembly and bone resorption. *J. Cell Biol.* **111,** 2543–2552.

Morgan, S. J., Smith, A. D., and Parker, P. F. (1990). Purification and characterization of bovine brain type I phosphatidylinositol kinase. *Eur. J. Biochem.* **191,** 761–767.

Morla, A., Schreurs, J., Miyajima, A., Wang, J. Y. (1988). Hematoporetic growth factors activate the tyrosine phosphorylation of distinct sets of proteins in interleukin-3-dependent murine cell lines. *Mol. Cell Biol.* **8,** 2214–2218.

Murrills, R. J., Stein, L. S., Horbert, W. R., Dempster, D. W. (1992). Effects of phrobol Myristate acetate on rat and chick osteoclasts. *J. Bone Miner. Res.* **7,** 415–423.

Oldberg, A., Franzen, A., and Heinegard, D. (1986). Cloning and sequence of analysis of rat bone sialoprotein (osteopontin) cDNA reveals an ARg–Gly–Asp cell-binding sequence. *Proc. Natl. Acad. Sci. U.S.A.* **93,** 8819–8823.

Oldberg, A., Franzen, A., Heinegard, D., Pierschbacher, M., and Ruoslahti, E. (1988). Identification of a bone sialoprotein receptor in osteosarcoma cells. *J. Biol. Chem.* **263,** 19433–19436.

Orloff, J. J., Wu, T. L., Stewart, A. F. (1989). Parathyroid hormone-like proteins: biochemical responses and receptor interactions. *Endocr. Rev.* **10,** 476–495.

Orti, E., Bodwell, J. E., and Munck, A. (1992). Phosphorylation of steroid hormone receptors. *End. Rev.* **13,** 105–128.

Otsu, M., Hiles, I., Gout, I., Fry, M. J., Ruiz-Larrea, F., Panayotou, G., Thompson, A., Dhand, R., Hsuan, J., Totty, N. (1991). Characterization of two 85 kd proteins that associate with receptor tyrosine kinases, middle-T/pp60c-src complexes and PLI3-kinase. *Cell* **65,** 91–104.

Pacifici, R., Brown, C., Puscheck, E., Friedrich, E., Slatopolsky, E., Maggio, D., McCracken, R., Avioli, L. V. (1991a). Effect of surgical menopause and estrogen replacement on cytokine release from human blood mononuclear cells. *Proc. Natl. Acad. Sci. USA* **88,** 5134–5138.

Pacifici, R., Carano, A., Santoro, S. A., Rifas, L., Jeffrey, J. J., Malone, J. D., McCracken, R., Avioli, L. V. (1991b). Bone matrix constituents stimulate interleukin-1 release from human blood mononuclear cells. *J. Clin. Invest.* **87,** 221–228.

Potier, M., Huot, C., Koch, C., Hamet, P., Tremblay, J. (1991). Radiation-inactivation analysis of multidomain proteins: the case of particulate guanylyl cyclase. *In* "Methods in Enzomology" (R. Johnson and J. Corbin eds.), **195,** 423–435.

Rees Smith, B., Buckland, P. R. (1982). Structure-function relations of the thyrotropin receptor. *In* "Receptors, Antibodies and Disease, Ciba Foundation Symposium 90 (D. Evered, J. Whelan, eds.) pp. 114–132. Pitman, London.

Reid, I. A., Civitelli, R., Halstead, L. R., Avioli, L. V., Hruska, K. A. (1987). Parathyroid hormone acutely elevates intracellular calcium in osteoblast-like cells. *Am. J. Physiol.* **253,** E45–E51.

Reinholt, F. P., Hultenby, K., Oldberg, A., and Heinegard, D. (1990). Osteopontin—a possible anchor of osteoclasts to bone. *Proc. Natl. Acad. Sci. U.S.A.* **87,** 4473–4475.

Reshkin, S. J., and Murer, H. (1992). Involvement of C_3 exotoxin-sensitive G proteins (*rho*/*rac*) in PTH signal transduction in OK cells. *Am. J. Physiol.* **262,** F572–F577.

Rosenblatt, M., Kronenberg, H. M., Potts, Jr., J. T. (1989). Parathyroid hormone: physiology, chemistry, biosynthesis, secretion, metabolism, and mode of action. *In* "Endocrinology" (L. J. DeGroot ed.) pp. 848–891. Saunders, Philadelphia.

Ross, F. P., Chappel, J., Alvarez, J. I., Sander, D., Butler, W. T., Farach-Carson, M. C., Mintz, K. A., Robey, P. G., Teitelbaum, S. L., and Cheresh, D. A. (1993). Interactions between the bone matrix proteins osteopontin and bone sialoprotein and the osteoclast integrin $\alpha_v\beta_3$ potentiate bone resorption. *J. Biol. Chem.* **268,** 9901–9907.

Ryu, S. H., Suh, P.-G., Cho, K. S., Lee, K.-Y., Rhee, S. G. (1987). Bovine brain cytosol contains three immunologically distinct forms of inositol phospholipid-specific phospholipase C. *Proc. Natl. Acad. Sci.* **84,** 6649.

Satoh, T., Endo, M., Nakafuku, M., Nakamura, S., Kaziro, Y. (1990a). Platelet-derived growth factor stimulates formation of active $p21^{ras}$. GTP complex in Swiss mouse 3T3 cells. *Proc. Natl. Acad. Sci.* **87,** 5993–5997.

Satoh, T., Endo, M., Nakafuku, M., Akiyama, T., Yamamoto, T., Kaziro, Y. (1990b). Accumulation of p21[ras]. GTP in response to stimulation with epidermal growth factor and oncogene products with tyrosine kinase activity. *Proc. Natl. Acad. Sci.* **87,** 7926–7929.

Satoh, T., Uehara, Y., and Kaziro, Y. (1992). Inhibition of interleukin 3 and granulocyte-macrophage colony-stimulating factor stimulated increase of active ras-GTP by herbimycin A, a specific inhibitor of tyrosine kinases. *J. Biol. Chem.* **267,** 2537–2541.

Shibasaki, F., Homma, Y., and Takenawa, T. (1991). Two types of phosphoinositol 3-kinase from bovine thymus. Monomer and heterodimer form. *J. Biol. Chem.* **266,** 8108–8114.

Shigeno, C., Hiroki, Y., Westerberg, D. P., Potts, J. T., Jr., Serge, G. V. (1988). Photoaffinity labeling of parathyroid hormone receptors in clonal rat osteosarcoma cells. *J. Biol. Chem.* **263,** 3864–3871.

Silver, I. A., Murrills, R. J., and Etherington, D. J. (1988). Microelectrode studies on the acid microenvironment beneath adherent macrophages and osteoclasts. *Exp. Cell Res.* **175,** 266–276.

Snyder, S. H. (1985). The molecular basis of communication between cells. *Sci. Am.* **253**(4), 132–140.

Soriano, P., C. Montgomery, R. Geske, and A. Bradley. (1991). Targeted disruption of the c-*src* proto-oncogene leads to osteopetrosis in mice. *Cell* **64,** 693–702.

Stephens, L., Hawkins, P. T., and Downes, C. P. (1989). Metabolic and structural evidence for the existence of a third species of polyphosphoinositide in cells: D-phosphatidyl-myo-inositol 3-phosphate. *Biochem. J.* **259,** 267–276.

Suda, T., Takahashi, N., and Martin, T. J. (1992). Modulation of osteoclast differentiation. *Endo. Rev.* **13,** 66–80.

Suzuki, Y., Hruska, K. A., Reid, L., Alvarez, U., Avioli, L. V. (1989). Characterization of phospholipase C activity of the plasma membrane and cytosol of an osteoblast-like cell line. *Am. J. Med. Sci.* **296,** 135–144.

Takenawa, T., Nagai, Y. (1981). Purification of phosphatidylinositol-specific phospholipase C from rat liver. *J. Biol. Chem.* **256,** 6769–6775.

Teti, A., Blair, H. C., Schlesinger, P., Grano, M., Zambonin-Zallone, A., Kahn, A., Teitelbaum, S. L., Hruska, K. A. (1989). Extracellular protons acidify osteoclasts, reduce cytosolic calcium and promote expression of cell-matrix attachment structures. *J. Clin. Invest.* **84,** 773–789.

Teti, A., Colucci, S., Grano, M., Argentino, L., Zambonin-Zallone, A. (1990). A protein kinase C regulates the organization of the cytoskeleton in osteoclasts (abstract). *J. Bone Miner. Res.* **5,** S213.

Ullrich, A., and Schlessinger, J. (1990). Signal transduction by receptors with tyrosine kinase activity. *Cell* **61,** 203–212.

Van-Leeuwen, J. P., Bos, M. P., Lowik, C. W., Herrmann-Erlee, M. P. (1988). Effect of parathyroid hormone and parathyroid hormone fragments on the intracellular ionized calcium concentration in an osteoblast cell line. *Bone & Mineral* **4,** 177–188.

Waldman, S. A., Leitman, D. C., Murad, F. (1991). Immunoaffinity purication of soluble guanylyl cyclase. *In* "Methods in Enzymology" (R. Johnson and J. Corbin eds.) **195,** 391–404. Academic Press.

Whitman, M., Downes, C. P., Keeler, C. P., Keller, T., Cantley, L. (1988). Type I phosphatidylinositol kinase mades a novel inositol phospholipid, phosphatidylinositol-3-phosphate. *Nature* 332, 644–646.

Yamaguchi, D. T., Hahn, T. J., Iida-Klein, A., Kleeman, C. R., Muallem, S. (1987). Parathyroid hormone-activated calcium channels in an osteoblasts-like clonal osteosarcoma cell line. *J. Biol. Chem.* **262,** 7711–7718.

Yatani, A., J. Codina, Y. Imoto, J. P. Reeves, L. Birnbaumer, and A. M. Brown. (1987). A G protein directly regulates mammalian cardiac calcium channels. *Science Wash. DC* **238,** 1288–1292.

Yoon, K., Buenaga, R., and Rodan, G. A. (1987). Tissue specificity and developmental expression of rat osteopontin. *Biochem. Biophys. Res. Commun.* **148,** 1129–1136.

Zaidi, M., Chambers, T. J., Bevis, P. J. R., Beacham, J. L., GainesDas, R. E., and MacIntyre, I. (1988). Effects of peptides from the calcitonin genes on bone and bone cells. *Q. J. Exp. Physiol.* **73,** 471–485.

Zaidi, M. (1990). Modularity of osteoclast behavior and "mode-specific" inhibition of osteoclast function. *Biosci. Rep.* **10,** 547–556.

Zambonin-Zallone, A., Teti, A., Grano, M., et al. (1989). Immunocytochemical distribution of extracellular matrix receptors in human osteoclasts: A B3 integrin is colocalized with vinculin and talin in the podosomes of osteoclastoma giant cells. *Exp. Cell Res.* **182,** 645–652.

Zambonin-Zallone, A., Teti, A., Carano, A., and Marchisio, P. C. (1988). The distribution of podosomes in osteoclasts cultured on bone laminae: Effect of retinol. *J. Bone Min. Res.* **3,** 517–523.

Zhang, J., Fry, M. J., Waterfield, M. D., Jaken, S., Liao, L., Fox, J. E. B., Rittenhouse, S. E. (1992). Activated phosphoinositide 3-kinase associates with membrane skeleton in thrombin-exposed platelets. *J. Biol. Chem.* **267,** 4686–4692.

14

CELLULAR AND MOLECULAR BIOLOGY OF THE OSTEOCLAST

ROLAND BARON, JAN-HINDRIK RAVESLOOT, LYNN NEFF, MUNMUN CHAKRABORTY, DIPTENDU CHATTERJEE, ABDERRAHIM LOMRI, and WILLIAM HORNE

Cellular and Molecular Biology of Bone

I. INTRODUCTION

The osteoclast is the cell responsible for the resorption of the bone matrix. Bone resorption is a necessary process for the normal development of the skeleton, for its adaptability, and for its maintenance. This cellular process is essential in the growth, remodeling, and repair of bone and is, under normal conditions, tightly coupled to the process of bone formation by the osteoblast. It is the balance between these two cellular activities that determines skeletal mass and shape at any point in time.

Before discussing the cellular and molecular biology of the osteoclast in greater detail, we first summarize its essential features. The osteoclast is a highly motile cell that attaches to, and migrates along, the surface of bone, mostly composed of the interface between bone and bone marrow (endosteum) but also including interfaces with fibrous connective tissues such as the periosteum at the periphery of bones and the periodontal ligament, which serves to attach the teeth to their bony support. The osteoclast is a multinucleated cell (although mononuclear osteoclasts are also encountered) that is formed by the asynchronous fusion of mononuclear precursors derived from the bone marrow and differentiating within the granulocyte–macrophage lineage. The osteoclast attaches to the mineralized bone matrix that it is going to resorb by forming a tight ringlike zone of adhesion—the sealing zone. This attachment involves the specific interaction between adhesion molecules in the cell's membrane and some specific proteins found in the bone matrix or at the surface of bone. The space contained inside this ring of attachment and between the osteoclast and the bone matrix constitutes the bone resorbing compartment. The osteoclast synthesizes several proteolytic enzymes, which are then vectorially transported and secreted into this extracellular bone resorbing compartment. Simultaneously, the osteoclast lowers the pH of this compartment by extruding protons across its apical membrane (facing the bone matrix). The concerted action of the enzymes and the low pH in the bone resorbing compartment leads to the extracellular digestion of the mineral and organic phases of the bone matrix. After resorbing to a certain depth, determined by mechanisms that remain to be elucidated, the osteoclast detaches and moves along the bone surface before reattaching and forming another resorption lacuna, usually in close proximity to the first one. In the process, a certain volume of bone matrix has been removed, only to be replaced, under normal circumstances, by newly formed matrix a few days later. Calcium, phosphate, and other components of the matrix, most of which have been completely digested but some of which may have been only mobilized during this process, will either be eliminated, serve locally as messengers, be reutilized at sites where bone formation and mineralization occur, or be used to maintain the proper ionic concentrations in the extracellular fluids.

The osteoclast is therefore a morphologically and functionally polarized cell (Fig. 1), with a pole facing the bone matrix, where attachment occurs and toward which most of the secretion is targeted (the apical pole), and a pole facing the soft tissues in the local microenvironment (bone marrow or periosteum), which provides mostly, but not exclusively, regulatory functions (the basolateral pole).

A. Morphological Features of the Osteoclast

At the light microscopic (LM) level, the osteoclast is characterized by its size (50–100 μm average), its multinucleation (usually 2–10 nuclei), and its presence within a resorptive (Howship's) lacuna along the edge of the calcified matrix–bone marrow interface. The osteoclast is most often found in close apposition to the calcified matrix. The apical area of the cell, closest to the matrix, is characterized by densely stained patches at the periphery (attachment apparatus) and a lightly stained, highly vacuolated, and striated center area corresponding to the ruffled border (Fig. 1). The cytoplasm is usually strongly basophilic, granular, and foamy, with vacuoles of varying sizes located mostly between the nuclei and the ruffled border area. The nuclei are characteristically heterogeneous in size, shape, and basophilia, a possible reflection of the asynchronous fusion of mononuclear precursors (Nijweide *et al.*, 1986).

Ultrastructural analysis (Holtrop and King, 1977; Baron *et al.*, 1985b, 1988) of the osteoclast (Fig. 1) confirms and extends the light microscopic observations. First, the morphological polarity of the cell is evident, with a marked contrast existing between the apical and basolateral domains. The peripheral zone of the apical domain of the plasma membrane of the osteoclast is very closely apposed to the extracellular matrix. This so-called *sealing zone* (Schenk *et al.*, 1967) is characterized by a very narrow space (0.2–0.5 nm) between the plasma membrane of the cell and the calcified matrix and by the presence of an organelle-free area in the adjacent cytoplasm, the *clear zone* (Fig. 1), which is characteristically enriched in contractile proteins (King and Holtrop, 1975; Marchisio *et al.*, 1984). Toward the center of the apical domain, the plasma membrane of

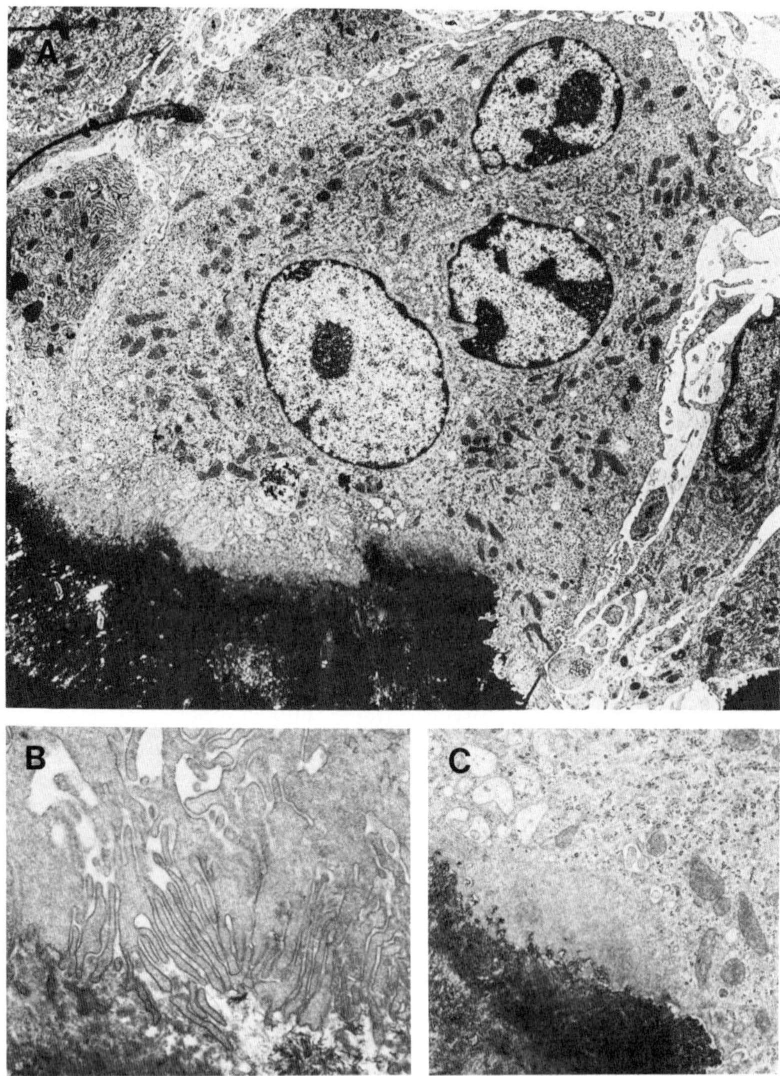

FIGURE 1 Morphological features of the osteoclast. Three electron micrographs show an overview of the osteoclast morphology (A) and higher magnification of the ruffled border area (B) and the sealing zone area (C). Note the multinucleation, the large number of mitochondriae, and the close apposition to the mineralized bone surface (bottom, black hydoxyapatite crystals in these undecalcified preparations). The ruffled border area (B) shows the deep infoldings of the plasma membrane and the dissolution of apatite crystals in the underlying area. In the Higher magnification of the sealing zone (C) of the osteoclast shown at the top, note the very tight apposition of the membrane to the bone matrix and the organelle-free and filamentous clear zone in the cytoplasm adjacent to the sealing zone. (D) Graphic representation of the morphology of an osteoclast, as seen in the electron micrographs. ER, endoplasmic reticulum; Go, Golgi complexes; mi, mitochondria; n, nuclei. Electron micrographs from P. Van Tran and R. Baron and from L. Neff and R. Baron, unpublished.

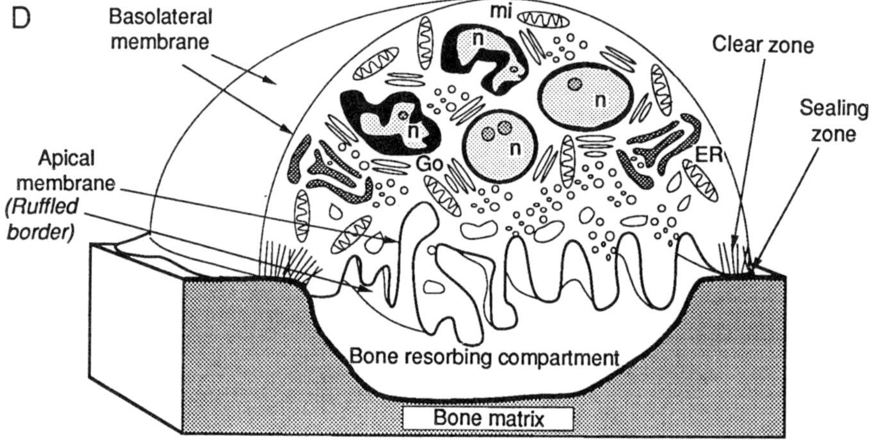

FIGURE 1 (*continued*)

the osteoclast develops progressively deeper infoldings, which reach a high degree of geometric complexity. Numerous folds and ampullar spaces are present that, due to the random orientation of the sections through the cells, often appear as "vacuoles" in the cell's cytoplasm. The cytoplasmic side of the membrane forming the ruffled border is lined by small, regularly spaced studs (Kallio *et al.*, 1971). In contrast, the basolateral domain of the plasma membrane is relatively smooth and is not lined on its cytoplasmic face by any defined structure.

In all species examined so far, one of the most consistent morphological features of the cytoplasmic organization of the osteoclast is the perinuclear distribution of multiple Golgi complexes, which entirely surround each of the cell's nuclei (see Figs. 1 and 2) (Baron *et al.*, 1985b). All of these Golgi stacks are functionally oriented with the trans side (site of exit of recently synthesized proteins), facing away from the nuclei, as demonstrated by the immunolocalization of sialyltransferase (Baron *et al.*, 1988). The osteoclast is not, as often thought, poor in rough endoplasmic reticulum: This organelle is quite well developed but remains limited in extent relative to the large size of the cell. The endoplasmic reticulum is usually concentrated in the basal portion of the cytoplasm of the osteoclast. The perinuclear envelopes, which are an extension of the endoplasmic reticulum, are in close association with the cis side of the multiple Golgi complexes and are most often also found to be actively engaged in the biosynthesis of secretory proteins and enzymes (Baron *et al.*, 1985b, 1988). In addition, the osteoclast is characteristically rich in mitochondria, free polysomes, and coated transport vesicles in the Golgi areas as well as in multiple vacuolar structures of heterogeneous size and shape, concentrated mostly between the nuclei and the ruffled border area. As already mentioned, many, but not all, of

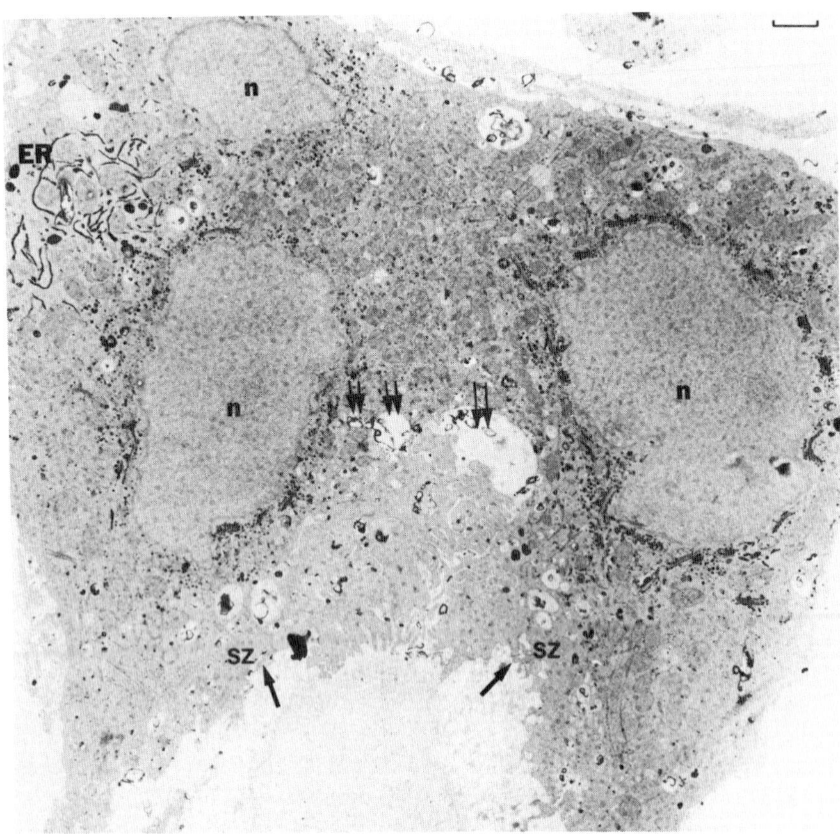

FIGURE 2 Localization of lysosomal enzymes in the osteoclast. Section of an osteoclast reacted for the lysosomal enzyme arylsulfatase. The osteoclast contains multiple nuclei (n), an endoplasmic reticulum (ER) where lysosomal enzymes are synthesized, and prominent Golgi stacks around each nucleus, which are also loaded with enzymes. The cell is attached to bone matrix (bottom) and forms a separate compartment underneath itself, limited by the sealing zone (SZ, single arrows). The plasma membrane of the cell facing this compartment is extensively folded and forms the ruffled border, with pockets of extracellular space between the folds (double arrows). Multiple small vesicles transporting enzymes toward the bone matrix can be seen in the cytoplasm. 9000×. Reproduced from the *Journal of Cell Biology*, 1985, 101, 2210–2222 by copyright permission of the Rockefeller University Press.

these vacuoles are pockets of extracellular space cut obliquely through the folds of the plasma membrane.

B. Structure–Function Relationship

The structural features of the osteoclast that have already been described are all the reflection of specific functions. The clear zone and the sealing zone are responsible for the attachment of the osteoclast to the bone matrix; the ruffled border corresponds to the area of ion transport and protein secretion; the basolateral membrane is a major site for regulatory influences; the endoplasmic reticulum and the Golgi complexes are responsible for the synthesis of both secretory and membrane proteins; and the cytoskeleton is involved in the cell's motility, its attachment to bone matrix, and the intracellular transport of membrane vesicles for secretion and membrane trafficking. Mitochondria provide the adenosine triphosphate (ATP) required for the various energy-dependent systems within the cell, especially the several ion-transporting ATPases, and are a source of CO_2, which is used by carbonic anhydrase to produce protons for the acidification of the bone resorbing compartment. Finally, the abundant free polysomes reflect the fact that the osteoclast, like any other cell, needs to synthesize numerous soluble proteins for its own cytoplasmic use. Because the structure–function relationships are the most important relationships in trying to understand the biology of the osteoclast, we now describe the various functions that this cell must perform to resorb bone, adding further details on the morphological organization of the cell as needed.

II. MOTILITY, ATTACHMENT, AND ESTABLISHMENT OF THE BONE RESORBING COMPARTMENT

A. Cytoskeletal Organization

All three types of filaments that have been demonstrated in the cytoskeleton of other cells, actin microfilaments, intermediate filaments, and microtubules are found in osteoclasts. Both the intermediate filaments, composed of vimentin, and the microtubules are radially organized in osteoclasts (Marchisio *et al.*, 1984; Turksen *et al.*, 1988; Warshafsky *et al.*, 1985). Individual cells often exhibit multiple microtubule organizing centers, and there may be a one-to-one correspondence between the microtubule organizing centers and the nuclei (Matthew *et al.*, 1967; Marchisio *et al.*, 1984; Turksen *et al.*, 1988). Inhibition of microtubule assembly interferes with the cell's functions, possibly because of the role played by microtubules in intracellular translocation of membrane vesicles between the plasma membrane and various intracellular compartments. Disruption of microtubules could interfere with the synthesis, processing, or secretion of hydrolytic enzymes into the resorption compartment or with the translocation of ion pumps or channels to the plasma membrane of the cell, resulting in failure of the cell to efficiently acidify the resorbing compartment (Hunter *et al.*, 1989; Baron *et al.*, 1990b).

Cytoskeletal structures composed of actin filaments and various actin-binding proteins are involved in osteoclast migration and in adhesion of osteoclasts to bone and other surfaces. Osteoclasts that are migrating across a surface (bone or glass) have a pattern of actin filaments that is similar to what is seen in other motile cells in culture (Turksen *et al.*, 1988; Zambonin-Zallone *et al.*, 1988; Lakkakorpi *et al.*, 1989). While actin filaments are present throughout the cell, the leading edge of the cell, or lamellipodium, which has an irregular, coarsely ruffled appearance, contains a prominent network of relatively disorganized actin filaments. Stress fibers, which are commonly seen in, for example, cultured fibroblasts, are seldom observed in osteoclasts. Myosin is also found throughout the cell, but, in contrast to actin, it is more concentrated in the central region of the cell and relatively less abundant in the lamellipodia (Warshafsky *et al.*, 1985; Turksen *et al.*, 1988).

B. Attachment Apparatus

1. The Clear Zone

The most striking and unique feature of the osteoclast actin cytoskeleton is at the site of cell contact with the substratum. In osteoclasts observed under a variety of conditions, there is a prominent peripheral band of F-actin, which contains actin filaments oriented parallel to the plane of the underlying substrate and running around the cell periphery as well as numerous punctate structures where the actin filaments are organized in bundles perpendicular to the plane of the substratum (Marchisio *et al.*, 1984, 1987; Zambonin-Zallone *et al.*, 1989; Lakkakorpi and Vaananen, 1991; Lakkakorpi *et al.*, 1989; Kanehisa *et al.*, 1990; Teti *et al.*, 1991). When osteoclasts are cultured on bone, the band circumscribes the area of active bone resorption (Turksen *et al.*, 1988; Lakkakorpi *et al.*, 1989; Kanehisa *et al.*, 1990) and, thus, presumably corresponds to the clear zone, where the high density of cytoskeletal elements excludes organelles from the region of the cytoplasm immediately adjacent to the plasma membrane. In electron micrographs, bundles of actin filaments can be observed oriented perpendicular to the bone surface and extending into short cell processes that enter irregularities of the bone surface (King and Holtrop, 1975; Zambonin-Zallone *et al*, 1988). These bundles of actin filaments apparently correspond to the punctate F-actin structures in isolated osteoclasts cultured on bone slices or on glass. Confocal microscopy has demonstrated that in actively resorbing osteoclasts these sites of close cell–substratum contacts may even extend into and across the resorption pits, forming multilacunar resorption areas under the same osteoclast (Taylor *et al.*, 1989; Lakkakorpi and Vaananen, 1991).

2. The Podosomes and the Sealing Zone

The punctate actin structures in the peripheral band, termed *podosomes* (Marchisio *et al.*, 1984; Teti *et al.*, 1991), apparently occur only in cells of monocytic origin (osteoclasts and monocytes) and in cells that have been transformed by the *src*, *fps*, and *abl* oncogenes (Marchisio *et al.*, 1987). In addition to the bundles of actin filaments, the podosomes contain a number of other proteins that have been reported to occur at sites of cell–substratum or cell–cell interaction (for review, see Teti *et al.*, 1991). These include fimbrin, α-actinin, and gelsolin, which are closely associated with the actin filaments in the core of the podosome, as well as vinculin and talin, which appear to form rosette structures surrounding the podosome cores (Marchisio *et al.*, 1984; Teti *et al*, 1991) (Figs. 3 and 4).

In terms of composition and function, podosomes and focal adhesion plaques are clearly related (Teti *et al.*, 1991). There are, however, important functional differences. In contrast to focal adhesion plaques,

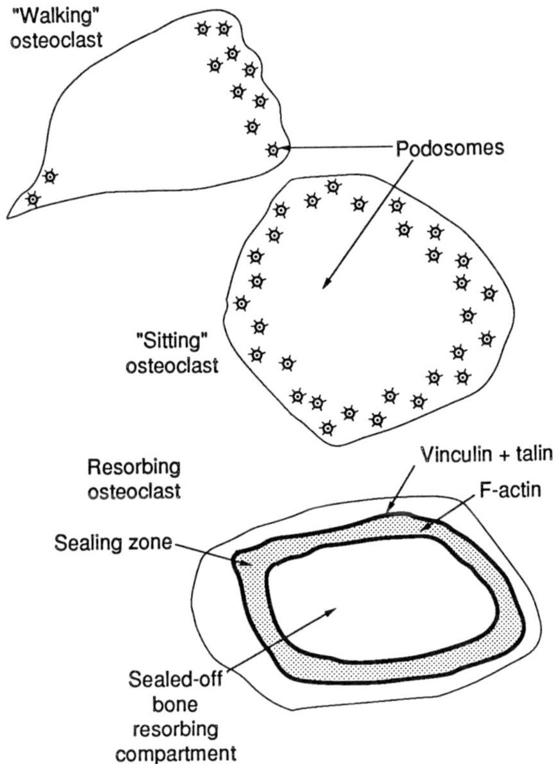

FIGURE 3 Schematic representation of three different functional conformations of the attachment apparatus of the osteoclast. In osteoclasts that are moving along bone surfaces ("walking"), the attachment apparatus is organized in focal points of adhesion (podosomes) at the front and tail of the cell; when the cell reaches the area to be resorbed, more podosomes are formed and they organize in a peripheral position ("sitting"); when the osteoclast is actively resorbing a bone substrate, there is rearrangement of the attachment structures and, possibly, loss of the podosomes: A double vinculin ring is found with F-actin inside.

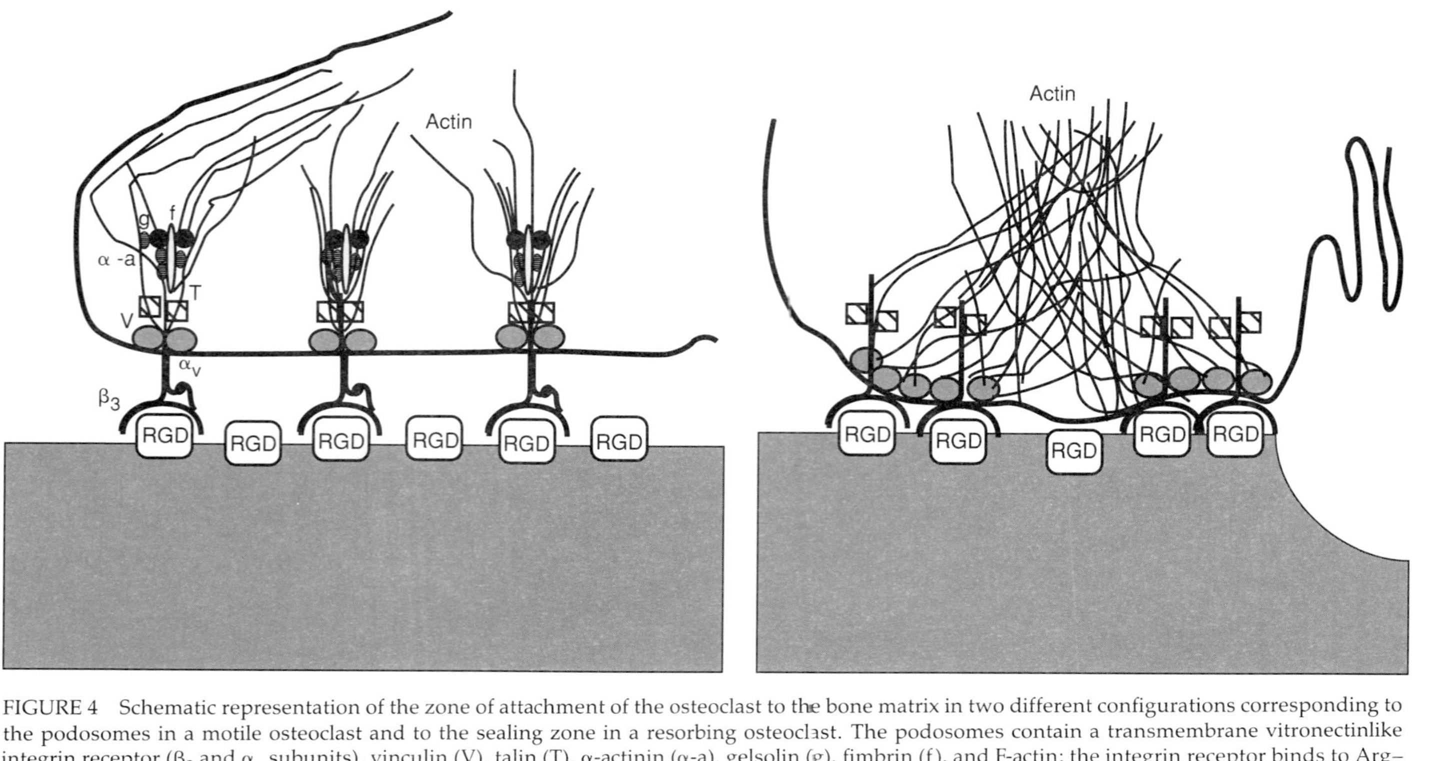

FIGURE 4 Schematic representation of the zone of attachment of the osteoclast to the bone matrix in two different configurations corresponding to the podosomes in a motile osteoclast and to the sealing zone in a resorbing osteoclast. The podosomes contain a transmembrane vitronectinlike integrin receptor (β_3 and α_v subunits), vinculin (V), talin (T), α-actinin (α-a), gelsolin (g), fimbrin (f), and F-actin; the integrin receptor binds to Arg–Gly–Asp sequences (RGD) present in extracellular matrix proteins (osteopontin, bone sialoprotein, and collagen type I for the α_1/β_2 receptor). In a resorbing osteoclast, the attachment forms the sealing zone, bringing the membrane of the cell in closer interaction with the matrix, sealing off the bone resorbing compartment (right); the presence of integrin receptors and the reorganization of vinculin, talin, and F-actin, however, are still controversial. See Lakkakorpi *et al.*, (1991).

which are relatively stable and involve very close association of the cell membrane and the substratum (10–15 nm), podosomes are less tightly associated with the substratum (30 nm) and are highly dynamic, changing size and location and appearing and disappearing with life-spans of 2–12 min (Kanehisa *et al.*, 1990; Lakkakorpi and Vaananen, 1991). It has been suggested that these properties of podosomes may be related to the fact that cells expressing podosomes are both highly motile and able to interact with and degrade extracellular matrix proteins (Teti *et al.*, 1991).

As discussed later, this difference in "tightness" of podosomes versus focal adhesion plaques may parallel a functional difference between attachment points and a sealing zone, thereby explaining the apparent reorganization of the attachment apparatus at the time of active resorption by the osteoclast.

Recent observations by Ali *et al.* (1984), Lakkakorpi *et al.* (1989; Lakkakorpi and Vaananen, 1991), and Kanehisa *et al.* (1990; Kanehisa and Heersche, 1988) have provided us with a dynamic view of the attachment of the osteoclast to the bone matrix (Fig. 3). In highly motile ("walking") osteoclasts, few podosomes are observed and seem to be confined to the irregularly shaped leading edge of the cell, or lamellipodium. Upon arrest and attachment, numerous podosomes are formed and organized in a peripheral ring, as described in cells attached on glass. Rapidly thereafter, the seal is established; then the punctate "podosome" structures either are replaced by two concentric rings of vinculin and talin circumscribing a broad central zone of F-actin (Kanehisa *et al.*, 1990; Lakkakorpi and Vaananen, 1991) or reach such density that they cannot be individually observed when they arrange into two concentric rings. These changes in the organization of the attachment structures could lead to the establishment of a tighter sealing zone by bringing the plasma membrane of the cell closer to the matrix than in the podosome mode. These observations therefore suggest a distinction in time and in specific cell–matrix interactions between the motile cell ("walking"), the cell recently arrested at a future resorbing site ("sitting"), forming a first ringlike structure of punctate attachment sites, and the resorbing cell, with a functionally tight sealing zone (Figs. 3 and 4).

3. The Role of Integrins

The cytoskeletal complexes already described provide an anchoring structure necessary for stabilizing the interaction of the osteoclast with the bone surface, but because they are limited to the cytoplasmic side of the plasma membrane they are not directly responsible for that interaction. This role is filled by integral membrane proteins whose cytoplasmic domains interact with the cytoskeleton while their extracellular domains bind to bone matrix proteins. These transmembrane proteins are members of the integrin family of adhesion molecules, which mediate cell–substratum and cell–cell interactions (Hynes, 1987, 1992). Integrins are heterodimeric molecular assemblies of an α subunit and a β subunit with specific, receptorlike, extracellular binding sites that recognize specific sequences in matrix proteins such as the Arg–Gly–Asp (RGD) sequence, a motif known to represent the core ligand for most members of the integrin family (Ruoslahti and Pierschbacher, 1987) (Fig. 5). The amino acid sequence surrounding the RGD motif determines the specificity and the affinity with which integrins will recognize and bind a specific matrix protein (Ruoslahti and Pierschbacher, 1987; Horton and Davies, 1989).

Osteoclasts express at least two α subunits, α_1 and α_v, and at least two β subunits, β_1 and β_3, implying that multiple integrins may be involved in osteoclast adhesion to the bone matrix. The α_v and β_3 proteins form a dimer that is closely related, if not identical, to the vitronectin receptor (VNR) and is expressed at high levels in osteoclast membranes (Davies *et al.*, 1989). Prior to attachment of the osteoclast to the bone surface, the VNR is distributed over the entire surface of the cell (L.

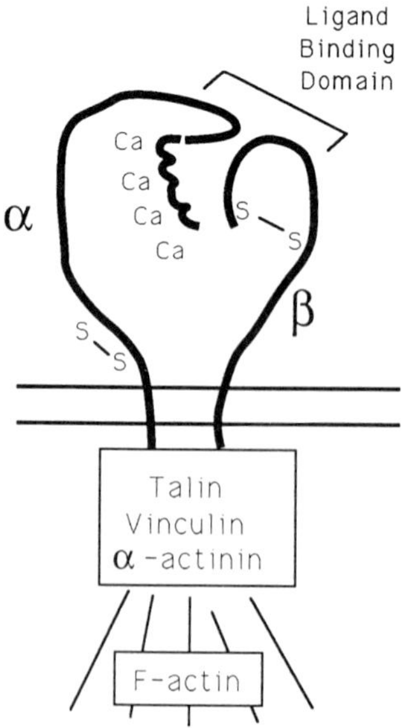

FIGURE 5 Schematic representation of an integrin receptor. The receptor is composed of one α subunit with an N-terminal calcium-binding region and a β subunit with an N-terminal cysteine bond. The C-termini of both subunits are cytoplasmic, and the ligand-binding domain, involving both subunits, is at the N-termini. The cytoplasmic tails are linking to various cytoskeletal proteins and to F-actin (see text).

Neff, A. Billecocq, B. Prallet, and R. Baron, unpublished work, 1991). In attached and polarized osteoclasts, it is restricted largely to the basolateral membrane (Neff *et al.*, unpublished work, 1991), although it may be expressed, albeit at a much lower level, in the apical (i.e., ruffled border) membrane (Lakkakorpi *et al.*, 1991). The VNR has been demonstrated in podosomes by immunofluorescence (Zambonin-Zallone *et al*,. 1989), but it has not been detected in the established sealing zone (Neff *et al.*, unpublished work, 1991; Lakkakorpi *et al.*, 1991), raising the question of whether the integrins are indeed involved in the formation of the seal at the periphery of the resorbing compartment (Lakkakorpi *et al.*, 1991). These discrepancies could, however, be due to technical difficulties and to the limited access of epitopes that would be present in the tight sealing zone area.

Although the detailed receptor–ligand interactions at the attachment site are only beginning to be elucidated, several RGD-containing matrix proteins have been identified (Teti *et al.*, 1991). Of these, collagen type I, osteopontin, and bone sialoprotein II (BSPII) are the most likely candidates to fill the role of integrin-binding proteins in bone (Teti *et al.*, 1991). Most interestingly, data is now accumulating that suggest that the osteoclast synthesizes and secretes both osteopontin and BSPII, raising the possibility that the osteoclast itself deposits the adhesion molecules required for its attachment to the bone surface and for establishing the sealed-off bone resorbing compartment.

The important role of integrins in bone resorption, whether it is in the sealing zone, the podosomes, and/or the motility of the osteoclast, is nevertheless well demonstrated. Monoclonal antibodies raised against the osteoclast VNR (Davies *et al.*, 1989) inhibit bone resorption and spontaneous lamellipodial motility and induce the retraction of the osteoclast, much like the effect of calcitonin (Chambers *et al.*, 1986). Furthermore, synthetic or natural peptides containing the Arg–Gly–Asp (RGD) sequence inhibit bone resorption by isolated osteoclasts in culture (Sato *et al.*, 1990). Finally, binding of RGD-containing peptides to osteoclasts induces changes in intracellular calcium and other signal transduction events including tyrosine-phosphorylation (Neff *et al.*, 1992), leading to an activation of the cell (Miyauchi *et al.*, 1991b). These results indicate that the integrins play an important role in the osteoclast function and, despite the controversy regarding their presence at the sealing zone membrane, most likely mediate the attachment of the osteoclast to the bone surface.

4. Regulation of Bone Resorption and the Attachment Apparatus

Calcitonin, which directly inhibits bone resorption after binding to receptors present on the osteoclast basolateral membrane (Nicholson *et al.*, 1986), causes the concurrent loss of the peripheral band of podosomes and other actin filaments and the appearance of a concentration of unorganized actin filaments in the center of the cell with a distribution very similar to that of myosin (Warshafsky *et al.*, 1985; Hunter *et al.*, 1989; Lakkakorpi and Vaananen, 1990). Formation of such a centrally located actomyosin network is likely to be functionally related to the calcitonin-induced retraction of the osteoclast (Chambers and Magnus, 1982) and to the arrested secretion associated with the internalization of the apical membrane (Baron *et al.*, 1990b). In contrast, retinol and retinoic acid, which have been reported to stimulate bone resorption, affect the organization of both microfilaments and microtubules, promoting podosome formation and inducing the reversible depolymerization of microtubules (Zambonin-Zallone *et al.*, 1988; Oreffo *et al.*, 1988).

Thus, when osteoclasts are activated or inhibited, rapid and dramatic changes occur in their cytoskeleton and attachment structures, further demonstrating the functional importance of these structures in bone resorption (Teti *et al.*, 1991).

Several of these cytoskeletal changes might be associated with the regulation of the osteoclast's intracellular calcium levels and/or pH (Teti *et al.*, 1991). Inhibition of bone resorption is associated with an elevation of cytoplasmic Ca^{2+} levels after calcitonin treatment (Malgaroli *et al.*, 1989), and similar cytoskeletal changes are seen when cytoplasmic Ca^{2+} is increased by membrane depolarization, elevating extracellular Ca^{2+} levels, or treating osteoclasts with a calcium channel agonist (Miyauchi *et al.*, 1990). Reducing intracellular calcium, in contrast, promotes the formation of podosomes (Teti *et al.*, 1989a). Podosome assembly is also induced by decreased cytosolic pH, but this effect is mediated via a decreased cytoplasmic Ca^{2+} (Teti *et al.*, 1989a). Furthermore, the binding of RGD-containing matrix proteins to their integrin receptors induces a decrease in intracellular Ca^{2+} and activates bone resorption (Miyauchi *et al.*, 1991b). One likely candidate for mediating these effects of calcium on actin filament organization is gelsolin, which has been identified in the podosome (Marchisio *et al.*, 1987) and is known to affect actin filaments in ways that are differentially activated as Ca^{2+} concentrations increase from nanomolar to micromolar.

C. Summary

Hence, the cytoskeleton and integral membrane receptors of the integrin family play essential roles in osteoclast motility, in the specific attachment on the bone surface, in establishing the seal at the periphery of the extracellular bone resorbing compartment, and in regulating the activity of the osteoclast. All these are essential components of the integrated function of the osteoclast (i.e., bone resorption), which, as discussed later, also involves the biosynthesis and secretion of several enzymes and the acidification of the bone resorbing compartment.

III. PROTEINS DESTINED FOR EXPORT: BIOSYNTHETIC AND SECRETORY FUNCTIONS OF THE OSTEOCLAST

The morphological and cytoskeletal polarity of the osteoclast is only a reflection of the functional polarity of the secretory and transport components of this cell. The osteoclast is actively engaged in the biosynthesis and vectorial secretion of proteolytic enzymes and other proteins.

A. Acid Phosphatase and Other Lysosomal Enzymes Found in the Secretory Pathway

One of the main cytochemical characteristics of the osteoclast is its enrichment in lysosomal enzymes. Ultrastructural studies have shown, however, that this high concentration of lysosomal enzymes in the osteoclast is not due to the presence of phagocytic structures such as secondary lysosomes. Instead, these enzymes are found, for the most part, in elements of the exocytic pathway (Baron *et al.*, 1985b, 1988). Using a variety of techniques to localize multiple enzymes, their presence was demonstrated in the endoplasmic reticulum, in the Golgi complex, and in numerous transport vesicles of the osteoclast. Thus, localization of arylsulfatase, β-glycerophosphatase, and acid phosphatase by enzyme cytochemistry and localization of β-glucuronidase, cathepsin C, and tartrate-resistant acid phosphatase by immunocytochemistry (Baron *et al.*, 1985b, 1988; Andersson *et al.*, 1986; Reinholt *et al.*, 1990b) have shown the abundant concentration of these enzymes in the lumen of the endoplasmic reticulum cisternae, including the perinuclear envelopes, in the cisternae of the Golgi complexes and in numerous small (50–75 nm), coated vesicles in the Golgi complex areas and throughout the cytoplasm. The latter are particularly abundant between the nuclei and the deep portions of the ruffled border membrane folds (Baron *et al.*, 1985b, 1988). In addition, some typical secondary lysosomal vacuoles were found, which are filled also with enzymes but are uncoated, are larger, and have a heterogeneous content. These classical lysosomes are, however, relatively few and concentrated mostly in the basal portion of the cell, facing the bone marrow compartment. Consequently, this enrichment in lysosomal enzymes does not reflect a high phagocytic activity but, rather, a high biosynthetic activity.

The newly synthesized lysosomal enzymes are vectorially secreted into the bone resorbing compartment (Fig. 6). Although many early studies (Vaes, 1968, 1988; Lucht, 1971; Doty and Schofield, 1972), in particular reports of the *in vitro* correlation between bone resorption and the release of tartrate resistant acid phosphatase (TRAP) into the medium (Vaes, 1968; Chambers *et al.*, 1987), strongly suggested that the osteoclast secreted lysosomal enzymes into the sealed-off extracellular

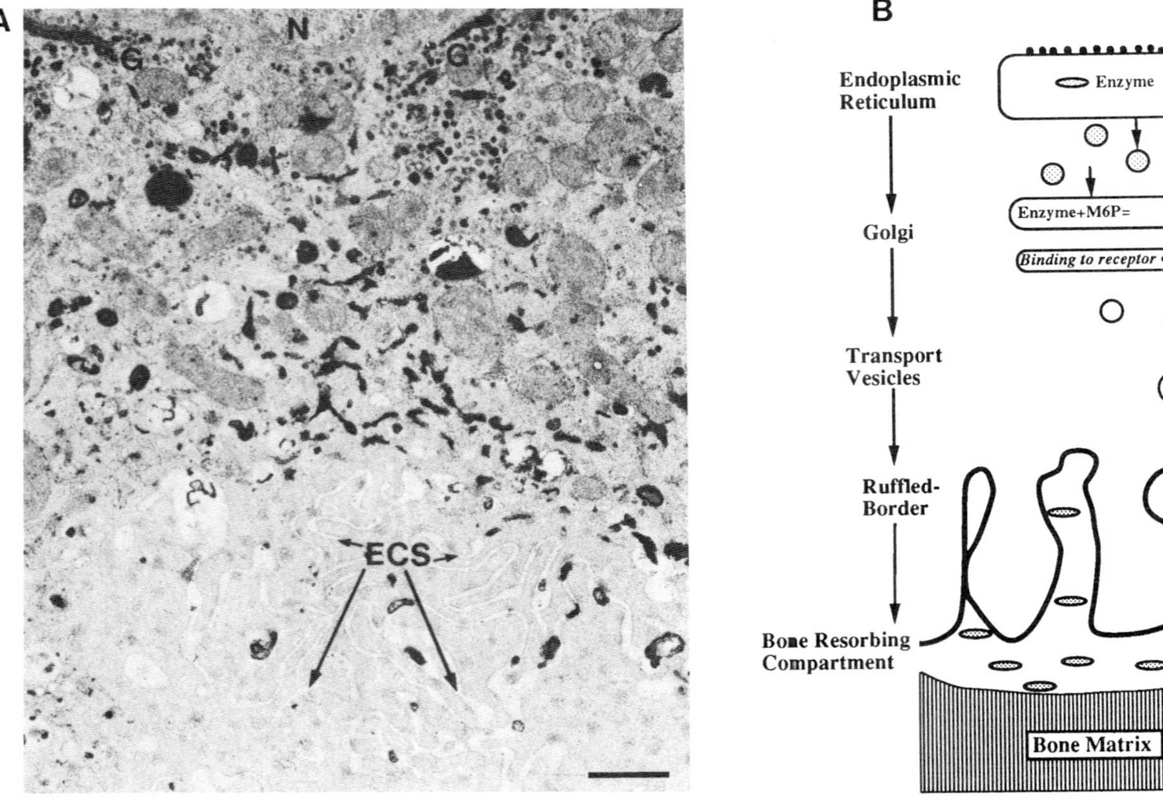

FIGURE 6 Vectorial transport and mannose-6-phosphate and pH-dependent secretion of lysosomal enzymes in the osteoclast. (A) From the perinuclear (nucleus [N]) Golgi complexes (G), where they are found in the cisternae, the lysosomal enzymes are transported in numerous transport vesicles (primary lysosomes) toward the ruffled border (bottom) delimiting extracellular space (ECS, arrows) where the enzymes are secreted. Scale, 1000 nm. From Baron *et al*. (1986, Fig. 1). (B) A schematic representation of the vectorial and mannose-6-phosphate (M6P)-dependent transport and release of the lysosomal enzymes (see text for detailed explanations). Rec, receptor.

bone resorbing compartment, this conclusion remained controversial until immunocytochemical studies were possible. Antibodies to TRAP and cathepsin C have, however, allowed the localization of these enzymes in the extracellular bone resorbing compartment (Baron *et al.*, 1988; Andersson *et al.*, 1986; Reinholt *et al.*, 1990b), hereby demonstrating that the osteoclast is actively engaged in the synthesis and secretion of lysosomal enzymes. These enzymes proceed through the Golgi and are vectorially transported from the trans-Golgi region to the ruffled border apical membrane in coated transport vesicles, which fuse exclusively with the ruffled border's plasma membrane and release their content into the bone resorbing compartment (Baron *et al.*, 1985b, 1988).

This specific release of the enzymes after targeting of secretory transport vesicles to the apical domain of the cell's membrane is accomplished by their association with the mannose-6-phosphate receptor (Baron *et al.*, 1988) (Fig. 6). In most cells, lysosomal enzymes are sorted out of the main flow of secretory and membrane proteins at the level of the Golgi apparatus (Kornfeld, 1986; Von Figura and Hasilik, 1986) and specifically targeted to late endosomes along the endocytic pathway (Brown *et al.*, 1986). The newly synthesized enzymes are recognized by a specific binding protein, the mannose-6-phosphate receptor (Creek and Sly, 1984), which is highly concentrated in membranes of the Golgi compartment (Brown and Farquhar, 1984, 1987) and binds mannose-6-phosphate, a specific marker added to the lysosomal enzymes in the *cis*-Golgi elements. This mechanism ensures that the enzymes are sorted out of the flow of other proteins synthesized by the cells. The enzymes remain bound to the mannose-6-phosphate receptor until they reach an acidified compartment where the low pH induces their dissociation from the receptor. In this manner, enzymes are released in the proper compartment, which, in all cells, is the late endosome (Brown *et al.*, 1986).

Despite the fact that the situation is quite different in the osteoclast, which secretes the newly synthesized enzymes instead of targeting them to an intracellular organelle, high concentrations of the cation-independent mannose-6-phosphate receptor are found, co-localized with lysosomal enzymes along the exocytic pathway (Baron *et al.*, 1988, 1990b). Both the enzymes and the mannose-6-phosphate receptors are present in the transport vesicles and, being in a post-Golgi compartment and at neutral pH, the enzymes must be bound to their receptor during transport toward the ruffled border. These transport vesicles then fuse with the ruffled border membrane, probably via specific interactions, and the enzymes dissociate from their receptors upon exposure to the low pH in the bone resorbing compartment. If this process is similar in the osteoclast and in other cells, the receptors are most likely recycled to the Golgi for other rounds of transport after delivery.

This could represent the mechanism by which the osteoclast ensures that the lysosomal enzymes are not secreted toward the surrounding tissues at the basolateral pole of the cell: A transport vesicle that would reach the basolateral membrane would not release the enzymes in the microenvironment because the ligands do not dissociate from the receptors at neutral pH. It is also possible that mannose-6 phosphate receptors present at the basolateral domain of the cell play a role in recapturing enzymes that would leak from the resorbing compartment, particularly when the cell detaches and enters in a motile phase (Baron *et al.*, 1988; Blair *et al.*, 1988).

B. Nature and Specificity of the Secreted Enzymes

The nature and specificity of the enzymes that are secreted by the osteoclast into the bone resorbing compartment and that participate in the degradation of the extracellular matrix, both collagen and noncollagenous proteins, have been the subject of intense discussion over the years (Table I). Although the best characterized osteoclast enzyme is the tartrate-resistant acid phosphatase type 5 (Hammarstrom *et al.*, 1971), which has been recently cloned and sequenced (Ek-Rylander *et al.*, 1991), other acid phosphatases as well as aryl-sulfatase, β-glucuronidase, and β-glycerophosphatase have been localized in the osteoclast's biosynthetic pathway (Lucht, 1971; Doty and Schofield, 1972; Baron *et al.*, 1985b, 1988; Andersson *et al.*, 1986; Reinholt *et al.*, 1990b). More recently, based on localization and functional studies, it has become apparent that the osteoclast also synthesizes and secretes several

TABLE I Secretory Products of the Osteoclast

Lysosomal enzymes
Tartrate-resistant acid phosphatase
β-glycerophosphatase
Arylsulfatase
β-glucuronidase
Cysteine-proteinases (cathepsin B, C, and L)
Non-lysosomal enzymes
Collagenase
Stromelysin
Tissue plasminogen activator
Lysozyme
Other (matrix) proteins
Bone sialoprotein
Osteopontin
Transforming growth factor β
Cytokines
Interleukin 6

cysteine-proteinases, among which are the cathepsins B, L (Delaisse and Vaes, 1991; Delaisse *et al.*, 1991), and C (R. Baron, L. Neff, and F. Mainferme, unpublished results, 1991). The importance of these enzymes is that they are capable of degrading collagen in an acidic environment (Delaisse *et al.*, 1991; Delaisse and Vaes, 1991), such as the one encountered in the bone resorbing compartment. As such, the presence of these cysteine-proteinases resolved the apparent contradiction between the low pH at the resorbing site and the inability of neutral collagenase to degrade collagen at such a pH *in vitro*. Furthermore, several recent studies have directly demonstrated the importance of lysosomal cysteine-proteinases in bone resorption: Inhibition of these enzymes led to a decrease in bone resorption (Delaisse and Vaes, 1991) and to the accumulation of partially digested collagen in the resorbing lacunae (Everts *et al.*, 1992).

Although the cathepsins can digest collagen at acidic pH, making collagenase theoretically unnecessary, the possibility that this enzyme still plays an important role in bone resorption has been raised by three recent studies. First, and despite the fact that the exact source of the collagenase is not yet firmly established, collagenase has been detected in the bone matrix underlying resorbing osteoclasts (Baron *et al.*, 1990a), and within the cell itself both collagenase (Delaisse *et al.*, 1992) and its messenger RNA (Okamura, 1992) have been localized. Second, inhibition of matrix metalloproteinases leads to an accumulation of partially digested collagen in the bone resorbing compartment (Everts *et al.*, 1992), an effect similar to inhibiting cysteine-proteinases. Finally, tissue plasminogen activator (Grills *et al.*, 1990) as well as stromelysin (Delaisse *et al.*, 1992), two enzymes that can activate latent collagenase, have also been found in osteoclasts, further suggesting a role for matrix metalloproteinases in bone resorption. Hence the osteoclast synthesizes and secretes several classes of enzymes: phosphatases, sulfatases, cysteine-proteinases, and matrix metalloproteinases.

Although the pH optimum of the cysteine- and metalloproteinase classes of enzymes in their ability to degrade bone matrix collagen are very different (4.0 to 5.0 and 7.5, respectively), it is possible to envision cooperation between the two classes of enzymes during the resorption of bone, particularly because collagenase still has an important activity around pH 6 (Delaisse and Vaes, 1991) and stromelysin can still degrade proteoglycans at pH 5. To integrate these new data, we have been led to speculate elsewhere, in collaboration with others (Baron *et al.*, 1990a; Delaisse *et al.*, 1992; Everts *et al.*, 1992), about possible modes in which the two classes of enzymes would cooperate in osteoclastic bone resorption. Besides their direct collagenolytic action, lysosomal cysteine-proteinases can generate active collagenase (Eeckhout and Vaes, 1977). The actions of the lysosomal acid cysteine-proteinases and of collagenase may therefore be exerted sequentially (Baron *et al.*, 1990). The first phase of the osteoclastic resorption process would involve mainly the cysteine-proteinases, degrading the bulk of the collagen as soon as the matrix is demineralized and in the fully sealed-off and acidified bone resorbing compartment. In contrast, collagenase action would occur at more neutral pH and in two circumstances—during active bone resorption and after displacement of the cell from its previous resorbing site.

During active osteoclastic bone resorption, the buffering capacity exerted by the solubilized bone salts is likely to cause a gradient of pH extending from the most acidic zone, in the immediate vicinity of the ruffled border and its proton pumps, to a more neutral zone, deeper in the resorbing lacuna and toward the interface between mineralized and demineralized matrix. The higher pH in these regions would both favor the activation of procollagenase by the lysosomal cysteine-proteinases, because this process is more efficient around pH 6 than at lower pH (Eeckhout and Vaes, 1977), and render the collagenolytic action of collagenase predominant, because cysteine-proteinases are quite inefficient near neutral pH.

Second, as already discussed, the osteoclast moves along the bone surface, successively walking, sitting, resorbing, and walking again (see Fig. 3). The detachment of the cell at the end of a resorptive phase causes the sudden neutralization of the pH of the resorbing compartment, resulting from its opening to outside extracellular fluids. This would immediately prevent further actions of the cysteine-proteinases, leaving a fringe of already demineralized but as yet undegraded collagen, as seen at the bottom of resorption pits eroded by isolated osteoclasts (Ali *et al.*, 1984; Murrills *et al.*, 1989). The role of collagenase could then be envisioned as removing that collagen fringe at the neutral pH present after displacement of the osteoclast, so as to allow the completion of the resorbing process despite the absence of a sealed-off subosteoclastic acidic microenvironment. Thus, it would then be the cooperative action of a set of acidic and neutral pH classes of enzymes that leads to a complete degradation of the extracellular matrix at the resorbing site.

C. Synthesis and Secretion of Other Proteins by the Osteoclast

It is becoming increasingly apparent that the osteoclast does not restrict its biosynthetic and secretory activity to making enzymes that are directly involved in bone matrix degradation, albeit most of these secretory products are probably important in bone resorption (Table I). As briefly mentioned earlier, there is immunocytochemical and/or *in situ* hybridization information strongly suggesting that the osteoclast is synthesizing two of the RGD-containing matrix proteins involved in its own adhesion. Both osteopontin, which has been demonstrated at the site of attachment of the osteoclast sealing zone (Reinholt *et al.*, 1990a), and BSP have been shown to be synthesized by the osteoclast or potential precursors (Masi *et al.*, 1991). Although these findings can be considered surprising because these molecules are also, and mostly, synthesized by the osteoblast, they are just another illustration of the fact that many cells synthesize and secrete the proteins that they need to attach on substrates. There is also preliminary evidence suggesting that the osteoclast might produce transforming growth factor β (Qi *et al.*, 1991; Hosoi *et al.*, 1991), which the acid environment could activate from its latent form, potentially leading to a local coupling message for the osteoblasts (Oreffo *et al.*, 1989; Pfeilschifter *et al.*, 1990).

D. Generation of Oxygen-Derived Free Radicals

In addition to the secretion of enzymes and protons, osteoclasts, like other cells in the monocyte–macrophage lineage, synthesize and secrete lysozyme (Hilliard *et al.*, 1990), an enzyme involved in the cleavage of certain polysaccharides, and may locally generate oxygen-derived free radicals to resorb bone. It has been shown that superoxide anions, one in a class of oxygen-derived free radicals, are present both intracellularly and under the osteoclast, in the bone resorbing compartment (Garrett *et al.*, 1990; Key *et al.*, 1990). Furthermore, patients with the autosomal recessive form of osteopetrosis show both a defect in superoxide generation by leukocytes and a defect in osteoclastic bone resorption, thereby suggesting that superoxide generation may be involved in normal bone resorption (Beard *et al.*, 1986). This possibility is further supported by the facts that the generation of free radicals is regulated by agents regulating bone resorption (Garrett *et al.*, 1990; Key *et al.*, 1990), that inhibition of free-radical generation by superoxide dismutase blocks the stimulation of bone resorption by a number of agents (Garrett *et al.*, 1990), and that H_2O_2 stimulates bone resorption (Bax *et al.*, 1992). This hypothesis has been further supported by the fact that an antigen highly enriched in the osteoclast plasma membrane and recognized by a monoclonal antibody with high specificity for this cell (121F [Oursler *et al.*, 1985]) may be a superoxide dismutase (Oursler *et al.*, 1989). Although the exact importance of this pathway in bone resorption remains to be determined, recent reports that the free radical nitric oxide inhibits and hydrogen peroxide stimulates bone resorption by isolated osteoclasts (MacIntyre *et al.*, 1991; Bax *et al.*, 1992) lends further support for the concept that free radicals play a role in bone resorption.

E. Summary

The motile osteoclast is thus capable of attachment to the bone surface, sealing off an extracellular compartment and secreting into this compartment several enzymes and proteins necessary for bone resorption. The morphological polarity mentioned earlier therefore parallels a functional polarity. Because acidification of the bone resorbing compartment is required and is also a polarized function of the cell, ionic transport at the basolateral and apical domains of the plasma membrane of the osteoclast has emerged as a major element in the biology of the osteoclast. These processes are discussed in the following section.

IV. CYTOSOLIC AND MEMBRANE PROTEINS: MEMBRANE COMPOSITION AND ION TRANSPORT

Many proteins made by the osteoclast are, unlike the molecules discussed in the previous section, not destined for export but will remain in the cell's cytosol or be inserted in specific domains of the membrane, such as the numerous integral-membrane proteins involved in ion transport and attachment (proton pumps, Na,K-ATPase, integrins, Ca-ATPase, anion exchangers, etc.).

Acidification of the extracellular bone resorbing compartment has emerged over the last few years as one of the most important features of the biology of the osteoclast (Vaes, 1968, 1988; Baron *et al.*, 1985b, 1986a; Blair *et al.*, 1989; Baron, 1989; Arnett and Dempster, 1986), often determining the requirements for other specific features (sealing zone, enzymes, ruffled border, etc.). The description of the organization and function of the osteoclast membrane domains needs, therefore, to be organized around the molecular mechanisms involved in acid secretion by this cell. Briefly, the polarized extrusion of protons across the apical membrane of the osteoclast imposes several ionic and charge constraints on the cell that require several ion-generating and ion-transport systems in the cytosol and in both domains of the plasma membrane (Baron, 1989; Sims *et al.*, 1991a). The identification and characterization of these systems has been greatly facilitated by the extended knowledge of the ion-transport properties of other cells involved in acid secretion, particularly the kidney tubule intercalated cell and the gastric parietal or oxyntic cell. It had been known for several years that the osteoclast is highly enriched in carbonic anhydrase (Gay and Mueller, 1974), an enzyme that plays an essential role in acid-secreting cells such as the kidney tubule intercalated cell and the gastric mucosa oxyntic cell. In humans, defective carbonic anhydrase II (CAII) is associated with cerebral calcification, tubular acidosis, and osteopetrosis (Sly *et al.*, 1983), providing direct genetic evidence for CAII involvement in bone resorption. Similarly, inhibition of carbonic anhydrase activity with acetazolamide blocks the ability of the osteoclast to efficiently resorb bone (Minkin and Jennings, 1972; Waite *et al.*, 1970). As discussed later, carbonic anhydrase generates protons and bicarbonate from carbon dioxide and water, providing the cells with protons for acidification at their apical membranes.

A. The Apical Membrane and the Process of Acidification

The apical membrane of the osteoclast, or ruffled border, is directly involved in the molecular mechanisms of bone resorption (Fig. 7). It is the target for the specific delivery of the newly synthesized secretory enzymes and the site of proton extrusion for acidification of the bone resorbing compartment. The extensive folding of this domain of the plasma membrane is probably due both to the intense vesicular traffic associated with secretion and to the need for increasing the number of proton pumps via an amplification of the apical membrane.

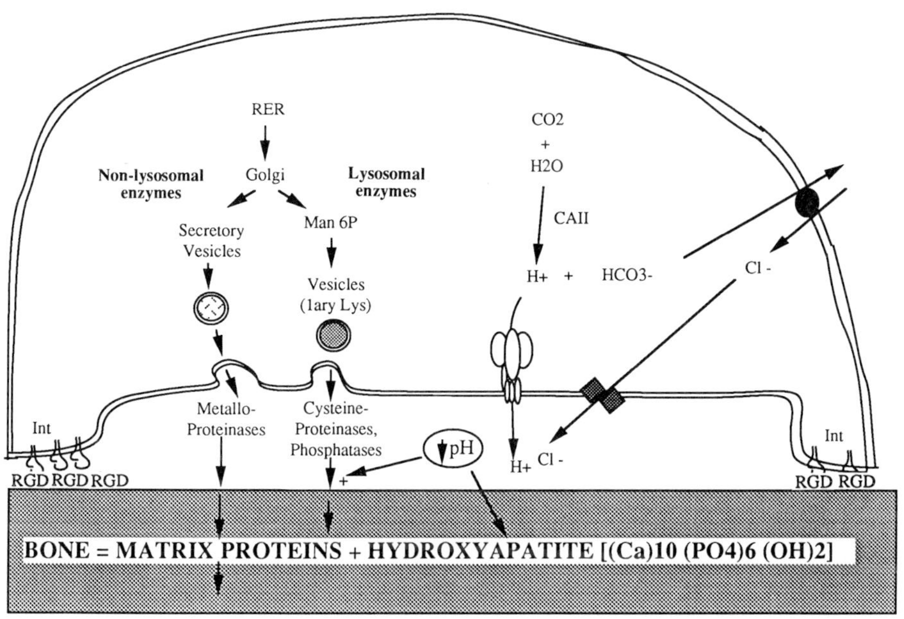

FIGURE 7 Schematic representation of the relationship of enzyme secretion, attachment, and acidification in the mechanisms of bone resorption. Sealing at the periphery of the resorbing compartment is assured by binding of integrin receptors to Arg–Gly–Asp (RGD)-containing extracellular proteins. Cysteine-proteinases and metalloproteases are secreted into this sealed-off compartment, possibly transported in two separate secretory vesicles. Acidification is ensured by the vacuolarlike proton adenine triphosphatase in the ruffled border membrane. The protons are generated intracellularly by carbonic anhydrase, and the by-product of the reaction, bicarbonate, is secreted basolaterally in exchange for chloride by a band 3-like anion exchanger. The chloride is then transported into the resorbing compartment by chloride channels, thereby avoiding hyperpolarization of the osteoclast's membrane that would result from proton transport. The low pH favors the action of lysosomal enzymes and dissolves the hydroxyapatite, and the presence of metalloproteases broadens the pH range at which resorption can occur. CAII, carbonic anhydrase II; Int, integrins; Man 6P, mannose-6-phosphate; RER, rough endoplasmic reticulum.

There are several reasons why the apical ruffled border membrane of the osteoclast could be expected to resemble the limiting membrane of vesicular elements of the endocytic pathway in other cells. First, this membrane limits a compartment that is the extracellular functional equivalent of a lysosome, with high concentrations of enzymes and a low pH (Mellman *et al.*, 1986). Second, the ruffled border membrane is the target for delivery of newly synthesized lysosomal enzymes by post-Golgi transport vesicles, as are late endosomes in other cells (Brown *et al.*, 1986). Finally, this membrane is responsible for establishing and maintaining the pH gradient necessary for bone resorption, as are the lysosomal and endosomal membranes for the degradation of internalized material. Immunocytochemical localization of lysosomal and/or endosomal integral membrane proteins has revealed that the ruffled border expresses several endosomal membrane proteins (Baron *et al.*, 1985a,b, 1988) but not the proteins that are normally restricted to mature secondary lysosomes (Baron *et al.*, 1985a). These lysosomal/endosomal proteins are restricted to the apical domain of the osteoclast's plasma membrane and cannot be demonstrated in the basolateral domain. These observations both established that some analogy exists between endosomal membranes and the ruffled border and demonstrated that the polarity of the osteoclast that is observed morphologically and in the trafficking of secretory proteins is also present at the level of plasma membrane composition, with some integral membrane proteins being specifically segregated into one or the other domains of the plasma membrane. The attachment apparatus of the cell very likely plays here a role similar to that of the tight junctions in epithelial cells in preventing the lateral diffusion of these restricted proteins from one membrane domain to the other. This could explain why the detachment of the osteoclast after calcitonin treatment leads to a loss of membrane polarity (Baron *et al.*, 1990b).

This parallel with endosomes, however, is limited because, as described later, these two membranes differ in other respects: One is intracellular when the other is not, at least in actively resorbing osteoclasts. They also differ in the number and, possibly, the specific nature of the proton pumps.

Proton pumps serve a number of important functions in eukaryotic cells (Mellman *et al.*, 1986; Forgac, 1989). In intracellular organelles of the endocytic or exocytic pathways, luminal acidification allows the dissociation of ligands from their receptors, the proteolytic processing of peptides, and the entry of viruses into cells (Mellman *et al.*, 1986) and energizes the limiting membranes of secretory vesicles, allowing the parallel uptake and accumulation of small molecules and amino acids (Hell *et al.*, 1988; Maycox *et al.*, 1988). Proton pumps are also present at the plasma membrane of cells that are specialized in proton secretion, such as the gastric parietal cell, cells lining the distal and collecting kidney tubules (Al-Awqati, 1986), and the osteoclast in bone (Baron *et al.*, 1985b; Blair *et al.*, 1989), where they allow acidification of extracellular fluids. During bone resorption, osteoclast-mediated acidification is required for the dissolution of the mineral phase and the enzymatic degradation of the organic phase of the extracellular matrix (Baron *et al.*, 1985b; Vaes, 1988; Blair *et al.*, 1986, 1989).

The H^+-ATPases responsible for acidification in these various cells and organelles are distinguished from each other on the basis of their structure and their sensitivity to various specific inhibitors. The gastric proton pump (P-ATPase), which exchanges H^+ for K^+, is composed of one α (100 kDa) and one β (55 kDa) subunit. The α subunit forms a phosphorylated intermediate and, hence, is inhibited by vanadate (Al-Awqati, 1986; Nelson, 1991). In contrast, the mitochondrial F_0-F_1 (F-ATPase) and the vacuolar H^+-ATPase (V-ATPase) (Fig. 8), which share a common evolutionary origin (Nelson and Taiz, 1989), are multisubunit structures (8–10 subunits) with a proton-conducting channel buried in the membrane (F_0 and V_0) and a large cytoplasmic catalytic portion (F_1 and V_1). They are resistant to vanadate and differ in their sensitivity to other specific inhibitors: The V-ATPase is sensitive to *N*-ethylmaleimide (NEM) and to bafilomycin A1 (Bowman *et al.*, 1988) but not to oligomycin, whereas the reverse is true for the F-ATPase (Al-Awqati, 1986; Nelson, 1991).

In an early report, based on the cross-reactivity of an antibody to a lysosomal membrane protein with the H^+/K^+-ATPase, the possibility that the osteoclast proton pump could be of the P type was suggested (Baron *et al.*, 1985b). Further studies with antibodies to the gastric proton pump, however, failed to demonstrate its presence in osteoclast membranes (R. Baron, L. Neff, and M. Caplan, unpublished, 1991). More recently, based on pharmacological and immunochemical data, it has been suggested that the H^+-ATPase present in osteoclast membranes is of the V type, closely resembling the pumps present in kidney membranes (Blair *et al.*, 1989; Bekker and Gay, 1990b; Vaananen *et al.*, 1990). More specifically, proton transport in inside-out vesicles prepared from osteoclast-enriched cell fractions has been reported to be sensitive to NEM and other V-ATPase inhibitors but not to vanadate or oligomycin. By Western blot, these preparations have been shown to contain the 70-, 60-, and 31-kDa subunits of the V-ATPases (Fig. 8), and cross-reacting antigens have been immunolocalized at the apical ruffled border membrane of the osteoclast (Blair *et al.*, 1989; Vaananen *et al.*, 1990) (Fig. 9).

There is, however, evidence in the literature showing subtle differences in the properties and structure of mammalian V-ATPases (Forgac, 1989; Nelson, 1991; Wang and Gluck, 1990). The vacuolar proton pumps contained in coated vesicles, endosomes, chromaffin granules, and kid-

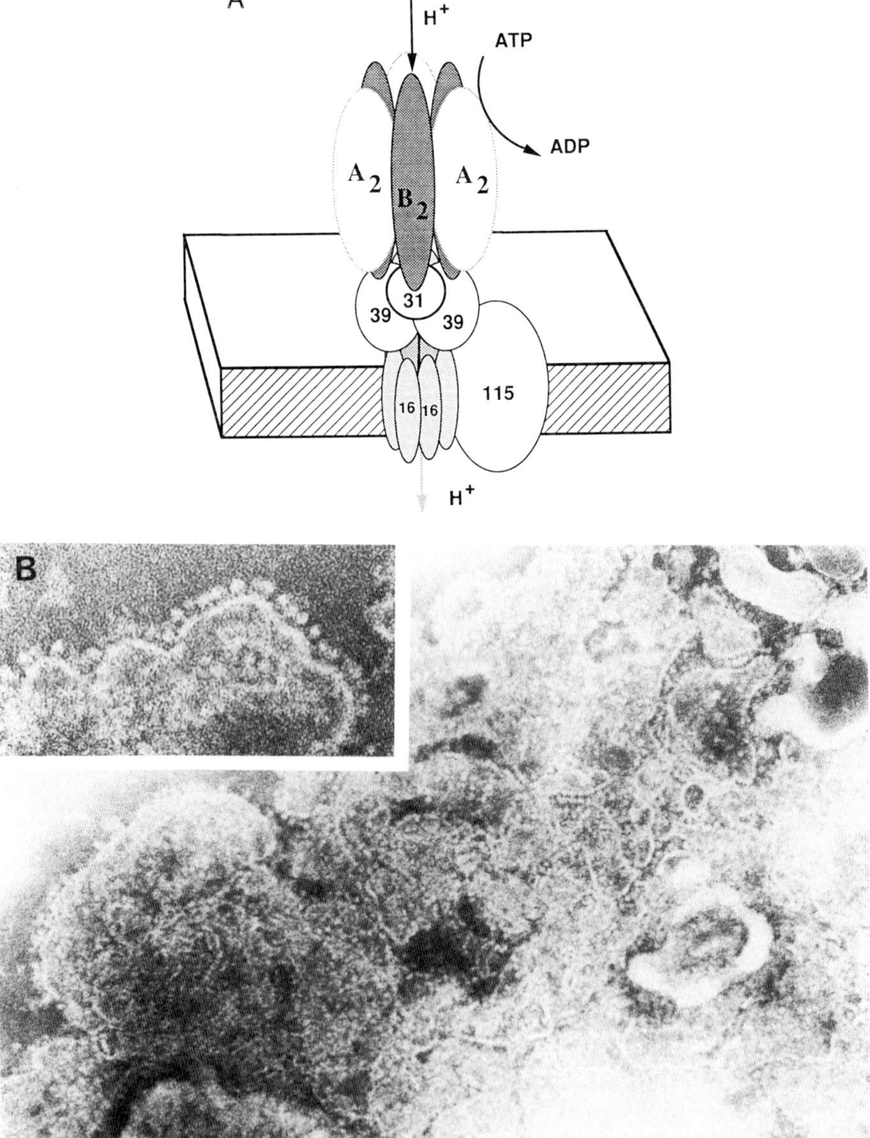

FIGURE 8 (A) Schematic representation of the prototype vacuolar adenosine triphosphatase H^+-ATPase (A) and electron microscopy of ruffled border membrane vesicles (B). The 70-kDa subunit A and the 60-kDa subunit B, each in three copies, together form the catalytic portion of the enzyme, which binds ATP on the cytoplasmic side of the membrane. In the osteoclast, two isoforms (A_2 and B_2) are expressed, conferring to the proton ATPase unique pharmacological properties (Chatterjee *et al.*, 1992, Fig. 1). Other subunits comprise 31- and 39–42-kDa subunits forming the stem and 16-kDa subunits forming the DCCD-binding proton pore through the membrane; the 100–115-kDa subunits are thought to be regulatory polypeptides. (B) The electron microscopy of negatively stained osteoclast-derived microsomal vesicles demonstrates the presence of a high density of characteristic V_0V_1 ball-and-stalk multisubunit H^+ pump structures, visible only in

ney tubule plasma membranes differ in their subunit composition, and it has been suggested that some of these differences may be due to the existence of isoforms of one or more subunit, with only small differences in their apparent molecular weight (Forgac, 1989; Wang and Gluck, 1990). These observations have led to the hypothesis that variations in isoforms of the multisubunit vacuolar proton pump(s) may constitute the basis for the differential targeting, properties, and regulation of H^+-ATPases present in different organelles in the same cell, in different cells, or in different organs. Recently, two isoforms of one of the subunits forming the catalytic domain (subunit B) have indeed been cloned and sequenced (Puopolo *et al.*, 1992; R. D. Nelson *et al.*, 1992).

In our hands (Chatterjee *et al.*, 1992), H^+-transport by inside-out vesicles derived from osteoclast-enriched cell preparations was, unlike other known proton-transport systems, not only sensitive to the V-ATPase inhibitors NEM and bafilomycin A1 but also to the P-ATPase inhibitor vanadate. The ability of osteoclast-derived vesicles to transport protons is about 50-fold higher than that of microsomes from the other cells present in these preparations (Chatterjee *et al.*, 1992) and at least 2-fold greater than that of highly purified endosomes (Schmid *et al.*, 1989; Fuchs *et al.*, 1989). This very efficient H^+ transport suggests that a high concentration of pumps is present in the osteoclast apical membrane. High magnification electron microscopy on negatively stained osteoclast microsomes showed the presence of high densities of characteristic ball-and-stalk structures (Fig. 8), compatible with the presence of a high number of copies of multisubunit H^+-ATPases (Chatterjee *et al.*, 1992), and similar to the V-type ATPases observed in kidney tubule apical membranes (D. Brown *et al.*, 1987).

These results, taken together with several earlier observations (Baron *et al.*, 1985b; Anderson *et al.*, 1986; Tuukkanen and Vaananen, 1986), suggest the possibility that the osteoclast proton pump has properties of both V-type and P-type ATPases, essentially making it a novel class of proton-transport system.

B. Role of the Basolateral Membrane and Ion Channels in Acidification, Intracellular pH, and Membrane Potential Regulation

The apparently simple process of pumping protons across the ruffled border membrane to acidify the resorption lacuna imposes, in fact, a complex and demanding set of ionic requirements to the cell, essentially

---

the inside-out configuration, in 30–40% of the microsomal vesicles. In these whole-mount preparations, the V_0V_1 structures appear as lightly stained dots over the vesicle membranes (28,400×). Characteristic ball-and-stalk structures (the diameter of the "ball" is ~ 10 nm) can be observed in an area where the membrane is seen sideways Inset, 68,600×. ADP, adenosin diphosphate. From Chatterjee *et al.* (1992).

designed to maintain the electrochemical balance of the osteoclast during bone resorption (Fig. 7 and 10). This ionic balancing act requires the coordinated activity of electrogenic ion pumps, ion channels, and electroneutral ion exchangers in order to maintain the cytoplasmic pH and the transmembrane electrical potential within narrow physiological ranges. The two components that are directly involved in the acidification process and dictate the other requirements are the apical electrogenic proton pump itself and CAII (Fig. 9).

Briefly, the protons that the H^+-ATPase transports across the apical membrane are generated in the cytoplasm by the reversible hydration of CO_2 to produce carbonic acid, which ionizes to form protons and HCO_3^- (Gay and Mueller, 1974; Maren, 1967) (Fig. 7 and 10). The generation of bicarbonate and the transfer of the protons out of the cell by the pump alter both the cytoplasmic pH, which becomes more alkaline, and

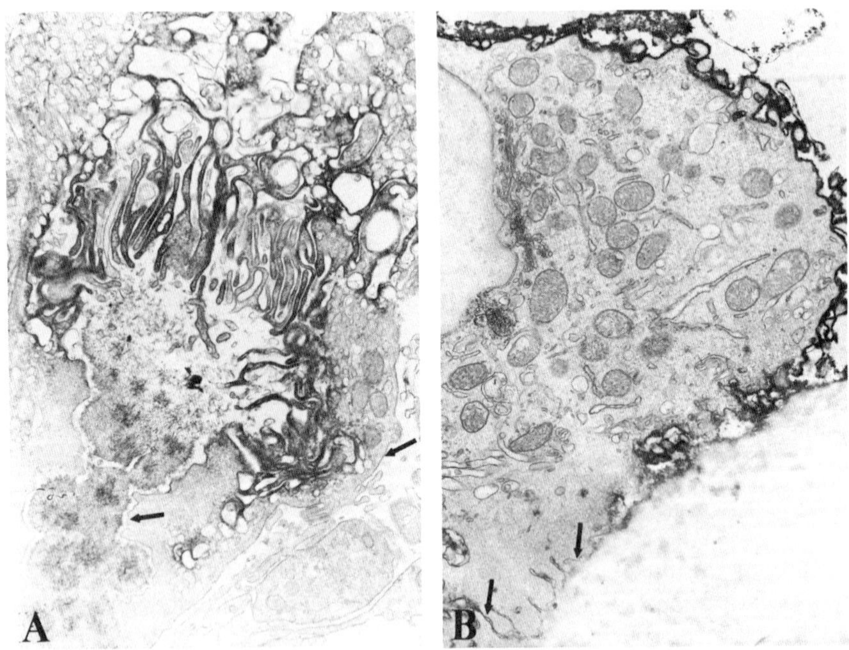

FIGURE 9 Molecular evidence for the polarization of the osteoclast's plasma membrane. The polarity of the osteoclast is not only morphological and secretory (see Figs. 1 and 5) but also molecular, with some membrane molecules being restricted to the ruffled border apical membrane (here, in (A), the 31-kDa subunit of the proton pump; L. Neff, D. Chatterjee, W. Horne, and R. Baron, unpublished) or to the basolateral membrane (here, in (B), the b subunit of the sodium pump; from Baron *et al.*, 1990b). In (A) note the unlabeled basolateral and sealing zone membranes (arrow) and, in (B) the unlabeled sealing zone and ruffled border membranes (arrows).

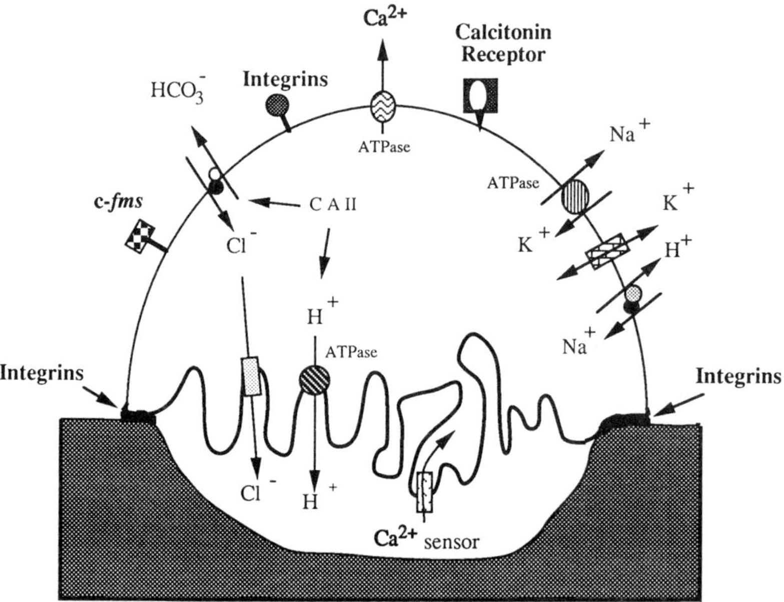

FIGURE 10 Major ion-transport systems that have been identified in the osteoclast. Single circles represent ion-motive adenine triphosphatases (ATPases), double circles represent ion exchangers, and rectangles represent ion channels (see text for detailed explanations). CAII, carbonic anhydrase II.

the membrane potential, which becomes hyperpolarized as progressively more positive charges move out across the membrane and negative charges accumulate in the cytosol. Cytoplasmic pH and membrane potential therefore must be tightly controlled.

Cytoplasmic pH is maintained near neutrality by the action of electroneutral ion exchangers present in the basolateral membrane (Fig. 5), probably at higher number of copies in the osteoclast than in other cell types due to its very active proton transport. Both an acid extruder (the Na^+/H^+ antiporter) and an acid loader (i.e., base extruder, the HCO_3^-/Cl^- exchanger) are present in the osteoclast. When the proton pump is active, the increasing alkalinity in the cytosol activates the HCO_3^-/Cl^- exchanger, which extrudes the excess HCO_3^- in a one-to-one exchange for extracellular Cl^-, preventing the intracellular pH from rising excessively (Teti *et al.*, 1989a; Hall and Chambers, 1989; Horne *et al.*, 1991). If, on the other hand, intracellular pH should fall for some reason—for example, if the pump should acutely stop extruding the protons generated by carbonic anhydrase—then the Na^+/H^+ antiporter is activated and extrudes protons across the basolateral membrane in exchange for extracellular Na^+ (Hall and Chambers, 1990; Chakraborty *et al.*, 1991). It is, in fact, the gradient of Na^+ across the membrane, established by the activity of the sodium pumps found in high numbers in the basolateral membrane of the osteoclast (Baron *et al.*, 1986a), that drives the antiporter. These two exchangers therefore keep the intracellular pH within physiologic range but at the cost of an increasing intracellular Cl^- or Na^+ concentration. The activity of the Na^+/K^+-ATPase (Fig. 9) will allow the sodium that enters the cell to be transported out in a process that exchanges 3 Na^+ for 2 K^+, thereby resulting, like the activity of the proton pump, in hyperpolarizing the membrane potential.

The hyperpolarization resulting from the activities of the H^+pump and the Na^+/K^+-ATPase, to which the activity of the Ca^{2+}-ATPase also contributes (Bekker and Gay, 1990a), would rapidly reach a level that makes it nearly impossible to transport any more positively charged ions out of the cell. This would eventually hamper proton generation and electrogenic proton transport, if they were not alleviated by the activity of voltage-sensitive ion channels, which transport positively charged ions into the cell and/or negatively charged ions out of the cell and, thus, discharge the electrical potential. Two such ionic conductances have been recently described.

First, whole-cell patch-clamp studies have shown that both freshly isolated neonatal rat and embryonic chick osteoclasts possess inwardly rectifying K^+, or G_{ki}, channels (Ravesloot *et al.*, 1989a,b; Sims and Dixon, 1989; Sims *et al.*, 1991b). These G_{ki} channels are activated (opened) by a hyperpolarizing E_m. If E_m becomes more negative than the K^+ equilibrium potential (E_k) of −80 mV, then K^+ ions flow through G_{ki} channels into the cell, reducing the hyperpolarization. In addition to G_{ki} channels, embryonic chick osteoclasts express two other types of K^+ channels that are both activated by depolarizing E_m values. The first channel type is present in about 70% of the osteoclasts and is activated transiently when E_m becomes less negative than −20 mV. At those E_m values, the channels give rise to outward directed K^+ currents that decrease in time (inactivate). The second type of K^+ channel activates at E_m values more positive than +30 mV and gives rise to outward K^+ currents in a sustained fashion. As a result of the activity of these K^+ channels, E_m is clamped at values close to E_k. Moreover, G_{ki} channels allow the osteoclast to switch rapidly between this value of E_m and −15 mV and vice versa (Sims *et al.*, 1991a).

Second, there is evidence indicating the presence of chloride conductances in osteoclast membranes. Using on-cell and cell-free patch-clamp techniques, high-conductance anion channels were identified in neonatal rat osteoclasts (Schoppa *et al.*, 1990; Sims *et al.*, 1991b). A chloride channel with a very high conductance of about 400 pS and voltage sensitivity (opened only at potentials between −30 and +30 mV) was observed (Schoppa *et al.*, 1990). Chloride conductances have also been identified in inside-out vesicles (Blair *et al.*, 1991), and the presence of Cl^- in the acidification buffers is necessary for H^+ transport (Blair *et al.*, 1989; Bekker and Gay, 1990b; Vaananen *et al.*, 1990; Chatterjee *et al.*, 1991), but it is not known whether the implied chloride conductance in vesicles corresponds to the conductances that were electrophysiologically characterized. Chloride as well as potassium channels may therefore be present in both the apical and basolateral domains of the plasma membrane of the osteoclast.

It is therefore apparent that the regulation of H^+ transport at the apical surface of the osteoclast is tightly linked to the regulation of intracellular pH and membrane potential, mostly accomplished by exchangers, pumps, and channels present in the basolateral membrane of the cell. These various elements are closely coupled and also depend on the intracellular and extracellular concentrations of various ions, among which calcium is the most prominent.

C. Role of Calcium in Osteoclast Regulation

There are several obvious reasons for calcium to be of major importance in the regulation of the osteoclast's function. First, the cell's activity is regulated by several calciotropic hormones, whether directly or indirectly, whose main function is the maintenance of calcium homeostasis. Second, the osteoclast is most probably the cell of the body that is exposed to the highest local concentrations of calcium, resulting from the dissolution of hydroxyapatite crystals in the acidic microenvironment of the bone resorbing compartment. Third, bone resorption being a means by which the organism mobilizes calcium from the skeleton, the calcium that is generated by the action of the osteoclast on bone matrix must find its way to the extracellular fluids of the body.

Recent studies with isolated osteoclasts have shed some light on these issues. Two groups (Zaidi, 1990; Malgaroli *et al.*, 1989; Miyauchi *et al.*, 1990) have independently demonstrated that the incubation of the cells in the presence of calcitonin or in the presence of high extracellular calcium (2–4 m*M*) led to an increase in intracellular calcium concentrations and an inhibition of bone resorption.

Based on pharmacological studies, the effect of calcitonin on intracellular Ca^{2+} has been attributed mostly to the plasma membrane and intracellular Ca^{2+}-ATPases (Malgaroli *et al.*, 1989). The basolateral membrane of the osteoclast has indeed been shown to contain high levels of a Ca^{2+}-ATPase, which could play a major role in these movements of Ca^{2+} (Bekker and Gay, 1990a).

Although in whole-cell patch-clamp experiments no Ca^{2+} (or Na^+) channels were found (Sims and Dixon, 1989; Ravesloot *et al.*, 1989b), experiments with fluorescent Ca^{2+} dyes indicate that osteoclasts present on glass coverslips, but not on bone, do express voltage-dependent Ca^{2+} channels (Miyauchi *et al.*, 1990). In addition, the osteoclast was found to express a novel type of Ca^{2+} channel that opens upon a rise in the extracellular Ca^{2+} concentration (Zaidi, 1990; Miyauchi *et al.*, 1990). This "calcium sensor," or calcium-activated calcium channel, is strongly reminiscent of the regulatory system found in the principal cells of the parathyroid gland, which themselves respond to elevations in serum calcium by an inhibition of parathyroid hormone secretion (Zaidi, 1990). Interestingly, more recent data from these two groups indicate that the threshold for opening of the osteoclast calcium sensor would be at much higher concentrations of Ca than the parathyroid cells, fitting with the very different microenvironmental calcium concentrations seen by these two cell types.

Because a rise in the intracellular Ca^{2+} concentration has been shown to inhibit the bone resorbing activity of osteoclasts (Zaidi, 1990; Miyauchi *et al.*, 1990) as well as to affect the number and distribution of podosomes (Miyauchi *et al.*, 1990), the current hypothesis is that elevations in Ca_i might primarily affect elements of the cytoskeleton and only secondarily the resorbing activity of the cells. This is most interesting in view of the fact that calcitonin, which has been shown to induce an increase in intracellular calcium (Malgaroli *et al.*, 1989) also disorganizes the cytoskeletal elements involved in the motility, attachment, and secretory organization of the osteoclast (Warshafsky *et al.*, 1985; Baron *et al.*, 1990b; Lakkakorpi and Vaananen, 1990). Other examples of the involvement of Ca_i in the regulation of osteoclastic activity are provided by platelet-activating factor (PAF), which elevates Ca_i and mimicks the other effects of calcitonin (Wood *et al.*, 1991) and the effects of RGD-containing peptides and pH, which may increase (Shankar *et al.*, 1992; Argentino *et al.*, 1992) or decrease Ca_i and activate bone resorption (Teti *et al.*, 1989a; Miyauchi *et al.*, 1991b).

Taken together, these results (Teti *et al.*, 1991; Zaidi, 1990) have suggested a model (Fig. 11) by which (1) elevations in Ca_i lead to inactivation of the osteoclast and, conversely, activation of the cell may, in some circumstances, be associated with a decrease in Ca_i, (2) elevation of Ca_i may be achieved by ligand-induced (calcitonin, platelet activating factor, RGD proteins) or ion-induced mechanisms (Ca^{2+}, H^+, K^+), and (3) elevation of Ca concentrations in the sealed-off extracellular bone resorbing compartment would lead to opening of the Ca sensor, elevation of Ca_i, inactivation and detachment of the osteoclast, and thereby diffusion of the mobilized extracellular calcium into the extracellular fluids. The osteoclast would then reattach and go through a second cycle of resorbing activity, thereby explaining the cyclic motility of the cell as well as the multilacunar nature of resorption sites *in vivo* and *in vitro*.

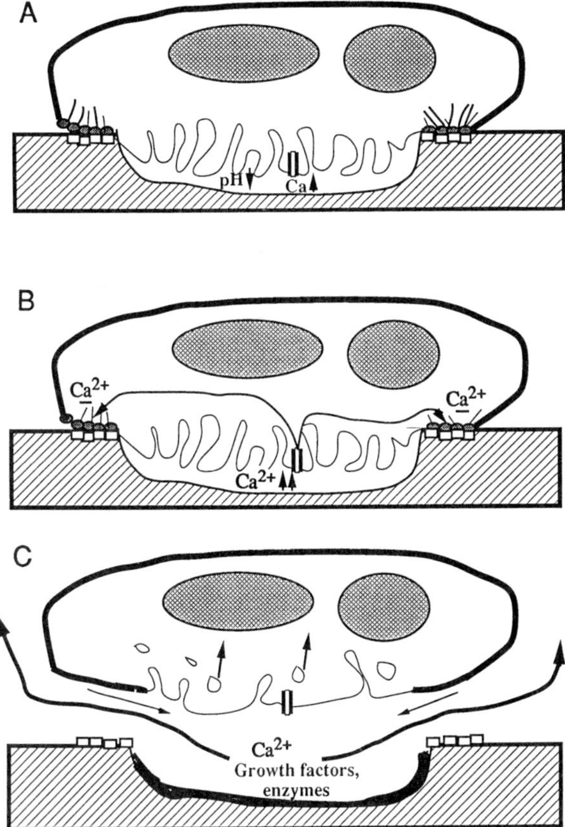

FIGURE 11 Schematic representation of the putative role of a calcium sensor in arresting bone resorption and inducing the detachment and mobility of the osteoclast. (A) Active resorption increases calcium concentration in the resorbing compartment. (B) Calcium sensor in apical membrane opens, increasing intracellular calcium and disrupting the attachment apparatus. (C) The osteoclast detaches, calcium and other factors and enzymes are released in the extracellular fluids, and the apical membrane is internalized. (D) The osteoclast (with only basolateral markers) moves and reattaches nearby to form a second lacuna (seen from above in lower part of figure); the factors and/or matrix at the former resorption lacuna now recruit locally the osteoblast precursors (coupling).

Because clear evidence indicates that the cytoskeleton of the osteoclast can be found in different arrangements (Lakkakorpi and Vaananen, 1991), it is most likely that these correspond to the various steps of the cyclic activity of the osteoclast (Teti *et al.*, 1991; Lakkakorpi and Vaananen, 1991). It is possible that these various functional conformations also correspond to the two levels of membrane potential measured electrophysiologically (Sims and Dixon, 1989).

V. REGULATION OF THE MATURE OSTEOCLAST

In addition to regulating the differentiation of precursors in the monocyte–macrophage lineage into osteoclasts, several cytokines and osteotropic hormones also regulate the activity of the mature osteoclast. Because this question is addressed in detail elsewhere in this volume (Chapter 13), we only mention some of the most important aspects of the hormonal and local regulation of osteoclast function.

The mature osteoclast is directly and negatively regulated by calcitonin, for which the cell expresses a high number of receptors in most species (Warshawsky *et al.*, 1980; Nicholson *et al.*, 1986, 1988). Receptors and/or direct responses to other peptide or steroid hormones involved in calcium or skeletal metabolism, such as parathyroid hormone or vitamin D_3, for instance, have not yet been demonstrated in these cells (Chambers *et al.*, 1985; Merke *et al.*, 1986), although precursors clearly respond to vitamin D_3 (Miyaura *et al.*, 1981; Roodman *et al.*, 1985; Billecocq *et al.*, 1990b) and, possibly, to parathyroid hormone. Recent evidence also suggests that the osteoclast expresses receptors for estrogen (Oursler *et al.*, 1991), potentially linking directly the activity of this cell to postmenopausal osteoporosis. Finally, receptor and nonreceptor tyrosine kinases have been shown to be functionally important in the osteoclast through the observations that osteopetrotic mutants express a defective colony-stimulating factor (CSF) (Yoshida *et al.*, 1990; Felix *et al.*, 1990; Wiktor-Jedrzejczak *et al.*, 1990) and that knock-out of the protooncogene c-*src* leads to an osteopetrotic phenotype (Soriano *et al.*, 1991).

From the point of view of osteoclast biology, while we still know very little about hormone and cytokine receptors in osteoclasts and their effects on bone resorption, it is already known that some of these regulatory events affect the functional determinants described earlier. Thus, calcitonin affects the cytoskeleton and attachment structures (Chambers and Magnus, 1982; Dempster *et al.*, 1987; Lakkakorpi and Vaananen, 1990; Warshafsky *et al.*, 1985), the secretion of enzymes (Vaes, 1972; Baron *et al.*, 1990b), the motility and volume of the osteoclast (Chambers *et al.*, 1984; Chambers and Magnus, 1982; Chambers and Moore, 1983), and some ion-transport systems (Hunter *et al.*, 1988; Chakraborty *et al.*, 1991; Malgaroli *et al.*, 1989). 1,25-Dihydroxyvitamin D_3 ($1,25(OH)_2D_3$), which has a profound influence on the differentiation of bone marrow precursors into osteoclasts (Miyaura *et al.*, 1981; Roodman *et al.*, 1985; Billecocq *et al.*, 1990b; Takahashi *et al.*, 1988c, 1989), induces the expression of several of the phenotypic and functional markers of the osteoclast: multinucleation and TRAP (Miyaura *et al.*, 1981; Roodman *et al.*, 1985), calcitonin receptors and VNRs (Takahashi *et al.*, 1988a; Billecocq *et al.*, 1990a), CAII and the sodium pump (Billecocq *et al.*, 1990b; Billecocq *et al.*, 1990a), and proton pumps (Kurihara *et al.*, 1990).

Similarly, several natural or synthetic compounds that inhibit bone resorption interfere with these functions and vice versa. Hence, inhibition of attachment with RGD peptides (Horton and Davies, 1989; Sato *et al.*, 1990), inhibition of enzymatic degradation with inhibitors of cysteine-proteinases or metalloproteinases (Delaisse and Vaes, 1991; Everts *et al.*, 1992) or with antibodies to TRAP (Andersson *et al.*, 1991), inhibition of carbonic anhydrase (Waite *et al.*, 1970), sodium pump (Prallet *et al.*, 1988), or proton pump activity (Sundquist *et al.*, 1990), and inhibition of bicarbonate–chloride (Hall and Chambers, 1989) or sodium–proton exchange (Hall and Chambers, 1990) all lead to an inhibition of bone resorption.

The predominant feature of hormonal regulation of the osteoclast is the possibility that many regulatory loops are indirect, involving a different target cell, which, in turn, is supposedly secreting or expressing a factor(s) that activates osteoclasts (Rodan and Martin, 1981). Two potential examples of such indirect loops, although this might be more controversial than before, are the effects of both parathyroid hormone (PTH) and $1,25(OH)_2D_3$ on bone resorption, which, based on the evidence that osteoclasts may not express receptors for these hormones, were thought to be mediated via the osteoblasts and/or stromal cells. A few recent reviews have, however, suggested that osteoclasts and/or their precursors may express receptors for and be activated by PTH/parathyroid hormone-related peptide (PTHrP), at least its N-terminal region (Agarwala and Gay, 1992), and be directly and profoundly inhibited by a short sequence (amino acids 107–111) of PTHrP in its C-terminal, non-PTH like region (Fenton *et al.*, 1991), suggesting the existence of a different receptor for this region of the molecule. Early hematopoietic precursors of the osteoclast may also express receptors for vitamin D_3, raising questions as to whether the effects of these hormones on bone resorption are, at least in part, direct. In the same category (i.e., receptors that may be present in the osteoclast but that require further evidence) are nuclear receptors for estrogens (Oursler *et al.*, 1991) and membrane receptors for PAF (Wood *et al.*, 1991) and CSF-1, recently suggested to be present and to mediate biological responses. Osteoclasts may also express receptors for interleukins 4 and 6 (Roodman *et al.*, 1992), but these still remain to be demonstrated.

In contrast, the effects of calcitonin are known to be direct, and the osteoclast expresses large numbers of calcitonin receptors at its surface (Warshawsky *et al.*, 1980; Nicholson *et al.*, 1986, 1988) as well as integrin receptors (Davies *et al.*, 1989; Horton and Davies, 1989) ($\alpha_v\beta_3$ and $\alpha_2\beta_1$), which now emerge as being not only attachment molecules but also receptors for ligand-induced regulation of cellular activity (Miyauchi *et al.*, 1991a). The mechanisms involved in mediating the effects of these agents are discussed elsewhere in this volume (Chapter 13).

VI. ORIGIN AND DIFFERENTIATION OF THE OSTEOCLAST

Finally, a description of the cellular and molecular biology of the osteoclast should include some current indication as to the origin of this cell, because many of the osteoclast's functional characteristics might indeed belong to the family to which it belongs (i.e., the monocyte–macrophage family of cells). As discussed in a recent review by Suda *et al.* (1992), osteoclasts are now widely accepted to be of hematopoietic origin and probably of the macrophage colony-forming unit (CFU-M)-derived monocyte–macrophage family. This conclusion is based on a large body of evidence from the literature, starting with the pioneering experiments of Walker (1973, 1975a,b), who demonstrated that transplants of spleen or bone marrow hematopoietic cells could restore bone resorption in irradiated osteopetrotic mutants. Direct cellular filiation was then demonstrated with the use of quail-chick chimerae (Kahn and Simmons, 1975; Jotereau and LeDouarin, 1978). More recently, *in vitro* experiments have shown that unselected bone marrow cells, colonies of the granulocyte–macrophage colony-forming unit or CFU-M lineage, or, even, mature macrophages can lead to the formation of osteoclasts under adequate culture conditions (Udagawa *et al.*, 1990; Scheven *et al.*, 1986; Nijweide *et al.*, 1986; Burger *et al.*, 1984; Hagenaars *et al.*, 1991). It therefore seems probable that cells of the monocyte–macrophage lineage can form osteoclasts at most stages of their differentiation.

Work done in the last few years has also clearly demonstrated that, to differentiate fully, osteoclasts require interactions with cells of the stromal lineage, to which the osteoblast belongs, and/or their soluble products (Burger *et al.*, 1984; Takahashi *et al*,. 1988b; Udagawa *et al.*, 1989). Interestingly, however, soluble CSF is not sufficient to support osteoclast formation *in vitro*, and direct cell–cell interactions between osteoclast precursors and stromal cells may be necessary (Udagawa *et al.*, 1990; Takahashi *et al.*, 1991). These conclusions, made from *in vitro* experiments, are further supported by the observation that one osteopetrotic mutation, namely the *op/op* mouse, is due to a defect in stroma-derived colony-stimulating activity (Wiktor-Jedrzejczak *et al.*, 1982). In these mutants, a single point mutation in the CSF molecule has been found, and this genetic defect is sufficient to prevent osteoclast formation (Yoshida *et al.*, 1990), which can consequently be restored by treatment with normal CSF (Felix *et al.*, 1990; Wiktor-Jedrzejczak *et al.*, 1990). Similar results have recently been reported in the *tl/tl* rat (Marks *et al.*, 1992).

Besides CSF, a cytokine locally secreted and possibly directly presented by stromal cells, osteoclast differentiation also require osteotropic hormones (Suda *et al.*, 1992). Of particular importance is the active metabolite of vitamin D_3, $1,25(OH)_2D_3$, which seems to represent an absolute requirement for osteoclast differentiation (Miyaura *et al.*, 1981; Takahashi *et al.*, 1988c; Billecocq *et al.*, 1990a). Other osteotropic hormones, such as calcitonin, PTH/PTHrP and estrogens, can modulate osteoclast differentiation but do not seem to be required *in vitro* for the formation of osteoclasts from bone marrow precursors (Miyaura *et al.*, 1986; Takahashi *et al.*, 1988; Akatsu *et al.*, 1989; Suda *et al.*, 1992).

It therefore appears that osteoclasts develop from hematopoietic cells in the monocyte–macrophage lineage and that the formation of the mature resorbing cell requires stromal cell CSF, which possibly needs to be presented by cell–cell interactions, and $1,25(OH)_2D_3$. Mononuclear precursors progressively express all of the phenotypic markers of the mature cell, migrate, and attach to bone, even before fusing into multinucleated osteoclasts (Suda *et al*,. 1992; Baron *et al.*, 1986b).

VII. A NEW CHALLENGE: THE PROTOONCOGENE *c-src* IS REQUIRED FOR NORMAL OSTEOCLAST FUNCTION

Recently, a very striking discovery was made in osteoclast biology, relating the protooncogene c-*src*, a nonreceptor tyrosine kinase that is involved in signal transduction and expressed at high levels in neurons and platelets, to bone resorption; deletion of the gene encoding c-*src* led to osteopetrosis, thereby suggesting that it plays a critical role in bone resorption (Soriano *et al.*, 1991).

The c-*src* protooncogene is highly conserved throughout evolution and widely expressed (Cooper, 1989; Eiseman and Bolen, 1990). Although the level of expression is low in most cells, some cell types, particularly neurons (Cotton and Brugge, 1983) and blood platelets (Varshney *et al.*, 1986; Golden *et al.*, 1986), express the gene product ($pp60^{c\text{-}src}$) and the c-*src* kinase activity at high levels. Although the physiological role of $pp60^{c\text{-}src}$ is not fully understood (Parsons and Weber, 1989; Cooper, 1989; Eiseman and Bolen, 1990), it is known that $pp60^{c\text{-}src}$ and the other members of the Src family, which share highly conserved sequences both within and outside the kinase catalytic domain, play important roles in signal transduction mechanisms that contribute to the regulation of cell growth and development (Cantley *et al.*, 1991; Cooper, 1989). They can be activated by various transmembrane tyrosine kinase receptors, such as the platelet-derived growth factor receptor (Kypta *et al.*, 1990), and, in turn, phosphorylate various substrate proteins on tyrosine residues, including the ras guanosine triphosphatease-activating protein (Ellis *et al.*, 1990; Cooper, 1989), one of the subunits of the IP3-kinase (p85) (Gutkind *et al.*, 1990; Fukui *et al.*, 1988) and two proteins (M_r = 80 and 85 kDa) that co-localize with F-actin in peripheral extensions of normal cells and in matrix attachment structures (podosomes) in v-*src*-transformed cells (Wu *et al.*, 1991).

To better understand the physiological role of pp60$^{c\text{-}src}$, Soriano *et al.* (1991) performed the targeted disruption of its encoding gene by homologous recombination in mouse embryos. Surprisingly, cell proliferation and other basal functions did not seem to be altered in the mutant c-*src*$^{-}$ mice, and no obvious phenotypic or functional abnormalities were noticed in brain and other neuronal tissues or in platelets, possibly due to the presence in these cells of other src-like tyrosine kinases (e.g., c-*fyn*, c-*yes*), which might impart a degree of functional redundancy with pp60$^{c\text{-}src}$ (Soriano *et al.*, 1991). Unexpectedly, striking skeletal abnormalities with a phenotype of osteopetrosis were observed in the recombinant mice. These included a failure of the incisors to erupt, a slower growth, shorter and abnormally shaped long bones, and a decreased bone marrow cavity. All these changes are characteristic of impaired osteoclast activity (Marks, 1984). The presence of osteoclasts in apparently normal numbers in the c-*src*$^{-}$ animals suggested that the defect was with the function rather than the differentiation of the cells. These data therefore suggested a critical role of normal c-*src* expression in osteoclast function, and the lack of other phenotypic changes in these mutant mice suggested that the requirement for normal c-*src* expression in bone was more stringent than in other tissues. The failure of osteoclasts of c-*src*$^{-}$ mice to resorb bone at a normal rate could, however, be due either to alterations within the osteoclasts or to alterations in other cells, which would be necessary for the complete differentiation or activation of the osteoclast (Suda *et al.*, 1992). Recent data strongly point to c-src playing its critical role within the osteoclast itself. First, pp60$^{c\text{-}src}$ is expressed at high levels in osteoclasts, where it is associated predominantly with intracellular organelles but not in other bone cells (Horne *et al.*, 1992). Osteoclasts were also found to express c-*fyn*, c-*lyn*, and c-*yes*, three other members of the Src family of proteins that, in the osteoclast, therefore cannot compensate for the role of c-*src* (Horne *et al.*, 1992). Culturing bone marrow cells with $1,25(OH)_2D_3$, which induces differentiation into osteoclastlike cells (Billecocq *et al.*, 1990a,b), resulted in increased expression of pp60$^{c\text{-}src}$. These results, taken together with the observation of Soriano *et al.* (1991), suggest that pp60$^{c\text{-}src}$ is part of the osteoclast phenotype and may perform a function associated with intracellular vesicles in the osteoclast that is essential for normal bone resorption. These results clearly demonstrate a direct link between the osteoclast and c-*src* expression and strongly suggest, but by no means prove, that the osteoclast may be the site of the specific defect in c-*src*$^{-}$ mice that leads to impaired osteoclast function and osteopetrosis. This possibility is further supported by the ability of Soriano, Mundy, and colleagues to reverse the osteopetrotic condition of the *src*$^{-}$ mice by transplanting hematopoietic tissues (liver or bone marrow) from normal mice (G. Mundy and P. Soriano, personal communication).

While these initial studies did not attempt to identify specific function(s) of pp60$^{c\text{-}src}$ in osteoclastic bone resorption, some possibilities are suggested by considering the immunolocalization results (Horne *et al.*, 1992) and the known characteristics of the protein in other cells. The higher concentration of the protein was not found at the plasma membrane but, rather, at the limiting membrane of intracellular vesicles. This is in contrast to the situation in platelets, where the majority of the pp60$^{c\text{-}src}$ is associated with the plasma membrane and the membrane of the surface-connected canalicular system (Ferrell *et al.*, 1990). The association of pp60$^{c\text{-}src}$ with secretory granules and vesicles has also been reported in platelets (Rendu *et al.*, 1989), chromaffin cells (Parsons and Creutz, 1986; Grandori and Hanafusa, 1988), and neurons (Hirano *et al.*, 1988; Barnekow *et al.*, 1990), further suggesting that the protein may contribute to vesicle targeting or membrane fusion. In the osteoclast, targeted delivery of intracellular vesicles to the apical surface of the cell is required in order to secrete enzymes into the resorption compartment and to insert the proton pump and other transporters involved in acid secretion at the apical ruffled border membrane (Baron *et al.*, 1992). Although the apparent association of pp60$^{c\text{-}src}$ with intracellular vesicles in the osteoclast is consistent with this possible function, further work will be required to clarify this issue.

It has also been suggested that pp60$^{c\text{-}src}$ may be involved in integrin-related attachment and signal transduction mechanisms, another potential role that is of particular interest with regard to osteoclasts. In platelets, inhibiting the binding of the integrin GP IIb-IIIa to fibrinogen blocks agonist-induced tyrosine kinase activity (Golden *et al.*, 1990), whereas in epidermal carcinoma cells antibody-mediated clustering of integrin $\alpha_5\beta_1$ activates tyrosine kinases (Kornberg *et al.*, 1991). In osteoclasts, integrins, specifically $\alpha_v\beta_3$, are involved in bone resorption (Davies *et al.*, 1989), probably mediating the attachment of the osteoclast to the bone matrix (Zambonin-Zallone *et al.*, 1989) and participating in the sealing off of the extracellular bone resorbing compartment. Although pp60$^{v\text{-}src}$ is known to associate with focal points of adhesion in v-*src*-transformed fibroblasts (Rohrschneider, 1980), pp60$^{c\text{-}src}$ never appears in focal adhesion plaques or podosomes in NIH overexpresser cells (David-Pfeuty and Nouvian-Dooghe, 1990). In agreement with these results, confocal microscopic results fail to show a specific association of pp60$^{c\text{-}src}$ with adhesion structures (Horne *et al.*, 1992). Integrin binding to bone matrix might nevertheless involve pp60$^{c\text{-}src}$ and provide a signal leading to the development of cell polarity and the formation of the ruffled border at the site of bone resorption. Consistent with such a possible involvement of protein tyrosine kinases in integrin function in osteoclasts, preliminary studies in our laboratory (Neff *et al.*, 1992) indeed suggest that a rapid, transient wave of tyrosine phosphorylation is triggered in isolated osteoclasts upon exposure to RGD-containing peptides that bind to integrin.

Further elucidation of the functional role of c-*src* in osteoclasts will definitely be one of the major themes of research in this field in the next few years and will undoubtedly lead to major advances in our understanding of the mechanisms of bone resorption and even, possibly, to unraveling the physiological role that c-*src* plays in all cells.

VIII. SUMMARY AND CONCLUSION

The osteoclast is a multinucleated giant cell formed by the fusion of mononuclear precursors differentiated from cells of the monocyte–macrophage hematopoietic lineage. The main determinants of the biology of the osteoclast are first its attachment to the bone matrix, leading to the formation of the sealed-off bone resorbing compartment and, second, the polarized acidification of, and secretion of enzymes into, this compartment. These activities require tight control of different membrane domains both in terms of their composition and targeting and in terms of ion transport. The biology and activity of these membrane domains are strongly interdependent. The differentiation of the osteoclast is regulated by stromal cells and by several calciotropic hormones and cytokines; of particular importance are $1,25(OH)_2D_3$ and CSF-1. The activity of the fully mature osteoclast is, directly or indirectly, regulated by soluble factors and hormones, by molecules of the extracellular matrix, and by its ionic environment. Finally, the coupling of bone formation to bone resorption (i.e., the local recruitment of osteoblasts at sites where the osteoclast has resorbed bone), may depend on the activity of the osteoclast itself. The maintenance of a normal skeletal shape and mass therefore closely depends on a normal osteoclastic activity, as do bone growth and repair. Furthermore, the improvement of our means of therapeutic intervention with antiresorptive agents will depend on our better understanding of the molecular basis of bone resorption.

REFERENCES

Agarwala, N., and Gay, C. V. (1992). Specific binding of parathyroid hormone to living osteoclasts. *J. Bone Miner. Res.* **7,** 531–539.

Akatsu, T., Takahashi, N., Udagawa, N., Sato, K., Nagata, N., Moseley, J. M., Martin, T. J., and Suda, T. (1989). Parathyroid hormone (PTH)-related protein is a potent stimulator of osteoclast-like multinucleated cell formation to the same extent as PTH in mouse marrow cultures. *Endocrinology* **125,** 20–27.

Al-Awqati, Q. (1986). Proton-translocating ATPases. *Cell Biol.* **2,** 179–199.

Ali, N. N., Boyde, A., and Jones, S. J. (1984). Motility and resorption: Osteoclastic activity in vitro. *Anat. Embryol.* **170,** 51–56.

Anderson, R. E., Woodbury, D. M., and Jee, W. S. S. (1986). Humoral and ionic regulation of osteoclast acidity. *Calcif. Tissue Int.* **39,** 252–258.

Andersson, G., Ek-Rylander, B., Hammarstrom, L. E., Lindskog, S., and Toverud, S. U. (1986). Immunocytochemical localization of a tartrate-resistant and vanadate-sensitive acid nucleotide tri- and diphosphatase. *J. Histochem. Cytochem.* **34,** 293.

Andersson, G., Ek-Rylander, B., and Minkin, C. (1992). Acid phosphatases. *In* "The Biology and Physiology of the Osteoclast" (B. R. Rifkin and C. V. Gay, eds.), in press. CRC Press, Boca Raton, Florida.

Argentino, L., Colucci, S., Grano, M., Barattolo, R., Zambonin-Zallone, A., and Teti, A. (1992). Phosphorylation pathways regulate extracellular calcium-induced signal transduction mechanisms in mammalian osteoclasts. *Bone Min.* **17**(*Suppl. 1*), 82.

Barnekow, A., Jahn, R., and Schartl, M. (1990). Synaptophysin: A substrate for the protein tyrosine kinase pp60$^{c\text{-}src}$ in intact synaptic vesicles. *Oncogene* **5,** 1019–1024.

Baron, R. (1989). Molecular mechanisms of bone resorption by the osteoclast. *Anat. Rec.* **224,** 317–324.

Baron, R., Neff, L., Lippincott-Schwartz, J., Louvard, D., Mellman, I., Helenius, A., and Marsh, M. (1985a). Distribution of lysosomal membrane proteins in the osteoclast and their relationship to acidic compartments. *J. Cell Biol.* **101,** 53a.

Baron, R., Neff, L., Louvard, D., and Courtoy, P. J. (1985b). Cell-mediated extracellular acidification and bone resorption: Evidence for a low pH in resorbing lacunae and localization of a 100 kD lysosomal membrane protein at the osteoclast ruffled border. *J. Cell Biol,* **101,** 2210–2222.

Baron, R., Neff, L., Roy, C., Boisvert, A., and Caplan, M. (1986a). Evidence for a high and specific concentration of (Na+,K+)ATPase in the plasma membrane of the osteoclast. *Cell* **46,** 311–320.

Baron, R., Neff, L., Tran Van, P., Nefussi, J.-R., and Vignery, A. (1986b). Kinetic and cytochemical identification of osteoclast precursors and their differentiation into multinucleated osteoclasts. *Am. J. Pathol.* **122,** 363–378.

Baron, R., Neff, L., Brown, W., Courtoy, P. J., Louvard, D., and Farquhar, M. G. (1988). Polarized secretion of lysosomal enzymes: Co-distribution of cation-independent mannose-6-phosphate receptors and lysosomal enzymes along the osteoclast exocytic pathway. *J. Cell Biol.* **106,** 1863–1872.

Baron, R., Eeckhout, Y., Neff, L., Francois-Gillet, C., Henriet, P., Delaisse, J. M., and Vaes, G. (1990a). Affinity purified antibodies reveal the presence of (pro)collagenase in the subosteoclastic bone resorbing compartment. *J. Bone Miner. Res.* **5,** S203.

Baron, R., Neff, L., Brown, W., Louvard, D., and Courtoy, P. J. (1990b). Selective internalization of the apical plasma membrane and rapid redistribution of lysosomal enzymes and mannose 6-phosphate receptors during osteoclast inactivation by calcitonin. *J. Cell Sci.* **97,** 439–447.

Baron, R., Chakraborty, M., Chatterjee, D., Horne, W., Lomri, A., and Ravesloot, J.-H. (1993). Biology of the osteoclast. *In* "Physiology and Pharmacology of Bone" (G. R. Mundy and T. J. Martin, eds.). Springer, New York. (In press).

Bax, B. E., Towhidul Alam, A. S. M., Banerji, B., Bax, C. M. R., Bevis, P. J. R., Stevens, C. R., Moonga, B. S., Blake, D. R., and Zaidi, M. (1992). Stimulation of osteoclastic bone resorption by hydrogen peroxide. *Biochem. Biophys. Res. Commun.* **183,** 1153–1158.

Beard, C. J., Key, L. L., Newburger, P. E., Ezekowitz, A. B., Arceci, R., Miller, B., Proto, P., Ryan, T., Anast, C. S., and Simons, E. R. (1986). Neutrophil defect associated with malignant osteopetrosis. *J. Lab. Clin. Med.* **109,** 498–505.

Bekker, P. J., and Gay, C. V. (1990a). Characterization of a calcium ATPase in osteoclast plasma membrane. *J. Bone Miner. Res.* **5,** 557–567.

Bekker, P. J., and Gay, C. V. (1990b). Biochemical characterization of an electrogenic vacuolar proton pump in purified chicken osteoclast plasma membrane vesicles. *J. Bone Miner. Res.* **5,** 569–579.

Billecocq, A., Emanuel, J., Jamsa-Kellokumpu, S., Prallet, B., Levenson, R., and Baron, R. (1990a). 1,25(OH)2D3 induces the concomittant expression of the vitronectin receptor, sodium pumps and carbonic anhydrase II in avian bone marrow cells. *In* "Calcium Regulation and Bone Metabolism" (D. V. Cohn, F. H. Glorieux, and T. J. Martin, eds.), pp. 152–156. Elsevier, Amsterdam.

Billecocq, A., Rettig-Emanuel, J., Levenson, R., and Baron, R. (1990b). 1a,25-Dihydroxyvitamin D_3 regulates the expression of carbonic anhydrase II in non-erythroid avian bone marrow cells. *Proc. Natl. Acad. Sci. USA* **87,** 6470–6474.

Blair, H. C., Kahn, A. J., Crouch, E. C., Jeffrey, J. J., and Teitelbaum, S. L. (1986). Isolated osteoclasts resorb the organic and inorganic components of bone. *J. Cell Biol.* **102,** 1164–1172.

Blair, H. C., Teitelbaum, S. L., Schimke, P. A., Konsek, J. D., Koziol, C. M., and Schlesinger, P. H. (1988). Receptor-mediated uptake of a mannose-6-phosphate bearing glycoprotein by isolated chicken osteoclasts. *J. Cell. Physiol.* **137,** 476–482.

Blair, H. C., Teitelbaum, S. L., Ghiselli, R., and Gluck, S. (1989). Osteoclastic bone resorption by a polarized vacuolar proton pump. *Science* **245,** 855–857.

Blair, H. C., Teitelbaum, S. L., Tan, H. L., Koziol, C. M., and Schlesinger, P. H. (1991). Passive chloride permeability charge coupled to H+ ATPase of avian osteoclast ruffled membrane. *Am. J. Physiol.* **260,** C1315–C1324.

Bowman, E. J., Siebers, A., and Altendorf, K. (1988). Bafilomycins: A class of inhibitors of membrane ATPases from microorganisms, animal cells, and plant cells: (Membrane ATPase/vacuolar ATPase/macrolide antibiotics). *Proc. Natl. Acad. Sci. USA* **85,** 7972–7976.

Brown, D., Gluck, S., and Hartwig, J. (1987). Structure of the novel membrane-coating material in proton-secreting epithelial cells and identification as an H+ ATPase. *J. Cell Biol.* **105,** 1637–1648.

Brown, W. J., and Farquhar, M. G. (1984). The mannose-6-phosphate receptor for lysosomal enzymes is concentrated in cis Golgi cisternae. *Cell* **36,** 295–307.

Brown, W. J., and Farquhar, M. G. (1987). The distribution of 215 kD mannose-6-phosphate receptors within cis (heavy) and trans (light) Golgi subfractions varies in different cell types. *Proc. Natl. Acad. Sci. USA* **84,** 9001–9005.

Brown, W. J., Goodhouse, J., and Farquhar, M. G. (1986). Mannose-6-phosphate receptors for lysosomal enzymes cycle between the Golgi complex and endosomes. *J. Cell Biol.* **103,** 1235–1247.

Burger, H. C., van der Meer, J. W. M., and Nijweide, P. J. (1984). Osteoclast formation from mononuclear phagocytes: Role of bone-forming cells. *J. Cell Biol.* **99,** 1901–1906.

Cantley, L. C., Auger, K. R., Carpenter, C., Duckworth, B., Graziani, A., Kapeller, R., and Soltoff, S. (1991). Oncogenes and signal transduction. *Cell* **64,** 281–302.

Chakraborty, M., Su, Y., Nathanson, M., Rubega-Male, A., Slayman, C., and Baron, R. (1991). The effects of calcitonin in rat osteoclasts and in a kidney cell line (LLC-PK1) are mediated via an inhibition of the Na+/H+ antiporter. *J. Bone Miner. Res.* **6,** S134.

Chambers, T. J., and Magnus, C. J. (1982). Calcitonin alters behavior of isolated osteoclasts. *J. Pathol.* **136,** 27–39.

Chambers, T. J., and Moore, A. (1983). The sensitivity of isolated osteoclasts to morphological transformation by calcitonin. *J. Clin. Endocrinol. Metab.* **57,** 819–824.

Chambers, T. J., Athanasou, N. A., and Fuller, K. (1984). Effect of parathyroid hormone and calcitonin on the cytoplasmic spreading of isolated osteoclasts. *J. Endocrinol.* **102,** 281–286.

Chambers, T. J., McSheehy, P. M., Thomson, B. M., and Fuller, K. (1985). The effect of calcium regulating hormones and prostaglandins on bone resorption by osteoclasts disaggregated from neonatal rabbit bones. *Endocrinology* **116,** 224–239.

Chambers, T. J., Fuller, K., Carby, J. A., Pringle, J. A. S., and Horton, M. A. (1986). Monoclonal antibodies against osteoclasts inhibit bone resorption in vitro. *Bone Miner.* **1,** 127–135.

Chambers, T. J., Fuller, K., and Darby, J. A. (1987). Hormonal regulation of acid phosphatase release by osteoclasts disaggregated from neonatal rat bone. *J. Cell. Physiol.* **132,** 90–96.

Chatterjee, D., Leit, M., Neff, L., Chakraborty, M., Jamsa-Kellokumpu, S., Fuchs, R., and Baron, R. (1991). A new and specific type of vacuolar proton pump is present at the osteoclast ruffled-border membrane. *J. Bone Miner. Res.* **6,** S197.

Chatterjee, D., Neff, L., Chakraborty, M., Leit, M., Jamsa-Kellokumpu, S., Fuchs, R., and Baron, R. (1992). Sensitivity to vanadate and isoforms of subunits A and B distinguish the osteoclast proton-pump from other vacuolar H^+ ATPases *Proc. Natl. Acad. Sci. USA* **89,** 6257–62.

Cooper, J. A. (1989). The src family of protein-tyrosine kinases. *In* "Peptides and Protein Phosphorylation" (B. Kemp and P. F. Alewood, eds.), pp. 85–113. CRC Press, Boca Raton, Florida.

Cotton, P. C., and Brugge, J. S. (1983). Neural tissues express high levels of the cellular src gene product pp60[super]c-src. *Mol. Cell. Biol.* **3,** 1157–1162.

Creek, K. E., and Sly, W. S. (1984). The role of the phosphomannosyl receptor in the transport of acid hydrolases to lysosomes. *In* "Lysosomes in Biology and Pathology" (J. T. Dingle, R. T. Dean, and W. S. Sly, eds.) pp. 63–82. Elsevier, Amsterdam.

David-Pfeuty, T., and Nouvian-Dooghe, Y. (1990). Immunolocalization of the cellular src protein in interphase and mitotic c-src overexpressor cells. *J. Cell Biol.* **111,** 3097–3116.

Davies, J., Warwick, J., Totty, N., Philp, R., Helfrich, M., and Horton, M. (1989). The osteoclast functional antigen, implicated in the regulation of bone resorption, is biochemically related to the vitronectin receptor. *J. Cell. Biol.* **109,** 1817–1826.

Delaisse, J. M., and Vaes, G. (1992). Mechanism of mineral solubilization and matrix degradation in osteoclastic bone resorption. *In* "The Biology and Physiology of the Osteoclast" (B. R. Rifkin and C. V. Gay, eds.), in press. CRC Press, Boca Raton, Florida.

Delaisse, J. M., Ledent, P., and Vaes, G. (1991). Collagenolytic cysteine proteinases of bone tissue. *Biochem. J.* **279,** 167–174.

Delaisse, J. M., Neff, L., Eeckhout, Y., Su, T., Vaes, G., and Baron, R. (1992). Evidence for the presence of (pro)collagenase in osteoclasts. *Bone Miner.* **17**(Suppl.), 82 (abstract).

Dempster, D. W., Murrills, R. J., Herbert, W. R., and Arnett, T. R. (1987). Biological activity of chicken calcitonin: Effects on neonatal rat and embryonic chicks osteoclasts. *J. Bone Miner. Res.* **2,** 443–448.

Doty, S. B., and Schofield, B. H. (1972). Electron microscopic localization of hydrolytic enzymes in osteoclasts. *Histochem. J.* **4,** 245–258.

Eeckhout, Y., and Vaes, G. (1977). Further studies on the activation of procollagenase, the latent precursor of bone collagenase. Effects of lysosomal cathepsin B, plasmin and kallikrein, and spontaneous activation. *Biochem. J.* **166,** 21–31.

Eiseman, E., and Bolen, J. B. (1990). src-related tyrosine protein kinases as signaling components in hematopoietic cells. *Cancer Cells* **2,** 303–310.

Ek-Rylander, B., Bill, P., Norgard, M., Nilsson, S., and Andersson, G. (1991). Cloning, sequence and developmental expression of a type 5, tartrate-resistant, acid phosphatase of rat bone. *J. Biol. Chem.* **266,** 24684–24689.

Ellis, C., Moran, M., McCormick, F., and Pawson, T. (1990). Phosphorylation of GAP and GAP-associated proteins by transforming and mitogenic tyrosine kinases. *Nature (London)* **343,** 377–381.

Everts, V., Delaisse, J. M., Korper, W., Niehof, A., Vaes, G., and Beertsen, W. (1992). The degradation of collagen in the bone-resorbing compartment underlying the osteoclast involves both cysteine-proteinases and matrix metalloproteinases. *J. Cell. Physiol.* **150,** 221–231.

Felix, R., Cecchini, M. G., and Fleisch, H. (1990). Macrophage colony stimulating factor restores in vitro bone resorption in the op/op osteopetrotic mouse. *Endocrinology* **127,** 2592–2594.

Fenton, A. J., Kemp, B. E., Kent, G. N., Moseley, J. M., Zheng, M.-H., Rowe, D. J., Britto, J. M., Martin, T. J., and Nicholson, G. C. (1991). A carboxyl-terminal peptide from the parathyroid hormone-related protein inhibits bone resorption by osteoclasts. *Endocrinology* **129,** 1762–1768.

Ferrell, J. E., Jr., Noble, J. A., Martin, G. S., Jacques, Y. V., and Bainton, D. F. (1990). Intracellular localization of pp60$^{c\text{-}src}$ in human platelets. *Oncogene* **5,** 1033–1036.

Forgac, M. (1989). Structure and function of vacuolar class of ATP-driven proton pumps. *Physiol. Rev.* **69,** 765–796.

Fuchs, R., Male, P., and Mellman, I. (1989). Acidification and ion permeabilities of highly purified rat liver endosomes. *J. Biol. Chem.* **264,** 2212–2220.

Fukui, Y., Kornbluth, S., Jong, S. M., Wang, L. H., and Hanafusa, H. (1988). Phosphatidylinositol kinase type I activity associates with various oncogene products. *Oncogene Res.* **4,** 283–292.

Garrett, I. R., Boyce, B. F., Oreffo, R. O. C., Bonewald, L., Poser, J., and Mundy, G. R. (1990). Oxygen-derived free radicals stimulate osteoclastic bone resorption in rodent bone in vitro and in vivo. *J. Clin. Invest.* **85,** 632–639.

Gay, C. V., and Mueller, W. J. (1974). Carbonic anhydrase and osteoclasts: Localization by labelled inhibitor autoradiography. *Science* **183,** 432–434.

Golden, A., Nemeth, S. P., and Brugge, J. S. (1986). Blood platelets express high levels of the pp60c-src-specific tyrosine kinase activity. *Proc. Natl. Acad. Sci. USA* **83,** 852–856.

Golden, A., Brugge, J. S., and Shattil, S. J. (1990). Role of platelet membrane glycoprotein IIb-IIIa in agonist-induced tyrosine phosphorylation of platelet proteins. *J. Cell Biol.* **111,** 3117–3127.

Grandori, C., and Hanafusa, H. (1988). P60$^{c\text{-}src}$ is complexed with a cellular protein in subcellular compartments involved in exocytosis. *J. Cell Biol.* **107,** 2125–2135.

Grills, B. L., Gallagher, J. A., Allan, E. H., Yumita, S., and Martin, T. J. (1990). Identification of plasminogen activator in osteoclasts. *J. Bone Miner. Res.* **5,** 499–505.

Gutkind, J. S., Lacal, P. M., and Robbins, K. C. (1990). Thrombin-dependent association of phosphatidylinositol-3 kinase with p60c-src and p59c-fyn in human platelets. *Mol. Cell. Biol.* **10,** 3806–3809.

Hagenaars, C. E., Kawilarang-de Haas, E. W. M., van der Kraan, A. A. M., Spooncer, E., Dexter, T. M., and Nijweide, P. J. (1991). Interleukin-3-dependent hematopoietic stem cell lines capable of osteoclast formation in vitro. *J. Bone Miner. Res.* **6,** 947–954.

Hall, T. J., and Chambers, T. J. (1989). Optimal bone resorption by isolated rat osteoclasts requires chloride/bicarbonate exchange. *Calcif. Tissue Int.,* **45,** 378–380.

Hall, T. J., and Chambers, T. J. (1990). Na+/H+ antiporter is the primary proton transport system used by osteoclasts during bone resorption. *J. Cell Physiol.* **142,** 420–424.

Hammarstrom, L. E., Hanker, J. S., and Toverud, S. U. (1971). Cellular differences in acid phosphatase isoenzymes in bone and teeth. *Clin. Orthop.* **78,** 151.

Hell, J. W., Maycox, P. R., Stadler, H., and Jahn, R. (1988). Uptake of GABA by rat brain synaptic vesicles isolated by a new procedure. *EMBO J.* **7,** 3023–3029.

Hilliard, T. J., Meadows, G., and Kahn, A. J. (1990). Lysozyme synthesis in osteoclasts. *J. Bone Miner. Res.* **5,** 1217–1222.

Hirano, A. A., Greengard, P., and Huganir, R. L. (1988). Protein tyrosine kinase activity and its endogenous substrates in rat brain: A subcellular and regional survey. *J. Neurochem.* **50,** 1447–1455.

Holtrop, M. E., and King, G. J. (1977). The ultrastructure of the osteoclast and its functional implications. *Clin. Orthop. Relat. Res.* **123,** 177–196.

Horne, W., Moya, M., Neff, L., Chatterjee, D., Kellokumpu, S., Kopito, R., Alper, S., and Baron, R. (1991). A band 3-related chloride/bicarbonate exchanger is highly expressed at the basolateral membrane of osteoclasts. *J. Bone Miner. Res.* **6,** S95.

Horne, W., Neff, L., Chatterjee, D., Lomri, A., Levy, J. B., and Baron, R. (1992). Osteoclasts express high levels of pp60 *c-src* in association with intracellular organelles. *J. Cell Biol.* **119,** 1003–1013.

Horton, M. A., and Davies, J. (1989). Perspectives: Adhesion receptors in bone. *Bone Miner.* **4,** 803–807.

Hosoi, T., Asaka, T., Tomita, T., Takeda, J., Ouchi, Y., and Orimo, H. (1991). Demonstration of local delivery and activation of TGF-beta by osteoclasts in the resorption lacunae. *J. Bone Miner. Res.* **6,** S264.

Hunter, S. J., Schraer, H., and Gay, C. V. (1988). Characterization of isolated and cultured chick osteoclasts: The effects of acetazolamide, calcitonin and PTH on acid production. *J. Bone Miner. Res.* **3,** 297–303.

Hunter, S. J., Schraer, H., and Gay, C. V. (1989). Characterization of the cytoskeleton in isolated chick osteoclasts: Effects of calcitonin. *J. Histochem. Cytochem.* **37,** 1529–1537.

Hynes, R. O. (1987). Integrins: A family of cell surface receptors. *Cell* **48,** 549–554.

Hynes, R. O. (1992). Integrins: Versatility, modulation, and signaling in cell adhesion. *Cell* **69,** 11–25.

Jotereau, F. V., and LeDouarin, N. M. (1978). The developmental relationships between osteocytes and osteoclasts. A study using the quail-chick nuclear markers in endochondral ossification. *Dev. Biol.* **63,** 253–265.

Kahn, A. J., and Simmons, D. J. (1975). Investigation of cell lineage in bone using a chimera of chick and quail embryonic tissue. *Nature (London)* **258,** 325–327.

Kallio, D. M., Garant, P. R., and Minkin, D. (1971). Evidence of coated membranes in the ruffled border of the osteoclast. *J. Ultrastruct. Res.* **37,** 169–177.

Kanehisa, J., and Heersche, J. N. M. (1988). Osteoclastic bone resorption: In vitro analysis of the rate of resorption and migration of individual osteoclasts. *Bone,* **9,** 73–79.

Kanehisa, J., Yamanaka, T., Doi, S., Turksen, K., Heersche, J. N. M., Aubin, J. E., and Takeuchi, H. (1990). A band of F-actin containing podosomes is involved in bone resorption by osteoclasts. *Bone* **11,** 287–293.

Key, L. L., Ries, W. L., Taylor, R. G., Hays, B. D., and Pitzer, B. L. (1990). Oxygen-derived free radicals in osteoclasts: The specificity and location of the nitroblue tetrazolium reaction. *Bone* **11,** 115–119.

King, G. J., and Holtrop, M. E. (1975). Actin-like filaments in bone cells of cultured calvaria as demonstrated by binding to heavy meromyosin. *J. Cell Biol.* **66,** 445–451.

Kornberg, L. J., Earp, H. S., Turner, C. E., Prockop, C., and Juliano, R. L. (1991). Signal transduction by integrins: Increased protein tyrosine phosphorylation caused by clustering of integrins. *Proc. Natl. Acad. Sci. USA* **88,** 8392–8396.

Kornfeld, S. (1986). Trafficking of lysosomal enzymes in normal and disease states. *J. Clin. Invest.* **77,** 1–6.

Kurihara, N., Gluck, S., and Roodman, G. D. (1990). Sequential expression of phenotype markers for osteoclasts during differentiation of precursors for multinucleated cells formed in long term human marrow cultures. *Endocrinology* **127,** 3215–3221.

Kypta, R. M., Goldberg, W., Ulug, E. T., and Courtneidge, S. A. (1990). Association between the PDGF receptor and members of the src family of tyrosine kinases. *Cell* **62,** 481–492.

Lakkakorpi, p., and Vaananen, H. K. (1990). Calcitonin, PGE2 and dibutyril-cAMP disperse the specific microfilament structure in resorbing osteoclasts. *J. Histochem. Cytochem.* **38,** 1487–1493.

Lakkakorpi, P. T., and Vaananen, H. K. (1991). Kinetics of the osteoclast cytoskeleton during the resorption cycle in vitro. *J. Bone Miner. Res.* **6,** 817–826.

Lakkakorpi, p., Tuukkanen, J., Hentunen, T., Jarvelin, K., and Vaananen, K. (1989). Organization of osteoclast microfilaments during the attachment to bone surface in vitro. *J. Bone Miner. Res.* **4,** 817–825.

Lakkakorpi, p., Horton, M. A., Helfrich, M. H., Karhukorpi, E.-K., and Vaananen, H. K. (1991). Kinetic and confocal microscopic studies of the microfilaments and vitronectin receptor in osteoclasts. *J. Bone Miner. Res.* **6,** S149.

Lucht, U. (1971). Acid phosphatase of osteoclasts demonstrated by electron microscopic histochemistry. *Histochemie* **28,** 103–117.

MacIntyre, I., Zaidi, M., Towhidul-Alam, A. S. M., Datta, H. K., Moonga, B. S., Lidbury, P. S., Hecker, M., and Vane, J. R. (1991). Osteoclastic inhibition: An action of nitric oxide not mediated by cyclic GMP. *Proc. Natl. Acad. Sci. USA* **88,** 2936–2940.

Malgaroli, A., Meldolesi, J., Zallone, A. Z., and Teti, A. (1989). Control of cytosolic free calcium in rat and chicken osteoclasts: The role of extracellular calcium and calcitonin. *J. Biol. Chem.* **264,** 14342–14347.

Marchisio, P. C., Naldini, L., Cirillo, D., Primavera, M. V., Teti, A., and Zambonin-Zallone, A. (1984). Cell-substratum interactions of cultured avian osteoclasts is mediated by specific adhesion structures. *J. Cell Biol.* **99,** 1696–1705.

Marchisio, P. C., Cirillo, D., Teti, A., Zambonin-Zallone, A., and Tarone, G. (1987). Rous sarcoma virus transformed fibroblasts and cells of monocytic origin display a peculiar dot-like organization of cytoskeletal proteins involved in microfilament-membrane interactions. *Exp. Cell Res.* **169,** 202–214.

Maren, T. H. (1967). Carbonic anhydrase in the animal kingdom: Chemistry, physiology and inhibition. *Physiol. Rev.* **47,** 595–781.

Marks, S. C., Jr. (1984). Congenital osteopetrotic mutations as probes of the origin, structure, and function of osteoclasts. *Clin. Orthop.* **189,** 239–263.

Marks, S. J., Jr., Wojtowicz, A., Szperl, M., Urbanowska, E., Mackay, C. A., Wiktor-Jedrzejczak, W., Stanley, E. R., and Aukerman, S. L. (1992). Administration of colony stimulating factor-1 corrects some macrophage, dental, and skeletal defects in an osteopetrotic mutation (toothless, tl) in the rat. *Bone* **13,** 89–93.

Masi, L., Brandi, M. L., Gehron-Robey, P., Kerr, J. M., Young, M. F., Bernabei, P. A., and Yanagishita, M. (1991). Bone sialoprotein (BSP) expression in human monoblastic cell line (FLG 29.1). *J. Bone Miner. Res.* **6,** S262.

Matthew, J. L., Martin, J. H., and Race, G. J. (1967). Giant-cell centrioles. *Science* **155,** 1423.

Maycox, P. R., Deckwerth, T., Hell, J. W., and Jahn, R. (1988). Glutamate uptake by brain synaptic vesicles: Energy dependence of transport and functional reconstitution in proteoliposomes. *J. Biol. Chem.* **263,** 15423–15428.

Mellman, I., Fuchs, R., and Helenius, A. (1986). Acidification of the endocytic and exocytic pathway. *Annu. Rev. Biochem.* **55,** 663–700.

Merke, J., Klaus, G., Waldherr, R., and Ritz, E. (1986). No 1,25-dihydroxyvitamin D3 receptors on osteoclasts of calcium-deficient chicken despite demonstrable receptors on circulating monocytes. *J. Clin. Invest.* **77,** 312–314.

Minkin, C., and Jennings, J. J. (1972). Carbonic anhydrase and bone remodeling: Sulfonamide inhibition of bone resorption in organ culture. *Science* **176,** 1031–1033.

Miyauchi, A., Hruska, K. A., Greenfield, E. M., Duncan, R., Alvarez, J. I., Barattolo, R., Colucci, S., Zambonin-Zallone, A., Teitelbaum, S. L., and Teti, A. (1990). Osteoclast cytosolic calcium, regulated by voltage-gated calcium channels and extracellular calcium, controls podosome assembly and bone resorption. *J. Cell Biol.* **111,** 2543–2552.

Miyauchi, A., Alvarez, J., Greenfield, E. M., Teti, A., Grano, M., Colucci, S., Zambonin-Zallone, A., Ross, F. P., Teitelbaum, S. L., Cheresh, D., and Hruska, K. A. (1991a). Recognition of osteopontin and related peptides by an $\alpha_v\beta_3$ integrin stimulates immediate cell signals in osteoclasts. *J. Biol. Chem.* **266,** 20369–20374.

Miyauchi, A., Alvarez, J., Greenfield, E., Teti, A., Zambonin-Zallone, A., Ross, F. P., Teitelbaum, S. L., Cheresh, D., and Hruska, K. (1991b). Matrix protein binding to the osteoclast adhesion integrin mediates a reduction in Ca_i. *J. Bone Miner. Res.* **6,** S96.

Miyaura, C., Abe, E., Kuribayasha, T., Tanaka, H., Konno, K., Nishii, Y., and Suda, T. (1981). 1a,25-dihydroxyvitamin D3 induces differentiation of myeloid leukaemia cells. *Biochem. Biophys. Res. Commun.* **102,** 937–943.

Miyaura, C., Segawa, A., Nagasawa, H., Abe, E., and Suda, T. (1986). Effects of retinoic acid on the activation and fusion of mouse alveolar macrophages induced by 1a,dihydroxyvitamin D3. *J. Bone Miner. Res.* **1,** 359–368.

Murrills, R. J., Shane, R., Lindsay, R., and Dempster, D. W. (1989). Bone resorption by isolated human osteoclasts in vitro: Effects of calcitonin. *J. Bone Miner. Res.* **4,** 259–268.

Neff, L., Horne, W., Male, P., Stadel, J. M., Samanen, J., Ali, F., Levy, J. B., and Baron, R. (1992). A cyclic RGD peptide induces a wave of tyrosine phosphorylation and the translocation of a c-src substrate, (p85) in isolated rat osteoclasts. *J. Bone Miner. Res.* **7,** S106 (abstract).

Nelson, N. (1991). Structure and pharmacology of the proton-ATPases. *Trends Pharm. Sci.* **12,** 71–75.

Nelson, N., and Taiz, L. (1989). The evolution of H+-ATPases. *TIBS* **14,** 113–116.

Nelson, R. D., Guo, X.-L., Masood, K., Brown, D., Kalkbrenner, M., and Gluck, S. (1992). Selectively amplified expression of an isoform of the vacuolar H^+-ATPase 56-kilodalton subunit in renal intercalated cells. *Proc. Natl. Acad. Sci. USA* **89,** 3541–3545.

Nicholson, G. C., Moseley, J. M., Sexton, P. M., Mendelsohn, F. A. O., and Martin, T. J. (1986). Abundant calcitonin receptors in isolated rat osteoclasts. Biochemical and autoradiographic characterization. *J. Clin. Invest.* **78,** 355–360.

Nicholson, G. C., D'Snatos, C. S., Evans, T., Moseley, J. M., Kemp, B. E., and Martin, T. J. (1988). Solubilization of functional calcitonin receptors. *Biochem. J.* **253,** 505–510.

Nijweide, P. J., Burger, E. H., and Feyen, J. H. (1986). Cells of bone: Proliferation, differentiation and hormonal regulation. *Physiol. Rev.* **66,** 855–886.

Okamura, T. (1992). Detection of collagenase mRNA in bovine root resorbing tissue by in situ hybridization. *Jpn. J. Oral Biol.* **34,** 95–111.

Oreffo, R. O. C., Teti, A., Francis, M. J. O., Triffitt, J. T., Carano, A., and Zambonin-Zallone, A. (1988). Effect of vitamin A on bone resorption: Evidence for a direct stimulation of isolated chicken osteoclasts by retinol and retinoic acid. *J. Bone Miner. Res.* **3,** 203–210.

Oreffo, R. O. C., Mundy, G. R., Seyedin, S. M., and Bonewald, L. F. (1989). Activation of the bone-derived latent TGF beta complex by isolated osteoclasts. *Biochem. Biophys. Res. Commun.* **158,** 817–823.

Oursler, M. J., Bell, L. V., Clevinger, B., and Osdoby, P. (1985). Identification of osteoclast-specific monoclonal antibodies. *J. Cell Biol.* **100,** 1592–1600.

Oursler, M. J., Li, L., and Osdoby, P. (1989). Characterization of an osteoclast membrane protein related to superoxide dismutase. *J. Bone Miner. Res.* **4,** 591A.

Oursler, M. J., Osdoby, P., Pyfferoen, J., Riggs., B. L., and Spelsberg, T. C. (1991). Avian osteoclasts as estrogen target cells. *Proc. Natl. Acad. Sci. USA* **88,** 6613–6617.

Parsons, J. T., and Weber, M. J. (1989). Genetics of src: Structure and functional organization of a protein tyrosine kinase. *Curr. Topics Microbiol. Immunol.* **147,** 79–127.

Parsons, S. J., and Creutz, C. E. (1986). $p60^{c\text{-}src}$ activity detected in the chromaffin granule membrane. *Biochem. Biophys. Res. Commun.* **134,** 736–742.

Pfeilschifter, J., Bonewald, L., and Mundy, G. R. (1990). Characterization of the latent transforming growth factor b complex in bone. *J. Bone Miner. Res.* **5,** 49–58.

Prallet, B., Beresford, J., Neff, L., and Baron, R. (1988). Ouabain inhibits bone resorption in organ culture and in isolated rat osteoclasts. *Calcif. Tissue Int.* **38,** 3902 (*Abstract).*

Puopolo, K., Kumamoto, C., Adachi, I., Magner, R., and Forgac, M. (1992). Differential expression of the "B" subunit of the vacuolar H^+-ATPase in bovine tissues. *J. Biol. Chem.* **267,** 3696–3706.

Qi, D. Y., Oreffo, R. O. C., Symons, G. A., DiGiovine, F. S., Seid, J., Duff, G. W., and Russell, R. G. G. (1991). TNF-alfa and TGF-beta gene expression in day 14 GM-CFC-derived osteoclasts detected by in situ hybridization. *J. Bone Miner. Res.* **6,** S263.

Ravesloot, J. H., Ypey, D. L., Nijweide, P. J., Buisman, H. P., and Vrijheid-Lammers, T. (1989a). Three voltage-activated K+ cinductances and an ATP activated conductance in freshly isolated embrynic chick osteoclasts. *Pfluegers Arch.* **414,** S166–S167.

Ravesloot, J. H., Ypey, D. L., Vrijheid-Lammers, T., and Nijweide, P. J. (1989b). Voltage-activated K+ conductances in freshly isolated embryonic chicken osteoclasts. *Proc. Natl. Acad. Sci. USA* **86,** 6821–6825.

Reinholt, F. P., Hultenby, K., Oldberg, A., and Heinegard, D. (1990a). Osteopontin: A possible anchor of osteoclasts to bone. *Proc. Natl. Acad. Sci. USA* **87,** 4473–4475.

Reinholt, F. P., Mengarelli-Wildhom, S., Ek-Rylander, B., and Andersson, G. (1990b). Ultrastructural localization of a tartrate-resistant acid ATPase in bone. *J. Bone Miner. Res.* **5,** 1055.

Rendu, F., Lebret, M., Danielian, S., Fagard, R., Levy-Toledano, S., and Fischer, S. (1989). High pp60c-src level in human platelet dense bodies. *Blood* **73,** 1545–1551.

Rodan, G. A., and Martin, T. J. (1981). Role of osteoblasts in hormonal control of bone resorption: A hypothesis. *Calcif. Tissue Int.* **33,** 349–351.

Rohrschneider, L. R. (1980). Adhesion plaques of Rous sarcoma virus-transformed cells contain the src gene product. *Proc. Natl. Acad. Sci. USA* **77,** 3514–3518.

Roodman, G. D., Ibbotson, K. J., MacDonald, B. R., Kuehl, T. J., and Mundy, G. R. (1985). 1,25-Dihydroxyvitamin D3 causes formation of multinucleated cells with several osteoclast characteristics in cultures of primate marrow. *Proc. Natl. Acad. Sci. USA* **82,** 8213–8217.

Roodman, G. D., Kurihara, N., Ohsaki, Y., Kukita, A., Hosking, D., Demulder, A., Smith, J. F., and Singer, F. R. (1992). Interleukin 6—A potential autocrine/paracrine factor in Paget's disease of bone. *J. Clin. Invest.* **89,** 46–52.

Ruoslahti, E., and Pierschbacher, M. D. (1987). New perspectives in cell adhesion: RGD and integrins. *Nature (London)* **238,** 491–497.

Sato, M., Sardana, M. K., Grasser, W. A., Garsky, V. M., Murray, J. M., and Gould, R. J. (1990). Echistatin is a potent inhibitor of bone resorption in culture. *J. Cell Biol.* **111,** 1713–1723.

Schenk, R., Spiro, D., and Wiener, J. (1967). Cartilage resorption in tibial epiphyseal plate of growing rats. *J. Cell Biol.* **34,** 275–291.

Scheven, B. A. A., Visser, J. W. M., and Nijweide, P. J. (1986). In vitro osteoclast generation from different bone marrow fractions, including a highly enriched haematopoietic stem cell population. *Nature (London)* **321,** 79–81.

Schmid, S., Fuchs, R., Kielian, M., Helenius, A., and Mellman, I. (1989). Acidification of endosome subpopulations in wild-type Chinese hamster ovary cells and temperature-sensitive acidification-defective mutants. *Cell. Biol.* **108,** 1291–1300.

Schoppa, N. E., Su, Y., Baron, R., and Boulpaep, E. L. (1990). Identification of single ion channels in neanatal rat osteoclasts. *J. Bone Miner. Res.* **5,** S204.

Shankar, G., Helfrich, M. H., and Horton, M. A. (1992). Effect of RGD-peptides on intracellular calcium and retractile responses in rat osteoclasts. *Bone Miner,* **17,**(Suppl. 1), 188.

Sims, S. M., and Dixon, S. J. (1989). Inwardly rectifying K+ current in osteoclasts. *Am. J. Physiol.* **256,** C1277–C1282.

Sims, S. M., Kelly, M. E. M., Arkett, S. A., and Dixon, S. J. (1992). Electrophysiology of osteoclasts. *In* "Biology and Physiology of the Osteoclast" (B. R. Rifkin and C. V. Gay, eds.), in press. CRC Press, Boca Raton, Florida.

Sims, S. M., Kelly, M. E. M., and Dixon, S. J. (1991b). K+ and Cl− currents in freshly isolated rat osteoclasts. *Eur. J. Physiol.* (in press).

Sly, W. S., Hewett-Emmett, D., Whyte, M. P., Yu, Y. S., and Tashian, R. E. (1983). Carbonic anhydrase II deficiency identified as the primary defect in the autosomal recessive syndrome of osteopetrosis with renal tubular acidosis and cerebral calcification. *Proc. Natl. Acad. Sci. USA* **80,** 2752–2756.

Soriano, P., Montgomery, C., Geske, R., and Bradley, A. (1991). Targeted disruption of the c-src proto-oncogene leads to osteopetrosis in mice. *Cell* **64,** 693–702.

Suda, T., Takahashi, N., and Martin, T. J. (1992). Modulation of osteoclast differentiation. *Endo. Rev.* **13,** 66–80.

Sundquist, K., Lakkakorpi, p., Wallmark, B., and Vaananen, K. (1990). Inhibition of osteoclast proton transport by bafilomycin-A1 abolishes bone resorption. *Biochem. Biophys. Res. Commun.* **168,** 309–313.

Takahashi, N., Akatsu, T., Sasaki, T., Nicholson, G. C., Mosley, J. M., Martin, T. J., and Suda, T. (1988a). Induction of calcitonin receptors by 1,25-dihydroxyvitamin D3 in osteoclast-like multinucleated cells formed from mouse bone marrow cells. *Endocrinology* **123,** 1504–1510.

Takahashi, N., Akatsu, T., Udagawa, N., Sasaki, T., Yamaguchi, A., Moseley, J. M., Martin, T. J., and Suda, T. (1988b). Osteoblastic cells are involved in osteoclast formation. *Endocrinology* **123,** 2600–2602.

Takahashi, N., Yamana, H., Yoshiki, S., Roodman, G. D., Mundy, G. R., Jones, S. J., Boyde, A., and Suda, T. (1988c). Osteoclast-like cell formation and its regulation by osteotropic hormones in mouse bone marrow cultures. *Endocrinology* **122,** 1373–1382.

Takahashi, N., Kukita, T., MacDonald. B. R., Bird, A., Mundy, G. R., McManus, L. M., Miller, M., Boyde, A., Jones, S. J., and Roodman, G. D. (1989). Osteoclast-like cells form in long-term human bone marrow but not in peripheral blood cultures. *J. Clin. Invest.* **83,** 543–550.

Takahashi, N., Udagawa, N., Akatsu, T., Tanaka, H., Shionome, M., and Suda, T. (1991). Role of colony-stimulating factors in osteoclast development *J. Bone Miner. Res.* **6,** 977–986.

Taylor, M. L., Boyde, A., and Jones, S. J. (1989). The effect of fluoride on the patterns of adherence of osteoclasts cultured on and resorbing dentine: A 3D assessment of vinculin-labelled cells using confocal microscopy. *Anat. Embryol.* **180,** 427–435.

Teti, A., Blair, H. C., Teitelbaum, S. L., Kahn, A. J., Koziol, C., Konsek, J., Zambonin-Zallone, A., and Schlesinger, P. H. (1989a). Cytoplasmic pH regulation and chloride bicarbonate exchange in avian osteoclasts. *J. Clin. Invest.* **84,** 227–233.

Teti, A., Blair, H. C., Schlesinger, P., Grano, M., Zambonin-Zallone, A., Kahn, A. J., Teitelbaum, S. L., and Hruska, K. A. (1989a). Extracellular protons acidify osteoclasts, reduce cytosolic calcium, and promote expression of cell–matrix attachment structures. *J. Clin. Invest.* **84,** 773–780.

Teti, A., Marchisio, P. C., and Zambonin-Zallone, A. (1991). Clear zone in osteoclast function: Role of podosomes in regulation of bone-resorbing activity. *Am. J. Physiol.* **261,** C1–C7.

Turksen, K., Kanehisa, J., Opas, M., and Heersche, J. N. M. (1988). Adhesion patterns and cytoskeleton of rabbit osteoclasts on bone slices and glass. *J. Bone Miner. Res.* **3,** 389–399.

Tuukkanen, J., and Vaananen, H. K. (1986). Omeprazole, a specific inhibitor of H+/K+-ATPase, inhibits bone resorption in vitro. *Calcif. Tissue Int.* **38,** 123–125.

Udagawa, N., Takahashi, N., Akatsu, T., Sasaki, T., Yamaguchi, A., Kadama, H., Martin, T. J., and Suda, T. (1989). The bone marrow derived stromal cell lines MC3T3-G2/PA6 and ST2 support osteoclast-like cell differentiation in co-cultures with spleen cells. *Endocrinology,* **125,** 1805–1813.

Udagawa, N., Takahashi, N., Akatsu, T., Tanaka, H., Sasaki, T., Nishihara, T., Koga, T.,

Martin, T. J., and Suda, T. (1990). Origin of osteoclasts: Mature monocytes and macrophages are capable of differentiating into osteoclasts under a suitable microenvironment prepared by bone marrow-derived stromal cells. *Proc. Natl. Acad. Sci. USA* **87,** 7260–7264.

Vaananen, H. K., Karhukorpi, E. K., Sundquist, K., Roininen, I., Hentunen, T., Tuukkanen, J., and Lakkakorpi, p. (1990). Evidence for the presence of a proton pump of the vacuolar H+-ATPase type in the ruffled border of osteoclasts. *J. Cell Biol.* **111,** 1305–1311.

Vaes, G. (1968). On the mechanisms of bone resorption: The action of parathyroid hormone on the excretion and synthesis of lysosomal enzymes and on the extracellular release of acid by bone cells. *J. Cell Biol.* **39,** 676–697.

Vaes, G. (1972). Inhibitory actions of calcitonin on resorbing bone explants in culture and on their release of lysosomal hydrolases. *J. Dent. Res.* **51**(Suppl.), 362–366.

Vaes, G. (1988). Cellular biology and biochemical mechanism of bone resorption. *Clin Orthop.* **231,** 239–271.

Varshney, G. C., Henry, J., Kahn, A., and Phan-Dinh-Tuy, F. (1986). Tyrosine kinases in normal human blood cells. Platelet but not erythrocyte band 3 tyrosine kinase is p60c-src. *FEBS Lett.* **205,** 97–103.

Von Figura, K., and Hasilik, A. (1986). Lysosomal enzymes and their receptors. *Annu. Rev. Biochem.* **55,** 167–193.

Waite, L. C., Volkert, W. A., and Kenny, A. D. (1970). Inhibition of bone resorption by acetazolamide in the rat. *Endocrinology* **87,** 1129–1139.

Walker, D. G. (1973). Osteopetrosis in mice cured by temporary parabiosis *Science* **180,** 875.

Walker, D. G. (1975a). Bone resorption restored in osteopetrotic mice by transplants of normal bone marrow and spleen cells. *Science* **190,** 784.

Walker, D. G. (1975b). Spleen cells transmit osteopetrosis in mice. *Science* **190,** 785.

Wang, Z.-Q., and Gluck, S. (1990). Isolation and properties of bovine kidney brush border vacuolar H+-ATPase. *J. Biol. Chem.* **265,** 21957–21965.

Warshafsky, B., Aubin, J. E., and Heersche, J. N. M. (1985). Cytoskeleton rearrangements during calcitonin-induced changes in osteoclast motility in vitro. *Bone* **6,** 179–185.

Warshawsky, H., Goltzman, D., Rouleau, M. F., and Bergeron, J. J. M. (1980). Direct in vivo demonstration by radioautography of specific binding sites for calcitonin in skeletal and renal tissues of the rat. *J. Cell Biol.* **85,** 682–694.

Wiktor-Jedrzejczak, W., Ahmed, A., Szczylik, C., and Skelly, R. R. (1982). Hematological characterization of congenital osteopetrosis in op/op mouse. *J. Exp. Med.* **156,** 1516.

Wiktor-Jedrzejczak, W., Bartocci, A., Ferrante, A. W., Ahmed-Ansari, A., Sell, K. W., Pollard, J. W., and Stanley, E. R. (1990). Total absence of CSF 1 in the macrophage deficient osteopetrotic (op/op) mouse. *Proc. Natl. Acad. Sci. USA* **87,** 4828–4832.

Wood, D. A., Hapak, L. K., Sims, S. M., and Dixon, S. J. (1991). Direct effects of platelet-activating factor on isolated rat osteoclasts. *J. Biol. Chem.* **266,** 15369–15376.

Wu, H., Reynolds, A. B., Kanner, S. B., Vines, R. R., and Parsons, J. T. (1991). Identification and characterization of a novel cytoskeleton-associated pp60^{c-src} substrate. *Mol. Cell. Biol.* **11,** 5113–5124.

Yoshida, H., Hayashi, S. I., Kunisada, T., Ogawa, M., Nishikawa, S., Okamura, H., Sudo, T., Shultz, L. D., and Nishikawa, S. I. (1990). The murine mutation osteopetrosis is in the coding region of the macrophage colony stimulating factor gene. *Nature* **345,** 442–444.

Zaidi, M. (1990). "Calcium receptors" on eukaryotic cells with special reference to the osteoclast. *Biosci. Rep.* **10,** 493–507.

Zambonin-Zallone, A., Teti, A., Carano, A., and Marchisio, P. C. (1988). The distribution of podosomes in osteoclasts cultured on bone laminae: Effects of retinol. *J. Bone Miner. Res.* **3,** 517–523.

Zambonin-Zallone, A., Teti, A., Grano, M., Rubinacci, A., Abbadini, M., Gaboli, M., and Marchisio, P. C. (1989). Immunocytochemical distribution of extracellular matrix receptors in human osteoclasts: A beta 3 integrin is colocalized with vinculin and talin in the podosomes of osteoclastoma giant cells. *Exp. Cell. Res.* **182,** 645–652.

15

c-*fos* ONCOGENE EXPRESSION IN CARTILAGE AND BONE TISSUES OF TRANSGENIC AND CHIMERIC MICE

AGAMEMNON E. GRIGORIADIS, ZHAO-QI WANG, and ERWIN F. WAGNER

Cellular and Molecular Biology of the Bone

I. INTRODUCTION

The study of oncogene function is a fascinating and rapidly growing area of research that has only recently become amenable to biochemical and molecular biological analyses. Attempts are being made not only to delineate the potential role of oncogenes in human disease and neoplasia, but also to understand their normal roles in the regulation of mammalian embryonic development and cellular differentiation. In this chapter, we focus on a specific oncogene, namely, the c-*fos* protooncogene, and attempt to describe its role in (1) the regulation of normal bone and cartilage cell differentiation and (2) the development of osteogenic and chondrogenic tumors, using mouse model systems.

A. Bone, Cartilage, and Oncogenes

The formation of bone and cartilage tissues during skeletogenesis is under the control of hormonal and local regulatory factors (for review, see Rodan and Rodan, 1984; Nijweide *et al.*, 1986; Marks and Popoff, 1988; Heersche and Aubin, 1990). The formation of bone occurs in two different ways: (1) endochondral bone formation, which is thought to involve a cartilage intermediate and (2) intramembranous bone formation, in which bone cell progenitors differentiate directly into bone-synthesizing osteoblasts (Hall, 1978). However, it is not known which factors and genes regulate these processes and how many steps are required for activating chondrogenesis and osteogenesis. At the cellular level, it is generally accepted that the cells comprising bone and cartilage tissues are related to each other and to other mesenchymal cells at the level of earlier undifferentiated mesenchymal progenitor cells (for review, see Owen, 1985; Aubin *et al.*, 1990). Thus, osteogenic, chondrogenic, myogenic, adipogenic, and fibroblastic lineages have a putative common origin (see also Grigoriadis *et al.*, 1988, 1990). Precisely where in such a hierarchy lineage restriction occurs, and what factors determine this lineage commitment, is not entirely known, although the identification of a family of related myogenic regulatory molecules (e.g., MyoD, myogenin) has shed some light on the mechanisms of muscle determination (for review, see Weintraub *et al.*, 1991).

Regulatory molecules for osteogenesis and chondrogenesis do exist, as evidenced by the marked effects on bone and cartilage differentiation of both systemic factors (e.g., steroid and polypeptide hormones) and local regulatory factors such as the bone morphogenetic proteins, transforming growth factors (e.g., TGF-β), cytokines, and prostaglandins (for review, see Wozney *et al.*, 1988; Heersche and Aubin, 1990). Despite the large number of *in vitro* and *in vivo* studies describing the regulation of bone and cartilage differentiation (Caplan and Pechak, 1987; Wozney, 1989), the specific cell types upon which these molecules act, and whether they act directly or indirectly to regulate commitment of progenitor cells to osteogenic and chondrogenic lineages, is not clear. Identification of such factors is therefore of central importance not only for investigating normal bone and cartilage development but also for understanding the cellular and molecular mechanisms underlying the perturbation of cell growth and differentiation in metabolic bone diseases and bone and cartilage neoplasias.

Oncogenes are dominant transforming genes that were first discovered through the identification of DNA and RNA tumor viruses that caused tumors after injection into animals. Together with the discovery that certain RNA tumor viruses (retroviruses) could capture and transduce mutated versions of normal cellular genes (protooncogenes), these observations confirmed that these normal genes were causally related to the process of tumorigenesis (Bishop, 1987). Furthermore, the association of some protooncogenes with specific chromosomal translocations and other genetic abnormalities (Haluska *et al.*, 1987; Klein, 1987; Solomon *et al.*, 1991), together with the ability of oncogenes transfected into tissue culture cells to cause morphological transformation (Weinberg, 1985), suggested that protooncogenes were involved in tumorigenesis.

In recent years, major research efforts have focused on trying to elucidate the role of different protooncogenes in the genetic control of cellular growth and differentiation (Wagner and Müller, 1986). Oncogene products are thought to play an important role in these normal cellular processes, as evidenced by the many observations that the products of oncogenes are involved in many cellular functions. For example, protooncogenes and their products can be grouped functionally into different categories: They can encode growth factors (e.g., c-*sis*), growth factor receptors (e.g., c-*erb*B, c-*fms*), and regulators of gene expression, that is, nuclear transcription factors (e.g., c-*fos*, c-*jun*, c-*erbA*) (for review, see Varmus, 1984; Bishop, 1987; Ransone and Verma, 1990; Aaronson, 1991). Whether some of these oncogenes and transcription factors also have a role in osteogenesis and chondrogenesis is supported by only a few studies. Specifically, amplification and overexpression of several protooncogenes such as c-*myc*, c-*ras*, c-*sis*, c-*bos*, and c-*abl* have been observed in some spontaneous and radiation-induced osteosarcomas and osteosarcoma cell lines (Schön *et al.*, 1986; Bogenmann *et al.*, 1987; Nardeux *et al.*, 1987). However, no consistent correlation was observed between oncogene expression and bone tumor formation, perhaps due to tumor and cellular heterogeneity. With respect to other oncogenes, it has recently been shown that mice that lack the product of the c-*src* oncogene develop a phenotype reminiscent of osteopetrosis (Soriano *et al.*, 1991). Finally, there has been recent interest in the group of tumor suppressor genes (e.g., retinoblastoma, p53) because mutations in, or inactivation of, these genes are associated with deregulated cell growth and human neoplasias (for review, see Weinberg, 1991). Specifically, p53 has gained much attention since rearrangements in this gene have been observed in many different human cancers including osteosarcomas (Masuda *et al.*, 1987; Nigro *et al.*, 1989; Levine *et al.*, 1991). Further support for a role for p53 in osteosarcoma formation comes from studies using transgenic mice that overexpress mutant alleles of the p53 oncogene and that develop, among other neoplasias, osteosarcomas (Lavigueur *et al.*, 1989).

Perhaps the most studied oncogene with respect to bone and cartilage differentiation and osteosarcoma formation is the c-*fos* protooncogene (reviewed in detail in Section II). Although many researchers have proposed that c-*fos* can act as a "master switch" for cell proliferation and differentiation (Verma and Sassone-Corsi, 1987), its functional role in bone and cartilage cells is not clear. With regard to our own studies with c-*fos*, we will describe an *in vivo* approach using gene transfer into mouse embryos in order to assess the potential biological role of this protooncogene and transcription factor in normal bone and cartilage formation and tumorigenesis (see Section III).

B. Oncogene Function Studied in Transgenic and Chimeric Mice

Many studies aimed at investigating the potential function of oncogenes in the control of growth and differentiation of normal and malignant cells have been performed *in vitro* using defined cell culture systems. However, the information obtained is limiting with respect to gaining insights into potential functions of oncogenes in the context of the development of the whole organism. At present, there are two well-established techniques for analyzing oncogene function *in vivo:* gene transfer either directly into mouse embryos (transgenic mice) or into embryonic stem (ES) cells, resulting in the generation of chimeric mice (see later). Specifically, two different experimental approaches have been used to address specific questions about oncogene function: first, a *gain-of-function* approach, in which one assesses the consequences of overexpression of specific oncogenes on the growth and development of an organism and second, a *loss-of-function* approach, in which the coding sequences of specific oncogenes are "inactivated" on either one or both alleles, leading to the lack of a functional oncoprotein in the organism and a potential lethal or pathological phenotype (for review, see Rossant, 1991; Wagner *et al.*, 1991).

With regard to the gain-of-function approach, it has become clear from numerous studies using transgenic and chimeric mice (see later) that oncogene expression can predispose mice to cellular alterations resulting in the predictable development of tumors that would not arise spontaneously. This is a direct confirmation for the causal role of oncoproteins in cancer; thus, these mice provide ideal model systems for unraveling the role of specific oncogenes in the normal development of specific tissues and for understanding the roles of different oncogenes in multistep tumorigenesis (Knudson, 1986; Hanahan, 1988; Adams and Cory, 1991). Thus, the use of transgenic and chimeric mice allows one to approach problems that cannot be addressed adequately using cell culture techniques—for example, identifying target tissues that may be susceptible to aberrant levels of a particular oncogene, analyzing oncogene cooperativity in relation to transformation and tumor formation, and identifying target cells and normal cellular genes that may be targets for specific oncogenes.

Although the characteristics of transgenic and chimeric mice and the respective methodologies for generating them have been discussed previously in great depth (for review, see Hanahan, 1988; Wagner, 1990; Wagner and Keller, 1992), the basic methods and the rationales for choosing the appropriate system are outlined briefly in the following.

1. DNA Injection into Fertilized Eggs

The method of choice for transferring genetic information stably into the germ line of mice involves the injection of recombinant DNA into one of the pronuclei of a fertilized mouse egg (one-cell embryo) (Fig. 1A). Injected eggs are then transferred to pseudopregnant females and allowed to develop to term. On average, approximately 10–20% of the mice born will carry the injected DNA (the *transgene*). These positive "founder" animals are then bred to obtain offspring that also carry the transgene, thereby establishing a unique family of mice. Expression analysis can then be performed to determine if, and in which tissues, the transgene is expressed. Expression of the transgene is normally stable over many generations, although the site of integration of the transgene in the chromosome often influences the pattern of expression (Palmiter and Brinster, 1986). In addition, the type of DNA vector used can affect expression of the transgene. Generally, three types of vector constructions are commonly used in dissecting specific aspects of oncogene function. First, constructs with oncogenes under the control of their own regulatory elements permit both stage- and tissue-specific expression of the oncogene. Second, hybrid constructs whereby specific oncogenes are fused to regulatory elements of broad specificity allow ectopic expression of the oncogene in a broad range of tissues. Third, hybrid constructs in which oncogenes are fused to cell- and/or tissue-specific

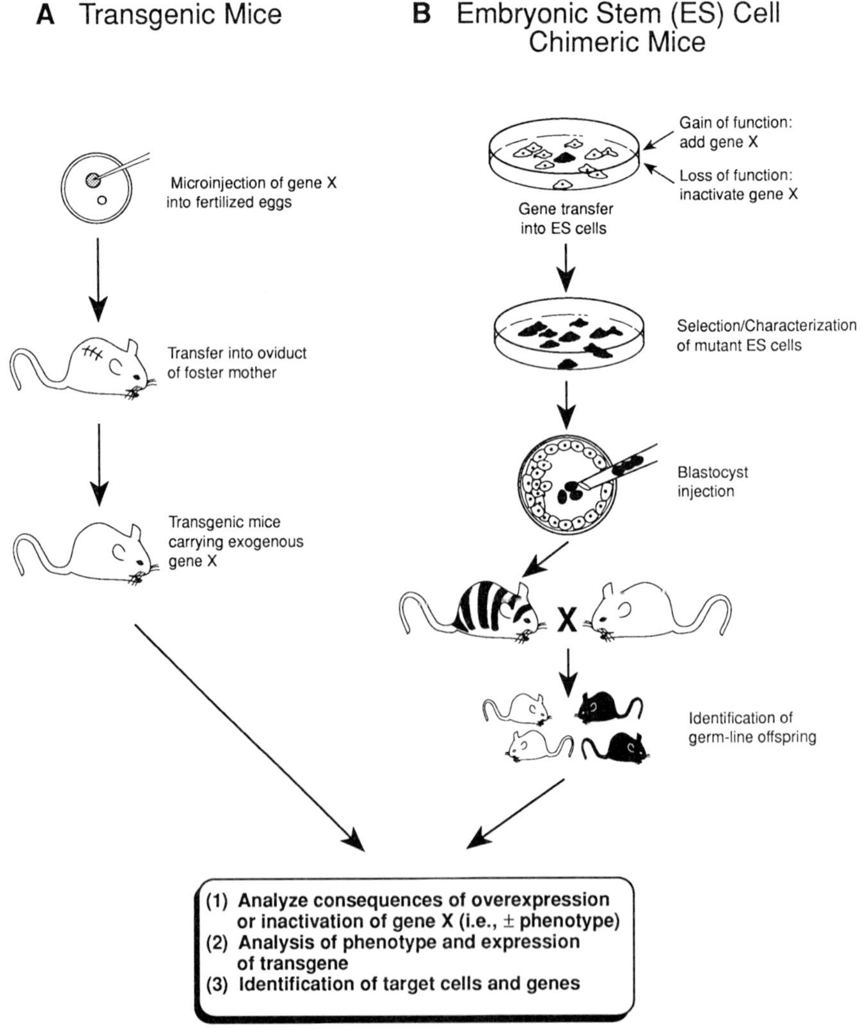

FIGURE 1 Two experimental approaches for introducing genes into mice. (A) Generation of transgenic mice. A DNA construct containing the gene of interest (e.g., gene X) is injected into the male pronucleus of a fertilized mouse egg. Injected eggs are transferred into the oviducts of foster mothers, and the resulting progeny are analyzed by Southern blot analysis for the presence of the transgene. (B) Generation of embryonic stem (ES) cell chimeric mice. ES cells are isolated from blastocysts and maintained *in vitro* in an undifferentiated state. For the gain-of-function approach, the gene of interest (e.g., gene X) is introduced into ES cells, resulting in overexpression of the encoded protein. For the loss-of-function approach, a targeting construct containing gene X is introduced into ES cells, resulting in the disruption of the endogenous locus on one allele as a result of homologous recombination. The second allele can be targeted using a similar strategy. Individual ES clones are selected *in vitro* and injected into blastocysts for generation of chimeric mice. Chimeras are identified and bred to obtain germ-line offspring. The consequences of overexpression or inactivation of specific genes can subsequently be analyzed.

promoter elements allow targeted expression of oncogenes to specific tissues (for a detailed review, see Hanahan, 1988; Wagner, 1990).

2. Gene Transfer Using Embryonic Stem Cells and Chimeras

A second method for introducing foreign genes into mouse embryos has been the use of ES cells (Fig. 1B). ES cells are derived from mouse blastocysts [day 3 post coitus, (p.c.), preimplantation embryos], specifically from the inner cell mass from which the embryo develops. Blastocysts can be explanted *in vitro* and ES cells established as permanent cell lines under well-defined culture conditions (Robertson, 1987; Wagner *et al.*, 1991). Furthermore, ES cells can be introduced into blastocysts, where they participate in the development of all somatic as well as germ cells (Bradley *et al.*, 1984; Robertson, 1987). The resulting mice are chimeric, and the extent of chimerism, that is, the extent to which derivatives of the injected ES cells contributed to the development of the mouse, is routinely estimated by examining specific genetic markers in the chimeric mouse. Thus, individual clones of ES cells are first selected *in vitro* for a specific genetic modification and then are introduced into blastocysts by microinjection for colonization of somatic and germ cells. Modifications include either ectopic expression of exogenous genes (gain-of-function) or inactivation of genes via homologous recombination (loss-of-function) (for review, see Capecchi, 1989; Wagner *et al.*, 1991). The consequences of altered gene expression can be embryonic lethality (Williams *et al.*, 1988) or a pathological phenotype that may be manifested at a later point in development or in postnatal life (Wang *et al.*, 1991). Because ES cells are first selected *in vitro* either for high expression or for absence of specific gene products, it is possible to study the effects of a specific alteration throughout embryogenesis. This offers one potential advantage in using the ES cell system versus the conventional DNA injection route, in which the expression of transgenes is commonly observed after birth (Palmiter and Brinster, 1986).

II. THE *fos* ONCOGENE

The protooncogene c-*fos* is the normal cellular homolog of the v-*fos* oncogene. v-*fos* was first detected as the transforming gene present in both the FBJ- and the FBR-murine sarcoma viruses (MSVs) isolated from spontaneous and radiation-induced osteosarcomas, respectively (Finkel *et al.*, 1966, 1975; Finkel and Biskis, 1968). Both viruses induce at a high frequency osteogenic sarcomas when injected into neonatal mice. Specifically, virus-induced tumors arise on several bones of the body and histological analyses characterized the tumors as a unique type of chondroosseous neoplasm (Ward and Young, 1976). Although the precise cellular origin of the tumors is not entirely clear, it is thought that they originate from periosteal cells (Ward and Young, 1976). Interestingly, in contrast to spontaneous osteosarcomas very few, if any, metastases are seen in virally induced tumors. In addition to the effects *in vivo*, both viruses have also been shown to transform fibroblasts *in vitro* that subsequently have the capacity to produce tumors when injected into syngeneic or nude mice (Jenuwein *et al.*, 1985). Further confirmation of the pathology induced by v-*fos* was recently shown by Schmidt *et al.* (1986), who injected FBR-MSV into newborn mouse mandibular condyles at a stage where the principal cell type is cartilage, and showed that the virus induced osseous lesions. Taken together, therefore, these findings implicated the *fos* oncogene in cellular transformation and in the development of a specific type of neoplasia of mesenchymal cells.

A. Structure and Regulation of Expression

The complete nucleotide sequences of both v-*fos* and c-*fos* oncogenes have been determined (van Beveren *et al.*, 1983, 1984; van Straaten *et al.*, 1983), and the organization of the murine genomic c-*fos* and v-*fos* genes is shown in Figure 2. c-*fos*, which is comprised of four exons encoding a mature mRNA transcript of 2.2 kilobases (van Beveren *et al.*, 1983; Renz *et al.*, 1985) encodes a protein (Fos) containing 380 amino acids with an apparent molecular mass of 55-kDa (van Beveren *et al.*, 1983; Curran *et al.*, 1984). In contrast, the v-*fos* protein encoded by the FBJ-MSV is 381 amino acids long, and the FBR-MSV-encoded protein is a 75-kDa *gag-fos* fusion protein containing 554 amino acids, the differences residing in the carboxy-terminal regions of the molecules (for review, see Müller, 1986; Ovitt and Rüther, 1989). Despite these slight differences, all Fos proteins are nuclear phosphoproteins, although c-Fos undergoes more extensive posttranslational modification (Curran *et al.*, 1984).

The potential role of c-*fos* as an important regulatory molecule is inferred in part by its high conservative between different species. Murine Fos exhibits approximately 97 and 94% homology with the rat and human Fos proteins, respectively (Curran *et al.*, 1987) and approximately 79% homology with the chicken Fos protein (Fujiwara *et al.*, 1987; Mölders *et al.*, 1987). More significant, however, is the fact that there is virtually complete amino acid identity in all these species in a specific region of the protein that has been shown to be responsible for the transforming activity of the v-*fos* protein (Jenuwein and Müller, 1987). This stretch of 88 amino acids contains a highly basic domain that is responsible for binding to DNA (Lucibello *et al.*, 1989) and a region consisting of periodic repeats of leucine residues every seven amino acids. This region, termed the *leucine zipper* (Landschultz *et al.*, 1988), is found in a variety of transcription factors (for review, see Kouzarides and Ziff, 1989; Ziff, 1990), notably c-*jun* (see also Section II.C). Many

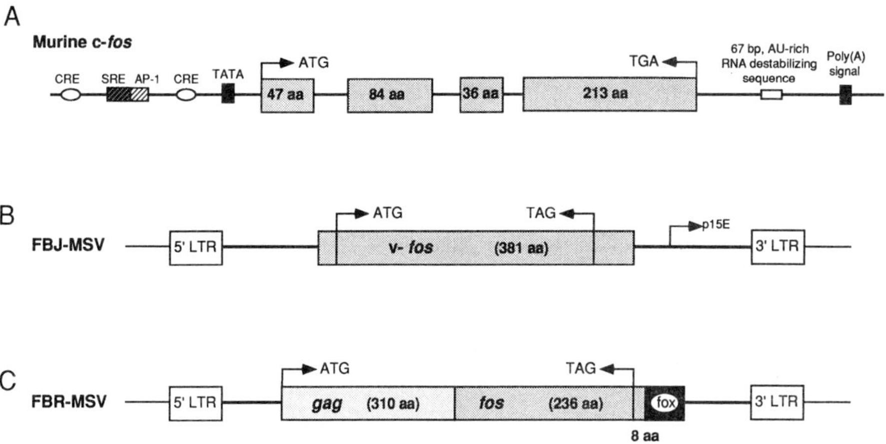

FIGURE 2 Structure of the murine genomic c-*fos* gene (A) and the FBJ- (B) and FBR-MSV (C) v-*fos* genes. The stippled boxes in (A) represent the exons and the number of amino acids encoded by each exon is indicated. The introns and the 5′ and 3′ unstranslated sequences are represented by a solid line. The v-*fos* and *gag–fos* fusion genes present in the FBJ- and FBR-MSVs are also shown with the corresponding number of amino acids that they encode. Also included are the initiation and termination codons for each gene and the approximate locations of the cyclic adenosine monophosphate response element (CRE), serum response element (SRE), AP-1 consensus sequence, TATA box, RNA destabilizing sequence, and polyadenylation (Poly(A)) signal present in the c-*fos* gene. Open boxes represent the 5′ and 3′ long terminal repeat (LTR) sequences of the FBJ- and FBR-MSVs. The coding region for the viral envelope protein p15E is also indicated in (B). The FBR-MSV sequence terminates in sequences designated *fox; fox* genes are an abundant class of RNA present in mouse tissue at loci not related to c-*fos*. The data in this figure were compiled from van Beveren *et al.* (1983, 1984) and Ransone and Verma (1990). MSV, murine sarcoma virus.

studies have shown using mutation and deletion analyses that the leucine zipper is necessary for the formation of heterodimers between Fos and Jun (Kouzarides and Ziff, 1988; Sassone-Corsi *et al.*, 1988; Gentz *et al.*, 1989; Turner and Tjian, 1989). Besides being the essential dimerization domain, the leucine zipper motif of Fos and Jun is also necessary for transcriptional activation and transformation (Schuermann *et al.*, 1989). Different experimental approaches, including competition studies, mutagenesis, and DNA-affinity studies, identified the specific DNA sequence to which Fos–Jun complexes bind as the consensus sequence of transcription factor AP-1 (activator protein 1). This AP-1 site is also referred to as the TPA (12-*O*-tetradecanoyl-phorphol 13-acetate)-responsive element (TRE) and has a consensus sequence of TGA(C/G)TCA (Angel *et al.*, 1987; Lee *et al.*, 1987; Chiu *et al.*, 1988). Thus, genes that contain AP-1 sites in their regulatory regions are believed to be targets of transcriptional regulation by Fos–Jun complexes. To date, only a few genes have been identified and shown in transient transfection experiments to be target genes for Fos–Jun complexes (e.g., osteocalcin [Schüle *et al.*, 1990], the matrix metalloproteinases collagenase [Schönthal *et al.*, 1988] and transin/stromelysin [Kerr *et al.*, 1988], preproenkephalin [Sonnenberg *et al.*, 1990], nerve growth factor [Hengerer *et al.*, 1990]). Interestingly, some of these genes are also thought to have distinct roles in normal bone metabolism and in bone remodeling, further implicating c-*fos* in the regulation of bone cell function.

Two other conserved domains in the c-*fos* gene should also be mentioned because they are, in fact, regions that are shared among most nuclear oncogenes. First, the rapid and transient induction of c-*fos* mRNA following cell stimulation by different external signals (e.g., serum) has established c-*fos* as a key member of the immediate early gene family. These molecules are implicated in signal transduction and the control of cell proliferation and include members of the *fos* and *jun* gene families (see Section II.C), c-*myc* and *egr*-1, to mention but a few (Curran, 1988; Sukhatme *et al.*, 1988; Bravo, 1990). The sequence responsible for this activation lies in the promoter region of the c-*fos* gene and is termed the serum response element (SRE; Treisman, 1985) although multiple control sequences have been identified (Fig. 2; see also Shaw *et al.*, 1989, for review). Finally, the short half-life of the c-*fos* mRNA is due, in part, to the presence of a 67-basepair sequence in the 3′ nontranslated region that is responsible for the rapid degradation of c-*fos* mRNA (Miller *et al.*, 1984; Meijlink *et al.*, 1985). Removal of these sequences stabilizes message levels (Rahmsdorf *et al.*, 1987; Wilson and Treisman, 1988), and, as described later, its removal is important in ensuring high, stable expression of c-*fos* mRNA and Fos protein to elicit biological effects.

B. Biological Role of c-*fos*

Despite the many efforts and investigations into the biochemical and molecular aspects of Fos function, the biological role of c-*fos* is not entirely clear. Gene transfer studies *in vitro* have implicated a role for c-*fos* in cell differentiation. For example, overexpression of c-*fos* can stimulate the expression of differentiation markers in F9 embryonal carcinoma cells (Müller and Wagner, 1984; Rüther *et al.*, 1985) but not in the monocyte–macrophage lineage (Mitchell *et al.*, 1986; see also Wagner and Müller, 1986, for review). With regard to cell proliferation, many experiments using c-*fos* anti-sense constructs and specific antibodies supported a role for c-*fos* in cell-cycle regulation (Nishikura and Murray, 1987; Riabowol *et al.*, 1988; Kovary and Bravo, 1991).

Although c-*fos* is present and can be induced in many different cell types *in vitro*, the evidence *in vivo* suggests that the expression of c-*fos* during embryonic development of both mice and humans, and in the adult organism, is in fact quite restricted. During early development, c-*fos* is expressed entirely in extraembryonic tissues (Müller *et al.*, 1982; Mason *et al.*, 1985), while embryonic expression during late development is restricted to the growth regions of fetal bones (Dony and Gruss, 1987; De Togni *et al.*, 1988; Sandberg *et al.*, 1988; Closs *et al.*, 1990). The function of c-*fos* at these sites of bone and cartilage formation is not known; however, it has been suggested recently that the phenotypic change that occurs in mouse mandibular condyles from cartilage to bone is preceded by a burst of c-*fos* expression (Closs *et al.*, 1990). Moreover, in isolated chick periosteal organ cultures in which bone formation can be regulated by glucocorticoids, it was shown that dexamethasone induces transient expression of c-*fos* specifically in osteogenic cells (Birek *et al.*, 1991). In the adult organism, c-*fos* is expressed in certain hematopoietic cell types, notably in macrophages, granulocytes, and mast cells (Müller, 1986; Curran, 1988). It has also been observed in other mouse tissues (e.g., salivary glands, intestine [our unpublished observations]). Thus, although expression in several tissues has been observed, the function of c-*fos* in the development of these tissues and cell types is still not clear.

Important in delineating a biological role for c-*fos* was the demonstration that c-*fos* itself can transform mesenchymal cells *in vitro* if expressed at high levels (Miller *et al.*, 1984). Indeed, in some murine osteosarcomas (Schön *et al.*, 1986) and in most human osteosarcomas (Wu *et al.*, 1990), c-*fos* has been shown to be expressed at high levels. These data suggest that Fos protein may function in osteogenic cells although its causal role cannot be established by these studies. That overexpression of c-Fos is indeed causally related to transformation of osteogenic tissues was only shown in studies involving gene transfer into mice (Rüther *et al.*, 1987, 1989; Wang *et al.*, 1991) and analyzing the consequences of deregulated c-*fos* expression in an *in vivo* setting. These studies are reviewed in detail in Section III below.

C. Fos-Related Genes

c-*fos* belongs to a multigene family that includes the oncogenes *fos*B, *fra*-1, and *fra*-2 (Cohen and Curran, 1988; Zerial *et al.*, 1989; Nishina *et al.*, 1990). Similarly, the other component of the AP-1 complex, c-*jun*, is also a member of a larger family including *junB* and *junD* oncogenes (Ryder *et al.*, 1988; Hirai *et al.*, 1989). All Fos- and Jun-related proteins are extremely well conserved in the DNA binding and leucine zipper regions (see Section II.A) but exhibit some differences with respect to basal levels of expression and kinetics of serum inducibility (for review, see Kouzarides and Ziff, 1989). More importantly for the purposes of gene regulation is the fact that all combinations of heterodimers between Fos family members and Jun family members are possible and have similar affinities for DNA binding. In addition, Jun family members can form homodimers with each other but with significantly lower affinities for DNA, and Fos family members cannot form homodimers (for review, see Kouzarides and Ziff, 1989; Ransone and Verma, 1990) (see also Fig. 3). The fact that such a large number of Fos-Jun-related heterodimers exist has enormous implications for the regulation of specific genes. Thus, AP-1-dependent transcription of specific genes may depend on specific combinations of Fos–Jun or Jun–Jun heterodimers in particular target cells.

That the different members of the Fos–Jun families have different functions in the developing organism has been inferred by analysis of their expression during development. In contrast to c-*fos*, *fos*B is not expressed in the developing limbs of the mouse embryo, but in the nervous system, and exhibits a different pattern of expression in the

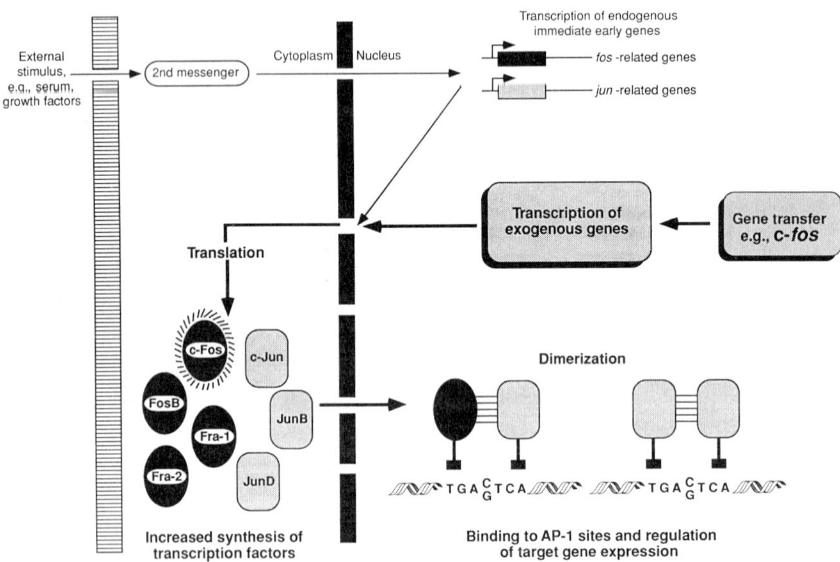

FIGURE 3 Schematic representation of the role of Fos and Jun in signal transduction and regulation of gene expression. Different external stimuli can induce transient expression of endogenous immediate early-gene messenger RNA (mRNA) and protein—for example, *fos*- and *jun*-related genes. Alternatively, gene transfer methods can be utilized to express constitutively high mRNA levels of exogenously added oncogenes, (e.g., c-*fos*), leading to increased protein levels. Proteins are rapidly translocated to the nucleus, where they can form heterodimers (e.g., between Fos family and Jun family members) or homodimers (e.g., within the Jun family). Thus, altered levels of transcription factor complexes can bind to the activator protein 1 (AP-1) DNA consensus sequence (TGA[C/G]TCA) and regulate target gene expression.

adult organism (Redemann-Fibi *et al.*, 1991). Perhaps a more significant observation is that the expression of c-*jun* is high in similar tissues as c-*fos* during development, namely, in the chondrogenic regions surrounding long bones (Wilkinson *et al.*, 1989). (The potential biological implications of this co-expression are discussed in Section IV.) Finally, the pattern of *jun*B expression in the developing embryo differs from that of c-*jun* (Wilkinson *et al.*, 1989). Although these observations suggest different functions for each member of the Fos–Jun family, they emphasize the potential complexity of AP-1-dependent transcriptional regulation. As we describe in the following sections, our approaches at addressing this complexity are to modulate *in vivo* the levels of one component of this complex system (i.e., c-*fos*) and assess the phenotypic consequences of this deregulated expression on the whole organism (Fig. 3).

III. OVEREXPRESSION OF c-*fos* LEADS TO SKELETAL DEFECTS IN TRANSGENIC AND CHIMERIC MICE

A. Rationale

To study the role of c-*fos* during development and to understand the specificity of *fos*-induced tumorigenesis, we have made several DNA constructs in which the murine genomic c-*fos* gene was fused to different ubiquitous promoter elements in order to enable ectopic expression in a wide variety of tissues. Here, we focus on the studies using the inducible human metallothionine promoter (hMT) and the promoter region of the heavy chain of the major histocompatibility complex (MHC) class I antigen, H-2K^b (H2) (Fig. 4 and Table I). Both promoters have been shown to direct expression of linked oncogenes to a wide variety of cell types (for review, see Wagner, 1990). The constructs also contained one additional modification: The mRNA destabilizing sequence that is present in the 3′ untranslated region of c-*fos* (see Section II.A) was removed and replaced with the 3′ long terminal repeat (LTR) from the FBJ-MSV. The rationale for this was twofold. First, the LTR provides a polyadenylation signal for termination of the mRNA molecule, and, second, it ensures the stability of c-*fos* message levels. Indeed, in cell transfection experiments using NIH 3T3 fibroblasts, this modification resulted in the stable expression of c-*fos* mRNA (Rüther *et al.*, 1985). These constructs have been designated MT-c-*fos*LTR and H2-c-*fos*LTR. In addition, the MT-c-*fos*LTR construct also contains the neomycin-resistance gene (*neo*) fused to the SV40 promoter to enable selection of cells in culture (Fig. 4).

In this review, we focus our discussion on the results obtained with transgenic mice harboring the H2-c-*fos*LTR construct and on chimeric mice harboring the MT-c-*fos*LTR construct. It should be noted that trans-

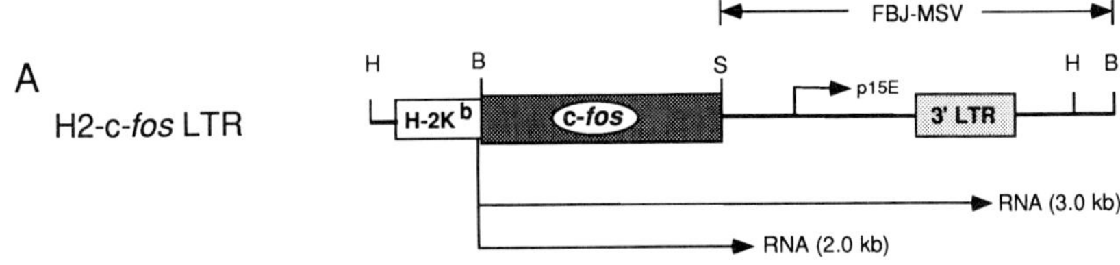

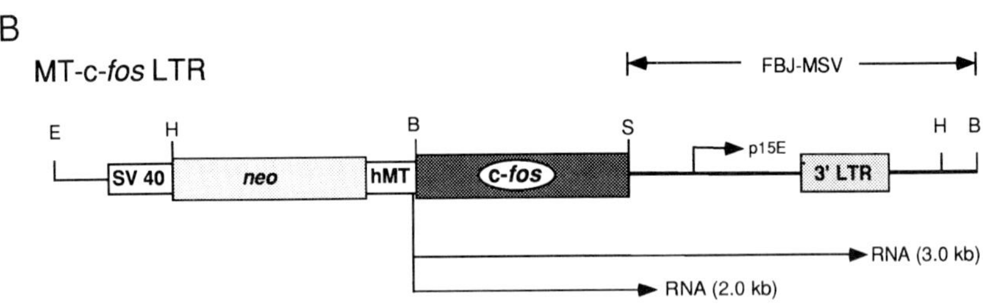

FIGURE 4 DNA constructs used for ectopic expression of c-*fos* in transgenic (H2-c-*fos*LTR) and in embryonic stem cell chimeric (MT-c-*fos*LTR) mice. In (A) the genomic c-*fos* gene is expressed from the major histocompatibility complex class I antigen H-2k^b promoter and in (B) from the human metallothionine promoter (hMT). In addition, the construct in (B) also contains the gene encoding neomycin resistance (*neo*) fused to the SV40 promoter to enable selection of c-*fos*-expressing clones *in vitro*. In both constructs, sequences 3′ to the *Sal*I site are derived from the FBJ-MSV, as indicated, and contain the 3′ FBJ LTR (3′ LTR) and the coding sequence from the retroviral envelope protein p15E. From this construct, two c-*fos* messenger RNA transcripts are synthesized: A 3.0-kilobase (kb) transcript terminating at the poly(A) site in the 3′ LTR and a second 2.0-kb transcript terminating at a cryptic poly(A) site present in the FBJ-derived sequence 3′ to the *Sal*I site. E, *Eco*RI; H, *Hin*dIII; B, *Bam*HI; S, *Sal*I; X, *Xba*I; LTR, long terminal repeat; MSV, murine sarcoma virus.

genic mice generated by using a MT-c-*fos*LTR construct develop a phenotype identical to the H2-c-*fos*LTR mice, but at a much lower penetrance (only 15%; see Rüther *et al.*, 1987, 1989). Therefore, the latter family is more suitable for analysis of c-*fos* function. With regard to the ES cell experiments, only constructs with the hMT promoter have been used; the H2-c-*fos*LTR construct has not been used in ES cells because the promoters of MHC class I antigens are apparently not active at such early stages of development (David-Watine *et al.*, 1990), thus precluding selection of c-*fos*-overexpressing cells.

B. Different Skeletal Tissues Are Affected

In H2-c-*fos*LTR transgenic mice as early as 4 weeks after birth, noticeable swellings were observed on the long bones, specifically in the areas of the distal femur and proximal tibia. These lesions increased significantly in size after only a few months and progressed to large calcified tumors in virtually all bones of the body (Fig. 5B). The penetrance of osteosar-

TABLE I Expression of Different *fos* Constructs in Transgenic Mice[a]

Construct	No.[b]	S	P	L	T	H	Lu	G	I	K	M	B	SG	Bone		Tu
														LB	Ca	
MT-c-*fos*	7	–	±	–	–	–	–	±	ND	–	–	–	–	ND	ND	0
H2-c-*fos*	5	+++	±	+	+++	+	+++	+	ND	±	±	–	+++	+	ND	0
MT-c-*fos*LTR	2	–	++	–	–	+++	+	+	ND	++	++	+++	+	++	ND	+++
H2-c-*fos*LTR	4	+	–	±	+	+++	+++	–	–	++	++	+++	+	++	++	++++
HMG-c-*fos*LTR	5[c]	–	ND	±	+	+	++	±	ND	+	++	–	–	ND	ND	ND[c]
MMTV-c-*fos*Δ	6[d]	–	ND	++	–	ND	+++	++	±	+++	ND	+	++	ND	ND	0
H2-*fos*B LTR	5	+++	ND	++	+++	–	–	ND	+++	ND	ND	–	–	+++	+	0

[a]Shown are the relative abundancies of exogenous c-*fos* messenger RNA in mouse tissues as estimated by Northern blot analysis: S, spleen; P, pancreas; L, liver; T, thymus; H, heart, Lu, lungs; G, gonads; I, intestine; K, kidneys; M, skeletal muscle; B, brain; SG, salivary glands; LB, long bones; Ca, calvaria; Tu, bone tumor; ND, not determined. Corresponding expression of Fos and FosB proteins was confirmed in H2-c-*fos*LTR, H2-*fos*BLTR, and MMTV-c-*fos*Δ families. MT, human metallothionine promoter; H2, murine major histocompatibility complex class I H-2K^b promoter; MMTV, mouse mammary tumor virus promoter; HMG, 3-hydroxy-3-methylglutaryl-CoA-reductase promoter; LTR, 3′LTR from the FBJ-MSV; Δ, deletion of the 3′ RNA destabilizing sequences in the c-*fos* gene. The following constructs are able to transform fibroblasts *in vitro:* MT-c-*fos*LTR, H2-c-*fos*LTR, MMTV-c-*fos*Δ (only in the presence of dexamethasone), and H2-*fos*BLTR (data not shown).

[b]Number of independent mouse lines. Data shown for each construct represent a summary of expression from all mouse lines.

[c]Data are from only one family, which expressed the transgene at very low levels; however, osteosarcomas developed at a low frequency after at least 1 year and were not further analyzed.

[d]Data represent levels of exogenous c-*fos* RNA after induction *in vivo* with dexamethasone.

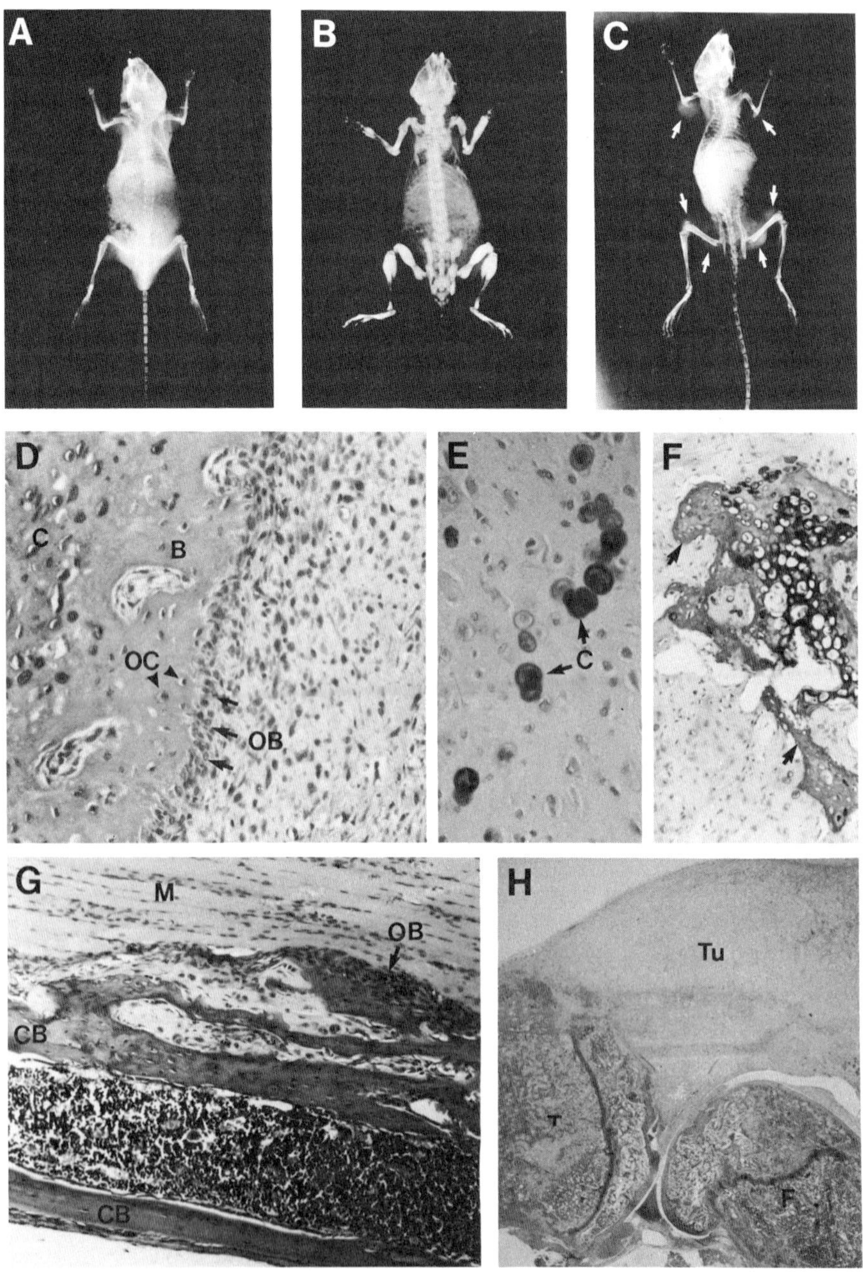

FIGURE 5 X-ray and histological analyses of the pathology observed in c-*fos* transgenic and chimeric mice. X rays of a normal mouse (A), an H2-c-*fos*LTR transgenic mouse (B), and an MT-c-*fos*LTR chimeric mouse (C). Note the large calcified tumor present throughout the skeleton of the transgenic mouse compared to those present in the joints of the chimeric mouse (arrows). (D) A chondroblastic osteosarcoma from an H2-c-*fos*LTR trans-

coma formation was 100%, because all mice carrying the transgene (i.e., both heterozygotes and homozygotes) developed these tumors (Grigoriadis *et al.*, 1993). The high penetrance and severity of the phenotype in H2-c-*fos*LTR mice enabled us to address specific questions on a background that was highly uniform. This phenotype was similar in different families of transgenic mice, suggesting that the phenotype observed is apparently independent of the site of integration of the transgene (see also Table I).

To address the consequences of ectopic c-*fos* expression during development, where endogenous levels show a very restricted pattern of expression (see Section II.B), we isolated different ES cell clones that were selected for high c-*fos* expression and injected them into blastocysts. The resulting chimeric mice were all born healthy with no apparent abnormalities, suggesting that high levels of Fos protein during early embryonic development can be tolerated. However, as early as 3–4 weeks of age the chimeric mice developed palpable lesions in the areas of the spine and joints of the long bones (Wang *et al.*, 1991), reminiscent of the phenotype observed in the transgenic mice. Upon further examination, however, it was evident that these lesions were different from the osteosarcomas observed in transgenic mice, in that the lesions were generally not mineralized, although some contained sites of ossification (Fig. 5C). Examination of the affected mice at autopsy confirmed the presence of large tumors associated with the spine, limbs, ribs, and shoulders. The same phenotype was observed in chimeric mice generated from several different c-*fos*-expressing ES cell clones and the frequency of tumor formation ranged from 60 to 100%, depending on which *fos*-ES clone was used. These data suggest that the phenotype did not depend on the integration site of the DNA construct but, rather, was a result of c-*fos* overexpression during development. It should be mentioned that germ-line transmission has not yet been obtained; that is, we

genic mouse showing large areas of mineralized bone containing osteocytes (arrowheads) and lined with cuboidal cells resembling osteoblasts (arrows). In addition, areas of chondrocyte differentiation are also present. Magnification, ×120. (E) Tumor from an MT-c-*fos*LTR chimeric mouse showing a typical region containing chondrocytes at different stages of differentiation. Magnification, ×230. Some of the chimeric tumors also contained areas of ossification surrounding the chondrocytes (arrows in F). Magnification, ×100. (G) Pretumor lesion present in the distal tibia of a 3-week-old H2-c-*fos*LTR transgenic mouse showing a region of new bone formation adjacent to the existing cortical bone and lined by osteoblastlike cells. Magnification, ×100. (H) Hind limb of a 6-week-old MT-c-*fos*LTR chimeric mouse showing a large tumor present around and inside the knee joint. Magnification, ×50. B, bone; BM, bone marrow; C, chondrocytes; CB, cortical bone; F, femur; OB, osteoblasts; OC, osteocytes; M, muscle; T, tibia; Tu, tumor. Sections in D–H are 4- to 6-μm paraffin sections and stained with hematoxylin and eosin.

have not yet been able to show that the *fos*-ES cell derivatives have differentiated into functional germ cells. Thus, expression of exogenous c-*fos* in these chimeras can only occur in those tissues in which the *fos*-ES cells randomly contributed, and this contribution is different in each chimera. This is in contrast to the *fos* transgenic mice, where all tissues in all animals contain the transgene and each generation of mice develops the identical phenotype. Nevertheless, it is interesting that despite the random ES contribution in the *fos* chimeras, all mice develop the same phenotype (Wang *et al.*, 1991).

Histological analysis was performed on both transgenic and chimeric tumors in order to characterize in greater detail the nature of the observed pathology. As shown in Fig. 5, the characteristics of each tumor are related but display fundamentally different properties. c-*fos* transgenic tumors resembled typical osteosarcomas, containing vast areas of osteoid and mineralized bone lined by cuboidal osteoblastlike cells (Fig. 5D) expressing high levels of alkaline phosphatase when stained histochemically (data not shown). The tumors were vascularized and also contained areas of chondrocyte differentiation, which is typical of chondroblastic osteosarcomas (see also Rüther *et al.*, 1989). In contrast, the hallmark of c-*fos* chimeric tumors was not bone formation but the presence of distinct foci of chondrogenic cells and differentiated chondrocytes surrounded by abundant extracellular matrix (Fig. 5E). The matrix stained intensely with Alcian blue and exhibited metachromasia after Toluidine blue staining confirming the presence of sulfated proteoglycans, which are abundant in cartilage matrix. Although the majority of tumors were unmineralized, a small number of tumors also contained areas of ossification (Fig. 5F). In transgenic osteosarcomas, the tumors appeared to originate from regions corresponding to the periosteum of the long bones and consisted of areas of active bone formation (Fig. 5G). In contrast, the tumors in the *fos* chimeras were always associated with the joints and appeared to destroy the articular surfaces of the long bones (Fig. 5H). Since chondrocytes were the predominant cell type present in the *fos* chimera tumors, we have designated them as chondrogenic or chondrosarcomalike tumors to distinguish them from the osteogenic and osteosarcomalike tumors observed in the *fos* transgenic mice. In both cases, the tumors surrounded virtually the entire vertebral column and the long bones and were highly invasive, as evidenced by the presence of ectopic bone and cartilage formation in the bone marrow spaces of transgenic and chimeric bones, respectively (data not shown). Thus, despite some similarities in skeletal localization, the tumors induced in the *fos* transgenic and chimeric mice are clearly different, suggesting that the affected cells in each tumor type may be different. Nevertheless, the cellular specificity of c-*fos* was reminiscent of the chondroosseous neoplasms induced by the v-*fos*-containing FBJ- and FBR-MSVs (Ward and Young, 1976).

C. Osteogenic and Chondrogenic Tumors Are the Result of c-*fos* Overexpression

To investigate exogenous c-*fos* expression in different transgenic and chimeric mouse tissues and to determine the tissue specificity of the transgene, we performed Northern blot analyses on tumor tissues as well as on unaffected tissues. In both transgenic and chimeric mice, the highest levels of transgene expression were observed in the respective tumor tissues. In addition, stable expression of exogenous c-*fos* was detected in a wide variety of other tissues (Rüther *et al.*, 1987; Wang *et al.*, 1991; Grigoriadis *et al.*, 1993), some at very high levels (summarized in Table I) and in which no abnormalities were ever observed. Thus, despite widespread and efficient expression in different tissues, the generated phenotypes were restricted to bone in the *fos* transgenics and to cartilage in the *fos* chimeras.

The complete penetrance of osteosarcoma formation in the transgenic mice enabled us to correlate the timing of exogenous c-*fos* expression with the onset of the phenotype, thereby addressing the causal role of c-*fos* in tumor formation. Time course experiments using Northern blot and X-ray analyses indicated that the transgene is first expressed in the bones of transgenic mice at approximately 2–3 weeks after birth, whereas the first evidence of pathological lesions was at 4 weeks of age (Grigoriadis *et al.*, 1993). Therefore, expression of the transgene preceded the onset of the phenotype, implicating c-*fos* in the development of osteosarcomas in transgenic mice. With regard to chimeric mice, we have confirmed that exogenous Fos protein is expressed during embryogenesis when the endogenous gene is normally not expressed. Specifically, expression was observed in several tissues of mesodermal and ectodermal origins, for example, in the developing prevertebral regions, thyroid gland, muscle, and cervical ganglia (Wang *et al.*, 1991). Taken together, the widespread tissue expression and the timing of expression suggest that high levels of c-*fos* can confer a specific growth advantage to osteogenic and chondrogenic cells.

Since all primary chondro- and osteosarcomas expressed high levels of exogenous c-*fos*, it was of interest to assess whether the expression of other genes was affected. To this end, we screened several tumors from different transgenic and chimeric mice for expression of AP-1-associated genes (e.g., endogenous c-*fos*, *fos*B, c-*jun*, *jun*B, *jun*D) and bone- and cartilage-associated genes (e.g., type I collagen, type II collagen, alkaline phosphatase, osteopontin, osteocalcin). The data, summarized in Table II, show that primary transgenic osteosarcomas expressed high but variable levels of all osteoblast-associated markers tested, but not all AP-1 genes were expressed. One major difference between the osteogenic and chondrogenic tumors was in the expression of endogenous c-*fos*: Generally, the bone tumors did not express endogenous c-*fos*, whereas

TABLE II Expression of AP-1 and Bone- and Cartilage-Associated Genes in Tumors Isolated from c-*fos* Transgenic and Chimeric Mice[a]

Tumor	AP-1 genes					Bone and cartilage genes				
	c-*fos*									
	Exogenous	Endogenous	c-*jun*	*jun*B	*jun*D	Type I collagen	Type II collagen	AP	OP	OC
Bone[b]										
1	++++	−	+	+++	+++	+	−	+++	+++	++++
2	++++	−	++	+++	+++	+	±	+++	++	++++
3	+++	−	+	++++	+++	+	±	++++	+++	+++
4	++	−	++	+++	+++	+	−	++++	+++	++++
Cartilage[c]										
1	++++	+	++	ND	ND	+	+	ND	+++	+
2	+++	+	+++	ND	ND	+	+	ND	+++	+++++
3	++++	+	++++	ND	ND	+	+	ND	+++++	++
4	++++	++	++++	ND	ND	++	±	ND	++++	++

[a] Data represent the relative abundancies of gene expression as estimated by Northern blot analysis. All blots were hybridized with specific probes against c-*fos*, c-*jun*, *jun*B, *jun*D, type I collagen, type II collagen, alkaline phosphatase (AP), osteopontin/2ar (OP), and osteocalcin (OC). ND = not determined.

[b] Primary bone tumors isolated from H2-c-*fos*LTR transgenic mice.

[c] Primary cartilage tumors isolated from MT-c-*fos*LTR chimeric mice.

the cartilage tumors co-expressed both endogenous and exogenous c-*fos* RNA. With respect to other AP-1 genes, it is of interest that c-*jun* levels in osteosarcomas were only moderately high while the chondrosarcomas expressed high levels of c-*jun*, which appeared to correlate with the exogenous c-*fos* levels (see also Section III.D). In addition, the chondrosarcomas expressed the cartilage marker gene type II collagen and, interestingly, some bone markers as well (Table II). It should be noted here, however, that histological analysis showed that all these tumors were quite heterogeneous. Thus, the information gained from the primary tumors is limiting with respect to the specific cell types that are responsible for the expression of these specific markers.

D. Targets for Ectopic c-*fos* Expression Are Osteogenic and Chondrogenic Cells

To characterize the potential target cells for the c-*fos* oncoprotein and to confirm that these target cells are indeed different in the transgenic and chimeric mice, we isolated several cell populations from both osteogenic and chondrogenic tumors. All primary and clonal tumor-derived cell lines were characterized with respect to morphology, doubling time, gene expression, differentiation capacity, and tumorigenicity (see also Wang *et al.*, 1991; 1993; Grigoriadis *et al.*, 1993). Cell lines generally have a fibroblastic morphology typical of transformed cells and their doubling times ranged from 18 to 48 hours. All the cell lines tested were tumorigenic *in vivo* when injected into syngeneic or nude mice (Table III). Of greater significance, however, was the observation that the cell-induced tumors had similar morphological and histological characteristics to the original parent tumors. Thus, while the cell lines isolated from the osteosarcomas induced tumors containing osteoid and bone, the chondrosarcoma-derived cells induced tumors containing cartilage (Figs. 6A and B). In addition, one cartilage tumor-derived clonal cell line also induced tumors that were mostly cartilagenous but also contained areas of ossification (wT2-1; Wang *et al.*, 1991). Because the tumors induced by these different cell lines were clearly different, we have designated the cell lines isolated from the *fos* chimeras and transgenics as chondrogenic (i.e., they form tumors containing cartilage) and osteogenic (i.e., they form tumors containing osteoid and bone), respectively. It should be emphasized, however, that there was considerable heterogeneity among the osteogenic cell lines; while some clearly formed bone (e.g., Fig. 6A), others were anaplastic and poorly differentiated, containing some osteoid, numerous tumor giant cells, and many mitoses (Table III).

Analysis of exogenous c-*fos* expression in the chondrogenic and osteogenic tumors induced by the different cell lines indicated that all

TABLE III Characteristics of Cell Lines Isolated from c-*fos*-Induced Bone and Cartilage Tumors[a]

	AP-1 genes					Bone and cartilage genes					
	c-*fos*										
Cell line	Exogenous	Endogenous	c-*jun*	*jun*B	*fra*-1	Type I collagen	Type II collagen	AP	OP	OC	Tumorigenicity (phenotype)[b]
Bone[c]											
R3	+++	−	+	+++	++	−++	−	±	+++	−	+ (anaplastic sarcoma + osteoid)
C1	++++	−	+	+++	++	+++	−	++	+	−	+ (ND)
P1	++	−	±	+++	+	++++	−	++++	++	−	+ (osteosarcoma)
K6	++++	−	++	++++	+++	+++	−	++	++	−	+ (anaplastic sarcoma + osteoid)
Cartilage[d]											
wT2-1	++++	++++	++++	+++	++++	+++	+	−	+++	−	+ (chondrosarcoma + ossification)
wT2-7	++++	+	+++	+++	++++	+++	+++	−	+++	−	+ (chondrosarcoma)
wT2-8	+++	+	+++	+++	ND	+++	++	±	+++	−	+ (chondrosarcoma)
wT2-9	++	±	+++	+++	+++	+++	++++	+	+++	−	+ (chondrosarcoma)

[a]Data represent the relative abundancies of gene expression as estimated by Northern blot analysis. All blots were hybridized with specific probes against c-*fos*, c-*jun*, *jun*B, *fra*-1, type I collagen, type II collagen, alkaline phosphatase (AP), osteopontin/2ar (OP), and osteocalcin (OC). ND = not determined.

[b]Tumorigenicity studies performed by injecting 1–2 $\times$ 10^6 cells into syngeneic or nude mice. Tumor diagnosis was based on morphological and histological identification (see text for details).

[c]Primary cell lines established from bone tumors induced in H2-c-*fos*LTR transgenic mice (see also Grigoriadis *et al.*, 1993).

[d]Clonal cell lines established from cartilage tumors induced in MT-c-*fos*LTR chimeric mice (see also Wang *et al.*, 1991).

tumors expressed high levels of exogenous c-*fos* (data not shown). Specifically, as demonstrated using RNA *in situ* hybridization and immunostaining analyses, high expression of exogenous c-*fos* RNA and protein was localized predominantly in the chondrogenic cells of the cartilage tumors and in the osteogenic cells of the bone tumors (Figs. 6C–6F). These data further support the notion that c-*fos* has affected osteogenic cells in the transgenic mice and chondrogenic cells in the chimeric mice. However, we have not ruled out the possibility that earlier progenitor cells are also targets for Fos overexpression (e.g., wT2-1 cells; Wang *et al.*, 1991).

A comparison of gene expression in the bone- and cartilage tumor-derived cell lines revealed some important differences. Despite high expression of exogenous c-*fos* in all cell lines tested, the expression of the endogenous gene was not detectable in the osteogenic cell lines but was present in the chondrogenic cell lines concomitantly with the exogenous c-*fos* (Fig. 7). Whether this suggests that there are differences between the cell lines in the ability of Fos protein to repress its own promoter (for review, see Shaw *et al.*, 1989) is not known at this time. Similar to the primary tumors, expression of c-*jun* was higher in chondrogenic cells and appeared to correlate with levels of exogenous c-*fos* (e.g., Fig. 6G). In contrast, c-*jun* expression levels in osteogenic cells were only moderately high and were independent of c-*fos*. Not all Jun family members showed the same expression pattern, as *jun*B was expressed ubiquitously in all cell lines tested (Fig. 7). Expression of the *fos*-related gene *fra*-1 also appeared to correlate with exogenous levels of c-*fos* in both osteogenic and chondrogenic cells (Table III). With regard to bone- and cartilage-associated genes, the most significant differences were in the expression of cartilage-specific type II collagen, which was present in all chondrogenic cell lines but not in any of the osteogenic cells. This is consistent with the notion that these cells are indeed chondrogenic. Regarding osteoblastic marker genes, the most notable observation was that osteocalcin expression was not detectable in any cell line expressing c-*fos* (Table III). This was also observed in cell lines isolated previously from osteosarcomas in MT-c-*fos*LTR transgenic mice (Goralczyk *et al.*, 1990). This is interesting in view of the recent findings by Schüle *et al.* (1990), who identified an AP-1 site in the promoter region of the human osteocalcin gene and demonstrated in transient transfection assays that AP-1 activity was associated with decreased transcriptional activation of the osteocalcin gene. Thus, this gene may represent one example of a *fos*/*jun*/AP-1-regulated gene in osteoblastic cells. Whether other osteoblastic genes are also regulated by AP-1 activity, or certain parameters associated with osteoblastic cells are affected in high c-*fos*-expressing cell lines, is not known. To this end, we are currently investigating the cyclic

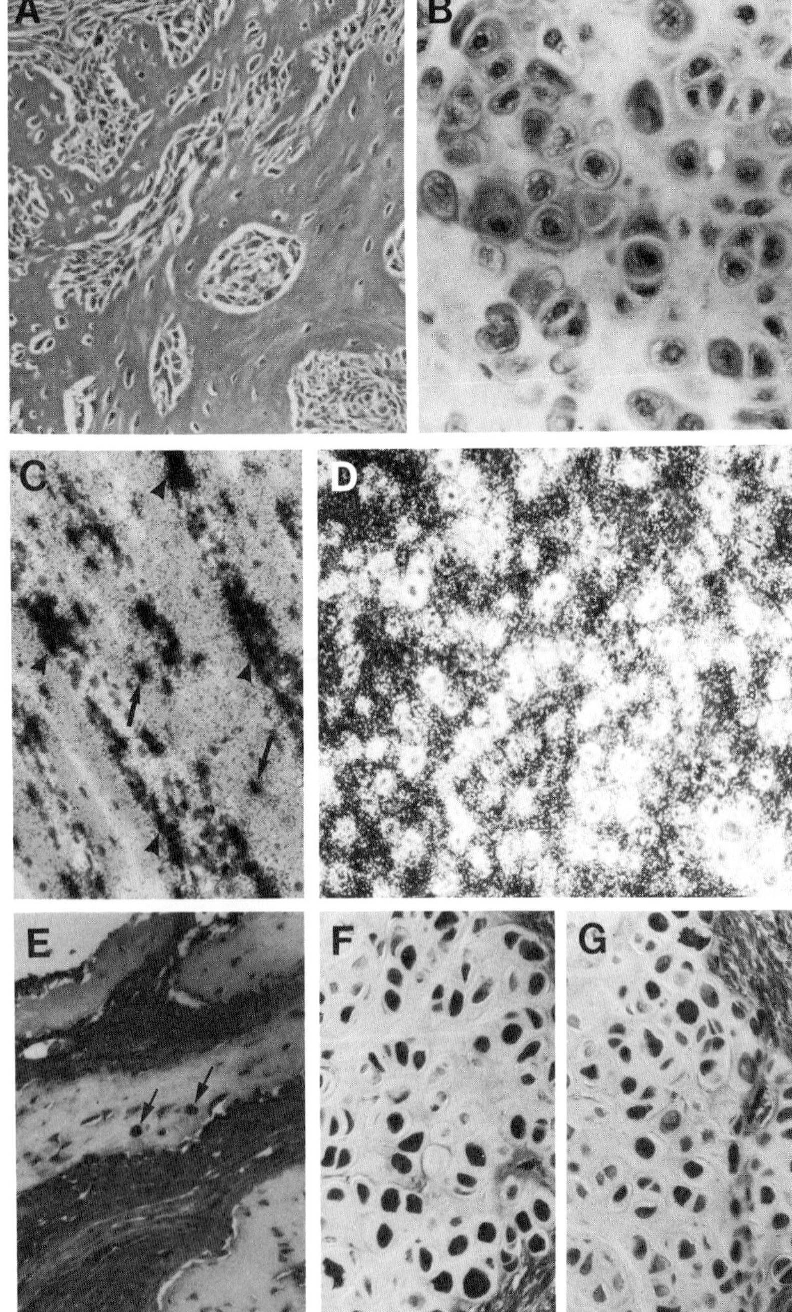
A
B
C
D
E
F
G

adenosine monophosphate (cAMP) responses of isolated clonal cell lines to different hormones (e.g., parathyroid hormone, prostaglandin E_2) and also the steroid hormone regulation of collagen synthesis and alkaline phosphatase activity, markers that are commonly associated with osteoblastic cells (see Rodan and Rodan, 1984; Heersche and Aubin, 1990). In this regard, it is interesting that the cell lines isolated by Goralczyk *et al.*, (1990) exhibited a reduced cAMP response to parathyroid hormone but retained high alkaline phosphatase activity. Such studies are necessary to further characterize these different cell populations. A summary of the results obtained in the cell lines derived from c-*fos* transgenic and chimeric mice described here is summarized in Table III.

Finally, it is evident that the chondrogenic cell lines isolated from the *fos* chimeric tumors are quite stable; some of these clones have expressed type II collagen even after 1 year in continuous culture (Wang *et al.*, 1993). This is of interest because of the well-known difficulty in establishing cartilage cell lines *in vitro,* since they tend to lose rapidly certain phenotypic properties (e.g., type II collagen; for review, see von der Mark, 1986). Although some stable cartilage-forming cell lines have been described (Grigoriadis *et al.*, 1989), the c-*fos*-transformed cell lines described here should enable the further study of the role of c-*fos* in cartilage cell transformation and regulation of chondrogenic cells.

E. Summary

In the preceding studies, we utilized two experimental systems to express high levels of exogenous Fos protein in mice. Both routes enabled

←

FIGURE 6 Analysis of tumors induced by cell lines derived from c-*fos* transgenic and chimeric mice. (A) Histological section through an osteosarcoma induced by P1 cells derived from an H2-c*fos*LTR transgenic bone tumor showing abundant bone formation. No chondrocytes are present in this tumor. Magnification, ×150. (B) Section through a chondrosarcoma induced by wT2 cells derived from a MT-c-*fos*LTR chimeric cartilage tumor showing abundant cartilage formation. Magnification ×460. (C) *In situ* hybridization for c-*fos* in a P1-induced osteosarcoma showing localization of the transgene to the osteoblasts (arrowheads) and osteocytes (arrows) (bright field). (D) *In situ* hybridization for c-*fos* in a wT2-induced chondrosarcoma showing expression of c-*fos* in chondrogenic cells (dark field). For both (C) and (D), an anti-sense riboprobe was used that was specific for the exogenous c-*fos* transcripts. Immunocytochemistry for Fos protein demonstrated specific nuclear staining for Fos protein in osteoblasts (arrowheads) and osteocytes (arrows) of P1-induced bone tumors (E), and in chondrocytes of wT2-induced cartilage tumors (F). In addition, chondrocytes in wT2-induced tumors also expressed high levels of Jun protein (G). All sections are 4- to 6-μm paraffin sections. Magnification for C, D, F and G, ×230; magnification for E, × 100.

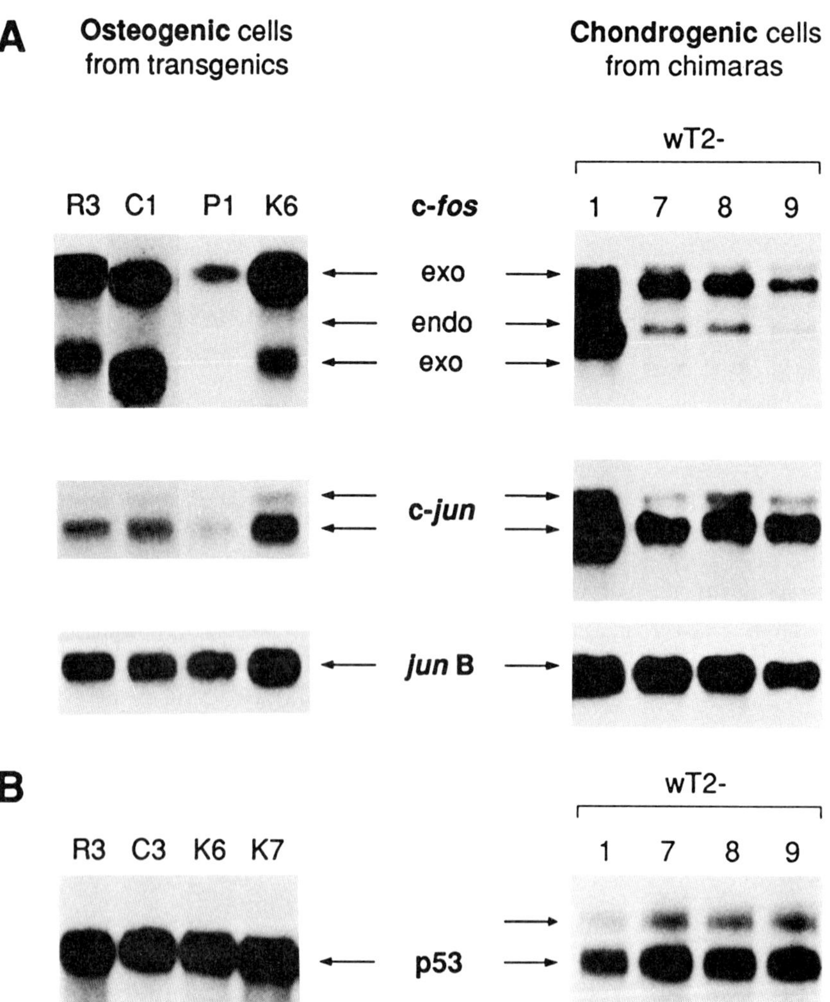

FIGURE 7 Northern blot analysis of AP-1 (A) and (B) p53 gene expression in cell lines isolated from c-*fos* transgenic (osteogenic cells) and chimeric (chondrogenic cells) mice. Poly(A)$^+$ messenger RNA from osteogenic cell lines and clonal chondrogenic cell lines was hybridized first with a c-*fos* probe followed by probes for c-*jun*, *jun*B, and p53. Approximately equal amounts of RNA are loaded in each lane (data not shown). A complete summary of gene expression in several osteogenic and chondrogenic cell lines is shown in Table III.

us to disrupt the normal physiological processes of mesenchymal cell development leading to specific tumors. The causal role of Fos in generating these phenotypes was demonstrated by several lines of evidence. First, exogenous c-*fos* was expressed in many different organs, yet only bone and cartilage tissues were affected, tissues that are also natural targets for the v-*fos* oncogene. Second, the timing of Fos expression indicated that it preceded the first morphological changes. In fact, the observed time difference between expression of the transgene and appearance of early lesions in both transgenic and chimeric mice suggested that other events were required to elicit the observed biological effect. Third, the expression of the introduced gene in both transgenic and chimeric mice was highest in pathological tissues, specifically in different cell types, that is, in osteogenic cells in transgenic mice and chondrogenic cells in chimeric mice. Finally, and perhaps most importantly, cell lines isolated from each tumor type retained the ability to induce tumors in nude mice with morphological features similar to those of the original primary *fos*-induced tumors. Thus, osteosarcoma-derived cell lines gave rise to osteogenic tumors and cartilage tumor-derived cell lines induced cartilage tumors. Taken together, we have developed an experimental basis for postulating that there are two distinct but developmentally related target cell populations that are sensitive to high levels of Fos.

IV. SPECIFICITY OF c-*fos* ACTION

A. Timing of c-*fos* Expression

Although the phenotypes observed in both transgenic and chimeric mice are somewhat related, we propose that the observed phenotypes reflect different target cell specificities for Fos. Two major lines of evidence could account for this altered specificity. First, the expression of c-*fos* in the chimeras occurs earlier than it does in the transgenic mice by virtue of the fact that they were generated using high c-*fos*-expressing ES cells; exogenous Fos is also expressed earlier than the endogenous gene. In the case of transgenic mice, we have shown that expression of the transgene occurs at birth or shortly thereafter (Heckl and Wagner, 1989; Grigoriadis *et al.*, 1993). One possible consequence of the different timing of c-*fos* expression is that high levels of Fos protein in chimeric embryos would be able to act on a different set of target cells, which may not be present after birth, when exogenous Fos protein is expressed in transgenic mice. Examples of "early" target cells may include mesenchymal progenitor cells, which have not yet committed to chondrogenic and/or osteogenic lineages and which may become transformed by high levels of Fos (Miller *et al.*, 1984). That fibroblastic cells isolated from the *fos* chimeras can also be transformed by high Fos levels is supported by the observations that fibroblasts isolated from unaffected tissues expressing the transgene (e.g., lungs) give rise to fibrosarcomas when injected into syngeneic or nude mice (Wang *et al.*, 1991). Alternatively, the appropriate secondary signals and cooperating factors necessary for tumor induction may also be developmentally regulated, being present during development and not in the postnatal animal. Clearly, other events must be occurring in the chimeric mice as evidenced by the fact that the early lesions and subsequent tumor growth were observed only after birth and in young mice. In this regard, the location of the early lesions also supports the notion of different cellular specificities of Fos action. In the *fos* chimeras, we have observed early lesions originating along the vertebral column, specifically from the areas adjacent to the intervertebral disc (Wang *et al.*, 1991). These lesions contained abundant fibroblastic cells in addition to several foci containing chondrocytes. In contrast, early lesions in the *fos* transgenics were usually first observed along the long bones, specifically in the periosteal regions of the distal femur and proximal tibia. In addition, the transgenic mice always develop large lesions in the calvaria, specifically in the areas of the parietooccipital sutures (Grigoriadis *et al.*, 1993), whereas calvaria lesions have never been observed in the *fos* chimeras.

The second piece of evidence supporting a different target cell specificity for Fos relates to cooperating factors that are necessary for transformation and initiation of tumor progression. In addition to the absolute levels of Fos protein, it has been shown that Fos can only exert its effects on transcriptional activation and transformation in cells that also contain the necessary cooperating genes (e.g., *jun*-related genes) (Schuermann *et al.*, 1989). In this regard, it is interesting that the cells that appear to be targets for Fos in the chimeric mice (i.e., chondrogenic cells) also express very high levels of c-*jun* (Wilkinson *et al.*, 1989). Together with the observations that endogenous c-*jun* RNA appears to be coordinately regulated with levels of exogenous c-*fos* in all chimeric tissues and cell lines analyzed (Wang *et al.*, 1991), the data suggest that high levels of both Fos and Jun in chondrogenic cells during development are responsible for initiation of proliferation/transformation. In contrast, the osteosarcomas and osteosarcoma-derived cell lines from the *fos* transgenics do not show a similar pattern of c-*jun* expression; therefore, c-*jun* does not appear to be a limiting factor in the action of c-*fos* (see also Rüther and Wagner, 1989; Grigoriadis *et al.*, 1993). In addition to the actual formation of Fos–Jun complexes, it is known that specific modifications of these complexes, either via phosphorylation or via interactions with other proteins, can affect DNA binding and subsequent transactivation of responsive genes (e.g., see Ofir *et al.*, 1990; Pulverer *et al.*, 1991) and that these modifications can also occur in a cell type-specific fashion (Baichwal *et al.*, 1991). Thus, it is possible that chondrocytes at a specific time point in development contain the necessary factors to interact with the high levels of Fos–Jun complexes, thereby conferring a specific proliferative effect on these cells. In contrast, osteogenic cells and other tissues that also express high levels of Fos may not contain these specific factors. That overexpression alone is not enough to elicit transformation *in vivo* is best illustrated in the *fos* transgenic mice that express high levels of Fos in tissues that never develop any pathology (e.g., heart, lung in H2-c-*fos*LTR mice; see Table I). In addition, transgenic mice harboring an MMTV-c-*fos*Δ construct can be induced by dexamethasone to express transiently high levels of exogenous c-*fos* RNA and protein in several tissues with no resulting phenotype, despite the fact that the same construct can conditionally transform fibroblasts *in vitro* (Table I). Why transformation has not been observed in these mice is not entirely clear but may reflect the fact that the levels of Fos present after induction do not persist long enough to trigger subsequent events.

Other factors that may be implicated in the specificity of Fos action in transgenic and chimeric mice are tumor suppressor genes, specifically p53, which is a negative regulator of the cell cycle (for review, see Levine *et al.*, 1991; Weinberg, 1991). Mutations in p53, either by alteration, inactivation, or interaction with specific oncogene products can lead to distinct neoplasias, including osteosarcoma (Masuda *et al.*, 1987). It is tempting to speculate that p53, via interaction with Fos–Jun complexes, may affect the ultimate specificity of Fos action. Because exogenous Fos is expressed in the chimeras during embryonic development, one can argue that there is a greater potential for interaction with p53 proteins resulting in the alteration of cell-cycle control. In contrast, these processes would not occur in the transgenic animals that do not express exogenous Fos before birth. Indeed, that p53 may be important in the genesis of the different tumor types observed in the transgenic and chimeric mice is supported by preliminary observations showing differences in p53 RNA expression between the chondrogenic and osteogenic cell lines described here (Fig. 7). Whether these differences are due to rearrangements in p53 is currently being analyzed.

Finally, we have also observed that in both transgenic and chimeric mice the *fos*-related gene *fra-1* is also expressed at high levels in tumors and cell lines that express high levels of exogenous c-*fos*. Although the significance of this is not yet known, the data imply that *fra-1* may also be a cooperating factor in the cascade of events induced by increased Fos levels. In this regard, recent evidence indicating that *fra-1* expression can be regulated by c-*fos* is of interest (Braselmann *et al.*, 1992).

Taken together, our results regarding the timing of c-*fos* expression and the potential involvement of cooperating oncogenes (e.g., c-*jun*, *fra*-1, p53) may explain the specificity of Fos action but may also suggest why we do not observe osteosarcomas as a primary phenotype in the *fos* chimeras. It should be mentioned that we cannot rule out the possibility that osteosarcomas may develop in older chimeras, since the rapid growth and aggressiveness of the cartilage tumors did not enable us to examine older mice. These speculations should become clearer as we assess the phenotype in the chimeras after passage through the germ line. Although the ES cells that were used can colonize the germ line (Pease and Williams, 1990), it is possible that the *fos*-expressing ES cells described here may be restricted in their developmental potential and fail to differentiate into functional germ cells and subsequently into the target cells responsible for osteosarcoma formation. Thus, it is necessary to ensure that the transgene is present in all skeletal compartments and then ascertain whether the cartilage phenotype in the chimeras is altered in any way. Experiments at achieving germ-like transmission in the *fos* chimeras are currently in progress.

B. Does the 3′ FBJ-Long Terminal Repeat Affect the Target Cell Specificity of c-*fos*?

As described earlier in Section III.A, the DNA constructs that we have used contain a modification whereby the 3′ LTR from the FBJ-MSV replaces the 3′ destabilizing sequences in the c-*fos* gene. Several lines of evidence suggest that this LTR is necessary for the generation of both the osteogenic and chondrogenic phenotypes (see also Wang *et al.*, 1991). Briefly, the function of the 3′ LTR is twofold. First, it provides a polyadenylation signal for the termination of c-*fos* mRNA, thereby ensuring high expression levels. Second, it is possible that the 3′ LTR, which contains both viral promoter and enhancer elements, imparts specificity of expression to bone and cartilage tissues. The enhancer may function in influencing the expression specificity from the MT and H2 promoters. Further evidence comes from the fact that transgenic and chimeric mice harboring MT-c-*fos* constructs without the 3′ LTR modification do not express efficiently exogenous c-*fos*, especially in bone tissues (for summary, see Table I and Wang *et al.*, 1991). In addition, that the 3′ LTR is the sole viral element necessary for stable c-*fos* expression is supported by the observations that chimeric mice harboring an MT-c-*fos*(Δp15E)LTR construct develop identical cartilage tumors to those already described (Wang *et al.*, 1991). Finally, constructs containing H2-c-*fos* without the 3′ LTR do not express efficient levels of the transgene in bone tissues. Rather, the expression profile of exogenous c-*fos* resembles more the expression of endogenous class I H-$2K^b$ genes: Highest expression is detected in hematopoietic organs such as the thymus and spleen (see also Table I) and these mice develop hyperplastic thymuses (increased numbers of thymic epithelial cells) and enlarged spleens (Rüther *et al.*, 1988; see also Rüther and Wagner, 1989, for review). Thus, the presence of the 3′ FBJ-LTR may well influence the tissue specificity of a strong promoter that can ensure high expression levels (Table I).

The specificity of c-*fos* action in the context of the H2 promoter and the 3′ LTR was further investigated by replacing the c-*fos* gene with the *fos*-related gene *fos*B (see also Section II.C). Transgenic mice harboring an H2-*fos*BLTR construct express *fos*B RNA and FosB protein in several different tissues, including high levels in bone (Grigoriadis *et al.*, 1993; see also Table I). However, no phenotype has yet been observed in mice older than 12 months of age. These results further support the notion that it is the combination of c-*fos* and the possible specificity imparted by the 3′ LTR that is responsible for the effects seen on target cells. Interestingly, endogenous *fos*B is not expressed in developing bones (Redemann-Fibi *et al.*, 1991) as are, for example, c-*fos* and c-*jun*; thus, it is also possible that there is no obvious role for this oncogene in the regulation of bone and cartilage cell differentiation. Furthermore, FosB/Jun heterodimers may not transactivate the relevant AP-1-regulated genes to elicit a phenotype.

C. Model for c-*fos* Action in Osteogenic and Chondrogenic Cells

It is generally accepted that overexpression of oncoproteins like c-Fos would, in combination with the appropriate member of the Jun family, result in the transactivation of genes that contain AP-1-responsive elements. This alone may not be enough to lead to the observed phenotype (e.g., a mitogenic or differentiation response), but it may trigger the expression of other so-called secondary factors that are necessary to elicit further events. These secondary factors could include "master" regulatory genes (e.g., "MyoD-like") that are involved in the determination and commitment of cells to specific lineages. These factors are likely to be cell- and/or tissue-specific and would therefore transduce the initial signal (in this case, overexpressed Fos) to specific cell types. We postulate that mesenchymal cells, specifically osteogenic and chondrogenic cells and perhaps earlier uncommitted cells, contain the necessary factors to carry out these functions. Other mesenchyme-derived cells such as fibroblasts (Miller *et al.*, 1984), muscle cells (Lassar *et al.*, 1989) and adipocytes (Distel *et al.*, 1987) have been shown *in vitro* to be sensitive to altered levels of Fos protein, although we have not observed any effects in these cell compartments *in vivo*, which also express high levels of the introduced gene.

Based on the evidence summarized in this review, we propose a model that postulates that high levels of Fos protein in transgenic and chimeric mice affects specifically the osteogenic and chondrogenic lineages and that the ultimate response of each cell type to these high levels is determined, at least in part, by cooperating oncogenes like c-Jun and perhaps by other cell-specific factors that can modify the Fos–Jun–AP-1 transcription factor complex (Fig. 8). Thus, high levels of Fos and Jun during embryogenesis of the *fos* chimeric mice may affect specifically chondrogenic cells that have the appropriate cooperating factors (e.g., factor "X") to trigger the appropriate downstream genes necessary for cartilage tumor formation (e.g., gene "A"). Examples of cooperating factors may be proteins that can modify the AP-1 complex (e.g., via phosphorylation) or other genes (e.g., *fra*-1, p53). A similar mechanism may also occur in earlier progenitors such as bipotential chondroprogenitor and osteoprogenitor cells (e.g., wT2-1 cells). Since exogenous Fos is not expressed during embryogenesis in the *fos* transgenic mice, such an effect would not be observed. Instead, Fos is expressed postnatally and acts specifically in osteogenic cells that have the necessary factors (e.g., factor "Y") for activation of bone tumor formation (e.g., via gene "B"). Further experiments, such as those described in the following section, will be necessary to test this hypothesis and understand the molecular basis for these differences.

V. CONCLUSIONS AND PERSPECTIVES

Gaining a molecular understanding of growth control, cellular differentiation, and the multistep nature of tumorigenesis awaits further experimentation. The amplification of individual oncogenes—for example, via point mutations or chromosomal rearrangements such as deletions or translocations, may represent the initial step in the activation of a tumorigenic phenotype that triggers a cascade of subsequent events. As we

FIGURE 8 A proposed model for the specific action of c-*fos* in chondrogenic and osteogenic cells based on studies from transgenic and chimeric mice. In c-*fos* chimeric mice (top panel), high levels of exogenous Fos protein (F) are present throughout embryogenesis in different tissues, including chondrogenic cells. In addition to also expressing Jun protein, chondrogenic cells before birth also contain cell type-specific factors (factor "X"), which can cooperate with or modify the Fos–Jun–AP-1 complexes. This specific interaction results in the activation of a multistep cascade (via gene "A"), resulting in the formation of chondrosarcomalike tumors. In contrast, exogenous Fos (F) is not expressed during embryogenesis in c-*fos* transgenic mice (bottom panel); therefore, a similar effect on those chondrogenic cells would not be observed. Instead, Fos acts on osteogenic cells present in postnatal transgenic mice that contain cell type-specific factors (factor "Y") for activation of osteosarcoma formation in a multistep fashion (via gene "B"). Shaded area, exogenous Fos expression; mpc, mesenchymal progenitor cell; c/opc, putative chondro-osteoprogenitor cell.

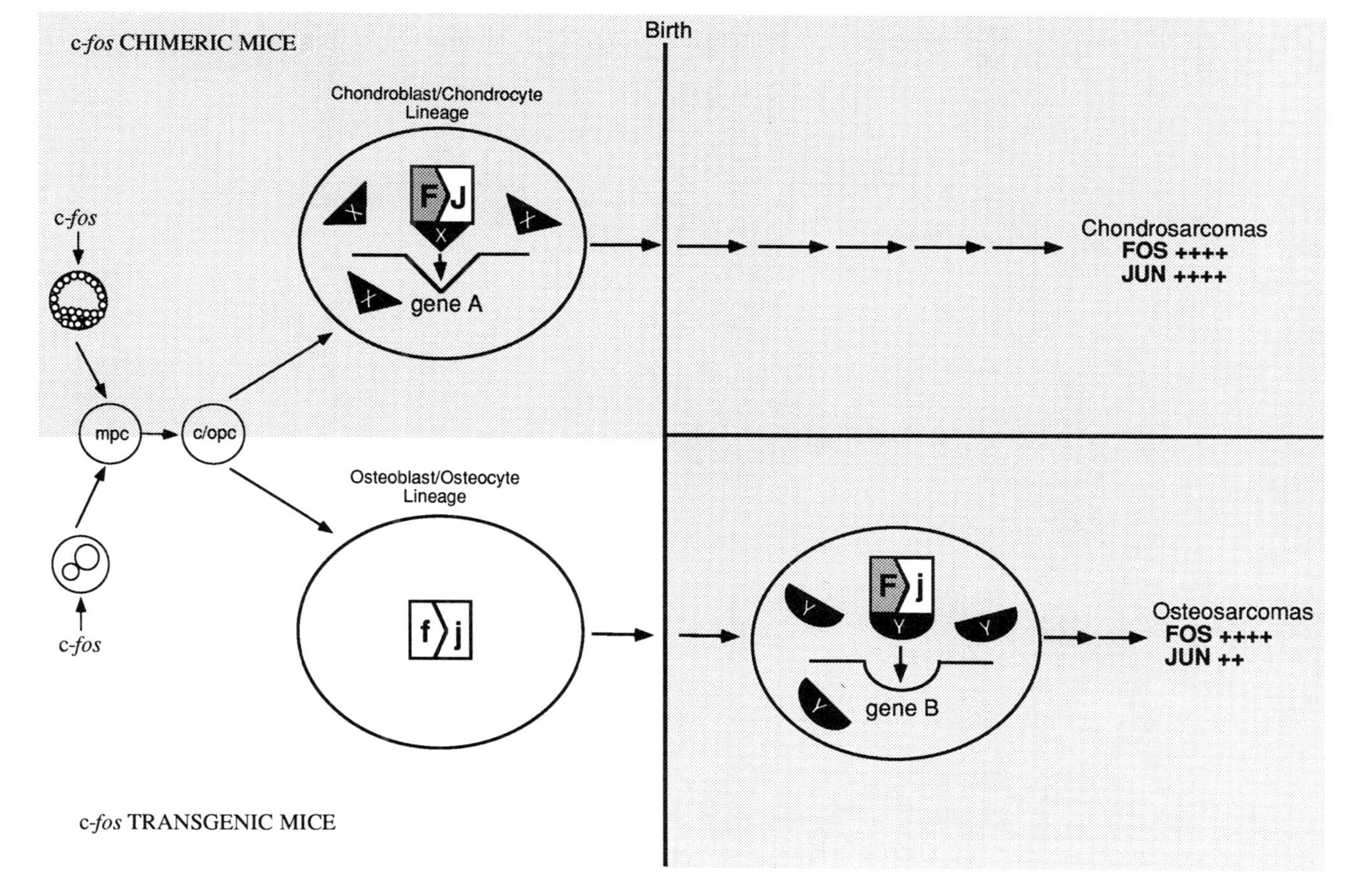
c-fos CHIMERIC MICE
Birth
Chondroblast/Chondrocyte Lineage
c-fos
F
J
X
gene A
Chondrosarcomas
FOS ++++
JUN ++++
mpc
c/opc
Osteoblast/Osteocyte Lineage
f
j
c-fos
F
j
Y
gene B
Osteosarcomas
FOS ++++
JUN ++
c-fos TRANSGENIC MICE

have shown, some of these events may be the association with, or activation of, other oncogenes (e.g., c-*jun*, *fra*-1). Indeed, oncogenes can cooperate in transformation events, as demonstrated by the fact that the Ha-*ras* oncogene can potentiate c-*jun*-mediated transformation via a mechanism that may involve in part, enhanced phosphorylation of the c-Jun N-terminal activation domain (Schütte *et al.*, 1989; Binétruy *et al.*, 1991). Moreover, the power of using transgenic mouse models to assess cooperativity between nuclear oncogenes and other "downstream" genes in neoplasia has recently been shown. Using transgenic mice overexpressing the c-*myc* oncogene from an immunoglobulin heavy-chain enhancer (Eμ), new genes were identified that cooperated with c-*myc* and were obligatory for the development of B-cell lymphomas (for review, see Berns, 1991; Adams and Cory, 1991). Such studies confirm the potential usefulness of mouse models that develop a reproducible and predictable phenotype, such as the *fos* transgenic and chimeric mice.

To address the importance of the Jun protein in discriminating between bone and cartilage tumors, we are generating H2-c-*jun*LTR transgenic mice in an attempt to attain high levels of Jun in osteogenic and chondrogenic tissues. Interestingly, transgenic mice harboring an H2-c-*jun* construct without the 3′ LTR do not develop any phenotype and do not express in bone or cartilage tissues (F. Hilberg, personal communication). However, transgenic mice expressing the v-*jun* oncogene from an H2 promoter also do not express the transgene in bone tissue but develop fibrosarcomas upon wounding (Schuh *et al.*, 1990). Finally, approaches aimed at expressing protooncogenes in selective cell compartments are being considered whereby, for example, c-*fos* is fused to the promoters for type I or type II collagen. Such experiments should result in the attainment of high levels of Fos protein in bone and cartilage tissues, respectively, and thus will help to address the lineage specificity observed for Fos action.

To define unequivocally the roles of Fos and Jun in the regulation of osteogenic and chondrogenic cell development, we have recently used the technology of gene targeting by homologous recombination to generate mouse lines that are deficient in either Fos or Jun. These genes were inactivated at the level of ES cells, and the phenotypic consequences in mice lacking functional oncoproteins were determined. ES cells lacking Fos are viable and exhibit no deficiencies in growth and differentiation *in vitro* (Wang *et al.*, 1992; Field *et al.*, 1992). However, mice lacking Fos are growth-retarded and develop osteopetrosis, with severe deficiencies in bone and cartilage remodeling (Wang *et al.*, 1992; Johnson *et al.*, 1992). ES cells lacking c-*jun* are also unaffected *in vitro;* however, mice lacking functional Jun die during embryogenesis (Hilberg and Wagner, 1992; F. Hilberg, personal communication). Thus, these data provide unequivocal evidence that Fos is an important molecule for normal skeletogenesis and, furthermore, that different AP-1 transcription factors have distinct roles during development.

Finally, the fact that the cell lines that we have isolated from the osteogenic and chondrogenic tumors are distinctly different suggests that they are important tools for identifying novel Fos-regulated genes acting in the cartilage and bone cell lineages. Together with cell lines isolated from Fos- and Jun-deficient mice, these cell lines provide an excellent system for biochemical studies aimed at identifying the specific AP-1 complexes and the possible modifications of these proteins that occur in the different cell types. These studies will eventually lead to a greater understanding of the regulation of expression of cartilage- and bone-associated genes and will better define the role of transcription factors in chondrogenic and osteogenic cell differentiation.

ACKNOWLEDGMENTS

We thank Uta Möhle-Steinlein for expert technical assistance, Dr. Karl Schellander for generating the transgenic mice, Norma Howells for maintaining our mouse colony, and Hannes Tkadletz for photography. We are also grateful to Dr. Ulrich Rüther for helpful suggestions and critical reading of the manuscript.

REFERENCES

Aaronson, S. A. (1991). Growth factors and cancer. *Science* **254,** 1146–1153.

Adams, J. M., and Cory, S. (1991). Transgenic models of tumor development. *Science* **254,** 1161–1167.

Angel, P., Imagawa, M., Chiu, R., Stein, B., Imbra, R. J., Rahmsdorf, H. J., Jonat, C., Herrlich, P., and Karin, M. (1987). Phorbol ester-inducible genes contain a common *cis* element recognized by a TPA-modulated transacting factor. *Cell* **49,** 729–739.

Aubin, J. E., Heersche, J. N. M., Bellows, C. G., and Grigoriadis, A. E. (1990). Osteoblastic lineage analysis in fetal rat calvaria cells. *In* "Calcium Regulation and Bone Metabolism" (D. V. Cohn, F. H. Glorieux, and T. J. Martin, eds.), pp. 362–370. Elsevier, Amsterdam.

Baichwal, V. R., Park, A., and Tjian, R. (1991). v-Src and EJ Ras alleviate repression of c-Jun by a cell-specific inhibitor. *Nature (London)* **352,** 165–168.

Berns, A. (1991). Separating the wheat from the chaff. *Curr. Biol.* **1,** 28–29.

Binétruy, B., Smeal, T., and Karin, M. (1991). Ha-Ras augments c-Jun activity and stimulates phosphorylation of its activation domain. *Nature (London)* **351,** 122–127.

Birek, C., Huang, H. Z., Birek, P., and Tenenbaum, H. C. (1991). c-*fos* oncogene expression in dexamethasone stimulated osteogenic cells in chick embryo periosteal cultures. *Bone Miner.* **15,** 193–208.

Bishop, J. M. (1987). The molecular genetics of cancer. *Science* **235,** 305–311.

Bogenmann, E., Moghadam, H., DeClerck, Y. A., and Mock, A. (1987). c-*myc* amplification and expression in newly established human osteosarcoma cell lines. *Cancer Res.* **47,** 3808–3814.

Bradley, A., Evans, M., Kaufmann, M. H., and Robertson, E. (1984). Formation of germline chimeras from embryo-derived teratocarcinoma cell lines. *Nature (London)* **309,** 255–256.

Braselmann, S., Bergers, G., Wrighton, C., Graninger, P., Superti-Furga, G., and Busslinger, M. (1992). Identification of Fos target genes by the use of selective induction systems. *J. Cell Sci. Suppl.* **16,** 97–109.

Bravo, R. (1990). Growth factor inducible genes in fibroblasts. *In* "Growth Factors, Differentiation Factors, and Cytokines (A. Habenicht, ed.), pp. 324–343. Springer-Verlag, Berlin and Heidelberg.

Capecchi, M. R. (1989). Altering the genome by homologous recombination. *Science* **244,** 1288–1292.

Caplan, A. I., and Pechak, D. G. (1987). The cellular and molecular embryology of bone formation. *In* "Bone and Mineral Research" (W. A. Peck, ed.), Vol. 5, pp. 117–183. Elsevier, Amsterdam.

Chiu, R., Boyle, W. J., Meek, J., Smeal, T., Hunter, T., and Karin, M. (1988). The c-*fos* protein interacts with c-*jun*/AP-1 to stimulate transcription of AP-1-responsive genes. *Cell* **54,** 541–552.

Closs, E. I., Murray, A. B., Schmidt, J., Schön, A., Erfle, V., and Strauss, P. G. (1990). c-*fos* expression precedes osteogenic differentiation of cartilage cells *in vitro*. *J. Cell Biol.* **111,** 1313–1323.

Cohen, D. R., and Curran, T. (1988). *Fra*-1: A serum-inducible, cellular immediate–early gene that encodes a *fos*-related antigen. *Mol. Cell. Biol.* **8,** 2063–2069.

Curran, T. (1988). The *fos* oncogene. *In* "The Oncogene Handbook" (E. P. Reddy, A. M. Skalka, and T. Curran, eds.), pp. 307–325. Elsevier, Amsterdam.

Curran, T., Miller, A. D., Zokas, L., and Verma, I. M. (1984). Viral and cellular *fos* proteins: A comparative analysis. *Cell* **36,** 259–268.

Curran, T., Gordon, M. B., Rubino, K. L., and Sambucetti, L. C. (1987). Isolation and characterization of the c-*fos* (rat) cDNA and analysis of post-translational modification *in vitro*. *Oncogene* **2,** 79–84.

David-Watine, B., Israël, A., and Kourilsky, P. (1990). The regulation and expression of MHC class I genes. *Immunol. Today* **11,** 286–292.

De Togni, P., Niman, H., Raymond, V., Sawchenko, P., and Verma, I. M. (1988). Detection of *fos* protein during osteogenesis by monoclonal antibodies. *Mol. Cell. Biol.* **8,** 2251–2256.

Distel, R. J., Ro, H.-S., Rosen, B. S., Groves, D. L., and Spiegelman, B. M. (1987). Nucleoprotein complexes that regulate gene expression in adipocyte differentiation: Direct participation of c-*fos*. *Cell* **49,** 835–844.

Dony, C., and Gruss, P. (1987). Proto-oncogene c-*fos* expression in growth regions of fetal bone and mesodermal web tissue. *Nature (London)* **328,** 711–714.

Field, S. J., Johnson, R. S., Mortensen, R. M., Papaioannou, V. E., Spiegelman, B. M., and Greenberg, M. E. (1992). Growth and differentiation of embryonic stem cells that lack an intact c-*fos* gene. *Proc. Natl. Acad. Sci. USA* **89,** 9306–9310.

Finkel, J. P., and Biskis, B. O. (1968). Experimental induction of osteosarcomas. *Prog. Exp. Tumor Res.* **10,** 72–111.

Finkel, J. P., Biskis, B. O., and Jinkins, P. B. (1966). Virus induction of osteosarcoma in mice. *Science* **151,** 698–701.

Finkel, J. P., Reilly, C. A., Jr., and Biskis, B. O. (1975). Viral etiology of bone cancer. *Front. Radiat. Ther. Oncol.* **10,** 28–39.

Fujiwara, K. T., Ashida, K., Nishina, H., Iba, H., Miyajima, N., Nishizawa, M., and Kawai, S. (1987). The chicken c-*fos* gene: Cloning and nucleotide sequence analysis. *J. Virol.* **61,** 4012–4018.

Gentz, R., Rauscher III., F. J., Abate, C., and Curran, T. (1989). Parallel association of Fos and Jun leucine zippers juxtaposes DNA binding domains. *Science* **243,** 1695–1699.

Goralczyk, R., Closs, E. I., Rüther, U., Wagner, E. F., Strauss, P. G., Erfle, V., and Schmidt, J. (1990). Characterization of *fos*-induced osteogenic tumours and tumour-derived murine cell lines. *Differentiation* **44,** 122–131.

Grigoriadis, A. E., Heersche, J. N. M., and Aubin, J. E. (1988). Differentiation of muscle, fat, cartilage and bone from progenitor cells present in a bone-derived clonal cell population: Effect of dexamethasone. *J. Cell Biol.* **106,** 2139–2151.

Grigoriadis, A. E., Aubin, J. E., and Heersche, J. N. M. (1989). Effects of dexamethasone and vitamin D_3 on cartilage differentiation in a clonal chondrogenic cell population. *Endocrinology* **125,** 2103–2110.

Grigoriadis, A. E., Heersche, J. N. M., and Aubin, J. E. (1990). Continuously growing bipotential and monopotential myogenic, adipogenic and chondrogenic subclones isolated form the multipotential RCJ 3.1 clonal cell line. *Dev. Biol.* **142,** 313–318.

Grigoriadis, A. E., Scheller, K., Wang, Z.-Q., and Wagner, E. F. (1993). Osteoblasts are target cells for transformation in c-fos transgenic mice. *J Cell. Biol.* In Press.

Hall, B. K. (1978). "Developmental and Cellular Skeletal Biology." Academic Press, New York.

Haluska, F. G., Tsujimoto, Y., and Croce, C. M. (1987). Oncogene activation by chromosome translocation in human malignancy. *Annu. Rev. Genet.* **21,** 321–346.

Hanahan, D. (1988). Dissecting multistep tumorigenesis in transgenic mice. *Annu. Rev. Genet.* **22,** 479–519.

Heckl, K., and Wagner, E. F. (1989). *In situ* analysis of c-*fos* expression in transgenic mice. *In* "Molecular Genetics of Early *Drosophila* and Mouse Development" (M. R. Cappechi, ed.), pp. 117–129. *Current Communications in Molecular Biology.* Cold Spring Harbor Laboratory Press, Cold Spring Harbor, New York.

Heersche, J. N. M., and Aubin, J. E. (1990). Regulation of cellular activity of osteoblasts. *In* "Bone: A Treatise. Vol. 1: The Osteoblast and Osteocyte" (B. K. Hall, ed.), pp. 327–349. The Telford Press, Caldwell, New Jersey.

Hengerer, B., Lindholm, D., Heumann, R., Rüther, U., Wagner, E. F., and Thoenen, H. (1990). Lesion-induced increase in nerve growth factor mRNA is mediated by c-*fos.* *Proc. Natl. Acad. Sci. USA* **87,** 3899–3903.

Hilberg, F., and Wagner, E. F. (1992). Embryonic stem (ES) cells lacking functional c-*jun.* Consequences for growth and differentiation, AP-1 activity and tumorigenicity. *Oncogene* **7,** 2371–2380.

Hirai, S., Ryseck, R.-P., Mechta, F., and Bravo, R. (1989). Characterization of *jun*D: A new member of the *jun* protooncogene family. *EMBO J.* **8,** 1433–1439.

Jenuwein, T., and Müller, R. (1987). Structure-function analysis of *fos* protein: A single amino acid change activates the immortalizing potential of v-*fos.* *Cell* **48,** 647–657.

Jenuwein, T., Müller, D., Curran, T., and Müller, R. (1985). Extended life span and tumorigenicity of nonestablished mouse connective tissue cells transformed by the *fos* oncogene of FBR-MuSV. *Cell* **41,** 629–637.

Johnson, R. S., Spiegelman, B. M., and Papaioannou, V. (1992). Pleiotropic effects of a null mutation in the c-*fos* proto-oncogene. *Cell* **71,** 577–586.

Kerr, L. D., Holt, J. T., and Matrisian, L. M. (1988). Growth factors regulate transin gene expression by c-*fos*-dependent and c-*fos*-independent pathways. *Science* **242,** 1424–1427.

Klein, G. (1987). The approaching era of the tumor suppressor genes. *Science* **238,** 1539–1545.

Knudson, A. G. (1986). Genetics of human cancer. *Annu. Rev. Genet.* **20,** 231–252.

Kouzarides, T., and Ziff, E. (1988). The role of the leucine zipper in the fos–jun interaction. *Nature (London)* **336,** 646–651.

Kouzarides, T., and Ziff, E. (1989). Behind the Fos and Jun leucine zipper. *Cancer Cells* **1,** 71–76.

Kovary, K., and Bravo, R. (1991). The Jun and Fos protein families are both required for cell cycle progression in fibroblasts. *Mol. Cell Biol.* **11,** 4466–4472.

Landschultz, W. H., Johnson, P. F., and McKnight, S. L. (1988). The leucine zipper protein:

A hypothetical structure common to a new class of DNA binding proteins. *Science* **240,** 1759–1764.

Lassar, A. B., Thayer, M. J., Overell, R. W., and Weintraub, H. (1989). Transformation by activated *ras* or *fos* prevents myogenesis by inhibiting expression of MyoD1. *Cell* **58,** 659–667.

Lavigueur, A., Maltby, V., Mock, D., Rossant, J., Pawson, T., and Bernstein, A. (1989). High incidence of lung, bone, and lymphoid tumors in transgenic mice overexpressing mutant alleles of the p53 oncogene. *Mol. Cell. Biol.* **9,** 3982–3991.

Lee, W., Mitchell, P., and Tjian, R. (1987). Purified transcription factor AP-1 interacts with TPA-inducible enhancer elements. *Cell* **49,** 741–752.

Levine, A. J., Momand, J., and Finlay, C. A. (1991). The p53 tumour suppressor gene. *Nature (London)* **351,** 453–456.

Lucibello, F. C., Lowag, C., Neuberg, M., and Müller, R. (1989). Trans-repression of the mouse c-*fos* promoter: A novel mechanism of *fos*-mediated trans-regulation. *Cell* **59,** 999–1007.

Marks, S. C., and Popoff, S. N. (1988). Bone cell biology: The regulation of development, structure and function in the skeleton. *Am. J. Anat.* **183,** 1–44.

Mason, J., Murphy, D., and Hogan, B. (1985). Expression of c-*fos* in parietal endoderm, amnion and differentiating F9 teratocarcinoma cells. *Differentiation* **30,** 76–81.

Masuda, H., Miller, C., Koeffler, H. P., Battifora, H., and Cline, M. J. (1987). Rearrangement of the p53 gene in human osteogenic sarcomas. *Proc. Natl. Acad. Sci. USA* **84,** 7716–7719.

Meijlink, F., Curran, T., Miller, A. D., and Verma, I. M. (1985). Removal of a 67-base pair sequence in the non-coding region of proto-oncogene *fos* converts it to a transforming gene. *Proc. Natl. Acad. Sci. USA* **82,** 4987–4991.

Miller, A. D., Curran, T., and Verma, I. M. (1984). c-*fos* protein can induce cellular transformation: A novel mechanism of activation of cellular oncogene. *Cell* **36,** 51–60.

Mitchell, R. L., Henning-Chubb, C., Huberman, E., and Verma, I. M. (1986). c-*fos* expression is neither sufficient nor obligatory for differentiation of monomyelocytes to macrophages. *Cell* **45,** 497–504.

Mölders, H., Jenuwein, T., Adamkiewicz, J., and Müller, R. (1987). Isolation and structural analysis of a biologically active chicken c-*fos* cDNA: Identification of evolutionarily conserved domains in *fos* protein. *Oncogene* **1,** 377–385.

Müller, R. (1986). Cellular and viral *fos* genes: Structure, regulation of expression and biological properties of their encoded products. *Biochim. Biophys. Acta* **823,** 207–225.

Müller, R., and Wagner, E. F. (1984). Differentiation of F9 teratocarcinoma stem cells after transfer of c-*fos* proto-oncogenes. *Nature (London)* **311,** 438–442.

Müller, R., Slamon, D. J., Tremblay, J. M., Cline, M. J., and Verma, I. M. (1982). Differential expression of cellular oncogenes during pre- and postnatal development of the mouse. *Nature (London)* **299,** 640–644.

Nardeux, P. C., Daya-Grosjean, L., Landin, R. M., Andéol, Y., and Suàrez, H. G. (1987). A c-*ras*-Ki oncogene is activated, amplified and overexpressed in a human osteosarcoma cell line. *Biochem. Biophys. Res. Commun.* **146,** 395–402.

Nigro, J. M., Baker, S. J., Preisinger, A. C., Jessup, J. M., Hostetter, R., Cleary, K., Bigner, S. H., Davidson, N., Baylin, S., Devilee, P., Glover, T., Collins, F. S., Weston, A., Modali, R., Harris, C. C., and Vogelstein, B. (1989). Mutations in the *p53* gene occur in diverse human tumour types. *Nature (London)* **342,** 705–708.

Nijweide, P., Berger, E. H., and Feyden, J. H. M. (1986). Cells of bone: Proliferation, differentiation and hormonal regulation. *Physiol. Rev.* **66,** 855–886.

Nishikura, K., and Murray, J. M. (1987). Antisense RNA of proto-oncogene c-*fos* blocks renewed growth of quiescent 3T3 cells. *Mol. Cell. Biol.* **7,** 639–649.

Nishina, H., Sato, T., Suzuke, T., Sato, M., and Iba, H. (1990). Isolation and characterization of *fra*-2, an additional member of the *fos* gene family. *Proc. Natl. Acad. Sci. USA* **87,** 3619–3623.

Ofir, R., Dwarki, V. J., Rashid, D., and Verma, I. M. (1990). Phosphorylation of the C terminus of Fos protein is required for transcriptional transrepression of the c-*fos* promoter. *Nature (London)* **348,** 80–82.

Ovitt, C. E., and Rüther, U. (1989). The proto-oncogene c-*fos:* Structure, expression, and functional aspects. *Oxford Surveys on Eukaryotic Genes* **6,** 33–51.

Owen, M. (1985). Lineage of osteogenic cells and their relationship to the stromal system. *In* "Bone and Mineral Research" (W. A. Peck, ed.), Vol. 3, pp. 1–25. Elsevier, Amsterdam.

Palmiter, R. D., and Brinster, R. L. (1986). Germ-line transformation of mice. *Annu. Rev. Genet.* **20,** 465–499.

Pease, S., and Williams, R. L. (1990). Formation of germ-line chimeras from embryonic stem cells maintained with recombinant leukemia inhibitory factor. *Exp. Cell Res.* **190,** 209–211.

Pulverer, B. J., Kyriakis, J. M., Avruch, J., Nikolakaki, E., and Woodgett, J. R. (1991). Phosphorylation of c-*jun* mediated by MAP kinases. *Nature (London)* **353,** 670–674.

Rahmsdorf, H. J., Schönthal, A., Angel, P., Litfin, M., Rüther, U., and Herrlich, P. (1987). Posttranscriptional regulation of c-*fos* mRNA expression. *Nucleic Acids Res.* **15,** 1643–1659.

Ransone, L. J., and Verma, I. M. (1990). Nuclear proto-oncogenes *fos* and *jun*. *Annu. Rev. Cell Biol.* **6,** 539–557.

Redemann-Fibi, B., Schuermann, M., and Müller, R. (1991). Stage and tissue-specific expression of *fos*B during mouse development. *Differentiation* **46,** 43–49.

Renz, M., Neuberg, M., Kurz, C., Bravo, R., and Müller, R. (1985). Regulation of c-*fos* transcription in mouse fibroblasts: Identification of DNase I-hypersensitive sites and regulatory upstream sequences. *EMBO J.* **4,** 3711–3716.

Riabowol, K. T., Vosatka, R. J., Ziff, E. B., Lamb, N. J., and Feramisco, J. R. (1988). Microinjection of *fos*-specific antibodies blocks DNA synthesis in fibroblast cells. *Mol. Cell. Biol.* **8,** 1670–1676.

Robertson, E. J. (1987). Embryo-derived stem cell lines. *In* "Teratocarcinomas and Embryonic Stem Cells: A Practical Approach" (E. J. Robertson, ed.), pp. 71–112. IRL Press, Oxford.

Rodan, G. A., and Rodan, S. B. (1984). Expression of the osteoblastic phenotype. *In* "Bone and Mineral Research" (W. A. Peck, ed.), Vol. 2, pp. 244–285. Elsevier, Amsterdam.

Rossant, J. (1991). Gene disruption in mammals. *Curr. Opinion Genet. Devel.* **1,** 236–240.

Rüther, U., and Wagner, E. F. (1989). The specific consequences of c-*fos* expression in transgenic mice. *Prog. Nucleic Acid Res. Mol. Biol.* **36,** 235–245.

Rüther, U., Wagner, E. F., and Müller, R. (1985). Analysis of the differentiation-promoting potential of inducible c-*fos* genes introduced into embryonal cells. *EMBO J.* **4,** 1775–1781.

Rüther, U., Garber, C., Komitowski, D., Müller, R., and Wagner, E. F. (1987). Deregulated c-*fos* expression interferes with normal bone development in transgenic mice. *Nature (London)* **325,** 412–416.

Rüther, U., Müller, W., Sumida, T., Tokuhisa, T., Rajewsky, K., and Wagner, E. F. (1988). c-*fos* expression interferes with thymus development in transgenic mice. *Cell* **53,** 847–856.

Rüther, U., Komotowski, D., Schubert, F. R., and Wagner, E. F. (1989). c-*fos* expression induces bone tumors in transgenic mice. *Oncogene* **4,** 861–865.

Ryder, K., Lau, L. F., and Nathans, D. (1988). A gene activated by growth factors is related to the oncogene v-*jun*. *Proc. Natl. Acad. Sci. USA* **85,** 1487–1491.

Sandberg, M., Vuorio, T., Hirvonen, H., Alitalo, K., and Vuorio, E. (1988). Enhanced expression of TGF-β and c-*fos* mRNA in the growth plates of developing human long bones. *Development* **102,** 461–470.

Sassone-Corsi, P., Ranson, L. K., Lamp, W. W., and Verma, I. M. (1988). Direct interaction between Fos and Jun nuclear oncoproteins: Role of leucine zipper domain. *Nature (London)* **336,** 692–695.

Schmidt, J., Livne, E., Erfle, V., Gössner, W., and Silbermann, M. (1986). Morphology and *in vivo* growth characteristics of an atypical murine proliferative osseous lesion induced *in vitro*. *Cancer Res.* **46,** 3090–3098.

Schön, A., Michiels, L., Janowski, M., Merregaert, J., and Erfle, V. (1986). Expression of proto-oncogenes in murine osteosarcomas. *Int. J. Cancer* **38,** 67–74.

Schönthal, A., Herrlich, P., Rahmsdorf, H. J., and Ponta, H. (1988). Requirement for *fos* gene expression in the transcriptional activation of collagenase by other oncogenes and phorbol esters. *Cell* **54,** 325–334.

Schuermann, M., Neuberg, M., Hunter, J. B., Jenuwein, T., Ryseck, R.-P., Bravo, R., and Müller, R. (1989). The leucine repeat motif in *fos* protein mediates complex formation with *jun*/AP-1 and is required for transformation. *Cell* **56,** 507–516.

Schuh, A. C., Keating, S. J., Monteclaro, F. S., Vogt, P. K., and Breitman, M. L. (1990). Obligatory wounding requirement for tumorigenesis in v-*jun* transgenic mice. *Nature (London)* **346,** 756–760.

Schüle, R., Umesono, K., Mangelsdorf, D., Bolado, J., Pike, J. W., and Evans, R. M. (1990). Jun-Fos and receptors for vitamins A and D recognize a common response element in the human osteocalcin gene. *Cell* **61,** 497–504.

Schütte, J., Minna, J. D., and Birrer, M. J. (1989). Deregulated expression of human c-*jun* transforms primary rat embryo cells in cooperation with an activated c-Ha-*ras* gene and transforms Rat-1a cells as a single gene. *Proc. Natl. Acad. Sci. USA* **86,** 2257–2261.

Shaw, P. E., Hipskind, R. A., Schröter, H., and Nordheim, A. (1989). Transcriptional regulation of proto-oncogene c-*fos*. *In* "Nucleic Acids and Molecular Biology" (F. Eckstein and D. M. J. Lilley, eds.), Vol. 3, pp. 120–132. Springer-Verlag, Berlin and Heidelberg.

Solomon, E., Borrow, J., and Goddard, A. D. (1991). Chromosome aberrations and cancer. *Science* **254,** 1153–1160.

Sonnenberg, J. L., Rauscher III, F. J., Morgan, J. I., and Curran, T. (1990). Regulation of proenkephalin by Fos and Jun. *Science* **246,** 1622–1625.

Soriano, P., Montgomery, C., Geske, R., and Bradley, A. (1991). Targeted disruption of the c-*src* proto-oncogene leads to osteopetrosis in mice. *Cell* **64,** 693–702.

Sukhatme, V. P., Cao, X., Chang, L. C., Tsai-Morris, C.-H., Stamenkovich, D., Ferreira, P. C. P., Cohen, D. R., Edwards, S. A., Shows, T. B., Curran, T., Le Beau, M. M., and Adamson, E. D. (1988). A zinc finger-encoding gene coregulated with c-*fos* during growth and differentiation, and after cellular depolarization. *Cell* **53,** 37–43.

Treisman, R. (1985). Transient accumulation of c-*fos* RNA following serum stimulation requires a conserved 5′ element and c-*fos* 3′ sequences. *Cell* **42,** 889–902.

Turner, R., and Tjian, R. (1989). Leucine repeats and an adjacent DNA binding domain mediate the formation of functional cFos–cJun heterodimers. *Science* **243,** 1689–1694.

van Beveren, C., van Straaten, F., Curran, T., Müller, R., and Verma, I. M. (1983). Analysis of FBJ-MuSV provirus and c-*fos* (mouse) gene reveals that viral and cellular *fos* gene products have different carboxy termini. *Cell* **32,** 1241–1255.

van Beveren, C., Enami, S., Curran, T., and Verma, I. M. (1984). FBR osteosarcoma virus. II. Nucleotide sequence of the provirus reveals that the genome contains sequences acquired from two cellular genes. *Virology* **135,** 229–243.

van Straaten, F., Müller, R., Curran, T., van Beveren, C., and Verma, I. M. (1983). Complete nucleotide sequence of human c-*onc* gene: Deduced amino acid sequence of the human c-*fos* protein. *Proc. Natl. Acad. Sci. USA* **80,** 3183–3187.

Varmus, H. E. (1984). The molecular genetics of cellular oncogenes. *Annu. Rev. Genet.* **18,** 553–612.

Verma, I. M., and Sassone-Corsi, P. (1987). Proto-oncogene *fos:* Complex but versatile regulation. *Cell* **51,** 513–514.

16

MOLECULAR BIOLOGY OF CARTILAGE MATRIX

SERGIO LINE*, CRAIG RHODES, and YOSHIHIKO YAMADA

**Present address:* Faculdade de Odontologia de Piracicaba–UNICAMP. Av. Limeira s/n. Caixa Postal 52. 13400 Piracicaba-SP. Brazil.

Cellular and Molecular Biology of the Bone

I. INTRODUCTION

Cartilage is a highly specialized connective tissue with distinct morphological and biochemical characteristics. It is characterized by isolated round-shaped chondrocytes surrounded by extensive areas of extracellular matrix. Chondrocytes are differentiated from other mesenchymal cells by the production and secretion of large amounts of cartilage-specific matrix components. Among the major products are the fibril-forming collagen type II, hyaluronic acid, the large core proteoglycan (aggrecan), and link protein. These components have been extensively studied through biochemical, histological, and immunochemical methods. Other minor collagens as types IX, X, and XI are also found in the cartilage matrix (Vuorio and de Crombrugghe, 1990). Collagen types IX and XI are present in all cartilage tissues, whereas collagen type X is produced only by hypertrophic chondrocytes. Collagen type IX is found associated with collagen type II fibrils, where it is thought to mediate the interaction with proteoglycans. Collagen type XI can polymerize to form fibrils and is also found in association with collagen type II, where it may play a role in determining fibril diameter.

Hormones and vitamins have a role in cartilage expression and maturation. Interaction of these factors with specific receptors can mediate transcriptional activity through direct binding of the complex to regulatory sequences of cartilage genes. For example, vitamin D is known to be important for skeletal development (Suda *et al.*, 1985), and when added to mesenchymal cells it causes a dramatic increase in the transcription of collagen type II and aggrecan (Tsonis, 1991). The involvement of the synthetic glucocorticoid dexamethasone in the maintenance of chondrocyte phenotype is controversial, and both chondroprotective and destructive influences have been reported (Pelletier and Martel-Pelletier, 1989).

The synthesis of cartilage components is coordinated during development, and disregulation may result in disease states. Co-expression of collagen type II, aggrecan, and link protein is characteristic of hyaline cartilage. The isolation of complementary DNA (cDNA) and genomic clones for the major cartilage proteins have allowed the determination of their primary structure, genomic arrangement, and identification of transcriptional regulatory sites in these genes. This has allowed not only a better understanding of the basic mechanisms of cartilage function, but also elucidation of the role of these molecules in pathological processes, such as rheumatoid arthritis and genetic anomalies.

The purpose of this chapter is to review the molecular biology of the three main protein components of the cartilage matrix (collagen type II, aggrecan, and link protein). Special emphasis will be given to recently published data on gene structure and on transcriptional regulation of these genes.

II. COLLAGEN TYPE II

Collagens are a family of proteins that form the major structural components of the extracellular matrix. These molecules are formed by three polypeptide chains that interact with one another to form a right-handed helical structure (Prockop and Kivirikko, 1984). The most common and well studied are collagen types I, II, and III, whose molecules can coordinately polymerize to form the collagen fibrils, which possess a typical cross-banded structure when observed by electron microscopy. In these three collagens, the helical domain is formed by a long repeat of about 1000 residues of the sequence Gly–X–Y.

Unlike collagen types I and III, which are ubiquitously distributed, collagen type II has a relatively restricted localization. In the adult, collagen type II is the major structural component of the hyaline cartilage of the articular surfaces, and it is also found in other tissues such as the nucleous pulposo of the intervertebral disc and the retina, sclera, and lens of the eye. More recently, collagen type II has been shown to be expressed in a variety of subepithelial locations in developing chicken and mouse embryos (Kosher and Solursh, 1989; Cheah *et al.*, 1991; Wood *et al.*, 1991).

A. cDNA and Exon–Intron Structure

Collagen type II is a homotrimer formed by α1(ll) chains. The human collagen type II gene is located in chromosome 12 in region 12q13 (Solomon *et al.*, 1985). It is formed by 54 exons in both the human and mouse and spans 28.9 kilobases (kb) (Metsaranta *et al.*, 1991).

The triple-helical domain consists of an uninterrupted sequence of 1014 amino acids that are coded by 44 exons. The majority of exons coding for this domain are formed by 54 basepairs (bp) or multiples of it (108, 162). Due to this characteristic, it has been suggested that the collagen genes originated by duplication of a common precursor exon containing 54 bp (Yamada *et al.*, 1980). A comparison between human and mouse sequences in the triple-helical domain shows a 96.7% identity at the amino acid level, including the two lysines at positions 87 and 930, which participate in the inter- and intramolecular cross-link of collagen type II. The two other lysines located at position 122 of the N-propeptide and position 17 of the C-propeptide domain are also conserved between these species.

The N-propeptide domain exhibits the highest degree of divergence among the fibrillar collagens. It is composed by eight exons (Su *et al.*, 1989) as opposed to the six exons of the genes coding for collagen type I (α1(1), α2(1)) (D'Allesio *et al.*, 1988; de Wet *et al.*, 1987; Tate *et al.*, 1983) and five exons for collagen type III (α1(3)) (Benson-Chandra *et al.*, 1989) (Fig. 1). The first exon contains an untranslated segment of approximately 150

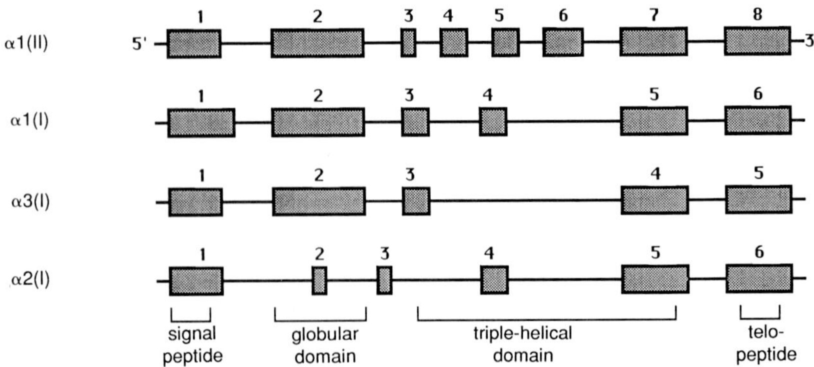

FIGURE 1 Genomic structure and subdomain organization of the exons coding for the N-propeptide of the collagen types I, II, and III. Exons (shaded boxes) are drawn approximately to scale, whereas the introns (lines) do not appear to scale.

bp and an 85-bp sequence, mostly coding for the signal peptide (Khono *et al.*, 1985; Ryan *et al.*, 1990; Metsaranta *et al.*, 1991). The second exon codes for a cysteine-rich globular domain of 69 amino acids. This exon was absent in the early cDNA clones published in rat and in human (Khono *et al.*, 1985; Baldwin *et al.*, 1989b). A more recent analysis of the human N-propeptide domain, however, revealed an additional exon located between the first and second exons (Ryan *et al.*, 1990). The presence of an alternatively spliced second exon in the collagen type II gene was further demonstrated by polymerase chain reaction analysis from messenger RNA (mRNA) prepared from various cartilage specimens (Ryan and Sandell, 1990). In the chick external cartilage from Day 14 embryos, collagen type II mRNA without the second exon is the predominant form (about 90%). In precartilage from limb mesenchyme, however, only the gene containing the second exon is detected. The spliced form occurs also in the mouse gene (Metsaranta *et al.*, 1991). The alternative splice is another distinctive characteristic of the N-propeptide domain of the collagen type II gene, because it does not occur in the other fibrillar collagen types. Exons 3–7 code for a triple-helical domain that contains 24 Gly–X–Y repeats, and exon 8 codes for the 19 amino acids of the N-telopeptide and part of the triple-helical domain.

The C-propeptide domain is coded by the last four exons. In the collagen type II gene, this domain codes for 273 amino acids. The terminal portion containing 267 amino acids (C-propeptide) is cleaved extracellularly, leaving a short telopeptide of 27 amino acids. The C-propeptide is important for the alignment of the collagen molecules, permitting assembly of the triple helix (Bateman *et al.*, 1989). The 3′ untranslated sequence of the type II procollagen mRNA is about 430 bp in mammals and 513 bp in the chicken (Sandel *et al.*, 1984; Sangiorgi *et al.*, 1985; Elima *et al.*, 1987; Metsaranta *et al.*, 1991).

B. Genetic Defects

The important role of the collagen type II gene in development is dramatically exemplified by certain genetic diseases that are caused by structural defects in the collagen type II coding sequence. For example, a single basepair mutation in exon 46 at position 943 that converts the amino acid glycine (GGC) to serine (AGC) was associated with a lethal perinatal form of short-limb dwarfism (Vissing *et al.*, 1989). A single basepair mutation that converts the codon CGT for arginine at position 519 to TGT for cysteine was also found in individuals with progressive osteoarthritis associated with mild chondroplasia (Ala-Kokko *et al.*, 1990). The mutation was found in all nine affected individuals of a family but not in unaffected members or in unrelated individuals. A 390-bp deletion was found in the collagen type II gene of affected members of a family with spondyloepiphyseal dysplasia, which is characterized by disproportionate short stature and pleiotropic involvement of the skeletal and ocular systems. As a consequence of the deletion, the entire exon 48 was eliminated from the coding sequence, resulting in a 36-amino acid loss in the triple-helix domain (Lee *et al.*, 1989). A premature stop codon in the collagen type II gene was found in affected members of a family with Stickler syndrome, which is an autosomal dominant disorder that effects the eye, ears, joints, and skeleton (Ahmad *et al.*, 1991).

C. Transcriptional Regulation

The transcription of the collagen type II gene is regulated by multiple *cis*-acting elements (Fig. 2). The promoter of this gene contains a TATA box between −20 and −30. This TATA element as well as it localization is conserved among the rat, human, and mouse sequences (Khono *et al.*, 1985; Ryan *et al.*, 1990; Metsaranta *et al.*, 1991). Another feature that is conserved in the promoter element is the sequence GGGCGG, which is repeated several times between −50 and −300. These regions are poten-

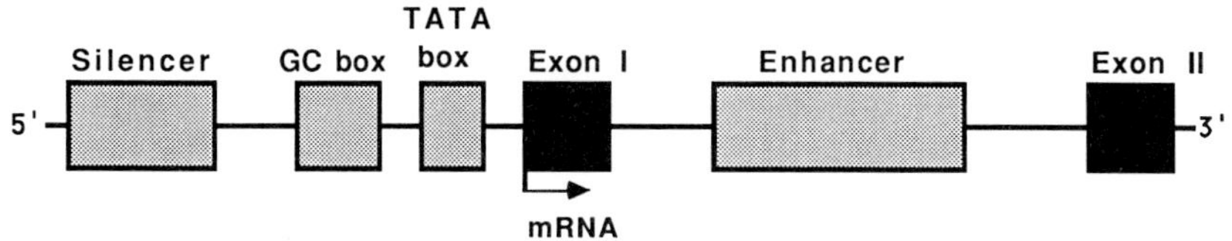

FIGURE 2 *Cis*-acting elements of the collagen type II gene.

tial binding sites for several zinc-finger proteins that are known transcriptional activators that include SP1, NGFI-A and NGFC-C (Crosby *et al.*, 1991). Transfection of constructs containing the promoter region driving the expression of the enzyme chloramphenical acetyltransferase (CAT) shows that the promoter activity is weak in chondrocytes. An explanation to these results came with the identification of an enhancer element. Like some other collagen genes, the enhancer element is located in the first intron of the gene (Horton *et al.*, 1987). To identify the *cis*-acting elements present in the first intron, several restriction fragments were subcloned in the promoter–CAT construct and transfected into chondrocytes. These studies revealed that 500 bp is required for full enhancer activity. This relatively large-sized collagen type II enhancer suggests that multiple sites within the 550-bp fragment are required for the full enhancer activity.

The enhancer element is also responsible for the tissue specificity of the collagen type II gene expression. Such specificity was determined by transfecting the collagen type II promoter–CAT–enhancer plasmid in various cell types as well as by the generation of transgenic animals (Yamada *et al.*, 1990). The results showed that high levels of the expression of the CAT gene occurred only in the tissues and in cell types that normally express the collagen type II gene (Fig. 3). Because the enhancer is preferentially active in chondrocytes, it is likely that specific nuclear factors bind to this region to modulate the transcription of the collagen type II gene.

To identify the *trans*-acting elements that bind to the collagen type II enhancer, we used the southwestern method to screen a human chondrocyte cDNA expression library (Vinson *et al.*, 1988). Using a 30-bp double-stranded oligonucleotide probe contained in the 550-bp enhancer, we cloned a protein that showed strong and specific binding to the enhancer region. Sequence analysis revealed that this protein is a new member of the zinc-finger family of transcription factors. We also showed that a 4-bp deletion in the 30-bp probe inhibited the formation of the complex with the *in vitro* translated protein and affected CAT activity in chondrocytes.

Silencer elements were also identified upstream from the promoter region, between −360 and −460 and between −620 and −700 (Savagner *et al.*, 1990). Constructs containing the silencer elements were shown to inhibit CAT activity in both fibroblasts and HeLa cells while not affecting the CAT gene expression in chondrocytes. In addition, gel retardation experiments showed that nuclear factors from HeLa cells bind specifically to a DNA fragment containing the silencer element, whereas condrocyte nuclear factors did not show any activity.

Transcription of the collagen type II gene can be modulated by exogenous factors that are present in serum. Three steroid hormones were

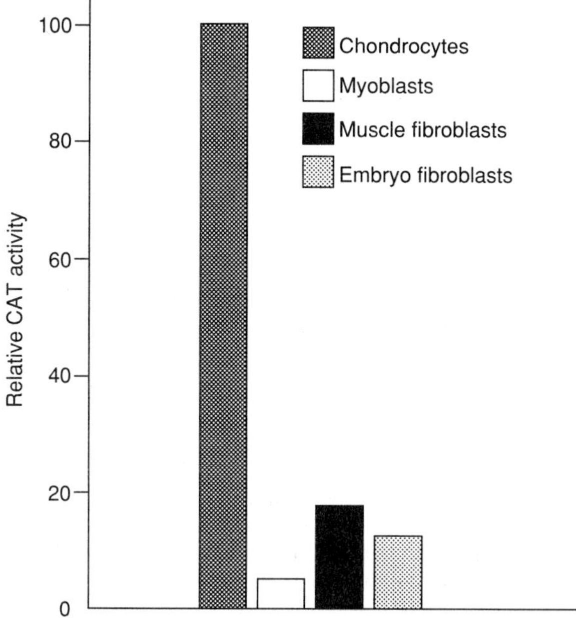

FIGURE 3 Relative chloramphenicol acetyltransferase (CAT) activity of construct containing the *cis*-acting regulatory regions of the collagen type II gene. Note the high level of CAT expression in chondrocytes contrasting with the other cell types.

found to affect transcriptional activity of the collagen type II gene. Dexamethasone and retinoic acid inhibited the collagen type II promoter–enhancer-mediated CAT activity, whereas vitamin D increased the activity (Fig. 4). Because these steroid hormones modulate the transcriptional activity through receptor-mediated DNA binding, these receptors may be directly binding to the sequences of the collagen type II gene.

III. AGGRECAN

The large aggregating chondroitin sulfate proteoglycan (aggrecan) is the major proteoglycan in cartilage with a protein core M_r ~250,000. Aggrecan is highly substituted with chondroitin sulfate and keratan sulfate chains as well as N- and O-linked oligosaccharides. It has the unique property of forming aggregates with hyaluronic acid. A small glycoprotein called link protein, which is structurally and functionally similar to the N-terminus of aggrecan, serves to stabilize the hyaluronic acid–aggrecan complex. These aggregates function to hold water and to resist compression in the joint.

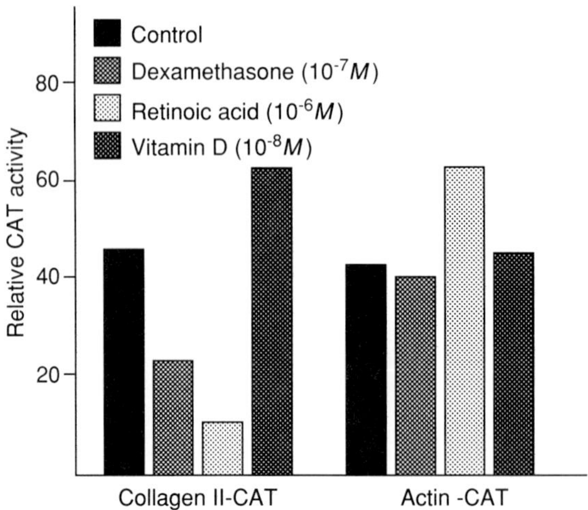

FIGURE 4 Effect of steroid hormones in the transcriptional activity of the collagen type II gene. Compare the difference between control and hormone-treated cells transfected with the collagen type II *cis*-acting elements and construct containing actin promoter and enhancer regions. CAT, chloramphenicol acetyltransferase.

A. Protein Structure

The complete cDNA sequence of both rat and human (Doege *et al.*, 1987, 1991) as well as partial sequences of chicken (Sai *et al.*, 1986) and bovine aggrecan (Oldberg *et al.*, 1987; Antonsson *et al.*, 1989) have been published. The composite rat sequence is about 6700 nucleotides long, whereas the human sequence contains 7137 nucleotides encoding for 2314 amino acids. Comparison between human and mouse sequences showed an overall identity of about 75% as well as a conservation of the internal domains. Aggrecan contains three cysteine-rich globular domains: G1 and G2, located in the N-terminus, and G3, located at the C-terminus. The three other domains that constitute this molecule are an interglobular domain (IGD) and two glycosaminoglycan attachment domains, a keratan sulfate-binding domain (KS) and a chondroitin sulfate-binding domain (CS), located in the large central portion (Fig. 5).

1. G1 Domain

The G1 domain interacts with hyaluronic acid and is therefore designated HABR for hyaluronic acid-binding domain (Mörgelin *et al.*, 1988). This domain is formed by three subregions: A, B, and B′. The globular structure of B and B′ are stabilized by two cystine bonds, while A contains two cysteine residues forming a single disulfide bond. Each of

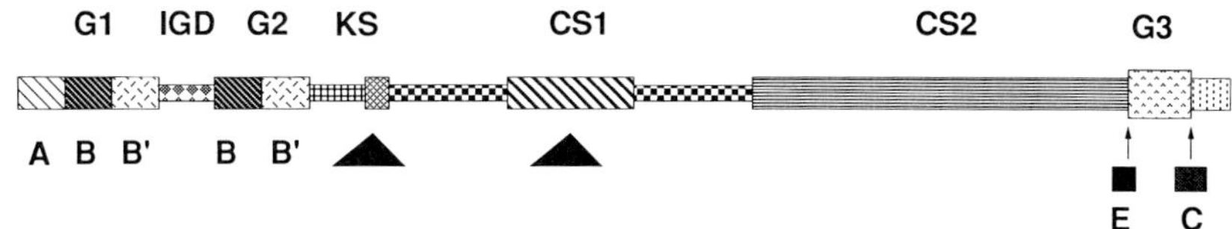

FIGURE 5 Schematic model of human aggrecan. Arrowheads show the localization of the human-specific repeat sequences. The points of insertion of the alternatively sliced epidermal growth factor-like (E) and complementary regulatory protein-like (C) globular domains in G3 are indicated with arrows. IGD, interglobular domain; KS, keratan sulfate-binding domain.

these subregions also contains a potential N-glycosylation site located near the C-terminus. Comparison between rat and human sequences showed a particularly high homology in this area, with 100% homology in the B subdomain.

2. G2 Domain

The G2 domain shows striking similarities to domain G1, not only in the structural organization but also in sequence. The G2 domain contains B and B′ segments but lacks the A subdomain. The B and B′ motifs are highly homologous to the corresponding parts in domain G1; B and B′ motifs exhibit 69 and 61% homology, respectively.

3. G3 Domain

The C-terminal G3 domain is also formed by three subregions. It contains a lectin-binding domain that has a 38% homology with chicken and rat hepatocyte lectins (Drickamer, 1988). Experiments using the *in vitro* expressed lectin-binding domain have confirmed the prediction from the cDNA sequence analysis, by showing a distinct specificity of the interaction between the peptide coding for this domain with both galactose and fucose residues (Halberg *et al.*, 1988). The two additional subdomains have homology with epidermal growth factor (EGF) (Baldwin *et al.*, 1989a) and with complement regulatory proteins (CRPs). These domains are generated by alternative splicing of its mRNA (Baldwin *et al.*, 1989a; Doege *et al.*, 1991).

4. Keratan Sulfate-Binding Domain

The KS domain is rich in proline–serine and proline–threonine sequences, which are putative keratan sulfate attachment sites. This domain is formed by 119 amino acids in the human gene and 113 in the rat.

The human sequence also contains 11 highly conserved repeats of the hexameric sequence E–E–P–(S,F)–P–S, which is not found in the rat sequences. Although there is no direct evidence, the prevalence of proline repeats in this domain supports the hypothesis that this amino acid mediates the aggrecan–keratan sulfate interaction.

5. Chondroitin Sulfate-Binding Domain

The CS domain is the largest domain of the aggrecan molecule; it extends through 1372 amino acids in the human gene and 1104 in the rat. This domain is characterized mainly by its high content in serine–glycine repeats, which function as attachment sites for the chondroitin sulfate glycosaminoglycan. This domain can be subdivided into two regions, CS1 and CS2, according to the arrangement of the repeats. In the CS1 domain, the serine–glycine sequences are uniformly spaced, separated by highly conserved amino acids, whereas in the CS2 domain the repeats are grouped in clusters without symmetry. A unique feature of the human aggrecan CS1 domain is the presence of a 19-amino acid sequence that is repeated 19 times. These repeats are not found in rat aggrecan.

6. Interglobular Domain

The IGD domain separates G1 from G2. It is formed by 137 amino acids in the rat gene and 129 in the human. The first half is formed by 60 amino acids and is 90% conserved between human and rat, including a potential N-glycosylation site at amino acid position 488.

Several other proteins share some of the structure features found in aggrecan (Fig. 6). Link protein (see later) consists of subdomain structure homologous to the A, B, and B′ subdomains of the G1 region of aggrecan. Versican, another class of proteoglycan that was originally identified in fibroblasts, shows strong similarity to both N- and C-terminal domains of aggrecan (Zimmerman and Ruoslahti, 1989). Versican lacks the domain G2 but has an extra repeat of the EGF-like domain. It shows 54% sequence homology with the aggrecan G1 domain, and 67% with the lectinlike and CRP-like domains. The EGF-like domains of G3 shares 44% identity between versican and aggrecan. Several other proteins including the lymphocyte homing receptors Hermes/CD44 (Goldstein *et al.*, 1989; Stamenkovic *et al.*, 1989) and MEL-14 (Lasky *et al.*, 1989; Siegelman *et al.*, 1989) and two activated endothelial cell membrane proteins, ELAM-1 (Bevilacqua *et al.*, 1989) and GMP-140 (Johnston *et al.*, 1989), also show some similarity to aggrecan. Hermes/CD44 has a single subdomain B at the N-terminus. While versican is not known to bind hy-

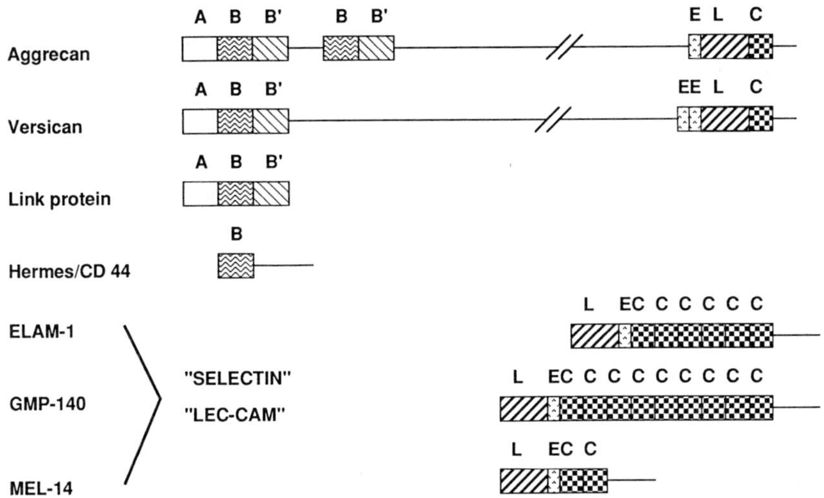

FIGURE 6 Schematic of related structural features among proteins sharing homologous domains with aggrecan. Related domains are shown as boxes and are labeled A, B, and B′ for the link motifs; E, L, and C for the epidermal growth factor-like, lectin-like, and complementary regulatory protein-like domains, respectively.

aluronic acid, Hermes/CD44 binds to tissues in a hyaluronic acid-dependent manner (Aruffo *et al.*, 1990). These results suggest that the B subunit may be sufficient for hyaluronic acid binding. ELAM-1, MEL-14, and GMP-140 are a family that has a unique structure consisting of a lectin-like domain, an EGF-like domain, and a CRC-like domain. Contrary to aggrecan and versican, the lectin-like domain precedes the EGF-like domain.

B. Gene Structure

Rat and human aggrecan contains 15 exons that span approximately 100 kb. The exon structure of this gene corresponds almost exactly to the structural domains predicted in the cDNA sequence. The G1 domain is encoded by exons 3, 4, and 5, which correspond to subdomains A, B, and B′, respectively. The IGD domain is coded by exon 6, whereas domains B and B′ correspond to exons 7 and 8, respectively. The two glycosaminoglycan attachment domains, the KS and CS domains, are encoded by two distinct exons: exon 9 for KS and the large exon 10 for CS. The exception is domain G3, where the three subdomains are coded by the last five exons. The 372-bp untranslated sequence is mostly contained in exon 1, whereas the small exon 2 contains the sequence for the signal peptide.

IV. LINK PROTEIN

Link protein occurs in various tissues such as sclera and tendon, although it has been best studied as a component of hyaline cartilage. Link protein stabilizes the interaction of the large sulfated proteoglycan, aggrecan, and hyaluronic acid.

A. Protein Structure

The entire coding sequence of the human (Rhodes *et al.*, 1991), rat (Doege *et al*, 1986; Rhodes *et al.*, 1988), and chick (Deak *et al.*, 1986) link protein has been determined by cDNA cloning. Some protein sequence has also been obtained (Neame *et al.*, 1987). The sequence consists of three domains—A, B, and B′—and shows high homology to G1 domain of aggrecan. Similar to aggrecan, B and B′ domains represent an internal repeat. The sequence of the B region is more strongly conserved in different species than the A domain. Proteolytic analysis suggests that the B motifs bind to the hyaluronic acid, whereas the A domain interacts with aggrecan. Complementary DNA analysis predicts an alternative splicing of link protein mRNA (Rhodes *et al.*, 1988). The insertion of a 58-amino acid sequence occurs in the A domain. However, protein with this inserted sequence has not yet been detected in tissues.

As predicted, the link and aggrecan genes share a common exon structure and are evolutionarily related. In both proteins, the domains are coded by single exons. A structural model for link protein is shown in Figure 7.

B. Transcriptional Regulation

Sequence analysis showed that the link protein promoter lacks a typical TATA box and contains an AP-1-like site and a cyclic adenosine monophosphate responsivelike element at rat sequences −315 and −192 that were conserved in human and rat. Like collagen type II the link protein gene contains an enhancer element that is localized in the first intron (Rhodes *et al.*, 1991). The enhancer functions in an orientation-independent manner and causes a severalfold increase in CAT activity when compared to constructs containing the promoter alone. The transcriptional activity of the link gene is also regulated by steroid hormones. Figure 8 shows the relative CAT activity of the promoter–Cat–enhancer construct when transfected in chick chondrocytes and cultured in the presence of certain hormones. As in the case of collagen type II regulatory regions, retinoic acid caused a dramatic decrease in CAT expression, whereas vitamin D promoted an increase in CAT activity. In contrast, dexamethasone caused an increase in CAT activity.

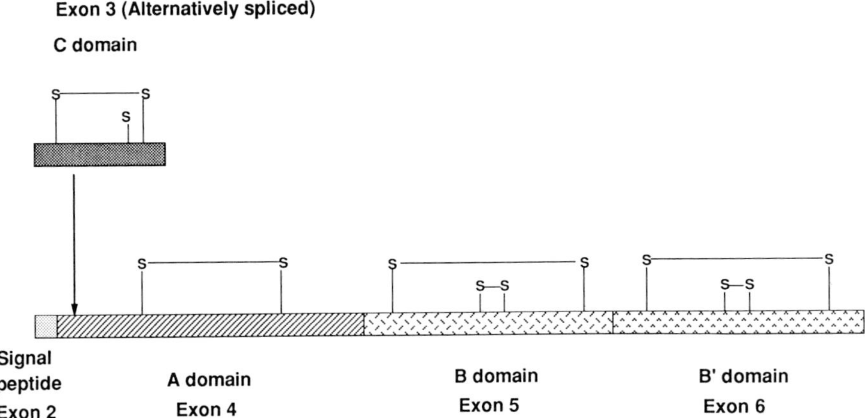

FIGURE 7 Schematic model of rat link protein. The predicted globular domains are shown with the corresponding exons. Exon 1, which codes for the 5′ untranslated sequence, does not appear in the scheme.

V. SUMMARY

Considerable progress has been made in understanding the structure of cartilage components by both molecular cloning and protein sequencing. These components include collagen type II, collagen types IX and XI, link protein, and large aggregating proteoglycan (aggrecan). The

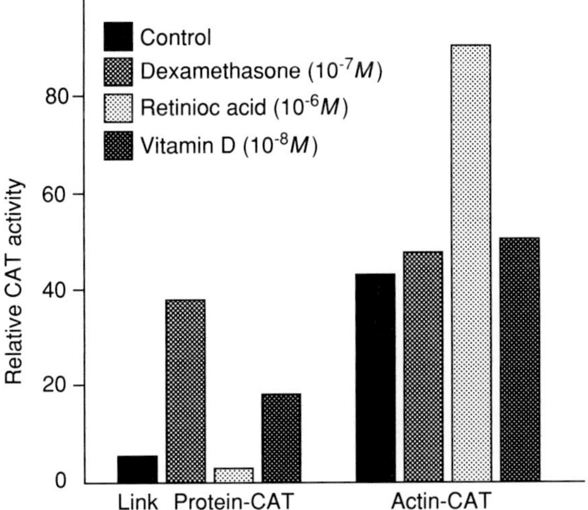

FIGURE 8 Effect of steroid hormones in the transcriptional activity of the link protein gene. Compare the difference between control and hormone-treated cells transfected to the collagen type II *cis*-acting elements and construct containing actin promoter and enhancer regions. CAT, chloramphenicol acetyltransferase.

function of these components has also been studied. Some evidence indicates a direct linkage between certain heritable disorders of cartilage and mutations of the collagen type II gene.

Gene regulation studies in cartilage have just begun. These studies indicate that the transcription of the collagen type II and link protein genes is regulated by multiple *cis*-acting sites, which are modulated by exogenous factors such as hormones and vitamins. Although these proteins are co-expressed in cartilage, their expression can be differentially regulated as well as co-regulated. The cloning and characterization of transcription factors that interact with the regulatory regions of these genes will be a key step for understanding the dynamic processes of cartilage expression and bone formation.

REFERENCES

Ahmad, N. N., Ala-Kokko, L., Knowlton, R. G., Jimenez, S. A., Weaver E. J., Maguire, J. I., Tasman, W., and Prockop, D. J. (1991). "Stop codon in the procollagen II gene (COL2A1) in a family with the Stickler syndrome (arthro-ophthalmopathy)." *Proc. Natl. Acad. Sci. USA* **88,** 6624–6627.

Ala-Kokko, L., Baldwin, C. T., Moskowitz, R. W., and Prockop, D. J. (1990). "Single base mutation in the type II procollagen gene (COL2A1) as a cause of primary osteoarthritis associated with a mild chondrodysplasia." *Proc. Natl. Acad. Sci. USA* **87,** 6565–6568.

Antonsson, P., Heinegard, D., and Oldberg, A. (1989). "The keratan sulfate-enriched region of bovine cartilage proteoglycan consists of a consecutively repeated hexapeptide motif." *J. Biol. Chem.* **264,** 16170–16173.

Aruffo, A., Stamenkovic, I., Melnick, M., Underhill, C., and Seed, B. (1990). "CD44 is the principal cell surface receptor for hyaluronate." *Cell* **61,** 1303–1313.

Baldwin, C. T., Reginato, A. M., and Prockop, D. J. (1989a). "A new epidermal growth factor-like domain in the human core protein for the large cartilage-specific proteoglycan. Evidence for alternative splicing of the domain." *J. Biol. Chem.* **264,** 15747–15750.

Baldwin, C. T., Reginato, A. M., Smith, C., Jimenez, S. A., and Prockop, D. J. (1989b). "Structure of cDNA clones coding for human type II procollagen. The alpha 1(II) chain is more similar to the alpha 1(I) chain than two other alpha chains of fibrillar collagens." *Biochem. J.* **262,** 521–528.

Bateman, J. F., Lamande, S. R., Dahl, H.-H. M., Chan, D., Mascara, T., and Cole, W. G. (1989). "A frameshift mutation results in a truncated nonfunctional carboxyl-terminal pro alpha 1(I) propeptide of type I collagen in osteogenesis imperfecta." *J. Biol. Chem.* **264,** 10960–10964.

Benson-Chandra, V., Su, M.-V., Weil, D., Chu, M.-L., and Ramirez, F. (1989). "Cloning and analysis of the 5′ portion of the human type-III procollagen gene (COL3A1)." *Gene* **78,** 255–265.

Bevilacqua, M. P., Stengelin, S., Gimbrone, M. A., Jr., and Seed, B. (1989). "Endothelial leukocyte adhesion molecule 1: an inducible receptor for neutrophils related to complement regulatory proteins and lectins." *Science* **243,** 1160–1165.

Cheah, K. S. E., Lau, E. T., Au, P. K. C., and Tam, P. P. L. (1991). "Expression of the mouse alpha 1(II) collagen gene is not restricted to cartilage during development." *Development* **111,** 945–953.

Crosby, S. D., Puetz, J. J., Simburger, K. S., Fahrner, T. J., and Milbrandt, T. J. (1991). "The early response gene NGFI-C encodes a zinc finger transcriptional activator and is a member of the GCGGGGGCG (GSG) element-binding protein family." *Mol. Cell Biol.* **11,** 3835–3841.

D'Allesio, M., Bernard, M., Pretorius, P. J., de Wet, W., and Ramirez, F. (1988). "Complete nucleotide sequence of the region encompassing the first twenty-five exons of the human pro alpha 1(I) collagen gene (COL1A1). *Gene* **67,** 105–115.

Deak, F., Kiss, I., Sparks, K. J., Argraves, W. S., Hampikian, G., and Goetinck, P. F. (1986). "Complete amino acid sequence of chicken cartilage link protein deduced from cDNA clones." *Proc. Nat'l. Acad. Sci. USA* **83,** 3766–3770.

de Wet, W., Bernard, M., Benson-Chanda, V., Chu, M. L., Dickson, L., Weil, D., and Ramirez, F. (1987). "Organization of the human pro-alpha 2(I) collagen gene." *J. Biol. Chem.* **262,** 16032–16036.

Doege, K., Hassell, J. R., Sasaki, M., and Yamada, Y. (1986). "Link protein cDNA sequence reveals a tandemly repeated protein structure." *Proc. Natl. Acad. Sci. USA* **83,** 3761–3765.

Doege, K., Sasaki, M., Horigan, E., Hassell, J. R., and Yamada, Y. (1987). "Complete primary structure of the rat cartilage proteoglycan core protein deduced from cDNA clones." *J. Biol. Chem.* **262,** 17757–17767.

Doege, K. J., Sasaki, M., Kimura, T., and Yamada, Y. (1991). "Complete coding sequence and deduced primary structure of the human cartilage large aggregating proteoglycan, aggrecan. Human-specific repeats, and additional alternatively spliced forms." *J. Biol. Chem.* **266,** 894–902.

Drickamer, K. (1988). "Two distinct classes of carbohydrate-recognition domains in animal lectins." *J. Biol. Chem.* **263,** 9557–9560.

Elima, K., Vuorio, T., and Vuorio, E. (1987). "Determination of the single polyadenylation site of the human pro alpha 1(II) collagen gene." *Nucleic Acids Res.* **15,** 9499–9504.

Goldstein, L. A., Zhou, D. F. H., Picker, L. J., Minty, C. N., Bargatze, R. F., and Butcher, E. C. (1989). "A human lymphocyte homing receptor, the hermes antigen, is related to cartilage proteoglycan core and link proteins." *Cell* **56,** 1063–1072.

Halberg, D. F., Proulx, G., Doege, K., Yamada, Y., and Drickamer, K. (1988). "A segment of the cartilage proteoglycan core protein has lectin-like activity." *J. Biol. Chem.* **263,** 9486–9490.

Horton, W., Miyashita, T., Khono, K., Hassell, J. R., and Yamada, Y. (1987). "Identification of a phenotype-specific enhancer in the first intron of the rat collagen II gene." *Proc. Natl. Acad. Sci. USA* **84,** 8864–8868.

Johnston, G. I., Cook, R. G., and McEver, R. P. (1989). "Cloning of GMP-140, a granule membrane protein of platelets and endothelium: sequence similarity to proteins involved in cell adhesion and inflammation." *Cell* **56,** 1033–1044.

Khono, K., Sullivan, M., and Yamada, Y. (1985). "Structure of the promoter of the rat type II procollagen gene." *J. Biol. Chem.* **260,** 4441–4447.

Kosher, R. A., and Solursh, M. (1989). "Widespread distribution of type II collagen during embryonic chick development." *Dev. Biol.* **131,** 558–566.

Lasky, L. A., Singer, M. S., Yednock, T. A., Dowbenko, D., Fennie, C., Rodriguez, H., Nguyen, T., Stachel, S., and Rosen, S. D. (1989). "Cloning of a lymphocyte homing receptor reveals a lectin domain." *Cell* **56,** 1045–1055.

Lee, B., Vissing, H., Ramirez, F., Rogers, D., and Rimoin, D. (1989). "Identification of the molecular defect in a family with spondyloepiphyseal dysplasia." *Science* **244,** 978–980.

Metsäranta, M., Toman, D., de Crombrugghe, B., and Vuorio, E. (1991). Mouse type II collagen gene. Complete nucleotide sequence, exon structure, and alternative splicing. *J. Biol. Chem.* **266,** 16862–16869.

Mörgelin, M., Paulsson, M., Hardingham, T. E. Heinegard, D., and Engel, J. (1988).

"Cartilage proteoglycans. Assembly with hyaluronate and link protein as studied by electron microscopy." *Biochem. J.* **253,** 175–185.

Neame, P. J., Christner, J. E., and Baker, J. R. (1987). "Cartilage proteoglycan aggregates. The link protein and proteoglycan amino-terminal globular domains have similar structures." *J. Biol. Chem.* **262,** 17768–17778.

Oldberg, A., Antonsson, P., and Heinegard, D. (1987). "The partial amino acid sequence of bovine cartilage proteoglycan, deduced from a cDNA clone, contains numerous Ser-Gly sequences arranged in homologous repeats." *Biochem. J.* **243,** 255–259.

Pelletier, J. P., and Martel-Pelletier, J. (1989). "The therapeutic effects of NSAID and corticosteroids in osteoarthritis: to be or not to be." *J. Rheumatol.* **16,** 266–269.

Rhodes, C., Doege, K., Sasaki, M., and Yamada, Y. (1988). "Alternative splicing generates two different mRNA species for rat link protein." *J. Biol. Chem.* **263,** 6063–6067.

Rhodes, C., Savagner, P., Line, S., Sasaki, M., Chirigos, M., Doege, K., and Yamada, Y. (1991). Characterization of the promoter for the rat and human-link protein gene. *Nucleic Acids Res.* **19,** 1933–1939.

Ryan, M. C., and Sandell, L. J. (1990). "Differential expression of a cysteine-rich domain in the amino-terminal propeptide of type II (cartilage) procollagen by alternative splicing of mRNA." *J. Biol. Chem.* **265,** 10334–10339.

Ryan, M. C., Sieraski, M., and Sandell, L. J. (1990). "The human type II procollagen gene: identification of an additional protein-coding domain and location of potential regulatory sequences in the promoter and first intron." *Genomics* **8,** 41–48.

Sai, S., Tanaka, T., Kosher, R. A., and Tanzer, M. L. (1986). "Cloning and sequence analysis of a partial cDNA for chicken cartilage proteoglycan core protein." *Proc. Natl. Acad. Sci. USA* **83,** 5081–5085.

Sandell, L. J., Prentice, H. L., Kravis, D., and Upholt, W. B. (1984). "Structure and sequence of the chicken type II procollagen gene. Characterization of the region encoding the carboxyl-terminal telopeptide and propeptide." *J. Biol. Chem.* **259,** 7826–7834.

Sangiorgi, F. O., Benson-Chanda, V., de Wet, W. J., Sobel, M. E., and Ramirez, F. (1985). "Analysis of cDNA and genomic clones coding for the pro alpha 1 chain of calf type II collagen." *Nucleic Acids Res.* **12,** 1025–1038.

Savagner, P., Miyashita, T., and Yamada, Y. (1990). "Two silencers regulate the tissue-specific expression of the collagen II gene." *J. Biol. Chem.* **265,** 6669–6674.

Siegelman, M. H., van der Rijn, M., and Weissman, I. L. (1989). "Mouse lymph node homing receptor cDNA clone encodes a glycoprotein revealing tandem interaction domains." *Science* **243,** 1165–1172.

Solomon, E., Hiorns, L. R., Spurr, M., Kurkinen, D., Barlow, D., Hogan, B. L. M., and Dalgleish, R. (1985). "Chromosomal assignments of the genes coding for human types II, III, and IV collagen: a dispersed gene family." *Proc. Natl. Acad. Sci. USA* **82,** 3330–3334.

Stamenkovic, I., Amiot, M., Pesando, J. M., and Seed, B. (1989). "A lymphocyte molecule implicated in lymph node homing is a member of the cartilage link protein family." *Cell* **56,** 1062–1075.

Su, M. W., Benson-Chandra, V., Vissing, H., and Ramirez, F. (1989). "Organization of the exons coding for pro alpha 1(II) collagen N-propeptide confirms a distinct evolutionary history of this domain of the fibrillar collagen genes." *Genomics* **4,** 438–41.

Suda, S., Takahashi, N., Shinki, T., Houriuchi, N., Yamaguchi, A., Yoshiki, S., Enomoto, S., and Suda, T. (1985). "1 alpha,25-dihydroxyvitamin D3 receptors and their action in embryonic chick chondrocytes." *Calcif. Tissue Int.* **37,** 82–90.

Tate, V. E., Finer, M. H., Boedtker, H., and Doty, P. (1983). "Chick pro alpha 2(I) collagen gene: exon location and coding potential for the prepropeptide." *Nucleic Acids Res.* **11,** 91–104.

Tsonis, P. A. (1991). "1,25-Dihydroxyvitamin D3 stimulates chondrogenesis of the chick limb bud mesenchymal cells." *Dev. Biol.* **143,** 130–134.

Vinson, C. R., LaMarco, K. L., Johnson, P. F., Landschulz, W. H., and McNight, S. L. (1988). "In situ detection of sequence-specific DNA binding activity specified by a recombinant bacteriophage." *Genes Dev.* **2,** 801–806.

Vissing, H., D'Alessio, M., Lee, B., Ramirez, F., Godfrey, M., and Hollister, D. W. (1989). "Glycine to serine substitution in the triple helical domain of pro-alpha 1 (II) collagen results in a lethal perinatal form of short-limbed dwarfism." *J. Biol. Chem.* **264,** 18265–18267.

Vuorio, E., and de Crombrugghe, B. (1990). "The family of collagen genes." *Annu. Rev. Biochem.* **59,** 837–872.

Wood, A., Ashhurst, D. E., Corbett, A., and Thorogood, P. (1991). "The transient expression of type II collagen at tissue interfaces during mammalian craniofacial development." *Development* **111,** 955–968.

Yamada, Y., Avvedimento, V. E., Mudryj, M., Ohkubo, H., Vogeli, G., Irani, M., Pastan, I., and de Crombrugghe, B. (1980). "The collagen gene: evidence for its evolutionary assembly by amplification of a DNA segment containing an exon of 54 bp." *Cell* **22,** 877–892.

Yamada, Y., Miyashita, T., Savagner, P., Horton, W., Brown, K. S., Abramczuk, J., Hou-Xiang, X., Khono, K., Bolander, M., and Bruggeman, L. (1990). Regulation of the collagen II gene *in vitro* and in transgenic mice. *In* "Structure, Molecular Biology and Pathology of Collagen" (R. Fleischmajer, B. Olsen, and K. Kuhn, eds.), Vol. 580, pp. 81–87. *Ann. N.Y. Acad. Sci.*

Zimmerman, D. R., and Ruoslahti, E. (1989). *EMBO J.* **8,** 2975–2981.

INDEX

H

R

S

T

U

V